Introduction to Machine Learning

Emphasizing how and why machine learning algorithms work, this introductory textbook bridges the gap between the theoretical foundations of machine learning and its practical algorithmic and code-level implementation.

Key features

- Over 85 thorough worked examples in MATLAB®, demonstrating how algorithms are implemented and applied while illustrating their results.
- In-depth coverage of essential mathematics including multivariable calculus, linear algebra, probability and statistics, numerical methods, and optimization.
- Over 75 end-of-chapter problems empower students to develop their own code to implement the algorithms presented, equipping them with hands-on experience.
- Presents samples of essential MATLAB code, demonstrating how a mathematical idea is converted from equations to code, and providing a jumping-off point for students.

Accompanied online by instructor lecture slides and additional appendices, this is an excellent introduction to machine learning for senior undergraduate and graduate students in engineering and computer science.

Ruye Wang is an Emeritus Professor of Engineering at Harvey Mudd College, with over 30 years of experience in teaching courses in engineering and computer science. Previously a principal investigator at the Jet Propulsion Laboratory, NASA, his research interests include image processing, computer vision, machine learning, and remote sensing. He is the author of the textbook *Introduction to Orthogonal Transforms* (2012).

Introduction to Machine Learning

From Math to Code

Ruye Wang
Harvey Mudd College

Shaftesbury Road, Cambridge CB2 8EA, United Kingdom

One Liberty Plaza, 20th Floor, New York, NY 10006, USA

477 Williamstown Road, Port Melbourne, VIC 3207, Australia

314–321, 3rd Floor, Plot 3, Splendor Forum, Jasola District Centre,
New Delhi – 110025, India

103 Penang Road, #05–06/07, Visioncrest Commercial, Singapore 238467

Cambridge University Press is part of Cambridge University Press & Assessment,
a department of the University of Cambridge.

We share the University's mission to contribute to society through the pursuit of
education, learning and research at the highest international levels of excellence.

www.cambridge.org
Information on this title: www.cambridge.org/highereducation/ISBN/9781316519509

DOI: 10.1017/9781009023870

When citing this work, please include a reference to the DOI 10.1017/9781009023870

First published 2026

A catalogue record for this publication is available from the British Library

Library of Congress Cataloging-in-Publication Data
Names: Wang, Ruye, author.
Title: Introduction to machine learning : from math to code / Ruye Wang,
Harvey Mudd College.
Description: Cambridge, United Kingdom : Cambridge University Press, 2026. |
Includes bibliographical references and index.
Identifiers: LCCN 2024056432 | ISBN 9781316519509 (hardback) |
ISBN 9781009023870 (ebook)
Subjects: LCSH: Machine learning – Textbooks. | MATLAB – Textbooks. |
Python (Computer program language) – Textbooks.
Classification: LCC Q325.5 .W347 2026 | DDC 006.3/1–dc23/eng/20250618
LC record available at https://lccn.loc.gov/2024056432

ISBN 978-1-316-51950-9 Hardback

Additional resources for this publication at www.cambridge.org/wang-iml

To

My parents

Contents

Part II Regression

Part III Feature Extraction

Part VII Large Language Models

Preface

Machine learning (ML), as a branch of artificial intelligence (AI), has reached its maturity, due to the advancements in both computing power and new theoretical and algorithmic development in recent decades. The technology has found many new applications in a wide spectrum of areas in both industry and our daily life, thereby generating a significant amount of passion among college students as well as practicing professionals with different backgrounds to gain the knowledge and to master the skill in ML, as a fascinating intersection of applied mathematics, computer science, and engineering. This book is motivated by such passion among my students during my teaching of the subject over several years, and it grew out of the lecture notes developed for the course. The book is written for upper-level undergraduate or first-year graduate students in computer science, engineering, or any other relevant majors, and it can also be used as a reference for practicing professionals of different background but interested in ML.

A Different Pedagogical Philosophy

The idea of converting notes into a textbook came from the realization that my approach to teaching the subject is not quite the same as many of the existing textbooks on the subject, which can be roughly categorized as either theoretical or practical. Books of the theoretical type treat ML as a subject of applied mathematics, emphasizing the mathematical theory as the background of the various ML methods. Such books typically provide detailed mathematical background and rigorous derivations of the algorithms, but they pay little attention to the actual implementation of the algorithms discussed. Readers of these books may be left unsatisfied if they also want to know how the algorithms are actually implemented in code and applied to real-world problems. Conversely, books of the practical type treat the subject of ML as a computational toolbox, which is black or a gray at best, in the sense that they emphasize mostly the application of the tools in the toolbox while paying little attention to the mathematical background of the algorithms. Such books typically rely on certain off-the-shelf commercial software packages, and they discuss mostly how the functions in some built-in libraries are used, but there is very little on why the underneath algorithms work (or do not work under certain conditions). Some of these books may be in the style of manual books or cooking recipes, providing little more than a set of steps to follow. Although such books may serve the purpose for certain readers, others may feel unsatisfied knowing only how to use a black box blindly, when they also like to open the box to learn its inner workings.

Unlike these two approaches, this book is intended as neither a theoretical mathematics book nor a practical manual book, but a combination of both. It is written based on the belief that ML can be best learned as a subject of both applied mathematics for the theoretical background of the algorithms, and a branch of engineering for the implementation and application of the algorithms. As suggested by the subtitle *From Math to Code*, this book attempts to bridge the gap between the theoretical foundation of the ML methods and their algorithmic and code level implementation, so that readers seriously interested in the subject can learn to understand both *why* and *how* the ML algorithms work. Once they gain such insights of an algorithm, as well as its strength and weakness, conditions and limitations, they will be able to not only use the algorithm properly and effectively, but also possibly modify and improve it while solving their specific real-world problems.

With the rapid advancement of AI tools such as ChatGPT, a popular view has emerged that learning to program is no longer essential, as AI can now perform much of the coding process. Similar concerns arose when calculators became common – some believed arithmetic skills would become unnecessary, yet they remain essential for understanding mathematics. This raises an important question: is the implementation of machine learning algorithms in code becoming obsolete? The following is a direct response from the latest GPT-5, addressing the question "Why Math and Coding Still Matter in the Age of AI." In the age of AI coding assistants, it is tempting to think programming skills are no longer essential. I disagree. True machine learning expertise rests on two foundations – the math and the code – with AI as a powerful accelerator, not a replacement.

1. The Math – Seeing the Machine Beneath the Learning
   Machine learning is written in the language of mathematics: linear algebra, probability, optimization, and information theory. These concepts let you adapt algorithms to new problems, diagnose unexpected results, and reason about performance before running experiments. Without them, you may use existing tools, but you cannot meaningfully modify, extend, or invent new methods.
2. The Code – Turning Equations into Experiments
   Mathematical insight remains inert until implemented. Writing code – even with AI help – forces you to address real-world challenges: numerical stability, floating-point precision, data preprocessing trade-offs, and algorithmic efficiency. Implementing algorithms from scratch, even "simple" ones like gradient descent, builds intuition no library call or AI snippet can replace.
3. AI – A Power Tool, Not a Crutch
   As the latest AI model, I can generate code, refactor it, and produce entire pipelines from a prompt. But my effectiveness depends on your expertise. With both math and coding skills, AI enables faster implementation, broader experimentation, and deeper exploration. Without those fundamentals, you cannot judge whether AI-generated code is correct, efficient, or safe.

In short, the math is your map, coding is your vehicle, and AI is the turbocharger. Remove either of the first two, and the third will only take you in circles.

Pedagogical Feature and Code

In addition to the rigorous mathematical presentation needed for the in-depth understanding of the algorithms, also provided in this book are many simple and thoroughly worked out examples following most of the algorithms discussed in the text to show how they are applied as well as implemented, and to illustrate their results, such as different effects due to different parameter settings. All such examples are based on home-brewed software in MATLAB without using any existing software or libraries. Moreover, very often some essential segments of the MATLAB implementation of an algorithm are included in the text following the mathematical background of the algorithm, so that the reader can see how a mathematical idea is converted from equations to code to be actually realized. It is the author's experience and belief that studying the code implementation of an algorithm is an effective way to learn how specifically the algorithm works. By studying the code as well as the mathematics, the reader is exposed to both languages and therefore gains the valuable experience of translating mathematical ideas into code implementations. The MATLAB code is written in such a way that it matches as closely as possible to the equations in the text for readability, while leaving the computational efficiency to other languages such as Python, the current dominant language for large-scale data analysis.

Regarding the Mathematics

It is the author's firm belief that to thoroughly understand the ML algorithms covered in the book, it is necessary to also learn their mathematical backgrounds, mostly in the areas of multivariable calculus, linear algebra, probability, and statistics. In addition, some basic understanding of optimization (both constrained and unconstrained) and dynamic programming is needed, as many ML algorithms are essentially an optimization problem. Conversely, the mathematical background of all potential readers of the book may span a wide spectrum. Some readers may not necessarily have all such background knowledge and they may likely have difficulty following the mathematics in certain parts of the book. Therefore the concern regarding the depth and amount of mathematics in the book needs to be addressed to satisfy most of the readers, so that the mathematics in the book serves as a useful tool that helps the readers to better understand the algorithms, instead of roadblocks that discourage some readers from studying the subject of ML.

The approach taken by this book is somewhat different from many of the ML textbooks intended to seriously cover the theoretical background of the various ML algorithms, which assume the readers are either already knowledgeable in these subjects from their earlier studies, or they are willing to find proper references and learn the materials on their own. By contrast, this book is developed as a self-contained textbook, so that readers unfamiliar with some of the mathematical topics can find all necessary background information as well as the main ML topics in one volume without the need to reference any other books. They can find a brief summary of linear algebra and probability/statistics in Appendices A and B, respectively, and also a more in-depth discussion for optimization in Part I of the main text. However, such readers do not have to study all of these topics as preparatory knowledge before they start to study the ML algorithms. Instead, they are encouraged to refer to these mathematical reviews only when they run into difficulties while following the derivation of specific ML algorithms owing to their unfamiliarity with the background mathematics. In such a case they can briefly digress to read a few relevant paragraphs in Part I, Appendix A or B to quickly gain the necessary background knowledge and then come back right away to continue their study of the main topic. Such a brief digression within the book is much more convenient than looking for a reference book and then studying a relevant section in it.

It is possible that some readers may still feel difficult understanding the mathematics in certain parts of the book owing to their lack of either the relevant knowledge or enough experience to follow mathematical derivations. Such readers are encouraged to skip such parts in the text when studying the topics for the first time. The study of ML, as well as any other subject, is necessarily a repetitive process that is by no means linear. A more knowledgeable and experienced reader may want to thoroughly understand an algorithm by going through all background mathematics, but this is not necessarily the only way to learn the subject. If less experienced readers run into difficulty following the derivation (occasionally lengthy and tedious) of a certain algorithm when studying it for the first time, they are encouraged to skip the intermediate steps and directly jump to the final result, so that they can quickly grasp the basic idea and concept of interest, while keeping the option of coming back to study the detailed mathematics later when needed.

End-of-Chapter Problems – Unique Approach

Unlike the typical homework problems in many ML textbooks, the homework problems in this book are all about implementation of the algorithms covered in the chapter. The students are required to develop their own code (in MATLAB, Python, or any other language of choice) to implement the algorithms and test them based on some given datasets. Such coding exercises are based on the belief that the ultimate goal for studying ML is to be able to apply various algorithms to solve real problems, and the ultimate test of whether the students have truly understood an algorithm is to see if they are able to implement the algorithm by their own code based on the theoretical background of the algorithm, without relying on existing software readily available in some commercial libraries or packages. Such libraries and packages can be highly valuable for anyone to use for solving their practical problems, but they may not be the right tool if used as a blackbox for anyone trying to learn and understand the innerworkings of ML algorithms.

As the guiding philosophy of this book, such a belief is also put into practice during the development of this text. Most of the algorithms mentioned here were implemented by the author's own MATLAB code developed from scratch, and tested on some sample datasets as shown in the many examples in the text. Moreover, some code segments and functions reflecting the essential parts of each of the algorithms are also provided in the text to serve two purposes. First, according to the author's own experience, seeing how an algorithm is actually implemented in code makes it more efficient to thoroughly learn the algorithm. Second, the provided code segments make the required coding homework less challenging, so that the students do not have to spend too much time on developing their code from scratch. However, a proper tradeoff needs to be made in terms of how much code to provide. It is hoped that the amount of code in the text achieves a reasonable balance between the two extremes of providing complete code and no code at all.

Most homework problems require students to develop their own code to implement the algorithms presented in the chapters. For those with less programming experience, MATLAB is recommended, as it allows them to focus on the conceptual and mathematical aspects of the algorithms without being distracted by issues such as computational efficiency or numerical precision. More experienced students, however, are encouraged to use Python, the general-purpose language most widely adopted for machine learning in industry. Through this process the students should gain not only the in-depth understanding of the theoretical background of the algorithms in terms of why and how they work, but also first-hand experiences of the implementation of these algorithms in terms of what type of problems an algorithm is designed to solve (or unable to solve), the pros and cons of a specific algorithm in comparison to other similar algorithms, and how to choose the proper values for the parameters of the algorithm for it to achieve the best performance. Code of inexperienced students may not be professionally developed with high computational efficiency and accuracy and good coding style, but once they have gained the insights of the algorithms, they can always take full advantage of the professionally developed commercial software packages and apply them properly to solve real problems more effectively. Moreover, they may potentially also be able to develop their own algorithms in the future to best solve specific problems they may have at hand.

For the students to test their own code by implementing various algorithms discussed in the text, they need to apply their code to some datasets, such as those used for the examples in the book. Some of the datasets can actually be generated by the code provided in the homework, while a few others are provided on a CUP website for the book at www.cambridge.org/wang-iml. The size and scale of such datasets may be too small to be practical, but they serve the purpose of debugging the code while avoiding a long execution time if the code is applied to a large-scale dataset. Once a piece of code has been thoroughly debugged and tested, it can be applied to more realistic datasets of much larger scale, such as many of those publically available on the internet for testing and benchmarking ML algorithms. The readers are highly encouraged to do so.

Structure and Organization of the Book

The book is organized in seven parts as listed in the following, although not all of them need to or can be covered in a typical one-semester course. Depending on the students' background, the chapters in Part I can be covered in the first few lectures dedicated to the background mathematics for the later chapters if the students are not already familiar with the topics, or, alternatively, these topics may only need to be quickly reviewed whenever they are needed while discussing the various ML algorithms if the students are knowledgeable enough in these subjects.

Part II on regression needs to be taught early on before other topics, not only because the mathematics is relatively more straightforward than that in the later chapters, but also some basic issues discussed in this part (such as bias-variance tradeoff and regularization in Sections 4.2 and 4.3) are relevant in general to most learning methods covered in later chapters.

Part III on feature extraction can be treated as the preprocessing stage for the main learning algorithms to be discussed in Parts IV and V. Finally, Parts VI and VII as two independent and slightly more advanced units can be covered in the later part of the course after the previous chapters are covered as their background.

- Part I Mathematical foundations

  This part is mostly dedicated to optimization, covering constrained as well as unconstrained methods, both of essential importance in ML, as many ML algorithms are formulated as an optimization problem to either minimize the difference between the output of a ML model and some observed training data, or to maximize the likelihood of the model parameters given the training data. Most ML textbooks do not discuss how such optimization methods are actually carried out, assuming the readers are already familiar with the subject or will pick up such knowledge on their own. However, optimization is a big subject of in its own right, and a reader who is unfamiliar with it may not even know where to start. For this reason, the necessary features of optimization that are relevant to ML algorithms are covered in Part I for the convenience of the reader.

  Also included in Part I is the topic of solving equation systems, which is one of the most fundamental computational operations for many algorithms, and it is also most closely related to optimization, in the simple sense that a variable x that maximizes function $f(x)$ can be found by solving the equation $f'(x) = 0$, by, for example, the most widely used Newton's method.
- Part II Regression

  This part covers the most important regression algorithms widely used in ML, including linear and nonlinear regression, logistic and softmax regression, and

Gaussian process regression. All such regression methods are closely related to classification, the core problem in ML, in the most straightforward sense that by simply thresholding the regression function, its domain is divided into two regions corresponding to two classes, that is, the regression method can be readily used as a binary classifier. Moreover, a more sophisticated classifier can be obtained if the regression function is converted into the probability for a data sample to belong to one of the classes. This way we can not only classify a data sample into a class but also gain the confidence or certainty of doing so.

- Part III Feature selection
  This part is about feature selection and dimensionality reduction, which can be considered as the preprocessing stage of the main ML operations of classification or clustering. The core method of this part is principal component analysis (PCA), by which the dimensionality of the dataset can be reduced significantly, while the information pertaining to classification or clustering contained in the data is mostly conserved. We also consider a set of variations of the PCA methods, such as kernel PCA and probabilistic PCA. In this part we also consider the method of independent component analysis (ICA), by which the independent signal components can be restored from the mixture of their linear combinations.
- Part IV Classification
  This part covers a set of core classification methods widely used in ML, including various supervised methods based on the training dataset containing a set of data samples each labeled by its class identity, such as the k-nearest neighbors (KNN) method and the minimum distance classifier, the adaptive boosting (AdaBoost), the naive Bayesian classifier, the support vector machine (SVM) with kernel mapping, and Gaussian process classifier, both binary and multiclass, and the decision tree method. In this part we also discuss unsupervised clustering methods without any training data, such as k-means method, and methods based on Gaussian and Bernoulli mixture models. All such classification and clustering can be considered as the partitioning of the multidimensional feature space into multiple regions, each corresponding to one of the classes or clusters.
- Part V Neural networks
  In Part V, we introduce a set of important algorithms based on artificial neural networks inspired by the biological neural network in the brain, including the Hebbian learning for association between two sets of patterns, the Hopfield network for self-association, the perceptron network based on kernel mapping for both binary and multiclass classification, and the most popular backpropagation method for multiclass classification. In this part we also briefly discuss the idea of deep learning based on multilayer backpropagation learning, typified by the convolutional neural network (CNN) for image object recognition. In addition, we also discuss unsupervised neural network algorithms such as competitive learning and the closely related self-organizing maps (SOM).

  We note that although the neural network methods covered in this part may seem conceptually very different from those regression and classification methods covered in previous parts, they may all look similar in the sense that the mathematical model of the learning taking place in a single layer of a neural network is actually the same as a linear regression followed by a nonlinear mapping function, called activation function in the context of neural networks, such as the logistic mapping considered in Part II.

- Part VI Reinforcement learning
  This part provides an introduction to reinforcement learning (RL), an big area of active and ongoing study that has attracted a lot of interest in both research and applications. RL is considered as the third type of learning methods fundamentally different from the two traditional of learning methods, the supervised learning (regression and classification) and unsupervised learning (clustering). Instead of assigning data points in the feature space into different classes or clusters as in supervised or unsupervised learning, RL aims at optimizing a sequence of actions for an agent to take for the purpose of receiving maximum returns from a given environment, based on its interaction with the environment. In this sense, RL can also be considered as a different type of optimization achieved by unsupervised interactive learning (without labeled training data). When combined with deep learning, RL has achieved amazing results in recent years, the most famous example being the software AlphaGo, which beat the best human player of the board game Go. Also, reinforcement learning is used in large language models (LLMs) primarily during **fine-tuning** to align the model's behavior with human preferences – a process called: **Reinforcement Learning from Human Feedback (RLHF)**
- Part VII Large language models
  This final part of the book discusses the latest advancements in the development of large language models in the past few years. All of these models are based on the fundamental Transformer architecture first introduced in 2017, which, as a major breakthrough, revolutionized the field of natural language processing. Central to this transformation is the principal idea that "attention is all you need," allowing models to dynamically focus on relevant parts of the input data. This chapter delves into the intricacies of LLMs, exploring the underlying principles of backpropagation learning that enable these models to optimize their performance effectively. We will examine various important concepts, such as self and cross-attention and transfer learning, and explore the issues of emergent properties and machine understanding. We will also briefly introduce several important generative neural network models developed in recent years, beyond the Transformer models, for the general task of generating novel data samples based on patterns learned from existing datasets. These models have expanded the capabilities of machine learning to create novel and realistic synthetic data, including images, text, audio, and video, and have found applications across a wide range of fields.

Online resources

In addition to all the materials in the book, there is a dedicated Cambridge University Press website at [www.cambridge.org/wang-iml] that provides students and instructors with supplementary materials and practical tools to enhance learning and instruction, thereby deepening their understanding of concepts and algorithms. These additional resources include the following:

- Two mathematical appendices that provide background support for the machine learning algorithms discussed in the main text: Appendix A, *A Review of Linear Algebra*, and Appendix B, *A Review of Probability and Statistics*. These appendices revisit essential concepts and formulae relevant to specific algorithms, while also offering a concise overview of foundational topics from these broader fields.
- PowerPoint lecture slides covering the most essential topics discussed in the chapters. Instructors adopting the book can use these slides as a foundation to further develop their own more comprehensive and detailed lecture materials.
- Sample datasets used in the book's examples, which students can use for the coding exercises at the end of each chapter. By comparing their results with those presented in the book, students can evaluate their implementations and gain deeper insights.

Acknowlegement

Finally, I would like to express my sincere gratitude to Hunter Whaples, a former student in my machine learning class at Harvey Mudd College, now a PhD student at MIT, who created a set of lecture slides for the main chapters of the book.

The cover image (by Daniel Berger) is a color-coded map of one cubic millimeter of human brain tissue, illustrating 4,000 axons connecting to a single neuron. This visualization was created through a collaboration that combined electron microscopy imaging by a Harvard research team led by Jeff Lichtman with AI algorithms developed by Google Research to color-code and reconstruct the brain's complex wiring.

Part I

Mathematical Foundations

Many machine learning (ML) problems can be formulated as an optimization problem to build a model of a set of observed data, called *training dataset* or simply *training set*, so that the model output matches the data in an optimal manner in a certain sense. Typically this is done by either minimizing or maximizing an *objective function* of a set of model parameters, which measures the goodness of the model. For example, in either regression or classification, the goal of a certain type of learning algorithms is to minimize the error of the model, the difference between the model prediction and the given observed data, measured in a certain way, such as the squared error. In other types of algorithms, the goal is to maximize the *likelihood* of the model parameters, defined as the probability of observing the given data conditioned on these parameters.

In the typical *supervised learning* paradigm, the modeling process can be formulated so as to construct a function $y = f(\mathbf{x}, \boldsymbol{\theta})$ of some d variables as the components of a vector $\mathbf{x} = [x_1, \cdots, x_d]^T$, parameterized by a set of M parameters as the components of a vector $\boldsymbol{\theta} = [\theta_1, \cdots, \theta_M]^T$. The goal is for this function to best fit a set of N data points in the training set, denoted by $\mathcal{D} = \{(\mathbf{x}_n, y_n), \quad n = 1, \cdots, N\}$. While the form of the model function is usually assumed to be known based on some prior knowledge, such as a polynomial (e.g., linear or quadratic) function, or certain *probability density distribution (pdf)* of the data, the values of the M parameters in $\boldsymbol{\theta}$ are unknown and to be estimated based on the training set $\mathcal{D}$. The estimation can be carried out by various methods such as the following two that are widely used in ML.

- *Least squares estimation (LSE)*
  The sum of squared error (SSE) between the model output and the desired output according to the training set is to be minimized

$$J_{sse}(\boldsymbol{\theta}) = \sum_{n=1}^{N} [y_n - f((\mathbf{x}_n, \boldsymbol{\theta}))]^2 \tag{0.1}$$

- *Maximum likelihood estimate (MLE)*
  The likelihood of the parameter $\boldsymbol{\theta}$, the conditional probability of observing the data in $\mathcal{D}$ given a specific parameter $\boldsymbol{\theta}$, is to be maximized

$$J_{mle}(\boldsymbol{\theta}) = L(\boldsymbol{\theta}|\mathcal{D}) \propto p(\mathcal{D}|\boldsymbol{\theta}) = \prod_{n=1}^{N} p(y_n|\mathbf{x}_n, \boldsymbol{\theta}) \tag{0.2}$$

Solving either the minimization or maximization problem we obtain the optimal model parameter $\boldsymbol{\theta}^*$ for the model to best fit the observed data.

To prepare for the discussions of various ML algorithms in future parts of this book, such as how to find specifically the optimal $\boldsymbol{\theta}^*$ that minimizes/maximizes function $J(\boldsymbol{\theta})$ just discussed, in this part we will consider a set of typical methods for the optimization problem in the most generic form, and find the optimal solution $\mathbf{x}^* = [x_1^2, \cdots, x_N^*]^T$ that either maximizes or minimizes a real valued multivariable function $y = f(\mathbf{x}) = f(x_1, \cdots, x_N)$. As maximizing $f(\mathbf{x})$ is equivalent to minimizing $-f(\mathbf{x})$, we typically only need to consider the minimization problem to find the optimal solution $\mathbf{x}^*$ that minimizes $f(\mathbf{x})$, denoted by

$$\mathbf{x}^* = \arg\min_{\mathbf{x}} f(\mathbf{x}), \qquad \text{i.e.,} \qquad f(\mathbf{x}^*) = \min_{\mathbf{x}} f(\mathbf{x}) \leq f(\mathbf{x}) \tag{0.3}$$

Here the function $f(\mathbf{x})$ can be either linear or nonlinear, and an optimization problem is either *unconstrained* if the independent variable $\mathbf{x}$ of the function is allowed to take any value in the function domain $\mathbb{R}^N$, or *constrained* if $\mathbf{x}$ must be inside a subset of $\mathbb{R}^N$, called the *feasible region*. In ML, both linear and nonlinear methods, either constrained or not, are widely used. For example, the linear least squares method (LLS) for regression (to minimize the squared error such as that given in Eq. (0.1), Section 5.1) is solved as a linear unconstrained optimization problem, whereas the method of support vector machine (SVM) for classification (to maximize the margin between two classes, Section 14.1) is solved as a nonlinear constrained optimization problem.

In the following, we will review some necessary mathematical topics as the foundation for many of the learning methods to be discussed in future parts, including the numerical methods for solving equations in Chapter 1, which are not only important in their own right, but also most closely related to both unconstrained and constrained optimization problems in Chapters 2 and 3, which in turn will be needed for many important ML algorithms to be discussed in the upcoming chapters.

1 Solving Equations

In this chapter, we consider some basic methods for solving an equation system in the most general form of M simultaneous equations of N variables

$$\begin{cases} f_1(x_1,\cdots,x_N) = f_1(\mathbf{x}) = 0 \\ \cdots\cdots\cdots \\ f_M(x_1,\cdots,x_N) = f_M(\mathbf{x}) = 0 \end{cases} \tag{1.1}$$

Here $\mathbf{x} = [x_1,\cdots,x_N]^T \in \mathbb{R}^N$ is an N–D column vector representing a point in the N–D vector space $\mathbb{R}^N$ spanned by $\{x_1,\cdots,x_N\}$, and $yf_m(\mathbf{x})$ $(m = 1,\cdots,M)$ is a scalar valued multivariate function that maps any point in its domain $\mathbf{x} \in \mathbb{R}^N$ to a scalar value y. This equation system can be represented more concisely in vector form

$$\mathbf{f}(\mathbf{x}) = \mathbf{0} \tag{1.2}$$

where $\mathbf{f} = [f_1,\cdots,f_M]^T$ is a column vector containing the M functions as components and $\mathbf{0} = [0,\cdots,0]^T$ is a vector containing M zeros. Solving this equation system means to finding a solution $\mathbf{x}^* \in \mathbb{R}^N$ that satisfies all M equations. If such a solution does not exist, we would still like to find an approximate solution by which certain error (e.g., the sum of squared error (SSE) $||\mathbf{f}(\mathbf{x})||^2$), is minimized.

Geometrically, a function $y = f(\mathbf{x})$ defined over domain $\mathbb{R}^N$ is a *hypersurface* in an $N+1$ dimensional space, of which the $N+1$st dimension represents the function value $y = f(\mathbf{x})$ corresponding to a point $\mathbf{x} \in \mathbb{R}^N$. Specially the hypersurface is a curve if $N = 1$ or a surface if $N = 2$. The *roots* or *zeros* of function $f(\mathbf{x})$, that is, the solutions of equation $f(\mathbf{x}) = 0$, are composed of all points on the intersection, if it exists, of the hypersurface $y = f(\mathbf{x})$ and the hyperplane $y = 0$.

For example, in the special case of $M = N = 2$, functions $f_1(x_1,x_2)$ and $f_2(x_1,x_2)$ are two surfaces defined over the 2-D space spanned by x_1 and x_2, and the roots of each of the two functions are on a curve as the intersection of the corresponding surface and the 2-D space. The solutions of the system composed of the two equations $f_1 = 0$ and $f_2 = 0$ are the intersection of these two curves in the 2-D space if they do intersect, otherwise no solution exists.

Example 1.1

Consider a simultaneous equation system with $M = N = 2$

$$\begin{cases} f_1(x_1, x_2) = x_1^2 + x_2^2 - 2 = 0 \\ f_2(x_1, x_2) = x_1 + x_2 - C = 0 \end{cases}$$

The first function $f_1(x_1, x_2)$ is a parabolic cone in the $N + 1 = 3$ dimensional space centrally symmetric to the vertical axis, and its roots form a circle $x_1^2 + x_2^2 = 2$ on the 2-D plane spanned by x_1 and x_2 centered at the origin (0, 0) with radius 1; the second function $f_2(x_1, x_2)$ is a plane in the 3-D space through the origin, and its roots form a straight line $x_2 = C - x_1$ on the 2-D plane. The solutions of the equation system are where the two curves intersect:

- if $C = 0$, there are two solutions $(1, -1)$ and $(-1, 1)$;
- if $C = 2$, there is only one solution $(1, 1)$;
- if $C = 3$, the two curves do not intersect, that is, no solution exists.

1.1 Linear Equation Systems

Consider a linear equation system of M equations and N variables

$$\begin{cases} f_1(\mathbf{x}) & = f_1(x_1, \cdots, x_N) & = \sum_{j=1}^N a_{1j}x_j - b_1 = 0 \\ \cdots & \cdots & \cdots \\ f_M(\mathbf{x}) & = f_1(x_1, \cdots, x_N) & = \sum_{j=1}^N a_{Mj}x_j - b_M = 0 \end{cases} \tag{1.3}$$

which can be expressed in vector form

$$\mathbf{f}(\mathbf{x}) = \mathbf{A}\mathbf{x} - \mathbf{b} = \mathbf{0} \tag{1.4}$$

where

$$\mathbf{f}(\mathbf{x}) = \begin{bmatrix} f_1(\mathbf{x}) \\ \vdots \\ f_M(\mathbf{x}) \end{bmatrix}, \quad \mathbf{A} = \begin{bmatrix} a_{11} & \cdots & a_{1N} \\ \vdots & \ddots & \vdots \\ a_{M1} & \cdots & a_{MN} \end{bmatrix}, \quad \mathbf{x} = \begin{bmatrix} x_1 \\ \vdots \\ x_N \end{bmatrix}, \quad \mathbf{b} = \begin{bmatrix} b_1 \\ \vdots \\ b_M \end{bmatrix} \tag{1.5}$$

In general, the existence and uniqueness of the solution can be determined based on the *fundamental theorem of linear algebra* (A.349 in Section A.5), depending on the rank R of the coefficient matrix $\mathbf{A}$, and whether the system is underdetermined (underconstrained) if $M < N$, or overdetermined (overconstrained) if $M > N$.

If $\mathbf{A}$ is a full rank square matrix with $R = M = N$, then its inverse $\mathbf{A}^{-1}$ exists and the system has a unique solution $\mathbf{x} = \mathbf{A}^{-1}\mathbf{b}$. But if $M > N = R$ and $\mathbf{b}$ is not in the column space of $\mathbf{A}$, the system is overconstrained with no solutions. In this case we can still find an optimal approximate solution that minimizes the SSE, called *objective function* in general optimization problems, defined as

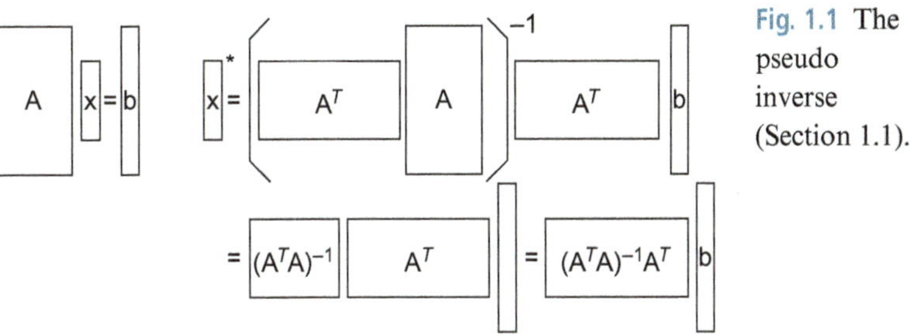

Fig. 1.1 The pseudo inverse (Section 1.1).

$$\begin{aligned}\varepsilon(\mathbf{x}) &= \frac{1}{2}||\mathbf{r}||^2 = \frac{1}{2}\mathbf{r}^T\mathbf{r} = \frac{1}{2}(\mathbf{b}-\mathbf{A}\mathbf{x})^T(\mathbf{b}-\mathbf{A}\mathbf{x}) \\ &= \frac{1}{2}\left(\mathbf{b}^T\mathbf{b} - \mathbf{x}^T\mathbf{A}^T\mathbf{b} - \mathbf{b}^T\mathbf{A}\mathbf{x} + \mathbf{x}^T\mathbf{A}^T\mathbf{A}\mathbf{x}\right) \end{aligned} \tag{1.6}$$

where $\mathbf{r} = \mathbf{b}-\mathbf{A}\mathbf{x}$ is the *residual* of the system, and $||\mathbf{r}||^2 = \sum_{m=1}^M r_m^2$ (thereby the term sum of squares error). The optimal solution $\mathbf{x}^*$ that minimizes $\varepsilon(\mathbf{x})$ can be found by setting its derivative with respect to $\mathbf{x}$ to zero (see Section A.4.3 for differentiation with respect to a vector variable)

$$\begin{aligned}\frac{d}{d\mathbf{x}}\varepsilon(\mathbf{x}) &= \frac{1}{2}\frac{d}{d\mathbf{x}}||\mathbf{r}||^2 = \frac{1}{2}\frac{d}{d\mathbf{x}}\left(\mathbf{b}^T\mathbf{b} - \mathbf{x}^T\mathbf{A}^T\mathbf{b} - \mathbf{b}^T\mathbf{A}\mathbf{x} + \mathbf{x}^T\mathbf{A}^T\mathbf{A}\mathbf{x}\right) \\ &= -\mathbf{A}^T\mathbf{b} + \mathbf{A}^T\mathbf{A}\mathbf{x} = \mathbf{0} \end{aligned} \tag{1.7}$$

Solving this matrix equation we get the optimal approximate solution, as illustrated in Fig. 1.1:

$$\mathbf{x}^* = \left(\mathbf{A}^T\mathbf{A}\right)^{-1}\mathbf{A}^T\mathbf{b} = \mathbf{A}^-\mathbf{b} \tag{1.8}$$

where $\mathbf{A}^- = \left(\mathbf{A}^T\mathbf{A}\right)^{-1}\mathbf{A}^T$ is the *pseudo-inverse* (see Section A.6.1) of the non-square matrix $\mathbf{A}$. Here we assume the M equations are independent, that is, the $N \times N$ square matrix $\mathbf{A}^T\mathbf{A}$ has a full rank $R = N$ (i.e., it is nonsingular and invertible with all eigenvalues being non-zero).

If the M equations are barely independent, that is, some eigenvalues of $\mathbf{A}^T\mathbf{A}$ are very close to zero, then it is near-singular but still invertible, and some eigenvalues of its inverse $(\mathbf{A}^T\mathbf{A})^{-1}$ may take huge values. Consequently the result in Eq. (1.8) may vary greatly due to some small random fluctuation in either $\mathbf{A}$ or $\mathbf{b}$. In this case, the system is said to be *ill-conditioned* or *ill-posed*, as it is prone to noise, that is, its solution is highly sensitive to noise and therefore unreliable and unstable.

Such an ill-posed problem can be addressed by *regularization*, so that the solution $\mathbf{x}$ is forced to avoid taking huge values and is therefore less sensitive to noise. This can be done by including in the objective function an additional penalty term for large values of $\mathbf{x}$ scaled by a parameter λ

$$J(\mathbf{x}) = \frac{1}{2}||\mathbf{r}||^2 + \frac{\lambda}{2}||\mathbf{x}||^2 \tag{1.9}$$

By minimizing $J(\mathbf{x})$, we can obtain a solution $\mathbf{x}$ of small norm $||\mathbf{x}||$ as well as a small error $\varepsilon(\mathbf{x})$. Just as before, the solution $\mathbf{x}$ can be obtained by setting the derivative of $J(\mathbf{x})$ to zero

$$\frac{d}{d\mathbf{x}}J(\mathbf{x}) = -\mathbf{A}^T\mathbf{b} + \mathbf{A}^T\mathbf{A}\mathbf{x} + \lambda\mathbf{x} = \mathbf{0} \tag{1.10}$$

and solving the resulting equation to obtain

$$\mathbf{x}^* = (\mathbf{A}^T\mathbf{A} + \lambda\mathbf{I})^{-1}\mathbf{A}^T\mathbf{b} \tag{1.11}$$

We see that even if $\mathbf{A}^T\mathbf{A}$ is near singular, matrix $\mathbf{A}^T\mathbf{A}+\lambda\mathbf{I}$ is not due to the additional term $\lambda\mathbf{I}$.

Here λ is called the *hyperparameter*. By adjusting it we can make a proper tradeoff between accuracy and stability.

- Small λ: the solution is more accurate but less stable as it is more prone to noise, that is, the variance error may be large. This is called *overfitting*.
- Large λ: the solution is less accurate but more stable as it is less affected by noise. This is called *underfitting*.

The issue of overfitting versus underfitting is of essential importance in machine learning (ML) in general; it will appear repeatedly and be specifically addressed while discussing various regression and classification algorithms in the later chapters.

1.2 The Bisection and Secant Methods

Most equation systems of interest are nonlinear. Unlike the linear equation systems, which can be solved under the guidance of the fundamental theorem of linear algebra, for nonlinear equation systems, there exists neither a theory regarding the existence and uniqueness of their solutions, nor closed-form solutions in general. We therefore have to rely on numerical methods to find one or more solutions, if they do exist, in an iterative manner from certain initial guess of the solution, if the iteration converges, that is, the difference between two consecutive results eventually approach to zero.

We first consider some methods that find a solution of a single-variable nonlinear equation $f(x) = 0$, by searching iteratively through a neighborhood of the domain, in which a solution is known to be located.

1.2.1 The Bisection Search

This method requires two initial guesses $x_0 < x_1$ satisfying $f(x_0)f(x_1) < 0$. As $f(x_0)$ and $f(x_1)$ are on opposite sides of the x-axis $y = 0$, the solution x^* at which $f(x^*) = 0$ must reside somewhere in between these two guesses, that is, $x_0 < x^* < x_1$.

Given such two end points x_0 and x_1, we could find any point x_2 in between and use it to replace one of the end points at which the function value $f(x)$ has the same sign as that of $f(x_2)$. The new search interval is $a = x_2 - x_0$ if $f(x_2)f(x_1) > 0$, or $b = x_1 - x_2$ if $f(x_0)f(x_2) > 0$, either of which is smaller than the previous interval

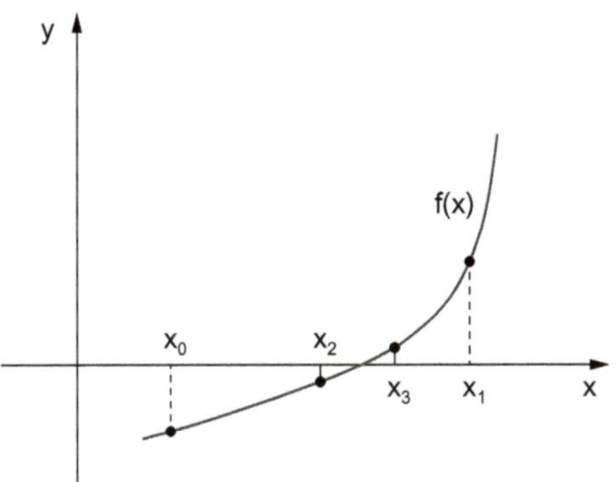

Fig. 1.2 Bisection search.

$a + b = x_1 - x_0$. This process is then carried out iteratively until the solution is eventually approached, with a guaranteed convergence.

To avoid the worst possible case in which the solution always happens to be in the larger of the two sections a and b, we typically choose x_2 to be the middle point $x_2 = (x_0 + x_1)/2$ between the two end points, so that $a = b = (x_1 - x_0)/2$, that is, the new search interval is always halved in each step of the iteration

$$x_{n+1} = \frac{x_n + x_{n-1}}{2} \tag{1.12}$$

as illustrated in Fig. 1.2. Here, without loss of generality, we have assumed $f(x_{n-1})f(x_{n+1}) > 0$ and x_{n-1} is replaced by x_{n+1}.

For example, if we assume that the root is located at $x^* = 2.2$ between the two initial values $x_0 = 0$ and $x_1 = 8$, then we get

Table 1.1

n	0	1	2	3	4	5	6	7	8
x_n	0	8	4	2	3	2.5	2.25	2.125	...
$\|e_n\|$	2.2	5.8	1.8	0.2	0.8	0.3	0.05	0.075	...

Note that the error $|e_n| = |x_n - x^*|$ does not necessarily always reduce monotonically. However, as in each iteration the search interval is always halved, the worst possible error is also halved

$$|e_{n+1}| = |x_{n+1} - x^*| \leq \frac{|x_n - x_{n-1}|}{2} = \frac{|x_0 - x_1|}{2^n} \approx \frac{|e_n|}{2} \tag{1.13}$$

In general, to measure how quickly or slowly an iterative method converges, we consider the following limit of the ratio of errors of two consecutive iterations

$$\lim_{n\to\infty} \frac{|e_{n+1}|}{|e_n|^q} = \lim_{n\to\infty} \frac{|e_{n+1}|}{|e_n|} \leq \mu \tag{1.14}$$

where q is the *order of convergence* and μ is the *rate of convergence*. The higher the order q, the more rapidly the iteration converges. For the bisection method we have

$$\lim_{n\to\infty} \frac{|e_{n+1}|}{|e_n|^q} = \lim_{n\to\infty} \frac{|e_{n+1}|}{|e_n|} \leq \mu = \frac{1}{2} \tag{1.15}$$

We see that its order of convergence is $q = 1$ (it converges linearly) with the rate of convergence $\mu = 1/2$. Compared to other methods to be considered later, the linear convergence of the bisection method is rather slow, but it has the advantage that no derivative $f'(x)$ of the given function is needed and the given function does not need to be differentiable.

1.2.2 The Secant Method

As in the bisection method, we assume there are two initial values x_0 and x_1 available, but they no longer have to satisfy $f(x_0)f(x_1) < 0$. Here the iteration is based on the zero-crossing of the secant line passing through the two points $f(x_0)$ and $f(x_1)$, instead of the middle point between them. The equation for the secant is

$$y = f(x_0) + \frac{x - x_0}{x_1 - x_0}(f(x_1) - f(x_0)) = f(x_0) + \hat{f}'(x_0)(x - x_0) \tag{1.16}$$

where $\hat{f}'(x_0)$ is the finite difference that approximates the derivative $f'(x_0)$

$$\hat{f}'(x_0) = \frac{f(x_1) - f(x_0)}{x_1 - x_0} = \frac{\Delta f(x_0)}{\Delta x_0} \overset{\Delta x_0 \to 0}{\Longrightarrow} f'(x_0) \tag{1.17}$$

Setting $y = 0$ in Eq. (1.17), we obtain the zero crossing point

$$x = x_0 - \frac{x_1 - x_0}{f(x_1) - f(x_0)} f(x_0) = x_0 - \frac{f(x_0)}{\hat{f}'(x_0)} \tag{1.18}$$

As shown in Fig. 1.3, the new value x just obtained is a better estimate of the root than either x_0 or x_1, and is used to replace one of them.

- If $f(x_0)f(x_1) < 0$, then x_i $(i = 0, 1)$ satisfying $f(x_i)f(x) > 0$ is replaced by $x_2 = x$ (same as the bisection method).
- If $f(x_0)f(x_1) > 0$, then x_i $(i = 0, 1)$ with greater $|f(x_i)|$ is replaced by $x_2 = x$.

This process is then carried out iteratively to approach the root

$$x_{n+1} = x_n - \frac{f(x_n)}{\hat{f}'(x_n)} \tag{1.19}$$

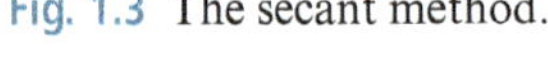
Fig. 1.3 The secant method.

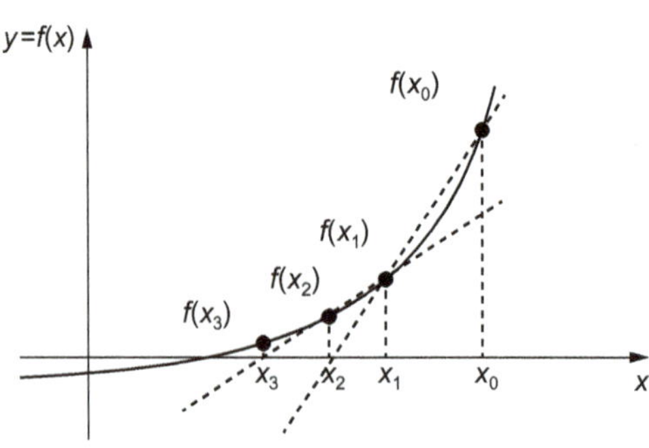

If the derivative $f'(x)$ is available, it can be used to replace its approximation $\hat{f}'(x_n)$ and the iteration becomes the Newton–Raphson method (Section 1.4)

$$x_{n+1} = x_n - \frac{f(x_n)}{f'(x_n)} \tag{1.20}$$

In case $f(x_n) = f(x_{n-1})$, the approximated derivative is zero $\hat{f}'(x_n) = 0$, and x_{n+1} cannot be found. In such a case, a root may exist between x_n and x_{n-1}. Therefore the way to resolve this problem is to combine the bisection search with the secant method so that when $\hat{f}' = 0$, the algorithm will switch to a bisection search. This is Dekker's method

$$x_{n+1} = \begin{cases} x_n - f(x_n)/\hat{f}'(x_n) & \text{if } f(x_n) \neq f(x_{n-1}) \\ (x_n + x_{n-1})/2 & \text{otherwise} \end{cases} \tag{1.21}$$

The order of convergence of the secant method is $q = 1.618$ as shown in Appendix 1 of this chapter, that is, the secant method converges more quickly than the bisection search method with $q = 1$ (which does not take advantage of the information of the specific function $f(x)$).

1.3 Fixed-Point Iteration

Fixed-point iteration is a method of finding the fixed point of a given function. It can be used to solve the generic equation system $\mathbf{f}(\mathbf{x}) = \mathbf{0}$, as seen in this section. In particular, it will be used to solve the Bellman equation in dynamic programming, as we will see in Chapter 20.

We first consider the special case of solving a single-variable nonlinear equation $f(x) = 0$ ($M = N = 1$). To find a solution x^* that satisfies the equation, we could first convert it into an equivalent equation $g(x) = x$, so that an x satisfying one of the equations will also satisfy the other. We then carry out an iteration $x_{n+1} = g(x_n)$ from some initial value x_0. If the iteration converges at a point x^*, that is, $g(x^*) = x^*$, then we also have $f(x^*) = 0$ (i.e., x^* is a solution of the equation $f(x) = 0$). Consider the following two examples.

Example 1.2

$$f(x) = \log(x) - 0.5 = 0$$

To solve the equation, we construct another function

$$g_1(x) = x - f(x) = x - \log(x) + 0.5$$

so that the equation $g(x) = x$ is equivalent to the given equation $f(x) = 0$ (left Fig. 1.4). We then carry out the iteration $x_{n+1} = g(x_n)$ from an initial value, such as $x_0 = 0.5$, and obtain

Table 1.2

n	x_n	$g(x_n)$	$f(x_n)$
0	0.50000	1.69315	−1.19315
1	1.69315	1.66656	0.02659
2	1.66656	1.65580	0.01076
3	1.65580	1.65152	0.00428
4	1.65152	1.64982	0.00169
5	1.64982	1.64915	0.00067
6	1.64916	1.64889	0.00026
7	1.64889	1.64879	0.00010
8	1.64879	1.64875	0.00004
9	1.64875	1.64873	0.00002
10	1.64873	1.64873	0.00001
11	1.64873	1.64872	0.00000
12	1.64872	1.64872	0.00000

We see that the iteration converges to $x^* = \lim_{n\to\infty} g(x_n) = 1.64872$ satisfying $g(x^*) = x^* - f(x^*) = x^*$, and, equivalently, x^* is also the solution of the given equation $f(x) = 0.5 - \log(x) = 0$

$$0.5 - \log(1.64872) = 0, \qquad \text{i.e.,} \qquad e^{0.5} = \sqrt{e} = 1.64872$$

Alternatively, we could construct a different function

$$g'(x) = x + f(x) = x + \log(x) - 0.5$$

also equivalent to $f(x) = 0$ (right Fig. 1.4). But this iteration no longer converges!

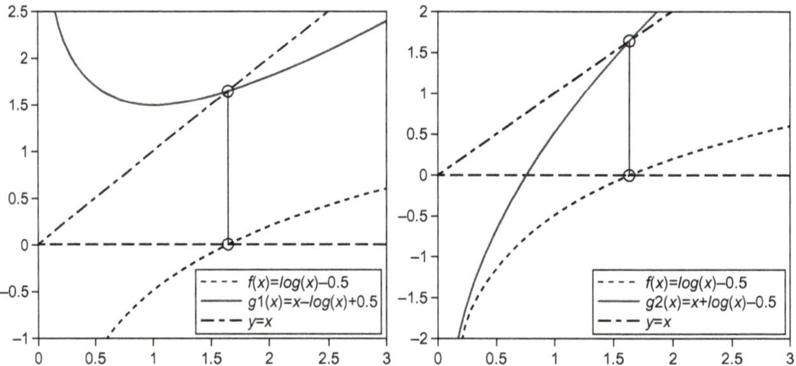

Fig. 1.4 Example 1.2: Iteration converges for $g(x)$ (left) but diverges for $g'(x)$ (right).

Example 1.3

$$f(x) = e^x - 1/x = 0$$

Taking a logarithm of the equation, we obtain an equivalent form (left in Fig. 1.5)

$$g(x) = x = e^{-x}$$

and an iteration

$$x_{n+1} = g(x_n) = e^{-x_n} \overset{n\to\infty}{\Longrightarrow} x^* = 0.5671$$

that converges to the root of the given equation $f(x) = e^x - 1/x = 0$, that is, $e^{-0.5671} = 0.5671$.

However, if the given equation is converted into an alternative form $g'(x) = x = -\ln(x)$ (right in Fig. 1.5), the iteration no longer converges.

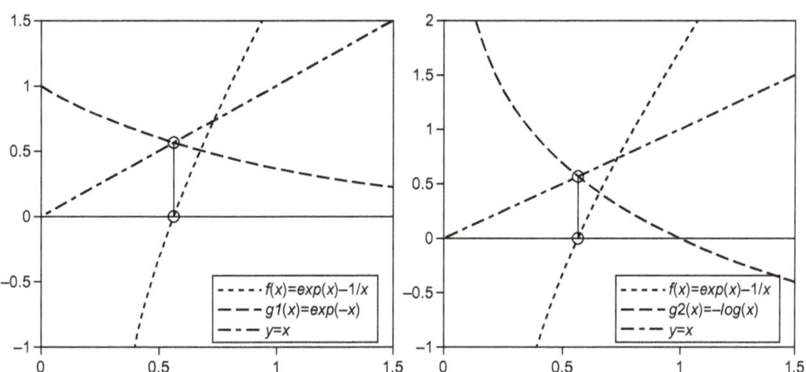

Fig. 1.5 Example 1.3: Iteration converges for $g(x)$ (left) but diverges for $g'(x)$ (right).

Why does this iterative method work in some cases but fail in others? To answer this we need to understand the theory behind the method, the *fixed point* of a *contraction function*.

If a single variable function $g(x)$ satisfies

$$|g(x_1) - g(x_2)| \leq k|x_1 - x_2|, \qquad k \geq 0 \tag{1.22}$$

it is said to be *Lipschitz continuous*, and k is a *Lipschitz constant*. If $k = 1$, then $g(x)$ is a *non-expansive* function, if $0 \leq k < 1$, then $g(x)$ is a *contraction function* or simply a *contraction*. These concepts can be generalized to solve multivariate equations $\mathbf{f}(\mathbf{x}) = \mathbf{0}$.

Definition: In a metric space V with certain distance $d(\mathbf{x}, \mathbf{y})$ defined between any two points $\mathbf{x}, \mathbf{y} \in V$, a function $\mathbf{g} : V \to V$ is a *contraction* if

$$d(\mathbf{g}(\mathbf{x}), \mathbf{g}(\mathbf{y})) \leq k\, d(\mathbf{x}, \mathbf{y}) \tag{1.23}$$

The smallest k value that satisfies this expression is called the *Lipschitz constant*.

Intuitively, a contraction reduces the distance between points in the space, that is, it brings them closer together. A function $\mathbf{g}(\mathbf{x})$ may not be a contraction throughout its entire domain, but it can be a contraction in the neighborhood of a certain point

$\mathbf{x}^* \in V$, composed of all points $\mathbf{x}$ satisfying $d(\mathbf{x},\mathbf{x}^*) < \epsilon > 0$ (i.e., $\mathbf{x}$ is a neighbor of $\mathbf{x}^*$ as defined by ϵ) so that

$$d(\mathbf{g}(\mathbf{x}),\mathbf{g}(\mathbf{x}^*)) \leq k\, d(\mathbf{x},\mathbf{x}^*), \qquad 0 \leq k < 1 \tag{1.24}$$

Definition: A *fixed point* $\mathbf{x}^*$ of a function $\mathbf{g}(\mathbf{x})$ is a point in its domain that is mapped to itself

$$\mathbf{x}^* = \mathbf{g}(\mathbf{x}^*) \tag{1.25}$$

We immediately have

$$\mathbf{g}(\mathbf{g}(\mathbf{x}^*)) = \mathbf{g}^2(\mathbf{x}^*) = \mathbf{x}^*, \cdots\cdots, \mathbf{g}(\mathbf{g}(\mathbf{g}(\cdots \mathbf{g}(\mathbf{x}^*)\cdots))) = \mathbf{g}^n(\mathbf{x}^*) = \mathbf{x}^* \tag{1.26}$$

A fixed point $\mathbf{x}^*$ is an *attractive fixed point* if any point $\mathbf{x}$ in its neighborhood converges to $\mathbf{x}^*$, that is, $\lim_{n\to\infty} \mathbf{g}^n(\mathbf{x}) = \mathbf{x}^*$.

We further consider two theorems regarding the fixed point and a contraction mapping.

Fixed-point theorem: *Let* $\mathbf{g}(\mathbf{x})$ *be a contraction satisfying*

$$d(\mathbf{g}(\mathbf{x}) - \mathbf{g}(\mathbf{y})) \leq k\; d(\mathbf{x} - \mathbf{y}) \qquad (0 \leq k < 1) \tag{1.27}$$

then there exists a unique fixed point $\mathbf{x}^* = \mathbf{g}(\mathbf{x}^*)$, *which can be found by an iteration from an arbitrary initial point* $\mathbf{x}_0$

$$\lim_{n\to\infty} \mathbf{x}_n = \lim_{n\to\infty} \mathbf{g}(\mathbf{x}_{n-1}) = \lim_{n\to\infty} \mathbf{g}^2(\mathbf{x}_{n-2}) = \cdots = \lim_{n\to\infty} \mathbf{g}^n(\mathbf{x}_0) = \mathbf{x}^* \tag{1.28}$$

Contraction mapping theorem: *Let* $\mathbf{x}^* = \mathbf{g}(\mathbf{x}^*)$ *be a fixed point of a differentiable function* $\mathbf{g}(\mathbf{x})$, *that is,* $\partial g_i/\partial x_j$ *exists for any* $1 \leq i,j \leq N$. *If the norm of the Jacobian matrix is smaller than 1,* $||\mathbf{J}_\mathbf{g}(\mathbf{x}^*)|| < 1$, *then* $\mathbf{g}(\mathbf{x})$ *is a contraction at* $\mathbf{x}^*$.

The Jacobian matrix of $\mathbf{g}(\mathbf{x})$ is defined as

$$\mathbf{g}'(\mathbf{x}) = \mathbf{J}_\mathbf{g}(\mathbf{x}) = \begin{bmatrix} \frac{\partial g_1}{\partial x_1} & \cdots & \frac{\partial g_1}{\partial x_N} \\ \vdots & \ddots & \vdots \\ \frac{\partial g_N}{\partial x_1} & \cdots & \frac{\partial g_N}{\partial x_N} \end{bmatrix} \tag{1.29}$$

The proofs of the two theorems just given are presented in Appendix 3 of the current chapter.

In the special case of a single-variable function $g(x)$ with $N = 1$, its Jacobian is simply its derivative $g'(x)$, and we have

$$g(x) - g(x^*) = g'(x^*)(x - x^*) + \frac{1}{2}g''(x^*)(x - x^*)^2 + R(x - x^*) \tag{1.30}$$

and

$$|g(x) - g(x^*)| \leq |g'(x^*)(x - x^*)| \leq |g'(x^*)|\;|(x - x^*)| \tag{1.31}$$

If $|g'(x)| < 1$, then $g(x)$ is a contraction at x^*.

Now we should understand why in the examples of the previous section the iteration leads to convergence in some cases but divergence in other cases: if $|g'(x)| < 1$,

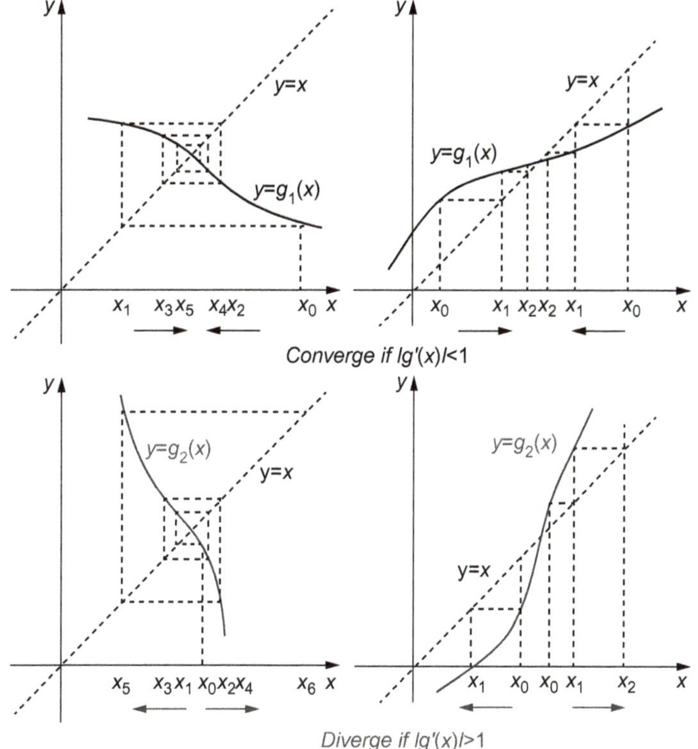

Fig. 1.6 Convergence (top) and divergence (bottom) of the fixed point method.

the iteration will converge to the root x^* of $f(x) = 0$, but if $|g'(x)| > 1$, it will never converge.

The iterative process $x_{n+1} = g(x_n)$ for finding the fixed point of a single-variable function $g(x)$ is shown in Fig. 1.6 as the intersections of the function $y = g(x)$ and the identity function $y = x$. If $|g'(x)| < 1$, the iteration converges as shown in the two cases on top, while if $|g'(x)| > 1$ it diverges as shown in the two cases at the bottom.

In general the fixed-point iteration converges linearly

$$\lim_{n\to\infty} \frac{|e_{n+1}|}{|e_n|} = \lim_{n\to\infty} \frac{|x_{n+1} - x^*|}{|x_n - x^*|} = \lim_{n\to\infty} \frac{|g(x_n) - g(x^*)|}{|x_n - x^*|} = |g'(x^*)| = \mu < 1 \tag{1.32}$$

However, if the iteration function $g(x)$ has zero derivative at the fixed point $g'(x^*) = 0$, we have

$$\begin{aligned} g(x) - g(x^*) &= g'(x^*)(x - x^*) + \frac{1}{2}g''(x^*)(x - x^*)^2 + R(x - x^*) \\ &= \frac{1}{2}g''(x^*)(x - x^*)^2 + R(x - x^*) \end{aligned} \tag{1.33}$$

and the iteration converges quadratically

$$\lim_{n\to\infty}\frac{|e_{n+1}|}{|e_n|^2}=\lim_{n\to\infty}\frac{|x_{n+1}-x^*|}{|x_n-x^*|^2}=\lim_{n\to\infty}\frac{|g(x_n)-g(x^*)|}{|x_n-x^*|^2}=\frac{1}{2}|g''(x^*)|=\text{constant} \tag{1.34}$$

Moreover, if $g''(x^*)=0$, then the iteration converges cubically.

Example 1.4

$$f(x)=x^3-x-2=0$$

This equation can converted into an equivalent form of $g(x)=x$ in two different ways

$$g_1(x)=x=\sqrt[3]{x+2},\qquad \text{and}\qquad g_2(x)=x=x^3-2$$

These two functions are plotted in Fig. 1.7, together with $f(x)$ and the identity function. The iteration based on $g_1(x)$ (left) converges to the solution $x^*=1.5214$ for any initial guess $-\infty<x_0<\infty$, as $g_1(x)$ is a contraction. However, the iteration based on $g_2(x)$ (right) does not converge to x^* as it is not a contraction in the neighborhood of x^*. In fact, the iteration will diverge toward either $-\infty$ if $x_0<x^*$ or ∞ if $x_0>x^*$.

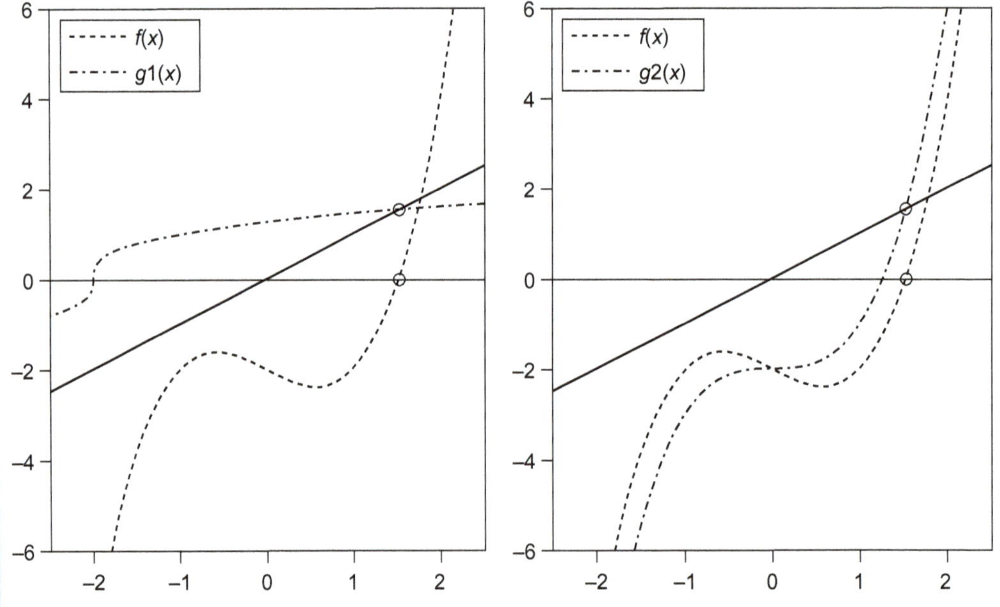

Fig. 1.7 Example 1.4, plots of $g_1(x)$ (left) and $g_2(x)$ (right).

Example 1.5

$$f(x)=x^3-3x+1=0$$

we first convert it into the form of $g(x)=x$ in two different ways

$$g_1(x) = x = \sqrt[3]{3x - 1}, \qquad g_2(x) = x = \frac{1}{3}(x^3 - 1)$$

As can be seen in Fig. 1.8, this equation has three solutions

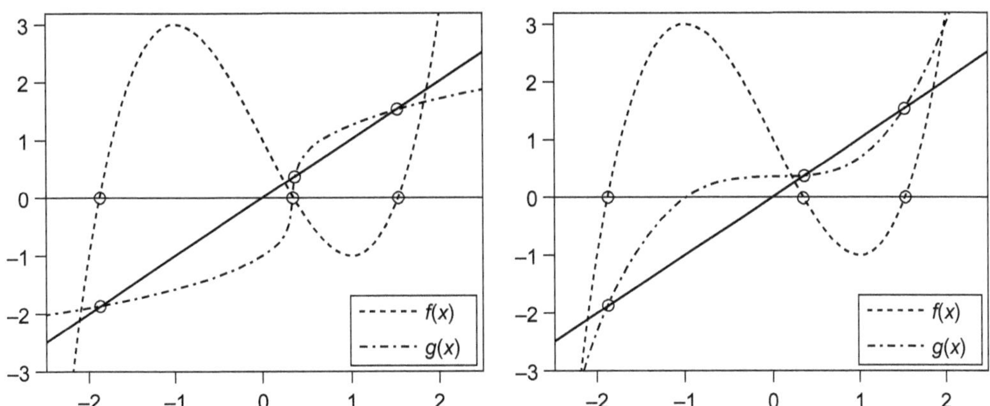

Fig. 1.8 Example 1.5, plots of $g_1(x)$ (left) and $g_2(x)$ (right).

$$x_1 = -1.8794, \qquad x_2 = 0.3473, \qquad x_3 = 1.5321$$

of which x_1 and x_3 can be obtained by the iteration based on $g_1(x)$ and x_2 can be obtained by the iteration based on $g_2(x)$. But neither of them can find all three roots.

- As shown in the plot on the left, $|g_1'(x)| < 1$ for all $-\infty < x < \infty$ except in the neighborhood of $x_2 = 0.3473$, that is, $g_1(x)$ is a contraction mapping everywhere except around x_2. Therefore the iteration starting from any initial guess x_0 will converge to either $x_1 = -1.8749$ if $x_0 < x_2$, or $x_3 = 1.5321$ if $x_0 > x_2$.

$$x_{n+1} = g_1(x_n) \overset{n\to\infty}{\Longrightarrow} \begin{cases} x_1 = -1.8749 & x_0 < x_2 \\ x_3 = 1.5321 & x_0 > x_2 \end{cases}$$

  However, as $g_1(x)$ is not a contraction mapping around $x = x_2$, the iteration will never converge to x_2.
- As shown in the plot on the right, $|g_2'(x)| > 1$ for all $-\infty < x < \infty$ except in the neighborhood of x_2, that is, $g_2(x)$ is not a contraction mapping around either x_1 or x_3. Therefore the iteration based on $g_2(x)$ will not converge to either x_1 or x_3, but it may converge to x_2, if the initial guess x_0 is in the range $x_1 < x_0 < x_3$. However, if x_0 is outside this range the iteration will diverge toward either $-\infty$ if $x_0 < x_1$ or ∞ if $x_0 > x_3$.

$$x_{n+1} = g_2(x_n) \overset{n\to\infty}{\Longrightarrow} \begin{cases} -\infty & x_0 < x_1 \\ x_2 = 0.3473 & x_1 < x_0 < x_3 \\ \infty & x_0 > x_3 \end{cases}$$

Table 1.3

n	x_n	$f(x_n)$
1	$-2.0408275e+00$	$-1.3775173e+00$
2	$-1.9240239e+00$	$-3.5041089e-01$
3	$-1.8919392e+00$	$-9.6254050e-02$
4	$-1.8829328e+00$	$-2.7019276e-02$
5	$-1.8803890e+00$	$-7.6311660e-03$
6	$-1.8796694e+00$	$-2.1590449e-03$
7	$-1.8794656e+00$	$-6.1114717e-04$
8	$-1.8794080e+00$	$-1.7301761e-04$
9	$-1.8793916e+00$	$-4.8983741e-05$
10	$-1.8793870e+00$	$-1.3868146e-05$
11	$-1.8793857e+00$	$-3.9263250e-06$
12	$-1.8793853e+00$	$-1.1116151e-06$

Example 1.6

$$f(x) = \sin(x) - x^3 = 0$$

This equation can be converted into the form $g(x) = x$ in different ways:

- $g_0(x) = x = \sqrt[3]{\sin(x)}, \quad g_0'(x) = \dfrac{\cos(x)}{3\,\sin(x)^{2/3}}$

Table 1.4

n	x_n	$f(x_n)$
0	$2.000000e+00$	-7.0907
1	$9.688027e-01$	-0.085089
2	$9.375886e-01$	-0.018075
3	$9.306842e-01$	-0.0041048
4	$9.291018e-01$	-0.00094611
5	$9.287364e-01$	-0.00021881
6	$9.286518e-01$	$-5.0647e-05$
7	$9.286322e-01$	$-1.1725e-05$
8	$9.286277e-01$	$-2.7144e-06$

- $g_1(x) = x = \sin(x)/x^2, \quad g_1'(x) = \dfrac{\cos(x)}{x^2} - \dfrac{2\,\sin(x)}{x^3}$
- $g_2(x) = x + \sin(x) - x^3, \quad g_2'(x) = \cos(x) - 3\,x^2 + 1$
- $g_3(x) = x - \dfrac{\sin(x) - x^3}{\cos(x) - 3x^2}, \quad g_3'(x) = -\dfrac{(6x + \sin(x))(\sin(x) - x^3)}{(\cos(x) - 3x^2)^2}$

Table 1.5

n	x_n	$f(x_n)$
0	$2.000000e+00$	-7.0907
1	$1.428913e+00$	-1.9276
2	$1.106787e+00$	-0.46152
3	$9.637850e-01$	-0.073886
4	$9.304466e-01$	-0.0036294
5	$9.286316e-01$	$-1.0509e-05$
6	$9.286263e-01$	$-8.9029e-11$

Example 1.7

- $f(x) = e^x - \dfrac{1}{x}$

  $$g_0(x) = x = e^{-x}, \quad g_0'(x) = e^{-x}$$

  The iteration from any initial guess $x_0 > 0$ will converge to $x^* = g_1(x^*) = 0.56714$.

- $g_1(x) = x = -\log(x), \quad g_1'(x) = 1/x$

  Around $x = x^* = 0.56714$, $g_1'(x) = 1/x > 1$, the iteration does not converge.

- $g_2(x) = x - \dfrac{e^x - 1/x}{e^x + 1/x^2}, \quad g_2'(x) = \dfrac{(e^x - 1/x)(e^x - 2/x^3)}{(e^x + 1/x^2)^2}$

Example 1.8

Consider a three-variable linear vector function $\mathbf{f}(\mathbf{x})$ of arguments $\mathbf{x} = [x,\ y,\ z]^T$

$$\mathbf{f}(\mathbf{x}) = \mathbf{0}, \qquad \begin{cases} f_1(\mathbf{x}) = 6x + 3y + 2z - 18 = 0 \\ f_2(\mathbf{x}) = 2x + 7y + 3z - 25 = 0 \\ f_3(\mathbf{x}) = x + 3y + 5z - 22 = 0 \end{cases}$$

from which the g-function can be obtained

$$\mathbf{g}(\mathbf{x}) = \mathbf{x}, \qquad \begin{cases} g_1(\mathbf{x}) = x = -(3y + 2z - 18)/6 \\ g_2(\mathbf{x}) = y = -(2x + 3z - 25)/7 \\ g_3(\mathbf{x}) = z = -(x + 3y - 22)/5 \end{cases}$$

The Jacobian $\mathbf{g}(\mathbf{x})$ of this linear system is a constant matrix

$$\mathbf{J}_g = \begin{bmatrix} 0 & -1/2 & -1/3 \\ -2/7 & 0 & -3/7 \\ -1/5 & -3/5 & 0 \end{bmatrix}$$

with the induced $p = 2$ norm (maximum singular value) $||\mathbf{J}_g|| = 0.851 < 1$. Consequently, the iteration $\mathbf{x}_{n+1} = \mathbf{g}(\mathbf{x}_n)$ converges from an initial guess $\mathbf{x}_0 = [1, 1, 1]^T$ to the solution $\mathbf{x} = [1, 2, 3]^T$.

Alternatively, the g-function can also be obtained as

$$\mathbf{g}'(\mathbf{x}) = \mathbf{x}, \qquad \begin{cases} g_1'(\mathbf{x}) = x = -(3y + 5z - 22) \\ g_2'(\mathbf{x}) = y = -(2x + 3z - 25)/7 \\ g_3'(\mathbf{x}) = z = -(6x + 3y - 18)/2 \end{cases}$$

The Jacobian is

$$\mathbf{J}_{g'} = \begin{bmatrix} 0 & -3 & -5 \\ -2/7 & 0 & -3/7 \\ -3 & -3/2 & 0 \end{bmatrix}$$

with the induced $p = 2$ norm $||\mathbf{J}_g|| = 5.917 > 1$. The iteration does not converge.

Example 1.9

Consider a three-variable nonlinear function $\mathbf{f}(\mathbf{x})$ of arguments $\mathbf{x} = [x,\ y,\ z]^T$

$$\mathbf{f}(\mathbf{x}) = \mathbf{0}, \qquad \begin{cases} f_1(\mathbf{x}) = x^2 - 2x + y^2 - z + 1 = 0 \\ f_2(\mathbf{x}) = xy^2 - x - 3y + yz + 2 = 0 \\ f_3(\mathbf{x}) = xz^2 - 3z + yz^2 + xy = 0 \end{cases}$$

The g-function can be obtained as

$$\mathbf{g}(\mathbf{x}) = \mathbf{x}, \qquad \begin{cases} g_1(\mathbf{x}) = x = (x^2 + y^2 - z + 1)/2 \\ g_2(\mathbf{x}) = y = (xy^2 - x + yz + 2)/3 \\ g_3(\mathbf{x}) = z = (xz^2 + yz^2 + xy)/3 \end{cases}$$

With $\mathbf{x}_0 = [0,\ 0,\ 0]^T$ and after $n > 170$ iterations $\mathbf{x}_{n+1} = \mathbf{g}(\mathbf{x}_n)$ converges to $\mathbf{x}_n = [1.098933,\ 0.367621,\ 0.144932]^T$, with error $||\mathbf{f}(\mathbf{x}_n)|| < 10^{-7}$. However, the iteration may not converge from other possible initial guesses.

Finally, we consider Aitken's method, by which the iteration $x_{n+1} = g(x_n)$ can be accelerated based on two consecutive points x_0 and x_1, as shown in Fig. 1.9.

The secant line of $g(x)$ that goes through the two points $P_0 = (x_0, g(x_0) = x_1)$ and $P_1 = (x_1, g(x_1) = x_2)$ is represented by the equation in terms of its slope

$$\frac{y - g(x_0)}{x - x_0} = \frac{g(x_1) - g(x_0)}{x_1 - x_0} \tag{1.35}$$

Solving for y, we get

$$y = g(x_0) + (x - x_0)\frac{g(x_1) - g(x_0)}{x_1 - x_0} = x_1 + (x - x_0)\frac{x_2 - x_1}{x_1 - x_0} \tag{1.36}$$

To accelerate, instead of moving from x_0 to x_1, we move to the point x at which this secant line intersects with the identity function $y = x$. We can therefore replace y in Eq. (1.36) by x and solve the resulting equation

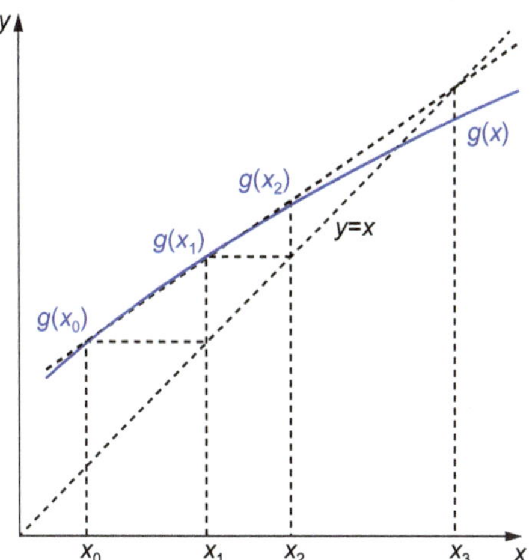

Fig. 1.9 Aitken's method.

$$x = x_1 + (x - x_0)\frac{x_2 - x_1}{x_1 - x_0} \tag{1.37}$$

to obtain

$$x = \frac{x_1(x_1 - x_0) - x_0(x_2 - x_1)}{(x_1 - x_0) - (x_2 - x_1)} = x_0 - \frac{(x_1 - x_0)^2}{x_2 - 2x_1 + x_0} = x_0 - \frac{(\Delta x_0)^2}{\Delta^2 x_0} \tag{1.38}$$

where Δx_0 and $\Delta^2 x_0$ are respectively the first- and second-order differences defined as follows:

$$\Delta x_0 = x_1 - x_0, \qquad \Delta x_1 = x_2 - x_1 \tag{1.39}$$

$$\Delta^2 x_0 = \Delta x_1 - \Delta x_0 = (x_2 - x_1) - (x_1 - x_0) = x_2 - 2x_1 + x_0 \tag{1.40}$$

This result can then be converted into an iterative process

$$\begin{cases} x_{n+1} = g(x_n) \\ x_{n+2} = g(x_{n+1}) \\ x_{n+3} = x_n - (x_{n+1} - x_n)^2/(x_{n+2} - 2x_{n+1} + x_n) \end{cases} \tag{1.41}$$

Given x_n, we skip $x_{n+1} = g(x_n)$ and $x_{n+2} = g(x_{n+1})$ but directly move to x_{n+3} computed based on x_{n+1} and x_{n+2}, thereby making a greater step toward the solution.

Example 1.10

Solve $x^3 - 3x + 1 = 0$. Construct $g(x) = (3x - 1)^{1/3}$. It takes 18 iterations for the regular fixed-point algorithm with initial guess $x_0 = 1$, to get $x_{18} = 1.5320887$ that satisfies $|f(x_n)| < 10^{-6}$, but it only takes three iterations for Aitken's method to converge to the same result:

Table 1.6

n	x_n	$f(x_n)$
0	1.0000	−1.0000
1	1.259921	$-7.797631e-01$
2	1.406056	$-4.384047e-01$
3	1.476396	$-2.110206e-01$
4	1.507985	$-9.476760e-02$
5	1.521751	$-4.129594e-02$
6	1.527672	$1.776362e-02$
7	1.530205	$-7.598914e-03$
8	1.531286	$3.242990e-03$
9	1.531747	$-1.382619e-03$
10	1.531943	$-5.892137e-04$
11	1.532027	$-2.510521e-04$
12	1.532062	$-1.069599e-04$
13	1.532078	$-4.556840e-05$
14	1.532084	$-1.941335e-05$
15	1.532087	$-8.270551e-06$
16	1.532088	$-3.523444e-06$
17	1.532089	$-1.501066e-06$
18	1.532089	$-6.394874e-07$

Table 1.7

n	x_n	$f(x_n)$
0	1.0000	−1.0000
1	1.5937361	$2.668730e-01$
2	1.5323992	$1.254749e-03$
3	1.5320889	$3.421160e-08$

1.4 Newton–Raphson Method (Univariate)

The Newton–Raphson method, or simply Newton's method, solves the general equation $f(x) = 0$ more efficiently as its converges more quickly (quadratic) than the secant method previously discussed, owing to the derivative $f'(x)$ assumed to be available, as an extra piece of information in addition to the function value $f(x)$ itself. Newton's method is based on the approximation of the function $f(x)$ by its Taylor series expansion at any x_0 as an initial guess of the root

$$f(x) = f(x_0)+f'(x_0)(x-x_0)+\frac{1}{2!}f''(x)(x-x_0)^2+\cdots+\frac{1}{n!}f^{(n)}(x_0)(x-x_0)^n+\cdots \quad (1.42)$$

If $f(x)$ is linear, that is, its slope $f'(x)$ is a constant at any point x, then all terms other than the first two are zero, and the equation becomes

$$f(x) = f(x_0) + f'(x)(x - x_0) = 0 \quad (1.43)$$

Solving this equation for x we get the root x^*

$$x^* = x_0 - \frac{f(x_0)}{f'(x_0)} = x_0 + \Delta x_0 \tag{1.44}$$

where $\Delta x_0 = -f(x_0)/f'(x_0)$ is the step increment needed to move from the initial guess x_0 to the root x^*

$$\Delta x_0 = -\frac{f(x_0)}{f'(x_0)} \begin{cases} < 0 & \text{if } f(x_0) \text{ and } f'(x_0) \text{ are of the same sign} \\ > 0 & \text{if } f(x_0) \text{ and } f'(x_0) \text{ are of different signs} \end{cases} \tag{1.45}$$

If $f(x)$ is nonlinear, the sum of the first two terms of its Taylor series is only an approximation of the function, and the result in Eq. (1.44) is no longer the root, but an approximation of the root, which can be improved iteratively to gradually approach the root x^*

$$x_{n+1} = x_n + \Delta x_n = x_n - \frac{f(x_n)}{f'(x_n)}, \qquad n = 0, 1, 2, \cdots \tag{1.46}$$

as illustrated in Fig. 1.10. Note that this iteration is the same as Eq. (1.19) for the secant method, with $\hat{f}'(x_n)$ replaced by the derivative $f'(x_n)$.

The Newton–Raphson method can be considered as the fixed-point iteration $x_{n+1} = g(x_n)$ based on

$$g(x) = x - f(x)/f'(x) \tag{1.47}$$

based on the assumption that $f'(x) \neq 0$. The root x^* at which $f(x^*) = 0$ is also the fixed point of $g(x)$, that is, $g(x^*) = x^*$. For the iteration to converge, $g(x)$ needs to be a contraction with $|g'(x)| < 1$. Consider

$$g'(x) = \left(x - \frac{f(x)}{f'(x)} \right)' = 1 - \frac{(f'(x))^2 - f(x)f''(x)}{(f'(x))^2} = \frac{f(x)f''(x)}{(f'(x))^2} \tag{1.48}$$

At the root $x = x^*$ where $f(x^*) = 0$, if $f'(x^*) \neq 0$, then $g'(x^*) = 0 < 1$, that is, $g(x)$ is a contraction and the iteration $x_{n+1} = g(x_n)$ converges quadratically when x_n is close enough to x^*.

Some additional considerations of the Newton–Raphson methods are outlined below.

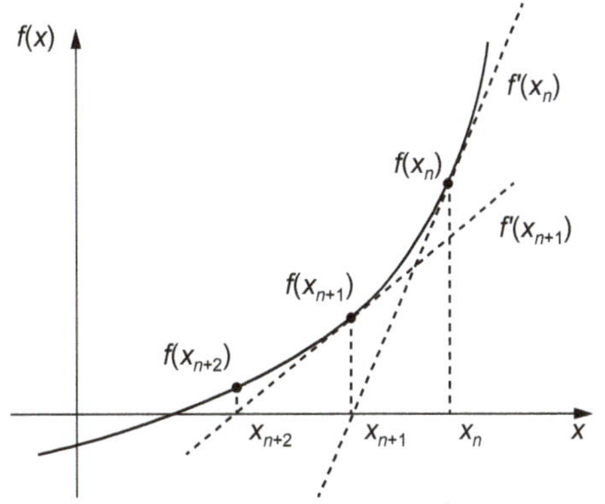

Fig. 1.10 Newton–Raphson method.

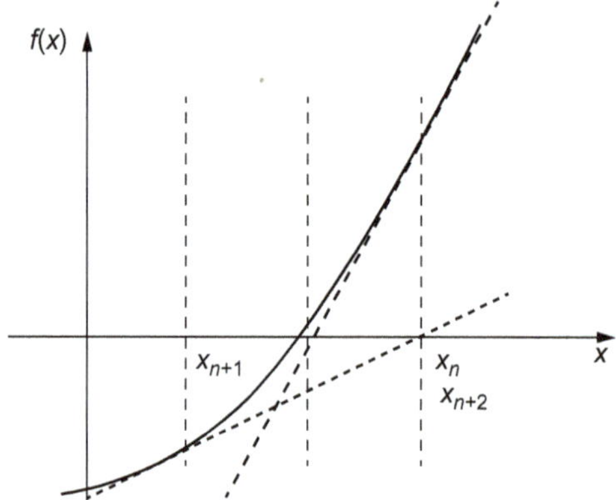

Fig. 1.11 Oscillation.

- If $f'(x_n) = 0$ (with a horizontal tangent line), the iteration cannot proceed, but we can modify x_n by adding a small value ϵ to x_n so that $f'(x + \epsilon) \neq 0$.
- It is difficult to know the number of roots of a nonlinear equation (unlike the case of a linear equation), which can vary from zero (e.g., $f(x) = x^2 + 1$) to infinity (e.g., $f(x) = \cos(x)$). One can try different initial guesses in the range of interest to see if different roots can be found.
- Sometimes a parameter δ can be used to control the step size of the iteration

$$x_{n+1} = x_n - \delta \frac{f(x_n)}{f'(x_n)} \tag{1.49}$$

  - If $\delta < 1$, the iteration is de-accelerated. Although the convergence becomes slower, this may be desirable if the function $f(x)$ is not smooth with many local variations.
  - If $\delta > 1$, the iteration is accelerated. The convergence may or may not be accelerated. Owing to the greater step size, the root may be skipped and missed. Sometimes the convergence may become significantly slowed or even oscillate around the true root, such as the example shown in Fig. 1.11 with $\delta = 2$.

The Newton–Raphson method converges quadratically

$$\lim_{n\to\infty} \frac{|e_{n+1}|}{|e_n|^2} \leq \mu \tag{1.50}$$

as shown in Appendix 2 of this chapter, that is, it converges more quickly than both the bisection method ($q = 1$) and the secant method ($q = 1.618$).

Example 1.11

$$f(x) = x^3 - 4x^2 + 5x - 2 = (x-1)^2(x-2) = 0 \tag{1.51}$$

Obviously this equation has a repeated root $x = 1$ as well as a single root $x = 2$. We have

$$f'(x) = 3x^2 - 8x + 5 = (3x-5)(x-1), \qquad f''(x) = 6x - 8 \tag{1.52}$$

Note that at the root $x^* = 1$ we have $f'(x^*) = f'(1) = 0$. We further find

$$g(x) = x - \frac{f(x)}{f'(x)} = x - \frac{(x-1)^2(x-2)}{(3x-5)(x-1)} = x - \frac{(x-1)(x-2)}{3x-5} \tag{1.53}$$

and

$$\begin{aligned} g'(x) &= \frac{f(x)f''(x)}{(f'(x))^2} = \frac{(x-1)^2(x-2)(6x-8)}{(3x-5)^2(x-1)^2} \\ &= \frac{(x-2)(6x-8)}{(3x-5)^2} = \begin{cases} 1/2 & x = 1 \\ 0 & x = 2 \end{cases} \end{aligned}$$

We therefore have

$$e_{n+1} = g'(x)e_n - \frac{1}{2}g''(x)e_n^2 + O(e_n^3) \overset{n\to\infty}{\Longrightarrow} \begin{cases} e_n/2 & x = 1 \\ -g''(x)e_n^2/2 & x = 2 \end{cases} \tag{1.54}$$

We see that the iteration converges quadratically to the single root $x = 2$, but only linearly to the repeated root $x = 1$.

We consider in general a function with a repeated root at $x = a$ of multiplicity k

$$f(x) = (x-a)^k h(x) \tag{1.55}$$

its derivative is

$$f'(x) = k(x-a)^{k-1}h(x) + (x-a)^k h'(x) = (x-a)^{k-1}[kh(x) + (x-a)h'(x)] \tag{1.56}$$

As $f'(a) = 0$, the convergence of the Newton–Raphson method to $x = a$ is linear, rather than quadratic.

In such a case, we can accelerate the iteration by using a step size $\delta = k > 1$

$$g(x) = x - k\,\frac{f(x)}{f'(x)} \tag{1.57}$$

and

$$g'(x) = 1 - k\,\frac{(f'(x))^2 - f(x)f''(x)}{(f'(x))^2} = 1 - k + k\,\frac{f(x)f''(x)}{(f'(x))^2} \tag{1.58}$$

Now we show that this $g'(x)$ is zero at the repeated root $x = a$, therefore the convergence to this root is still quadratic.

We substitute $f(x) = (x-a)^k h(x)$, $f'(x) = k(x-a)^{k-1}[kh + (x-a)h']$, and

$$\begin{aligned} f''(x) &= [(x-a)^{k-1}[kh + (x-a)h']]' \\ &= 1 - k + k\,(k-1)(x-a)^{k-2}[kh + (x-a)h'] \\ &\quad + (x-a)^{k-1}[(k+1)h' + (x-a)h''] \end{aligned} \tag{1.59}$$

into the expression just given for $g'(x)$ to get (after some algebra)

$$g'(x) = 1 - k + k\frac{f(x)f''(x)}{(f'(x))^2}$$
$$= 1 - k + k\,\frac{h(k-1)(kh + (x-a)h') + h(x-a)((k+1)h' + (x-a)h')}{(kh + (x-a)h')^2} \tag{1.60}$$

At $x = a$, we get

$$g'(x)\Big|_{x=a} = 1 - k + k\frac{(k-1)kh^2}{(kh)^2} = 0 \tag{1.61}$$

that is the convergence to the repeated root at $x = a$ is no longer linear but quadratic. The difficulty, however, is that the multiplicity k of a root is unknown ahead of time. If $\delta > 1$ is used blindly, some root may be skipped and the iteration may oscillate around the real root.

1.5 Newton–Raphson Method (Multivariate)

The Newton–Raphson method discussed in the previous section for solving a single-variable equation $f(x) = 0$ can be generalized to the case of multivariate equation systems containing M equations of N variables in $\mathbf{x} = [x_1, \cdots, x_N]^T$

$$\mathbf{f}(\mathbf{x}) = \begin{bmatrix} f_1(\mathbf{x}) \\ \vdots \\ f_M(\mathbf{x}) \end{bmatrix} = \begin{bmatrix} f_1(x_1, \cdots, x_N) \\ \vdots \\ f_M(x_1, \cdots, x_N) \end{bmatrix} = \begin{bmatrix} 0 \\ \vdots \\ 0 \end{bmatrix} = \mathbf{0} \tag{1.62}$$

Again, we first consider the Taylor series expansion (Eq. (A.388)) of each of the M functions in then neighborhood of an initial guess $\mathbf{x}_0 = [x_{01}, \cdots, x_{0N}]^T$

$$f_m(\mathbf{x}) = f_m(\mathbf{x}_0) + \mathbf{g}^T_{f_m}(\mathbf{x}_0)(\mathbf{x} - \mathbf{x}_0) + r_m(||\mathbf{x} - \mathbf{x}_0||^2)$$
$$\approx f_m(\mathbf{x}_0) + \sum_{n=1}^{N} \frac{\partial f_m(\mathbf{x}_0)}{\partial x_n}(x_n - x_{0n}) \qquad (m = 1, \cdots, M) \tag{1.63}$$

where $r_m(||\mathbf{x} - \mathbf{x}_0||^2)$ represents the second- and higher-order terms in the series beyond the linear term, which can be neglected if $||\mathbf{x} - \mathbf{x}_0||$ is small. These M equations can be expressed in matrix form

$$\begin{bmatrix} f_1(\mathbf{x}) \\ \vdots \\ f_M(\mathbf{x}) \end{bmatrix} \approx \begin{bmatrix} f_1(\mathbf{x}_0) \\ \vdots \\ f_M(\mathbf{x}_0) \end{bmatrix} + \begin{bmatrix} \frac{\partial f_1(\mathbf{x}_0)}{\partial x_1} & \cdots & \frac{\partial f_1(\mathbf{x}_0)}{\partial x_N} \\ \vdots & \ddots & \vdots \\ \frac{\partial f_M(\mathbf{x}_0)}{\partial x_1} & \cdots & \frac{\partial f_M(\mathbf{x}_0)}{\partial x_N} \end{bmatrix} \begin{bmatrix} x_1 - x_{01} \\ \vdots \\ x_N - x_{0N} \end{bmatrix} \tag{1.64}$$

or more concisely

$$\mathbf{f}(\mathbf{x}) \approx \mathbf{f}(\mathbf{x}_0) + \mathbf{J}(\mathbf{x}_0)\;(\mathbf{x} - \mathbf{x}_0) = \mathbf{f}_0 + \mathbf{J}_0\;\Delta\mathbf{x} \tag{1.65}$$

where $\Delta\mathbf{x} = \mathbf{x} - \mathbf{x}_0$, and $\mathbf{f}_0 = \mathbf{f}(\mathbf{x}_0)$ and $\mathbf{J}_0 = \mathbf{J}(\mathbf{x}_0)$ are respectively the function $\mathbf{f}(\mathbf{x})$ and its *Jacobian matrix* $\mathbf{J}_f(\mathbf{x})$ both evaluated at $\mathbf{x}_0$. We further consider solving the equation system $\mathbf{f}(\mathbf{x}) = \mathbf{0}$ in the following two cases.

- $M = N$: The number of equations is the same as the number of unknowns, the Jacobian $\mathbf{J}(\mathbf{x})$ is a square matrix and its inverse $\mathbf{J}^{-1}$ exists in general. In the special case where $\mathbf{f}(\mathbf{x}) = \mathbf{A}\mathbf{x} - \mathbf{b}$ is linear, the Taylor series contains only the first two terms while all higher-order terms are zero, and the approximation in Eq. (1.65) becomes exact with $\mathbf{J} = \mathbf{A}$. To find the root $\mathbf{x}^*$ of the equation, we set $\mathbf{f}(\mathbf{x})$ in the equation to zero and solve the resulting equation for $\mathbf{x}$ to obtain the solution

$$\mathbf{x}^* = \mathbf{x}_0 - \mathbf{J}_0^{-1}\mathbf{f}_0 = \mathbf{x}_0 - \mathbf{A}^{-1}\mathbf{f}_0 \tag{1.66}$$

For example, if $\mathbf{x}_0 = \mathbf{0}$ and $\mathbf{f}(\mathbf{x}_0) = -\mathbf{b}$, then $\mathbf{x}^* = \mathbf{A}^{-1}\mathbf{b}$.

If $\mathbf{f}(\mathbf{x})$ is nonlinear, the sum of the first two terms of the Taylor series is only an approximation, and the result in Eq. (1.66) is only an approximate root, which can be further improved iteratively to move from the initial guess $\mathbf{x}_0$ toward the root $\mathbf{x}^*$

$$\mathbf{x}_{n+1} = \mathbf{x}_n + \Delta\mathbf{x}_n = \mathbf{x}_n - \mathbf{J}_n^{-1}\,\mathbf{f}_n \tag{1.67}$$

where $\Delta\mathbf{x}_n = -\mathbf{J}_n^{-1}\,\mathbf{f}_n$ is the increment, representing the *search direction* at the nth step. The iteration moves $\mathbf{x}_n$ in the N–D space spanned by $\{x_1, \cdots, x_N\}$ from some initial guess $\mathbf{x}_0$ along such a path that all function values $f_m(\mathbf{x}),\ \ m = 1, \cdots, M)$ are reduced. As in the univariate case, a scaling factor δ_n can be used to control the step size of the iteration

$$\mathbf{x}_{n+1} = \mathbf{x}_n + \delta_n\,\Delta\mathbf{x}_n = \mathbf{x}_n - \delta_n\,\mathbf{J}_n^{-1}\,\mathbf{f}_n \tag{1.68}$$

The step size can be adjusted based on the specific function shape. When $\delta_n < 1$, the step size becomes smaller and the convergence of the iteration is slower, we will have a better chance not to skip a solution, which may happen if $\mathbf{f}(\mathbf{x})$ is not smooth and the step size is too big.

The algorithm is listed in the following (where the tolerance *tol* is preset to a small value):

  - Select $\mathbf{x}_0$
  - Obtain $\mathbf{f}_0 = \mathbf{f}(\mathbf{x}_0)$ and $\mathbf{J}_0$
  - Obtain $\mathbf{J}_0^{-1}$
  - $n = 0$
  - While $||\mathbf{x}_{n+1} - \mathbf{x}_n|| > tol$ do
    - $\mathbf{x}_{n+1} = \mathbf{x}_n - \mathbf{J}_n^{-1}\mathbf{f}_n$
    - Find $\mathbf{f}_{n+1}$ and $\mathbf{J}_{n+1}$
    - $n = n + 1$

- $M > N$: There are more equations than unknowns, that is, equation $\mathbf{f}(\mathbf{x}) = \mathbf{0}$ is an over-constrained system, and the Jacobian $\mathbf{J}(\mathbf{x})$ is an $M \times N$ non-square matrix with no inverse, that is, no solution exists for the equation $\mathbf{f}(\mathbf{x}) = \mathbf{0}$ in general. But we can still seek to find an optimal solution $\mathbf{x}^*$ that minimizes the following SSE

$$\varepsilon(\mathbf{x}) = \frac{1}{2}||\mathbf{f}(\mathbf{x})||^2 = \frac{1}{2}\mathbf{f}(\mathbf{x})^T\mathbf{f}(\mathbf{x}) = \frac{1}{2}\sum_{m=1}^{M} f_m^2(\mathbf{x}) \tag{1.69}$$

The gradient vector of this SSE is

$$\begin{aligned}\mathbf{g}_\varepsilon(\mathbf{x}) &= \frac{d}{d\mathbf{x}}\varepsilon(\mathbf{x}) = \frac{d}{d\mathbf{x}}\left(\frac{1}{2}\sum_{m=1}^{M} f_m^2(\mathbf{x})\right) = \sum_{m=1}^{M}\frac{d\,f_m(\mathbf{x})}{d\mathbf{x}}f_m(\mathbf{x}) \\ &= \mathbf{J}_f^T(\mathbf{x})\,\mathbf{f}(\mathbf{x})\end{aligned} \tag{1.70}$$

where $\mathbf{J}_f(\mathbf{x})$ is the Jacobian of $\mathbf{f}(\mathbf{x})$. We can also obtain the nth component ($n = 1, \cdots, N$) of $\mathbf{g}_\varepsilon(\mathbf{x})$ as

$$\frac{\partial\varepsilon(\mathbf{x})}{\partial x_n} = \sum_{m=1}^{M}\frac{\partial f_m(\mathbf{x})}{\partial x_n}f_m(\mathbf{x}) = \sum_{m=1}^{M} J_{mn}\,f_m(\mathbf{x}) \tag{1.71}$$

where $J_{mn} = \partial f_m(\mathbf{x})/\partial x_n$ is the component in the mth row and nth column of $\mathbf{J}_f(\mathbf{x})$. At the optimal solution $\mathbf{x}^*$ that minimizes this error $\varepsilon(\mathbf{x})$, we have $\mathbf{g}_\varepsilon(\mathbf{x}^*) = \mathbf{0}$.

In the special case when the function $\mathbf{f}(\mathbf{x}) = \mathbf{A}\mathbf{x} - \mathbf{b}$ is linear with $\mathbf{J}(\mathbf{x}) = \mathbf{A}$, it can be represented as the sum of the first two terms of its Taylor series in Eq. (1.65), and the gradient is

$$\mathbf{g}_\varepsilon(\mathbf{x}) = \mathbf{A}^T\,\mathbf{f}(\mathbf{x}) = \mathbf{A}^T\,[\mathbf{f}(\mathbf{x}_0) + \mathbf{A}\,\Delta\mathbf{x}] = \mathbf{A}^T(\mathbf{f}_0 + \mathbf{A}\Delta\mathbf{x}) \tag{1.72}$$

where $\mathbf{x}_0$ is any chosen initial guess. At the optimal solution $\mathbf{x}^*$ that minimizes $\varepsilon(\mathbf{x})$, this gradient is zero

$$\mathbf{g}_\varepsilon(\mathbf{x}^*) = \mathbf{A}^T(\mathbf{f}_0 + \mathbf{A}\Delta\mathbf{x}) = \mathbf{0} \tag{1.73}$$

Solving this equation for $\Delta\mathbf{x}$ we get the optimal increment

$$\Delta\mathbf{x} = -(\mathbf{A}^T\mathbf{A})^{-1}\mathbf{A}^T\mathbf{f}_0 = -\mathbf{A}^-\,\mathbf{f}_0 \tag{1.74}$$

Here $\mathbf{A}^- = (\mathbf{A}^T\mathbf{A})^{-1}\mathbf{A}^T$ is the *pseudo-inverse* (Section A.6.1) of the non-square matrix $\mathbf{A}$. Now the optimal solution can be found as

$$\mathbf{x}^* = \mathbf{x}_0 + \Delta\mathbf{x} = \mathbf{x}_0 - \mathbf{A}^-\,\mathbf{f}_0 \tag{1.75}$$

For example, if $\mathbf{x}_0 = \mathbf{0}$ and $\mathbf{f}(\mathbf{x}_0) = \mathbf{A}\mathbf{x}_0 - \mathbf{b} = -\mathbf{b}$, we have $\mathbf{x}^* = \mathbf{A}^-\mathbf{b}$, same as Eq. (1.8).

In the general case when $\mathbf{f}(\mathbf{x})$ is nonlinear, the sum of the first two terms of its Taylor series is its approximation, and the presented result is an approximation of the root, which can be further improved iteratively to move from the initial guess toward the root

$$\mathbf{x}_{n+1} = \mathbf{x}_n + \Delta\mathbf{x}_n = \mathbf{x}_n - \mathbf{J}_n^-\,\mathbf{f}_n \tag{1.76}$$

This iteration will converge to $\mathbf{x}^*$ at which $\mathbf{g}_\varepsilon(\mathbf{x}^*) = \mathbf{0}$, and the squared error $\varepsilon(\mathbf{x})$ is minimized.

Comparing Eqs (1.67) and (1.76), we see that they are essentially the same, with the only difference that the regular inverse $\mathbf{J}^{-1}$ is used when $M = N$, but the pseudoinverse $\mathbf{J}^{-}$ is used when $M > N$ and $\mathbf{J}^{-1}$ does not exist.

The Newton–Raphson method assumes the availability of the analytical expressions of all partial derivatives $J_{mn} = \partial f_m(\mathbf{x})/\partial x_n$ $(m = 1,\cdots,M,\ n = 1,\cdots,N)$ in the Jacobian matrix $\mathbf{J}$. However, when this is not the case, J_{mn} needs to be approximated by the forward or central difference (secant) method

$$J_{mn} = \frac{\partial f_m(x_1,\cdots,x_N)}{\partial x_n} \approx \frac{f_m(x_1,\cdots,x_n+h,\cdots,x_N) - f_m(x_1,\cdots,x_n,\cdots,x_N)}{h}$$
$$\approx \frac{f_m(x_1,\cdots,x_n+h,\cdots,x_N) - f_m(x_1,\cdots,x_n-h,\cdots,x_N)}{2h} \tag{1.77}$$

where h is a small increment.

Example 1.12

$$\begin{cases} 3\,x_1 - \cos(x_2x_3) - 3/2 = 0 \\ 4x_1^2 - 625\,x_2^2 + 2x_3 - 1 = 0 \\ 20\,x_3 + e^{-x_1x_2} + 9 = 0 \end{cases}$$

$$\mathbf{J} = \begin{bmatrix} 3 & x_3\sin(x_2x_3) & x_2\sin(x_2x_3) \\ 8\,x_1 & -1250\,x_2 & 2 \\ -x_2\,e^{-x_1x_2} & -x_1\,e^{-x_1x_2} & 20 \end{bmatrix}$$

Table 1.8

n	$\mathbf{x}$	*Error*
0	(1.000000, 1.000000, 1.000000)	$6.207e+02$
1	(1.232701, 0.503132, −0.473253)	$1.541e+02$
2	(0.832592, 0.251806, −0.490636)	$3.884e+01$
3	(0.833238, 0.128406, −0.494702)	$9.517e+00$
4	(0.833275, 0.069082, −0.497147)	$2.200e+00$
5	(0.833281, 0.043585, −0.498206)	$4.063e-01$
6	(0.833282, 0.036117, −0.498517)	$3.486e-02$
7	(0.833282, 0.035343, −0.498549)	$3.741e-04$
8	(0.833282, 0.035335, −0.498549)	$4.498e-08$
9	(0.833282, 0.035335, −0.498549)	$5.551e-16$

Example 1.13

$$\mathbf{f}(\mathbf{x}) = \mathbf{0}, \qquad \begin{cases} f_1(\mathbf{x}) = x_1^2 - 2x_1 + x_2^2 - x_3 + 1 = 0 \\ f_2(\mathbf{x}) = x_1x_2^2 - x_1 - 3x_2 + x_2x_3 + 2 = 0 \\ f_3(\mathbf{x}) = x_1x_3^2 - 3x_3 + x_2x_3^2 + x_1x_2 = 0 \end{cases}$$

$$\mathbf{J} = \begin{bmatrix} 2x_1 - 2 & 2x_2 & -1 \\ x_2^2 - 1 & 2x_1x_2 - 3 + x_3 & x_2 \\ x_3^2 + x_2 & x_3^2 + x_1 & 2x_1x_3 - 3 + 2x_2x_3 \end{bmatrix}$$

With $\mathbf{x}_0 = [1,\ 2,\ 3]^T$, we obtain a root:

Table 1.9

n	$\mathbf{x}$	*Error*
0	(1.00000, 2.00000, 3.00000)	$2.064e+01$
1	(0.10256, 1.64103, 2.56410)	$4.303e+00$
2	(1.52062, 1.41113, 0.19859)	$2.689e+00$
3	(1.94123, 0.77134, 0.89465)	$1.217e+00$
4	(1.06737, 1.19117, 0.48353)	$1.144e+00$
5	(1.26825, 0.95182, 0.88028)	$3.323e-01$
6	(0.95899, 1.03384, 0.96813)	$1.171e-01$
7	(1.00171, 1.00007, 0.99718)	$4.162e-03$
8	(1.00000, 1.00000, 1.00000)	$6.701e-06$

With $\mathbf{x}_0 = [0,\ 0,\ 0]^T$, we get another root:

Table 1.10

n	$\mathbf{x}$	*Error*
0	(0.00000, 0.00000, 0.00000)	$2.236e+00$
1	(0.50000, 0.50000, 0.00000)	$5.728e-01$
2	(0.83951, 0.47531, 0.13580)	$1.175e-01$
3	(0.98582, 0.41849, 0.15069)	$2.639e-02$
4	(1.05417, 0.38715, 0.14717)	$6.088e-03$
5	(1.08565, 0.37339, 0.14558)	$1.264e-03$
6	(1.09693, 0.36849, 0.14503)	$1.618e-04$
7	(1.09888, 0.36764, 0.14494)	$4.817e-06$

Broyden's method

In the Newton–Raphson method, two main operations are carried out in each iteration: (a) evaluate the Jacobian matrix $\mathbf{J_f}(\mathbf{x}_n)$ and (b) obtain its inverse $\mathbf{J_f}^{-1}(\mathbf{x}_n)$. To avoid the expensive computation for these operations, we can consider using Broyden's method, one of the so-called *quasi-Newton methods*, which approximates the inverse of the Jacobian $\mathbf{J}_{n+1}^{-1} = \mathbf{J}^{-1}(\mathbf{x}_{n+1})$ from the $\mathbf{J}_n^{-1} = \mathbf{J}^{-1}(\mathbf{x}_n)$ in the previous iteration step, so that it can be updated iteratively from the initial $\mathbf{J}_0^{-1} = \mathbf{J}^{-1}(\mathbf{x}_0)$.

We first consider the iteration step in the single-variable case to see how to estimate the next derivative $f'_{n+1} = f'(x_{n+1})$ from the current $f'_n = f'(x_n)$ by the secant method

$$f'_{n+1} \approx \hat{f}'_{n+1} = \frac{f_{n+1} - f_n}{x_{n+1} - x_n} = \frac{\delta f_n}{\delta x_n} = \frac{\hat{f}'_n \delta x_n - \hat{f}'_n \delta x_n + \delta f_n}{\delta x_n}$$
$$= \hat{f}'_n + \frac{\delta f_n - \hat{f}'_n \delta x_n}{\delta x_n} = \hat{f}'_n + \hat{\delta f}'_n \tag{1.78}$$

where

- $\delta f_n = f_{n+1} - f_n$ is the true increment of the function over the interval $\delta x_n = x_{n+1} - x_n$;
- $\hat{f}'_n \delta x_n$ is the estimated increment of the function based on the previous derivative $\hat{f}'_n$;
- $\hat{\delta f}'_n$ is the estimated increment of the derivative

$$\hat{\delta f}'_n = \frac{\delta f_n - \hat{f}'_n \delta x_n}{\delta x_n} \tag{1.79}$$

This equation indicates that the derivative f'_{n+1} in the $(n+1)th$ step can be estimated by adding the estimated increment $\hat{\delta f}'_n$ to the derivative $\hat{f}'n$ in the current nth step, as shown in Fig. 1.12.

Having obtained $\hat{f}'_{n+1}$, we can use the same iteration in the Newton–Raphson method to find x_{n+1}

$$x_{n+1} = x_n - \frac{f(x_n)}{\hat{f}'_n} \tag{1.80}$$

This method for a single-variable case can be generalized to the multivariable case for solving $\mathbf{f}(\mathbf{x}) = \mathbf{0}$. Following the way we estimate the increment of the derivative of a single-variable function in Eq. (1.79), here we can estimate the increment of the Jacobian of a multivariable function

$$\delta\hat{\mathbf{J}}_n = \frac{(\delta\mathbf{f}_n - \hat{\mathbf{J}}_n\delta\mathbf{x}_n)\ \delta\mathbf{x}_n^T}{\delta\mathbf{x}_n^T\ \delta\mathbf{x}_n} = \frac{(\delta\mathbf{f}_n - \hat{\mathbf{J}}_n\delta\mathbf{x}_n)\delta\mathbf{x}_n^T}{||\delta\mathbf{x}_n||^2} \tag{1.81}$$

where $\delta\mathbf{x}_n = \mathbf{x}_{n+1} - \mathbf{x}_n$ and $\delta\mathbf{f}_n = \mathbf{f}_{n+1} - \mathbf{f}_n = \mathbf{f}(\mathbf{x}_{n+1}) - \mathbf{f}(\mathbf{x}_n)$. Now in each iteration, we can update the estimated Jacobian as well as the estimated root

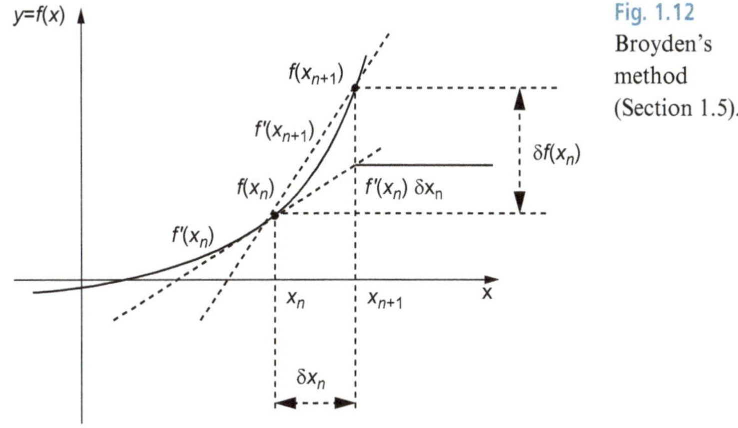

Fig. 1.12 Broyden's method (Section 1.5).

$$\hat{\mathbf{J}}_{n+1} = \hat{\mathbf{J}}_n + \delta\hat{\mathbf{J}}_n, \qquad \mathbf{x}_{n+1} = \mathbf{x}_n + \delta\mathbf{x}_n = \mathbf{x}_n - \hat{\mathbf{J}}_n^{-1}\mathbf{f}(\mathbf{x}_n) \tag{1.82}$$

The algorithm can be further improved so that the inverse of the Jacobian $\mathbf{J}_n$ is avoided. Specifically, consider the inverse Jacobian

$$\hat{\mathbf{J}}_{n+1}^{-1} = \left(\hat{\mathbf{J}}_n + \delta\hat{\mathbf{J}}_n\right)^{-1} = \left[\hat{\mathbf{J}}_n + \frac{(\delta\mathbf{f}_n - \hat{\mathbf{J}}_n\delta\mathbf{x}_n)\delta\mathbf{x}_n^T}{||\delta\mathbf{x}_n||^2}\right]^{-1} \tag{1.83}$$

We can apply the Sherman–Morrison formula (Section A.2.4)

$$(\mathbf{A} + \mathbf{u}\mathbf{v}^T)^{-1} = \mathbf{A}^{-1} - \frac{\mathbf{A}^{-1}\mathbf{u}\mathbf{v}^T\mathbf{A}^{-1}}{1 + \mathbf{v}^T\mathbf{A}^{-1}\mathbf{u}} \tag{1.84}$$

to the right-hand side of this equation by defining

$$\mathbf{A} = \hat{\mathbf{J}}_n, \qquad \mathbf{u} = \frac{\delta\mathbf{f}_n - \hat{\mathbf{J}}_n\delta\mathbf{x}_n}{||\delta\mathbf{x}_n||^2}, \qquad \mathbf{v} = \delta\mathbf{x}_n \tag{1.85}$$

and rewrite it as

$$\hat{\mathbf{J}}_{n+1}^{-1} = \hat{\mathbf{J}}_n^{-1} - \frac{\hat{\mathbf{J}}_n^{-1}(\delta\mathbf{f}_n - \hat{\mathbf{J}}_n\delta\mathbf{x}_n^T)/||\delta\mathbf{x}_n||^2\,\delta\mathbf{x}_n^T\,\hat{\mathbf{J}}_n^{-1}}{1 + \delta\mathbf{x}_n^T\hat{\mathbf{J}}_n^{-1}(\delta\mathbf{f}_n - \hat{\mathbf{J}}_n\delta\mathbf{x}_n)/||\delta\mathbf{x}_n||^2} = \hat{\mathbf{J}}_n^{-1} - \frac{(\hat{\mathbf{J}}_n^{-1}\delta\mathbf{f}_n - \delta\mathbf{x}_n)\,\delta\mathbf{x}_n^T\,\hat{\mathbf{J}}_n^{-1}}{\delta\mathbf{x}_n^T\hat{\mathbf{J}}_n^{-1}\delta\mathbf{f}_n} \tag{1.86}$$

We see that the next $\hat{\mathbf{J}}_{n+1}^{-1}$ can be iteratively estimated directly from the previous $\hat{\mathbf{J}}_n^{-1}$, thereby avoiding computing the inverse of $\hat{\mathbf{J}}_n$ altogether. The algorithm is listed as

- Select $\mathbf{x}_0$
- Find $\mathbf{f}_0 = \mathbf{f}(\mathbf{x}_0)$, $\mathbf{J}_0 = \mathbf{J}(\mathbf{x}_0)$, and $\mathbf{J}_0^{-1}$
- $\mathbf{x}_1 = \mathbf{x}_0 - \mathbf{J}_0^{-1}\mathbf{f}_0$
- $n = 0$
- While $||\mathbf{x}_{n+1} - \mathbf{x}_n|| > tol$ do
  - $\delta\mathbf{x}_n = \mathbf{x}_{n+1} - \mathbf{x}_n$
  - $\mathbf{f}_{n+1} = \mathbf{f}(\mathbf{x}_{n+1})$
  - $\delta\mathbf{f} = \mathbf{f}_{n+1} - \mathbf{f}_n$
  - $\mathbf{J}_n^{-1} = \mathbf{J}_n^{-1} - (\delta\mathbf{x}_n - \mathbf{J}_n^{-1}\delta\mathbf{f}_n)\delta\mathbf{x}_n^T\mathbf{J}_n^{-1}/(\delta\mathbf{x}_n^T\mathbf{J}_n^{-1}\delta\mathbf{f}_n)$
  - $\mathbf{x}_n = \mathbf{x}_{n+1}$, $\mathbf{f}_n = \mathbf{f}_{n+1}$
  - $\mathbf{x}_{n+1} = \mathbf{x}_n - \mathbf{J}_n^{-1}\mathbf{f}_n$
  - $n = n + 1$

PROBLEMS

Develop MATLAB code (or any other language) to implement the algorithms covered in this chapter to solve the equations in the examples in the text. Try a set of different initial guesses to see how they may affect how quickly the iteration converges.

1. Implement the bisection and secant methods and apply them to solve the single-variable equations considered in Examples 1.2 through 1.6 based on properly selected initial guesses.

2. Implement method of fixed-point iteration and apply it to solve the same single-variable equations in the previous problem based on properly selected initial guesses.
3. Implement the univariate Newton–Raphson method and apply it to solve the same single-variable equations discussed in this chapter.
4. Implement the multivariate Newton–Raphson method and apply it to solve the multivariable equations in Examples 1.12 and 1.13.

APPENDICES

1.A Order of Convergence of the Secant Method

Let x^* be the root at which $f(x^*) = 0$. The error of x_{n+1} is

$$\begin{aligned} e_{n+1} &= x_{n+1} - x^* = x_n - \frac{x_n - x_{n-1}}{f(x_n) - f(x_{n-1})} f(x_n) - x^* \\ &= \frac{(x_{n-1} - x^*) f(x_n) - (x_n - x^*) f(x_{n-1})}{f(x_n) - f(x_{n-1})} \\ &= \frac{e_{n-1} f(x_n) - e_n f(x_{n-1})}{f(x_n) - f(x_{n-1})} \end{aligned} \tag{1.87}$$

Consider the Taylor expansion of $f(x_n)$ around the root at which as $f(x^*) = 0$

$$\begin{aligned} f(x_n) &= f(x^* + e_n) = f(x^*) + f'(x^*) e_n + \frac{1}{2} f''(x^*) e_n^2 + O(e_n^3) \\ &= f'(x^*) e_n + \frac{1}{2} f''(x^*) e_n^2 + O(e_n^3) \end{aligned} \tag{1.88}$$

Similarly we also have

$$f(x_{n-1}) = f'(x^*) e_{n-1} + \frac{1}{2} f''(x^*) e_{n-1}^2 + O(e_{n-1}^3) \tag{1.89}$$

Substituting these into the expression for e_{n+1}, we get

$$\begin{aligned} e_{n+1} &= \frac{e_{n-1}[f'(x^*)e_n + \frac{1}{2} f''(x^*) e_n^2 + O(e_n^3)] - e_n[f'(x^*)e_{n-1} + \frac{1}{2} f''(x^*) e_{n-1}^2 + O(e_{n-1}^3)]}{[f'(x^*)e_n + \frac{1}{2} f''(x^*) e_n^2 + O(e_n^3)] - [f'(x^*)e_{n-1} + \frac{1}{2} f''(x^*) e_{n-1}^2 + O(e_{n-1}^3)]} \\ &= \frac{e_{n-1} e_n f''(x^*)(e_n - e_{n-1})/2 + O(e_{n-1}^4)}{(e_n - e_{n-1}) f'(x^*) + (e_n^2 - e_{n+1}^2) f''(x^*)/2 + O(e_{n-1}^3)} \\ &= \frac{e_{n-1} e_n f''(x^*)/2 + O(e_{n-1}^3)}{f'(x^*) + (e_n + e_{n+1}) f''(x^*)/2 + O(e_{n-1}^2)} \\ &= \frac{e_{n-1} e_n f''(x^*)/2 + O(e_{n-1}^3)}{f'(x^*) + O(e_n) + O(e_{n-1}^2)} \end{aligned} \tag{1.90}$$

When $n \to \infty$, the lowest order terms in both the numerator and denominator become the dominant terms as all other higher-order terms approach to zero, and we have

$$e_{n+1} = \frac{e_{n-1} e_n f''(x^*)}{2f'(x^*)} = C\, e_n e_{n-1} \tag{1.91}$$

where we have defined a constant $C = f''(x^*)/2f'(x^*)$. To find the order of convergence, we need to find q in

$$|e_{n+1}| \leq |C|\; |e_{n-1}|\; |e_n| = \mu |e_n|^q \tag{1.92}$$

Solving this equation for $|e_n|$ we get

$$|e_n| = \left(\frac{|C|}{\mu} |e_{n-1}| \right)^{1/(q-1)} = \left(\frac{|C|}{\mu} \right)^{1/(q-1)} |e_{n-1}|^{1/(q-1)} \tag{1.93}$$

Conversely, when $n \to \infty$ we also have

$$|e_n| = \mu |e_{n-1}|^q \tag{1.94}$$

Equating the right-hand sides of these two equations we get

$$\left(\frac{|C|}{\mu} \right)^{1/(q-1)} |e_{n-1}|^{1/(q-1)} = \mu |e_{n-1}|^q \tag{1.95}$$

which requires the following two equations to hold

$$q = \frac{1}{q-1}, \qquad \mu = \left(\frac{|C|}{\mu} \right)^{1/(q-1)} = \left(\frac{|C|}{\mu} \right)^q \tag{1.96}$$

These two equations can be solved separately to get

$$q = \frac{1+\sqrt{5}}{2} = 1.618, \qquad \mu = |C|^{q/(q+1)} = C^{(\sqrt{5}-1)/2} = |C|^{0.618} \tag{1.97}$$

that is

$$|e_{n+1}| = \mu |e_n|^q = |C|^{0.618} |e_n|^{1.618} = \left| \frac{f''(x)}{2f'(x)} \right|^{0.618} |e_n|^{1.618} \tag{1.98}$$

or

$$\lim_{n \to \infty} \frac{|e_{n+1}|}{|e_n|^{1.618}} = \mu = |C|^{0.618} \tag{1.99}$$

We see that the secant method has an order of convergence $q = 1.618$ with rate of convergence $\mu = |C|^{0.618}$.

1.B Order of Convergence of the Newton–Raphson Method

The order of convergence of the Newton–Raphson iteration can be found based on the Taylor expansion of $f(x)$ at the neighborhood of the root $x^* = x_n + e_n$ (Section A.7)

$$0 = f(x^*) = f(x_n) + f'(x_n) e_n + \frac{f''(x_n)}{2} e_n^2 + O(e_n^3) \tag{1.100}$$

where $e_n = x^* - x_n$ is the error at the nth step. Substituting the Newton–Raphson's iteration

$$x_{n+1} = x_n - \frac{f(x_n)}{f'(x_n)}, \quad \text{i.e.} \quad f(x_n) = f'(x_n)(x_n - x_{n+1}) \tag{1.101}$$

into this equation, we obtain

$$\begin{aligned} 0 &= f'(x_n)(x_n - x_{n+1}) + f'(x_n)(x^* - x_n) + \frac{f''(x_n)}{2}e_n^2 + O(e_n^3) \\ &= f'(x_n)(x^* - x_{n+1}) + \frac{f''(x)}{2}e_n^2 + O(e_n^3) \\ &= f'(x_n)e_{n+1} + \frac{f''(x_n)}{2}e_n^2 + O(e_n^3) \end{aligned} \tag{1.102}$$

that is

$$e_{n+1} = -\frac{f''(x_n)}{2f'(x_n)}e_n^2 + O(e_n^3) \tag{1.103}$$

When $n \to \infty$ all the higher-order terms disappear, and this equation can be written as

$$\lim_{n\to\infty} \frac{|e_{n+1}|}{|e_n|^2} \le \frac{|f''(x^*)|}{2|f'(x^*)|} = \mu \tag{1.104}$$

Alternatively, we can get the Taylor expansion in terms of $g(x)$

$$x_{n+1} = x^* - e_{n+1} = g(x_n) = g(x^* - e_n) = g(x^*) - g'(x^*)e_n + \frac{g''(x^*)}{2}e_n^2 + O(e_n^3) \tag{1.105}$$

Subtracting $g(x^*) = x^*$ from both sides we have

$$e_{n+1} = g'(x^*)e_n - \frac{g''(x^*)}{2}e_n^2 + O(e_n^3) \tag{1.106}$$

Now we find $g'(x)$ and $g''(x)$

$$g'(x) = \left(x - \frac{f(x)}{f'(x)}\right)' = 1 - \frac{(f'(x))^2 - f(x)f''(x)}{(f'(x))^2} = \frac{f(x)f''(x)}{(f'(x))^2} \tag{1.107}$$

and

$$\begin{aligned} g''(x) &= \left(\frac{f(x)f''(x)}{(f'(x))^2}\right)' \\ &= \frac{(f'(x)f''(x) + f(x)f'''(x))(f'(x))^2 - f(x)f''(x)2f'(x)f''(x)}{(f'(x))^4} \\ &= \frac{(f'(x))^3 f''(x)}{(f'(x))^4} = \frac{f''(x)}{f'(x)} \end{aligned} \tag{1.108}$$

Evaluating these at $x = x^*$ at which $f(x^*) = 0$, and substituting them back into the expression for e_{n+1} we see that the linear term is zero as $g'(x^*) = 0$, and obtain the same result as in Eq. (1.104)

$$\lim_{n\to\infty} \frac{|e_{n+1}|}{|e_n|^2} \le \frac{|f''(x^*)|}{2|f'(x^*)|} = \frac{|g''(x^*)|}{2} \tag{1.109}$$

We see that if $f'(x^*) \neq 0$, the order of convergence of the Newton–Raphson method is $q = 2$ with the rate of convergence $\mu = |f''(x)|/2|f'(x)|$. However, if $f'(x^*) = 0$, the convergence is linear rather than quadratic, as shown in some of the examples.

1.C Proofs of the Fixed-Point and Contraction Mapping Theorems

Proof of the fixed-point theorem

Here the distance $d(\mathbf{x}-\mathbf{y})$ is specifically defined as the *p-norm* (Section A.4.1) of the vector $\mathbf{x}-\mathbf{y}$

$$d(\mathbf{x},\mathbf{y}) = ||\mathbf{x}-\mathbf{y}||_p = \left(\sum_{n=1}^{N}(x_n - y_n)^p\right)^{1/p} \tag{1.110}$$

where $p \geq 1$, for example $p = 1, 2, \infty$. For convenience, we can drop p so that $d(\mathbf{x},\mathbf{y}) = ||\mathbf{x}-\mathbf{y}||$.

- We first prove constructively the existence of a fixed point. As $\mathbf{g}(\mathbf{x})$ is a contraction, we have

$$\begin{aligned}||\mathbf{x}_{n+1}-\mathbf{x}_n|| &= ||\mathbf{g}(\mathbf{x}_n)-\mathbf{g}(\mathbf{x}_{n-1})|| \leq k||\mathbf{x}_n-\mathbf{x}_{n-1}|| \leq k^2||\mathbf{x}_{n-1}-\mathbf{x}_{n-2}|| \leq \cdots \\ &\leq k^n||\mathbf{x}_1-\mathbf{x}_0||\end{aligned} \tag{1.111}$$

As $0 \leq k < 1$, we have $\lim_{n\to\infty}||\mathbf{x}_{n+1}-\mathbf{x}_n|| = 0$. This is a *Cauchy sequence* (Section A.1) that converges to some point $\lim_{n\to\infty}\mathbf{x}_n = \mathbf{x}^*$ also in the space. We further have

$$\mathbf{g}(\mathbf{x}^*) = \mathbf{g}(\lim_{n\to\infty}\mathbf{x}_n) = \lim_{n\to\infty}\mathbf{g}(\mathbf{x}_n) = \lim_{n\to\infty}\mathbf{x}_{n+1} = \mathbf{x}^* \tag{1.112}$$

that is the limit of the Cauchy sequence $\lim_{n\to\infty}\mathbf{x}_n = x^*$ is a fixed point.

- We next prove the uniqueness of the fixed point. Let $\mathbf{x}_1^*$ and $\mathbf{x}_2^*$ be two fixed points of $\mathbf{g}(\mathbf{x})$, then we have

$$||\mathbf{g}(\mathbf{x}_1^*)-\mathbf{g}(\mathbf{x}_2^*)|| \leq k||\mathbf{x}_1^*-\mathbf{x}_2^*|| = k||\mathbf{g}(\mathbf{x}_1^*)-\mathbf{g}(\mathbf{x}_2^*)|| \tag{1.113}$$

For any $k \neq 0$, this equation holds only if $||\mathbf{x}_1^*-\mathbf{x}_2^*|| = ||\mathbf{g}(\mathbf{x}_1^*)-f(\mathbf{x}_2^*)|| = 0$, that is, $\mathbf{x}_1^* = \mathbf{x}_2^*$ is the unique fixed point. □

Proof of the contraction mapping theorem

Consider the Taylor expansion (Section A.7) of the function $\mathbf{g}(\mathbf{x})$ in the neighborhood of $\mathbf{x}^*$

$$\mathbf{g}(\mathbf{x}) = \mathbf{g}(\mathbf{x}^*)+\mathbf{g}'(\mathbf{x}^*)(\mathbf{x}-\mathbf{x}^*)+R(\mathbf{x}-\mathbf{x}^*) = \mathbf{g}(\mathbf{x}^*)+\mathbf{J}_\mathbf{g}(\mathbf{x}^*)(\mathbf{x}-\mathbf{x}^*)+R(\mathbf{x}-\mathbf{x}^*) \tag{1.114}$$

where $R(\mathbf{x}-\mathbf{x}^*)$ is the remainder composed of second- and higher-order terms of $\delta = \mathbf{x}-\mathbf{x}^*$. Subtracting $\mathbf{g}(\mathbf{x}^*)$ and taking any p-norm on both sides, we get

$$||\mathbf{g}(\mathbf{x})-\mathbf{g}(\mathbf{x}^*)||_p = ||\mathbf{J}_\mathbf{g}(\mathbf{x}^*)(\mathbf{x}-\mathbf{x}^*)+R(\mathbf{x}-\mathbf{x}^*)||_p \tag{1.115}$$

When $\mathbf{x}\to\mathbf{x}^*$, the second- and higher-order terms of $\mathbf{x}-\mathbf{x}^*$ disappear and $R(\mathbf{x}-\mathbf{x}^*)\to 0$, we have

$$||\mathbf{g}(\mathbf{x})-\mathbf{g}(\mathbf{x}^*)||_p = ||\mathbf{J}_\mathbf{g}(\mathbf{x}^*)(\mathbf{x}-\mathbf{x}^*)||_p \leq ||\mathbf{J}_\mathbf{g}(\mathbf{x}^*)||_p \cdot ||\mathbf{x}-\mathbf{x}^*||_p, \tag{1.116}$$

The inequality is due to the *Cauchy–Schwarz inequality* (Section A.1) if $||\mathbf{J}_\mathbf{g}(\mathbf{x}^*)|| < 1$, the function $\mathbf{g}(\mathbf{x})$ is a contraction at $\mathbf{x}^*$. □

2 Unconstrained Optimization

Optimization, also known as *mathematical programming*, is a field of mathematical study with the general goal of obtaining the best solution to a given problem, and it finds a wide range of applications in many different areas, such as to minimize the error or cost, or to maximize the yields or profits. Mathematically an optimization problem can be formulated to find the values of a set of variables represented by a vector $\mathbf{x} = [x_1, \cdots, x_N]^T$ so that an the *objective function* $f(\mathbf{x}) = f(x_1, \cdots, x_N)$ is either maximized or minimized. The solution of this problem, called the optimal solution, is denoted by $\mathbf{x}^*$. The fact that the function is either maximized or minimized is expressed as

$$\mathbf{x}^* = \arg\max_{\mathbf{x}} f((\mathbf{x}) \qquad \text{i.e.,} \qquad f(\mathbf{x}^*) \geq f(\mathbf{x}) \tag{2.1}$$

or

$$\mathbf{x}^* = \arg\min_{\mathbf{x}} f((\mathbf{x}) \qquad \text{i.e.,} \qquad f(\mathbf{x}^*) \leq f(\mathbf{x}) \tag{2.2}$$

The optimization problem is *unconstrained* if the variable $\mathbf{x}$ can take any value in the domain of the function, the entire domain $\mathbb{R}^N$ of the objective function $f(\mathbf{x})$, otherwise it is *constrained* if $\mathbf{x}$ is constrained to take values only in a subset of the domain, such as $||\mathbf{x}|| < 5$.

Optimization plays an essential role in machine learning (ML) as many learning algorithms can be formulated as solving an optimization problem so that certain objectives or criterion functions can be either minimized or maximized. In general, we would like to minimize the difference between the observed data and the model prediction by the learning method measured in some way (e.g. the sum of squared error), or to maximize the similarity between them measured in some way (e.g., likelihood of the model parameters). This issue will be discussed first in Section 4.1, and then throughout the book chapters wherever optimization is involved in the specific algorithms.

We will consider unconstrained optimization in this chapter and constrained optimization problems in Chapter 3. We will discuss various methods used for solving optimization problems in the most abstract form of maximizing or minimizing a given objective function $f(\mathbf{x})$, which will take some specific meanings in the context of various learning algorithms in the future chapters.

Two main types of algorithms widely used in unconstrained optimization are the gradient descent method and Newton's method, both of which find the optimal solution iteratively based on certain approximation of the objective function. Specifically, the gradient descent method approximates the objective function by the first two terms for the constant and linear components of its Taylor expansion, while Newton's method approximates the function by the first three terms of its Taylor expansion, including the third term for the quadratic component as well (see Section A.7). In other words, the gradient method depends only on the gradient vector of the function representing the local slope of the corresponding hypersurface, while Newton's method depends on the second-order Hessian matrix describing more the global shape of the surface as well as the local slope. Consequently, as Newton's method approximates the function more accurately, it converges more quickly than the gradient method. However, this improved efficiency relies on the second-order derivatives of the function, which may not be as conveniently available as the first-order derivatives. In practice, both methods can be used depending on the tradeoff between algorithmic efficiency and the extra cost for computing the second-order derivatives.

2.1 Optimization by Solving Equations

Before discussing specific methods for optimization, we first show that an unconstrained minimization problem can be converted into a problem of equation solving and vice versa, that is, these two types of problems are equivalent in this sense.

- To minimize an objective function $f(\mathbf{x})$, we can set its gradient vector $\mathbf{g}_f(\mathbf{x})$ to zero and solve the resulting equation system

$$\mathbf{g}_f(\mathbf{x}) = \nabla_{\mathbf{x}} f(\mathbf{x}) = \frac{df(\mathbf{x})}{d\mathbf{x}} = \begin{bmatrix} \partial f/\partial x_1 \\ \vdots \\ \partial f/\partial x_N \end{bmatrix} = \mathbf{0} \tag{2.3}$$

The solution $\mathbf{x}^*$ of this of equation, known as a *critical or stationary point* of $f(\mathbf{x})$, is also a solution to the corresponding optimization problem, provided it is not a saddle point – and either maximizes or minimizes $f(\mathbf{x})$.

This equation system can be solved by any of the methods discussed in the previous chapter, such as the Newton–Raphson method, which finds the root of a general function $\mathbf{f}(\mathbf{x})$ iteratively: $\mathbf{x}_{n+1} = \mathbf{x}_n - \mathbf{J}_f^{-1}(\mathbf{x}_n)\,\mathbf{f}(\mathbf{x}_n)$. Here, specifically, to solve the equation $\mathbf{g}_f(\mathbf{x}) = \mathbf{0}$, we first obtain the Jacobian $\mathbf{J}_g(\mathbf{x}) = \mathbf{H}_f(\mathbf{x})$ of the gradient $\mathbf{g}_f(\mathbf{x})$ of $f(\mathbf{x})$, which is the Hessian of $f(\mathbf{x})$, and then carry out the following iteration to eventually find $\mathbf{x}^*$ that minimizes $f(\mathbf{x})$

$$\mathbf{x}_{n+1} = \mathbf{x}_n - \mathbf{J}_g^{-1}(\mathbf{x}_n)\,\mathbf{g}_f(\mathbf{x}_n) = \mathbf{x}_n - \mathbf{H}_f^{-1}(\mathbf{x}_n)\,\mathbf{g}_f(\mathbf{x}_n) \tag{2.4}$$

The second equality is due to the fact that the Jacobian of the gradient $\mathbf{g}_f$ of function $\mathbf{f}(\mathbf{x})$ is the Hessian of the function, that is, $\mathbf{J}_g = \mathbf{H}_f$.

- To solve an equation system $\mathbf{f}(\mathbf{x}) = [f_1(\mathbf{x}), \cdots, f_N(\mathbf{x})]^T = \mathbf{0}$, we can minimize an objective function defined as

$$J(\mathbf{x}) = \frac{1}{2}\mathbf{f}^T(\mathbf{x})\mathbf{f}(\mathbf{x}) = \frac{1}{2}||\mathbf{f}(\mathbf{x})||^2 = \frac{1}{2}\sum_{i=1}^{N} |f_i(\mathbf{x})|^2 \tag{2.5}$$

  To minimize $J(\mathbf{x})$, we set its gradient to zero:

$$\mathbf{g}_J(\mathbf{x}) = \nabla_{\mathbf{x}} J(\mathbf{x}) = \frac{d}{d\mathbf{x}}\left[\frac{1}{2}\mathbf{f}^T(\mathbf{x})\mathbf{f}(\mathbf{x})\right] = \mathbf{f}'(\mathbf{x})\ \ \mathbf{f}(\mathbf{x}) = \mathbf{J}_\mathbf{f}(\mathbf{x})\,\mathbf{f}(\mathbf{x}) = \mathbf{0} \tag{2.6}$$

  If $\mathbf{J}_\mathbf{f}(\mathbf{x})$ is full rank and therefore invertible, then $\mathbf{f}(\mathbf{x}) = \mathbf{J}_\mathbf{f}^{-1}(\mathbf{x})\mathbf{0} = \mathbf{0}$, i.e., $\mathbf{x}$ is a root of $\mathbf{f}(\mathbf{x})$. If $\mathbf{J}_\mathbf{f}(\mathbf{x})$ is not full rank, then in general $\mathbf{f}(\mathbf{x})$ does not have a root, except in the special case $\mathbf{f}(\mathbf{x})$ happens to be zero.

2.2 Newton's Method

We first consider Newton's method applied to minimizing a univariate function $f(x)$. The Taylor series expansion (Section A.7) of the function $f(x)$ around a point x_0 is

$$\begin{aligned} f(x) &= f(x_0) + f'(x_0)(x - x_0) + \frac{1}{2}f''(x_0)(x - x_0)^2 + \cdots + \frac{1}{n!}f^{(n)}(x_0)(x - x_0)^n + \cdots \\ &\approx f(x_0) + f'(x_0)(x - x_0) + \frac{1}{2}f''(x_0)(x - x_0)^2 = q(x) \end{aligned} \tag{2.7}$$

If $f(x)$ is a quadratic function denoted by $q(x)$, then its Taylor series contains only the first three terms (constant, linear, and quadratic). In this case, we can find the vertex point at which $f(x) = q(x)$ reaches its extremum, by first setting its derivative to zero

$$\begin{aligned} q'(x) &= \frac{d}{dx}q(x) = \frac{d}{dx}\left[f(x_0) + f'(x_0)(x - x_0) + \frac{1}{2}f''(x_0)(x - x_0)^2\right] \\ &= f'(x_0) + f''(x_0)(x - x_0) = 0 \end{aligned} \tag{2.8}$$

and then solving the resulting equation to obtain

$$x^* = x = x_0 - \frac{f'(x_0)}{f''(x_0)} = x_0 + \Delta x_0 \tag{2.9}$$

where $\Delta x_0 = -f'(x_0)/f''(x_0)$ is the step we need to take to move from any initial point x_0 to the solution x^* in a single step. Note that $f(x^*)$ is a minimum if $f''(x^*) > 0$ but a maximum if $f''(x^*) < 0$.

If $f(x) \neq q(x)$ is not quadratic, the result given here can still be considered as an approximation of the solution, which can be improved iteratively from the initial guess x_0 to eventually approach the solution

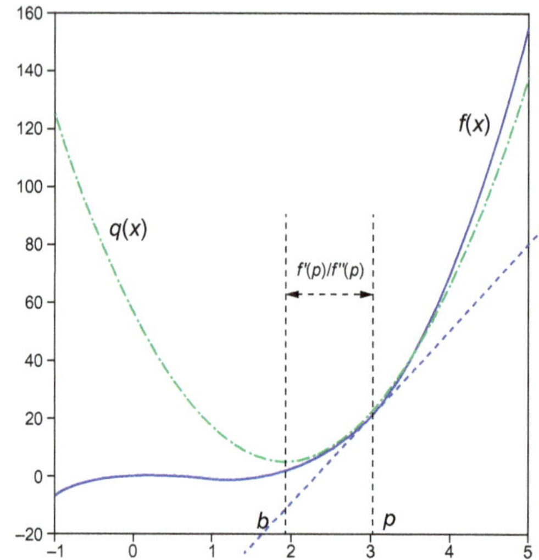

Fig. 2.1 Newton's method.

$$x_{n+1} = x_n + \Delta x_n = x_n - \frac{f'(x_n)}{f''(x_n)}, \qquad n = 0, 1, 2, \cdots \tag{2.10}$$

where $\Delta x_n = -f'(x_n)/f''(x_n)$ is the step size.

We see that in each iterative step, the function value $f(x_n)$ is fitted by a quadratic function $q(x_n)$ and its vertex at x_{n+1} is used as the updated approximated solution, at which the function is again fitted by another quadratic function $q(x_{n+1})$ for the next iteration, as shown in Fig. 2.1. Through this process the solution can be approached as $n \to \infty$.

We note that the iteration in Eq. (2.10) is the same as Eq. (1.46) in Chapter 1, if the function $f(x)$ is replaced by its derivative $f'(x)$. In other words, Newton's method for optimization is actually solving the equation $f'(x) = 0$, the necessary condition for an extremum of $f(x)$, by the Newton–Raphson method.

Newton's method for the minimization of a single-variable function $f(x)$ can be generalized for the minimization of a multivariable function $f(\mathbf{x}) = f(x_1, \cdots, x_N)$. Again, we approximate $f(\mathbf{x})$ by a quadratic function $q(\mathbf{x})$ containing only the first three terms of its Taylor series at some initial point $\mathbf{x}_0$

$$f(\mathbf{x}) \approx f(\mathbf{x}_0) + \mathbf{g}_0^T (\mathbf{x} - \mathbf{x}_0) + \frac{1}{2}(\mathbf{x} - \mathbf{x}_0)^T \mathbf{H}_0 (\mathbf{x} - \mathbf{x}_0) = q(\mathbf{x}) \tag{2.11}$$

where $\mathbf{g}_0$ and $\mathbf{H}_0$ are, respectively, the gradient vector and Hessian matrix of the function $f(\mathbf{x})$ at $\mathbf{x}_0$

$$\mathbf{g}_0 = \mathbf{g}_f(\mathbf{x}_0) = \frac{d}{d\mathbf{x}} f(\mathbf{x}_0) = \begin{bmatrix} \frac{\partial f(\mathbf{x}_0)}{\partial x_1} \\ \vdots \\ \frac{\partial f(\mathbf{x}_0)}{\partial x_N} \end{bmatrix},$$

$$\mathbf{H}_0 = \mathbf{H}_f(\mathbf{x}_0) = \frac{d}{d\mathbf{x}}\mathbf{g}(\mathbf{x}_0) = \frac{d^2}{d\mathbf{x}^2}f(\mathbf{x}_0) = \begin{bmatrix} \frac{\partial^2 f(\mathbf{x}_0)}{\partial x_1^2} & \cdots & \frac{\partial^2 f(\mathbf{x}_0)}{\partial x_1 \partial x_N} \\ \vdots & \ddots & \vdots \\ \frac{\partial^2 f(\mathbf{x}_0)}{\partial x_N \partial x_1} & \cdots & \frac{\partial^2 f(\mathbf{x}_0)}{\partial x_N^2} \end{bmatrix} \tag{2.12}$$

The stationary point of $f(\mathbf{x}) = q(\mathbf{x})$ can be found from any initial guess $\mathbf{x}_0$ by setting its derivative to zero

$$\begin{aligned} \frac{d}{d\mathbf{x}}q(\mathbf{x}) &= \frac{d}{d\mathbf{x}}\left[f(\mathbf{x}_0) + \mathbf{g}_0^T(\mathbf{x} - \mathbf{x}_0) + \frac{1}{2}(\mathbf{x} - \mathbf{x}_0)^T\mathbf{H}_0(\mathbf{x} - \mathbf{x}_0)\right] \\ &= \mathbf{g}_0 + \mathbf{H}_0(\mathbf{x} - \mathbf{x}_0) = \mathbf{0} \end{aligned} \tag{2.13}$$

and solving the resulting equation to obtain

$$\mathbf{x}^* = \mathbf{x} = \mathbf{x}_0 - \mathbf{H}(\mathbf{x}_0)^{-1}\mathbf{g}(\mathbf{x}_0) \tag{2.14}$$

Similar to the single-variable case in which $f(x^*)$ is either a minimum or maximum depends on whether the second-order derivative $f''(x^*)$ is greater or smaller than zero, here in multivariable case the function value $f(\mathbf{x}^*)$ at the stationary point is a maximum, minimum, or saddle point, depending on the curvature of the function represented by the Hessian matrix $\mathbf{H}^* = \mathbf{H}_f(\mathbf{x})|_{\mathbf{x}=\mathbf{x}^*}$ evaluated at $\mathbf{x}^*$.

- If $\mathbf{H}^* > 0$ is positive definite (all eigenvalues are positive), then $f(\mathbf{x})$ is *convex* and $f(\mathbf{x}^*)$ is the global minimum.
- If $\mathbf{H}^* < 0$ is negative definite (all eigenvalues are negative), then $f(\mathbf{x}$ is *concave* and $f(\mathbf{x}^*)$ is the global maximum.
- If $\mathbf{H}^*$ is indefinite (with both positive and negative eigenvalues), then $f(\mathbf{x})$ is neither convex nor concave, and $f(\mathbf{x}^*)$ is a saddle point (maximum in some directions, but minimum in others).

If $f(\mathbf{x})$ is quadratic and convex, the optimization problem is called *convex programming*, and its minimum can be easily obtained. (Equivalently, if $f(\mathbf{x})$ is quadratic and concave, its maximum can be easily obtained.)

But if $f(\mathbf{x}) \neq q(\mathbf{x})$ is not quadratic, the presented here is not a minimum or maximum, but it can still be used as an approximation of the solution, which can be improved iteratively to approach the solution

$$\mathbf{x}_{n+1} = \mathbf{x}_n + \Delta\mathbf{x}_n = \mathbf{x}_n - \mathbf{H}_n^{-1}\mathbf{g}_n = \mathbf{x}_n + \mathbf{d}_n, \qquad n = 0, 1, 2, \cdots \tag{2.15}$$

where $\mathbf{g}_n = \mathbf{g}(\mathbf{x}_n)$, $\mathbf{H}_n = \mathbf{H}(\mathbf{x}_n)$, and the increment $\mathbf{d}_n = \Delta\mathbf{x}_n = -\mathbf{H}_n^{-1}\mathbf{g}_n$ is called the *Newton search direction*. We note that this iteration is just a generalization of $x_{n+1} = x_n - f'(x_n)/f''(x_n)$ in 1-D case where $\mathbf{g}_n = f'(x_n)$ and $\mathbf{H}_n = f''(x_n)$, and the iteration $\mathbf{x}_{n+1} = \mathbf{x}_n - \mathbf{H}_n^{-1}\mathbf{g}_n$ is just Eq. (1.67) in Chapter 1, the Newton–Raphson method applied to solving equation $f'(\mathbf{x}) = \mathbf{g}_f(\mathbf{x}) = \mathbf{0}$, as the necessary condition for an extremum of $\mathbf{f}(\mathbf{x})$. The computational complexity for each iteration is $O(N^3)$ attributed to the inverse operation $\mathbf{H}_n^{-1}$.

For example, we can solve an over-constrained nonlinear equation system $\mathbf{f}(\mathbf{x}) = \mathbf{0}$ of M equations and $N < M$ unknowns by minimizing the following SSE

$$\varepsilon(\mathbf{x}) = \frac{1}{2}||\mathbf{f}(\mathbf{x})||^2 = \frac{1}{2}\sum_{m=1}^{M} f_m^2(\mathbf{x}) \tag{2.16}$$

To use the Newton–Raphson method, we first find the gradient of the error function

$$\mathbf{g}_\varepsilon(\mathbf{x}) = \frac{d}{d\mathbf{x}}\varepsilon(\mathbf{x}) = \frac{1}{2}\frac{d}{d\mathbf{x}}||\mathbf{f}(\mathbf{x})||^2 = \frac{d\mathbf{f}}{d\mathbf{x}}\,\mathbf{f} = \mathbf{J}_f^T\,\mathbf{f} \tag{2.17}$$

of which the ith component is

$$g_i = \frac{\partial}{\partial x_i}\left(\frac{1}{2}||\mathbf{f}||^2\right) = \frac{1}{2}\sum_{m=1}^{M}\frac{\partial}{\partial x_i}f_m^2 = \sum_{m=1}^{M}\frac{\partial f_m}{\partial x_i}\,f_m = \sum_{m=1}^{M} J_{mi}\,f_m \tag{2.18}$$

where $J_{mi} = \partial f_m/\partial x_i$ is the component of the function's Jacobian $\mathbf{J}_f(\mathbf{x})$ in the ith row and ith column.

We further find the component of the Hessian $\mathbf{H}_\varepsilon$ of the error function in the ith row and jth column

$$\begin{aligned} H_{ij} &= \frac{\partial^2}{\partial x_i \partial x_j}\left(\frac{1}{2}||\mathbf{f}||^2\right) = \frac{\partial}{\partial x_j}g_i = \sum_{m=1}^{M}\frac{\partial}{\partial x_j}\left[f_m\frac{\partial f_m}{\partial x_i}\right] \\ &= \sum_{m=1}^{M}\left[\frac{\partial f_m}{\partial x_i}\frac{\partial f_m}{\partial x_j} + f_m\frac{\partial^2 f_m}{\partial x_i \partial x_j}\right] \\ &\approx \sum_{m=1}^{M}\frac{\partial f_m}{\partial x_i}\frac{\partial f_m}{\partial x_j} = \sum_{m=1}^{M} J_{mi}J_{mj} \end{aligned} \tag{2.19}$$

Here J_{ij} is the component of the Jacobian $\mathbf{J}_f$ in the ith row and jth column, and we have dropped the second-order terms $f_m\,(\partial^2 f_m/\partial x_i \partial x_j)$. Now the Hessian matrix can be written as

$$\mathbf{H}_\varepsilon(\mathbf{x}) = \frac{d^2}{d\mathbf{x}^2}\varepsilon(\mathbf{x}) = \frac{d}{d\mathbf{x}}\mathbf{g}_\varepsilon(\mathbf{x}) \approx \mathbf{J}_f^T\mathbf{J}_f \tag{2.20}$$

Based on Newton–Raphson's method, the optimal solution $\mathbf{x}^*$ that minimizes $\varepsilon(\mathbf{x})$ can be found iteratively

$$\mathbf{x}_{n+1} = \mathbf{x}_n - \mathbf{H}_n^{-1}\mathbf{g}_n \approx \mathbf{x}_n - (\mathbf{J}_n^T\mathbf{J}_n)^{-1}\mathbf{J}_n^T\mathbf{f}_n \tag{2.21}$$

This is the same as pseudo-inverse method in Eq. (1.76) in Chapter 1.

Example 2.1

Consider the following quadratic function for example

$$q(\mathbf{x} = q(x_1,x_2)) = \frac{1}{2}\mathbf{x}^T\mathbf{A}\mathbf{x} = \frac{1}{2}[x_1\ x_2]\begin{bmatrix} a & b/2 \\ b/2 & c \end{bmatrix}\begin{bmatrix} x_1 \\ x_2 \end{bmatrix} = \frac{1}{2}(ax_1^2 + bx_1x_2 + cx_2^2)$$

The gradient vector and Hessian matrix of this function are, respectively

$$\mathbf{g} = \begin{bmatrix} ax_1 + bx_2/2 \\ bx_1/2 + cx_2 \end{bmatrix}, \qquad \mathbf{H} = \begin{bmatrix} a & b/2 \\ b/2 & c \end{bmatrix} = \mathbf{A}$$

The two eigenvalues of the symmetric matrix $\mathbf{H}$ can be found to be

$$\lambda_{1,2} = \frac{1}{2}[(a+c) \pm \sqrt{(a+c)^2 - 4ac + b^2}] \tag{2.22}$$

and the determinant of $\mathbf{H}$ is $\det \mathbf{H} = \lambda_1 \lambda_2 = ac - b^2/4$. The shape and extreme of the surface represented by $q(\mathbf{x})$ are determined by the coefficients a, b, c, which can take any values such as those discussed in the following and shown in Fig. 2.2.

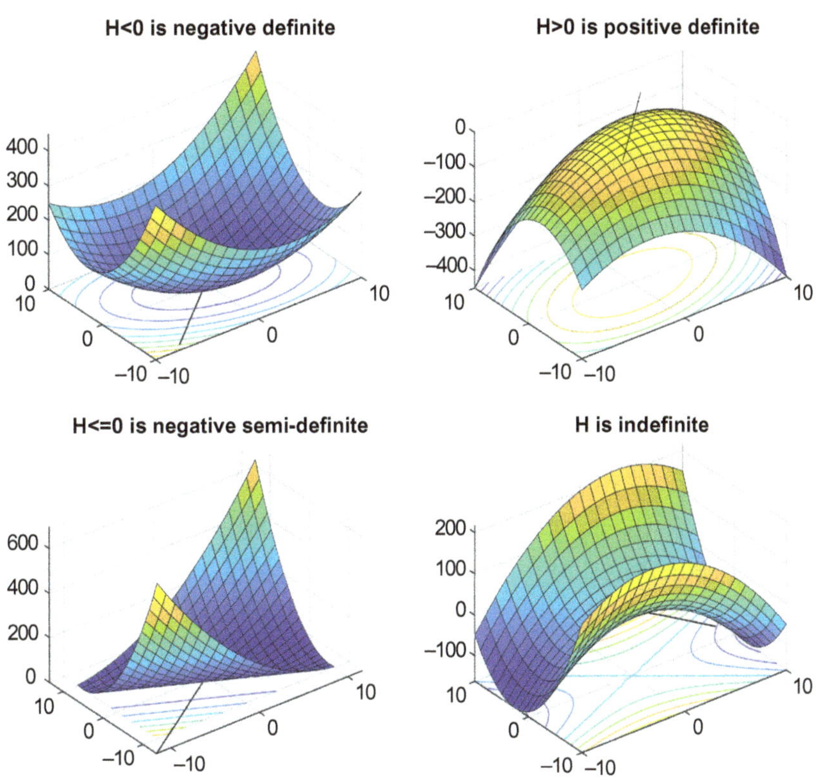

Fig. 2.2 Quadratic surfaces.
The contour lines of the quadratic function are shown in color-coded lines and the gradient vectors at the extreme/saddle points are shown in black.

- $a = 3$, $b = 2$, $c = 4$, $\lambda_1 = 2.382$, $\lambda_2 = 4.618$ are both positive, and $\mathbf{H}$ is positive definite. $q(\mathbf{x})$ has a minimum at $q(0, 0) = 0$.
- $a = -3$, $b = 2$, $c = -4$, $\lambda_1 = -4.618$ $\lambda_2 = -2.382$ are both negative, and $\mathbf{H}$ is negative definite. $q(\mathbf{x})$ has a maximum at $q(0, 0) = 0$.
- $a = 3$, $b = \sqrt{4ac}$, $c = 4$, $\lambda_1 = 7$, $\lambda_2 = 0$, and $\mathbf{H}$ is positive semi-definite. $q(0,0)$ is a minimum along one direction but constant along another.

- $a = -3,\ b = 2,\ c = 4,\ \lambda_1 = -3.140,\ \lambda_2 = 4.140$ are of different signs, and $\mathbf{H}$ is indefinite. $q(0,0) = 0$ is a saddle point, which is a maximum along one direction but a minimum along another.

These properties based on the (semi) definiteness of the Hessian $\mathbf{H}$ can be generalized to quadratic functions with more variables than two (see Section A.2.3).

We can speed up the convergence by a bigger step size $\delta > 1$. However, if δ is too big, the solution may be skipped and the iteration may not converge if it gets into an oscillation around the solution. Even worse, the iteration may become divergent. For such reasons, a smaller step size $\delta < 1$ may be preferred sometimes.

In summary, Newton's method approximates the function $f(\mathbf{x})$ at an estimated solution $\mathbf{x}_n$ by a quadratic equation (the first three terms of the Taylor's series) based on the gradient $\mathbf{g}_n$ and Hessian $\mathbf{H}$ of the function at $\mathbf{x}_n$, and treats the vertex of the quadratic equation as the updated estimate $\mathbf{x}_{n+1}$.

Newton's method requires the Hessian matrix as well as the gradient to be available. Moreover, it is necessary to calculate the inverse of the Hessian matrix in each iteration, which may be computationally expensive.

Example 2.2

The Newton's method is applied to solving the following non-linear equation system of $N = 3$ variables:

$$\begin{cases} f_1(x_1, x_2, x_3) = 3x_1 - (x_2 x_3)^2 - 3/2 \\ f_2(x_1, x_2, x_3) = 4x_1^2 - 625\,x_2^2 + 2x_2 - 1 \\ f_3(x_1, x_2, x_3) = exp(-x_1 x_2) + 20x_3 + 9 \end{cases}$$

with the exact solution ($x_1 = 0.5,\ x_2 = 0,\ x_3 = -0.5$). These equations can be expressed in vector form as $\mathbf{f}(\mathbf{x}) = \mathbf{0}$ and solved as an optimization problem with the objective function $J(\mathbf{x}) = \mathbf{f}^T(\mathbf{x})\mathbf{f}(\mathbf{x})$. The iteration from an initial guess $\mathbf{x}_0 = \mathbf{0}$ is shown in Table 2.1.

Table 2.1

n	$\mathbf{x} = [x_1, x_2, x_3]$	$\|\|\mathbf{f}(\mathbf{x})\|\|$
0	0.000000, 0.000000, 0.000000	$1.016120e+01$
1	0.500000, 0.500000, −0.500000	$1.552502e+02$
2	0.499550, 0.250800, −0.493801	$3.881300e+01$
3	0.500096, 0.126206, −0.496852	$9.702208e+00$
4	0.500025, 0.063914, −0.498405	$2.425198e+00$
5	0.500010, 0.032778, −0.499181	$6.059054e-01$
6	0.500005, 0.017231, −0.499570	$1.510777e-01$
7	0.500003, 0.009498, −0.499763	$3.737330e-02$
8	0.500002, 0.005712, −0.499857	$8.959365e-03$
9	0.500001, 0.003968, −0.499901	$1.900145e-03$

Table 2.1 **(cont.)**

n	$\mathbf{x} = [x_1, x_2, x_3]$	$\|\|\mathbf{f}(\mathbf{x})\|\|$
10	0.500001, 0.003326, −0.499917	$2.577603e-04$
11	0.500001, 0.003206, −0.499920	$8.932714e-06$
12	0.500001, 0.003202, −0.499920	$1.238536e-08$
13	0.500001, 0.003202, −0.499920	$2.371437e-14$

We see that after 13 iterations the algorithm converges to the following approximated solution with accuracy of $J(\mathbf{x}) = ||\mathbf{f}(\mathbf{x}^*)||^2 \approx 10^{-28}$

$$\mathbf{x}^* = \begin{bmatrix} 0.5000008539707297 \\ 0.0032017070323056 \\ -0.4999200212218281 \end{bmatrix} \tag{2.23}$$

2.3 Gradient Descent Method

Newton's method discussed in the previous section is based on the gradient $\mathbf{g}_f(\mathbf{x})$ and Hessian $\mathbf{H}_f(\mathbf{x})$, representing the local variation and more global curvature of the function $f(\mathbf{x})$ to be minimized. The method is not applicable if the Hessian $\mathbf{H}_f$ is not available, because either the analytical form of the function is not known or the cost for computing the inverse $\mathbf{H}_f^{-1}$ is too high. In such cases, we can use the gradient descent method based only on the gradient but no longer the Hessian.

We first consider the minimization of a single-variable function $f(x)$. From any initial point x_0, we can move to a nearby point x along the opposite direction of the gradient, so that the function value will be smaller than that at the current position x_0

$$x = x_0 - \delta f'(x_0) = x_0 + \Delta x_0, \quad \text{where} \quad \Delta x_0 = -\delta f'(x_0), \quad \delta > 0 \tag{2.24}$$

We see that regardless of whether $f'(x_0)$ is positive or negative, the function value $f(x)$ approximated by the first two terms of its Taylor series is always reduced if the positive step size $delta$ is small enough

$$f(x) \approx f(x_0) + f'(x_0)\Delta x_0 = f(x_0) - |f'(x_0)|^2\delta < f(x_0) \tag{2.25}$$

This process can be carried out iteratively

$$x_{n+1} = x_n - \delta_n\, f'(x_n), \qquad n = 0, 1, 2, \cdots \tag{2.26}$$

until eventually reaching a point x^* at which $f'(x^*) = 0$ and no further progress can be made, that is, a local minimum of the function is reached.

This simple method can be generalized to minimize a multivariable objective function $f(\mathbf{x}) = f(x_1, \cdots, x_N)$ in N–D space. The derivative of the 1-D case is generalized to the gradient vector $\mathbf{g}(\mathbf{x}) = df(\mathbf{x})/d\mathbf{x}$ of function $f(\mathbf{x})$, which is in the direction along which the function increases most rapidly with the steepest slope, perpendicular to the contour or isolines of the function $f(\mathbf{x})$. The fastest way to reduce $f(\mathbf{x})$ is to go downhill along the opposite direction of the gradient vector.

Specifically the gradient descent method (also called steepest descent or downhill method) carries out the iteration $\mathbf{x} = \mathbf{x}_0 + \Delta\mathbf{x}$, which can be considered as the first two terms of the Taylor series of $f(\mathbf{x})$ with $\Delta\mathbf{x} = -\delta\mathbf{g}$ ($\delta > 0$)

$$f(\mathbf{x}) \approx f(\mathbf{x}_0) + \mathbf{g}_0^T\Delta\mathbf{x} = f(\mathbf{x}_0) - \delta\mathbf{g}_0^T\mathbf{g}_0 = f(\mathbf{x}_0) - \delta||\mathbf{g}||^2 < f(\mathbf{x}_0) \tag{2.27}$$

As the function value is always reduced by this iteration, it can be carried out repeatedly

$$\mathbf{x}_{n+1} = \mathbf{x}_n - \delta_n\,\mathbf{g}_n = (\mathbf{x}_{n-1} - \delta_{n-1}\,\mathbf{g}_{n-1}) - \delta_n\,\mathbf{g}_n = \cdots = \mathbf{x}_0 - \sum_{i=0}^{n} \delta_n\,\mathbf{g}_i \tag{2.28}$$

until eventually reaching a point $\mathbf{x}^*$ at which $\mathbf{g}(\mathbf{x}^*) = \mathbf{0}$ and $f(\mathbf{x})$ is minimized.

Comparing Eq. (2.28) with Newton's method where the search direction is $\mathbf{d}_n = -\mathbf{H}_n^{-1}\mathbf{g}_n$ in Eq. (2.15) based on both $\mathbf{H}_n$ and $\mathbf{g}_n$, we see that here in the gradient descent method the search direction $\mathbf{d}_n = -\mathbf{g}_n$ relies solely on local information from the gradient $\mathbf{g}$, without incorporating the more global curvature information provided by the second-order Hessian $\mathbf{H}$. As a result, the gradient method typically has less information than Newton's method and may therefore be less efficient. For example, when the function is quadratic, as discussed before, Newton's method can find the solution in a single step from any initial guess, but it takes the gradient descent method multiple steps to reach the solution, as it always follows the negative direction of the local gradient, which does not lead toward the solution directly in general. However, the gradient descent method is computationally less expensive as the Hessian matrix is not needed.

Example 2.3

Consider a two-variable quadratic function

$$f(\mathbf{x}) = \mathbf{x}^T\mathbf{A}\mathbf{x} = [x_1\ x_2]\begin{bmatrix} a_{11} & a_{12} \\ a_{21} & a_{22} \end{bmatrix}\begin{bmatrix} x_1 \\ x_2 \end{bmatrix}$$
$$= [x_1\ x_2]\begin{bmatrix} 2 & 1 \\ 1 & 1 \end{bmatrix}\begin{bmatrix} x_1 \\ x_2 \end{bmatrix} = 2x_1^2 + 2x_1x_2 + x_2^2$$

where $\mathbf{A}$ is a symmetric positive definite matrix. This function has a minimum $f(x_1, x_2) = 0$ at $x_1 = x_2 = 0$.

We select an initial guess $\mathbf{x}_0 = [1,\ 2]^T$, at which the gradient is $\mathbf{g}_0 = [4,\ 3]^T$, as shown in Fig. 2.3. Now we compare the gradient method with Newton's method, in terms of the search direction $\mathbf{d}$ and progress of the iterations.

- Newton's method: Here we have

$$\mathbf{g} = \begin{bmatrix} 2x_1 + x_2 \\ x_1 + x_2 \end{bmatrix}, \quad \mathbf{H} = \mathbf{A} = \begin{bmatrix} 2 & 1 \\ 1 & 1 \end{bmatrix}, \quad \mathbf{H}^{-1} = \begin{bmatrix} 1 & -1 \\ -1 & 2 \end{bmatrix}$$

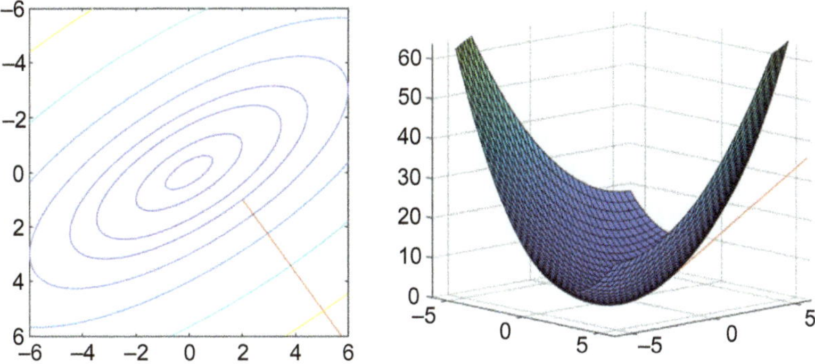

Fig. 2.3 Gradient of a quadratic surface.

and the search direction is $\mathbf{d}_0 = -\mathbf{H}^{-1}\mathbf{g}_0 = -[1,\ 2]^T$ (the red arrow in the figure). In a single iteration we reach

$$\mathbf{x}_1 = \mathbf{x}_0 - \mathbf{H}^{-1}\mathbf{g}_0 = \begin{bmatrix} 1 \\ 2 \end{bmatrix} - \begin{bmatrix} 1 & -1 \\ -1 & 2 \end{bmatrix} \begin{bmatrix} 4 \\ 3 \end{bmatrix} = \begin{bmatrix} 0 \\ 0 \end{bmatrix}$$

which is the minimum of the function.

- The gradient descent method: the search direction is $\mathbf{d}_0 = -\mathbf{g}_0 = -[4,\ 3]^T$, perpendicular to the contour of the function. The first iteration is

$$\mathbf{x}_1 = \mathbf{x}_0 - \delta\mathbf{g}_0 = \begin{bmatrix} 1 \\ 2 \end{bmatrix} - \delta \begin{bmatrix} 4 \\ 3 \end{bmatrix} = \begin{bmatrix} 1 - \delta 4 \\ 2 - \delta 3 \end{bmatrix}$$

We need to determine the step size δ (to be considered later) to find $\mathbf{x}_1$, and then continue the iteration.

Figure 2.4 compares the different search directions by Newton's method (red) and the gradient descent method (blue). We see that the solution is reached in a single step in Newton's method, based on the Hessian matrix $\mathbf{H}$ representing some global information of the elliptical shape of the contour line of the quadratic function, as well as the gradient $\mathbf{g}_0$ representing the local information at $\mathbf{x}_0$. In comparison, the search direction of the gradient descent method is based on only the local information in the gradient $\mathbf{g}_0$ without more global information in $\mathbf{H}$. Consequently, multiple steps are needed for the iteration to approach the solution. The gradient descent method is less effective than Newton's method owing to the lack of second-order derivatives carrying more global information.

In the gradient descent method, we also need to determine a proper step size δ. If δ is too small, the iteration may converge too slowly, especially when $f(x)$ reduces slowly toward its minimum. On the other hand, if δ is too large but $f(x)$ has some rapid variations in the local region, the minimum of the function may be skipped and the iteration may not converge. We will consider how to find the optimal step size in the next section.

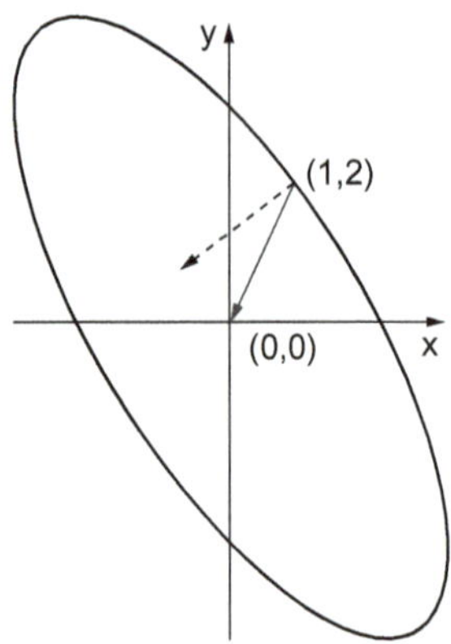

Fig. 2.4 Comparison of gradient descent and Newton's method.

2.4 Line Minimization

In general, in any iterative algorithm for minimizing a function $f(\mathbf{x})$, including Newton's method (Eq. (2.15)) and the gradient descent method (Eq. (2.28)), the variable $\mathbf{x}$ is updated iteratively $\mathbf{x}_{n+1} = \mathbf{x}_n + \delta_n \mathbf{d}_n$ based on both the search direction $\mathbf{d}_n$ and the step size δ, which need to be determined in such a way that the function value $f(\mathbf{x})$ is maximally reduced at each iteration.

First, the search direction $\mathbf{d}_n$ needs to point away from the gradient $\mathbf{g}_n$ along which $f(\mathbf{x}_n)$ increases most rapidly. In other words, the angle between $\mathbf{d}_n$ and $\mathbf{g}_n$ should be greater than $\pi/2$

$$\cos^{-1}\left(\frac{\mathbf{d}_n^T \mathbf{g}_n}{||\mathbf{d}_n||\ ||\mathbf{g}_n||}\right) > \frac{\pi}{2} \qquad \text{i.e.,} \qquad \mathbf{d}_n^T \mathbf{g}_n < 0 \tag{2.29}$$

This condition is indeed satisfied in both the gradient descent method with $\mathbf{d}_n = -\mathbf{g}_n$ and Newton's method with $\mathbf{d}_n = -\mathbf{H}_n^{-1}\mathbf{g}_n$

$$\mathbf{d}_n^T \mathbf{g}_n = -\mathbf{g}_n^T \mathbf{H}_n^{-1} \mathbf{g}_n < 0, \qquad \mathbf{d}_n^T \mathbf{g}_n = -\mathbf{g}_n^T \mathbf{g}_n = -||\mathbf{g}_n||^2 < 0 \tag{2.30}$$

Here $\mathbf{H}_n$ is assumed to be positive definite so that a minimum exists.

Next, the optimal step size δ_n needs to be such that the function value $f(\mathbf{x}_{n+1}) = f(\mathbf{x}_n + \delta_n \mathbf{d}_n)$ is minimized along the search direction $\mathbf{d}_n$. To do so, we set to zero the derivative of the function value with respect to δ_n, and obtain the *directional derivative* (Section A.4.3) along the direction of $\mathbf{d}_n$ based on the chain rule

$$\frac{d}{d\delta_n} f(\mathbf{x}_{n+1}) = \frac{d}{d\delta_n} f(\mathbf{x}_n + \delta_n \mathbf{d}_n) = \left(\frac{d\,f(\mathbf{x}_{n+1})}{d\mathbf{x}}\right)^T \frac{d(\mathbf{x}_n + \delta_n \mathbf{d}_n)}{d\delta_n} = \mathbf{g}_{n+1}^T \mathbf{d}_n = 0 \tag{2.31}$$

This result indicates that ideally the gradient $\mathbf{g}_{n+1} = f'(\mathbf{x}_{n+1})$ at the next point $\mathbf{x}_{n+1}$ should be perpendicular to the search direction $\mathbf{d}_n$. In other words, when traversing along $\mathbf{d}_n$, we should stop at the point $\mathbf{x}_{n+1} = \mathbf{x}_n + \delta \mathbf{d}_n$ at which the gradient $\mathbf{g}_{n+1}$ is perpendicular to $\mathbf{d}_n$, that is, it has zero component along $\mathbf{d}_n$, and the corresponding δ is the optimal step size.

To find such an optimal step size δ, we can consider $f(\mathbf{x}_{n+1}) = f(\mathbf{x}_n + \delta\mathbf{d}_n)$ as a function of δ treated as the single variable, and find the optimal value that minimizes $f(\mathbf{x}_{n+1})$. To do so, we assume that the value of δ is small enough so that this function can be approximated by the first three terms of its Taylor series at $\delta = 0$ (Maclaurin series of function $f(\delta)$)

$$\begin{aligned} f(\mathbf{x}_{n+1}) &= f(\mathbf{x}_n + \delta\mathbf{d}_n) \\ &\approx [f(\mathbf{x}_n + \delta\mathbf{d}_n)]_{\delta=0} + \delta \left[\frac{d}{d\delta} f(\mathbf{x}_n + \delta\mathbf{d}_n)\right]_{\delta=0} + \frac{\delta^2}{2}\left[\frac{d^2}{d\delta^2} f(\mathbf{x}_n + \delta\mathbf{d}_n)\right]_{\delta=0} \end{aligned} \tag{2.32}$$

The three terms of this expression can be further written as

$$[f(\mathbf{x}_n + \delta\mathbf{d}_n)]_{\delta=0} = f(\mathbf{x}_n) \tag{2.33}$$

$$\left[\frac{d}{d\delta} f(\mathbf{x}_n + \delta\mathbf{d}_n)\right]_{\delta=0} = \left[\mathbf{g}(\mathbf{x}_n + \delta\mathbf{d}_n)^T\mathbf{d}_n\right]_{\delta=0} = \mathbf{g}_n^T\mathbf{d}_n \tag{2.34}$$

$$\begin{aligned} \left[\frac{d^2}{d\delta^2} f(\mathbf{x}_n + \delta\mathbf{d}_n)\right]_{\delta=0} &= \left[\frac{d}{d\delta}\mathbf{g}(\mathbf{x}_n + \delta\mathbf{d}_n)^T\right]_{\delta=0} \frac{d}{d\delta}(\mathbf{x}_n + \delta\mathbf{d}_n) \\ &= \left[\frac{d}{d\mathbf{x}}\mathbf{g}(\mathbf{x})\ \frac{d}{d\delta}(\mathbf{x}_n + \delta\mathbf{d}_n)\right]_{\delta=0}^T \mathbf{d}_n \\ &= (\mathbf{H}_n\mathbf{d}_n)^T\mathbf{d}_n = \mathbf{d}_n^T\mathbf{H}_n\mathbf{d}_n \end{aligned} \tag{2.35}$$

where $\mathbf{H}_n = \mathbf{H}_n^T$ is the Hessian matrix of $f(\mathbf{x})$ at $\mathbf{x}_n$. Substituting these back into Eq. (2.32) we get:

$$f(\mathbf{x}_n + \delta\mathbf{d}_n) \approx f(\mathbf{x}_n) + \delta\,\mathbf{g}_n^T\,\mathbf{d}_n + \frac{\delta^2}{2}\,\mathbf{d}_n^T\mathbf{H}_n\,\mathbf{d}_n \tag{2.36}$$

To find the optimal δ that minimizes $f(\mathbf{x}_{n+1}) = f(\mathbf{x}_n + \delta\mathbf{d}_n)$, we first find its derivative with respect to δ and then set it to zero

$$\begin{aligned} \frac{d}{d\delta} f(\mathbf{x}_n + \delta\mathbf{d}_n) &\approx \frac{d}{d\delta}\left(f(\mathbf{x}_n) + \delta\mathbf{g}_n^T\,\mathbf{d}_n + \frac{\delta^2}{2}\mathbf{d}_n^T\mathbf{H}_n\,\mathbf{d}_n\right) \\ &= \mathbf{g}_n^T\,\mathbf{d}_n + \delta\,\mathbf{d}_n^T\mathbf{H}_n\,\mathbf{d}_n = 0 \end{aligned} \tag{2.37}$$

Solving this equation for δ we have

$$\delta_n = -\frac{\mathbf{g}_n^T\mathbf{d}_n}{\mathbf{d}_n^T\mathbf{H}_n\mathbf{d}_n} \tag{2.38}$$

Note that only if the value of this δ_n is small enough for the approximation in Eq. (2.32) to be good enough, will it be a good approximation of the true optimal step size (and each search direction is perpendicular to the previous one). Otherwise it may be very different from the true optimal step size.

Based on this result, we can get the optimal step size for both the Newton's method and gradient descent method as shown in the following:

- **Newton's method**
  The search direction is $\mathbf{d}_n = -\mathbf{H}_n^{-1}\mathbf{g}_n$, and the optimal step size is

$$\delta_n = -\frac{\mathbf{g}_n^T\mathbf{d}_n}{\mathbf{d}_n^T\mathbf{H}_n\mathbf{d}_n} = \frac{\mathbf{g}_n^T(\mathbf{H}_n^{-1}\mathbf{g}_n)}{(\mathbf{H}_n^{-1}\mathbf{g}_n)^T\mathbf{H}_n(\mathbf{H}_n^{-1}\mathbf{g}_n)} = \frac{\mathbf{g}_n^T(\mathbf{H}_n^{-1}\mathbf{g}_n)}{(\mathbf{H}_n^{-1}\mathbf{g}_n)^T\mathbf{g}_n} = 1 \tag{2.39}$$

  This confirms that the iteration in Eq. (2.15) is indeed optimal

$$\mathbf{x}_{n+1} = \mathbf{x}_n + \delta_n\mathbf{d}_n = x_n - \delta_n\mathbf{H}_n^{-1}\mathbf{g}_n = \mathbf{x}_n - \mathbf{H}_n^{-1}\mathbf{g}_n \tag{2.40}$$

- **Gradient descent method**
  The search direction is $\mathbf{d}_n = -\mathbf{g}_n$, and Eq. (2.31) becomes

$$\mathbf{g}_{n+1}^T\mathbf{d}_n = -\mathbf{g}_{n+1}^T\mathbf{g}_n = 0, \qquad \text{i.e.,} \qquad \mathbf{g}_{n+1} \perp \mathbf{g}_n \tag{2.41}$$

  indicating that the search direction $\mathbf{d}_{n+1} = -\mathbf{g}_{n+1}$ is always perpendicular to the previous one $\mathbf{d}_n = -\mathbf{g}_n$ (i.e., the iteration follows a zigzag path composed of a sequence of segments from the initial guess to the final solution). The optimal step size is

$$\delta_n = -\frac{\mathbf{g}_n^T\mathbf{d}_n}{\mathbf{d}_n^T\mathbf{H}_n\mathbf{d}_n} = \frac{\mathbf{g}_n^T\mathbf{g}_n}{\mathbf{g}_n^T\mathbf{H}_n\mathbf{g}_n} = \frac{||\mathbf{g}_n||^2}{\mathbf{g}_n^T\mathbf{H}_n\mathbf{g}_n} \tag{2.42}$$

  Given $\mathbf{H}_n$ as well as $\mathbf{g}_n$, we can find the optimal step size δ_n with complexity $O(N^2)$ for each iteration. However, as we assume the Hessian $\mathbf{H}_n$ is not available in the gradient descent method, the given optimal step size cannot be computed. We can instead approximate $f''(\delta)|_{\delta=0}$ in the third term of Eq. (2.32) at two nearby points at $\delta = 0$ and $\delta = \sigma$, where σ is a small value

$$\left[\frac{d^2}{d\delta^2}f(\mathbf{x}+\delta\mathbf{d})\right]_{\delta=0} = \left[\frac{d}{d\delta}f'(\mathbf{x}+\delta\mathbf{d})\right]_{\delta=0} = \lim_{\sigma\to 0}\frac{f'(\mathbf{x}+\sigma\mathbf{d}) - f'(\mathbf{x})}{\sigma}$$
$$\approx \frac{\mathbf{g}^T(\mathbf{x}+\sigma\mathbf{d})\,\mathbf{d} - \mathbf{g}^T(\mathbf{x})\,\mathbf{d}}{\sigma} = \frac{\mathbf{g}_\sigma^T\mathbf{d} - \mathbf{g}^T\mathbf{d}}{\sigma} \tag{2.43}$$

where $\mathbf{g} = \mathbf{g}(\mathbf{x})$ and $\mathbf{g}_\sigma = \mathbf{g}(\mathbf{x}+\sigma\mathbf{d})$. This approximation can be used to replace $\mathbf{d}_n^T\mathbf{H}_n\mathbf{d}_n$ in Eq. (2.37)

$$\frac{d}{d\delta}f(\mathbf{x}_n+\delta\mathbf{d}_n) \approx \mathbf{g}_n^T\mathbf{d}_n + \frac{\delta}{\sigma}\left(\mathbf{g}_{\sigma n}^T\mathbf{d}_n - \mathbf{g}_n^T\mathbf{d}_n\right) = 0 \tag{2.44}$$

Solving for δ we achieve the estimated optimal step size

$$\delta_n = -\frac{\sigma\,\mathbf{g}_n^T\mathbf{d}_n}{\mathbf{g}_{\sigma n}^T\mathbf{d}_n - \mathbf{g}_n^T\mathbf{d}_n} = -\frac{\sigma\,\mathbf{g}_n^T\mathbf{d}_n}{(\mathbf{g}_{\sigma n} - \mathbf{g}_n)^T\mathbf{d}_n} \tag{2.45}$$

Specifically in the gradient descent method with $\mathbf{d}_n = -\mathbf{g}_n$, this optimal step size becomes

$$\delta_n = -\frac{\sigma\,\mathbf{g}_n^T\mathbf{d}_n}{(\mathbf{g}_{\sigma n} - \mathbf{g}_n)^T\mathbf{d}_n} = -\frac{\sigma\,\mathbf{g}_n^T\mathbf{g}_n}{(\mathbf{g}_{\sigma n} - \mathbf{g}_n)^T\mathbf{g}_n} = \frac{\sigma\,||\mathbf{g}_n||^2}{||\mathbf{g}_n||^2 - \mathbf{g}_{\sigma n}^T\mathbf{g}_n} \tag{2.46}$$

and the iteration becomes

$$\mathbf{x}_{n+1} = \mathbf{x}_n - \delta_n\mathbf{g}_n = \mathbf{x}_n + \frac{\sigma\,||\mathbf{g}_n||^2}{\mathbf{g}_{\sigma n}^T\mathbf{g}_n - ||\mathbf{g}_n||^2}\,\mathbf{g}_n \tag{2.47}$$

Example 2.4

The gradient descent method applied to solve the same three-variable equation system previously solved by Newton's method

$$\begin{cases} f_1(x_1, x_2, x_3) = 3x_1 - (x_2x_3)^2 - 3/2 \\ f_2(x_1, x_2, x_3) = 4x_1^2 - 625\, x_2^2 + 2x_2 - 1 \\ f_3(x_1, x_2, x_3) = exp(-x_1x_2) + 20x_3 + 9 \end{cases}$$

The step size δ_n is determined by the secant method with $\sigma = 10^{-6}$. The iteration from an initial guess $\mathbf{x}_0 = \mathbf{0}$ is shown in the following table.

Table 2.2

n	$\mathbf{x} = [x_1, x_2, x_3]$	$\lvert\lvert\mathbf{f}(\mathbf{x})\rvert\rvert$
0	0.0000, 0.0000, 0.0000	$1.032500e+02$
10	0.4246, −0.0073, −0.5002	$1.535939e-01$
20	0.5015, 0.0064, −0.4998	$2.448241e-05$
30	0.5009, 0.0057, −0.4998	$9.178209e-06$
40	0.5006, 0.0052, −0.4998	$4.17587e-06$
50	0.5004, 0.0049, −0.4999	$2.122594e-06$
60	0.5003, 0.0047, −0.4999	$1.182466e-06$
100	0.5001, 0.0043, −0.4999	$1.805871e-07$
150	0.5000, 0.0041, −0.4999	$2.720744e-08$
200	0.5000, 0.0041, −0.4999	$4.934027e-09$
250	0.5000, 0.0040, −0.4999	$9.630085e-10$
300	0.5000, 0.0040, −0.4999	$1.939747e-10$
350	0.5000, 0.0040, −0.4999	$3.961646e-11$
400	0.5000, 0.0040, −0.4999	$8.140393e-12$
450	0.5000, 0.0040, −0.4999	$1.677968e-12$
500	0.5000, 0.0040, −0.4999	$3.462695e-13$
550	0.5000, 0.0040, −0.4999	$7.136628e-14$
600	0.5000, 0.0040, −0.4999	$1.474205e-14$

We see that after the first 100 iterations the error is reduced to about $J(\mathbf{x}) = ||\mathbf{f}(\mathbf{x}^*)||^2 \approx 10^{-14}$, and with 500 additional iterations the algorithm converges to the following approximated solution with accuracy of $J(\mathbf{x}) \approx 10^{-28}$

$$\mathbf{x}^* = \begin{bmatrix} 0.5000013623816102 \\ 0.0040027495837189 \\ -0.4999000311539049 \end{bmatrix} \tag{2.48}$$

Although the gradient descent method requires many more iterations than Newton's method to converge, the computational cost in each iteration is much reduced, as no more matrix inversion is needed.

When it is difficult or too computationally costly to find the optimal step size along the search direction, some suboptimal step size may be acceptable, such as in the quasi-Newton methods (Section 2.5) for minimization problems. In this case, although the step size δ is no longer required to be such that the function value at the new position $f(\mathbf{x}_{n+1}) = f(\mathbf{x}_n + \delta\mathbf{d}_n)$ is minimized along the search direction $\mathbf{d}_n$, the step size δ still has to satisfy the following *Wolfe conditions*:

- *Sufficient decrease (Armijo rule)*

$$f(\mathbf{x}_{n+1}) = f(\mathbf{x}_n + \delta\mathbf{d}_n) \leq f(\mathbf{x}_n) + c_1\,\delta\mathbf{d}^T\mathbf{g}_n \tag{2.49}$$

- *Curvature condition*

$$\mathbf{g}_{n+1}^T\mathbf{d}_n \geq c_2\,\mathbf{g}_n^T\mathbf{d}_n \tag{2.50}$$

As $\mathbf{g}_n^T\mathbf{d}_n < 0$, this condition can also be written in the following alternative form:

$$|\mathbf{g}_{n+1}^T\mathbf{d}_n| < |c_2\,\mathbf{g}_n^T\mathbf{d}_n| \tag{2.51}$$

Here the two constants c_1 and c_2 satisfy $0 < c_1 < c_2 < 1$.

In general, these conditions are motivated by the desired effect that after each iterative step, the function should have a shallower slope along $\mathbf{d}_n$, as well as a lower value, so that eventually the solution can be approached where $f(\mathbf{x})$ is minimum and the gradient is zero.

Specifically, to understand the first condition we represent the function to be minimized as a single-variable function of the step size $\phi(\delta) = f(\mathbf{x}_n + \delta\mathbf{d}_n)$, and its tangent line at the point $\delta = 0$ as a linear function $L_0(\delta) = a + b\delta$, where the intercept a can be found as $a = L_0(0) = \phi(\delta)|_{\delta=0} = f(\mathbf{x}_n)$, and the slope b can be found as the derivative of $\phi(\delta) = f(\mathbf{x}_n + \delta\mathbf{d}_n)$ at $\delta = 0$:

$$b = \frac{d}{d\delta}\,f(\mathbf{x}_n + \delta\mathbf{d}_n)\Big|_{\delta=0} = \mathbf{g}_n^T\mathbf{d}_n < 0 \tag{2.52}$$

which needs to be negative for the function value to be reduced, $f(\mathbf{x}_n + \delta\mathbf{d}_n) < f(\mathbf{x}_n)$. Now the function of the tangent line can be written as

$$L_0(\delta) = a + b\delta = f(\mathbf{x}_n) + \mathbf{g}_n^T\mathbf{d}_n\delta \tag{2.53}$$

Comparing this with a constant line $L_1(\delta) = f(\mathbf{x}_n)$ of slope zero, we see that any straight line between $L_0(\delta)$ and $L_1(\delta)$ can be described by $L(\delta) = f(\mathbf{x}_n) + c_1\,\mathbf{g}_n^T\mathbf{d}_n\delta$ with $0 < c_1 < 1$, with a slope $0 < c_1\mathbf{g}_n^T\mathbf{d}_n < \mathbf{g}_n^T\mathbf{d}_n$. The Armijo rule is to find any $\delta > 0$ that satisfies

$$f(\mathbf{x}_{n+1}) = f(\mathbf{x}_n + \delta\mathbf{d}_n) \leq f(\mathbf{x}_n) + c_1\,\mathbf{g}_n^T\mathbf{d}_n\delta < f(\mathbf{x}_n) \tag{2.54}$$

We see that the value of $f(\mathbf{x})$ is guaranteed to be reduced.

The second condition requires that at the new position $\mathbf{x}_{n+1}$ the slope of the gradient $\mathbf{g}_{n+1}$ along the search direction $\mathbf{d}_n$ be sufficiently reduced to be less than a specified value (determined by c_2), in comparison to the slope at the old position $\mathbf{x}_n$.

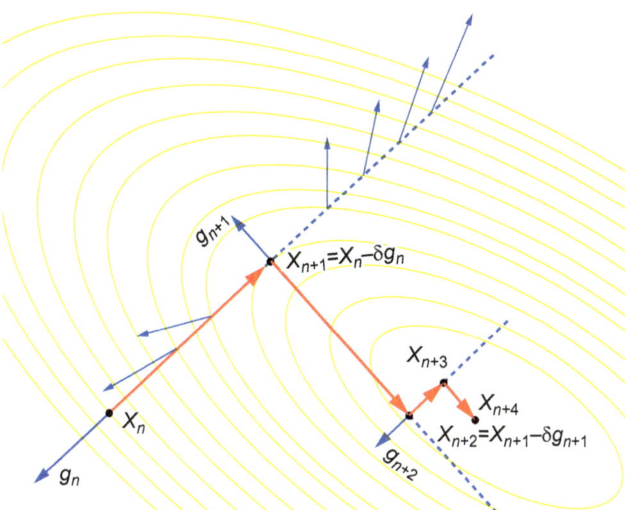

Fig. 2.5 Iterations of the gradient descent method.

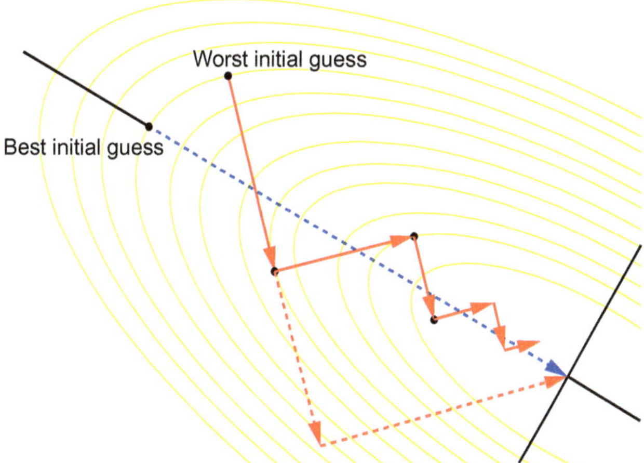

Fig. 2.6 Best (blue) and worst (red) initial guess.

The reason why $\mathbf{g}_{n+1} \perp \mathbf{g}_n$ can also be explained geometrically. As shown in Fig. 2.5, the gradient vectors at various points along the direction of $-\mathbf{g}_n$ are shown by the blue arrows, and their projections onto the direction represent the slopes of the function $f(\delta_n) = f(\mathbf{x}_n - \delta_n \mathbf{g}_n)$. Obviously at $\mathbf{x}_{n+1}$ where $f(\delta_n)$ reaches its minimum, its slope is zero. In other words, the projection of the gradient $\mathbf{g}_{n+1}$ onto the direction of $-\mathbf{g}_n$ is zero, that is, $\mathbf{g}_{n+1} \perp \mathbf{g}_n$ or $\mathbf{g}_{n+1}^T \mathbf{g}_n = 0$.

The gradient descent method gradually approaches a solution $\mathbf{x}$ of an N–D minimization problem by moving from the initial guess $\mathbf{x}_0$ along a zigzag path composed of a set of segments with any two consecutive segments perpendicular to each other. The number of steps depends greatly on the initial guess. As illustrated in Fig. 2.6 for a quadratic surface over a 2-D space, the best possible case (blue) is that the solution happens to be on the gradient direction of the initial guess, which could be reached in a single step, while the worst possible case (red) is that the gradient direction of the initial guess happens to be 45 degrees off from the gradient direction of the optimal case, and it takes many zigzag steps to go around the optimal path to reach the solution. Many of the steps are in the same direction as some of the previous steps.

To improve the performance of the gradient descent method we can include in the iteration a momentum term representing the search direction previously traversed

$$\mathbf{x}_{n+1} = \mathbf{x}_n - \delta_n \mathbf{g}_n + \alpha_n(\mathbf{x}_n - \mathbf{x}_{n-1}) \tag{2.55}$$

Now two consecutive search directions are no longer perpendicular to each other and the resulting search path is smoother than the zigzag path without the momentum term. The parameter α_m controls how much momentum is to be added.

Obviously it is most desirable not to repeat any of the previous directions traveled so that the solution can be reached in N steps, each in a unique direction in the N–D space. In other words, the subsequent steps are independent of each other, never interfering with the results achieved in the previous steps. Such a method will be discussed in the next section.

Example 2.5

The Rosenbrock function

$$f(x_1,x_2) = (a - x_1)^2 + b(x_2 - x_1^2)^2 \qquad \text{(e.g., } a = 1,\ b = 100)$$

is a two-variable non-convex function with a global minimum $f(x_1, x_2) = 0$ at the point $(1, 1)$, which is inside a long parabolic shaped valley as shown in the left panel of Fig. 2.7. As the slope along the valley is very shallow, it is difficult for an algorithm to converge quickly to the minimum. For this reason, the Rosenbrock function is often used to test various minimization algorithms.

Both the gradient method and Newton's method are used to find the minimum from an initial guess at $(-0.5, 1.5)$. While it takes only a few iteration steps for the Newton's method to converge to the minimum, following the search path shown in red in the right panel of Fig. 2.7, it takes many more steps for the gradient descent method to converge to the minimum, following the search path shown in blue in the right panel. Also, Fig. 2.8 shows a zoomed-in version of the last stage of the search path toward the final solution. We see that when the step size δ is small, the zigzag search path is composed of a long sequence of consecutive segments perpendicular to each other.

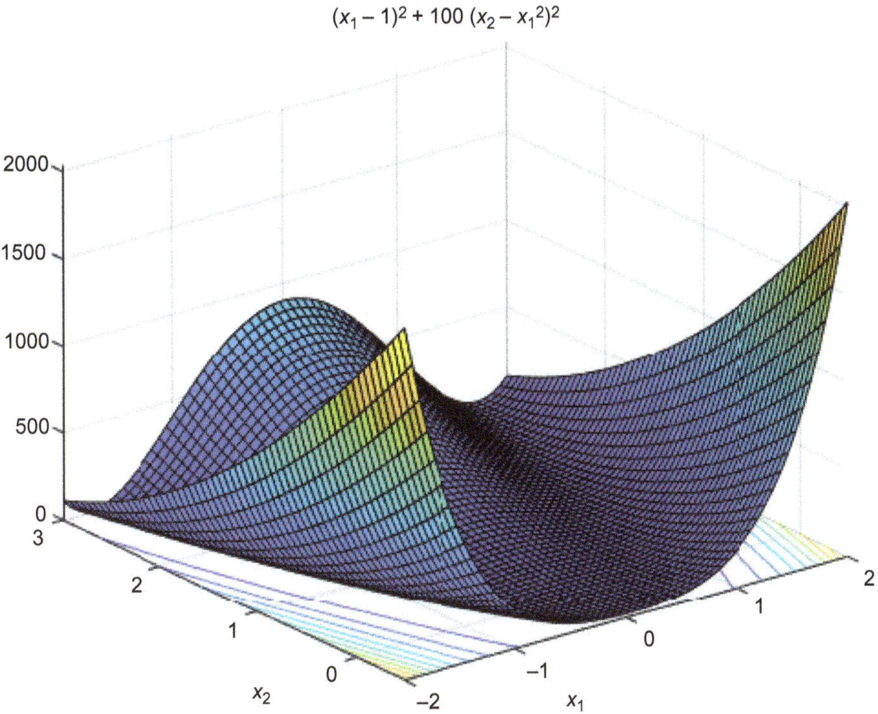

Fig. 2.7 Rosenbrock function (left) and the search paths to its minimum (right).
The search paths for Newton's and gradient descent methods are shown in red and blue, respectively.

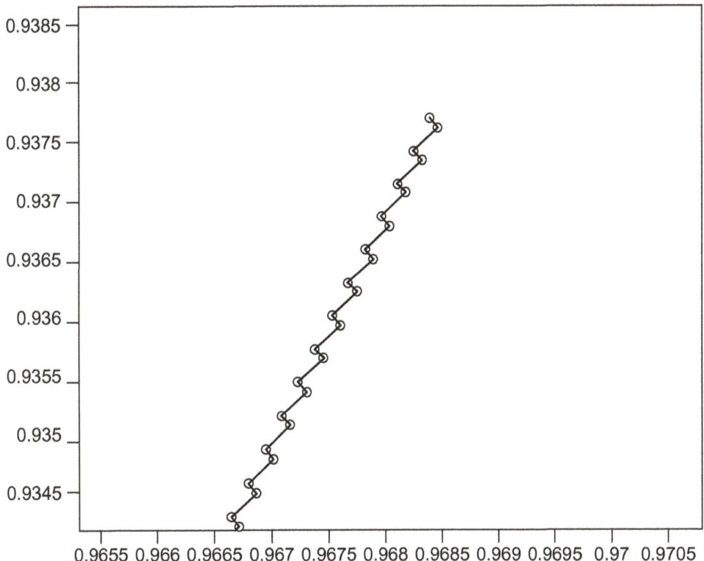

Fig. 2.8 Search path of the gradient descent (zoomed-in).

2.5 Quasi-Newton Methods

As we have seen, Newton's methods can be used to solve both nonlinear systems to find roots of a set of simultaneous equations, and optimization problems to minimize a scalar-valued objective function based on the iterations of the same form. Specifically,

- Solving equations $\mathbf{f}(\mathbf{x}) = \mathbf{0}$

$$\mathbf{x}_{n+1} = \mathbf{x}_n - \mathbf{J}(\mathbf{x}_n)^{-1}\mathbf{f}(\mathbf{x}_n) = \mathbf{x}_n - \mathbf{J}_n^{-1}\mathbf{f}_n \tag{2.56}$$

  where $\mathbf{J}_n = \mathbf{J}(\mathbf{x}_n)$ is the Jacobian matrix of $\mathbf{f}(\mathbf{x}_n)$,
- Minimizing $f(\mathbf{x})$

$$\mathbf{x}_{n+1} = \mathbf{x}_n - \mathbf{H}(\mathbf{x}_n)^{-1}\mathbf{g}(\mathbf{x}_n) = \mathbf{x}_n - \mathbf{H}_n^{-1}\mathbf{g}_n \tag{2.57}$$

  where $\mathbf{g}_n = \mathbf{g}(\mathbf{x}_n)$ and $\mathbf{H}_n = \mathbf{H}(\mathbf{x}_n)$ are the gradient vector and Hessian matrix of $f(\mathbf{x})$ at $\mathbf{x}_n$, respectively.

If the Jacobian $\mathbf{J}(\mathbf{x})$ or the Hessian matrix $\mathbf{H}(\mathbf{x})$ is not available, or if it is too computationally costly to calculate its inverse (with complexity $O(N^3)$), the quasi-Newton methods can be used to approximate the Hessian matrix or its inverse based only on the first-order derivative, the gradient $\mathbf{g}$ of $f(\mathbf{x})$ (with complexity $O(N^2)$), similar to Broyden's method previously considered (Section 1.5).

In the following, we consider the minimization of a function $f(\mathbf{x})$. Its Taylor expansion around point $\mathbf{x}_{n+1}$ is

$$f(\mathbf{x}) = f(\mathbf{x}_{n+1}) + (\mathbf{x} - \mathbf{x}_{n+1})^T\mathbf{g}_{n+1} + \frac{1}{2}(\mathbf{x} - \mathbf{x}_{n+1})^T\mathbf{H}_{n+1}(\mathbf{x} - \mathbf{x}_{n+1}) + O(||\mathbf{x} - \mathbf{x}_{n+1}||^3) \tag{2.58}$$

Taking the derivative with respect to $\mathbf{x}$, we obtain

$$\frac{d}{d\mathbf{x}}f(\mathbf{x}) = \mathbf{g}(\mathbf{x}) = \mathbf{g}_{n+1} + \mathbf{H}_{n+1}(\mathbf{x} - \mathbf{x}_{n+1}) + O(||\mathbf{x} - \mathbf{x}_{n+1}||^2) \tag{2.59}$$

Evaluating at $\mathbf{x} = \mathbf{x}_n$, we have $\mathbf{g}(\mathbf{x}_n) = \mathbf{g}_n$, and Eq. (2.59) can be written as

$$\begin{aligned}\mathbf{g}_{n+1} - \mathbf{g}_n &= \mathbf{H}_{n+1}(\mathbf{x}_{n+1} - \mathbf{x}_n) + O(||\mathbf{x}_{n+1} - \mathbf{x}_n||^2) \\ &= \mathbf{B}_{n+1}(\mathbf{x}_{n+1} - \mathbf{x}_n)\end{aligned} \tag{2.60}$$

where matrix $\mathbf{B}_n$ is the secant approximation of the Hessian matrix $\mathbf{H}_n$, and the last equality is called the *secant equation*. For convenience, we further define

$$\mathbf{s}_n = \mathbf{x}_{n+1} - \mathbf{x}_n, \qquad \mathbf{y}_n = \mathbf{g}_{n+1} - \mathbf{g}_n \tag{2.61}$$

so that this equation above can be written as

$$\mathbf{B}_{n+1}\mathbf{s}_n = \mathbf{y}_n, \quad \text{or} \quad \mathbf{B}_{n+1}^{-1}\mathbf{y}_n = \mathbf{s}_n \tag{2.62}$$

This is the *quasi-Newton equation*, which is the *secant condition* that must be satisfied by matrix $\mathbf{B}_n$, or its inverse $\mathbf{B}_n^{-1}$, in any of the quasi-Newton algorithms, all taking the follow general steps:

1. initialize $\mathbf{x}_0$ and $\mathbf{B}_0$, set $n = 0$;
2. compute gradient $\mathbf{g}_n$ and the search direction $\mathbf{d}_n = -\mathbf{B}_n^{-1}\mathbf{g}_n$;
3. get $\mathbf{x}_{n+1} = \mathbf{x}_n + \delta\mathbf{d}_n$ with step size δ satisfying the Wolfe conditions
4. update $\mathbf{B}_{n+1} = \mathbf{B}_n + \Delta\mathbf{B}_n$ or $\mathbf{B}_{n+1}^{-1} = \mathbf{B}_n^{-1} + \Delta\mathbf{B}_n^{-1}$, that satisfies the quasi-Newton equation;
5. if termination condition is not satisfied, $n = n + 1$, go back to second step.

For this iteration to converge to a local minimum of $f(\mathbf{x})$, $\mathbf{B}_n$ must be positive definite matrix, same as the Hessian matrix $\mathbf{H}$ it approximates. Specially, if $\mathbf{B}_n = \mathbf{I}$, then $\mathbf{d}_n = -\mathbf{g}_n$, the algorithm becomes the gradient descent method; also, if $\mathbf{B}_n = \mathbf{H}_n$ is the Hessian matrix, then the algorithm becomes the Newton's method.

In a quasi-Newton method, we can choose to update either $\mathbf{B}_n$ or its inverse $\mathbf{B}_n^{-1}$ based on one of the two forms of the quasi-Newton equation. We note that there is a dual relationship between the two ways to update, that is, by swapping $\mathbf{y}_n$ and $\mathbf{s}_m$, an updated formula for $\mathbf{B}_n$ can be directly applied to its inverse and vice versa.

- The *Symmetric Rank 1 (SR1)* algorithm
  Here matrix $\mathbf{B}_n$ is updated by an additional term:

$$\mathbf{B}_{n+1} = \mathbf{B}_n + \mathbf{u}\mathbf{u}^T \tag{2.63}$$

where $\mathbf{u}$ a vector and $\mathbf{u}\mathbf{u}^T$ is a matrix of rank 1. Imposing the secant condition, we get

$$\mathbf{B}_{n+1}\mathbf{s}_n = \mathbf{B}_n\mathbf{s}_n + \mathbf{u}\mathbf{u}^T\mathbf{s}_n = \mathbf{y}_n, \quad \text{i.e.,} \quad \mathbf{u}(\mathbf{u}^T\mathbf{s}_n) = \mathbf{y}_n - \mathbf{B}_n\mathbf{s}_n \tag{2.64}$$

This equation indicates $\mathbf{u}$ is in the same direction as $\mathbf{w} = \mathbf{y}_n - \mathbf{B}_n\mathbf{s}_n$, and can therefore be written as $\mathbf{u} = c\mathbf{w}$. Substituting this back we obtain

$$\mathbf{u}\mathbf{u}^T\mathbf{s}_n = c^2\mathbf{w}\mathbf{w}^T\mathbf{s}_n = \mathbf{w} \tag{2.65}$$

Solving this we get $c = 1/(\mathbf{w}^T\mathbf{s}_n)^{1/2}$

$$\mathbf{u} = c\mathbf{w} = \frac{\mathbf{y}_n - \mathbf{B}_n\mathbf{s}_n}{(\mathbf{w}^T\mathbf{s}_n)^{1/2}} = \frac{\mathbf{y}_n - \mathbf{B}_n\mathbf{s}_n}{((\mathbf{y}_n - \mathbf{B}_n\mathbf{s}_n)^T\mathbf{s}_n)^{1/2}} \tag{2.66}$$

and the iteration of $\mathbf{B}_n$ becomes

$$\mathbf{B}_{n+1} = \mathbf{B}_n + \mathbf{u}\mathbf{u}^T = \mathbf{B}_n + \frac{(\mathbf{y}_n - \mathbf{B}_n\mathbf{s}_n)(\mathbf{y}_n - \mathbf{B}_n\mathbf{s}_n)^T}{(\mathbf{y}_n - \mathbf{B}_n\mathbf{s}_n)^T\mathbf{s}_n} \tag{2.67}$$

Applying the Sherman–Morrison formula (Section A.2.4) we further get the inverse $\mathbf{B}_{n+1}$

$$\begin{aligned}\mathbf{B}_{n+1}^{-1} &= \left(\mathbf{B}_n + \mathbf{u}\mathbf{u}^T\right)^{-1} = \mathbf{B}_n^{-1} - \frac{\mathbf{B}_n^{-1}\mathbf{u}\mathbf{u}^T\mathbf{B}_n^{-1}}{1 + \mathbf{u}^T\mathbf{B}_n^{-1}\mathbf{u}} \\ &= \mathbf{B}_n^{-1} + \frac{(\mathbf{s}_n - \mathbf{B}_n^{-1}\mathbf{y}_n)(\mathbf{s}_n - \mathbf{B}_n^{-1}\mathbf{y}_n)^T}{(\mathbf{s}_n - \mathbf{B}_n^{-1}\mathbf{y}_n)^T\mathbf{y}_n}\end{aligned} \tag{2.68}$$

We note that this equation is dual to the previous one, and it can also be obtained based on the duality between the secant conditions for $\mathbf{B}_n$ and $\mathbf{B}_n^{-1}$. Specifically, as just discussed, we impose the secant condition for the inverse $\mathbf{B}_{n+1}^{-1}$ of the following form

$$\mathbf{B}_{n+1}^{-1} = \mathbf{B}_n^{-1} + \mathbf{u}\mathbf{u}^T \tag{2.69}$$

and obtain

$$\mathbf{B}_{n+1}^{-1}\mathbf{y}_n = \mathbf{B}_n^{-1}\mathbf{y}_n + \mathbf{u}\mathbf{u}^T\mathbf{y}_n = \mathbf{s}_n, \tag{2.70}$$

Then by repeating the same process, we get the same updated formula for $\mathbf{B}_n^{-1}$. This is the formula for directly updating $\mathbf{B}_n^{-1}$. For the iteration to converge to a local minimum, matrix $\mathbf{B}_{n+1}^{-1}$ needs to be positive definite as well as $\mathbf{B}_n^{-1}$, we therefore require

$$\mathbf{w}^T\mathbf{y}_n = (\mathbf{s}_n - \mathbf{B}_n^{-1}\mathbf{y}_n)^T\mathbf{y}_n = \mathbf{s}_n^T\mathbf{y}_n - \mathbf{y}_n^T\mathbf{B}_n^{-1}\mathbf{y}_n > 0 \tag{2.71}$$

It may be difficult to maintain this requirement throughout the iteration. This problem can be avoided in the following BFGS and DFP methods.

- The *BFGS (Broyden–Fletcher–Goldfarb–Shanno)* algorithm
This is a rank 2 algorithm in which matrix $\mathbf{B}_n$ is updated by two rank 1 terms

$$\mathbf{B}_{n+1} = \mathbf{B}_n + \alpha\mathbf{u}\mathbf{u}^T + \beta\mathbf{v}\mathbf{v}^T \tag{2.72}$$

Imposing the secant condition for $\mathbf{B}_n$, we get

$$\mathbf{B}_{n+1}\mathbf{s}_n = (\mathbf{B}_n + \alpha\mathbf{u}\mathbf{u}^T + \beta\mathbf{v}\mathbf{v}^T)\mathbf{s}_n = \mathbf{B}_n\mathbf{s}_n + \mathbf{u}(\alpha\mathbf{u}^T\mathbf{s}_n) + \mathbf{v}(\beta\mathbf{v}^T\mathbf{s}_n) = \mathbf{y}_n \tag{2.73}$$

or

$$\mathbf{u}(\alpha\mathbf{u}^T\mathbf{s}_n) + \mathbf{v}(\beta\mathbf{v}^T\mathbf{s}_n) = \mathbf{y}_n - \mathbf{B}_n\mathbf{s}_n \tag{2.74}$$

which can be satisfied if we let

$$\mathbf{u} = \mathbf{y}_n, \qquad \alpha = \frac{1}{\mathbf{s}_n^T\mathbf{y}_n}, \qquad \mathbf{v} = \mathbf{B}_n\mathbf{s}_n, \qquad \beta = -\frac{1}{\mathbf{v}^T\mathbf{s}_n} = -\frac{1}{\mathbf{s}_n^T\mathbf{B}_n\mathbf{s}_n} \tag{2.75}$$

Substituting these into $\mathbf{B}_{n+1} = \mathbf{B}_n + \alpha\mathbf{u}\mathbf{u}^T + \beta\mathbf{v}\mathbf{v}^T$ we get

$$\mathbf{B}_{n+1} = \mathbf{B}_n + \alpha\mathbf{u}\mathbf{u}^T + \beta\mathbf{v}\mathbf{v}^T = \mathbf{B}_n + \frac{\mathbf{y}_n\mathbf{y}_n^T}{\mathbf{y}_n^T\mathbf{s}_n} - \frac{\mathbf{B}_n\mathbf{s}_n\mathbf{s}_n^T\mathbf{B}_n}{\mathbf{s}_n^T\mathbf{B}_n\mathbf{s}_n} \tag{2.76}$$

Given this updated formula for $\mathbf{B}_n$, we can further find its inverse $\mathbf{B}_n^{-1}$ and then the search direction $\mathbf{d}_n = -\mathbf{B}_n^{-1}\mathbf{g}_n$.

Alternatively, given $\mathbf{B}_n$, we can further derive an updated formula directly for the inverse matrix $\mathbf{B}_n^{-1}$, so that the search direction $\mathbf{d}_n$ can be obtained without carrying out the inversion computation. Specifically, we first define $\mathbf{U} = [\mathbf{u}_1\ \mathbf{u}_2]$ and $\mathbf{V} = [\mathbf{v}_1\ \mathbf{v}_2]$, where

$$\mathbf{u}_1 = \mathbf{v}_1 = \frac{\mathbf{y}_n}{(\mathbf{s}_n^T\mathbf{y}_n)^{1/2}}, \qquad \mathbf{u}_2 = -\mathbf{v}_2 = \frac{\mathbf{B}_n\mathbf{s}_n}{(\mathbf{s}_n^T\mathbf{B}_n\mathbf{s}_n)^{1/2}} \tag{2.77}$$

so that the expression can be written as

$$\mathbf{B}_{n+1} = \mathbf{B}_n + \mathbf{u}_1\mathbf{v}_1^T + \mathbf{u}_2\mathbf{v}_2^T = (\mathbf{B}_n + \mathbf{U}\mathbf{V}^T)^{-1} \tag{2.78}$$

and then apply the Sherman–Morrison formula to get

$$\begin{aligned}\mathbf{B}_{n+1}^{-1} &= (\mathbf{B}_n + \mathbf{U}\mathbf{V}^T)^{-1} = \mathbf{B}_n^{-1} - \mathbf{B}_n^{-1}\mathbf{U}(\mathbf{I} + \mathbf{V}^T\mathbf{B}_n^{-1}\mathbf{U})^{-1}\mathbf{V}^T\mathbf{B}_n^{-1} \\ &= \mathbf{B}_n^{-1} - \mathbf{B}_n^{-1}\mathbf{U}\mathbf{C}^{-1}\mathbf{V}^T\mathbf{B}_n^{-1} \\ &= \mathbf{B}_n^{-1} - \mathbf{B}_n^{-1}[\mathbf{u}_1\ \mathbf{u}_2]\mathbf{C}^{-1}(\mathbf{B}_n^{-1}[\mathbf{v}_1\ \mathbf{v}_2])^T \end{aligned} \tag{2.79}$$

where we have defined

$$\mathbf{C} = \begin{bmatrix} c_{11} & c_{12} \\ c_{21} & c_{22} \end{bmatrix} = \mathbf{I} + \mathbf{V}^T\mathbf{B}_n^{-1}\mathbf{U} = \mathbf{I} + [\mathbf{v}_1\ \mathbf{v}_2]^T\mathbf{B}_n^{-1}[\mathbf{u}_1\ \mathbf{u}_2] \tag{2.80}$$

with

$$c_{11} = 1 + \mathbf{v}_1^T\mathbf{B}_n^{-1}\mathbf{u}_1 = 1 + \frac{\mathbf{y}_n^T\mathbf{B}_n^{-1}\mathbf{y}_n}{\mathbf{s}_n^T\mathbf{y}_n} \tag{2.81}$$

$$c_{22} = 1 + \mathbf{v}_2\mathbf{B}_n^{-1}\mathbf{u}_2 = 1 - \frac{\mathbf{s}_n^T\mathbf{B}_n\mathbf{B}_n^{-1}\mathbf{B}_n\mathbf{s}_n}{\mathbf{s}_n^T\mathbf{B}_n\mathbf{s}_n} = 0 \tag{2.82}$$

$$c_{12} = \mathbf{v}_1^T\mathbf{B}_n^{-1}\mathbf{u}_2 = \frac{\mathbf{y}_n^T\mathbf{B}_n^{-1}\mathbf{B}_n\mathbf{s}_n}{(\mathbf{s}_n^T\mathbf{y}_n)^{1/2}(\mathbf{s}_n^T\mathbf{B}_n\mathbf{s}_n)^{1/2}} = \frac{(\mathbf{s}_n^T\mathbf{y}_n)^{1/2}}{(\mathbf{s}_n^T\mathbf{B}_n\mathbf{s}_n)^{1/2}} \tag{2.83}$$

$$c_{21} = \mathbf{v}_2^T\mathbf{B}_n^{-1}\mathbf{u}_1 = -c_{12} \tag{2.84}$$

$$\mathbf{C} = \begin{bmatrix} c_{11} & c_{12} \\ -c_{12} & 0 \end{bmatrix}, \qquad \mathbf{C}^{-1} = \begin{bmatrix} 0 & -1/c_{12} \\ 1/c_{12} & c_{11}/c_{12}^2 \end{bmatrix} \tag{2.85}$$

Substituting this $\mathbf{C}^{-1}$ into the expression for $\mathbf{B}_{n+1}^{-1}$, we get:

$$\begin{aligned}\mathbf{B}_{n+1}^{-1} &= \mathbf{B}_n^{-1} - [\mathbf{B}_n^{-1}\mathbf{u}_1\ \mathbf{B}_n^{-1}\mathbf{u}_2]\begin{bmatrix} 0 & -1/c_{12} \\ 1/c_{12} & c_{11}/c_{12}^2 \end{bmatrix}[\mathbf{B}_n^{-1}\mathbf{v}_1\ \mathbf{B}_n^{-1}\mathbf{v}_2]^T \\ &= \mathbf{B}_n^{-1} - \frac{1}{c_{12}}[\mathbf{B}_n^{-1}\mathbf{u}_2\mathbf{v}_1^T\mathbf{B}_n^{-1} - \mathbf{B}_n^{-1}\mathbf{u}_1\mathbf{v}_2^T\mathbf{B}_n^{-1}] - \frac{c_{11}}{c_{12}^2}\mathbf{B}_n^{-1}\mathbf{u}_2\mathbf{v}_2^T\mathbf{B}_n^{-1} \end{aligned} \tag{2.86}$$

where

$$\frac{1}{c_{12}}\mathbf{B}_n^{-1}\mathbf{u}_2\mathbf{v}_1^T\mathbf{B}_n^{-1} = \frac{(\mathbf{s}_n^T\mathbf{B}_n\mathbf{s}_n)^{1/2}}{(\mathbf{s}_n^T\mathbf{y}_n)^{1/2}}\mathbf{B}_n^{-1}\frac{\mathbf{B}_n\mathbf{s}_n}{(\mathbf{s}_n^T\mathbf{B}_n\mathbf{s}_n)^{1/2}}\frac{\mathbf{y}_n^T}{(\mathbf{s}_n^T\mathbf{y}_n)^{1/2}}\mathbf{B}_n^{-1} = \frac{\mathbf{s}_n\mathbf{y}_n^T\mathbf{B}_n^{-1}}{\mathbf{s}_n^T\mathbf{y}_n} \tag{2.87}$$

$$-\frac{1}{c_{12}}\mathbf{B}_n^{-1}\mathbf{u}_1\mathbf{v}_2^T\mathbf{B}_n^{-1} = \frac{(\mathbf{s}_n^T\mathbf{B}_n\mathbf{s}_n)^{1/2}}{(\mathbf{s}_n^T\mathbf{y}_n)^{1/2}}\mathbf{B}_n^{-1}\frac{\mathbf{y}_n}{(\mathbf{s}_n^T\mathbf{y}_n)^{1/2}}\frac{\mathbf{s}_n^T\mathbf{B}_n}{(\mathbf{s}_n^T\mathbf{B}_n\mathbf{s}_n)^{1/2}}\mathbf{B}_n^{-1} = \frac{\mathbf{B}_n^{-1}\mathbf{y}_n\mathbf{s}_n^T}{\mathbf{s}_n^T\mathbf{y}_n} \tag{2.88}$$

$$\begin{aligned}\frac{c_{11}}{c_{12}^2}\mathbf{B}_n^{-1}\mathbf{u}_2\mathbf{v}_2^T\mathbf{B}_n^{-1} &= -\left(1 + \frac{\mathbf{y}_n^T\mathbf{B}_n^{-1}\mathbf{y}_n}{\mathbf{s}_n^T\mathbf{y}_n}\right)\frac{\mathbf{s}_n^T\mathbf{B}_n\mathbf{s}_n}{\mathbf{s}_n^T\mathbf{y}_n}\mathbf{B}_n^{-1}\frac{\mathbf{B}_n\mathbf{s}_n}{(\mathbf{s}_n^T\mathbf{B}_n\mathbf{s}_n)^{1/2}}\frac{\mathbf{s}_n^T\mathbf{B}_n}{(\mathbf{s}_n^T\mathbf{B}_n\mathbf{s}_n)^{1/2}}\mathbf{B}_n^{-1} \\ &= -\left(1 + \frac{\mathbf{y}_n^T\mathbf{B}_n^{-1}\mathbf{y}_n}{\mathbf{s}_n^T\mathbf{y}_n}\right)\frac{\mathbf{s}_n\mathbf{s}_n^T}{\mathbf{s}_n^T\mathbf{y}_n} \end{aligned} \tag{2.89}$$

Substituting these three terms back into Eq. (2.86), we obtain the updated formula for $\mathbf{B}_n^{-1}$

$$\mathbf{B}_{n+1}^{-1} = \mathbf{B}_n^{-1} - \frac{\mathbf{B}_n^{-1}\mathbf{y}_n\mathbf{s}_n^T + \mathbf{s}_n\mathbf{y}_n^T\mathbf{B}_n^{-1}}{\mathbf{s}_n^T\mathbf{y}_n} + \left(1 + \frac{\mathbf{y}_n^T\mathbf{B}_n^{-1}\mathbf{y}_n}{\mathbf{s}_n^T\mathbf{y}_n}\right)\frac{\mathbf{s}_n\mathbf{s}_n^T}{\mathbf{s}_n^T\mathbf{y}_n} \tag{2.90}$$

In summary, $\mathbf{B}_{n+1}^{-1}$ can be found by either Eq. (2.76) for $\mathbf{B}_{n+1}$ followed by an additional inversion operation, or Eq. (2.90) directly for $\mathbf{B}_{n+1}^{-1}$. The results obtained by these methods are equivalent.

- The *DFP (Davidon–Fletcher–Powell)* algorithm
  Same as the BFGS algorithm, the DFP algorithm is also a rank 2 algorithm, where instead of $\mathbf{B}_n$, the inverse matrix $\mathbf{B}_n^{-1}$ is updated by two rank 1 terms

$$\mathbf{B}_{n+1}^{-1} = \mathbf{B}_n^{-1} + \alpha\mathbf{u}\mathbf{u}^T + \beta\mathbf{v}\mathbf{v}^T \tag{2.91}$$

where α and β are real scalars, $\mathbf{u}$ and $\mathbf{v}$ are vectors. Imposing the secant condition for the inverse of $\mathbf{B}_n$, we get

$$\mathbf{B}_{n+1}^{-1}\mathbf{y}_n = (\mathbf{B}_n^{-1} + \alpha\mathbf{u}\mathbf{u}^T + \beta\mathbf{v}\mathbf{v}^T)\mathbf{y}_n = \mathbf{B}_n^{-1}\mathbf{y}_n + \mathbf{u}(\alpha\mathbf{u}^T\mathbf{y}_n) + \mathbf{v}(\beta\mathbf{v}^T\mathbf{y}_n) = \mathbf{s}_n \tag{2.92}$$

or

$$\mathbf{u}(\alpha\mathbf{u}^T\mathbf{y}_n) + \mathbf{v}(\beta\mathbf{v}^T\mathbf{y}_n) = \mathbf{s}_n - \mathbf{B}_n^{-1}\mathbf{y}_n \tag{2.93}$$

which can be satisfied if we let

$$\mathbf{u} = \mathbf{s}_n, \qquad \alpha = \frac{1}{\mathbf{s}^T\mathbf{y}_n}, \qquad \mathbf{v} = \mathbf{B}_n^{-1}\mathbf{y}_n, \qquad \beta = -\frac{1}{\mathbf{v}^T\mathbf{y}_n} = -\frac{1}{\mathbf{y}_n^T\mathbf{B}_n^{-1}\mathbf{y}_n} \tag{2.94}$$

Substituting these into $\mathbf{B}_{n+1}^{-1} = \mathbf{B}_n^{-1} + \alpha\mathbf{u}\mathbf{u}^T + \beta\mathbf{v}\mathbf{v}^T$ we have

$$\mathbf{B}_{n+1}^{-1} = \mathbf{B}_n^{-1} + \alpha\mathbf{u}\mathbf{u}^T + \beta\mathbf{v}\mathbf{v}^T = \mathbf{B}_n^{-1} + \frac{\mathbf{s}_n\mathbf{s}_n^T}{\mathbf{s}_n^T\mathbf{y}_n} - \frac{\mathbf{B}_n^{-1}\mathbf{y}_n\mathbf{y}_n^T\mathbf{B}_n^{-1}}{\mathbf{y}_n^T\mathbf{B}_n^{-1}\mathbf{y}_n} \tag{2.95}$$

Following the same procedure used for the BFGS method, we can also obtain the updated formula of matrix $\mathbf{B}_n$ for the DFP method

$$\mathbf{B}_{n+1} = \mathbf{B}_n - \frac{\mathbf{B}_n\mathbf{s}_n\mathbf{y}_n^T + \mathbf{y}_n\mathbf{s}_n^T\mathbf{B}_n}{\mathbf{y}_n^T\mathbf{s}_n} + \left(1 + \frac{\mathbf{s}_n^T\mathbf{B}_n\mathbf{s}_n}{\mathbf{y}_n^T\mathbf{s}_n}\right)\frac{\mathbf{y}_n\mathbf{y}_n^T}{\mathbf{y}_n^T\mathbf{s}_n} \tag{2.96}$$

We note that Eqs (2.95) and (2.76) form a duality pair, and Eqs (2.96) and (2.90) form another duality pair. In other words, the BFGS and FDP methods are dual of each other.

In both the DFP and BFGS methods, matrix $\mathbf{B}_n^{-1}$, as well as $\mathbf{B}_n$, must be positive definite, that is, $\mathbf{z}^T\mathbf{B}_n\mathbf{z} > 0$ must hold for any $\mathbf{z} \neq \mathbf{0}$. We now prove that this requirement is satisfied if the curvature condition of the Wolfe conditions is satisfied

$$\mathbf{g}_{n+1}^T\mathbf{d}_n \geq c_2\,\mathbf{g}_n^T\mathbf{d}_n \qquad \text{i.e.,} \qquad (\mathbf{g}_{n+1} - c_2\,\mathbf{g}_n)^T\,\mathbf{d}_n \geq 0 \tag{2.97}$$

Replacing the search direction $\mathbf{d}_n$ by $\delta\mathbf{d}_n = \mathbf{x}_{n+1} - \mathbf{x}_n = \mathbf{s}_n$, we get:

$$(\mathbf{g}_{n+1} - c_2\mathbf{g}_n)^T\mathbf{s}_n = \mathbf{g}_{n+1}^T\mathbf{s}_n - c_2\mathbf{g}_n^T\mathbf{s}_n \geq 0 \tag{2.98}$$

We also have:

$$\mathbf{y}_n^T \mathbf{s}_n = (\mathbf{g}_{n+1} - \mathbf{g}_n)^T \mathbf{s}_n = \mathbf{g}_{n+1}^T \mathbf{s}_n - \mathbf{g}_n^T \mathbf{s}_n \tag{2.99}$$

Subtracting the first equation from the second, we obtain

$$\mathbf{y}_n^T \mathbf{s}_n - (\mathbf{g}_{n+1}^T \mathbf{s}_n - c_2 \mathbf{g}_n^T \mathbf{s}_n) = (c_2 - 1)\mathbf{g}_n^T \mathbf{s}_n > 0 \tag{2.100}$$

The last inequality is attributable to the fact that $\mathbf{g}_n^T \mathbf{s}_n < 0$ and $c_2 < 1$. We therefore have

$$\mathbf{y}_n^T \mathbf{s}_n \geq \mathbf{g}_{n+1}^T \mathbf{s}_n - c_2 \mathbf{g}_n^T \mathbf{s}_n \geq 0 \tag{2.101}$$

Given $\mathbf{y}_n^T \mathbf{s}_n > 0$, we can further prove by induction that both $\mathbf{B}_{n+1}$ and $\mathbf{B}_{n+1}^{-1}$ based, respectively, on the updated formulae in Eqs (2.76) and (2.95) are positive definite. We first assume $\mathbf{B}_0$ and $\mathbf{B}_n$ are both positive definite, and express $\mathbf{B}_n$ in terms of its Cholesky decomposition, $\mathbf{B}_n = \mathbf{L}\mathbf{L}^T$. We further define $\mathbf{a} = \mathbf{L}^T \mathbf{z}$, $\mathbf{b} = \mathbf{L}^T \mathbf{s}$, and get

$$\mathbf{a}^T \mathbf{a} = \mathbf{z}^T \mathbf{B}_n \mathbf{z}, \quad \mathbf{b}^T \mathbf{b} = \mathbf{s}^T \mathbf{B}_n \mathbf{s}, \quad \mathbf{a}^T \mathbf{b} = \mathbf{z}^T \mathbf{B}_n \mathbf{s} \tag{2.102}$$

Now we can show that the sum of the first and third terms of Eq. (2.90) is positive definite

$$\mathbf{z}^T \left[\mathbf{B}_n - \frac{\mathbf{B}_n \mathbf{s}_n \mathbf{s}_n^T \mathbf{B}_n}{\mathbf{s}_n^T \mathbf{B}_n \mathbf{s}_n} \right] \mathbf{z} = \mathbf{z}^T \mathbf{B}_n \mathbf{z} - \frac{(\mathbf{z}^T \mathbf{B}_n \mathbf{s}_n)^2}{\mathbf{s}_n^T \mathbf{B}_n \mathbf{s}_n} = \mathbf{a}^T \mathbf{a} - \frac{(\mathbf{a}^T \mathbf{b})^2}{\mathbf{b}^T \mathbf{b}} \geq 0 \tag{2.103}$$

The last step is due to the Cauchy–Schwarz inequality. Also, as $\mathbf{s}_n^T \mathbf{y}_n \geq 0$, the second term of Eq. (2.76) is positive definite

$$\mathbf{z}^T \left(\frac{\mathbf{s}_n \mathbf{s}_n^T}{\mathbf{s}_n^T \mathbf{y}_n} \right) \mathbf{z} \geq 0 \tag{2.104}$$

Combining these two results, we obtain

$$\mathbf{z}^T \mathbf{B}_{n+1} \mathbf{z} = \mathbf{z}^T \left(\mathbf{B}_n - \frac{\mathbf{B}_n \mathbf{s}_n \mathbf{s}_n^T \mathbf{B}_n}{\mathbf{s}_n^T \mathbf{B}_n \mathbf{s}_n} \right) \mathbf{z} + \mathbf{z}^T \left(\frac{\mathbf{y}_n \mathbf{y}_n^T}{\mathbf{s}_n^T \mathbf{y}_n} \right) \mathbf{z} \geq 0 \tag{2.105}$$

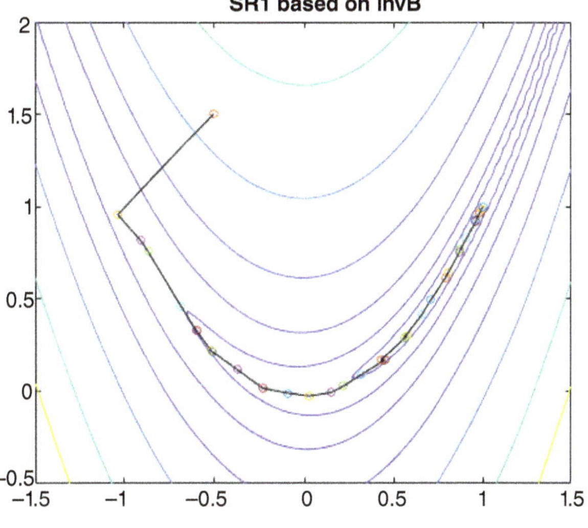

Fig. 2.9 Search path of SR1 based on $\mathbf{B}^{-1}$.

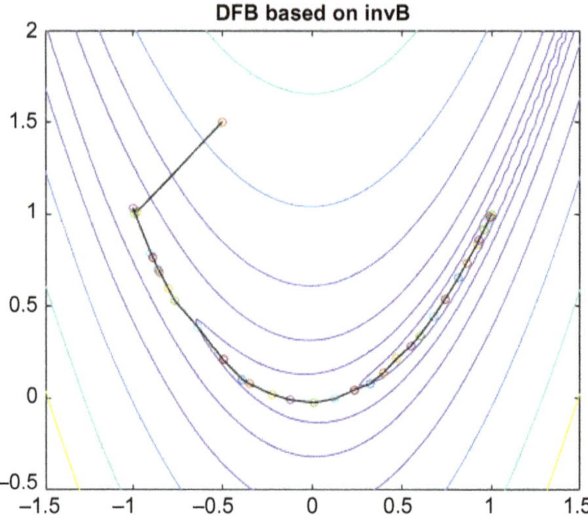

Fig. 2.10 Search path of DFP based on $\mathbf{B}^{-1}$.

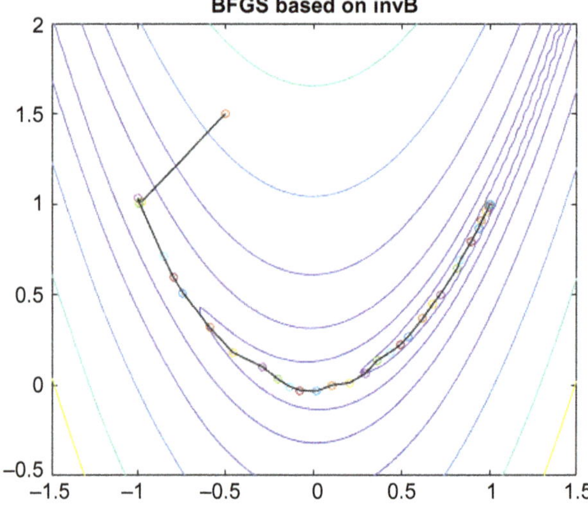

Fig. 2.11 Search path of BFGS based on $\mathbf{B}^{-1}$.

that is $\mathbf{B}_{n+1}$ based on the updated formula Eq. (2.76) is positive definite. Following the same steps, we can also show that $\mathbf{B}_{n+1}^{-1}$ based on Eq. (2.95) is positive definite.

As examples, Figs 2.9, 2.10. and 2.11, show, respectively, the search path of the SR1, DFP, and BFGS methods when applied to find the minimum of the Rosenbrock function.

2.6 Conjugate Gradient Method

The gradient descent method can be used to solve the minimization problem when the Hessian matrix of the objective function is not available. However, this method may be inefficient if it gets into a zigzag search pattern and repeats the same search directions many times. This problem can be avoided by the *conjugate gradient (CG)* method. If the objective function is quadratic, the CG method converges to the solution in N iterations without repeating any of the directions previously traversed. If the objective function is not quadratic, the CG method can still significantly improve the performance in comparison to the gradient descent method.

Again consider the approximation of the function $f(\mathbf{x})$ to be minimized by the first three terms of its Taylor series

$$f(\mathbf{x}) = f(\mathbf{x}_0 + \delta\mathbf{x}) \approx f(\mathbf{x}_0) + \mathbf{g}_0^T \delta\mathbf{x} + \frac{1}{2}\delta\mathbf{x}^T \mathbf{H}_0 \delta\mathbf{x} \tag{2.106}$$

If the Hessian matrix is positive definite, then $f(\mathbf{x})$ has a minimum. If function $f(\mathbf{x})$ is quadratic, this approximation becomes exact and the function can be written as

$$f(\mathbf{x}) = \frac{1}{2}\mathbf{x}^T \mathbf{A}\mathbf{x} - \mathbf{b}^T \mathbf{x} + c \tag{2.107}$$

This function's gradient and Hessian can be written as

$$\mathbf{g}(\mathbf{x}) = \frac{d}{d\mathbf{x}} f(\mathbf{x}) = \frac{d}{d\mathbf{x}}\left(\frac{1}{2}\mathbf{x}^T \mathbf{A}\mathbf{x} - \mathbf{b}^T \mathbf{x} + c\right) = \mathbf{A}\mathbf{x} - \mathbf{b} \tag{2.108}$$

and

$$\mathbf{H}(\mathbf{x}) = \frac{d^2}{d\mathbf{x}^2} f(\mathbf{x}) = \frac{d}{d\mathbf{x}}\mathbf{g} = \frac{d}{d\mathbf{x}}(\mathbf{A}\mathbf{x} - \mathbf{b}) = \mathbf{A} \tag{2.109}$$

Solving $\mathbf{g}(\mathbf{x}) = \mathbf{A}\mathbf{x} - \mathbf{b} = \mathbf{0}$, we obtain the solution $\mathbf{x}^* = \mathbf{A}^{-1}\mathbf{b}$, at which the function is minimized to

$$f(\mathbf{x}^*) = \frac{1}{2}(\mathbf{A}^{-1}\mathbf{b})^T \mathbf{A}(\mathbf{A}^{-1}\mathbf{b}) - \mathbf{b}^T(\mathbf{A}^{-1}\mathbf{b}) + c = -\frac{1}{2}\mathbf{b}^T \mathbf{A}^{-1}\mathbf{b} + c \tag{2.110}$$

At the solution $\mathbf{x}^*$ that minimizes $f(\mathbf{x})$, its gradient is

$$\mathbf{g}(\mathbf{x}^*) = \mathbf{A}\mathbf{x}^* - \mathbf{b} = \mathbf{0} \tag{2.111}$$

We also see that the minimization of the quadratic function $f(\mathbf{x})$ is equivalent to solving a linear equation $\mathbf{A}\mathbf{x} = \mathbf{b}$ with a symmetric positive definite coefficient matrix $\mathbf{A}$. The CG method considered here can therefore be used for solving both problems.

Conjugate basis vectors

We first review the concept of *conjugate vectors* (Section A.1.2), which is of essential importance in the CG method. Two vectors $\mathbf{u}$ and $\mathbf{v}$ are *mutually conjugate* or *A-conjugate* to each other with respect to a symmetric matrix $\mathbf{A} = \mathbf{A}^T$, if they satisfy

$$\mathbf{u}^T\mathbf{A}\mathbf{v} = (\mathbf{A}\mathbf{v})^T\mathbf{u} = \mathbf{v}^T\mathbf{A}\mathbf{u} = 0 \tag{2.112}$$

Specially, if $\mathbf{A} = \mathbf{I}$, the two conjugate vectors become orthogonal to each other, that is, $\mathbf{v}^T\mathbf{u} = 0$.

Similar to a set of N orthogonal vectors that can be used as the basis spanning an N–D space, a set of N mutually conjugate vectors $\{\mathbf{d}_0, \cdots, \mathbf{d}_{N-1}\}$ satisfying $\mathbf{d}_i^T\mathbf{A}\mathbf{d}_j = 0$ $(i \neq j)$ can also be used as a basis to span the N–D space. Any vector in the space can be expressed as a linear combination of these basis vectors.

Also we note that any set of N independent vectors can be converted by the Gram–Schmidt process (Section A.1.2) to a set of N basis vectors that are either orthogonal or A-conjugate to each other.

Example 2.6

Given two independent basis vectors of the 2-D space, and a positive-definite matrix

$$\mathbf{v}_1 = \begin{bmatrix} 1 \\ 0 \end{bmatrix}, \quad \mathbf{v}_2 = \begin{bmatrix} 0 \\ 1 \end{bmatrix}, \quad \mathbf{A} = \begin{bmatrix} 3 & 1 \\ 1 & 2 \end{bmatrix}$$

we can construct two A-orthogonal basis vectors by the Gram–Schmidt method, as shown in Fig. 2.12.

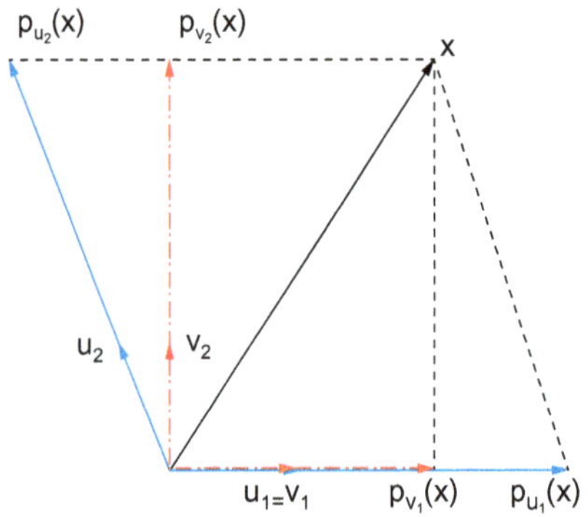

Fig. 2.12 A-orthogonality based on matrix **A**.

$$\mathbf{u}_1 = \mathbf{v}_1 = \begin{bmatrix} 1 \\ 0 \end{bmatrix}, \qquad \mathbf{u}_2 = \mathbf{v}_2 - \frac{\mathbf{u}_1^T\mathbf{A}\mathbf{v}_2}{\mathbf{u}_1^T\mathbf{A}\mathbf{u}_1}\mathbf{u}_1 = \begin{bmatrix} -1/3 \\ 1 \end{bmatrix}$$

The projections of a vector $\mathbf{x} = [2,\ 3]^T$ onto $\mathbf{v}_1$ and $\mathbf{v}_2$ are

$$\mathbf{p}_{\mathbf{v}_1}(\mathbf{x}) = \frac{\mathbf{v}_1^T\mathbf{x}}{\mathbf{v}_1^T\mathbf{v}_1}\mathbf{v}_1 = 2\begin{bmatrix} 1 \\ 0 \end{bmatrix} = \begin{bmatrix} 2 \\ 0 \end{bmatrix}, \quad \mathbf{p}_{\mathbf{v}_2}(\mathbf{x}) = \frac{\mathbf{v}_2^T\mathbf{x}}{\mathbf{v}_2^T\mathbf{v}_2}\mathbf{v}_2 = 3\begin{bmatrix} 0 \\ 1 \end{bmatrix} = \begin{bmatrix} 0 \\ 3 \end{bmatrix}$$

The A-projections of the same vector $\mathbf{x}$ onto $\mathbf{u}_1$ and $\mathbf{u}_2$ are

$$\mathbf{p}_{\mathbf{u}_1}(\mathbf{x}) = \frac{\mathbf{u}_1^T\mathbf{A}\mathbf{x}}{\mathbf{u}_1^T\mathbf{A}\mathbf{u}_1}\mathbf{u}_1 = 3\begin{bmatrix} 1 \\ 0 \end{bmatrix} = \begin{bmatrix} 3 \\ 0 \end{bmatrix}, \qquad \mathbf{p}_{\mathbf{u}_2}(\mathbf{x}) = \frac{\mathbf{u}_2^T\mathbf{A}\mathbf{x}}{\mathbf{u}_2^T\mathbf{A}\mathbf{u}_2}\mathbf{u}_2 = 3\begin{bmatrix} -1/3 \\ 1 \end{bmatrix} = \begin{bmatrix} -1 \\ 3 \end{bmatrix}$$

The original vector $\mathbf{x}$ can be represented in either of the two bases

$$\mathbf{x} = \begin{bmatrix} 2 \\ 3 \end{bmatrix} = 2\begin{bmatrix} 1 \\ 0 \end{bmatrix} + 3\begin{bmatrix} 0 \\ 1 \end{bmatrix} = 2\mathbf{v}_1 + 3\mathbf{v}_2 = \mathbf{p}_{\mathbf{v}_1}(\mathbf{x}) + \mathbf{p}_{\mathbf{v}_2}(\mathbf{x})$$
$$= 3\begin{bmatrix} 1 \\ 0 \end{bmatrix} + 3\begin{bmatrix} -1/3 \\ 1 \end{bmatrix} = 3\mathbf{u}_1 + 3\mathbf{u}_2 = \mathbf{p}_{\mathbf{u}_1}(\mathbf{x}) + \mathbf{p}_{\mathbf{u}_2}(\mathbf{x})$$

Search along a conjugate basis

Similar to the gradient descent method, which iteratively improves the estimated solution by following a sequence of orthogonal search directions $\{\mathbf{d}_0, \cdots, \mathbf{d}_n, \cdots\}$ with $\mathbf{d}_n = -\mathbf{g}_n$ satisfying $\mathbf{g}_n^T \mathbf{g}_{n+1} = 0$, the CG method also follows a sequence of search directions $\{\mathbf{d}_0, \cdots, \mathbf{d}_{N-1}\}$ A-orthogonal to each other, that is, $\mathbf{d}_i^T \mathbf{A} \mathbf{d}_j = 0 \; (i \neq j)$

$$\mathbf{x}_{n+1} = \mathbf{x}_n + \delta_n \mathbf{d}_n = \cdots = \mathbf{x}_0 + \sum_{i=0}^{n} \delta_i \mathbf{d}_i \tag{2.113}$$

We define the error at the nth step as $\mathbf{e}_n = \mathbf{x}_n - \mathbf{x}^*$. Then subtracting $\mathbf{x}^*$ from both sides, we get the iteration in terms of the errors

$$\mathbf{e}_{n+1} = \mathbf{e}_n + \delta_n \mathbf{d}_n = \cdots = \mathbf{e}_0 + \sum_{i=0}^{n} \delta_i \mathbf{d}_i \tag{2.114}$$

Owing to $\mathbf{g} = \mathbf{A}\mathbf{x} - \mathbf{b}$ in Eq. (2.111), we can find the gradient at the nth step $\mathbf{x}_n$ as

$$\mathbf{g}_n = \mathbf{A}\mathbf{x}_n - \mathbf{b} = \mathbf{A}(\mathbf{x}^* + \mathbf{e}_n) - \mathbf{b} = \mathbf{A}\mathbf{x}^* - \mathbf{b} + \mathbf{A}\mathbf{e}_n = \mathbf{A}\mathbf{e}_n \tag{2.115}$$

As the gradient at the solution is zero $\mathbf{g}(\mathbf{x}^*) = \mathbf{0}$, we can consider the gradient $\mathbf{g}_n$ at $\mathbf{x}_n$ as the residual of the nth iteration, and $\varepsilon = ||\mathbf{g}_n||^2$ an error measurement representing how close $\mathbf{x}_n$ is to the true solution $\mathbf{x}^*$.

The optimal step size given in Eq. (2.38) can now be written as

$$\delta_i = -\frac{\mathbf{d}_i^T \mathbf{g}_i}{\mathbf{d}_i^T \mathbf{A}\mathbf{d}_i} = -\frac{\mathbf{d}_i^T \mathbf{A}\mathbf{e}_i}{\mathbf{d}_i^T \mathbf{A}\mathbf{d}_i} = -\frac{\mathbf{d}_i^T \mathbf{A}\left(\mathbf{e}_0 + \sum_{j=0}^{i-1} \delta_j \mathbf{d}_n\right)}{\mathbf{d}_i^T \mathbf{A}\mathbf{d}_i}$$
$$= -\frac{\mathbf{d}_i^T \mathbf{A}\mathbf{e}_0 + \sum_{j=0}^{n-1} \delta_j \mathbf{d}_i^T \mathbf{A}\mathbf{d}_j}{\mathbf{d}_i^T \mathbf{A}\mathbf{d}_i} = -\frac{\mathbf{d}_i^T \mathbf{A}\mathbf{e}_0}{\mathbf{d}_i^T \mathbf{A}\mathbf{d}_i} \tag{2.116}$$

The last equality is attributed to the fact that $\mathbf{d}_i$ and $\mathbf{d}_j$ are A-orthogonal, $\mathbf{d}_i^T \mathbf{A}\mathbf{d}_j = 0$. Substituting this into Eq. (2.114) we obtain

$$\mathbf{e}_{n+1} = \mathbf{e}_n + \delta_n \mathbf{d}_n = \mathbf{e}_n - \left(\frac{\mathbf{d}_n^T \mathbf{A}\mathbf{e}_0}{\mathbf{d}_n^T \mathbf{A}\mathbf{d}_n}\right) \mathbf{d}_n \tag{2.117}$$

Conversely, we can represent the error $\mathbf{e}_0 = \mathbf{x}_0 - \mathbf{x}^*$ associated with the initial guess $\mathbf{x}_0$ as a linear combination of the A-orthogonal search vectors $\{\mathbf{d}_0, \cdots, \mathbf{d}_{N-1}\}$ as N basis vectors that span the N–D vector space

$$\mathbf{e}_0 = \sum_{i=0}^{N-1} c_i \mathbf{d}_i = \sum_{i=0}^{N-1} \mathbf{p}_{\mathbf{d}_i}(\mathbf{e}_0) = \sum_{i=0}^{N-1} \left(\frac{\mathbf{d}_i^T \mathbf{A} \mathbf{e}_0}{\mathbf{d}_i^T \mathbf{A} \mathbf{d}_i} \right) \mathbf{d}_i \tag{2.118}$$

where $\mathbf{p}_{\mathbf{d}_i}(\mathbf{e}_0)$ is the A-projection of $\mathbf{e}_0$ onto the ith basis vector $\mathbf{d}_i$

$$\mathbf{p}_{\mathbf{d}_i}(\mathbf{e}_0) = c_i \mathbf{d}_i = \left(\frac{\mathbf{d}_i^T \mathbf{A} \mathbf{e}_0}{\mathbf{d}_i^T \mathbf{A} \mathbf{d}_i} \right) \mathbf{d}_i \tag{2.119}$$

Note that the coefficient c_i happens to be the negative optimal step size in Eq. (2.116)

$$c_i = \frac{\mathbf{d}_i^T \mathbf{A} \mathbf{e}_0}{\mathbf{d}_i^T \mathbf{A} \mathbf{d}_i} = -\delta_i \tag{2.120}$$

Now the expression of $\mathbf{e}_{n+1}$ in Eq. (2.114) can be written as

$$\mathbf{e}_{n+1} = \mathbf{e}_0 + \sum_{i=0}^{n} \delta_i \mathbf{d}_i = \sum_{i=0}^{N-1} c_i \mathbf{d}_i - \sum_{i=0}^{n} c_i \mathbf{d}_i = \sum_{i=n+1}^{N-1} c_i \mathbf{d}_i = \sum_{i=n+1}^{N-1} \mathbf{p}_{\mathbf{d}_i}(\mathbf{e}_0) \tag{2.121}$$

We see that in each iteration, the number of terms in the summation for $\mathbf{e}_0$ is reduced by one, that is, the nth component $\mathbf{p}_{\mathbf{d}_n}(\mathbf{e}_0)$ of $\mathbf{e}_0$ along the direction of $\mathbf{d}_n$ is completely eliminated. After N such iterations, the error is reduced from $\mathbf{e}_0$ to $\mathbf{e}_N = \mathbf{0}$, and the true solution is obtained $\mathbf{x}_N = \mathbf{x}^* + \mathbf{e}_N = \mathbf{x}^*$.

Premultiplying $\mathbf{d}_k^T \mathbf{A}$ $(k \le n)$ on both sides of the Eq. (2.121), we get

$$\mathbf{d}_k^T \mathbf{A} \mathbf{e}_{n+1} = \sum_{i=n+1}^{N-1} c_i \mathbf{d}_k^T \mathbf{A} \mathbf{d}_j = 0 \tag{2.122}$$

We see that after $n+1$ iterations the remaining error $\mathbf{e}_{n+1}$ is A-orthogonal to all previous directions $\mathbf{d}_0, \cdots, \mathbf{d}_n$.

Owing to Eq. (2.115), Eq. (2.122) can also be written as

$$\mathbf{d}_k^T \mathbf{A} \mathbf{e}_{n+1} = \mathbf{d}_k^T \mathbf{g}_{n+1} = 0 \tag{2.123}$$

that is the gradient $\mathbf{g}_{n+1}$ is orthogonal to all previous search directions.

The conjugate gradient method is compared with the gradient descent method for the case of $N = 2$ in Fig. 2.13. We see that the first search direction is the same $-\mathbf{g}_0$ for both methods, but the next search direction $\mathbf{d}_1$ is A-orthogonal to $\mathbf{d}_0$, same as the next error $\mathbf{e}_1$, different from the search direction $-\mathbf{g}_1$ in gradient descent method. The conjugate gradient method finds the solution $\mathbf{x}$ in $N = 2$ steps, while the gradient descent method has to go through many more steps all orthogonal to each other before it finds the solution.

Find the A-orthogonal basis

The N A-orthogonal search directions $\{\mathbf{d}_0, \cdots, \mathbf{d}_{N-1}\}$ can be constructed based on any set of N independent vectors $\{\mathbf{v}_0, \cdots, \mathbf{v}_{N-1}\}$ by the Gram–Schmidt process

$$\mathbf{d}_n = \mathbf{v}_n - \sum_{j=0}^{n-1} \mathbf{p}_{\mathbf{d}_j}(\mathbf{v}_n) = \mathbf{v}_n - \sum_{m=0}^{n-1} \left(\frac{\mathbf{d}_m^T \mathbf{A} \mathbf{v}_n}{\mathbf{d}_m^T \mathbf{A} \mathbf{d}_m} \right) \mathbf{d}_m = \mathbf{v}_n - \sum_{m=0}^{n-1} \beta_{nm} \mathbf{d}_m \tag{2.124}$$

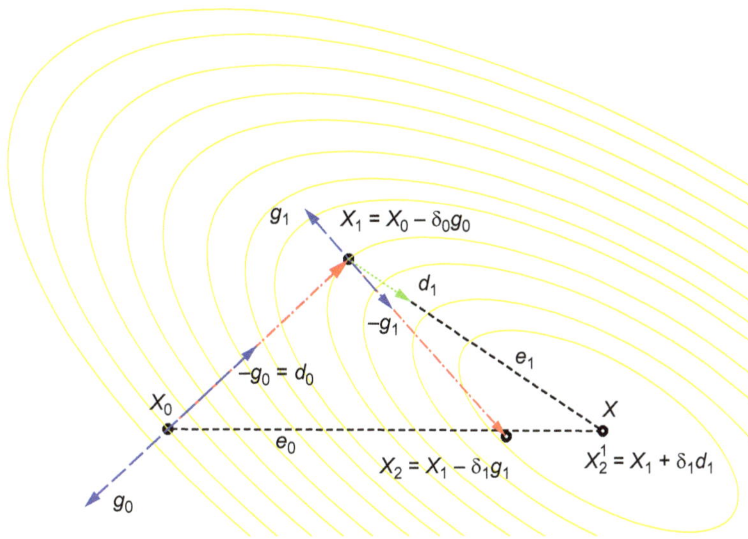

Fig. 2.13 Conjugate gradient descent.

where $\beta_{nm} = \mathbf{d}_m^T \mathbf{A} \mathbf{v}_n / (\mathbf{d}_m^T \mathbf{A} \mathbf{d}_m)$ and $\beta_{nm} \mathbf{d}_m$ is the A-projection of $\mathbf{v}_n$ onto each of the previous direction $\mathbf{d}_m$.

We will gain some significant computational advantage if we choose to use $\mathbf{v}_n = -\mathbf{g}_n$. Now the Gram–Schmidt process just discussed becomes

$$\mathbf{d}_n = -\mathbf{g}_n - \sum_{m=0}^{n-1} \beta_{nm} \mathbf{d}_m, \tag{2.125}$$

where

$$\beta_{nm} = \frac{\mathbf{d}_m^T \mathbf{A} \mathbf{v}_n}{\mathbf{d}_m^T \mathbf{A} \mathbf{d}_m} = -\frac{\mathbf{d}_m^T \mathbf{A} \mathbf{g}_n}{\mathbf{d}_m^T \mathbf{A} \mathbf{d}_m} \qquad (m < n) \tag{2.126}$$

This equation indicates that $\mathbf{g}_n$ can be written as a linear combination of all previous search directions $\mathbf{d}_0, \cdots, \mathbf{d}_n$

$$\mathbf{g}_n = \sum_{i=0}^{n} \alpha_i \mathbf{d}_i \tag{2.127}$$

Premultiplying $\mathbf{g}_{n+1}^T$ on both sides, we have

$$\mathbf{g}_{n+1}^T \mathbf{g}_k = \mathbf{g}_{n+1}^T \left(\sum_{i=0}^{k} \alpha_i \mathbf{d}_i \right) = \sum_{i=0}^{k} \alpha_i \, \mathbf{g}_{n+1}^T \mathbf{d}_i = 0 \tag{2.128}$$

The last equality is due to $\mathbf{g}_{n+1}^T \mathbf{d}_k = 0$ in Eq. (2.123). We see that $\mathbf{g}_{n+1}$ is also orthogonal to all previous gradients $\mathbf{g}_0, \cdots, \mathbf{g}_n$.

Premultiplying $\mathbf{g}_k^T$ $(k \geq n)$ on both sides of Eq. (2.125), we get

$$\mathbf{g}_k^T \mathbf{d}_n = -\mathbf{g}_k^T \mathbf{g}_n - \sum_{m=0}^{n-1} \beta_{mn} \mathbf{g}_k^T \mathbf{d}_m = -\mathbf{g}_k^T \mathbf{g}_n = \begin{cases} -||\mathbf{g}_n||^2 & n = k \\ 0 & n < k \end{cases} \tag{2.129}$$

Note that all terms in the summation are zero as $\mathbf{g}_k^T \mathbf{d}_m = 0$ for all $m < n \leq k$ (Eq. (2.122)). Substituting $\mathbf{g}_n^T \mathbf{d}_n = -||\mathbf{g}_n||^2$ into Eq. (2.116), we get

$$\delta_n = -\frac{\mathbf{g}_n^T \mathbf{d}_n}{\mathbf{d}_n^T \mathbf{A} \mathbf{d}_n} = \frac{||\mathbf{g}_n||^2}{\mathbf{d}_n^T \mathbf{A} \mathbf{d}_n} \tag{2.130}$$

Next we consider

$$\mathbf{g}_{m+1} = \mathbf{A}\mathbf{x}_{m+1} - \mathbf{b} = \mathbf{A}(\mathbf{x}_m + \delta_m \mathbf{d}_m) - \mathbf{b} = (\mathbf{A}\mathbf{x}_m - \mathbf{b}) + \delta_m \mathbf{A}\mathbf{d}_m = \mathbf{g}_m + \delta_m \mathbf{A}\mathbf{d}_m \tag{2.131}$$

Premultiplying $\mathbf{g}_n^T$ with $n > m$ on both sides we get

$$\mathbf{g}_n^T \mathbf{g}_{m+1} = \mathbf{g}_n^T \mathbf{g}_m + \delta_m \mathbf{g}_n^T \mathbf{A}\mathbf{d}_m = \delta_m \mathbf{g}_n^T \mathbf{A}\mathbf{d}_m \tag{2.132}$$

where $\mathbf{g}_n^T \mathbf{g}_m = 0$ $(m \neq n)$. Solving for $\mathbf{g}_n^T \mathbf{A}\mathbf{d}_m$ we have

$$\mathbf{g}_n^T \mathbf{A}\mathbf{d}_m = \frac{1}{\delta_m} \mathbf{g}_n^T \mathbf{g}_{m+1} = \begin{cases} ||\mathbf{g}_n||^2/\delta_{n-1} & m = n-1 \\ 0 & m < n-1 \end{cases} \tag{2.133}$$

Substituting this into Eq. (2.126) we get

$$\beta_{nm} = -\frac{\mathbf{d}_m^T \mathbf{A}\mathbf{g}_n}{\mathbf{d}_m^T \mathbf{A}\mathbf{d}_m} = \begin{cases} -||\mathbf{g}_n||^2/\delta_{n-1}\mathbf{d}_{n-1}^T \mathbf{A}\mathbf{d}_{n-1} & m = n-1 \\ 0 & m < n-1 \end{cases} \tag{2.134}$$

which is non-zero only when $m = n - 1$, that is, there is only one non-zero term in the summation of the Gram–Schmidt formula for $\mathbf{d}_n$. This is the reason why we choose $\mathbf{v}_n = -\mathbf{g}_n$. We can now drop the second subscript m in β_{nm}, and Eq. (2.124) becomes

$$\mathbf{d}_n = \mathbf{v}_n - \sum_{m=0}^{n-1} \beta_{nm} \mathbf{d}_m = -\mathbf{g}_n - \beta_n \mathbf{d}_{n-1} \tag{2.135}$$

Substituting the step size $\delta_{n-1} = ||\mathbf{g}_{n-1}||^2/\mathbf{d}_{n-1}^T \mathbf{A}\mathbf{d}_{n-1}$ (Eq. (2.130)) into the expression for $\beta_{nm} = \beta_m$, we get

$$\beta_n = -\frac{||\mathbf{g}_n||^2}{||\mathbf{g}_{n-1}||^2} \tag{2.136}$$

We note that matrix **A** no longer appears in the expression.

The CG algorithm

Summarizing, we finally obtain the conjugate gradient algorithm in the following steps.

1. Set $n = 0$ and initialize the search direction (same as gradient descent)

$$\mathbf{d}_0 = -\mathbf{g}_0 \tag{2.137}$$

2. Terminate if the error $\varepsilon = ||\mathbf{g}_n||^2$ is smaller than a preset threshold. Otherwise, continue with the following.
3. Find optimal step size (Eq. (2.130)) and step forward

$$\delta_n = \frac{||\mathbf{g}_n||^2}{\mathbf{d}_n^T \mathbf{A} \mathbf{d}_n}, \qquad \mathbf{x}_{n+1} = \mathbf{x}_n + \delta_n \mathbf{d}_n \tag{2.138}$$

4. Update gradient

$$\mathbf{g}_{n+1} = \frac{d}{d\mathbf{x}} f(\mathbf{x}_{n+1}) \tag{2.139}$$

5. Find coefficient for the Gram–Schmidt process (Eq. (2.136))

$$\beta_{n+1} = -\frac{||\mathbf{g}_{n+1}||^2}{||\mathbf{g}_n||^2} \tag{2.140}$$

6. Update search direction (Eq. (2.135))

$$\mathbf{d}_{n+1} = -\mathbf{g}_{n+1} - \beta_{n+1} \mathbf{d}_n \tag{2.141}$$

   Set $n = n + 1$ and go back to step 2.

This algorithm assumes the objective function $f(\mathbf{x})$ to be quadratic with known $\mathbf{A}$. But when $f(\mathbf{x})$ is not quadratic, $\mathbf{A}$ is no longer available, we can still approximate $f(\mathbf{x})$ as a quadratic function by the first three terms in its Taylor series in the neighborhood of its minimum. Also, we need to modify the algorithm so that it does not depend on $\mathbf{A}$. Specifically, the optimal step size δ_n calculated in step 3 based on $\mathbf{A}$ can also be alternatively found by line minimization based on any suitable algorithms for 1-D optimization.

The MATLAB code for the conjugate gradient algorithm is listed here.

```
function xn=myCG(j,tol)        % j is the objective function to minimize
    syms d;                    % variable for 1-D symbolic function f(d)
    x=symvar(j).';             % symbolic variables in objective function
    J=matlabFunction(j);       % the objective function
    G=jacobian(j).';           % symbolic gradient of objective function
    G=matlabFunction(G);       % the gradient function
    xn=zeros(length(x),1);     % initial guess of x
    xc=num2cell(xn);
    e=J(xc{:});                % initial error
    gn=G(xc{:});               % gradient at xn
    dn=-gn;                    % use negative gradient as search direction
    n=0;
    while e>tol
        n=n+1;
        f=subs(j,x,xn+d*dn);   % convert n-D f(x) to 1-D f(d)
        delta=Opt1d(f);        % find delta that minimizes 1D function f(d)
        xn=xn+delta*dn;        % updata variable x
        xc=num2cell(xn);
```

```
        e=J(xc{:});                % new error
        gn1=G(xc{:});              % new gradient
        bt=-(gn1.'*gn1)/(gn.'*gn);     % find beta
        dn=-gn1-bt*dn;             % new search direction
        gn=gn1;                    % update gradient
        fprintf('%d\t(%.4f, %.4f, %.4f)\t%e\n',n,xn(1),xn(2),xn(3),e)
    end
end
```

Here is the function that uses Newton's method to find the optimal step size δ that minimizes the objective function as a 1-D function of δ

```
function x=Opt1d(f)           % f is 1-D symbolic function to minimize
    syms x;
    tol=10^(-3);
    d1=diff(f);               % 1st order derivative
    d2=diff(d1);              % 2nd order derivative
    f=matlabFunction(f);
    d1=matlabFunction(d1);
    d2=matlabFunction(d2);
    x=0.0;                    % initial guess of delta
    if d2(x)<=0               % 2nd order derivative needs to be positive
        x=rand-0.5;e
    end
    y=x+1;
    while abs(x-y) > tol      % minimization
        y=x;
        x=y-d1(y)/d2(y);      % Newton iteration
    end
end
```

Example 2.7

To compare the conjugate method and the gradient descent method, consider a very simple 2-D quadratic function

$$f(x,y) = \mathbf{x}^T\mathbf{A}\mathbf{x} = [x_1, x_2]\begin{bmatrix} 3 & 1 \\ 1 & 2 \end{bmatrix}\begin{bmatrix} x_1 \\ x_2 \end{bmatrix}$$

The performance of the gradient descent method depends significantly on the initial guess. For the specific initial guess of $\mathbf{x}_0 = [1.5,\ -0.75]^T$, the iteration gets into a zigzag pattern and the convergence is very slow, as shown in the following table.

Table 2.3

n	$\mathbf{x} = [x_1, x_2]$	$f(\mathbf{x})$
0	1.500000, −0.750000	2.812500
1	0.250000, −0.750000	$0.468750e-01$
2	0.250000, −0.125000	$7.812500e-02$
3	0.041667, −0.125000	$1.302083e-02$
4	0.041667, −0.020833	$2.170139e-03$
5	0.006944, −0.020833	$3.616898e-04$
6	0.006944, −0.003472	$6.028164e-05$
7	0.001157, −0.003472	$1.004694e-05$
8	0.001157, −0.000579	$1.674490e-06$
9	0.000193, −0.000579	$2.790816e-07$
10	0.000193, −0.000096	$4.651361e-08$
11	0.000032, −0.000096	$7.752268e-09$
12	0.000032, −0.000016	$1.292045e-09$
13	0.000005, −0.000016	$2.153408e-10$

However, as expected, the conjugate gradient method takes exactly $N = 2$ steps from any initial guess to reach at the solution

Table 2.4

n	$\mathbf{x} = [x_1, x_2]$	$f(\mathbf{x})$
0	1.500000, −0.750000	$2.812500e+00$
1	0.250000, −0.750000	$4.687500e-01$
2	0.000000, −0.000000	$1.155558e-33$

In Fig. 2.14, the search paths of these two methods are compared.

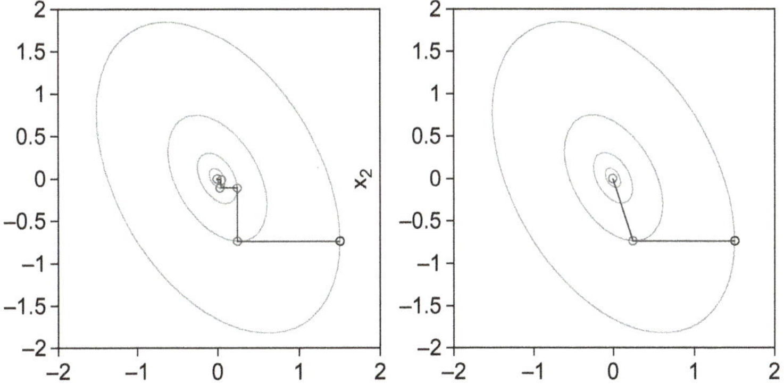

Fig. 2.14 Gradient (left) vs. conjugate gradient (right).

For an $N = 3$ example of $f(\mathbf{x}) = \mathbf{x}^T \mathbf{A} \mathbf{x}$ with

$$\mathbf{A} = \begin{bmatrix} 5 & 3 & 1 \\ 3 & 4 & 2 \\ 1 & 2 & 3 \end{bmatrix}$$

from an initial guess $\mathbf{x}_0 = [1,\ 2,\ 3]^T$, it takes the gradient descent method 41 iterations to reach $\mathbf{x}_{41} = [3.5486e-06,\ -7.4471e-06,\ 4.6180e-06]^T$ corresponding to $f(\mathbf{x}) = 8.5429e-11$. From the same initial guess, it takes the conjugate gradient method only $N = 3$ iterations to converge to the solution:

Table 2.5

n	$\mathbf{x} = [x_1, x_2, x_3]$	$f(\mathbf{x})$
0	1.000000, 2.000000, 3.000000	$4.500000e+01$
1	−0.734716, −0.106441, 1.265284	$2.809225e+00$
2	0.123437, −0.209498, 0.136074	$3.584736e-02$
3	−0.000000, 0.000000, 0.000000	$3.949119e-31$

For an $N = 9$ example, it takes over 4,000 iterations for the gradient descent method to converge with $||\mathbf{e}|| \approx 10^{-10}$, but exactly nine iterations for the CG method to converge with $||\mathbf{e}|| \approx 10^{-16}$.

Example 2.8

The search path of the conjugate gradient method applied to the minimization of the Rosenbrock function is shown in Fig. 2.15.

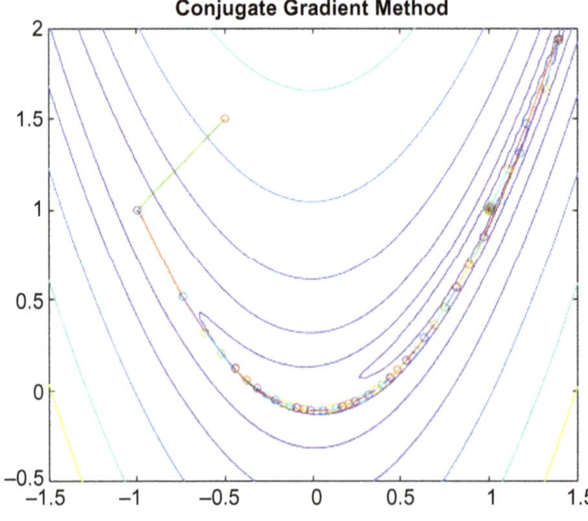

Fig. 2.15 Conjugate gradient applied to Rosenbrock surface.

Example 2.9

Solve the following $N = 3$ nonlinear equation system by the CG method

$$\begin{cases} f_1(x_1, x_2, x_3) = 3x_1 - (x_2x_3)^2 - 3/2 \\ f_2(x_1, x_2, x_3) = 4x_1^2 - 625\, x_2^2 + 2x_2 - 1 \\ f_3(x_1, x_2, x_3) = exp(-x_1x_2) + 20x_3 + 9 \end{cases}$$

The solution is known to be $x_1 = 0.5,\; x_2 = 0,\; x_3 = -0.5$.

This equation system can be represented in vector form as $\mathbf{f}(\mathbf{x}) = \mathbf{0}$ and the objective function is $J(\mathbf{x}) = \mathbf{f}^T(\mathbf{x})\mathbf{f}(\mathbf{x})$. The iteration of the CG method with an initial guess $\mathbf{x}_0 = \mathbf{0}$ is shown in the following table.

Table 2.6

n	$\mathbf{x} = [x_1, x_2, x_3]$	$J(\mathbf{x})$
0	0.0000,0.0000,0.0000	$1.032500e+02$
1	0.0113,0.0050,−0.5001	$3.160163e+00$
2	0.0188,−0.0021,−0.5004	$3.095894e+00$
3	0.5009,−0.0018,−0.5004	$7.268252e-05$
4	0.5009,−0.0017,−0.5000	$1.051537e-05$
5	0.5008,−0.0012,−0.5000	$6.511151e-06$
6	0.5001,−0.0005,−0.5000	$6.365321e-07$
7	0.5001,−0.0005,−0.5000	$5.667357e-07$
8	0.5002,−0.0004,−0.5000	$2.675128e-07$
9	0.5001,−0.0003,−0.5000	$1.344218e-07$
10	0.5001,−0.0002,−0.5000	$1.241196e-07$
11	0.5000,−0.0001,−0.5000	$2.120969e-08$
12	0.5000,−0.0001,−0.5000	$1.541814e-08$
13	0.5000,−0.0001,−0.5000	$7.282025e-09$
14	0.5000,−0.0001,−0.5000	$4.801781e-09$
15	0.5000,−0.0000,−0.5000	$4.463926e-09$

In comparison, the gradient descent method would need to take over 200 iterations (with much reduced complexity though) to reach this level of error.

Conjugate gradient method used for solving linear equation systems

As discussed before, if $\mathbf{x}$ is the solution that minimizes the quadratic function $f(\mathbf{x}) = \mathbf{x}^T\mathbf{A}\mathbf{x}/2 - \mathbf{b}^T\mathbf{x} + c$, with $\mathbf{A} = \mathbf{A}^T$ being symmetric and positive definite, it also satisfies $d\,f(\mathbf{x})/d\mathbf{x} = \mathbf{A}\mathbf{x} - \mathbf{b} = \mathbf{0}$. In other words, the optimization problem is equivalent to the problem of solving the linear system $\mathbf{A}\mathbf{x} - \mathbf{b} = \mathbf{0}$, both can be solved by the conjugate gradient method.

Now consider solving the linear system $\mathbf{A}\mathbf{x} = \mathbf{b}$ with $\mathbf{A} = \mathbf{A}^T$. Let $\mathbf{d}_i$, $(i = 1,\cdots,N)$ be a set of N A-orthogonal vectors satisfying $\mathbf{d}_i^T\mathbf{A}\mathbf{d}_j = 0$ $(i \neq j)$, which can be generated based on any N independent vectors, such as the standard basis vectors, by the Gram–Schmidt method. The solution $\mathbf{x}$ of the equation $\mathbf{Ax} = \mathbf{b}$ can be represented by these N vectors as

$$\mathbf{x} = \sum_{i=1}^{N} c_i \mathbf{d}_i \tag{2.142}$$

Now we have

$$\mathbf{b} = \mathbf{Ax} = \mathbf{A}\left[\sum_{i=1}^{N} c_i \mathbf{d}_i\right] = \sum_{i=1}^{N} c_i \mathbf{A}\mathbf{d}_i \tag{2.143}$$

Premultiplying $\mathbf{d}_j^T$ on both sides we get

$$\mathbf{d}_j^T \mathbf{b} = \sum_{i=1}^{N} c_i \mathbf{d}_j^T \mathbf{A}\mathbf{d}_i = c_j \mathbf{d}_j^T \mathbf{A}\mathbf{d}_j \tag{2.144}$$

Solving for c_j we have

$$c_j = \frac{\mathbf{d}_j^T \mathbf{b}}{\mathbf{d}_j^T \mathbf{A}\mathbf{d}_j} \tag{2.145}$$

Substituting this back into the expression for $\mathbf{x}$ we get the solution of the equation

$$\mathbf{x} = \sum_{i=1}^{N} c_i \mathbf{d}_i = \sum_{i=1}^{N} \left(\frac{\mathbf{d}_i^T \mathbf{b}}{\mathbf{d}_i^T \mathbf{A}\mathbf{d}_i} \right) \mathbf{d}_i \tag{2.146}$$

Also note that as $\mathbf{b} = \mathbf{Ax}$, the ith term of this summation is simply the A-projection of $\mathbf{x}$ onto the ith direction $\mathbf{d}_i$

$$\mathbf{p}_{\mathbf{d}_i}(\mathbf{x}) = \left(\frac{\mathbf{d}_i^T \mathbf{A}\mathbf{x}}{\mathbf{d}_i^T \mathbf{A}\mathbf{d}_i} \right) \mathbf{d}_i \tag{2.147}$$

One application of the conjugate gradient method is to solve the normal equation to find the least-square solution of an over-constrained equation system $\mathbf{Ax} = \mathbf{b}$, where the coefficient matrix $\mathbf{A}$ is M by N with rank $R = N < M$. As discussed previously, the normal equation of this system is

$$\mathbf{A}^T \mathbf{A}\mathbf{x} = \mathbf{A}^T \mathbf{b} \tag{2.148}$$

Here $\mathbf{A}^T\mathbf{A}$ is an N by N symmetric, positive definite matrix. This normal equation can be solved by the conjugate gradient method.

2.7 Issues of Local/Global Minimum

A main concern in optimization problems is whether the solution obtained by a certain algorithm is a local minimum or a global one. If function $f(\mathbf{x})$ to be optimized is quadratic and definite (either positive or negative definite), then any local extremum (either minimum or maximum) is the unique and global extremum. However, if $f(\mathbf{x})$ is not quadratic, then a local minimum found by the algorithm may not be global. In such a case, the *simulated annealing (SA)* algorithm can be used to search for the global minimum, by allowing uphill moves in the iteration to avoid getting stuck at a local minimum. SA mimics the annealing process in metallurgy, in which the temperature is carefully controlled to decrease slowly so that the material reaches a state with minimum free energy.

The way the SA method avoids becoming stuck at a local minimum is to allow the iteration steps to go uphill. Specifically, the objective function $f(\mathbf{x})$ evaluated at $\mathbf{x}_n$ is treated as the energy at $\mathbf{x}_n$, and an uphill move corresponding to an energy increase $\Delta E = f(\mathbf{x}_{n+1}) - f(\mathbf{x}_n)$ can be accepted if some probability function $P(\Delta E,T)$ defined in the following is greater than a threshold value Th within a certain range comparable with P

$$P(\Delta E,T) = e^{-\Delta E/T} > \text{Th} \tag{2.149}$$

where T represents the temperature. Obviously smaller ΔE and higher T will cause higher P, it is more probable for a move to be accepted.

During the entire search process, the temperature T gradually reduces so that the probability of accepting an uphill move also gradually reduces. In the early stage of the search, it is more probable to go uphill thereby getting out of a local minimum. However, toward the later part of the search process, such a move is less likely to happen, as presumably in this late stage of the search, the energy level is already low, not likely to be at a local minimum with high energy level. Throughout the search, the best result corresponding to the lowest energy level is always updated, to be returned to as the final solution at the end of the process.

PROBLEMS

Develop MATLAB code (or any other language) to implement the algorithms covered in this chapter to solve the equations in the examples in the text. Try at least three different initial guesses to see how they may affect how quickly the iteration converges.

1. Consider the linear over-constrained system in the form $\mathbf{Ax} = \mathbf{b}$, where $\mathbf{A}$ is a M by N matrix, $\mathbf{x}$ and $\mathbf{b}$ are, respectively, N–D and M–D vectors. Specifically, let $M = 9$ and $N = 5$, and generate in MATLAB by $rand('seed',1)$, $A = rand(M,N)$ and $b = rand(M,1)$, and then find the optimal solution of equation system so that the LS error $\varepsilon = ||\mathbf{Ax} - \mathbf{b}||^2$ is minimized.
2. Plot the following single-variable nonlinear functions. Make sure the domain of the function in the plot is wide enough to cover at least one root (zero-crossing):

$$f_1(x) = e^x - 1/x, \qquad f_2(x) = sin(x) - x^3, \qquad f_3(x) = x^3 + x^2 - 26\cos(x)$$

Then find their roots by solving the corresponding equations by the Newton–Raphson method.

3. Solve the same equations in the previous problem by minimizing the objective function defined as $o(x) = f^2(x)$ by Newton's method. Try at least three different initial guesses and compare results.
4. Solve the following equation systems by the multivariable Newton–Raphson method. Try at least three different initial guesses and compare results in terms of convergence of the iteration.

$$\begin{cases} f_1(\mathbf{x}) = x_1^2 - 2x_1 + x_2^2 - x_3 + 1 = 0 \\ f_2(\mathbf{x}) = x_1x_2^2 - x_1 - 3x_2 + x_2x_3 + 2 = 0 \\ f_3(\mathbf{x}) = x_1x_3^3 - 3x_3 + x_2x_3^2 + x_1x_2 = 0 \end{cases}$$

$$\begin{cases} f_1(\mathbf{x}) = 3x - 1 - (x_2x_3)^2 - 3/2 = 0 \\ f_2(\mathbf{x}) = 4x_1^2 - 500x_2^2 + 2x_2 - 1 = 0 \\ f_3(\mathbf{x}) = exp(-x_1x_2) + 20x_3 + 9 = 0 \end{cases}$$

5. Resolve the equation systems in the previous problem in vector form $\mathbf{f}(\mathbf{x}) = \mathbf{0}$, by minimizing the objective function defined as $o(\mathbf{x}) = ||\mathbf{f}(\mathbf{x})||^2) = \mathbf{f}^T\mathbf{f}$, using Newton's method for minimizing multivariable functions. Try at least three different initial guesses and compare results.
6. Make a 3-D surface plot and a 2-D contour plot of the Rosenbrock function

$$f(x, y) = (1 - x)^2 + 100(y - x^2)^2$$

Use both the gradient descent method (without second-order derivatives) and Newton's method to find the minimum of the function from different initial guesses such as: [−1, 2], [−1, 0], [0, 1], [2, 2]. Plot the search paths, same as in Fig. 2.7.

7. Resolve the same problem just discussed, but using various versions of quasi-Newton's method (SR1, BFGS, DFP). Try the same initial guesses and plot the search paths.
8. Resolve the same problem above but using the conjugate gradient method. Try the same initial guesses and plot the search paths.
9. Solve the following equation system by the conjugate gradient method

$$\begin{cases} f_1(x_1, x_2, x_3) = 3x_1 - (x_2x_3)^2 - 3/2 \\ f_2(x_1, x_2, x_3) = 4x_1^2 - 625\,x_2^2 + 2x_2 - 1 \\ f_3(x_1, x_2, x_3) = exp(-x_1x_2) + 20x_3 + 9 \end{cases}$$

3 Constrained Optimization

This chapter contains some theoretically intensive discussions about constrained optimization, including theories and methods for optimization with both equality and inequality constraints, and the Karush–Kahn–Tucker (KKT) conditions for both primal and dual problems of constrained optimization. Such discussions are necessary for the study of certain topics in future chapters, such as the method of support vector machines (SVMs), one of the most important classification method in machine learning (ML), to be considered in Chapter 14.

An optimization problem becomes more complicated if it is constrained, that is, the arguments $\mathbf{x}$ of the objective function $f(\mathbf{x})$ are subject to (abbreviated as s.t. in the following) certain equality and/or inequality constraints in terms of what values they can take. Such a generic constrained optimization problem can be formulated as

$$\begin{cases} \text{max/min:} & f(\mathbf{x}) = f(x_1, \cdots, x_N) \\ \text{s. t.:} & \begin{cases} h_i(\mathbf{x}) = 0, & (i = 1, \cdots m) \\ g_j(\mathbf{x}) \le 0, & (j = 1, \cdots n) \end{cases} \end{cases} \tag{3.1}$$

The set composed of all values of $\mathbf{x}$ that satisfy the $m + n$ constraints is called the *feasible region* of the problem. The goal is to find a point $\mathbf{x}^*$ in the feasible region at which $f(\mathbf{x}^*)$ is an extremum.

In the following we will first consider the most general case where the objective function $f(\mathbf{x})$ is nonlinear and so are the constraint functions $g_i(\mathbf{x})$ and $h_j(\mathbf{x})$. The processing of solving such a problem is called the *nonlinear programming (NLP)*. Here *programming* means the mathematical procedure for solving the optimization problems. We will consider the general method of Lagrange multiplier and a set of conditions that all solutions of such problems must satisfy.

We will then consider some specific algorithms for two special cases: *linear programming (LP)* in Section 3.4 for problems where the objective function and all constraint functions are linear (the feasible region is a polytope), and then *quadratic programming (QP)* in Section 3.6 for problems where the objective function is quadratic while the constraints are linear.

3.1 Optimization with Equality Constraints

The optimization problems subject to equality constraints can be generally formulated as

$$\begin{cases} \text{max/min:} & f(\mathbf{x}) = f(x_1, \cdots, x_N) \\ \text{s. t.:} & h_i(\mathbf{x}) = h_i(x_1, \cdots, x_N) = 0, \quad (i = 1, \cdots m) \end{cases} \tag{3.2}$$

Here we assume m is no bigger than N, as in general there does not exist a solution that satisfies more than N equations in the N–D space.

This problem can be visualized in the special case with $N = 2$ and $m = 1$, where both $f(\mathbf{x})$ and $h(\mathbf{x})$ are surfaces defined over the 2-D space spanned by x_1 and x_2, and $h(\mathbf{x}) = 0$ is the intersection line of $h(\mathbf{x})$ and the 2-D plane. The solution $\mathbf{x}^*$ must satisfy the following two conditions:

- $\mathbf{x}^*$ is on the intersection line satisfying $h(\mathbf{x}^*) = 0$;
- $\mathbf{x}^*$ is on the contour line $f(\mathbf{x}) = d$ (for some d as shown in Fig. 3.1, so that $f(\mathbf{x}^*)$ will neither increase nor decrease while moving along the contour line in the neighborhood of $\mathbf{x}^*$.

For both conditions to be satisfied, the two contours $h(\mathbf{x}) = 0$ and $f(\mathbf{x}) = d$ must coincide at $\mathbf{x}^*$, that is, they must have the same tangent line and therefore parallel gradients (perpendicular to the tangents)

$$\nabla_{\mathbf{x}} f(\mathbf{x}^*) = \lambda \ \nabla_{\mathbf{x}} \, h(\mathbf{x}^*) \tag{3.3}$$

where $\nabla_{\mathbf{x}} f(\mathbf{x}) = [\partial f/\partial x_1, \cdots, \partial f/\partial x_N]^T$ is the gradient of $f(\mathbf{x})$, and λ is a constant scaling factor which can be either positive if the two gradients are in the same direction or negative if they are in opposite directions. As here we are only interested in whether the two curves coincide or not, the sign of λ can be either positive or negative.

To find such a point $\mathbf{x}^*$, we first construct a new objective function, called the *Lagrange function* (or simply *Lagrangian*)

$$L(\mathbf{x}, \lambda) = f(\mathbf{x}) - \lambda \, h(\mathbf{x}) \tag{3.4}$$

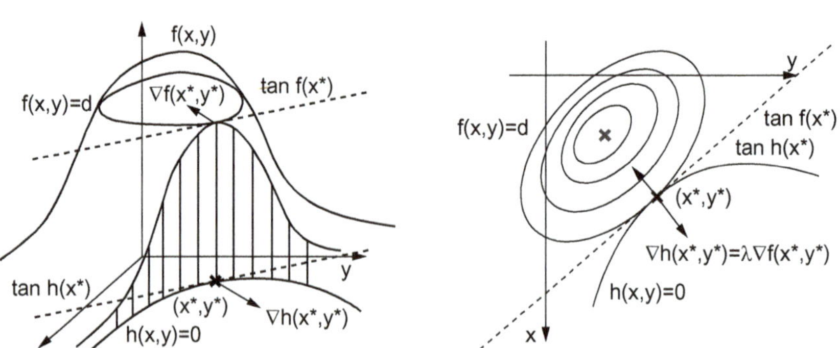

Fig. 3.1 Equality constrained optimization of function $f(x, y)$.

where λ is the *Lagrange multiplier*, which can be either positive or negative (i.e., the sign in front of λ can be either plus or minus), and then set the gradient of this objective function with respect to λ as well as $\mathbf{x}$ to zero

$$\nabla_{\mathbf{x},\lambda} L(\mathbf{x},\lambda) = \nabla_{\mathbf{x},\lambda}[f(\mathbf{x}) - \lambda\, h(\mathbf{x})] = \mathbf{0} \tag{3.5}$$

This equation contains two parts

$$\begin{cases} \nabla_{\mathbf{x}} f(\mathbf{x}) = \lambda\, \nabla_{\mathbf{x}}\, h(\mathbf{x}) \\ \nabla_{\lambda} L(\mathbf{x},\lambda) = \partial L(\mathbf{x},\lambda)/\partial\lambda = -h(\mathbf{x}) = 0 \end{cases} \tag{3.6}$$

This is an equation system containing $N + m = 2 + 1 = 3$ equations for the same number of unknowns x_1, x_2, λ

$$\frac{\partial f(\mathbf{x})}{\partial x_i} = \lambda \frac{\partial h(\mathbf{x})}{\partial x_i} \quad (i = 1,2), \qquad h(\mathbf{x}) = 0 \tag{3.7}$$

The solution $\mathbf{x}^* = [x_1^*, x_2^*]^T$ and λ^* of this equation system satisfies the two desired relationships: (a) $\nabla_{\mathbf{x}} f(\mathbf{x}) = \lambda\, \nabla_{\mathbf{x}}\, h(\mathbf{x})$, and (b) $h(\mathbf{x}^*) = 0$.

Specially, consider the following two possible cases.

- If $\lambda^* = 0$, then

$$\nabla_{\mathbf{x}} f(\mathbf{x}^*) = \lambda^*\, \nabla_{\mathbf{x}}\, h(\mathbf{x}^*) = 0 \tag{3.8}$$

  indicating $f(\mathbf{x}^*)$ is an extremum independent of the constraint, that is, the constraint $h(\mathbf{x}) = 0$ is *inactive*, and the optimization is unconstrained.
- If $\lambda^* \neq 0$, then $\nabla_{\mathbf{x}} f(\mathbf{x}^*) \neq 0$ (as in general $\nabla_{\mathbf{x}} h(\mathbf{x}^*) \neq \mathbf{0}$), indicating $f(\mathbf{x}^*)$ is not an extremum without the constraint, that is, the constraint is *active*, and the optimization is indeed constrained.

This discussion can be generalized from 2-D to any $N > 2$ dimensional space, in which the optimal solution $\mathbf{x}^*$ is to be found to extremize the objective $f(\mathbf{x})$ subject to m equality constraints $h_i(\mathbf{x}) = 0$ $(i = 1, \cdots, m)$, each representing a contour surface in the N–D space. The solution $\mathbf{x}^*$ must satisfy the following two conditions:

- $\mathbf{x}^*$ is on all m contour surfaces to satisfy $h_1(\mathbf{x}^*) = \cdots = h_m(\mathbf{x}^*) = 0$, that is, it is on the intersection of these surfaces, a curve in the space (e.g., the intersection of two surfaces in 3-D space is a curve);
- $\mathbf{x}^*$ is on a contour surface $f(\mathbf{x}^*) = d$ (for some value d) so that $f(\mathbf{x}^*)$ is an extremum, that is, it does not increase or decrease while moving along the curve in the neighborhood of $\mathbf{x}^*$.

Combining these conditions, we see that the intersection (a curve or a surface) of the m surfaces $h_i(\mathbf{x}) = 0$ must coincide with the contour surface $f(\mathbf{x}) = d$ at $\mathbf{x}^*$, that is, the intersection of the tangent surfaces of $h_i(\mathbf{x}) = 0$ through $\mathbf{x}^*$ must coincide

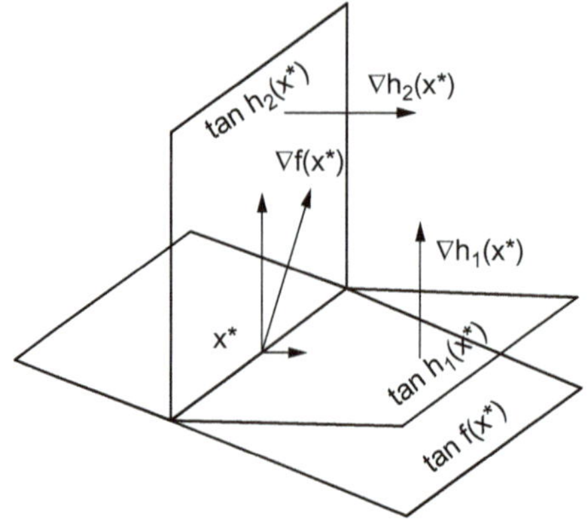

Fig. 3.2 Derivation of Lagrange function.

with the tangent surface of $f(\mathbf{x}) = d$ at $\mathbf{x}^*$, therefore the gradient $\triangledown f(\mathbf{x}^*)$ must be a linear combination of the m gradients $\triangledown h_i(\mathbf{x}^*)$

$$\triangledown_{\mathbf{x}} f(\mathbf{x}^*) = \sum_{i=1}^{m} \lambda_i^* \triangledown_{\mathbf{x}} h_i(\mathbf{x}^*) \tag{3.9}$$

as shown in Fig. 3.2.

To find such a solution $\mathbf{x}^*$ of the optimization problem, we first construct the Lagrange function

$$L(\mathbf{x},\boldsymbol{\lambda}) = f(\mathbf{x}) - \sum_{i=1}^{m} \lambda_i h_i(\mathbf{x}) \tag{3.10}$$

where $\boldsymbol{\lambda} = [\lambda_1, \cdots, \lambda_m]^T$ is a vector for m Lagrange multipliers. We then set its gradient to zero

$$\triangledown_{\mathbf{x},\lambda} L(\mathbf{x},\boldsymbol{\lambda}) = \triangledown_{\mathbf{x},\lambda} \left[f(\mathbf{x}) - \sum_{i=1}^{m} \lambda_i h_i(\mathbf{x}) \right] = \mathbf{0} \tag{3.11}$$

to get the following two equation systems of N and m equations, respectively

$$\triangledown_{\mathbf{x}} f(\mathbf{x}) = \sum_{i=1}^{m} \lambda_i \triangledown_{\mathbf{x}} h_i(\mathbf{x}) \quad \text{i.e.,} \quad \frac{\partial f(\mathbf{x})}{\partial x_j} = \sum_{i=1}^{m} \lambda_i \frac{\partial h_i(\mathbf{x})}{\partial x_j} \quad (j = 1, \cdots, N), \tag{3.12}$$

and

$$\frac{\partial L(\mathbf{x},\boldsymbol{\lambda})}{\partial \lambda_i} = h_i(\mathbf{x}) = 0 \quad (i = 1, \cdots, m) \tag{3.13}$$

The first set of N equations indicates that the gradient $\triangledown_{\mathbf{x}} f(\mathbf{x}^*)$ at $\mathbf{x}^*$ is a linear combination of the m gradients $\triangledown_{\mathbf{x}} h_i(\mathbf{x}^*)$, while the second set of m equations guarantees that $\mathbf{x}^*$ also satisfy the m equality constraints, both as desired. Solving these equations, we get the wished-for solution $\mathbf{x}^*$ together with the Lagrange multipliers $\lambda_1, \cdots, \lambda_m$. This is the *Lagrange multiplier method*, by which the constrained optimization problem is reformulated in such a way that the solution can be found by solving the equation system obtained by setting the gradient of the Lagrange function to zero.

Specially, if $\lambda_i = 0$, then the ith constraint is not active. More specially if $\lambda_i = 0$ for all $i = 1, \cdots, m$, then

$$\nabla_{\mathbf{x}} f(\mathbf{x}^*) = \sum_{i=1}^{m} \lambda_i \nabla_{\mathbf{x}} h_i(\mathbf{x}) = \mathbf{0} \tag{3.14}$$

indicating $f(\mathbf{x}^*)$ is an extremum independent of any of the constraints, that is, the optimization problem is unconstrained.

3.2 Optimization with Inequality Constraints

The optimization problems subject to inequality constraints can be generally formulated as

$$\begin{cases} \text{max/min:} & f(\mathbf{x}) = f(x_1, \cdots, x_N) \\ \text{s. t.:} & g_j(\mathbf{x}) = g_j(x_1, \cdots, x_N) \leq \text{ or } \geq 0, \qquad (j = 1, \cdots n) \end{cases} \tag{3.15}$$

Again, to visualize the problem we first consider an example with $N = 2$ and $n = 1$, as shown in Fig. 3.3 for the minimization (left) and maximization (right) of $f(\mathbf{x})$ subject to $g(\mathbf{x}) > 0$. Here the constrained solution $\mathbf{x}^* = [x_1^*, x_2^*]^T$ is on the boundary of the feasible region satisfying $g(x_1, x_2) = 0$, while the unconstrained extremum is outside the feasible region.

We consider the following two possible cases.

- First, if the unconstrained extremum is inside the feasible region, that is, the inequality constraint is inactive, then the solution $\mathbf{x}^*$ is the same as that of an unconstrained problem, not on the boundary of the feasible region, that is

$$\nabla_{\mathbf{x}} f(\mathbf{x}^*) = \mathbf{0}, \qquad g(\mathbf{x}^*) \neq 0 \tag{3.16}$$

Similar to the case of an inactive equality constraint, here the multiplier for an inactive inequality constraint is zero, that is, $\mu^* = 0$, and we get

$$\nabla_{\mathbf{x}} f(\mathbf{x}^*) = \mu^* \nabla_{\mathbf{x}} g(\mathbf{x}^*) = \mathbf{0} \tag{3.17}$$

the same as in Eq. (3.8).

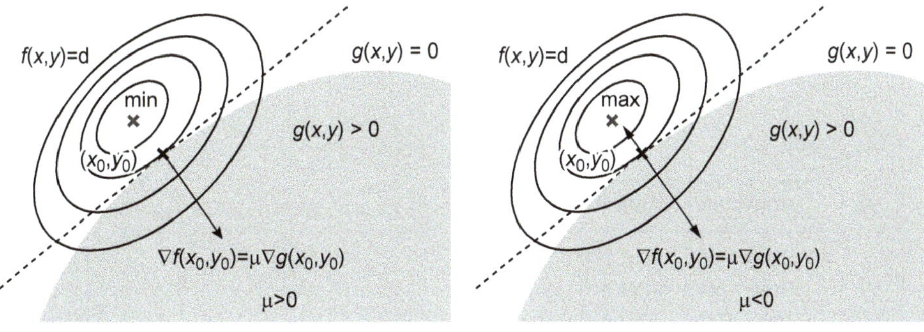

Fig. 3.3 Inequality constrained optimization of $f(x, y)$.

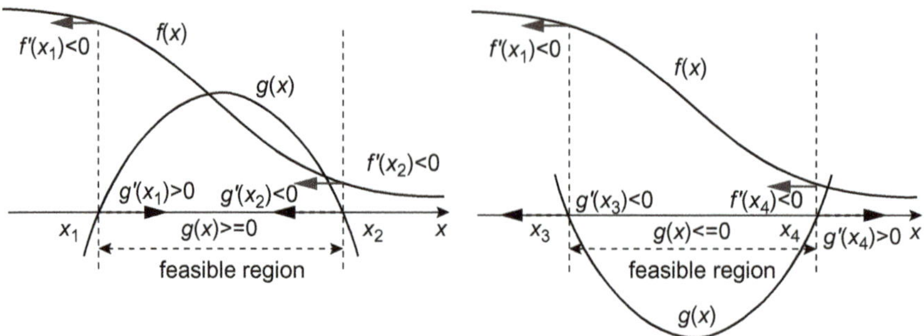

Fig. 3.4 Determination of signs of Lagrange multipliers.

- Second, if the unconstrained extremum at which $\triangledown_{\mathbf{x}} f(\mathbf{x}) = 0$ is outside the feasible region, that is, the inequality constraint is active, then the constrained solution $\mathbf{x}^*$ must be on the boundary of the feasible region, and it is different from the unconstrained solution

$$g(\mathbf{x}^*) = 0, \qquad \triangledown_{\mathbf{x}} f(\mathbf{x}^*) \neq \mathbf{0} \tag{3.18}$$

In this case, the problem becomes the same as an equality constrained problem considered before, that is, the objective function $f(\mathbf{x})$ and the constraint function $g(\mathbf{x})$ have the same tangent at $\mathbf{x}^*$ with parallel gradients

$$\triangledown_{\mathbf{x}} f(\mathbf{x}^*) = \mu^* \ \triangledown_{\mathbf{x}}\, g(\mathbf{x}^*) \quad \text{or} \quad \triangledown_{\mathbf{x}}\, f(\mathbf{x}^*) - \mu^* \ \triangledown_{\mathbf{x}}\, g(\mathbf{x}^*) = 0 \tag{3.19}$$

However, different from before, now we also need to consider whether the two gradients have the same or opposite directions, corresponding to $\mu > 0$ or $\mu < 0$, respectively. Specifically, there exist four possible cases in terms of the sign of μ, depending on whether $f(\mathbf{x})$ is to be maximized or minimized, and whether the constraint is $g(\mathbf{x}) \geq 0$ or $g(\mathbf{x}) \leq 0$, as the four points x_1 through x_4 illustrated in the 1-D examples in Fig. 3.4 and also summarized in the following tables:

Table 3.1

	$g(\mathbf{x}) \geq 0$	$g(\mathbf{x}) \leq 0$
max $f(\mathbf{x})$	(1) $\triangledown f(\mathbf{x}_1) = \mu \triangledown g(\mathbf{x}_1),\ \mu < 0$	(3) $\triangledown f(\mathbf{x}_3) = \mu \triangledown g(\mathbf{x}_3),\ \mu > 0$
min $f(\mathbf{x})$	(2) $\triangledown f(\mathbf{x}_2) = \mu \triangledown g(\mathbf{x}_2),\ \mu > 0$	(4) $\triangledown f(\mathbf{x}_4) = \mu \triangledown g(\mathbf{x}_4),\ \mu < 0$

Table 3.2

	max/min	subject to	μ
x_1	max $f(\mathbf{x})$	$g(\mathbf{x}) \geq 0$	$\mu < 0$
x_2	min $f(\mathbf{x})$	$g(\mathbf{x}) \geq 0$	$\mu > 0$
x_3	max $f(\mathbf{x})$	$g(\mathbf{x}) \leq 0$	$\mu > 0$
x_4	min $f(\mathbf{x})$	$g(\mathbf{x}) \leq 0$	$\mu < 0$

We can also consider the product of μ and $g(\mathbf{x})$ for either maximization or minimization

$$\mu\, g(\mathbf{x}) \begin{cases} \le 0 & \text{for maximization} \\ \ge 0 & \text{for minimization} \end{cases} \tag{3.20}$$

Summarizing these two cases, we see that $g(\mathbf{x}^*) \neq 0$ but $\mu^* = 0$ in the first case of an inactive constraint, and $g(\mathbf{x}^*) = 0$ but $\mu^* \neq 0$ in the second case of an active constraint, that is, the following holds in both cases

$$\mu^*\, g(\mathbf{x}^*) = 0 \tag{3.21}$$

This discussion can be generalized from 2-D to $N > 2$ dimensional space, in which the optimal solution $\mathbf{x}^*$ is to be found to extremize the objective $f(\mathbf{x})$ subject to n inequality constraints $g_i(\mathbf{x}) = 0$ $(i = 1, \cdots, n)$. To solve this inequality constrained optimization problem, we first construct the Lagrangian

$$L(\mathbf{x}, \boldsymbol{\mu}) = f(\mathbf{x}) - \sum_{i=1}^{n} \mu_i\, g_i(\mathbf{x}) \tag{3.22}$$

where $\boldsymbol{\mu} = [\mu_1, \cdots, \mu_n]^T$ is a vector for n Lagrange multipliers. (We note that in some literature, a plus sign is used in front of the summation of the second term. This is equivalent to our discussion here so long as the sign of μ indicated in Table 3.2 is negated.) Now Eq. (3.21) can be written as

$$(\boldsymbol{\mu}^*)^T\, g(\mathbf{x}^*) = 0 \tag{3.23}$$

We then set the gradient of the Lagrangian to zero

$$\nabla_{\mathbf{x},\boldsymbol{\mu}} L(\mathbf{x}, \boldsymbol{\mu}) = \nabla_{\mathbf{x},\boldsymbol{\mu}} \left[f(\mathbf{x}) - \sum_{i=1}^{n} \mu_i\, g_i(\mathbf{x}) \right] = \mathbf{0} \tag{3.24}$$

and get two equation systems of N and n equations, respectively

$$\nabla_{\mathbf{x}} f(\mathbf{x}) = \sum_{i=1}^{n} \mu_i\, \nabla_{\mathbf{x}}\, g_i(\mathbf{x}) \tag{3.25}$$

and

$$\nabla_{\boldsymbol{\mu}} \left[\sum_{i=1}^{n} \mu_i\, g_i(\mathbf{x}) \right] = \mathbf{0} \quad \text{i.e.,} \quad \frac{\partial L(\mathbf{x}, \boldsymbol{\mu})}{\partial \mu_i} = g_i(\mathbf{x}) = 0 \quad (i = 1, \cdots, n) \tag{3.26}$$

This result for the inequality constrained problems based on Eq. (3.24) looks the same as that for the equality constrained problems based on Eq. (3.11), but with an additional requirement regarding the sign of the Lagrange multipliers. For an equality constrained problem, the direction of the gradient $\nabla g(\mathbf{x})$ is of no concern, i.e., the sign of λ is unrestricted; but here for an inequality constrained problem, the sign of μ needs to be consistent with those shown in Table 3.2 unless when the constraint is inactive and $\mu = 0$.

As a simple example, consider the minimization of a function $f(\mathbf{x})$ by the method of gradient descent. To find the best increment $\Delta\mathbf{x}$ that minimizes $f(\mathbf{x} + \Delta\mathbf{x}) \approx$

$f(\mathbf{x}) + \mathbf{g}_f(\mathbf{x})^T \Delta\mathbf{x}$ subjected to the constraint of a limited length, i.e., $||\Delta\mathbf{x}||^2/2 \le C$, we first construct the Lagrangian

$$L(\mathbf{x}, \lambda) = f(\mathbf{x}) + \mathbf{g}_f(\mathbf{x})^T \Delta\mathbf{x} - \mu(||\Delta\mathbf{x}||^2/2 - C) \tag{3.27}$$

and set its derivative to zero

$$\frac{d}{d\Delta\mathbf{x}} L(\mathbf{x}, \lambda)\mathbf{g}_f(\mathbf{x}) - \mu\Delta\mathbf{x} = \mathbf{0} \tag{3.28}$$

where $\mu < 0$ according to Table 3.2. We then solve the equation to get

$$\Delta\mathbf{x} = \frac{1}{\mu}\mathbf{g}_f(\mathbf{x}) \tag{3.29}$$

and the iteration

$$\mathbf{x}_{n+1} = \mathbf{x}_n - \delta\mathbf{g}(\mathbf{x}_n) \tag{3.30}$$

same as the gradient descent method in Eq. (2.24).

We now consider the general optimization of an N–D objective function $f(\mathbf{x})$ subject to multiple constraints of qualities and inequalities

$$\begin{array}{ll} \text{max/min:} & f(\mathbf{x}) = f(x_1, \cdots, x_N) \\ \text{s. t.:} & \begin{cases} h_i(\mathbf{x}) = 0, & (i = 1, \cdots m) \\ g_j(\mathbf{x}) \le 0 \text{ or } g_j(\mathbf{x}) \ge 0, & (j = 1, \cdots n) \end{cases} \end{array} \tag{3.31}$$

For notational convenience, we denote the $m+n$ equality and inequality constraints in vector form as

$$\begin{array}{ll} \text{max/min:} & f(\mathbf{x}) \\ \text{s. t.:} & \begin{cases} \mathbf{h}(\mathbf{x}) = \mathbf{0} \\ \mathbf{g}(\mathbf{x}) \le \mathbf{0} \text{ or } \mathbf{g}(\mathbf{x}) \ge \mathbf{0} \end{cases} \end{array} \tag{3.32}$$

where $\mathbf{h}(\mathbf{x}) = [h_1(\mathbf{x}), \cdots, h_m(\mathbf{x})]^T$ and $\mathbf{g}(\mathbf{x}) = [g_1(\mathbf{x}), \cdots, g_n(\mathbf{x})]^T$.

To solve this optimization problem, we first construct the Lagrangian

$$L(\mathbf{x}, \boldsymbol{\lambda}, \boldsymbol{\mu}) = f(\mathbf{x}) - \sum_{i=1}^{m} \lambda_i h_i(\mathbf{x}) - \sum_{j=1}^{n} \mu_j g_j(\mathbf{x}) = f(\mathbf{x}) - \boldsymbol{\lambda}^T \mathbf{h}(\mathbf{x}) - \boldsymbol{\mu}^T \mathbf{g}(\mathbf{x}) \tag{3.33}$$

where the Lagrange multipliers in $\boldsymbol{\lambda} = [\lambda_1, \cdots, \lambda_m]^T$ and $\boldsymbol{\mu} = [\mu_1, \cdots, \mu_n]^T$ are for the m equality and n non-negative constraints, respectively. We then set to zero the gradient of $L(\mathbf{x}, \boldsymbol{\lambda}, \boldsymbol{\mu})$ with respect to both $\boldsymbol{\lambda}$ and $\boldsymbol{\mu}$ as well as $\mathbf{x}$.

$$\frac{\partial}{\partial\mathbf{x}} L(\mathbf{x}, \boldsymbol{\lambda}, \boldsymbol{\mu}) = \mathbf{0}, \quad \frac{\partial}{\partial\boldsymbol{\lambda}} L(\mathbf{x}, \boldsymbol{\lambda}, \boldsymbol{\mu}) = \mathbf{0}, \quad \frac{\partial}{\partial\boldsymbol{\mu}} L(\mathbf{x}, \boldsymbol{\lambda}, \boldsymbol{\mu}) = \mathbf{0} \tag{3.34}$$

Solving these equation systems, we get the solution $\mathbf{x}^*$. While λ_i can be either positive or negative, the sign of μ_j needs to be consistent with those specified in Table 3.2. Otherwise the inequality constraints is inactive.

Example 3.1

Find the extremum of $f(x_1, x_2) = x_1^2 + x_2^2$ subject to each of the three different constraints: $x_1 + x_2 = 1$, $x_1 + x_2 \le 1$, and $x_1 + x_2 \ge 1$, as illustrated in Fig. 3.5.

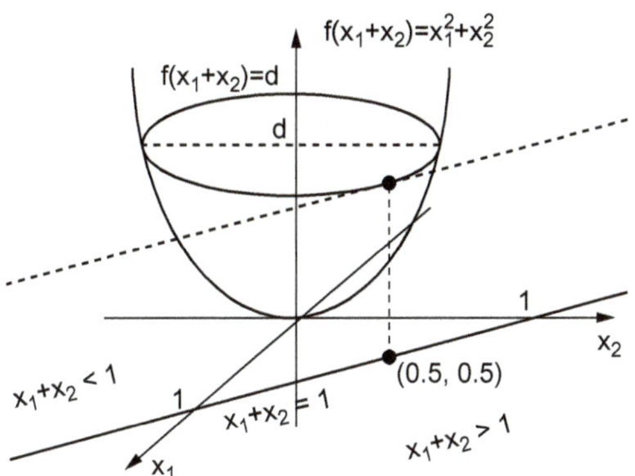

Fig. 3.5 Inequality constrained optimization.

- $\begin{cases} \text{min:} & f(x_1,x_2) = x_1^2 + x_2^2 \\ \text{s. t.:} & h(x_1,x_2) = x_1 + x_2 - 1 = 0 \end{cases}$

  The Lagrangian is

$$L(x_1,x_2,\lambda) = f(x_1,x_2) - \lambda g(x_1,x_2) = x_1^2 + x_2^2 - \lambda(x_1 + x_2 - 1)$$

  Solving the following equations

$$\frac{\partial L}{\partial x_1} = 2x_1 - \lambda = 0, \quad \frac{\partial L}{\partial x_2} = 2x_2 - \lambda = 0, \quad \frac{\partial L}{\partial \lambda} = x_1 + x_2 - 1 = 0$$

  we get $\lambda^* = 1$, $x_1^* = x_2^* = 0.5$, that is, the function is minimized at $f(0.5, 0.5) = 0.5$. We note that $f(x_1,x_2)$ and $g(x_1,x_2)$ have the same gradients at $x_1^* = x_2^* = 0.5$

$$\triangledown f(0.5,0.5) = \triangledown h(0.5,0.5) = [1,1]^T$$

- Consider these two problems

$$\begin{cases} \text{min:} & f(x_1,x_2) = x_1^2 + x_2^2 \\ \text{s. t.:} & g(x_1,x_2) = x_1 + x_2 - 1 \geq 0 \end{cases} \qquad \begin{cases} \text{max:} & f(x_1,x_2) = x_1^2 + x_2^2 \\ \text{s. t.:} & g(x_1,x_2) = x_1 + x_2 - 1 \leq 0 \end{cases}$$

  These problems are the two middle cases (2 and 3) in Table 3.2. They have the same Lagrangian as above with the same results $x_1^* = x_2^* = 0.5$ and $\mu^* = 1 > 0$. According to Table 3.2, $\mathbf{x}^*$ is the solution for the minimization problem subject to $g(\mathbf{x}) \geq 0$ (left), or the maximization problem subject to $g(\mathbf{x}) \leq 0$.
- Consider these two problems

$$\begin{cases} \text{max:} & f(x_1,x_2) = x_1^2 + x_2^2 \\ \text{s. t.:} & g(x_1,x_2) = x_1 + x_2 - 1 \geq 0 \end{cases} \qquad \begin{cases} \text{min:} & f(x_1,x_2) = x_1^2 + x_2^2 \\ \text{s. t.:} & g(x_1,x_2) = x_1 + x_2 - 1 \leq 0 \end{cases}$$

  These problems are cases 1 and 4 in Table 3.2. Again they have the same Lagrangian as in the previous cases and the same results $x_1^* = x_2^* = 0.5$, and $\mu^* = 1 > 0$. According to Table 3.2, $\mathbf{x}^*$ is *not* the solution of either of the two problems, that is, the constraint is inactive. We therefore need to assume $\mu^* = 0$ and solve the following equations for an unconstrained problem

$$\frac{\partial L}{\partial x_1} = 2x_1 = 0, \quad \frac{\partial L}{\partial x_2} = 2x_2 = 0$$

Now we get $x_1^* = x_2^* = 0$ and $\mathbf{x}^* = \mathbf{0}$, with minimum $f(\mathbf{x}^*) = 0$. This is the solution for the minimization problem (right), at which $g(\mathbf{x}^*) \neq 0$, that is, the constraint is inactive. However, this is not the solution for the maximization problem as the function is unbounded from above.

3.3 Duality and KKT Conditions

Previously we considered the constrained minimization problem

$$\begin{cases} \text{min:} & f_p(\mathbf{x}) \\ \text{s. t.:} & \mathbf{h}(\mathbf{x}) = \mathbf{0}, \qquad \mathbf{g}(\mathbf{x}) \leq \mathbf{0} \quad \text{or} \quad \mathbf{g}(\mathbf{x}) \geq \mathbf{0} \end{cases} \tag{3.35}$$

with its Lagrangian

$$L(\mathbf{x}, \boldsymbol{\lambda}, \boldsymbol{\mu}) = f_p(\mathbf{x}) - \boldsymbol{\lambda}^T \mathbf{h}(\mathbf{x}) - \boldsymbol{\mu}^T \mathbf{g}(\mathbf{x}) \tag{3.36}$$

The optimal solution $\mathbf{x}^*$ can be found by solving the following equation systems

$$\nabla_{\mathbf{x}} L(\mathbf{x}, \boldsymbol{\lambda}, \boldsymbol{\mu}) = \mathbf{0}, \quad \nabla_{\boldsymbol{\lambda}} L(\mathbf{x}, \boldsymbol{\lambda}, \boldsymbol{\mu}) = \mathbf{0}, \quad \nabla_{\boldsymbol{\mu}} L(\mathbf{x}, \boldsymbol{\lambda}, \boldsymbol{\mu}) = \mathbf{0} \tag{3.37}$$

at which the objective function reaches its minimum $p^* = f_p(\mathbf{x}^*)$.

In particular, if the objective function $f_p(\mathbf{x})$ and all inequality constraints $\mathbf{g}_i(\mathbf{x})$ are convex and the equality constraints $\mathbf{h}_i(\mathbf{x})$ are linear (affine), that is, the feasible region is convex (Section A.8.3), then the problem is a *convex optimization* with some nice properties as shown here.

This constrained minimization problem can be called the *primal problem*, as there also exists an equivalent *dual problem*, which may be easier to solve. To construct the dual problem, we first define the *dual function* as the lower bound of the Lagrangian in Eq. (3.36) to be minimized

$$f_d(\boldsymbol{\lambda}, \boldsymbol{\mu}) = \min_{\mathbf{x}} L(\mathbf{x}, \boldsymbol{\lambda}, \boldsymbol{\mu}) = \min_{\mathbf{x}} \left[f_p(\mathbf{x}) - \boldsymbol{\lambda}^T \mathbf{h}(\mathbf{x}) - \boldsymbol{\mu}^T \mathbf{g}(\mathbf{x}) \right] \tag{3.38}$$

We further find the tightest lower bound, the maximum of the dual function $f_d(\boldsymbol{\lambda}, \boldsymbol{\mu})$ over its variables $\boldsymbol{\lambda}$ and $\boldsymbol{\mu}$, by solving the following dual problem

$$\begin{cases} \text{max:} & f_d(\boldsymbol{\lambda}, \boldsymbol{\mu}) = \min_{\mathbf{x}} L(\mathbf{x}, \boldsymbol{\lambda}, \boldsymbol{\mu}) \\ \text{s. t.:} & \boldsymbol{\mu} \leq \mathbf{0} \quad \text{or} \quad \boldsymbol{\mu} \geq \mathbf{0} \end{cases} \tag{3.39}$$

Here $\boldsymbol{\lambda}$ for the equality constrains of the primal problem can be either positive or negative, that is, it is not constrained in the dual problem. However, the sign of $\boldsymbol{\mu}$ for the inequality constraints of the primal minimization problem needs to satisfy the following according to Eq. (3.20) (the second inequality for this minimization primal problem)

$$\boldsymbol{\mu}^T \mathbf{g}(\mathbf{x}) \geq 0 \tag{3.40}$$

that is $\boldsymbol{\mu}$ is subject to the inequality constraint of the dual problem. We denote the optimal solution of the dual problem by $\{\boldsymbol{\lambda}^*, \boldsymbol{\mu}^*\}$, and the maximum of the dual objective function by $d^* = f_d(\boldsymbol{\lambda}^*, \boldsymbol{\mu}^*)$.

Obviously the primal objective function of any feasible $\mathbf{x}$ is always no less than its lower bound, the dual objective function of any feasible $\{\boldsymbol{\lambda}, \boldsymbol{\mu}\}$

$$f_p(\mathbf{x}) \geq f_d(\boldsymbol{\lambda}, \boldsymbol{\mu}) \tag{3.41}$$

and this inequality also holds for the optimal solutions $\mathbf{x}^*$ and $\{\boldsymbol{\lambda}^*, \boldsymbol{\mu}^*\}$

$$p^* = f_p(\mathbf{x}) \geq f_d(\boldsymbol{\lambda}, \boldsymbol{\mu}) = d^* \tag{3.42}$$

The difference $p^* - d^*$ is defined as the *duality gap*. Obviously $p^* - d^* \geq 0$ always holds and the duality is called *weak duality*. In some problems where certain conditions (*Slater's conditions*) are satisfied, $p^* = d^*$, that is, d^* is the tightest lower bound, the reality is called *strong duality*.

We now consider the necessary and sufficient conditions for the solutions of both the primal and dual problems. We first introduce the *KKT* conditions:

- *Stationarity:* $\mathbf{x}^*$ maximizes or minimizes the Lagrangian, that is

$$\nabla_{\mathbf{x}} L(\mathbf{x}^*, \boldsymbol{\lambda}, \boldsymbol{\mu}) = \mathbf{0} \tag{3.43}$$

- *Primal feasibility:* $\mathbf{x}^*$ must be inside the feasible region

$$h_i(\mathbf{x}^*) = 0, \quad g_j(\mathbf{x}^*) \leq 0 \quad \text{or} \quad g_j(\mathbf{x}^*) \geq 0, \qquad (i = 1, \cdots, m, \; j = 1, \cdots, n) \tag{3.44}$$

- *Dual feasibility:* All multipliers in $\boldsymbol{\mu}$ for the inequalities must take the proper sign as specified by Table 3.2

$$\mu_j^* \leq 0 \quad \text{or} \quad \mu_j^* \geq 0, \quad (j = 1, \cdots, n) \tag{3.45}$$

- *Complementarity:* Either μ_j^* or $g_j(\mathbf{x}^*)$ must be zero depending on whether the jth constraint is inactive or not:
  - if inactive, then $g_j(\mathbf{x}^*) \neq 0$, but $\mu_j^* = 0$;
  - if active, then $\mu_j^* \neq 0$, but $g_j(\mathbf{x}^*) = 0$.

  Combining both cases, we get (Eq. (3.23))

$$(\boldsymbol{\mu}^*)^T \mathbf{g}(\mathbf{x}^*) = 0 \tag{3.46}$$

We now prove the following theorem stating that the KKT conditions and both necessary and sufficient for the solutions of the primal and dual problems to exist.

Theorem:

- *(Necessary) If $\mathbf{x}^*$ and $\{\boldsymbol{\lambda}^*, \boldsymbol{\mu}^*\}$ are solutions for the respective primal and dual problems with strong duality, then they satisfy all KKT conditions.*
- *(Sufficient) If $\mathbf{x}^*$ and $\{\boldsymbol{\lambda}^*, \boldsymbol{\mu}^*\}$ satisfy all KKT conditions, then they are solutions for the respective primal and dual problems, and strong duality holds, when the primal objective function $f_p(\mathbf{x}^*)$ and the inequality constraints $g_j(\mathbf{x})$ are convex, and equality constraints $h_j(\mathbf{x})$ are linear.*

Proof:

- Necessary conditions
  As $\mathbf{x}^*$ and $\{\lambda^*, \mu^*\}$ are, respectively, the primal and dual solutions, the primal and dual feasibilities must be satisfied, and so is stationarity as the optimal solution $\mathbf{x}^*$ satisfies Eq. (3.37). In addition, we also have $\mathbf{h}(\mathbf{x}^*) = \mathbf{0}$. To show complementarity also holds, consider

$$\begin{aligned} d^* &= f_d(\lambda^*, \mu^*) = \min_{\mathbf{x}} L(\mathbf{x}, \lambda^*, \mu^*) \\ &= \min_{\mathbf{x}} \left[f_p(\mathbf{x}) - (\lambda^*)^T \mathbf{h}(\mathbf{x}) - (\mu^*)^T \mathbf{g}(\mathbf{x}) \right] \\ &\le f_p(\mathbf{x}^*) - (\lambda^*)^T \mathbf{h}(\mathbf{x}^*) - (\mu^*)^T \mathbf{g}(\mathbf{x}^*) \\ &= f_p(\mathbf{x}^*) - (\mu^*)^T \mathbf{g}(\mathbf{x}^*) \\ &\le f_p(\mathbf{x}^*) = p^* \end{aligned} \tag{3.47}$$

  The last inequality is due to Eq. (3.40). But as $d^* = p^*$ owing to the strong duality, the inequalities need to be equalities, that is, $(\mu^*)^T \mathbf{g}(\mathbf{x}^*) = 0$ and complementarity holds.
- Sufficient conditions
  If $\mathbf{x}^*$ and $\{\lambda^*, \mu^*\}$ satisfy the KKT conditions, then they are feasible for the respective primal and dual problems. Also, as $f_p(\mathbf{x})$ and all $g_j(\mathbf{x})$ are convex and $h_j\mathbf{x})$ are linear, Lagrangian in Eq. (3.36) as a sum of convex functions is also convex (see Section A.8.3), and stationarity in Eq. (3.43) indicates $\mathbf{x}^*$ is a global minimum, therefore

$$\begin{aligned} d^* &= f_d(\lambda^*, \mu^*) = L(\mathbf{x}^*, \lambda^*, \mu^*) = f_p(\mathbf{x}^*) - (\lambda^*)^T \mathbf{h}(\mathbf{x}^*) - (\mu^*)^T \mathbf{g}(\mathbf{x}^*) \\ &= f_p(\mathbf{x}^*) = p^* \end{aligned} \tag{3.48}$$

  where the second line is due to the KKT conditions in Eqs (3.44) and (3.46). □

In summary, we see that the KKT conditions are necessary and satisfied by all solutions of constrained optimization problems, and they can also be sufficient for the solution of such problems if the primal problem further satisfies some additional conditions of convexity, such as in quadratic program if the quadratic objective function is also positive definite. In such a case, strong duality holds and the solution of the dual problem is the tightest lower bound of the primal, the same as the solution of the primal.

Consider, as an example of the inequality constrained minimization problems, the support vector machine for binary classification (Chapter 14). To find the decision boundary that optimally separates two classes, we need to solve the following minimization problem constrained by a set of inequalities imposed by the data samples $\{(\mathbf{x}_n, y_n),\ n = 1, \cdots, N\}$ in the training set

$$\begin{cases} \text{min:} & \frac{1}{2}||\mathbf{w}||^2 \\ \text{s. t.:} & y_n(\mathbf{x}_n^T\mathbf{w} + b) \geq 1 \quad (n = 1, \cdots, N) \end{cases} \tag{3.49}$$

where $\mathbf{w} = \sum_{n=1}^N \alpha_n y_n \mathbf{x}_n$. We note that the quadratic objective function is convex and so are the affine inequality constraints, that is, this is a convex problem. The solution $\mathbf{w}$ with minimum norm $||\mathbf{w}||^2$ subject to the inequality constraints can be found by solving the dual problem in terms of the Lagrange multipliers in $\boldsymbol{\alpha} = [\alpha_1, \cdots, \alpha_N]^T$, subject to certain inequality constraints (detailed derivation in Chapter 14)

$$\begin{cases} \text{max:} & \sum_{n=1}^N \alpha_n - \frac{1}{2}\sum_{n=1}^N \sum_{m=1}^N \alpha_n \alpha_m y_n y_m \left(\mathbf{x}_n^T \mathbf{x}_m\right) \\ \text{s. t.:} & \sum_{n=1}^N \alpha_n y_n = \mathbf{y}^T\boldsymbol{\alpha} = 0, \quad \alpha_n \geq 0 \quad (n = 1, \cdots, N) \end{cases} \tag{3.50}$$

As the dual function can be written a quadratic form

$$\boldsymbol{\alpha}^T\mathbf{1} - \frac{1}{2}\boldsymbol{\alpha}^T\mathbf{Q}\boldsymbol{\alpha} \tag{3.51}$$

where $\mathbf{Q}$ is an $N \times N$ symmetric matrix of which the component in the mth row and nth column is $Q(m,n) = y_m y_n \mathbf{x}_m^T \mathbf{x}_n$, this is a quadratic programming problem, which can be solved by the interior point algorithm to be discussed in Section 3.7.

3.4 Linear Programming (LP)

The basic problem in LP is to minimize/maximize a linear objective function $f(x_1, \cdots, x_N) = \sum_{i=1}^N c_i x_i$ of N variables $x_1, \cdots, x_N$ under a set of M linear constraints

$$\begin{array}{ll} \text{max:} & f(x_1, \cdots, x_N) = \sum_{i=1}^N c_i x_i \\ \text{s. t.:} & \begin{cases} \sum_{i=1}^N a_{1i} x_i \leq b_1 \\ \cdots\cdots\cdots\cdots \\ \sum_{i=1}^N a_{Mi} x_i \leq b_M \\ x_1 \geq 0, \cdots, x_N \geq 0 \end{cases} \end{array} \tag{3.52}$$

The LP problem can be more concisely represented in matrix form

$$\begin{array}{ll} \text{max:} & \mathbf{c}^T\mathbf{x} \\ \text{s. t.:} & \begin{cases} \mathbf{Ax} - \mathbf{b} \leq \mathbf{0} \\ \mathbf{x} \geq \mathbf{0} \end{cases} \end{array} \tag{3.53}$$

where

$$\mathbf{x} = \begin{bmatrix} x_1 \\ \vdots \\ x_N \end{bmatrix} \quad \mathbf{c} = \begin{bmatrix} c_1 \\ \vdots \\ c_N \end{bmatrix} \quad \mathbf{b} = \begin{bmatrix} b_1 \\ \vdots \\ b_M \end{bmatrix} \quad \mathbf{A} = \begin{bmatrix} a_{11} & \cdots & a_{1N} \\ \vdots & \ddots & \vdots \\ a_{M1} & \cdots & a_{MN} \end{bmatrix} \tag{3.54}$$

and the smaller or greater-than sign in a vector inequality should be understood as an element-wise relationship between the corresponding elements of the vectors on both sides.

For example, the objective function could be the total profit, the constraints could be some limited resources. Specifically, $x_1, \cdots, x_N$ may represent the quantities of N different types of products, $c_1, \cdots, c_N$ represent their unit prices, and a_{ij} represent the consumption of the ith resource by the jth product, the goal is to maximize the profit $f(x_1, \cdots, x_N)$ subject to the constraints imposed by the limited resources $b_1, \cdots, b_M$.

Here all variables are assumed to be non-negative. If there exists a variable that is not restricted, it can be eliminated. Specifically, we can solve one of the constraining equations for the variable and use the resulting expression of the variable to replace all its appearances in the problem. For example

$$\begin{array}{ll} \text{max:} & f(x_1, x_2, x_3) = 2x_1 - x_2 + 3x_3 \\ \text{s. t.:} & \begin{cases} x_1 - 2x_2 + x_3 = 3 \\ 3x_1 - x_2 + 4x_3 = 10 \\ x_2 \geq 0, \;\; x_3 \geq 0 \end{cases} \end{array} \tag{3.55}$$

Solving the first constraining equation for the free variable x_1, we get $x_1 = 3 + 2x_2 - x_3$. Substituting this into the objective function and the other constraint, we get $2x_1 - x_2 + x_3 = 3x_2 + x_3 + 6$ and $5x_2 + x_3 = 1$, the problem can be reformulated as

$$\begin{array}{ll} \text{max:} & f(x_2, x_3) = 3x_2 + x_3 + 6 \\ \text{s. t.:} & \begin{cases} 5x_2 + x_3 = 1 \\ x_2 \geq 0, \;\; x_3 \geq 0 \end{cases} \end{array} \tag{3.56}$$

Given the primal LP problem, we can further find its dual problem. We first construct the Lagrangian of primal problem

$$L(\mathbf{x}) = \mathbf{c}^T\mathbf{x} - \mathbf{y}^T(\mathbf{A}\mathbf{x} - \mathbf{b}) = (\mathbf{c} - \mathbf{A}^T\mathbf{y})^T\mathbf{x} + \mathbf{y}^T\mathbf{b} \tag{3.57}$$

where $\mathbf{y} = [y_1, \cdots, y_M]^T$ contains the Lagrange multipliers for the inequality constraints $\mathbf{A}\mathbf{x} - \mathbf{b} \leq \mathbf{0}$, and $\mathbf{y} \geq \mathbf{0}$ according to Table 3.2. We define the dual function as the maximum of $L(\mathbf{x})$ over $\mathbf{x}$

$$f_d(\mathbf{y}) = \max_{\mathbf{x}} L(\mathbf{x}) = \max_{\mathbf{x}}[\mathbf{c}^T\mathbf{x} - \mathbf{y}^T(\mathbf{A}\mathbf{x} - \mathbf{b})] = \max_{\mathbf{x}}[(\mathbf{c} - \mathbf{A}^T\mathbf{y})^T\mathbf{x} + \mathbf{y}^T\mathbf{b}] \tag{3.58}$$

To find the value of $\mathbf{x}$ that maximizes $L(\mathbf{x})$, we set its gradient to zero and get

$$\triangledown_{\mathbf{x}} L(\mathbf{x}) = \triangledown_{\mathbf{x}}[(\mathbf{c} - \mathbf{A}^T\mathbf{y})^T\mathbf{x} + \mathbf{y}^T\mathbf{b}] = \mathbf{c} - \mathbf{A}^T\mathbf{y} = \mathbf{0} \tag{3.59}$$

Substituting this into $L_d(\mathbf{y}) = \max_{\mathbf{x}} L(\mathbf{x})$ we get: $f_d(\mathbf{y}) = \mathbf{y}^T\mathbf{b}$, which is the upper bound of $f_p(\mathbf{x})$, under the constraint that $\mathbf{c} - \mathbf{A}^T\mathbf{y} \leq 0$ so that $(\mathbf{c} - \mathbf{A}^T\mathbf{y})^T\mathbf{x} \leq 0$ contributes negatively to $L(\mathbf{x})$ (as $\mathbf{x} \geq 0$)

$$f_d(\mathbf{y}) = \mathbf{b}^T\mathbf{y} \geq L(\mathbf{x}) = \mathbf{c}^T\mathbf{x} - \mathbf{y}^T(\mathbf{A}\mathbf{x} - \mathbf{b}) \geq \mathbf{c}^T\mathbf{x} = f_p(\mathbf{x}) \tag{3.60}$$

This upper bound needs to be minimized with respect to $\mathbf{y}$, to become the tightest upper bound, that is, the original primal problem of maximization in Eq. (3.53) has now been converted into the following dual problem of minimization

$$\begin{cases} \text{max}: & \mathbf{c}^T\mathbf{x} \\ \text{s.t.} & \mathbf{A}\mathbf{x} - \mathbf{b} \leq \mathbf{0}, \;\; \mathbf{x} \geq \mathbf{0} \end{cases} \Longleftrightarrow \begin{cases} \text{min} & \mathbf{b}^T\mathbf{y} \\ \text{s.t.} & \mathbf{A}^T\mathbf{y} - \mathbf{c} \geq \mathbf{0}, \;\; \mathbf{y} \geq \mathbf{0} \end{cases} \tag{3.61}$$

Following the same approach, we can also find the dual of a minimization primal problem

$$\begin{cases} \min: & \mathbf{c}^T\mathbf{x} \\ \text{s.t.} & \mathbf{A}\mathbf{x}-\mathbf{b} \leq \mathbf{0}, \;\; \mathbf{x} \geq \mathbf{0} \end{cases} \Longleftrightarrow \begin{cases} \max & \mathbf{b}^T\mathbf{y} \\ \text{s.t.} & \mathbf{A}^T\mathbf{y}-\mathbf{c} \leq \mathbf{0}, \;\; \mathbf{y} \leq \mathbf{0} \end{cases}$$
$$\Longleftrightarrow \begin{cases} \max & -\mathbf{b}^T\mathbf{y} \\ \text{s.t.} & \mathbf{A}^T\mathbf{y}+\mathbf{c} \geq \mathbf{0}, \;\; \mathbf{y} \geq \mathbf{0} \end{cases} \tag{3.62}$$

Moreover, we can show that the dual of a dual problem is the original primal problem.

If either the primal or the dual is feasible and bounded, so is the other, and they form a strong duality, the solution d^* of the dual problem is the same as the solution p^* of the primal problem.

An inequality constrained LP problem can be converted to an equality constrained problem by introducing *slack variables*

$$\sum_{i=1}^{N} a_i x_i \leq b \quad \Longrightarrow \quad \sum_{i=1}^{N} a_i x_i + s = b, \quad (s \geq 0) \tag{3.63}$$

so that it can be reformulated to take the *standard form*

$$\begin{array}{ll} \max: & z = f(\mathbf{x}) = \mathbf{c}^T\mathbf{x} = \sum_{j=1}^{N} c_j x_j \\ \text{s. t.:} & \begin{cases} \sum_{i=1}^{N} a_{1i}x_i & +s_1 & & & & = b_1 \\ \sum_{i=1}^{N} a_{2i}x_i & & +s_2 & & & = b_2 \\ \cdots & \cdots & \cdots & \cdots & \cdots & \cdots \\ \sum_{i=1}^{N} a_{Mi}x_i & & & & +s_M & = b_M \\ x_1 \geq 0, & \cdots, & x_N \geq 0, & s_1 \geq 0, & \cdots & s_M \geq 0 \end{cases} \end{array} \tag{3.64}$$

We further redefine the following:

- $\mathbf{x}$ is an $N+M$ dimensional augmented variable vector that includes the M slack variables $\{s_1, \cdots, s_M\}$ as well as the N original variables $\{x_1, \cdots, x_N\}$

$$\mathbf{x} = [x_1, \; x_2, \cdots, x_N, \; s_1, \; s_2, \cdots, s_M]^T \tag{3.65}$$

- $\mathbf{A}$ is redefined as an $M \times (N+M)$ augmented coefficient matrix that includes coefficients for both types of variables

$$\mathbf{A} = \left[\begin{array}{cccc|cccc} a_{11} & a_{12} & \cdots & a_{1N} & 1 & 0 & \cdots & 0 \\ a_{21} & a_{22} & \cdots & a_{1N} & 0 & 1 & \cdots & 0 \\ \vdots & \vdots & \ddots & \vdots & \vdots & \vdots & \ddots & \vdots \\ a_{M1} & a_{M2} & \cdots & a_{MN} & 0 & \cdots & 0 & 1 \end{array}\right] = [\mathbf{A}_{M\times N} \mid \mathbf{I}_{M\times M}] \tag{3.66}$$

  where $\mathbf{A}_{M\times N}$ are the coefficients for the N original variables $\{x_1, \cdots, x_N\}$, and the identity matrix $\mathbf{I}_{M\times M} = [\mathbf{e}_1, \cdots, \mathbf{e}_M]$ is for the unity coefficients of the M slack variables $\{s_1, \cdots, s_M\}$, each of which appears in the M constraints only once.

Now the LP problem can be expressed in the standard form (original form on the left, with $\mathbf{x}$ and $\mathbf{A}$ redefined on the right)

$$\begin{array}{ll} \text{max:} & \mathbf{c}^T\mathbf{x} \\ \text{s. t.:} & \begin{cases} \mathbf{Ax} + \mathbf{s} = \mathbf{b} \\ \mathbf{x} \geq \mathbf{0},\ \mathbf{s} \geq \mathbf{0} \end{cases} \end{array} \quad \text{or} \quad \begin{array}{ll} \text{max:} & \mathbf{c}^T\mathbf{x} \\ \text{s. t.:} & \begin{cases} \mathbf{Ax} = \mathbf{b} \\ \mathbf{x} \geq \mathbf{0} \end{cases} \end{array} \tag{3.67}$$

An LP problem can be viewed geometrically. If we normalize $\mathbf{c}$ so that $||\mathbf{c}|| = 1$, then the objective function $f(\mathbf{x}) = \mathbf{c}^T\mathbf{x}$ becomes the projection of $\mathbf{x}$ onto $\mathbf{c}$. Each of the M equality constraints in $\mathbf{Ax} = \mathbf{b}$ represents a hyper-plane in the N–D vector space perpendicular to its normal direction $\mathbf{a}_j = [a_{i1}, \cdots, a_{IN}]^T$

$$\sum_{j=1}^{N} a_{ij}x_j = \mathbf{a}_j^T\mathbf{x} = b_i, \qquad (i = 1, \cdots, M) \tag{3.68}$$

Also, each of the N non-negativity conditions $x_j \geq 0$ in $\mathbf{x} \geq \mathbf{0}$ corresponds to a hyper-plane perpendicular to the jth standard basis vector $\mathbf{e}_j$. In general, in an N–D space, if none of the hyper-planes is parallel to any others, then no more than N hyper-planes intersect at one point. For example, in a 2-D or 3-D space, two straight lines or three surfaces intersect at a point. Here for the LP problem in the N–D space, the total number of intersection points formed by the $N + M$ hyper-planes is

$$C_{M+N}^{N} = C_{M+N}^{M} = \frac{(M+N)!}{N!\ M!} \tag{3.69}$$

The polytope enclosed by these $N + M$ hyper-planes is called the feasible region, in which the optimal solution must lie. The vertices of the polytopic feasible region are a subset of the C_{M+N}^{N} intersections.

Fundamental theorem of linear programming

The optimal solution $\mathbf{x}^*$ of a linear programming problem formulated here is either a vertex of the polytopic feasible region P, or lies on a hyper-surface of the polytope, on which all points are optimal solutions.

Proof: Assume the optimal solution $\mathbf{x}^* \in P$ is interior to the polytopic feasible region P (as shown in Fig. 3.6), then there must exist some $\epsilon > 0$ such that the hyper-sphere of radius ϵ centered at $\mathbf{x}^*$ is inside P. Evaluating the objective function at a point on the hyper-sphere

$$\mathbf{x}' = \mathbf{x}^* + \epsilon\ \mathbf{c}/||\mathbf{c}|| \in P \tag{3.70}$$

we get

$$\begin{aligned} f(\mathbf{x}') &= \mathbf{c}^T\mathbf{x}' = \mathbf{c}^T(\mathbf{x}^* + \epsilon\ \mathbf{c}/||\mathbf{c}||) \\ &= \mathbf{c}^T\mathbf{x}^* + \epsilon\ \mathbf{c}^T\mathbf{c}/||\mathbf{c}|| = \mathbf{c}^T\mathbf{x}^* + \epsilon\ ||\mathbf{c}|| > \mathbf{c}^T\mathbf{x}^* = f(\mathbf{x}^*) \end{aligned} \tag{3.71}$$

that is $\mathbf{x}^*$ cannot be the optimal solution as assumed. This contradiction indicates that an optimal solution $\mathbf{x}^*$ must be on the surface of P, either at one of its vertices or on one of its surfaces. □

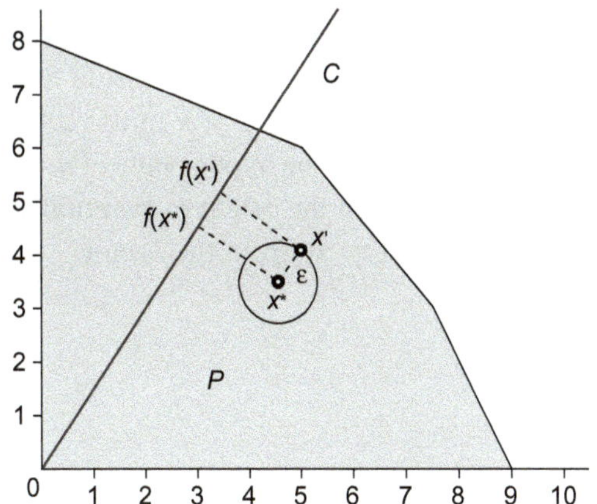

Fig. 3.6 Fundamental theorem of linear programming.

Based on this theorem, the optimization of a linear programming problem could be solved by exhaustively checking each of the C^N_{N+M} intersections formed by N of the $N + M$ hyper-planes to find (a) whether it is feasible (satisfying all $N + M$ constraints), and, if so, (b) whether the objective function $f(\mathbf{x})$ is maximized at the point. Specifically, we choose N of the $N + M$ equations for the constraints and solve this N-equation and N-unknown linear system to get the intersection point of N corresponding hyper-planes. This brute-force method is most straightforward, but the computational complexity is high when N and M, and therefore C^N_{N+M}, become large.

Example 3.2

$$\text{max:} \quad f_p(x_1, x_2) = 2x_1 + 3x_2 \qquad\qquad \text{max:} \quad f_p(\mathbf{x}) = \mathbf{c}^T\mathbf{x}$$

$$\text{s. t.:} \quad \begin{cases} 2x_1 + x_2 \le 18 \\ 6x_1 + 5x_2 \le 60 \\ 2x_1 + 5x_2 \le 40 \\ x_1 \ge 0, \quad x_2 \ge 0 \end{cases} \qquad \text{or} \qquad \text{s. t.:} \quad \begin{cases} \mathbf{Ax} - \mathbf{b} \le \mathbf{0} \\ \mathbf{x} \ge \mathbf{0} \end{cases}$$

where

$$\mathbf{x} = \begin{bmatrix} x_1 \\ x_2 \end{bmatrix}, \quad \mathbf{c} = \begin{bmatrix} 2 \\ 3 \end{bmatrix}, \quad \mathbf{b} = \begin{bmatrix} 18 \\ 60 \\ 40 \end{bmatrix}, \quad \mathbf{A} = \begin{bmatrix} 2 & 1 \\ 6 & 5 \\ 2 & 5 \end{bmatrix}$$

This problem has $N = 2$ variables with the same number of non-negativity constraints and $M = 3$ linear inequality constraints, as shown in Fig. 3.7. The $n + m = 5$ straight lines form $C^N_{N+M} = C^2_5 = 10$ intersections, out of which five are the vertices of the polygonal feasible region satisfying all constraints. The value of the objective function $f_p(x_1, x_2) = \mathbf{c}^T\mathbf{x} = c_1x_1 + c_2x_2$ is proportional to the projection of the 2-D variable vector $\mathbf{x} = [x_1, \ x_2]^T$ onto the coefficient vector $\mathbf{c} = [2, \ 3]^T$. The goal of this linear programming problem is to find a point $\mathbf{x}$ inside the polygonal feasible region with maximum $f(\mathbf{x}) = \mathbf{c}^T\mathbf{x}$ (the projection of $\mathbf{x}$ onto vector $\mathbf{c}$ if $||\mathbf{c}|| = 1$). Geometrically, this can be understood to push the hyper-plane $\mathbf{c}^T\mathbf{x} = z$, here a line in 2-D space, in its normal direction $\mathbf{c}$, away from the origin to eventually arrive at the vertex of the polytopic feasible region farthest away from the origin.

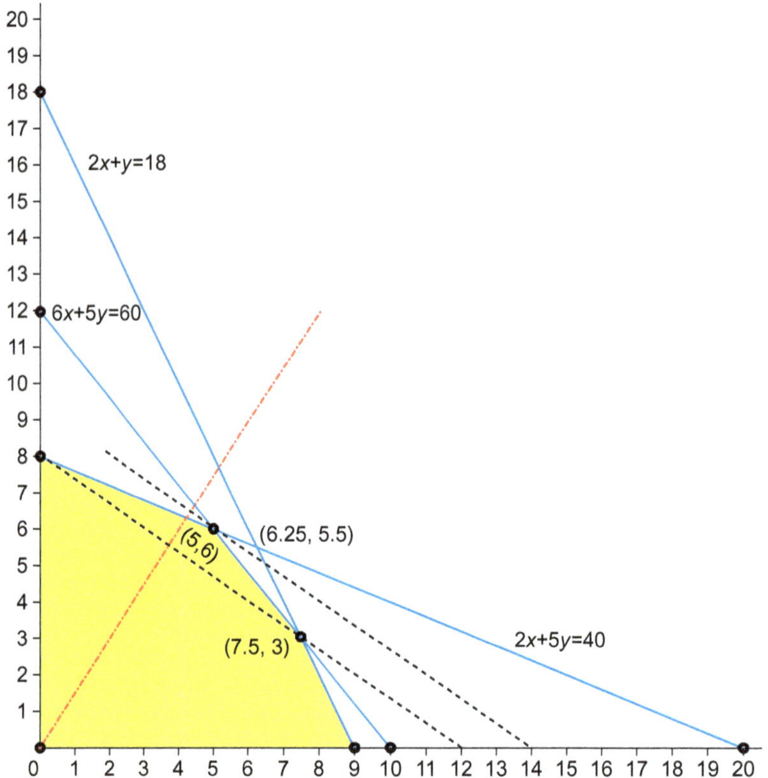

Fig. 3.7 Simplex algorithm example (original problem).

This LP problem can be converted into the standard form

$$\text{max:} \quad f_p(x_1,x_2) = 2x_1 + 3x_2 \qquad\qquad \text{max:} \quad f_p(x_1,x_2) = 2x_1 + 3x_2$$

$$\text{s. t.:} \begin{cases} 2x_1 + x_2 \leq 18 \\ 6x_1 + 5x_2 \leq 60 \\ 2x_1 + 5x_2 \leq 40 \\ x_1 \geq 0, \ \ x_2 \geq 0 \end{cases} \quad \Longrightarrow \quad \text{s. t.:} \begin{cases} 2x_1 + x_2 + s_1 = 18 \\ 6x_1 + 5x_2 + s_2 = 60 \\ 2x_1 + 5x_2 + s_3 = 40 \\ x_1 \geq 0, \ x_2 \geq 0, \ s_1 \geq 0, \ s_2 \geq 0, \ s_3 \geq 0 \end{cases}$$

Listed in the following table are the $C_{M+N}^N = C_5^2 = 10$ intersections $\mathbf{x}$, called basic solutions, together with the corresponding slack variables s_1, s_2, s_3 and the objective function value $f(\mathbf{x})$ (proportional to the projection of $\mathbf{x}$ onto $\mathbf{c}$). We see that out of the ten basic solutions, five are feasible (vertices of the polytopic feasible region) with all $M + N = 5$ variables taking non-negative values, while the remaining five are not feasible, as some of the slack variables are negative, and thereby violating the constraints. Out of the five feasible solutions, the one at $\mathbf{x}^* = (5,\ 6)$, is optimal with maximum objective function value $f_p(x_1, x_2) = 2x_1 + 3x_2 = 2 \times 5 + 3 \times 6 = 28$.

Table 3.3

	x_1	x_2	s_1	s_2	s_3	Feasibility	$\mathbf{c}^T\mathbf{x}/\|\|\mathbf{c}\|\| = 1$
1	5	6	2	0	0	feasible	28/7.77
2	6.25	5.5	0	−5	0	infeasible	29/8.04
3	7.5	3	0	0	10	feasible	24/6.66
4	20	0	−2	−60	20	infeasible	40/11.1
5	10	0	−2	0	20	infeasible	20/5.55
6	9	0	0	6	22	feasible	18/4.10
7	0	8	10	20	0	feasible	24/6.66
8	0	12	6	0	−20	infeasible	36/9.98
9	0	18	0	−30	−50	infeasible	54/14.99
10	0	0	18	60	40	feasible	0/0.00

This primal maximization problem can be converted into the dual minimization problem

$$\text{min:}\quad f_d(y_1, y_2, y_3) = 18y_1 + 60y_2 + 40y_3 \qquad\qquad \text{min:}\quad f_d(\mathbf{y}) = \mathbf{b}^T\mathbf{y}$$
$$\text{s. t.:}\quad \begin{cases} 2y_1 + 6y_2 + 2y_3 \ge 2 \\ y_1 + 5y_2 + 5y_3 \ge 3 \\ y_1 \ge 0,\ y_2 \ge 0,\ y_3 \ge 0 \end{cases} \qquad \text{or} \qquad \text{s. t.:}\quad \begin{cases} \mathbf{A}^T\mathbf{y} - \mathbf{c} \ge \mathbf{0} \\ \mathbf{y} \ge \mathbf{0} \end{cases}$$

As shown in Fig. 3.8, the five planes, including $m = 2$ from the constraining inequalities and $n = 3$ for the non-negative constraints $y_1 \le 0$, $y_2 \le 0$, and $y_3 \le 0$ for the three variables, form 10 intersection points:

Table 3.4

	(y_1, y_2, y_3)	Feasibility	Objective function
1	(1,0,0)	infeasible	18
2	(3,0,0)	feasible	54
3	(0,1/3,0)	infeasible	20
4	(0,3/5,0)	feasible	36
5	(0,0,1)	feasible	40
6	(0,0,3/5)	infeasible	24
7	(0,1/5,2/5)	feasible	28
8	(1/2,0,1/2)	feasible	29
9	(−2,1,0)	infeasible	24
10	(0,0,0)	infeasible	0

We see that at the feasible point $(0, 1/5, 2/5)$, the dual function $f_d(y_1, y_2, y_3) = 18y_1 + 60y_2 + 40y_3$ is minimized to be 28, the same as the maximum of the primal function $f_p(x_1, x_2) = 2x_1 + 3x_2$ at $(5,6)$.

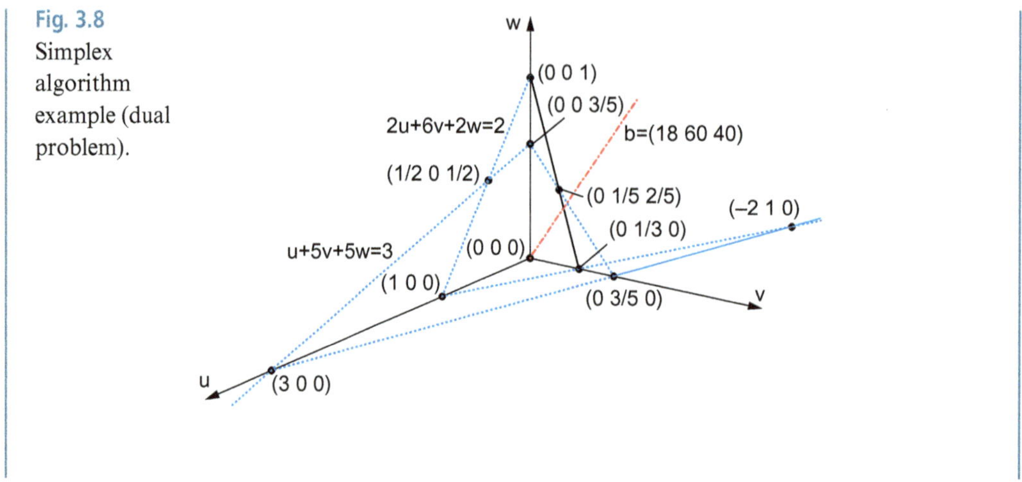

Fig. 3.8 Simplex algorithm example (dual problem).

3.5 The Simplex Algorithm

The simplex algorithm finds the optimal solution of a LP problem by an iterative process that traverses along a sequence of edges of the polytopic feasible region, starting at the origin and through a sequence of vertices $\mathbf{x}$ with progressively greater objective value $f(\mathbf{x})$, until eventually reaching the optimal solution. By doing so, it avoids checking exhaustively all vertices of the feasible region for optimality.

We first consider the equality constraint $\mathbf{Ax} = [\mathbf{A}_{M\times N} \mid \mathbf{I}_{M\times M}]\mathbf{x} = \mathbf{b}$ of the standard LP problem. This is an under-determined linear system of $N+M$ variables but only M equations, and only M of its $N+M$ columns of the coefficient matrix $\mathbf{A}$ are independent (assuming $\text{rank}(\mathbf{A}) = M$). Initially we choose the M columns of the identity coefficient matrix $\mathbf{I}$ as the independent columns, but subsequently we can choose any other M columns of $\mathbf{A}$ as the independent columns and convert them into a standard basis vector by Gauss–Jordan elimination together with the corresponding variables in $\mathbf{x}$. The resulting equation $\mathbf{Ax} = \mathbf{b}$ remains equivalent to the original one, that is, the M equality constraints are always preserved and never violated in the process.

For convenience of discussion and without loss of generality, we could reorder (not actually carried out) the $N+M$ columns in $\mathbf{A}$, together with the corresponding variables in $\mathbf{x}$, so that the constraining equation would always take the following form

$$\mathbf{Ax} = [\mathbf{A}_n \;\; \mathbf{I}]\begin{bmatrix} \mathbf{x}_n \\ \mathbf{x}_b \end{bmatrix} = \mathbf{A}_n\mathbf{x}_n + \mathbf{I}\,\mathbf{x}_b = \mathbf{b} \tag{3.72}$$

Now the variables in the M–D vector $\mathbf{x}_b$ corresponding to the M independent columns in $\mathbf{I}$ are the *basic variables*, while the N–D vector $\mathbf{x}_n$ corresponding to the remaining N dependent columns, now denoted by $\mathbf{A}_n$, are the *non-basic variables*. This equation will always hold if $\mathbf{x}_n = \mathbf{0}$ and $\mathbf{x}_b = \mathbf{b}$. Such a solution is called a *basic solution* of the linear system $\mathbf{Ax} = \mathbf{b}$

$$\mathbf{x} = \begin{bmatrix} \mathbf{x}_n \\ \mathbf{x}_b \end{bmatrix} = \begin{bmatrix} \mathbf{0} \\ \mathbf{b} \end{bmatrix} \tag{3.73}$$

Initially, $\mathbf{A}_n$ is the coefficient matrix in the inequality constraint $\mathbf{Ax} \leq \mathbf{b}$, and the corresponding non-basic variables $\mathbf{x}_n = \mathbf{0}$ is actually the origin. But through the iteration, we will keep converting some other columns of $\mathbf{A}$ into standard basis vectors $\{\mathbf{e}_1, \cdots, \mathbf{e}_M\}$, the column vectors of $\mathbf{I}$, by Gauss–Jordan elimination. The variables in $\mathbf{x}$ corresponding to the new standard basis vectors become the basic variables, while those corresponding to columns that become non-standard basis vectors become non-basic variables.

The basic solutions corresponding to the C_{N+M}^N ways to choose M of the $N + M$ columns of $\mathbf{A}$ as independent are actually the intersections formed by any N of the $N + M$ constraining hyper-planes in the N–D space. As noted before, only a subset of these C_{N+M}^N basic solutions satisfy the M constraints, and they are called *basic feasible solutions (BFS)*. The goal of the iterative process of the simplex algorithm is to find the optimal BFS that maximizes $f(\mathbf{x})$ without exhausting all C_{N+M}^N possibilities. This is done by selecting the columns in such a way that the value of the objective function will always be maximally increased, until eventually we find the optimal solution $\mathbf{x}^*$ at one of the vertexes of the polytopic feasible region.

Specially, the implementation of the simplex algorithm is based on a tableau with $N + M + 1$ columns and $M + 1$ rows, initialized as follows:

- The first M rows are for the coefficients in the equality constraints $\sum_{j=1}^N a_{ij}x_j + s_i = b_i$ $(i = 1, \cdots, M)$. An additional $(M + 1)$st row at the bottom is for the coefficients in the objective function $f(\mathbf{x}) = \sum_{j=1}^N c_j x_j$, which is zero initially at the origin $x_1 = \cdots = x_N = 0$, that is, $z - c_1x_1 - \cdots - c_N x_N = 0$.
- The first $N + M$ columns contain the coefficients for the N original variables $\{x_1, \cdots, x_N\}$ and the M slack variables $\{s_1, \cdots, s_M\}$, and an additional $(N + M + 1)$st column is for the constants $\mathbf{b} = [b_1, \cdots, b_M]^T$ on the right-hand side of the M constraint equations. The last element of the column is the right-hand side of the objective equation $z = f(\mathbf{x})$, which is initially zero.

Table 3.5

Basic variables	x_1	x_2	$\cdots$	x_N	s_1	s_2	$\cdots$	s_M	Basic solution
s_1	a_{11}	a_{12}	$\cdots$	a_{1N}	1	0	$\cdots$	0	b_1
s_2	a_{21}	a_{22}	$\cdots$	a_{2N}	0	1	$\cdots$	0	b_2
$\vdots$	$\vdots$	$\vdots$	$\ddots$	$\vdots$	$\vdots$	$\vdots$	$\ddots$	$\vdots$	$\vdots$
s_M	a_{M1}	a_{M2}	$\cdots$	a_{MN}	0	0	$\cdots$	1	b_M
z	$-c_1$	$-c_2$	$\cdots$	$-c_M$	0	0	$\cdots$	0	0

At this initial stage of the iteration, the N original variables $\mathbf{x}_n = [x_1, \cdots, x_N]^T$ are the non-basic variables, while the M slack variables $\mathbf{x}_b = [s_1, \cdots, s_M]^T$ are the basic variables and their coefficients form an $M \times M$ identity matrix $\mathbf{I}$, composed of the standard basis vectors. The corresponding feasible basic solution is simply

$$\mathbf{x}_b = [s_1, \cdots, s_M]^T = \mathbf{b}, \qquad \mathbf{x}_n = [x_1, \cdots, x_N]^T = \mathbf{0} \tag{3.74}$$

that satisfies all the constraints $\mathbf{A}\mathbf{x} = \mathbf{A}_n\mathbf{x}_n + \mathbf{I}\mathbf{x}_b = \mathbf{b}$.

In each of the subsequent iterations, we will select one of the non-basic variables, called the *entering variable*, to replace one of the basic variables, called the *leaving variable*, in such a way that the value of the objective function value $z = f(\mathbf{x})$ in the last row will be maximally increased, while all constraints remain satisfied. Here are the steps in each iteration.

- **Selection of the entering variable**
  This section is based on the maximization of $z = f(\mathbf{x})$. We select x_j in the jth column of the tableau if $-c_j$ is most negative, that is, x_j is most heavily weighted by $c_j = \max\{c_1, \cdots, c_N\}$, so that it will increase $z = \sum_{j=1}^N c_j x_j$ more than any other $x_k \ (k \neq j)$.
- **Selection of the leaving variable**
  This selection is based on the constraints imposed on the selected entering variable x_j. In general, the restriction on x_j set by the kth constraint $\sum_{j=1}^N a_{kj}x_j \leq b_k$ is $x_j \leq b_k/a_{kj}$ when $a_{kj} > 0$. If $b_i/a_{ij} = \min\{b_k/a_{1j}, \cdots, b_k/a_{Mj}\}$, then the ith constraint is most restrictive on x_j, we will therefore select the corresponding basic variable s_i as the leaving variable, that is, it becomes a non-basic variable to be set to zero, so that x_j can be maximally increased without violating the constraints. If all $a_{kj} < 0$ for all $k = 1, \cdots, M$, variable x_j is not bounded.
- **Gauss–Jordan elimination based on pivoting**
  To convert the entering variable x_j to a basic variable to replace the leaving variable s_i, we need to turn the corresponding jth column into a standard basis vector $\mathbf{e}_i$. This is realized by pivoting on a_{ij} in the following steps:

  1. Divide the pivot row $\mathbf{r}_i$ by a_{ij}: $\mathbf{r}_i \leftarrow \mathbf{r}_i/a_{ij}$. Now $a_{ij} = 1$.
  2. Subtract $a_{kj}\mathbf{r}_i$ from the kth row: $\mathbf{r}_k \leftarrow \mathbf{r}_k - a_{kj}\mathbf{r}_i$ so that $a_{kj} = 0$ for all $k = 1, \cdots, M+1$, $k \neq i$, including the last row $\mathbf{r}_{M+1}$.

  Now the jth column corresponding to the entering variable x_j becomes a standard basis vector $\mathbf{e}_i$, that is, x_j becomes a basic variable that takes the value of b_i, and the column corresponding to the leaving variable s_i is no longer a standard basis vector, and $s_i = 0$ as it is now a non-basic variable.

As Gauss–Jordan elimination converts the linear equations to a set of equivalent equations, the constraints remain satisfied throughout the process. Although the membership of the basic and non-basic variable groups keeps changing in the iteration, so long as $\mathbf{x}_b = \mathbf{b}$ and $\mathbf{x}_n = \mathbf{0}$, the M constraint equations $\mathbf{A}\mathbf{x} = \mathbf{A}_n\mathbf{x}_n + \mathbf{I}\mathbf{x}_b = \mathbf{b}$ always holds.

This iterative process keeps replacing the slack variables $\{s_1, \cdots, s_M\}$ (the basic variables initially) by the original variables $\{x_1, \cdots, x_N\}$ (the non-basic variables initially), one at a time, until all original variables become basic variables and all elements in the last row are non-negative.

The final result can be read out directly from the tableau. The variables corresponding to the M standard basis vectors $\mathbf{e}_i$ $(i = 1, \cdots, M)$ are the final basic variables that take the values in $\mathbf{b}$ in the right-most column of the tableau. They form the optimal basic solution, with the maximum of the objective function $z = f(\mathbf{x})$ given by the last element also in the last column. The remaining n variables corresponding to non-standard columns are non-basic variables that take the value zero. When $M > N$ (more constraints than variables), some of the slack variables may remain in the basic variable group taking non-zero values; when $M < N$, some of the original variable may be in the non-basic group taking the value zero.

Example 3.3

We re-solve the LP problem considered in the previous examples, now in the standard form

$$\text{max:} \quad f(\mathbf{x}) = 2x_1 + 3x_2$$
$$\text{s. t.:} \quad \begin{cases} 2x_1 + x_2 + s_1 = 18 \\ 6x_1 + 5x_2 + s_2 = 60 \\ 2x_1 + 5x_2 + s_3 = 40 \\ x_1 \geq 0,\ x_2 \geq 0,\ s_1 \geq 0,\ s_2 \geq 0,\ s_3 \geq 0 \end{cases}$$

The standard form is further converted to a tableau. The left-most column indicates the basic variables, the next $N = 2$ columns are for the variables x and y, the next $M = 3$ columns are for the slack variables u, v, and w. The right-most column is for the constants b_1, b_2, and b_3 on the right-hand side of the equations.

Table 3.6

	x_1	x_2	s_1	s_2	s_3	b
s_1	2	1	1	0	0	18
s_2	6	5	0	1	0	60
s_3	2	5	0	0	1	40
z	−2	−3	0	0	0	0

In this initial state, the basic variables are $s_1 = 18$, $s_2 = 60$, and $s_3 = 40$, the non-basic variables are $x_1 = x_2 = 0$. The corresponding BFS is at the origin.

- Select column $j = 2$ as the pivot column, as $-c_2 = -3 < -c_1$ is the most negative coefficient.
- Select $\mathbf{r}_3$ as the pivot row, as $b_3/a_{32} = 40/5 = 8 = \min\{18/1 = 18,\ 60/5 = 12,\ 40/5 = 8\}$.
- Divide pivot row $\mathbf{r}_i = \mathbf{r}_3 = [2,\ 5,\ 0,\ 0,\ 1,\ 40]$ by the pivot element $a_{ij} = 32 = 5$ to get $\mathbf{r}_i = \mathbf{r}_3 = [0.4,\ 1,\ 0,\ 0,\ 0.2,\ 8]$.
- Subtract $a_{kj}\mathbf{r}_k = a_{k2}\mathbf{r}_k$ from row $\mathbf{r}_k$, so that $a_{kj} = a_{k2} = 0$, $(k = 1, \cdots, M+1 = 4,\ k \neq i)$, including the last row $\mathbf{r}_4$.

Table 3.7

	x_1	x_2	s_1	s_2	s_3	b
s_1	1.6	0	1	0	−0.2	10
s_2	4	0	0	1	−1	20
x_2	0.4	1.0	0	0	0.2	8
z	−0.8	0	0	0	0.6	24

The entering variable $x_j = x_2$ becomes a basic variable, and the leaving variable $s_i = s_3$ becomes a non-basic variable, replaced by new basic variable x_2. The corresponding BFS is at $x_1 = 0$, $x_2 = b_3 = 8$, with $s_1 = 10$, $s_2 = 20$, $s_3 = 0$. The objective function value is $z = \mathbf{c}^T \mathbf{x} = 24$.

- Select column $j = 1$ as the pivot column, as $-c_1 = -0.8$ is the most (only) negative coefficient in the last row.
- Select row $i = 2$ as the pivot row, as the ratio $b_2/a_{21} = 20/4 = 5 = \min\{10/1.6 = 6.25,\ 20/4 = 5,\ 8/0.4 = 20\}$.
- Divide pivot row $\mathbf{r}_i = \mathbf{r}_2 = [4,\ 0,\ 0,\ 1,\ -1,\ 20]$ by the pivot element $a_{ij} = a_{21} = 4$ to get $\mathbf{r}_i = \mathbf{r}_2 = [1,\ 0,\ 0,\ 0.25,\ -0.25,\ 5]$.
- Subtract $a_{kj}\mathbf{r}_k = a_{k1}\mathbf{r}_k$ from row $\mathbf{r}_k$, so that $a_{kj} = a_{k1} = 0, (k = 1, \cdots, M + 1 = 4)$, including the last row $\mathbf{r}_4$.

Table 3.8

	x_1	x_2	s_1	s_2	s_3	b
s_1	0	0	1	−0.4	0.2	2
x_1	1	0	0	0.25	−0.25	5
x_2	0	1	0	−0,1	0.3	6
z	0	0	0	0.2	0.4	28

The entering variable x_1 becomes a basic variable, and the leaving variable s_1 becomes a non-basic variable, replaced by the new basic variable x_1. The corresponding BFS is at $x_1 = b_2 = 5$, $x_2 = b_3 = 6$, with $s_1 = 2$, $s_2 = s_3 = 0$. The objective function value is $z = \mathbf{c}^T \mathbf{x} = 28$. Now that both x_1 and x_2 have become basic variables, the optimal solution has been obtained.

This problem has $C_{N+M}^N = C_5^2 = 10$ basic solutions, corresponding to the same number of intersections, out of which five are feasible, as previously obtained. The simplex method finds three of these feasible solutions, starting from $x_1 = x_2 = 0$ at the origin with $z = 0$, through the vertex at $x_1 = 0$ and $x_2 = 8$ with $z = 24$, to the optimal solution at $x_1 = 5$ and $x_2 = 6$ with $z = 28$.

The MATLAB code for implementing the simplex algorithm is listed here. The function takes input including matrix $\mathbf{A}$ as well as the vectors $\mathbf{b}$ and $\mathbf{c}$, and generates the optimal solution $\mathbf{x}^*$, the corresponding maximum value of the objective function $f(\mathbf{x}) = \mathbf{c}^T \mathbf{x}^*$, together with the values of the slack variable $\mathbf{s}$.

```
function [x s z] = MySimplex(A,b,c)
    [m n]=size(A)           % coefficient matrix for variables x
    x=zeros(n,1);           % initial zero values for x
    s=zeros(m,1);
    A=[A eye(m) b; -c zeros(1,m+1)]; % initialization of tableau

    [cmin, pc]=min( A(m+1,1:n+m) );  % pc, index of first pivot column
    it=0;
    while cmin<0            % iteration until all c's in last row are zero
        it=it+1;
        rmin=9e9;
        for i=1:m           % find pivot row
            if A(i,n+m+1)~=0 & A(i,pc)>0
                w=A(i,n+m+1)./A(i,pc);
                if w<rmin
                    pr=i;  % pr, index of the pivot row
                    rmin=w;
                end
            end
        end
        p=A(pr,pc);         % get the pivot element
        A(pr,:)=A(pr,:)/p;  % modify pivot row
        for i=1:m+1         % modify all k+1 rows
            if i~=pr
                A(i,:)=A(i,:)-A(pr,:)*A(i,pc);
            end
        end
        z=A(m+1,n+m+1);     % maximum value of objective function so far

        [cmin, pc]=min( A(m+1,1:n+m) ); % pivot column of next iteration
    end
end
```

3.6 Quadratic Programming (QP)

A QP problem is to minimize a quadratic function subject to some equality and/or inequality constraints

$$\begin{array}{ll} \text{min:} & f(\mathbf{x}) = \frac{1}{2}[\mathbf{x}-\mathbf{m}]^T\mathbf{Q}[\mathbf{x}-\mathbf{m}] = \frac{1}{2}\mathbf{x}^T\mathbf{Q}\mathbf{x} + \mathbf{c}^T\mathbf{x} + c \\ \text{s. t.:} & \mathbf{A}\mathbf{x} \le \mathbf{b} \end{array} \tag{3.75}$$

where $\mathbf{Q} = \mathbf{Q}^T$ is an $N \times N$ positive definite matrix, $\mathbf{A}$ an $M \times N$ matrix, and $\mathbf{b}$ is an M–D vector. Also, $\mathbf{c} = -\mathbf{Q}\mathbf{m}$, $c = \mathbf{m}^T\mathbf{Q}\mathbf{m}/2$. Here the scalar constant c can be dropped as it does not play any role in the optimization. Note that $\mathbf{x}^T\mathbf{Q}\mathbf{x}$ is a hyper elliptic paraboloid with the minimum at point $\mathbf{m}$ in the N–D space.

We first consider the special case where the QP problem is only subject to equality constraints and we assume $m \leq N$, that is, the number of constraints is smaller than the number of unknowns in $\mathbf{x}$. Then the solution $\mathbf{x}$ has to satisfy $\mathbf{Ax} - \mathbf{b} = \mathbf{0}$, that is, it has to be on M hyper planes in the N–D space.

The Lagrangian function of the QP problem is

$$L(\mathbf{x}, \lambda) = f(\mathbf{x}) + \lambda^T(\mathbf{Ax} - \mathbf{b}) = \frac{1}{2}\mathbf{x}^T\mathbf{Qx} + \mathbf{c}^T\mathbf{x} + \lambda^T(\mathbf{Ax} - \mathbf{b}) \tag{3.76}$$

To find the optimal solution, we first equate the derivatives of the Lagrangian with respect to both $\mathbf{x}$ and λ to zero:

$$\begin{aligned} \nabla_{\mathbf{x}} L(\mathbf{x}, \lambda) &= \mathbf{Qx} + \mathbf{c} + \mathbf{A}^T\lambda = \mathbf{0} \\ \nabla_{\lambda} L(\mathbf{x}, \lambda) &= \mathbf{Ax} - \mathbf{b} = \mathbf{0} \end{aligned} \tag{3.77}$$

These two equations can be combined and expressed in matrix form as

$$\begin{bmatrix} \mathbf{Q} & \mathbf{A}^T \\ \mathbf{A} & \mathbf{0} \end{bmatrix} \begin{bmatrix} \mathbf{x} \\ \lambda \end{bmatrix} = \begin{bmatrix} -\mathbf{c} \\ \mathbf{b} \end{bmatrix} \tag{3.78}$$

We then solve this system of $m + N$ equations to get the optimal solution $\mathbf{x}^*$ and the corresponding λ^*.

Example 3.4

$$\begin{aligned} \text{min:} \quad & f(x_1, x_2) = x_1^2 + x_2^2 = \tfrac{1}{2}[x_1,\ x_2] \begin{bmatrix} 2 & 0 \\ 0 & 2 \end{bmatrix} \begin{bmatrix} x_1 \\ x_2 \end{bmatrix} = \tfrac{1}{2}\mathbf{x}^T\mathbf{Qx} \\ \text{s. t.:} \quad & \mathbf{Ax} = [1,\ 1] \begin{bmatrix} x_1 \\ x_2 \end{bmatrix} = x_1 + x_2 = \mathbf{b} = 1 \end{aligned}$$

where

$$\mathbf{Q} = \begin{bmatrix} 2 & 0 \\ 0 & 2 \end{bmatrix}, \quad \mathbf{A} = [1,\ 1], \quad \mathbf{c} = \begin{bmatrix} 0 \\ 0 \end{bmatrix}, \quad \mathbf{b} = 1;$$

$$\begin{bmatrix} 2 & 0 & 1 \\ 0 & 2 & 1 \\ 1 & 1 & 0 \end{bmatrix} \begin{bmatrix} x_1 \\ x_2 \\ \lambda \end{bmatrix} = \begin{bmatrix} 0 \\ 0 \\ 1 \end{bmatrix}$$

Solving this equation we get the solution $x_1^* = x_2^* = 0.5$ and $\lambda^* = -1$, at which the function $f(x_1^*, x_2^*) = 0.5$ is minimized subject to $x_1 + x_2 = 1$.

If $M = N$, that is, the number of equality constraints is the same as the number of variables, then the variable $\mathbf{x}$ is uniquely determined by the linear system $\mathbf{Ax} = \mathbf{b}$, as the intersect of $M = N$ hyper planes, independent of the objective function $f(\mathbf{x})$. Further if $M > N$, that is, the system $\mathbf{Ax} = \mathbf{b}$ is over constrained, and its solution does not exist in general. It is therefore more interesting to consider QP problems subject to both equality and inequality constraints

$$\text{min:} \qquad f(\mathbf{x}) = \tfrac{1}{2}\mathbf{x}^T\mathbf{Q}\mathbf{x} + \mathbf{c}^T\mathbf{x}$$

$$\text{s. t.:} \qquad \mathbf{A}\mathbf{x} \leq \mathbf{b}$$

3.7 Interior Point Methods

Consider the generic constrained optimization problem in the following form

$$\begin{aligned} \text{min:} \quad & f(\mathbf{x}) \\ \text{s. t.:} \quad & \begin{cases} \mathbf{h}(\mathbf{x}) = \mathbf{0} \\ \mathbf{g}(\mathbf{x}) \leq \mathbf{0} \quad \text{or} \quad \mathbf{g}(\mathbf{x}) \geq \mathbf{0} \\ \mathbf{x} \geq \mathbf{0} \end{cases} \end{aligned} \tag{3.79}$$

By introducing slack variables $\mathbf{s} \geq \mathbf{0}$, all inequality constraints can be converted into equality constraints

$$\mathbf{g}(\mathbf{x}) \leq \mathbf{0} \implies \mathbf{g}(\mathbf{x}) + \mathbf{s} = \mathbf{0}, \qquad \mathbf{g}(\mathbf{x}) \geq \mathbf{0} \implies \mathbf{g}(\mathbf{x}) - \mathbf{s} = \mathbf{0} \tag{3.80}$$

which can then be combined with the equality constraints and still denoted by $\mathbf{h}(\mathbf{x}) = \mathbf{0}$, while the slack variables $\mathbf{s}$ are also combined with the original variables and still denoted by $\mathbf{x}$. We assume there are in total M equality constraints and N variables. Now the optimization problem can be reformulated as

$$\begin{aligned} \text{min:} \quad & f(\mathbf{x}) \\ \text{s. t.:} \quad & \begin{cases} \mathbf{h}(\mathbf{x}) = \mathbf{0} \\ \mathbf{x} \geq \mathbf{0} \end{cases} \end{aligned} \tag{3.81}$$

The Lagrangian is

$$L(\mathbf{x}, \boldsymbol{\lambda}, \boldsymbol{\mu}) = f(\mathbf{x}) + \boldsymbol{\lambda}^T\mathbf{h}(\mathbf{x}) - \boldsymbol{\mu}^T\mathbf{x} \tag{3.82}$$

The KKT conditions are

$$\begin{cases} \nabla_{\mathbf{x}} L(\mathbf{x}, \boldsymbol{\lambda}, \boldsymbol{\mu}) = \mathbf{g}_f(\mathbf{x}) + \mathbf{J}_{\mathbf{h}}^T(\mathbf{x})\boldsymbol{\lambda} - \boldsymbol{\mu} = \mathbf{0} & \text{(stationarity)} \\ \mathbf{h}(\mathbf{x}) = \mathbf{0}, \quad \mathbf{x} \geq \mathbf{0} & \text{(primal feasibility)} \\ \boldsymbol{\mu} \geq \mathbf{0} & \text{(dual feasibility)} \\ \mu_j x_j = 0, \quad (j = 1, \cdots, N) \quad \text{or} \quad \mathbf{XM1} = \mathbf{0} & \text{(complementarity)} \end{cases} \tag{3.83}$$

where

$$\mathbf{X} = \begin{bmatrix} x_1 & \cdots & 0 \\ \vdots & \ddots & \vdots \\ 0 & \cdots & x_n \end{bmatrix}_{N\times N}, \qquad \mathbf{M} = \begin{bmatrix} \mu_1 & \cdots & 0 \\ \vdots & \ddots & \vdots \\ 0 & \cdots & \mu_N \end{bmatrix}_{N\times N} \tag{3.84}$$

and $\mathbf{J_f}(\mathbf{x})$ is the Jacobian matrix of $\mathbf{h}(\mathbf{x})$

$$\mathbf{J_h}(\mathbf{x}) = \begin{bmatrix} \frac{\partial h_1}{\partial x_1} & \cdots & \frac{\partial h_1}{\partial x_n} \\ \vdots & \ddots & \vdots \\ \frac{\partial h_m}{\partial x_1} & \cdots & \frac{\partial h_m}{\partial x_n} \end{bmatrix}_{M \times N} \tag{3.85}$$

To simplify the problem, the non-negativity constraints can be removed by introducing an indicator function

$$I(x) = \begin{cases} 0 & x \geq 0 \\ \infty & x < 0 \end{cases} \tag{3.86}$$

which, when used as a cost function, penalizes any violation of the non-negativity constraint $x \geq 0$. Now the optimization problem can be reformulated without the non-negative constants $\mathbf{x} \geq \mathbf{0}$

$$\begin{array}{ll} \text{min:} & f(\mathbf{x}) + \sum_{i=1}^N I(x_i) \\ \text{s. t.:} & \mathbf{h}(\mathbf{x}) = \mathbf{0} \end{array} \tag{3.87}$$

However, as the indicator function is not smooth and therefore not differentiable, it is approximated by the *logarithmic barrier function* which approaches $I(x)$ when the parameter $t > 0$ approaches infinity:

$$-\frac{1}{t}\ln(x) \stackrel{t\to\infty}{\Longrightarrow} I(x) \tag{3.88}$$

Now the optimization problem can be written as

$$\begin{array}{ll} \text{min:} & f(\mathbf{x}) - \frac{1}{t}\sum_{i=1}^N \ln x_i \\ \text{s. t.:} & \mathbf{h}(\mathbf{x}) = \mathbf{0} \end{array} \tag{3.89}$$

The Lagrangian is

$$L(\mathbf{x}, \boldsymbol{\lambda}, \boldsymbol{\mu}) = f(\mathbf{x}) + \boldsymbol{\lambda}^T \mathbf{h}(\mathbf{x}) - \frac{1}{t}\sum_{j=1}^N \ln x_j \tag{3.90}$$

and its gradient is

$$\nabla_{\mathbf{x}} L(\mathbf{x}, \boldsymbol{\lambda}, \boldsymbol{\mu}) = \mathbf{g}_f(\mathbf{x}) + \mathbf{J}_\mathbf{h}^T(\mathbf{x})\boldsymbol{\lambda} - \frac{1}{t}\mathbf{X}^{-1}\mathbf{1} \tag{3.91}$$

We define

$$\boldsymbol{\mu} = \frac{\mathbf{x}}{t} = \frac{1}{t}\mathbf{X}^{-1}\mathbf{1}, \quad \text{i.e.,} \quad \mathbf{X}\boldsymbol{\mu} = \mathbf{XM1} = \frac{\mathbf{1}}{t} \tag{3.92}$$

and combine this equation with the KKT conditions of the new optimization problem to get

$$\begin{cases} \mathbf{g}_f(\mathbf{x}) + \mathbf{J}_\mathbf{h}^T(\mathbf{x})\boldsymbol{\lambda} - \boldsymbol{\mu} = \mathbf{0} & \text{(stationarity)} \\ \mathbf{h}(\mathbf{x}) = \mathbf{0} & \text{(primal feasibility)} \\ \mathbf{XM1} - \mathbf{1}/t = \mathbf{0} & \text{(complementarity)} \end{cases} \tag{3.93}$$

The third condition is actually from the definition of $\boldsymbol{\mu}$ given here, but here it is labeled as complementarity, so that these conditions can be compared with the KKT conditions of the original problem. We see that the two sets of KKT conditions are similar to each other, but with the following differences:

- the non-negativity $\mathbf{x} \geq \mathbf{0}$ in the primal feasibility of the original KKT conditions is dropped;
- consequently the inequality $\boldsymbol{\mu} \geq \mathbf{0}$ in the dual feasibility of the original KKT conditions is also dropped;
- there is an extra term $\mathbf{1}/t$ in complementarity, which will vanish when $t \to \infty$.

The optimal solution that minimizes the objective function $f(\mathbf{x})$ can be found by solving the simultaneous equations in the modified KKT conditions (with no inequalities) given. To do so, we first express the equations in the KKT given conditions above as a nonlinear equation system $\mathbf{F}(\mathbf{x}, \boldsymbol{\lambda}, \boldsymbol{\mu}) = \mathbf{0}$, and find its Jacobian matrix $\mathbf{J_F}$ composed of the partial derivatives of the function $\mathbf{F}(\mathbf{x}, \boldsymbol{\lambda}, \boldsymbol{\mu})$ with respect to each of the three variables $\mathbf{x}$, $\boldsymbol{\lambda}$, and $\boldsymbol{\mu}$

$$\mathbf{F}(\mathbf{x}, \boldsymbol{\lambda}, \boldsymbol{\mu}) = \begin{bmatrix} \mathbf{g}_f(\mathbf{x}) + \mathbf{J}_\mathbf{h}^T(\mathbf{x})\boldsymbol{\lambda} - \boldsymbol{\mu} \\ \mathbf{h}(\mathbf{x}) \\ \mathbf{XM1} - \mathbf{1}/t \end{bmatrix}, \quad \mathbf{J_F} = \begin{bmatrix} \mathbf{W}(\mathbf{x}) & \mathbf{J}_\mathbf{h}^T(\mathbf{x}) & -\mathbf{I} \\ \mathbf{J_h}(\mathbf{x}) & \mathbf{0} & \mathbf{0} \\ \mathbf{M} & \mathbf{0} & \mathbf{X} \end{bmatrix} \tag{3.94}$$

where

$$\begin{aligned} \mathbf{W}(\mathbf{x}) &= \nabla_\mathbf{x} \left[\mathbf{g}_f(\mathbf{x}) + \mathbf{J}_\mathbf{h}^T(\mathbf{x})\boldsymbol{\lambda} - \boldsymbol{\mu} \right] \\ &= \nabla_\mathbf{x} \mathbf{g}_f(\mathbf{x}) - \nabla_\mathbf{x} \left[\sum_{i=1}^M \lambda_i \frac{\partial h_i}{\partial x_1}, \cdots, \sum_{i=1}^M \lambda_i \frac{\partial h_i}{\partial x_N} \right]^T \\ &= \mathbf{H}_f(\mathbf{x}) - \sum_{i=1}^M \lambda_i \mathbf{H}_{h_i}(\mathbf{x}) \end{aligned} \tag{3.95}$$

is symmetric as the Hessian matrices $\mathbf{H}_f$ and $\mathbf{H}_{h_i}$, $(i = 1, \cdots, M)$ are symmetric.

We can now use the Newton–Raphson method to find the solution of the nonlinear equation $\mathbf{F}(\mathbf{x}, \boldsymbol{\lambda}, \boldsymbol{\mu}) = \mathbf{0}$ iteratively

$$\begin{bmatrix} \mathbf{x}_{n+1} \\ \boldsymbol{\lambda}_{n+1} \\ \boldsymbol{\mu}_{n+1} \end{bmatrix} = \begin{bmatrix} \mathbf{x}_n \\ \boldsymbol{\lambda}_n \\ \boldsymbol{\mu}_n \end{bmatrix} + \alpha \begin{bmatrix} \delta\mathbf{x}_n \\ \delta\boldsymbol{\lambda}_n \\ \delta\boldsymbol{\mu}_n \end{bmatrix} \tag{3.96}$$

where α is a parameter that controls the step size and $[\delta\mathbf{x}_n, \delta\boldsymbol{\lambda}_n, \delta\boldsymbol{\mu}_n]^T$ is the search direction (*Newton direction*), which can be found by solving the following equation

$$\begin{aligned} \mathbf{J_F}(\mathbf{x}_n, \boldsymbol{\lambda}_n, \boldsymbol{\mu}_n) \begin{bmatrix} \delta\mathbf{x}_n \\ \delta\boldsymbol{\lambda}_n \\ \delta\boldsymbol{\mu}_n \end{bmatrix} &= \begin{bmatrix} \mathbf{W}(\mathbf{x}_n) & \mathbf{J}_\mathbf{h}^T(\mathbf{x}_n) & -\mathbf{I} \\ \mathbf{J_h}(\mathbf{x}_n) & \mathbf{0} & \mathbf{0} \\ \mathbf{M}_n & \mathbf{0} & \mathbf{X}_n \end{bmatrix} \begin{bmatrix} \delta\mathbf{x}_n \\ \delta\boldsymbol{\lambda}_n \\ \delta\boldsymbol{\mu}_n \end{bmatrix} \\ = -\mathbf{F}(\mathbf{x}_n, \boldsymbol{\lambda}_n, \boldsymbol{\mu}_n) &= - \begin{bmatrix} \mathbf{g}_f(\mathbf{x}_n) + \mathbf{J}_\mathbf{h}^T(\mathbf{x}_n)\boldsymbol{\lambda}_n - \boldsymbol{\mu}_n \\ \mathbf{h}(\mathbf{x}_n) \\ \mathbf{X_n M_n 1} - \mathbf{1}/t \end{bmatrix} \end{aligned} \tag{3.97}$$

To get the initial values for the iteration, we first find any point inside the feasible region $\mathbf{x}_0$ and then $\boldsymbol{\mu}_0 = \mathbf{x}_0/t$, based on which we further find $\boldsymbol{\lambda}_0$ by solving the first equation in the KKT conditions in Eq. (3.93).

Actually, this equation system can be separated into two subsystems, which are easier to solve. Premultiplying $\mathbf{X}^{-1}$ to the third equation

$$\mathbf{M}\delta\mathbf{x} + \mathbf{X}\delta\boldsymbol{\mu} = -\mathbf{XM1} + \mathbf{1}/t \tag{3.98}$$

we get

$$\mathbf{X}^{-1}\mathbf{M}\delta\mathbf{x} + \delta\boldsymbol{\mu} = -\mathbf{M1} + \mathbf{X}^{-1}\mathbf{1}/t = -\boldsymbol{\mu} + \mathbf{X}^{-1}\mathbf{1}/t \tag{3.99}$$

Adding this to the first equation

$$\mathbf{W}\delta\mathbf{x} + \mathbf{J}_\mathbf{h}^T(\mathbf{x})\delta\boldsymbol{\lambda} - \delta\boldsymbol{\mu} = -\mathbf{g}_f(\mathbf{x}) - \mathbf{J}_\mathbf{h}^T(\mathbf{x})\boldsymbol{\lambda} + \boldsymbol{\mu} \tag{3.100}$$

we get

$$(\mathbf{W} + \mathbf{X}^{-1}\mathbf{M})\delta\mathbf{x} + \mathbf{J}_\mathbf{h}^T(\mathbf{x})\delta\boldsymbol{\lambda} = -\mathbf{g}_f(\mathbf{x}) - \mathbf{J}_\mathbf{h}^T(\mathbf{x})\boldsymbol{\lambda} + \mathbf{X}^{-1}\mathbf{1}/t = -\nabla_\mathbf{x} L(\mathbf{x}, \boldsymbol{\lambda}, \boldsymbol{\mu}) \tag{3.101}$$

Combining this equation with the second one, we get an equation system of two vector variables $\delta\mathbf{x}$ and $\delta\boldsymbol{\lambda}$ with symmetric coefficient matrix

$$\begin{bmatrix} \mathbf{W} + \mathbf{X}^{-1}\mathbf{M} & \mathbf{J}_\mathbf{h}^T(\mathbf{x}) \\ \mathbf{J}_\mathbf{h}(\mathbf{x}) & \mathbf{0} \end{bmatrix} \begin{bmatrix} \delta\mathbf{x} \\ \delta\boldsymbol{\lambda} \end{bmatrix} = -\begin{bmatrix} \nabla_\mathbf{x} L(\mathbf{x}, \boldsymbol{\lambda}, \boldsymbol{\mu}) \\ \mathbf{h}(\mathbf{x}) \end{bmatrix} \tag{3.102}$$

Solving this equation system we get $\delta\mathbf{x}$ and $\delta\boldsymbol{\lambda}$, and we can further find $\delta\boldsymbol{\mu}$ by solving the third equation

$$\mathbf{M}\delta\mathbf{x} + \mathbf{X}\delta\boldsymbol{\mu} = -\mathbf{XM1} - \mathbf{1}/t \tag{3.103}$$

to get

$$\delta\boldsymbol{\mu} = \mathbf{X}^{-1}\mathbf{1}/t - \mathbf{X}^{-1}\mathbf{M}\delta\mathbf{x} - \boldsymbol{\mu} \tag{3.104}$$

We now consider the two special cases of linear and quadratic programming.

- **Linear programming**

$$\begin{aligned} \text{min:} \quad & f(\mathbf{x}) = \mathbf{c}^T\mathbf{x} \\ \text{s. t.:} \quad & \mathbf{h}(\mathbf{x}) = \mathbf{Ax} - \mathbf{b} = \mathbf{0}, \quad \mathbf{x} \geq \mathbf{0} \end{aligned} \tag{3.105}$$

  We have
  - $\mathbf{g}_f(\mathbf{x}) = \nabla_\mathbf{x} f(\mathbf{x}) = \mathbf{c}$
  - $\mathbf{J}_\mathbf{h}(\mathbf{x}) = \nabla_\mathbf{x}(\mathbf{Ax} - \mathbf{b}) = \mathbf{A}$.
  - $\mathbf{W}(\mathbf{x}) = \nabla_\mathbf{x}^2 L(\mathbf{x}, \boldsymbol{\lambda}, \boldsymbol{\mu}) = \mathbf{0}$.

  The Lagrangian is

$$L(\mathbf{x}, \boldsymbol{\lambda}, \boldsymbol{\mu}) = \mathbf{c}^T\mathbf{x} + \boldsymbol{\lambda}^T(\mathbf{Ax} - \mathbf{b}) - \boldsymbol{\mu}^T\mathbf{x} \tag{3.106}$$

The KKT conditions are

$$\begin{cases} \nabla_{\mathbf{x}} L(\mathbf{x}, \lambda, \mu) = \mathbf{g}_f(\mathbf{x}) + \mathbf{J}_{\mathbf{h}}^T(\mathbf{x})\lambda - \mu = \mathbf{c} + \mathbf{A}^T\lambda - \mu = \mathbf{0} \\ \mathbf{h}(\mathbf{x}) = \mathbf{A}\mathbf{x} - \mathbf{b} = \mathbf{0} \\ \mathbf{X}\mathbf{M}\mathbf{1} - \mathbf{1}/t = \mathbf{0} \end{cases}$$

The search direction of the iteration can be found by solving the this equation system

$$\begin{bmatrix} \mathbf{0} & \mathbf{A}^T & -\mathbf{I} \\ \mathbf{A} & \mathbf{0} & \mathbf{0} \\ \mathbf{M} & \mathbf{0} & \mathbf{X} \end{bmatrix} \begin{bmatrix} \delta\mathbf{x} \\ \delta\lambda \\ \delta\mu \end{bmatrix} = - \begin{bmatrix} \mathbf{c} + \mathbf{A}^T\lambda - \mu \\ \mathbf{A}\mathbf{x} - \mathbf{b} \\ \mathbf{X}\mathbf{M}\mathbf{1} - \mathbf{1}/t \end{bmatrix} \tag{3.107}$$

By solving the equation $\mathbf{c} + \mathbf{A}^T\lambda - \mu = \mathbf{0}$ we get the initial value λ. The MATLAB code for the interior point method for the LP problem is listed here

```
function x=InterierPointLP(A,c,b,x)
    % given A,b,c of the LP problem and initial value of x,
    % find optimal solution x that minimizes c'*x
    [m n]=size(A);        % m constraints, n variables
    z1=zeros(n);
    z2=zeros(m);          % zero matrices in coefficient matrix
    z3=zeros(m,n);
    I=eye(n);             % identity matrix in coefficient matrix
    y=ones(n,1);
    t=9;                  % initial value for parameter t
    alpha=0.3;            % small enough not to exceed boundary
    mu=x./t;
    lambda=pinv(A')*(mu-c); % initial value of lambda
    w=[x; lambda; mu];        % initial values for all three
    B=[c+A'*lambda-mu; A*x-b; x.*mu-y/t];
    p=[x(1) x(2)];        % initial guess of solution
    while norm(B)>10^(-7)
        t=t*9;            % increase parameter t
        X=diag(x);
        M=diag(mu);
        C=[z1 A' -I; A z2 z3; M z3' X];
        B=[c+A'*lambda-mu; A*x-b; x.*mu-y/t];
        dw=-inv(C)*B;     % find search direction
        w=w+alpha*dw;     % step forward
        x=w(1:n);
        lambda=w(n+1:n+m);
        mu=w(n+m+1:length(w));
        p=[p; x(1), x(2)];
    end
    scatter(p(:,1),p(:,2)); % plot trajectory of solution
end
```

Example 3.5

A given linear programming problem is first converted into the standard form

$$\text{max:}\quad f(x_1,x_2) = 2x_1 + 3x_2 \qquad\qquad \text{min:}\quad f(x_1,x_2,x_3,x_4,x_5) = -2x_1 - 3x_2$$

$$\text{s. t.:}\begin{cases} 2x_1 + x_2 \le 18 \\ 6x_1 + 5x_2 \le 60 \\ 2x_1 + 5x_2 \le 40 \\ x_1 \ge 0,\ \ x_2 \ge 0 \end{cases} \Longrightarrow \quad \text{s. t,:}\begin{cases} 2x_1 + x_2 + x_3 = 18 \\ 6x_1 + 5x_2 + x_4 = 60 \\ 2x_1 + 5x_2 + x_5 = 40 \\ x_i \ge 0,\ \ (i = 1,\cdots,5) \end{cases}$$

or in matrix form

$$\begin{aligned} \text{min:}\quad & f(\mathbf{x}) = \mathbf{c}^T\mathbf{x} \\ \text{s. t. :}\quad & \mathbf{h}(\mathbf{x}) = \mathbf{A}\mathbf{x} - \mathbf{b} = \mathbf{0},\ \ \mathbf{x} \ge \mathbf{0} \end{aligned}$$

with

$$\mathbf{x} = \begin{bmatrix} x_1 \\ x_2 \\ x_3 \\ x_4 \\ x_5 \end{bmatrix}, \quad \mathbf{c} = \begin{bmatrix} -2 \\ -3 \\ 0 \\ 0 \\ 0 \end{bmatrix}, \quad \mathbf{A} = \begin{bmatrix} 2 & 1 & 1 & 0 & 0 \\ 6 & 5 & 0 & 1 & 0 \\ 2 & 5 & 0 & 0 & 1 \end{bmatrix}, \quad \mathbf{b} = \begin{bmatrix} 18 \\ 60 \\ 40 \end{bmatrix}$$

where x_3, x_4, x_5 are the slack variables.

We choose an initial value $\mathbf{x}_0 = [1, 2, 1, 1, 1]^T$ and an initial parameter $t = 9$, which is scaled up by a factor of 9 in each iteration, We also used full Newton step size of $\alpha = 1$. As shown in Fig. 3.9, after eight iterations, the algorithm converged to the optimal solution $\mathbf{x}^* = [5,\ 6]^T$ corresponding to the maximum function value $f(\mathbf{x}^*) = 2x_1 + 3x_2 = 28$:

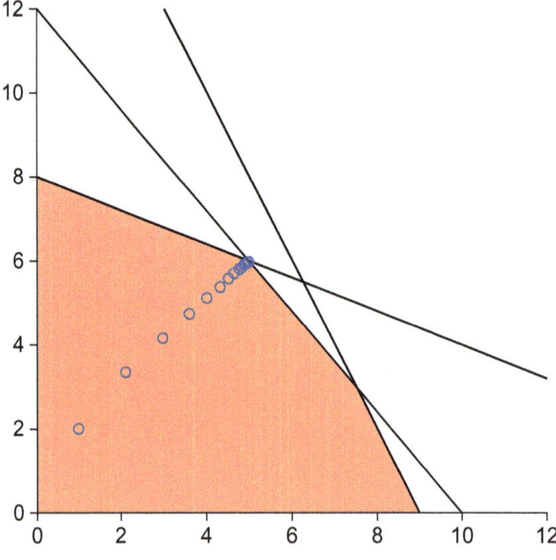

Fig. 3.9 Interior point method for linear programming.

Table 3.9

	$(x_1$	$x_2)$	$f(\mathbf{x})$	Error
1	$(1.000000e+00$	$2.000000e+00)$	−8.000000	52.413951
2	$(4.654514e+00$	$6.346354e+00)$	−28.348090	3.080204
3	$(5.040828e+00$	$5.946973e+00)$	−27.922575	0.213509
4	$(4.997282e+00$	$6.001764e+00)$	−27.999856	0.004262
5	$(4.999906e+00$	$5.999962e+00)$	−27.999697	0.000303
6	$(4.999989e+00$	$5.999996e+00)$	−27.999966	0.000034
7	$(4.999999e+00$	$6.000000e+00)$	−27.999996	0.000004
8	$(5.000000e+00$	$6.000000e+00)$	−28.000000	0.000000

At the end of the iteration, we also get the values of the slack variables: $x_3 = 2,\ x_4 = x_5 = 0$.

- **Quadratic programming**

$$\begin{array}{ll} \text{min:} & f(\mathbf{x}) = \frac{1}{2}\mathbf{x}^T\mathbf{Q}\mathbf{x} + \mathbf{c}^T\mathbf{x} \\ \text{s. t.:} & \mathbf{h}(\mathbf{x}) = \mathbf{A}\mathbf{x} - \mathbf{b} = \mathbf{0}, \quad \mathbf{x} \geq \mathbf{0} \end{array} \tag{3.108}$$

We have

- $\mathbf{g}_f(\mathbf{x}) = \nabla_{\mathbf{x}} f(\mathbf{x}) = \mathbf{Q}\mathbf{x} + \mathbf{c}$
- $\mathbf{J}_{\mathbf{h}}(\mathbf{x}) = \nabla_{\mathbf{x}}(\mathbf{A}\mathbf{x} - \mathbf{b}) = \mathbf{A}$.
- $\mathbf{W}(\mathbf{x}) = \nabla_{\mathbf{x}}^2 L(\mathbf{x}, \boldsymbol{\lambda}, \boldsymbol{\mu}) = \mathbf{Q}$.

The Lagrangian is

$$L(\mathbf{x}, \boldsymbol{\lambda}, \boldsymbol{\mu}) = \frac{1}{2}\mathbf{x}^T\mathbf{Q}\mathbf{x} + \mathbf{c}^T\mathbf{x} + \boldsymbol{\lambda}^T(\mathbf{A}\mathbf{x} - \mathbf{b}) - \boldsymbol{\mu}^T\mathbf{x} \tag{3.109}$$

The KKT conditions are

$$\begin{cases} \nabla_{\mathbf{x}} L(\mathbf{x}, \boldsymbol{\lambda}, \boldsymbol{\mu}) = \mathbf{g}_f(\mathbf{x}) + \mathbf{J}_{\mathbf{h}}^T(\mathbf{x})\boldsymbol{\lambda} - \boldsymbol{\mu} = \mathbf{Q}\mathbf{x} + \mathbf{c} + \mathbf{A}^T\boldsymbol{\lambda} - \boldsymbol{\mu} = \mathbf{0} \\ \mathbf{h}(\mathbf{x}) = \mathbf{A}\mathbf{x} - \mathbf{b} = \mathbf{0} \\ \mathbf{XM1} - \mathbf{1}/t = \mathbf{0} \end{cases}$$

The equation system for the Newton–Raphson method is

$$\begin{bmatrix} \mathbf{Q} & \mathbf{A}^T & -\mathbf{I} \\ \mathbf{A} & \mathbf{0} & \mathbf{0} \\ \mathbf{M} & \mathbf{0} & \mathbf{X} \end{bmatrix} \begin{bmatrix} \delta\mathbf{x} \\ \delta\boldsymbol{\lambda} \\ \delta\boldsymbol{\mu} \end{bmatrix} = - \begin{bmatrix} \mathbf{Q}\mathbf{x} + \mathbf{c} + \mathbf{A}^T\boldsymbol{\lambda} - \boldsymbol{\mu} \\ \mathbf{A}\mathbf{x} - \mathbf{b} \\ \mathbf{XM1} - \mathbf{1}/t \end{bmatrix} \tag{3.110}$$

The MATLAB code for the interior point method for the QP problem is listed in the following

```
function [x mu lambda]=InteriorPointQP(Q,A,c,b,x)
    n=length(c);         % n variables
    m=length(b);         % m constraints
    z2=zeros(m);
    z3=zeros(m,n);
    I=eye(n);
```

```
    y=ones(n,1);
    t=9;                  % initial value for parameter t
    alpha=0.1;            % small stepsize not to exceed boundary
    mu=x./t;
    lambda=pinv(A')*(mu-c-Q*x); % initial value of lambda
    w=[x; lambda; mu];     % initial value
    B=[Q*x+c+A'*lambda-mu; A*x-b; x.*mu-y/t];
    while norm(B)>10^(-7)
        t=t*9;            & increase parameter t
        X=diag(x);
        M=diag(mu);
        C=[Q A' -I; A z2 z3; M z3' X];
        B=[Q*x+c+A'*lambda-mu; A*x-b; x.*mu-y/t];
        dw=-inv(C)*B;  % find search direction
        w=w+alpha*dw;  % step forward
        x=w(1:n);
        lambda=w(n+1:n+m);
        mu=w(n+m+1:length(w));
    end
end
```

Example 3.6

A 2-D quadratic programming problem is to minimize four different quadratic functions under the same linear constraints as the linear programming problem in the previous example. In general, the quadratic function is given in the matrix form

$$\begin{aligned} \text{min:} \quad & f(\mathbf{x}) = [\mathbf{x}-\mathbf{m}]^T\mathbf{Q}[\mathbf{x}-\mathbf{m}] = \mathbf{x}^T\mathbf{Q}\mathbf{x} + \mathbf{c}^T\mathbf{x} + d \\ \text{s. t.:} \quad & \mathbf{h}(\mathbf{x}) = \mathbf{A}\mathbf{x} - \mathbf{b} = \mathbf{0}, \quad \mathbf{x} \geq \mathbf{0} \end{aligned}$$

with following four different sets of function parameters

$$\mathbf{Q} = \begin{bmatrix} 4 & -1 \\ -1 & 4 \end{bmatrix}, \begin{bmatrix} 2 & 1 \\ 1 & 2 \end{bmatrix}, \begin{bmatrix} 2 & 1 \\ 1 & 2 \end{bmatrix}, \begin{bmatrix} 3 & -1 \\ -1 & 3 \end{bmatrix}$$

$$\mathbf{m} = \begin{bmatrix} 4 \\ 10 \end{bmatrix}, \begin{bmatrix} 9 \\ 5 \end{bmatrix}, \begin{bmatrix} 7 \\ 2 \end{bmatrix}, \begin{bmatrix} -1 \\ 6 \end{bmatrix}$$

As shown in the four panels in Fig. 3.10 the iteration of the interior point algorithm brings the solution from the same initial position at $\mathbf{x}_0 = [2, \ 1]^T$ to the final position for each of the four objective functions $f(\mathbf{x})$, the optimal solution, which is on the boundary of the feasible region in all cases except the third one, where the optimal solution is inside the feasible region, that is, the optimization problem is not constrained. Also note that the trajectories of the solution during the iteration go straight from the initial guess to the final optimal solution by following the negative gradient direction of the quadratic function in all cases except in the last one, where it makes a turn while approaching the vertical boundary, obviously owing to the significantly greater value of the log barrier function close to the boundary, as shown in Fig. 3.11.

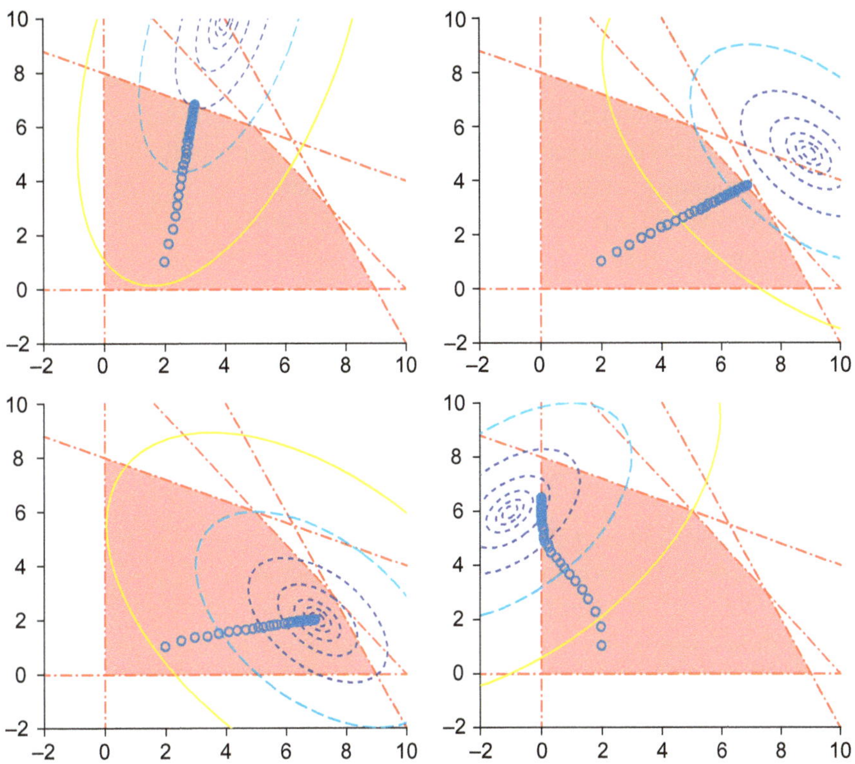

Fig. 3.10 Interior point method for quadratic programming.

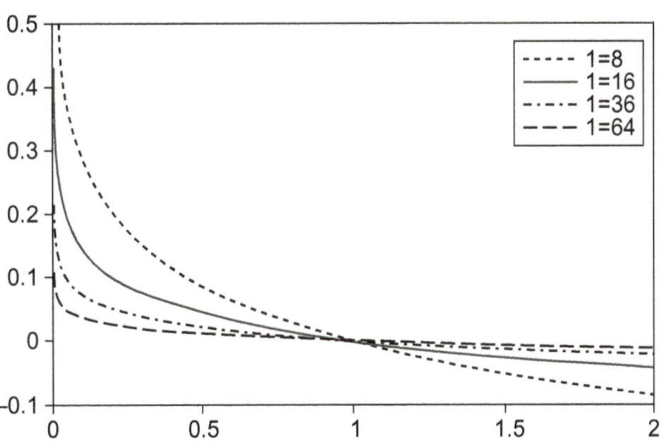

Fig. 3.11 Logarithmic barrier function.

We also note that the dual problem in the SVM algorithm in Eq. (3.50) is in the same form as in Eq. (3.108), that is, it is a quadratic programming problem that can be solved by the interior point algorithm.

PROBLEMS

1. Study carefully the first three sections of the chapter to understand the theoretical background of the constrained optimization problems. Then implement the algorithms discussed in the following sections for both linear and quadratic programming as stated in the following problems.
2. Develop MATLAB code for the brute force method for solving an LP problem in the generic form

$$\text{maximize} \quad f(x_1,\cdots,x_N) = c_1x_1 + \cdots + c_Nx_N = \textstyle\sum_{i=1}^N c_ix_i$$

$$\text{subject to:} \quad \begin{cases} a_{11}x_1 + \cdots + a_{1N}x_N = \sum_{i=1}^n a_{1i}x_i \le b_1 \\ \cdots\cdots\cdots\cdots\cdots\cdots\cdots\cdots\cdots \\ a_{M1}x_1 + \cdots + a_{MN}x_N = \sum_{i=1}^N a_{Mi}x_i \le b_M \\ x_1 \ge 0,\cdots,x_N \ge 0 \end{cases}$$

Specifically, find all C_{N+M}^N intersections formed by $N + M$ hyperplanes in the N–D space corresponding to the M constraints $\mathbf{A}_{M\times N}\mathbf{x}_{N\times 1} \le \mathbf{b}_{M\times 1}$ and the N conditions $\mathbf{x} \ge \mathbf{0}$, identify which of them are vertices of the polytope surrounded by these hyperplane, and find the vertex corresponding to the optimal solution that maximizes $f(\mathbf{x}) = \mathbf{c}^T\mathbf{x}$ for the given $\mathbf{c}$.
Run your code to solve the LP problem in the example in Section 3.4.
3. Develop MATLAB code to implement the simplex algorithm, then apply it to the LP problem in the previous problem, and the maximization problems

$$\text{maximize} \quad f(x_1, x_2, x_3) = 2x_1 - 3x_2 + 4x_3$$

$$\text{subject to:} \quad \begin{cases} 4x_1 - 3x_2 + x_3 \le 3 \\ x_1 + x_2 + x_3 \le 10 \\ 2x_1 + x_2 - x_3 \le 10 \\ x_1 \ge 0,\ x_2 \ge 0,\ x_3 \ge 0 \end{cases}$$

$$\text{maximize} \quad f(x_1, x_2, x_3) = x_1 + 2x_2 - x_3$$

$$\text{subject to:} \quad \begin{cases} 2x_1 + x_2 + x_3 \le 14 \\ 4x_1 + 2x_2 + 3x_3 \le 28 \\ 2x_1 + 5x_2 + 5x_3 \le 30 \\ x_1 \ge 0,\ x_2 \ge 0,\ x_3 \ge 0 \end{cases}$$

4. Develop code to implement the interior point method, and apply it to both the LP and QP problems in the two examples in Section 3.7.

Part II

Regression

Equipped with the mathematical tools covered in Part I, we are now ready to discuss various machine learning (ML) methods and algorithms in the rest of the book, starting with Part II for regression analysis. The general learning processes of all these learning methods may seem very different from each other in terms of their specific goals and mathematical setups, but they all fit into a generic model that captures the essential mechanism of the learning process.

In general all learning processes can be modeled as a black box, either a brain in human learning or a computer in machine learning. That takes some observed datasets as its input and generates the corresponding response as the output, representing some results learned from the input, depending on the specific learning task. As a simple ML example, the input data can be a set of numerical numbers, and the output a mathematical function that best characterizes certain relationship among these numbers. In a more complicated case, the input can be a set of images and the output some recognized objects that appear in the images. More generally, no matter how simple or complicated the learning problem is, the black box for the learning process can be represented symbolically in the simple form of a function $y = F(x,\theta)$, where the input x and output y of the model function may represent some large-scale dataset such as images and text files as well as some numbers. Here θ denotes a set of parameters of the model function. In most ML methods, the specific form of the model function is presumed to be known or given, while the main task is to find the optimal parameters in such a way that the output represents the best results learned from the data in some sense.

All learning methods covered in the book fit into the generic black box model. In Part II for regression analysis, the model function is simply a regression function $y = f(x,\theta)$ that characterizes the relationship between two sets of observed sample values of the input and output variables x and y. In Part IV, the model functions of the various classification methods are some specific algorithms (naive Bayes, support vector machine, etc.), and the goal of the learning process is to find the parameters used in the algorithm that best fit the given data. In Part V discussing the general method of artificial neural networks (ANN), the model functions are always a network composed of two or more layers of neurons or nodes, and the parameters are the connectivities or weights between the nodes of any two neighboring layers. While the number of model parameters in all these learning methods may vary tremendously, from two in the linear regression function

$y = f(x,\theta) = ax + b$, to many millions of weights in some deep learning neural networks or even billions of weights in some Large language models, they all fit into the black box learning model with the common goal of determining the model parameters represented by θ. Here the model parameters are not to be confused with some *hyperparameters* in many ML algorithms, which, different from the model parameters, may be used to control the learning process, such as the size of an ANN, or how closely the model needs to fit the observed data contaminated with noise.

In Part II, we will consider various methods for regression analysis, which is a process of *supervised learning*, with the general goal of establishing a model for the relationship between a set of independent variables (stimuli) and another set of dependent variables (responses) of a generic system. Such a model can be symbolically represented as $y = f(\mathbf{x},\boldsymbol{\theta})$, where $\mathbf{x} = [x_1,\cdots,x_d]^T$ is a column vector containing d independent variables, y is a scalar for the dependent variable, and column vector $\boldsymbol{\theta} = [\theta_1,\cdots,\theta_M]^T$ contains a set of M parameters of the model. The main task of a supervised learning process is to obtain the unknown model parameters in $\boldsymbol{\theta}$ by learning from a set of observed data, called the *training set*, represented by $\mathcal{D} = \{(\mathbf{x}_n, y_n), (n = 1,\cdots,N)\}$, containing a set N data samples $\mathbf{x}_n$ each labeled in some way by the corresponding y_n.

Specially, in regression, as one of the supervised learning methods, the model $y = f(\mathbf{x},\boldsymbol{\theta})$ is a specific *regression function* that maps a data point $\mathbf{x}$, a point in the d-dimensional space, to a scalar output y, represented by a hypersurface in a $d + 1$ dimensional space spanned by y as well as the d independent variables in $\mathbf{x}$. In particular, the hypersurface is a curve if $d = 1$, or a surface if $d = 2$.

The regression function can be assumed to be either deterministic or probabilistic depending on the specific algorithm. Typically the regression function is hypothesized to take a certain form (e.g., linear, higher-order polynomial, or exponential) based on some prior knowledge, while the parameters in $\boldsymbol{\theta}$ are to be determined by the regression algorithm based on the training set $\mathcal{D}$, which can also be denoted by $\mathcal{D} = \{\mathbf{X}, \mathbf{y}\}$ in terms of a $d \times N$ matrix $\mathbf{X} = [\mathbf{x}_1,\cdots,\mathbf{x}_N]$ and an N–D vector $\mathbf{y} = [y_1,\cdots,y_N]^T$. Throughout the discussion in all the following chapters, we will always assume the data samples in the training set are *independently and identically distributed (i.i.d.)*.

The goal of regression is to obtain the parameters in $\boldsymbol{\theta}$ so that the model prediction $\hat{y}_n = f(\mathbf{x}_n,\boldsymbol{\theta})$ matches optimally the corresponding y_n as the ground truth, that is, the hypersurface of the model best fits all training samples in the $d + 1$ dimensional space.

In this part, we will first in Chapter 4 treat regression as a simple and typical supervised learning process and illustrate some important issues such as underfitting versus overfitting commonly faced in general by all supervised learning methods including not only regression algorithms considered in this part, but also the classification algorithms to be considered in Parts IV and V. We will consider first various linear and nonlinear regression methods based on deterministic models in Chapters 5 and 6, and then some additional regression methods based on probabilistic models in Chapters 7 and 8, which can also be applied to classification problems.

4 Bias–Variance Tradeoff and Overfitting vs. Underfitting

Before discussing various regression algorithms, we first in this chapter consider some guiding philosophies and methodologies of supervised learning in general, such as frequentist versus Bayesian approach; the tradeoff and balance between bias and variance errors, or similarly, between overfitting and underfitting of the data; and some general approaches commonly used to address these issues, such as cross-validation and ensemble learning. Although here we consider such issues mostly in the context of regression analysis, we will come back to them over and over again when we discuss other supervised learning algorithms throughout the book, all of which need to address the same general issues by the same general approaches, each in its own specific way.

4.1 Supervised Learning as Optimization

In general, supervised learning, including both regression and classification as the two typical types of problems, can be formulated as a system modeling process based on the given training set $\mathcal{D} = \{\mathbf{X}, \mathbf{y}\}$, when the independent variable $\mathbf{x} \in \mathbf{X}$ and the dependent variable $y \in \mathbf{y}$ are treated, respectively, as the input (stimuli) and output (responses) of a system, of which the behavior in terms of the relationship between such input and output is to be modeled by a generic function $y = f(\mathbf{x}, \boldsymbol{\theta})$ parameterized by $\boldsymbol{\theta}$.

In general, a supervised learning problem can be formulated as the framework illustrated in Fig. 4.1, with the goal of constructing a mathematical model, symbolized by function $f(\mathbf{x}, \boldsymbol{\theta})$ parameterized by $\boldsymbol{\theta}$, in such a way that its output $\hat{\mathbf{y}}$ matches the labeling y corresponding to the input $\mathbf{x}$ in the training set $\mathcal{D}$ in some sense. Typically the form of the model $f(\mathbf{x}, \boldsymbol{\theta})$ is assumed to be known, which is either a specific function in regression, or certain specific classification algorithm (e.g., a statistic model or a neural network). In any case, the model parameters in $\boldsymbol{\theta}$ are to be determined optimally in some sense, so that the discrepancy between the actual output $\hat{\mathbf{y}}$, the prediction of the model, and the desired output $\mathbf{y}$, the ground truth given in the training set is minimized. In other words, a learning problem is typically solved as an optimization problem.

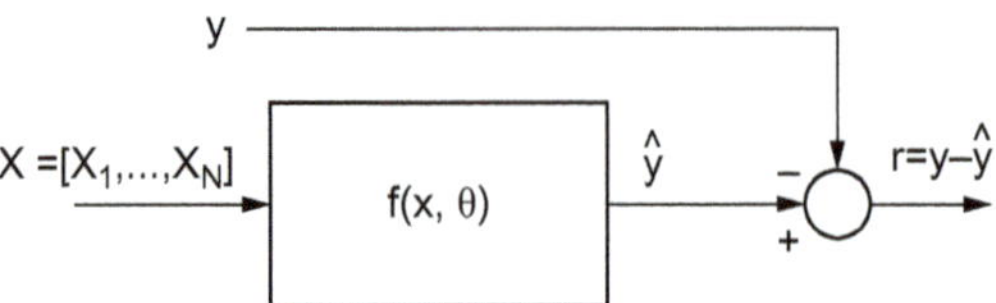

Fig. 4.1 A general model for Supervised learning.

Specifically, in regression analysis, the model $y = f(\mathbf{x}, \boldsymbol{\theta})$ is simply a function with independent variables $\mathbf{x}$ and dependent variable y, while in pattern classification (or object recognition) (to be discussed in Part IV), the model $y = f(\mathbf{x}, \boldsymbol{\theta})$ represents a specific classification method, by which a pattern described by the independent variable $\mathbf{x}$ as its *features* is assigned to one of the classes $\{C_1, \cdots, C_K\}$ indicated by the corresponding dependent variable y. As both regression and classification can be considered as a supervised learning process, much of the discussion in this chapter for various regression algorithms is also highly relevant to the classification algorithms to be considered in the future chapters.

The goal of a regression algorithm is for the regression model to optimally fit the observed data, so that ideally it reflects all variations owing to the valid *signal* in the data, but at the same time remains unaffected by the inevitable observation *noise* also existing in the data. In general, all learning algorithms aiming at modeling a set of observed data face the common dilemma between over and under trusting the observed data contaminated by noise, and they all need to address the same issue of how to best distinguish between signal and noise in the data, so that they neither *overfit* the data, that is, overly affected by noise, nor *underfit* data, that is, not matching the valid signal variation sufficiently closely. This issue of making a proper tradeoff between overfitting and underfitting will need to be addressed by various means in different learning algorithms, as we will see in this and future chapters.

The problem of regression analysis can be addressed from different philosophical viewpoints. First, from the *frequentist* point of view, the unknown model parameters in $\boldsymbol{\theta}$ are fixed deterministic variables and their values are to be estimated based on the training set. A typical method based on this viewpoint is the *least squares (LS)* method for both linear and nonlinear regression methods. Alternatively, from the point of view of *Bayesian inference*, the model parameters in $\boldsymbol{\theta}$ are random variables. We can estimate their *prior probability distribution* $p(\boldsymbol{\theta})$ based on some prior knowledge without observing any data. If no such prior knowledge is available, we can simply assume $p(\boldsymbol{\theta})$ is a uniform distribution, that is, all possible values of $\boldsymbol{\theta}$ are equally likely. Once the training data $\mathbf{X}$ are available and observed, we can estimate the *posterior probability* of the model parameters based on Bayes' theorem

$$p(\boldsymbol{\theta}|\mathbf{X}) = \frac{p(\mathbf{X}|\boldsymbol{\theta})p(\boldsymbol{\theta})}{p(\mathbf{X})} \propto p(\mathbf{X}|\boldsymbol{\theta})p(\boldsymbol{\theta}) \tag{4.1}$$

In general, the posterior $p(\boldsymbol{\theta}|\mathbf{X})$ is a narrower distribution than the prior $p(\boldsymbol{\theta})$, and therefore a more accurate description of the model parameters, owing to the reduced uncertainty of the parameters based on additional information gained from the observed data. The output $\hat{y} = f(\mathbf{x}, \boldsymbol{\theta})$ of such a regression model based on Bayesian inference is no longer a deterministic value, but a random variable described by its probability distribution in terms of its mean and variance. Typical methods based on this point of view include Bayesian linear regression, logistic regression and Gaussian process regression.

Based on either of these two viewpoints, different regression methods can be used to find the parameters $\boldsymbol{\theta}$ for the model function $\hat{y} = f(\mathbf{x}, \boldsymbol{\theta})$, such as those listed here.

- *Least squares (LS) method*

  This frequentist method measures how well the regression function models the training data by the *residual* defined as the difference between the model prediction $\hat{y}(\boldsymbol{\theta}) = f(\mathbf{x}, \boldsymbol{\theta})$ and the ground truth labeling value y corresponding to $\mathbf{x}$, for each of the N data points in the observed data

$$\mathbf{r}(\boldsymbol{\theta}) = \begin{bmatrix} r_1 \\ \vdots \\ r_N \end{bmatrix} = \begin{bmatrix} y_1 - \hat{y}_1 \\ \vdots \\ y_N - \hat{y}_N \end{bmatrix} = \begin{bmatrix} y_1 - f(\mathbf{x}_1, \boldsymbol{\theta}) \\ \vdots \\ y_N - f(\mathbf{x}_N, \boldsymbol{\theta}) \end{bmatrix} = \mathbf{y} - \hat{\mathbf{y}} \tag{4.2}$$

  Ideally, for the residual to be zero, $\mathbf{r}(\boldsymbol{\theta}) = \mathbf{0}$, we find the M model parameters as components of $\boldsymbol{\theta}$ by solving this nonlinear equation system of N equations of M unknowns in $\boldsymbol{\theta}$. However, as in general $N \gg M$, this system is overconstrained without a solution. We therefore can only find the optimal solution that minimizes the *sum of squared error (SSE)*:

$$\varepsilon(\theta) = \frac{1}{2}||\mathbf{r}||^2 = \frac{1}{2}\mathbf{r}^T\mathbf{r} = \frac{1}{2}\sum_{n=1}^{N} r_n^2 = \frac{1}{2}\sum_{n=1}^{N}(y_n - f(\mathbf{x}_n, \boldsymbol{\theta}))^2 \tag{4.3}$$

  Here the coefficient $1/2$ is included for mathematical convenience. When divided by the total number of samples N, this SSE becomes the average of all N squared errors, and the error becomes the *mean squared error (MSE)*. Minimizing either SSE or MSE produces the same results.

- *Maximum likelihood estimate (MLE)*

  This Bayesian inference method measures how well the regression function models the training data in terms of the likelihood $L(\boldsymbol{\theta}|\mathcal{D})$ of the model parameter $\boldsymbol{\theta}$ based on the observed dataset $\mathcal{D} = \{\mathbf{X}, \mathbf{y}\}$, which is proportional to the conditional probability of $\mathcal{D}$ given $\boldsymbol{\theta}$:

$$L(\boldsymbol{\theta}|\mathcal{D}) \propto p(\mathcal{D}|\boldsymbol{\theta}) = p(\mathbf{y}|\mathbf{X}, \theta) \tag{4.4}$$

  The optimal model parameters in $\boldsymbol{\theta}$ can be found as those that maximize $L(\boldsymbol{\theta}|\mathcal{D})$ or equivalently $p(\mathbf{y}|\mathbf{X}, \theta)$.

  We assume the residual $r = y - f(\mathbf{x}, \boldsymbol{\theta})$ is a zero-mean random variable (without loss of generality) with a normal probability density function (pdf)

$$\begin{aligned} p(r) = p(y - f(\mathbf{x}, \boldsymbol{\theta})) = \mathcal{N}(r, 0, \sigma^2) &= \frac{1}{\sqrt{2\pi\sigma^2}} \exp\left(-\frac{(y - f(\mathbf{x}, \boldsymbol{\theta}))^2}{2\sigma^2}\right) \\ &= \mathcal{N}(y, f(\mathbf{x}, \boldsymbol{\theta}), \sigma^2) = p(y|\mathbf{x}, \boldsymbol{\theta}) \end{aligned} \tag{4.5}$$

The justification for this assumed normal pdf is that it has the greatest entropy (Section B.2.1), or maximum uncertainty among all possible pdfs with the same variance, that is, given σ^2, the assumed normal pdf imposes least amount of constraint and thereby minimum bias.

The zero-mean pdf of $r = y - f(\mathbf{x}, \boldsymbol{\theta})$ can also be considered as the pdf of y with mean $f(\mathbf{x}, \boldsymbol{\theta})$, the conditional pdf $p(y|\mathbf{x}, \boldsymbol{\theta})$ of y given $\boldsymbol{\theta}$ as well as $\mathbf{x}$, based on the assumption that y and $\mathbf{x}$ are related by $y = f(\mathbf{x}, \boldsymbol{\theta})$. Then we can find the likelihood of $\boldsymbol{\theta}$ given the data samples in the training set, all assumed to be independently and identically distributed

$$\begin{aligned} L(\theta|\mathcal{D}) &= p(\mathbf{y}|\mathbf{X}, \theta) = \prod_{n=1}^{N} p(y = y_n|\mathbf{x}_n, \theta) \\ &= \prod_{n=1}^{N} \frac{1}{\sqrt{2\pi\sigma^2}} \exp\left(-\frac{(y_n - f(\mathbf{x}_n, \boldsymbol{\theta}))^2}{2\sigma^2}\right) \end{aligned} \tag{4.6}$$

We need to maximize this likelihood, or, equivalently and for mathematical convenience, minimize the negative log likelihood function

$$-l(\boldsymbol{\theta}|\mathcal{D}) = -\log L(\boldsymbol{\theta}|\mathbf{X}, \mathbf{y}) = -\sum_{n=1}^{N} \log p(y = y_n|\mathbf{x}_n, \theta) \tag{4.7}$$

$$\begin{aligned} &= -\sum_{n=1}^{N} \log\left[\frac{1}{\sqrt{2\pi\sigma^2}} \exp\left(-\frac{(y_n - f(\mathbf{x}, \boldsymbol{\theta}))^2}{2\sigma^2}\right)\right] \\ &= \frac{N}{2} \log(2\pi\sigma^2) + \frac{1}{2\sigma^2} \sum_{n=1}^{N} ||\mathbf{r}||^2 \end{aligned} \tag{4.8}$$

The first term and the coefficient $1/\sigma^2$ in the second term are both independent of θ and can therefore be dropped. We note that this result is equivalent to that in Eq. (4.3), that is, we can find the optimal $\boldsymbol{\theta}$ by either maximizing the likelihood $L(\boldsymbol{\theta}|\mathcal{D})$, or equivalently minimizing the SSE $\varepsilon(\boldsymbol{\theta})$.

We further note that negative log likelihood $-l(\boldsymbol{\theta}|\mathcal{D}) = -\log L(\boldsymbol{\theta}|\mathbf{X}, \mathbf{y})$ in Eq. (4.7) can also be interpreted differently as the cross entropy (Section B.2.2) between an unknown ground truth probability P and the model probability Q

$$H(P, Q) = -\sum_{i} P_i \log Q_i = -E_P[\log Q] \tag{4.9}$$

We rewrite Eq. (4.7) as the average of the logarithm of the model distribution $Q = p(y|\mathbf{x}, \theta)$ over all N observed samples in the dataset $\mathcal{D} = \{\mathbf{X}, \mathbf{y}\}$

$$-l(\boldsymbol{\theta}|\mathcal{D}) = -\sum_{n=1}^{N} \log p(y = y_n|\mathbf{x}_n, \theta) = -E_{(X,y)}[\log p(y = y_n|\mathbf{x}_n, \theta)] \tag{4.10}$$

This is the cross entropy $H(P, Q)$ estimated based on the N samples, which can be minimized for model Q to match the ground truth P as closely as possible.

- *Maximum a posteriori (MAP)*
  This is also a Bayesian inference method that measures how well the regression function models the training data by the posterior probability of the parameter $\boldsymbol{\theta}$ given in Eq. (4.1), proportional to the product of the likelihood $L(\boldsymbol{\theta}|\mathcal{D}) \propto p(\mathcal{D}|\boldsymbol{\theta})$ and the prior $p(\boldsymbol{\theta})$. If no prior knowledge about $\boldsymbol{\theta}$ is available, then $p(\boldsymbol{\theta})$ is a uniform distribution, and MAP is equivalent to MLE. However, if certain prior knowledge regarding $\boldsymbol{\theta}$ does exist, and the prior is not uniform, then the posterior better characterizes $\boldsymbol{\theta}$ than the likelihood and MAP may produce better results than MLE.

We conclude that regression analysis, and in general many supervised learning algorithms to be considered in the following chapters, can be treated as an over-constrained optimization problem with many more observed data points in the training set than the number of unknown parameters in the model. Following different philosophical view points, this optimization problem can be addressed by different methods, such as minimizing the sum of squared errors, maximizing the likelihood of the model parameters, or minimizing the negative likelihood, that is, the cross entropy between the model and unknown ground truth distributions of the given dataset, all of these methods are to be discussed in detail in the following sections.

More generally, some of the concepts and methods in regression analysis discussed in the following sections are also shared by many supervised learning algorithms for classification. For example, the general method of neural network is conceptually based on some ideas very different from regression, but both of these methods can be described mathematically as a function $\mathbf{y} = f(\mathbf{x}, \boldsymbol{\theta})$ that models the relationship between the independent variable $\mathbf{x}$ and dependent variable $\mathbf{y}$. While this function is called the regress function in this chapter, it is also a way to describe the complex behavior of a multilayer learning network to be discussed in Section 18.2

4.2 Bias–Variance Tradeoff

Before discussing specifically various regression algorithms, we first address an important issue that commonly appears in not only regression problems as discussed in the following, but in general also all supervised learning problems based on training dataset (or simply training set) such as classification.

In a generic regression problem, the goal is to model the unknown relationship between a dependent variable $y = f(\mathbf{x})$ and d independent variables in $\mathbf{x} = [x_1, \cdots, x_d]^T$, by a hypothesis regression function $\hat{f}(\mathbf{x})$, based on a set of training data $\mathcal{D} = \{(\mathbf{x}_n, y_n)\ n = 1, \cdots, N\}$ containing N observed data samples. As the observation is inevitably contaminated by some random noise e, both the observed value modeled by $y = f(\mathbf{x}) + e$ and the regression model $\hat{f}(\mathbf{x})$ need to be treated as random variables. We need to consider all possible errors of different types in the regression analysis as the consequence of such noisy data.

We first assume without loss of generality that the random noise e has a zero mean $E[e] = 0$ and variance σ^2, and it is independent of both f and $\hat{f}$

$$\begin{aligned}\text{Var}[e] &= E[(e - E[e])^2] = E[e^2] = \sigma^2 \\ E[e\, f] &= E[e]\, E[f], \qquad E[e\, \hat{f}] = E[e]\, E[\hat{f}]\end{aligned} \tag{4.11}$$

To measure how well the model $\hat{f}(\mathbf{x})$ fits the noisy data $y = f(\mathbf{x}) + e$, we define the *MSE* between them (with $\mathbf{x}$ dropped for simplicity)

$$\begin{aligned}\text{E}[(y - \hat{f})^2] &= \text{E}[(f + e - \hat{f})^2] = \text{E}\left[[(f - E\hat{f}) + (E\hat{f} - \hat{f}) + e]^2\right] \\ &= E\left[(f - E\hat{f})^2 + (E\hat{f} - \hat{f})^2 + e^2 + 2e(f - E\hat{f}) + 2e(E\hat{f} - \hat{f})\right. \\ &\quad \left. +2(f - E\hat{f})(E\hat{f} - \hat{f})\right] \\ &= \text{E}[(f - E\hat{f})^2] + \text{E}[(E\hat{f} - \hat{f})^2] + E(e^2) \\ &\quad +2\,\text{E}[e(f - E\hat{f})] + 2\,\text{E}[e(E\hat{f} - \hat{f})] + 2\,\text{E}[(E\hat{f} - \hat{f})(f - E\hat{f})] \\ &= \text{E}[(f - E\hat{f})^2] + \text{E}[(E\hat{f} - \hat{f})^2] + \text{E}[e^2]\end{aligned} \tag{4.12}$$

The last equality is due to the fact that all three cross terms are zero

$$\begin{cases} \text{E}[e(f - E\hat{f})] = \text{E}[e]\ \ \text{E}[f - E\hat{f}] = 0 \\ \text{E}[e(E\hat{f} - \hat{f})] = \text{E}[e]\ \ \text{E}[E\hat{f} - \hat{f}] = 0 \\ \text{E}[(E\hat{f} - \hat{f})(f - E\hat{f})] = (E\hat{f} - E\hat{f})(f - E\hat{f}) = 0 \end{cases} \tag{4.13}$$

Now the total mean squared error in Eq. (4.12) can be expressed in terms of three different types of errors

$$\text{Mean Squared Error} = \text{Variance Error} + \text{Bias Error} + \text{Irreducible Error} \tag{4.14}$$

where:

- *Bias Error:* $\text{E}[(f - E\hat{f})^2] = (f - E\hat{f})^2$ measures how well the estimated function $\hat{f}$ models the ground truth function f;
- *Variance Error:* $\text{E}[\hat{f} - E\hat{f}]^2 = \text{Var}[\hat{f}]$ measures the variation of the model;
- *Irreducible Error:* $\text{E}[e^2] = \sigma^2$ measures the inevitable observation noise.

The variance error and bias error are illustrated in Fig. 4.2. Comparing the top right and the bottom left cases we see that under the same total MSE and irreducible error σ^2, the bias error and the variance error are complementary to each other, that is, if one is higher, the other is lower, depending on the specific model function $\hat{f}$. This relationship is called the *bias–variance tradeoff*, which is related to the common issue of overfitting versus underfitting in all supervised learning problems in general as we will see in many of the learning algorithms throughout the book, such as support vector machine in Chapter 14, backpropagation network in Chapter 18, and reinforcement learning in Chapter 20.

- *Overfitting*
  To match the training data closely, a complex model may be needed to take into consideration all detailed variations in the data which may be caused by the noise

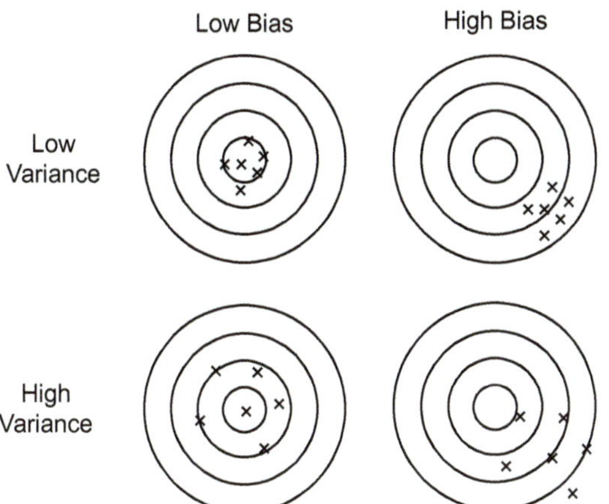

Fig. 4.2 Bias–variance tradeoff.

in the data. Such a model tends to have a low variance error, but may suffer from high bias error, as it may be overly affected by the random noise such as some outliers, as shown in the top-right case in Fig. 4.2. Such a model does not generalize well across multiple datasets contaminated differently by noise.

- *Underfitting*
  Conversely, to avoid being overly affected by the noise, a simple model may be used with the possible consequence that it may not match the detailed variation due to the valid signal in the data sufficiently closely. Such a model tends to have low bias error as it is insensitive to the varying noise across datasets, but it may suffer from high variance error as it may not match all valid signal variations. This is the lower-left case in Fig. 4.2.

Table 4.1

	Overfit	Underfit
Variance error	High	Low
Bias error	Low	High

These two possibilities are summarized in Table 4.1 and Fig. 4.3.

As both overfitting and underfitting result in poor prediction capability, they need to be avoided by making a proper tradeoff between bias-error/underfitting and variance-error/overfitting. This is one of the most important issues in general in all supervised learning algorithms aiming at modeling a system and predicting its behaviors, including regression for predicting the output value $\hat{y} = f(\mathbf{x})$ as a function of a given data point $\mathbf{x}$, and classification for providing a categorical class labeling of a given point $\mathbf{x}$.

The issue of overfitting versus underfitting is further illustrated in the examples in Fig. 4.4. In the case of regression (left panel), the data points are modeled by two different regression functions, and in classification, the 2-D space is partitioned

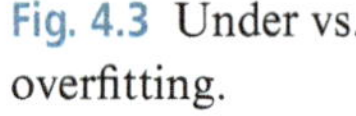

Fig. 4.3 Under vs. overfitting.

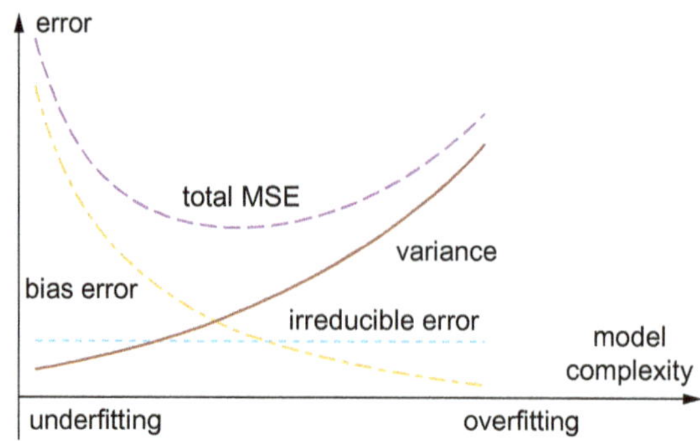

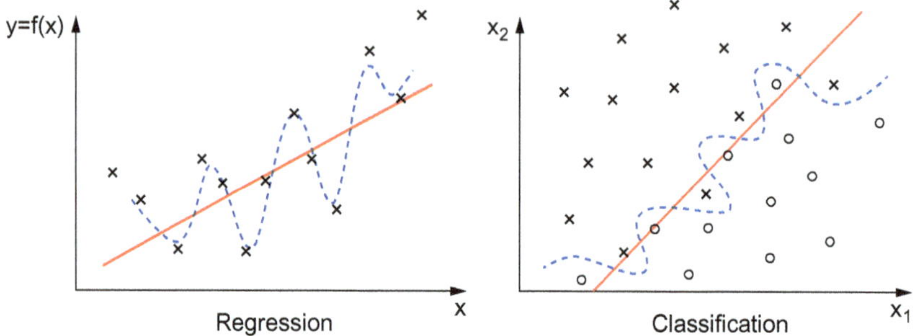

Fig. 4.4 Under and overfitting in regression and classification.

into two regions corresponding to two classes by two different *decision boundaries*. In both cases, the given datasets are overfitted by the blue curve underfitted by the red curve.

- Underfitting
  The simple linear regression function or decision boundary (red) may underfit the data as some legitimate variations in the data may be missed. This is a problem if all data points reliably reflect some complex variations owing to valid signal in the data.
- Overfitting
  The complex regression model (e.g., a high-order polynomial) or decision boundary (blue) may overfit the data if it is heavily affected by the variation due to observation noise instead of valid signal in the data. This is a problem if the data samples are taken from a simple process, but are fitted by an unnecessarily complex model owing to the noise, that is, a violation of the principle of Occam's razor.

Therefore we face the challenge of how to make a proper tradeoff between over and underfittings, so that the algorithm can produce an optimal model that fits the variations in the given data, in the sense that it maximally fits the valid signal in the data, but at the same time is minimally affected by the noise also in the data. Various methods to address this challenge will be considered in the following section and throughout the discussion of many supervised learning algorithms in the following chapters.

4.3 Cross-Validation

In the case when the entire training set is used for training to take full advantage of the available data which may be limited to start with, it is impossible to distinguish between variations owing to valid signal and those owing to random observation noise, and to judge whether the model produced by the learning algorithm in question either overfits or underfits the given data. This problem can be addressed by the method of cross-validation. If the dataset available for training the algorithm is of sufficient size, it can be partitioned into two or more subsets each used to either train or test the algorithm. This method is based on the simple idea that the valid signal variation should be shared by all sub-datasets, while they are contaminated differently by the noise due to its random nature. If the algorithm is trained by one subset of data but tested by another, then its performance measured by certain errors (calculated differently depending on the specific learning problem and algorithm used, e.g., residual for regression, or rate of error for classification) can be used to assess the algorithm in terms of whether it suffers from the overfitting problem.

Cross-validation has different variations depending on how the available dataset is partitioned.

- Exhaustive methods: the algorithm is trained and tested based on the available dataset divided in all possible ways into multiple subsets. Typically, the leave-p-out cross-validation uses p out of the total n sample points in the dataset to test the algorithm after it is trained by the remaining $n - p$ samples. To achieve unbiased result, we would need to repeat this process C_p^n times to exhaust all possible ways to choose p samples out of n. To avoid repeating too many times when n is large, we may need to consider the special case of either leave-pair-out ($p = 2$) or leave-one-out ($p = 1$).
- Non-exhaustive methods: Typically the K-fold cross-validation divides the dataset into K (e.g., $K = 4$ or 5) equal-sized subsets, of which $K - 1$ subsets are used for training, while the remaining one is reserved for testing. This process is repeated K times each based on a different testing subset and then the average of the K errors is used to check if the algorithm suffers overfitting. Specially we can simply let $K = 2$, or we can also randomly divide the available data into two subsets each for either training or testing, and calculate the average of the errors obtained while the process is repeated multiple times.

The method of cross-validation method has certain limitations. First, the method requires much increased computational complexity. Second, the method is effective only when the size of the training set available is sufficiently large and can therefore be divided into multiple subsets of meaningful sizes. If the size of the available dataset is small, as may be the case in some applications, the results generated by the cross-validation may not be reliable and meaningful.

4.4 Regularization and Ensemble Learning

The method of cross-validation can detect and assess overfitting problems, but it does not correct the problem by itself. Some additional approaches can be considered to address this problem.

First, we consider *regularization*, which is a method widely used in many supervised learning algorithms (including both regression and classification) to deal with the issue of overfitting versus underfitting. Typically this is done by adjusting certain hyperparameters of the model to make a proper tradeoff between the two extremes of overfitting and underfittings. In particular, if the problem at hand is ill-conditioned (see Section A.8.5), then regularization can be highly effective so that the algorithm is less sensitive to noise in the dataset and therefore less overfitting, that is, it is more generalizable while still able to capture the essential nature and behavior of the valid signal. We will show how regularization is actually implemented, such as the ridge regression in Section 5.2, and many other learning algorithm to be considered in future chapters.

Second, we consider the general method of *ensemble learning*, a general approach that takes advantage of multiple learning algorithms, each considered as a hypothesis, trained on multiple datasets, so that the resulting model can best fits the given data. This method is based on the idea that a single learning algorithm trained on a single dataset, here called a learner, may be *weak* as it may suffer from the problem of underfitting vs. overfitting, but a set of multiple such weak learners (either identical or different) can be trained on multiple datasets and combined in some way to become a *strong* learner. In general such an ensemble learning method comes in two different forms depending on how the weak learners are combined to form a strong learner.

- *Bootstrap aggregation (bagging)*
  This is a two-step process including *bootstrapping* in which a set of independent weak learners are trained in parallel by different subsets as random subsamples from the total dataset (with replacement), and *aggregation* in which a strong learner combines the results by the individual weak learners with equal weights, such as their average for regression, or their majority vote for classification. The weak learners can be either identical (homogeneous) or different (heterogeneous). In the latter case, the method is also called *stacking*, as the collection of all such weak learners based on different algorithms learning in parallel are treated as a stack. As we will see in Chapter 16, the popular method of *random forest* is a bagging algorithm that combines multiple decision tree classifiers for better classification performance.

- *Boosting*
  This is a multistep sequential process in which the weak learners are trained sequentially, in the sense that each learner is trained on a different subset of data, based on the learner in the previous step, thereby continuously improving the overall performance of the strong learner. As we will see in Section 13.4, the method of *adaptive boosting or AdaBoost* is such a boosting algorithm for classification.

5 Linear Regression

5.1 Linear Least Squares (LLS) Regression

In linear regression, the relationship between the dependent variable y and the d independent variables $\{x_1, \cdots, x_d\}$ is modeled by a linear hypothesis function

$$\hat{y} = f(\mathbf{x}, \boldsymbol{\theta}) = f(\mathbf{x}, \mathbf{w}, b) = b + \sum_{i=1}^{d} w_i x_i = b + \mathbf{x}^T \mathbf{w} \tag{5.1}$$

parameterized by $\boldsymbol{\theta}$ containing b and $\mathbf{w} = [w_1, \cdots, w_d]^T$ for the coefficients or weights in the linear function. The goal of a linear regression algorithm is to estimated the model parameters in $\boldsymbol{\theta}$ based on the training set $\mathcal{D} = \{\mathbf{X}, \mathbf{y}\}$ containing a set of N given data points in $\mathbf{X} = [\mathbf{x}_1, \cdots, \mathbf{x}_N]$ and the corresponding labelings in $\mathbf{y} = [y_1, \cdots, y_N]^T$.

For notational convenience, we redefine both $\mathbf{x}$ and $\mathbf{w}$ so that they become augmented $d + 1$ dimensional vectors: $\mathbf{x} = [x_0, x_1, \cdots, x_d]^T$ with an extra component $x_0 = 1$ and $\mathbf{w} = [w_0, w_1, \cdots, w_d]^T$ with an extra component $w_0 = b$. Now this linear model can be expressed more concisely as

$$\hat{y} = f(\mathbf{x}, \mathbf{w}) = \mathbf{x}^T \mathbf{w} \tag{5.2}$$

Geometrically, the linear regression function $f(\mathbf{x}) = \mathbf{w}^T \mathbf{x}$ is a straight line if $d = 1$, a plane if $d = 2$, or a hyperplane if $d > 2$, in the $d + 1$ dimensional space spanned by y as well as $\{x_1, \cdots, x_d\}$. The function has an intercept $f(\mathbf{0}) = w_0 = b$ along the y dimension (when $x_1 = \cdots = x_d = 0$), and a normal vector $[w_1, \cdots, w_d, -1]^T$. When the regression function is set to zero, that is, the regression line or plane is thresholded by the plane $y = 0$, it becomes an equation $f(\mathbf{x}) = \mathbf{x}^T \mathbf{w} = 0$, representing a point if $d = 1$, a straight line if $d = 2$, a plane or hyperplane if $d \geq 3$, in the d-dimensional space spanned by $\{x_1, \cdots, x_d\}$, as shown in Fig. 5.1 for $d = 2$.

We denote by $\mathbf{X}$ the matrix containing the N augmented data points $\mathbf{x}_n = [x_{0n} = 1, x_{1n}, \cdots, x_{dn}]^T$ $(n = 1, \cdots, N)$ as its column vectors

$$\mathbf{X} = \begin{bmatrix} 1 & \cdots & 1 \\ x_{11} & \cdots & x_{1N} \\ \vdots & \ddots & \vdots \\ x_{d1} & \cdots & x_{dN} \end{bmatrix}_{(d+1)\times N} = [\mathbf{x}_1, \cdots, \mathbf{x}_N] \tag{5.3}$$

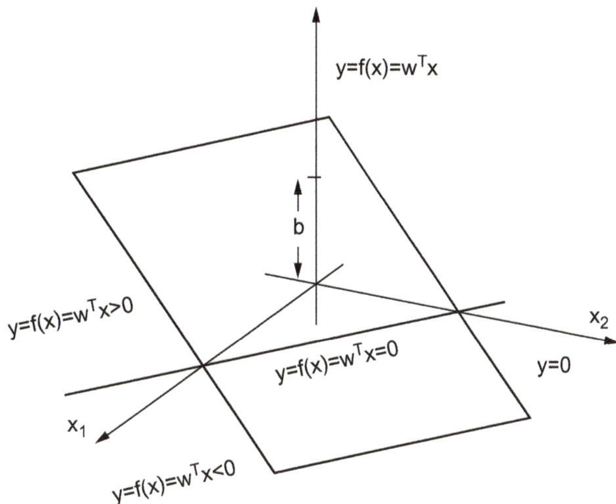

Fig. 5.1 Linear regression as a binary classifier.

and its transpose can be written as

$$\mathbf{X}^T = \begin{bmatrix} 1 & x_{11} & \cdots & x_{d1} \\ \vdots & \vdots & \ddots & \vdots \\ 1 & x_{1N} & \cdots & x_{dN} \end{bmatrix}_{N\times(d+1)} = \left[\mathbf{1}, \underline{\mathbf{x}}_1, \cdots, \underline{\mathbf{x}}_d\right] \tag{5.4}$$

where $\mathbf{1} = [1, \cdots, 1]^T$, and $\underline{\mathbf{x}}_i = [x_{i1}, \cdots, x_{iN}]^T \quad (i = 1, \cdots, d)$ is an N-dimensional vectors containing the ith components of all N data vectors in the dataset. Now the linear regression problem is simply to find the model parameter **w**, so that the model prediction

$$\hat{\mathbf{y}} = \mathbf{X}^T\mathbf{w} \tag{5.5}$$

matches the ground truth labeling **y**, and the residual of the linear model $\mathbf{r} = \mathbf{y} - \mathbf{X}^T\mathbf{w}$ is zero, that is, we need to solve the following linear system containing N equations of $d + 1$ unknowns

$$\mathbf{r} = \mathbf{y} - \hat{\mathbf{y}} = \mathbf{y} - \mathbf{X}^T\mathbf{w} = \mathbf{0} \tag{5.6}$$

If the number of data samples is equal to the number of unknown parameters, that is, $N = d + 1$, then **X** is square and invertible matrix (assuming all N samples are independent and therefore **X** has a full rank), and Eq. (5.6) can be solved to get the desired weight vector

$$\mathbf{w} = (\mathbf{X}^T)^{-1}\mathbf{y} \tag{5.7}$$

However, if there are many more data samples than the unknown parameters, that is, $N \gg d + 1$, then **X** is not square and invertible, then $\mathbf{X}^T\mathbf{w} = \mathbf{y}$ is an over-determined linear system without a solution and the prediction $\hat{y} = \mathbf{X}^T\mathbf{w}$ will not match the ground truth labeling **y** in general. In this case we need to find an optimal solution $\mathbf{w}^*$ that minimizes the sum of squared error (SSE) (proportional to mean squared error (MSE))

$$\varepsilon(\mathbf{w}) = \frac{1}{2}\sum_{n=1}^{N} r_n^2 = \frac{1}{2}||\mathbf{r}||^2 = \frac{1}{2}||\mathbf{y} - \mathbf{X}^T\mathbf{w}||^2$$
$$= \frac{1}{2}\mathbf{y}^T\mathbf{y} - \mathbf{y}^T\mathbf{X}^T\mathbf{w} + \frac{1}{2}\mathbf{w}^T\mathbf{X}\mathbf{X}^T\mathbf{w} \tag{5.8}$$

To find the optimal $\mathbf{w}$ that minimizes $\varepsilon(\mathbf{w})$ as the objective function, we set its gradient to zero and get the *normal equation*

$$\mathbf{g}_\varepsilon(\mathbf{w}) = \frac{d}{d\mathbf{w}}\varepsilon(\mathbf{w}) = \mathbf{X}\mathbf{X}^T\mathbf{w} - \mathbf{X}\mathbf{y} = \mathbf{0}, \quad \text{i.e.,} \quad \mathbf{X}\mathbf{X}^T\mathbf{w} = \mathbf{X}\mathbf{y} \tag{5.9}$$

Solving this equation for $\mathbf{w}$, we get the optimal weight vector

$$\mathbf{w}^* = (\mathbf{X}\mathbf{X}^T)^{-1}\mathbf{X}\mathbf{y} = (\mathbf{X}^T)^{-}\mathbf{y} \tag{5.10}$$

where $(\mathbf{X}^T)^- = (\mathbf{X}\mathbf{X}^T)^{-1}\mathbf{X}$ is the $(d+1) \times N$ pseudo-inverse (Section A.6.1) of the $N \times (d+1)$ matrix $\mathbf{X}^T$. This method for estimating the regression model parameters is called *ordinary least squares (OLS)*.

Having found the parameter $\mathbf{w}$ for the linear regression model $\hat{y} = \mathbf{x}^T\mathbf{w}$, we can predict the outputs corresponding to any unlabeled test dataset $\mathbf{X}_* = [\mathbf{x}_{1*}, \cdots, \mathbf{x}_{M*}]$

$$\hat{\mathbf{y}}_* = \mathbf{f}_* = \mathbf{X}_*^T\mathbf{w}^* = [\mathbf{1}, \underline{\mathbf{x}}_1, \cdots, \underline{\mathbf{x}}_N] \begin{bmatrix} w_0^* \\ w_1^* \\ \vdots \\ w_d^* \end{bmatrix} = w_0^*\mathbf{1} + \sum_{i=1}^{d} w_i^*\underline{\mathbf{x}}_i \tag{5.11}$$

as a linear combination of $\mathbf{1}$ and $\{\underline{\mathbf{x}}_1, \cdots, \underline{\mathbf{x}}_d\}$.

We can show geometrically that the estimated $\hat{\mathbf{y}} = \mathbf{x}^T\mathbf{w}^*$ based on the optimal $\mathbf{w}^*$ is indeed the best fit of the training data. We first rewrite the normal equation in Eq. (5.9) as

$$\mathbf{X}\mathbf{X}^T\mathbf{w}^* - \mathbf{X}\mathbf{y} = \mathbf{X}\left(\mathbf{X}^T\mathbf{w}^* - \mathbf{y}\right) = \mathbf{X}(\hat{\mathbf{y}} - \mathbf{y})$$
$$= \mathbf{X}\mathbf{r} = \begin{bmatrix} 1 & \cdots & 1 \\ x_{11} & \cdots & x_{1N} \\ \vdots & \ddots & \vdots \\ x_{d1} & \cdots & x_{dN} \end{bmatrix} \mathbf{r} = \begin{bmatrix} \mathbf{1}^T \\ \underline{\mathbf{x}}_1^T \\ \vdots \\ \underline{\mathbf{x}}_d^T \end{bmatrix} \mathbf{r} = \mathbf{0} \tag{5.12}$$

The last equality indicate that the residual vector $\mathbf{r}$ is perpendicular to each of the $d+1$ vectors of $\mathbf{X}^T = [\mathbf{1}, \underline{\mathbf{x}}_1, \cdots, \underline{\mathbf{x}}_d]$

$$\mathbf{1}^T\mathbf{r} = \sum_{n=1}^{N} r_n = 0, \quad \underline{\mathbf{x}}_1^T\mathbf{r} = \cdots = \underline{\mathbf{x}}_N^T\mathbf{r} = 0 \tag{5.13}$$

that is $\mathbf{r}$ is perpendicular to the space spanned by $\{\mathbf{1}, \underline{\mathbf{x}}_1, \cdots, \underline{\mathbf{x}}_d\}$, in which all possible estimated $\hat{\mathbf{y}} = \mathbf{x}^T\mathbf{w}$ based on any arbitrary $\mathbf{w}$ reside. We can therefore conclude that $||\mathbf{r}|| = ||\hat{\mathbf{y}} - \mathbf{x}^T\mathbf{w}^*||$ based on $\mathbf{w}^*$ is the smallest error among all possible $||\mathbf{r}'|| = ||\hat{\mathbf{y}} - \mathbf{x}^T\mathbf{w}||$ based on other $\mathbf{w} \neq \mathbf{w}^*$, that is, $\hat{\mathbf{y}} = \mathbf{x}^T\mathbf{w}^*$ is indeed the best with minimum residual $||\mathbf{r}||^2$. This is illustrated in Fig. 5.2 for $d = 1$.

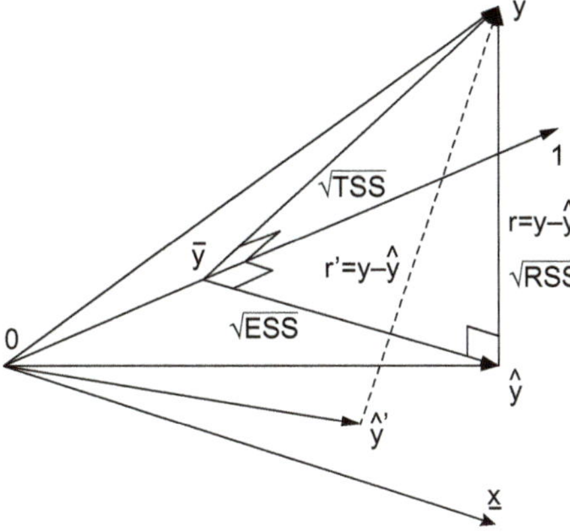

Fig. 5.2 Explained, residual and total sum of squares.

We first derive some properties of the model based on its output $\hat{\mathbf{y}} = \mathbf{X}^T\mathbf{w} = \mathbf{X}^T(\mathbf{X}\mathbf{X}^T)^{-1}\mathbf{X}$

$$||\hat{\mathbf{y}}||^2 = \hat{\mathbf{y}}^T\hat{\mathbf{y}} = \left(\mathbf{X}^T(\mathbf{X}\mathbf{X}^T)^{-1}\mathbf{X}\mathbf{y}\right)^T \left(\mathbf{X}^T(\mathbf{X}\mathbf{X}^T)^{-1}\mathbf{X}\mathbf{y}\right)$$
$$= \left(\mathbf{y}^T\mathbf{X}^T(\mathbf{X}\mathbf{X}^T)^{-1}\mathbf{X}\right)\left(\mathbf{X}^T(\mathbf{X}\mathbf{X}^T)^{-1}\mathbf{X}\mathbf{y}\right)$$
$$= \mathbf{y}^T\mathbf{X}^T(\mathbf{X}\mathbf{X}^T)^{-1}\mathbf{X}\mathbf{y} = \mathbf{y}^T\hat{\mathbf{y}} \tag{5.14}$$

based on which we also obtain the following results

$$\mathbf{r}^T\hat{\mathbf{y}} = (\mathbf{y}-\hat{\mathbf{y}})^T\hat{\mathbf{y}} = \mathbf{y}^T\hat{\mathbf{y}} - \hat{\mathbf{y}}^T\hat{\mathbf{y}} = 0 \tag{5.15}$$

$$||\mathbf{r}||^2 = \mathbf{r}^T\mathbf{r} = (\mathbf{y}-\hat{\mathbf{y}})^T(\mathbf{y}-\hat{\mathbf{y}}) = \mathbf{y}^T\mathbf{y} - 2\mathbf{y}^T\hat{\mathbf{y}} + \hat{\mathbf{y}}^T\hat{\mathbf{y}}$$
$$= \mathbf{y}^T\mathbf{y} - \mathbf{y}^T\hat{\mathbf{y}} = \mathbf{y}^T(\mathbf{y}-\hat{\mathbf{y}}) = \mathbf{y}^T\mathbf{r} \tag{5.16}$$

$$||\hat{\mathbf{y}}||^2 + ||\mathbf{r}||^2 = \mathbf{y}^T\hat{\mathbf{y}} + \mathbf{y}^T\mathbf{r} = \mathbf{y}^T(\hat{\mathbf{y}}+\mathbf{r}) = \mathbf{y}^T\mathbf{y} = ||\mathbf{y}||^2 \tag{5.17}$$

We next consider how to quantitatively measure the goodness of the linear regression model $y = \mathbf{w}^T\mathbf{x}$, that is, how well it fits the given dataset. To do so, we first define a vector $\bar{\mathbf{y}} = \bar{y}\,\mathbf{1}$ of which all components are the mean of $\{y_1, \cdots, y_N\}$

$$\bar{y} = \frac{1}{N}\sum_{n=1}^{N} y_n = \frac{1}{N}\mathbf{y}^T\mathbf{1} \tag{5.18}$$

and then define the following three *sums of squares* of the components (squared norm) of the three vectors $\hat{\mathbf{y}} - \bar{\mathbf{y}}$, $\mathbf{y} - \hat{\mathbf{y}}$, and $\mathbf{y} - \bar{\mathbf{y}}$

- *Explained sum of squares (ESS)* that measures the variation in the data (with respect to its mean) explained by the regression model

$$ESS = ||\hat{\mathbf{y}} - \bar{\mathbf{y}}||^2 = \sum_{n=1}^{N}(\hat{y}_n - \bar{y})^2 \tag{5.19}$$

- *Residual sum of squares (RSS)* that represents the variation in the data not explained by the data (due to noise or discrepancy between the data and the model)

$$RSS = ||\mathbf{r}||^2 = ||\mathbf{y} - \hat{\mathbf{y}}||^2 = \sum_{n=1}^{N} (y_n - \hat{y}_n)^2 \tag{5.20}$$

- *Total sum of squares* (TSS) for the total data variation

$$TSS = ||\mathbf{y} - \bar{\mathbf{y}}||^2 = \sum_{n=1}^{N} (y_n - \bar{y})^2 \tag{5.21}$$

We can show that the total sum of squares is the sum of the explained sum of squares and the residual sum of squares

$$\begin{aligned} TSS &= ||\mathbf{y} - \bar{\mathbf{y}}||^2 = ||\mathbf{y} - \bar{\mathbf{y}} + \hat{\mathbf{y}} - \hat{\mathbf{y}}||^2 = ||(\hat{\mathbf{y}} - \bar{\mathbf{y}}) + (\mathbf{y} - \hat{\mathbf{y}})||^2 \\ &= ||\hat{\mathbf{y}} - \bar{\mathbf{y}}||^2 + 2(\mathbf{y} - \hat{\mathbf{y}})^T(\hat{\mathbf{y}} - \bar{\mathbf{y}}) + ||\mathbf{y} - \hat{\mathbf{y}}||^2 \\ &= ||\hat{\mathbf{y}} - \bar{\mathbf{y}}||^2 + 2\mathbf{r}^T(\hat{\mathbf{y}} - \bar{\mathbf{y}}) + ||\mathbf{r}||^2 \\ &= ESS + 2\mathbf{r}^T\hat{\mathbf{y}} - 2\mathbf{r}^T\bar{\mathbf{y}} + RSS = ESS + RSS \end{aligned} \tag{5.22}$$

The last equality is due to the fact that the two middle terms are both zero (Eqs (5.13) and (5.15))

$$\begin{cases} \mathbf{r}^T\bar{\mathbf{y}} = \sum_{n=1}^{N} r_n\bar{y} = \bar{y}\sum_{n=1}^{N} r_n = 0 \\ \mathbf{r}^T\hat{\mathbf{y}} = (\mathbf{y} - \hat{\mathbf{y}})^T\hat{\mathbf{y}} = \mathbf{y}^T\hat{\mathbf{y}} - \hat{\mathbf{y}}^T\hat{\mathbf{y}} = 0 \end{cases}$$

We also see that all three vectors are perpendicular to vector $\mathbf{1}$

$$\begin{aligned} (\hat{\mathbf{y}} - \bar{\mathbf{y}})^T\mathbf{1} &= (\hat{\mathbf{y}} - \mathbf{y} + \mathbf{y} - \bar{\mathbf{y}})^T\mathbf{1} = (\hat{\mathbf{y}} - \mathbf{y})^T\mathbf{1} + (\mathbf{y} - \bar{\mathbf{y}})^T\mathbf{1} = 0 \\ (\mathbf{y} - \hat{\mathbf{y}})^T\mathbf{1} &= \sum_{n=1}^{N} (y_n - \hat{y}_n) = \sum_{n=1}^{N} r_n = 0 \\ (\mathbf{y} - \bar{\mathbf{y}})^T\mathbf{1} &= \sum_{n=1}^{N} (y_n - \bar{y}) = N(\bar{y} - \bar{y}) = 0 \end{aligned} \tag{5.23}$$

(The second equation here is attributed to Eq. (5.13).) The geometry of these three vectors and the simple relationship in Eq. (5.22) is illustrated in Fig. 5.2 for $d = 1$.

We can now measure the goodness of the regression model based on the three sums of squares just given. Given the total sum of squares TSS, if the model residual RSS is small, then ESS is large, indicating most of the variation in the data can be explained by the regression model, that is, it fits the data well and is therefore a good model. Otherwise if RSS is large, the model is not good. Specifically, the goodness of the model can be measured by the *coefficient of determination*, denoted by R^2 (*R-squared*), as the percentage of variation in the data that is explained by the model

$$R^2 = \frac{ESS}{TSS} = \frac{TSS - RSS}{TSS} = 1 - \frac{RSS}{TSS} \tag{5.24}$$

In general, $0 < R^2 < 1$. In the ideal case where the data are perfectly modeled by the linear regression function $f(\mathbf{x})$ with zero error $\mathbf{r} = \mathbf{y} - \hat{\mathbf{y}} = \mathbf{0}$, then $RSS = 0$, $TSS = ESS$, and $R^2 = 1$, that is, the data are completely explained by the model. In a nonideal case with $\mathbf{r} = \mathbf{y} - \hat{\mathbf{y}} \neq \mathbf{0}$, $R^2 < 1$.

How well a linear regression function models a given dataset depends on its nature such as how closely the variables in the dataset are related to each other and how noisy the dataset is, which can be measured in terms of the *correlation* between the variables. In general, the more correlated the variables, the better the dataset can be modeled, as shown in the following for the special case of $d = 1$. In this case how closely the two variables x and y in the 1-D dataset $\{(x_n, y_n) \ \ (n = 1, \cdots, N)\}$ are correlated can be measured by *correlation coefficient* defined as (Section B.1.3)

$$\rho = \frac{\sigma_{xy}^2}{\sqrt{\sigma_x^2 \, \sigma_y^2}} = \frac{\sigma_{xy}^2}{\sigma_x \, \sigma_y} \tag{5.25}$$

where σ_x^2 and σ_y^2 are the estimated variances of x and y

$$\sigma_x^2 = \frac{1}{N}\sum_{n=1}^{N}(x_n - \bar{x})^2 = \frac{1}{N}\sum_{n=1}^{N} x_n^2 - \bar{x}^2, \quad \sigma_y^2 = \frac{1}{N}\sum_{n=1}^{N}(y_n - \bar{y})^2 = \frac{1}{N}\sum_{n=1}^{N} y_n^2 - \bar{y}^2 \tag{5.26}$$

with

$$\bar{x} = \frac{1}{N}\sum_{n=1}^{N} x_n, \quad \bar{y} = \frac{1}{N}\sum_{n=1}^{N} y_n \tag{5.27}$$

and σ_{xy}^2 is their estimated covariance

$$\sigma_{xy}^2 = \frac{1}{N}\sum_{n=1}^{N}(x_n - \bar{x})(y_n - \bar{y}) = \frac{1}{N}\sum_{n=1}^{N} x_n y_n - \bar{x}\bar{y} \tag{5.28}$$

The correlation coefficient is a value between -1 and 1. It can be considered as the normalized covariance of the two variables x and y, indicating how much they are either positively or negatively correlated:

- $\rho = 1$: maximum positive correlation
- $0 < \rho < 1$: positive correlation
- $\rho = 0$: not correlated at all
- $-1 < \rho < 0$: negative correlation
- $\rho = -1$: maximum negative correlation.

The model parameters of this 1-D linear regression function $y = w_0 + w_1 x$ can be obtained based on Eq. (5.10), and expressed in terms of the variances and covariance of the two variables

$$
\begin{aligned}
\mathbf{w} &= \begin{bmatrix} w_0 \\ w_1 \end{bmatrix} = (\mathbf{X}\mathbf{X}^T)^{-1}\mathbf{X}\mathbf{y} \\
&= \left(\begin{bmatrix} 1 & \cdots & 1 \\ x_1 & \cdots & x_N \end{bmatrix} \begin{bmatrix} 1 & x_1 \\ \vdots & \vdots \\ 1 & x_N \end{bmatrix} \right)^{-1} \begin{bmatrix} 1 & \cdots & 1 \\ x_1 & \cdots & x_N \end{bmatrix} \begin{bmatrix} y_1 \\ \vdots \\ y_N \end{bmatrix} \\
&= \begin{bmatrix} N & \sum_{n=1}^N x_n \\ \sum_{n=1}^N x_n & \sum_{n=1}^N x_n^2 \end{bmatrix}^{-1} \begin{bmatrix} \sum_{n=1}^N y_n \\ \sum_{n=1}^N x_n y_n \end{bmatrix} = \begin{bmatrix} 1 & \bar{x} \\ \bar{x} & \sigma_x^2 + \bar{x}^2 \end{bmatrix}^{-1} \begin{bmatrix} \bar{y} \\ \sigma_{xy}^2 + \bar{x}\bar{y} \end{bmatrix} \\
&= \frac{1}{\sigma_x^2} \begin{bmatrix} \sigma_x^2 + \bar{x}^2 & -\bar{x} \\ -\bar{x} & 1 \end{bmatrix} \begin{bmatrix} \bar{y} \\ \sigma_{xy}^2 + \bar{x}\bar{y} \end{bmatrix} = \begin{bmatrix} \bar{y} - \sigma_{xy}^2/\sigma_x^2\, \bar{x} \\ \sigma_{xy}^2/\sigma_x^2 \end{bmatrix}
\end{aligned} \tag{5.29}
$$

Based on these model parameters we obtain the linear regression model

$$
\hat{y} = w_0 + w_1 x = \bar{y} - \frac{\sigma_{xy}^2}{\sigma_x^2}\bar{x} + \frac{\sigma_{xy}^2}{\sigma_x^2} x = \bar{y} + \frac{\sigma_{xy}^2}{\sigma_x^2}(x - \bar{x}) \tag{5.30}
$$

The coefficient of determination R^2 for the goodness of the regression model is closely related to the correlation coefficient ρ for correlation of the given data (independent of the model). Consider

$$
\begin{aligned}
\frac{1}{N} RSS &= \frac{1}{N}||y - \hat{y}||^2 = \frac{1}{N}\sum_{n=1}^N (y_n - w_0 - w_1 x_n)^2 \\
&= \frac{1}{N}\sum_{n=1}^N (y_n - (\bar{y} - \bar{x}w_1) - w_1 x_n)^2 = \frac{1}{N}\sum_{n=1}^N [(y_n - \bar{y}) - w_1(x_n - \bar{x})]^2 \\
&= \frac{1}{N}\sum_{n=1}^N [(y_n - \bar{y})^2 - 2w_1(y_n - \bar{y})(x_n - \bar{x}) + w_1^2(x_n - \bar{x})^2] \\
&= \sigma_y^2 - 2w_1\sigma_{xy}^2 + w_1^2\sigma_x^2 = \sigma_y^2 - 2\frac{\sigma_{xy}^2}{\sigma_x^2}\sigma_{xy}^2 + \frac{\sigma_{xy}^4}{\sigma_x^4}\sigma_x^2 = \sigma_y^2 - \frac{\sigma_{xy}^4}{\sigma_x^2} \\
&= \sigma_y^2\left(1 - \frac{\sigma_{xy}^4}{\sigma_x^2\sigma_y^2}\right) = \sigma_y^2(1 - \rho^2)
\end{aligned} \tag{5.31}
$$

We see that if the two variables x and y are more correlated with ρ closer to 1, then the residual error $RSS = ||\mathbf{r}||^2$ is smaller and R^2 is larger, indicating the data can be better fitted by the linear regression model.

Example 5.1

In 1929, Edwin Hubble published a paper "A relationship between distance and radial velocity among extra-galactic nebulae" (*Proceedings of the National Academy of Sciences of the United States of America (PANS)*, Vol. 15), where the linear regression method is used to model the relationship between the distance from Earth in megaparsec (Mpc = 3.26×10^6 light years) and the recessional velocity (in kilometer/second) of nebulae (galaxies), as shown in the left panel of Fig. 5.3. Based on this plot, Hubble estimated the slope of the linear regression, now called the *Hubble constant*

$$H_0 = \frac{\text{recessional velocity}}{\text{distance}} = \frac{v}{D}$$

to be 500 km/sec/Mpc. The estimated value of the Hubble constant has been continuously revised using different methods based on data with much wider range in distance collected by progressive technology. The more recent estimations of the Hubble constant are roughly in the range around 70 km/sec/Mpc, depending on the specific methods used (see `https://en.wikipedia.org/wiki/Hubble%27s_law`). One of such recent results is shown in the right panel of Fig. 5.3 ("Hubble's diagram and cosmic expansion," R. P. Kirshner, *PANS*, Vol. 101 (1), Dec. 2003, Copyright (2003) National Academy of Sciences, U.S.A.).

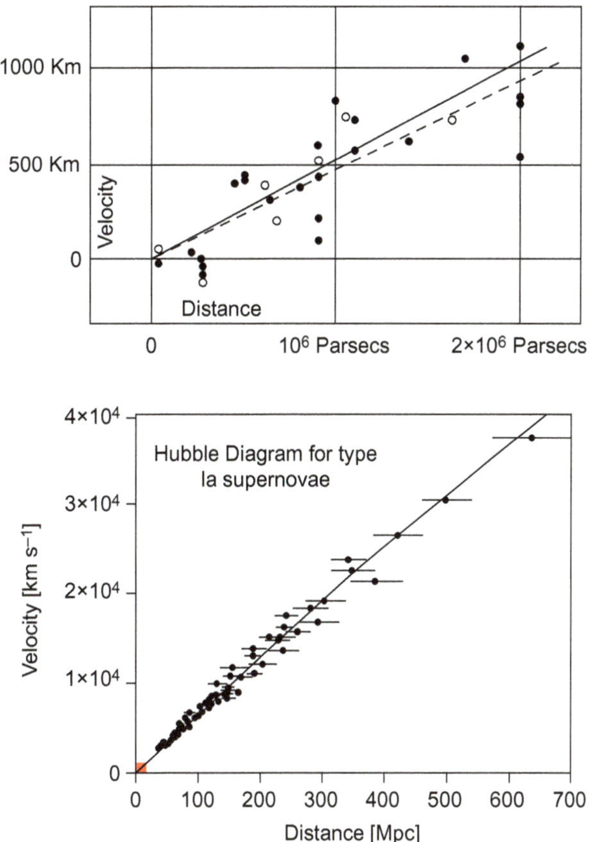

Fig. 5.3 Linear regression between recessional velocity and distance of nebulae, based on Hubble's original data (right) and modern data (left).

The simple linear relationship between the recessional velocity v and distance D has led to some very profound results such as the expansion of the universe, the *Big Bang* as the beginning of the expansion, and the age of the universe since the Big Bang, which can be estimated (assuming the expansion has been linear with $H_0 = 72$ since the Big Bang, and $c = 3 \times 10^5$ km/sec) to be

$$t_0 = \frac{1}{H_0} = \frac{D}{v} = \frac{1 \text{ Mpc}}{72 \text{ km/sec}} = \frac{3.26 \times 10^6 \times 3 \times 10^5}{72} = 13.6 \times 10^9 \text{years}$$

Example 5.2

Find a linear regression function $y = w_0 + w_1 x$ to fit the following $N = 6$ points:

Table 5.1

x	−3.4	−2.1	−0.8	0.3	1.7	2.5
y	−0.76	−1.04	1.75	1.82	3.17	3.15

Based on the data, we get

$$\bar{y} = 1.348, \quad \sigma_x^2 = 4.22, \quad \sigma_y^2 = 2.85, \quad \sigma_{xy}^2 = 3.27, \quad \rho = \frac{\sigma_{xy}^2}{\sigma_x \sigma_y} = 0.94$$

The augmented data array is

$$\mathbf{X} = \begin{bmatrix} 1 & 1 & 1 & 1 & 1 & 1 \\ -3.4 & -2.1 & -0.8 & 0.3 & 1.7 & 2.5 \end{bmatrix}$$

Solving the over-constrained equation system

$$\mathbf{y} = \begin{bmatrix} -0.76 \\ -1.04 \\ 1.75 \\ 1.82 \\ 3.17 \\ 3.15 \end{bmatrix} = \mathbf{X}^T \mathbf{w} = \begin{bmatrix} 1 & -3.4 \\ 1 & -2.1 \\ 1 & -0.8 \\ 1 & 0.3 \\ 1 & 1.7 \\ 1 & 2.5 \end{bmatrix} \begin{bmatrix} w_0 \\ w_1 \end{bmatrix}$$

we get

$$\mathbf{w} = \mathbf{X}^- \mathbf{y} = (\mathbf{X}\mathbf{X}^T)^{-1} \mathbf{X}\,\mathbf{y}$$

$$= \begin{bmatrix} 0.13 & 0.15 & 0.16 & 0.17 & 0.19 & 0.20 \\ -0.12 & -0.07 & -0.02 & 0.02 & 0.08 & 0.11 \end{bmatrix} \begin{bmatrix} -0.76 \\ -1.04 \\ 1.75 \\ 1.82 \\ 3.17 \\ 3.15 \end{bmatrix} = \begin{bmatrix} 1.58 \\ 0.78 \end{bmatrix}$$

and the linear regression model function $y = w_0 + w_1 x = 1.58 + 0.78\,x$, a straight line with intercept $w_0 = 1.58$, slope $w_1 = 0.78$, and normal direction $[0.78, -1]^T$, as shown in Fig. 5.4. The LS error is $||\mathbf{r}|| = ||\mathbf{y} - \hat{\mathbf{y}}|| = 1.38$. We further obtain

$$\bar{\mathbf{y}} = [1.348,\ 1.348,\ 1.348,\ 1.348,\ 1.348,\ 1.348]^T$$
$$\hat{\mathbf{y}} = \mathbf{X}^T \mathbf{w} = [-1.054,\ -0.047,\ 0.961,\ 1.813,\ 2.898,\ 3.518]^T$$

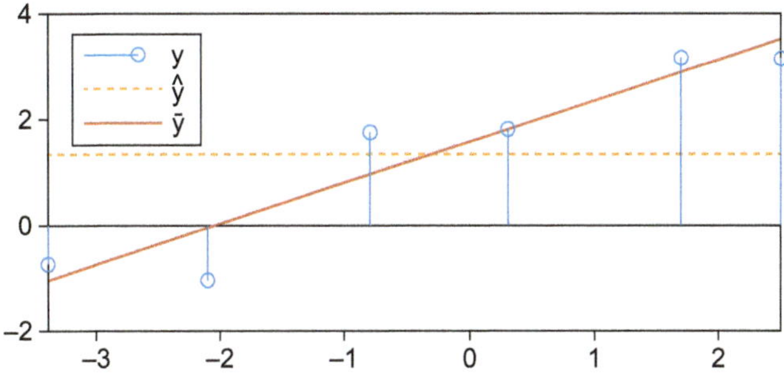

Fig. 5.4 Linear regression example.

and

$$ESS = 15.21, \quad RSS = 1.91, \quad TSS = 17.12, \quad R^2 = 0.89$$

We can also check that

$$\frac{1}{N} RSS = \sigma_y^2 (1 - \rho^2) = 0.318$$

Example 5.3

The two datasets given in the following table are fitted by two linear regression functions as shown in the left and right panels in Fig. 5.5.

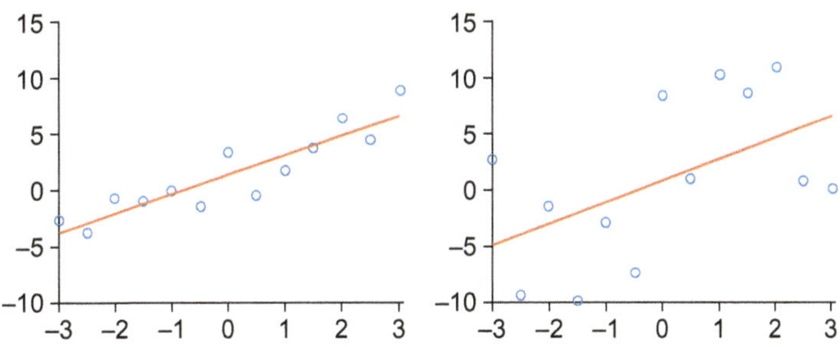

Fig. 5.5 Linear regressions with different R^2 and ρ.

Table 5.2

x	−3.0	−2.5	−2.0	−1.5	−1.0	−0.5	0.0	
y_1	−2.67	−3.76	−0.64	−0.93	−0.01	−1.42	3.38	......
y_2	2.73	−9.38	−1.44	−9.86	−2.82	−7.29	8.31	

Table 5.3

	0.5	1.0	1.5	2.0	2.5	3.0
······	−0.47	1.77	3.67	6.40	4.49	8.82
	0.98	10.18	8.58	10.86	0.77	0.20

Table 5.4

	w_0	w_1	ESS	RSS	TSS	R^2	ρ
*dataset*1	1.43	1.72	134.71	30.18	164.89	0.817	0.904
*dataset*2	0.91	1.92	166.98	443.92	610.89	0.273	0.523

Example 5.4

Figure 5.6 shows a set of $N = 30$ data points $\{(x_i, y_i),\ i = 1, \cdots, 30\}$ fitted by a 1-D linear regression function $y = w_0 + w_1 x$, a straight line, with $w_0 = 1.81$, $w_1 = 3.22$. The correlation coefficient is $\rho = 0.98$, and $ESS = 734.78$, $RSS = 32.98$, $TSS = 767.77$, $R^2 = 0.96$. Figure 5.7 shows three more examples with different correlation coefficient values $\rho = 0.982,\ 0.110,\ 0.989$, indicating the two variables are highly positively correlated, not correlated, and highly negatively correlated, respectively.

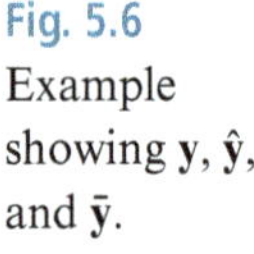

Fig. 5.6 Example showing $\mathbf{y}$, $\hat{\mathbf{y}}$, and $\bar{\mathbf{y}}$.

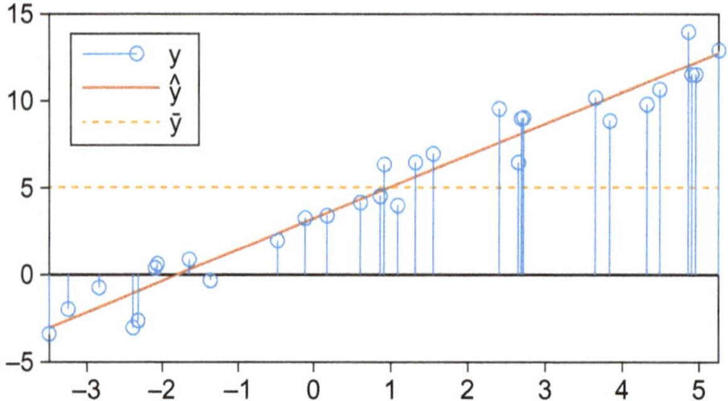

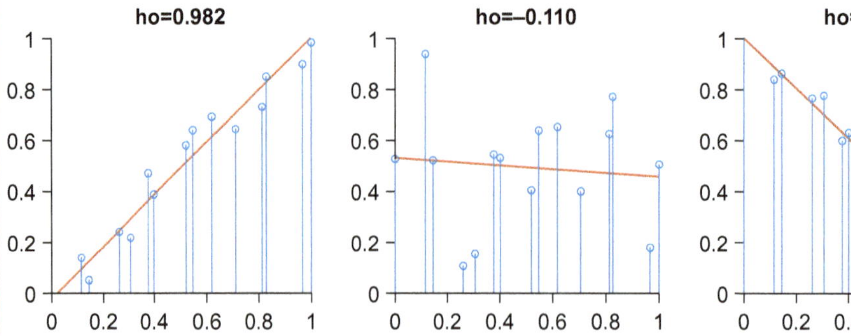

Fig. 5.7 Examples of different correlation coefficients.

Example 5.5

Figure 5.6 shows a set of $N = 90$ data points $\{(\mathbf{x}_i, y_i),\ i = 1, \cdots, 90\}$ fitted by a 2-D linear regression function $y = w_0 + w_1 x_1 + w_2 x_2$, with $w_0 = 2.82,\ w_1 = 1.85,\ w_2 = 4.31$. This is a plane as shown in Fig. 5.8 from two perspective view angles. Here $ESS = 677.00, RSS = 124.48,\ TSS = 802.47$ and $R^2 = 0.84$.

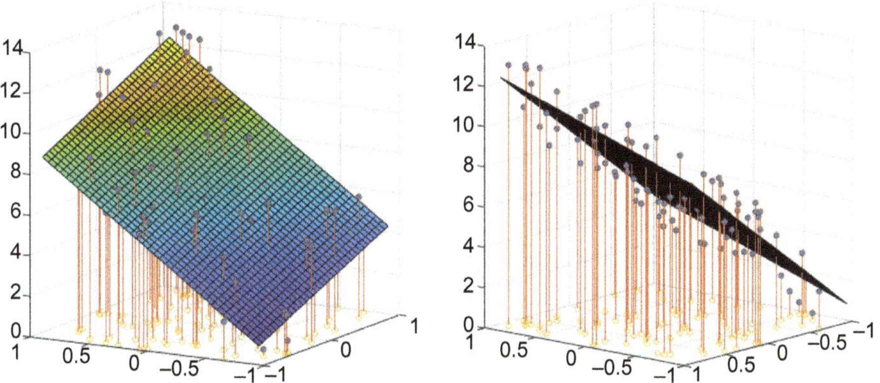

Fig. 5.8 Linear regression of a 2-D dataset.

Finally we note that the linear regression method can be used for binary classification, if the regression function $y = f(\mathbf{x})$, as a hyperplane in the $d + 1$ dimensional space spanned by $\{x_1, \cdots, x_d, y\}$, is thresholded by a constant C, for example $C = 0$. The resulting equation $f(\mathbf{x}) = C$ defines a hyperplane in the d-dimensional space spanned by $\{x_1, \cdots, x_d\}$, which partitions the space into two parts in such a way that all points on one side of the hyperplane satisfy $f(\mathbf{x}) > C$, while all points on the other side satisfy $f(\mathbf{x}) < C$. We see that the regression function is a binary classifier that separates all points $\mathbf{x}$ in the d-dimensional space into two classes C_+ and C_-, depending on whether $f(\mathbf{x})$ is greater or smaller than C. Now each y_n corresponding to $\mathbf{x}_n$ in the given dataset $\mathcal{D} = \{\mathbf{X}, \mathbf{y}\}$ can be treated as a label indicating $\mathbf{x}_n$ belong to class C_+ if $y_n > C$, or C_- if $y_n < C$, and the regression problem becomes a binary classification problem. This binary classifier is illustrated in Fig. 5.1 for $C = 0$.

5.2 Ridge Regression

The linear regression method considered in the previous section may be prone to noise and suffer from overfitting, if the problem is ill-conditioned, when the data samples in the dataset are barely independent and therefore ill-behaved. See more detailed discussion in Section A.8.5. In such a case, matrix $\mathbf{X}\mathbf{X}^T$ still has a full rank but it is close to singularity, that is some of its eigenvalues may be very close to zero, and the corresponding eigenvalues of its inverse $(\mathbf{X}\mathbf{X}^T)^{-1}$ may take very large values, causing some components of the model parameter $\mathbf{w} = (\mathbf{X}\mathbf{X}^T)^{-1}\mathbf{X}\mathbf{y}$ to also take very large values. Now any minor change in the dataset (in either $\mathbf{x}$ or y) owing to noise may be amplified by $\mathbf{w}$ to cause a major change in the model output $\hat{y} = \mathbf{w}^T\mathbf{x}$. In other words, the regression model becomes highly prone to noise, that is it may overfit a given dataset with a large variance error thereby causing poor generalizability.

This issue of ill-conditioning can be illustrated in the 2-D case, where a linear regression function $f(\mathbf{x}) = \mathbf{w}^T\mathbf{x}$, a 3-D plane, is to be determined to fit a set of given data points $\{(\mathbf{x}_n, y_n),\ n = 1, \cdots, N\}$. If two variables x_1 and x_2 in the dataset are highly correlated, that is they are narrowly distributed in the 2-D space around a straight line (similar to the case in either the first or third panel in Fig. 5.7), then the value of $|\rho|$ is large and close to 1, and some eigenvalues of $\mathbf{X}\mathbf{X}^T$ take small values close to 0, and the 2-D regression problem becomes ill-conditioned as it degenerates into a 1-D problem. If y is linearly related to x_1 and x_2, then all data points $\{(\mathbf{x}_n, y_n),\ n = 1, \cdots, N\}$ are closely distributed around a straight line in the 3-D space which can be fitted by any 3-D plane through the line with drastically different orientations (rotated differently around the 3-D line as an axis). Consequently it becomes difficult to determine a unique 3-D plane to fit the data as it may change drastically due to minor variations in y_n or $\mathbf{x}_n$.

Example 5.6

Figure 5.9 compares two different datasets: one well behaved, the other ill-behaved. In the first case shown in the left panel, the two independent variables x_1 and x_2 are widely distributed in the 2-D plane, that is they are uncorrelated with a small $\rho = -0.322$, and the matrix $\mathbf{X}\mathbf{X}^T$ has two large eigenvalues $\lambda_1 = 12.34$ and $\lambda_2 = 25.44$. This is a well-conditioned problem that results in a reliable regression plane as shown. In the second case shown in the right panel, x_1 and x_2 are narrowly distributed around a straight line in the 2-D plane, that is they are highly correlated with a large $\rho = 0.99$, and one of the two eigenvalues $\lambda_1 = 0.09$ and $\lambda_2 = 43.24$ of matrix $\mathbf{X}\mathbf{X}^T$ is close to zero. This is a ill-conditioned problem that results in an unstable regression plane that can be drastically affected (rotating around a straight line in the 3-D space) owing to noise in the data.

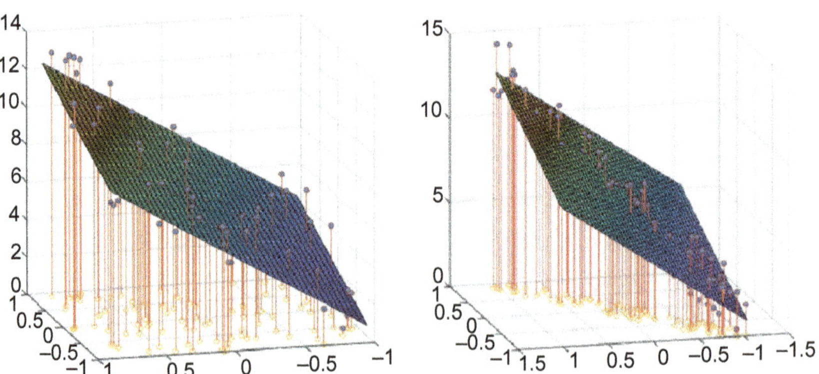

Fig. 5.9 2-D linear regression, well-conditioned (left) versus ill-conditioned (right).

To address the ill-conditioned problem in linear regression, the method of *ridge regression* can be used to regularize the problem by imposing some additional constraints to prevent $\mathbf{w}$ from taking large values while minimizing the squared error SSE in Eq. (5.8). To do so, we construct an objective function $J(\mathbf{w})$ to be minimized that contains not only the SSE term $\varepsilon = ||\mathbf{r}||^2/2$, but also a regularization or penalty term $\lambda||\mathbf{w}||^2/2$ to discourage large values in $\mathbf{w}$

$$J(\mathbf{w}) = \frac{1}{2}||\mathbf{y} - \mathbf{X}^T\mathbf{w}||^2 + \frac{\lambda}{2}||\mathbf{w}||^2 \tag{5.32}$$

so that the optimal $\mathbf{w}$ that minimizes this objective function has a small norm $||\mathbf{w}||$ as well as a small SSE. This method is called *weight decay* and the hyperparameter λ is called the *weight decay parameter*. Similar to Eq. (5.9), we set its gradient to zero

$$\frac{d}{d\mathbf{w}}J(\mathbf{w}) = \mathbf{X}\mathbf{X}^T\mathbf{w} - \mathbf{X}\mathbf{y} + \lambda\mathbf{w} = (\mathbf{X}\mathbf{X}^T + \lambda\mathbf{I})\mathbf{w} - \mathbf{X}\mathbf{y} = \mathbf{0} \tag{5.33}$$

and solve the resulting equation to get

$$\mathbf{w}^* = (\mathbf{X}\mathbf{X}^T + \lambda\mathbf{I})^{-1}\mathbf{X}\mathbf{y} \tag{5.34}$$

By adjusting the hyperparameter λ, we can make a proper tradeoff between stability of the result and its accuracy in terms of the goodness of the model measured by R^2).

- If λ is small, the result is more accurate but less stable as it is more prone to noise, that is it tends to overfit the data with a large bias error but small variance.
- If λ is large, the result is less affected by noise and therefore more stable, but it is less accurate, that is it tends to underfit the data with a small bias error but a large variance error.

In general, to prevent overfitting, regularization by means of including in the objective function a constraint term to impose some additional constraints over the model parameters is commonly used in many supervised learning algorithms. This is done for both regression and classification, including the classification methods of support vector machine in Section 14 and the backpropagation neural network in Section 18.2.

Example 5.7

The method of ridge regression is applied to the same ill-conditioned dataset in the right panel in Fig. 5.9, where the two independent variables x_1 and x_2 are highly correlated with $\rho = 0.99$. As the result, matrix $\mathbf{X}\mathbf{X}^T$ is close to singularity with one of its two eigenvalues $\lambda_1 = 43.24$ and $\lambda_2 = 0.09$ close to zero, that is the matrix is close to singularity and the problem is ill-conditioned.

When the regular linear regression ($\lambda = 0$) is applied, it produces the results shown in the top two panels in Fig. 5.10 for the dataset before and after the dependent variable y is slightly perturbed. We see that the orientations of the two resulting regression planes differ drastically, due to the perturbance in y. But when the ridge regression method (with $\lambda = 1$) is applied, it produces more stable results as shown in the bottom two panels of the figure. We see that the same perturbation in the dataset caused only a minor change in the orientation of the regression plane.

The cost of such more stable results is their accuracy, as the model goodness values R^2 for the dataset before and after the perturbance are reduced from 0.894 and 0.915 for the regular regression to 0.857 and 0.876, respectively, for the ridge regression.

This example shows that the stability of the solution of an ill-conditioned problem can be improved with the cost of lower accuracy (smaller R^2), and tradeoff between stability (low bias error) and accuracy (low variance error) can be made by adjusting parameter λ.

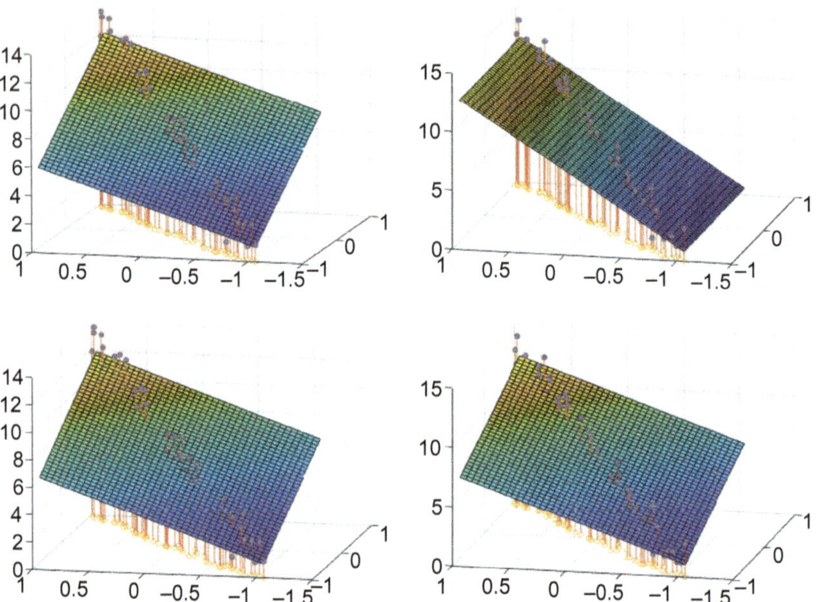

Fig. 5.10 Unstable results generated by regression applied to ill-conditioned dataset (top) and stable results generated by ridge regression (bottom).

5.3 Regression Based on Basis Functions

The method of linear regression considered previously can be generalized to model nonlinear relationships between the dependent variable y and the independent variables in $\mathbf{x} = [x_1, \cdots, x_d]^T$ by a regression function as a linear combination of K nonlinear basis functions of $\mathbf{x}$

$$\hat{y} = f(\mathbf{x}) = \sum_{k=1}^{K} w_k \, \varphi_k(\mathbf{x}, \boldsymbol{\theta}) = \boldsymbol{\varphi}(\mathbf{x})^T \mathbf{w} \tag{5.35}$$

where $\varphi_k(\mathbf{x}, \boldsymbol{\theta})$, $(k = 1, \cdots, K)$ is one of the basis functions parameterized by $\boldsymbol{\theta}$ that span the function space in which y resides, and $\boldsymbol{\varphi}(\mathbf{x}) = [\varphi_1(\mathbf{x}), \cdots, \varphi_K(\mathbf{x})]^T$ is a

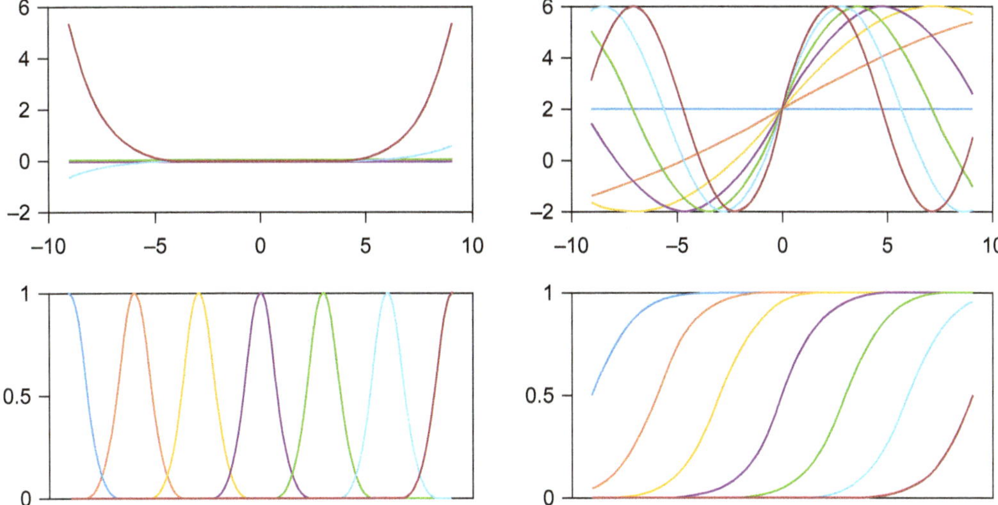

Fig. 5.11 Basis functions for regression.

K–D vector. Some of the commonly used basis functions are listed in the following and plotted in Fig. 5.11 for $K = 7$.

- Power function (top-left)

$$\varphi_k(x) = x^{k-1} \tag{5.36}$$

  This is the $(k-1)$th power of x, and the regression function $f(x)$ is $(K-1)$th order polynomial of x.
- Sinusoidal function (top-right)

$$\varphi_k(x,\omega) = \sin((k-1)\omega x), \quad \text{or} \quad \varphi_k(x,\omega) = \cos((k-1)\omega x) \tag{5.37}$$

  This is the $(k-1)$th harmonic of a sinusoidal function of x, and the regression function $f(x)$ is the first K terms of the sine or cosine series expansion of the given function.
- Gaussian function (bottom-left)

$$\varphi_k(\mathbf{x},\mathbf{c}_k,a) = \exp(a||\mathbf{x}-\mathbf{c}_k||^2) \tag{5.38}$$

  This is a member of a family of functions called *radial basis functions (RBFs)*, each centered at $\mathbf{c}_k$.
- Sigmoid function (bottom-right)

$$\varphi_k(\mathbf{x},\mathbf{c}_k,a) = \frac{1}{1+e^{-a||\mathbf{x}-\mathbf{c}_k||}} \tag{5.39}$$

In practice, choosing a proper type of basis functions may be a trial-and-error process depending on the specific dataset.

The original d-dimensional data space spanned by $x_1,\cdots,x_d$ is now converted to a K-dimensional space spanned by the K basis functions $\varphi_1(\mathbf{x}),\cdots,\varphi_K(\mathbf{x})$.

Applying such a regression model to each data point $\mathbf{x}_n$, we obtain

$$\hat{y}_n = \sum_{k=1}^{K} w_k \varphi_k(\mathbf{x}_n) = \boldsymbol{\varphi}(\mathbf{x}_n)^T \mathbf{w} \qquad (n = 1,\cdots,N) \tag{5.40}$$

These N equations can be written in matrix form

$$\hat{\mathbf{y}} = \begin{bmatrix} y_1 \\ \vdots \\ y_N \end{bmatrix} = \begin{bmatrix} \varphi^T(\mathbf{x}_1) \\ \vdots \\ \varphi_K^T(\mathbf{x}_N) \end{bmatrix} \mathbf{w} = \mathbf{\Phi}^T \mathbf{w} \tag{5.41}$$

where $\mathbf{\Phi}$ is a $K \times N$ matrix

$$\mathbf{\Phi} = [\varphi(\mathbf{x}_1),\cdots,\varphi(\mathbf{x}_N)] = \begin{bmatrix} \varphi_1(\mathbf{x}_1) & \cdots & \varphi_1(\mathbf{x}_N) \\ \vdots & \ddots & \vdots \\ \varphi_K(\mathbf{x}_1) & \cdots & \varphi_K(\mathbf{x}_N) \end{bmatrix} \tag{5.42}$$

Typically there are many more sample data points than the basis functions, that is $N \gg K$.

If, specially, $\varphi(u) = u$ is an identity function, then Eq. (5.35) becomes the same as Eq. (5.2) for linear regression. Also, as the nonlinear model $\hat{\mathbf{y}} = \mathbf{\Phi}^T\mathbf{w}$ in Eq. (5.41) is in the same form as the linear model $\hat{\mathbf{y}} = \mathbf{X}^T\mathbf{w}$ in Eq. (5.2), it can be solved the same way as in Eq. (5.10), by simply replacing $\mathbf{X}$ by $\mathbf{\Phi}$

$$\mathbf{w}^* = (\Phi\Phi^T)^{-1}\mathbf{\Phi}\mathbf{y} = (\mathbf{\Phi}^T)^{-}\mathbf{y} \tag{5.43}$$

where $(\mathbf{\Phi}^T)^- = (\Phi\Phi^T)^{-1}\mathbf{\Phi}$ is the $K \times N$ pseudo-inverse of the $N \times K$ matrix $\mathbf{\Phi}^T$. Now we get the output of the regression model

$$\hat{\mathbf{y}} = \mathbf{\Phi}^T\mathbf{w} = \mathbf{\Phi}^T\mathbf{\Phi}^-\mathbf{y} = \mathbf{\Phi}^T(\Phi\Phi^T)^{-1}\mathbf{\Phi}\mathbf{y} \tag{5.44}$$

The following MATLAB code segment shows the essential part of the algorithm, where $x(n)$ is the nth data sample $\mathbf{x}_n$ out of the total of N, $y(x)$ is the vector $\mathbf{y} = [y_1,\cdots,y_N]^T$ with $y_n = f(\mathbf{x}_n)$, and $c(k)$ contains the parameters for the kth basis function $\phi_k(\mathbf{x}) = \phi(\mathbf{x}, c(k))$ represented by `phi` in the code.

```
for n=1:N
    for k=1:K
        Phi(k,n)=phi(x(n),c(k));
    end
end
w=pinv(Phi')*y(x);        % weight vector by LS method
yhat=Phi'*w;              % reconstructed function
```

Example 5.8

The plots in Figs 5.12 through 5.15 show the approximations of a given function based on a set of N samples using linear regression based on the four different types of basis functions. We see that as the number of basis functions K increases, the approximation becomes more accurate.

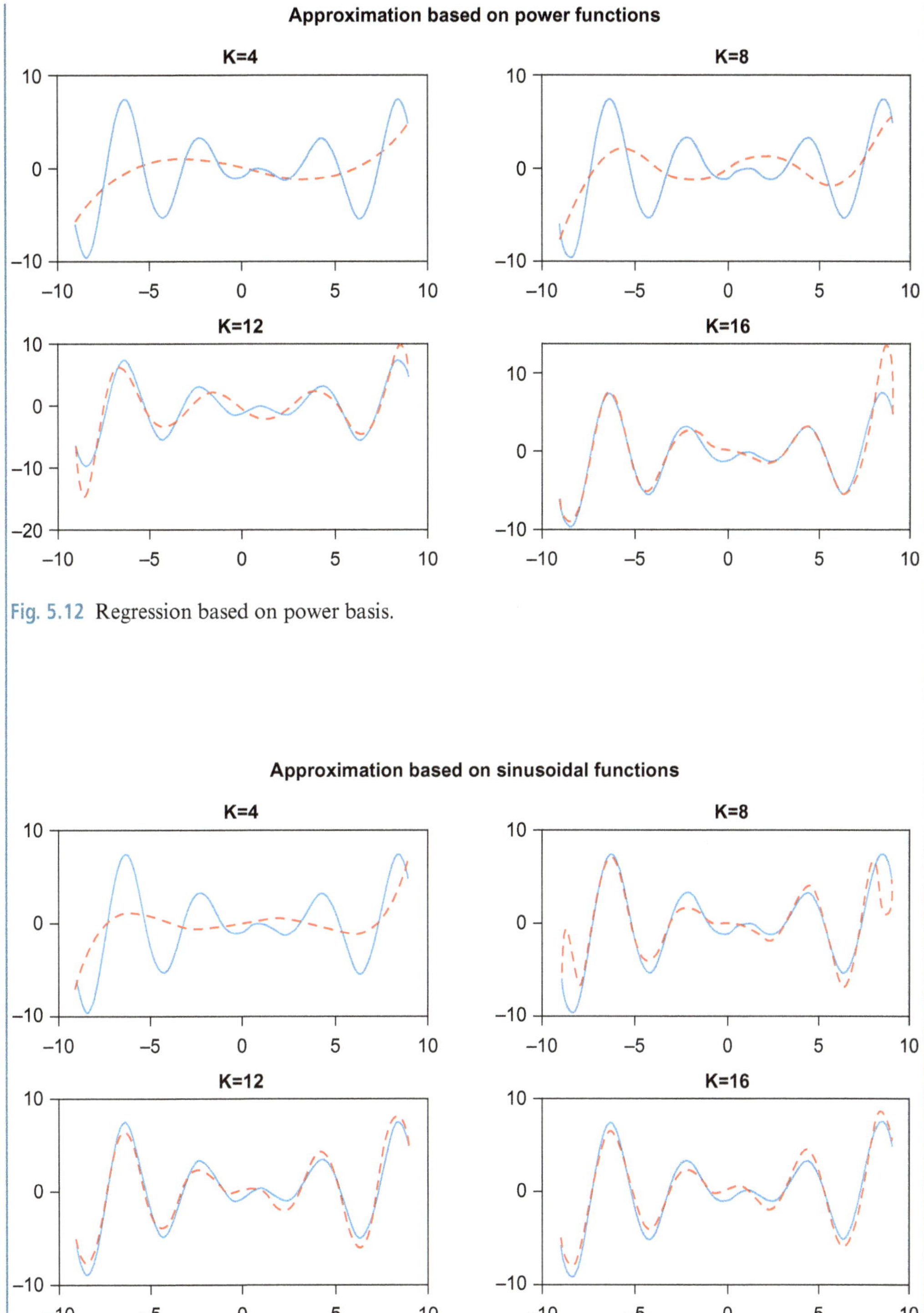

Fig. 5.12 Regression based on power basis.

Fig. 5.13 Regression based on sinusoidal basis.

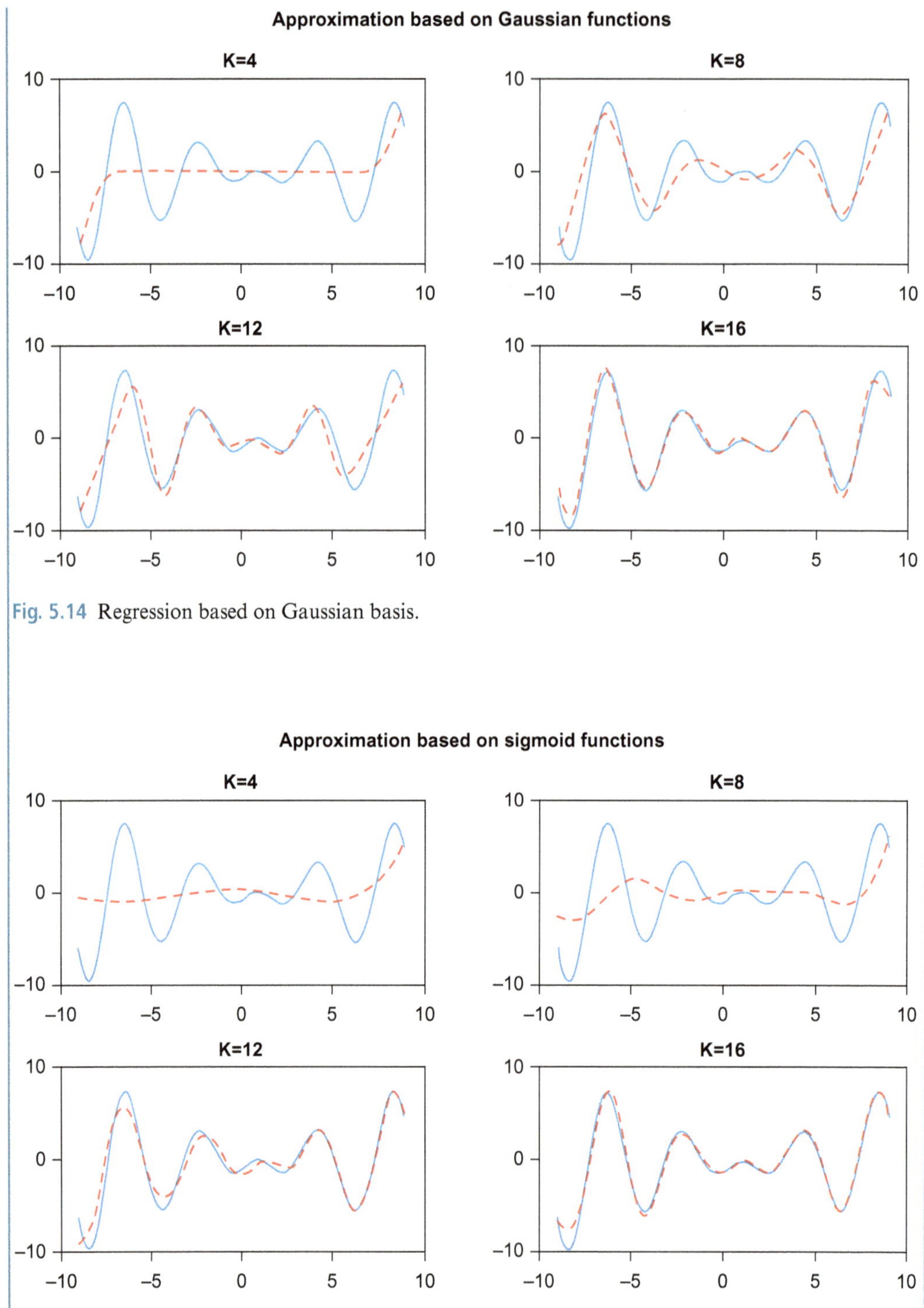

Fig. 5.14 Regression based on Gaussian basis.

Fig. 5.15 Regression based on sigmoid basis.

Example 5.9

The linear regression method based on basis functions is shown in Fig. 5.16, where a 2-D function $f(\mathbf{x}) = f(x_1, x_2)$ (left) is to be approximated based on the function values in $\{y_1, \cdots, y_N\}$ at a set N sample points $\{\mathbf{x}_1, \cdots, \mathbf{x}_N\}$ taken from a grid in the 2-D space (right). The regression is based on K 2-D Gaussian basis functions centered at different locations in the 2-D space shown in Fig. 5.17 for $K = 9$.

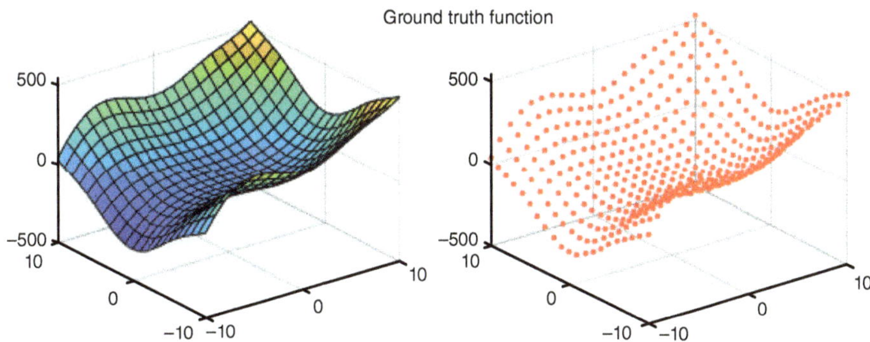

Fig. 5.16 A function and its samples in 2-D.

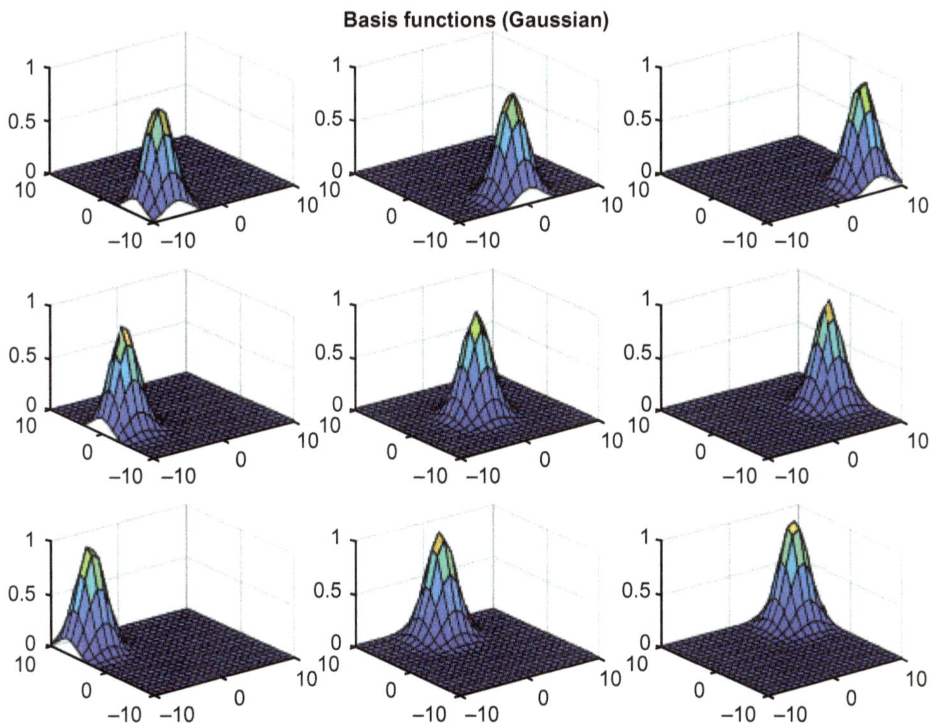

Fig. 5.17 Gaussian basis in 2-D.

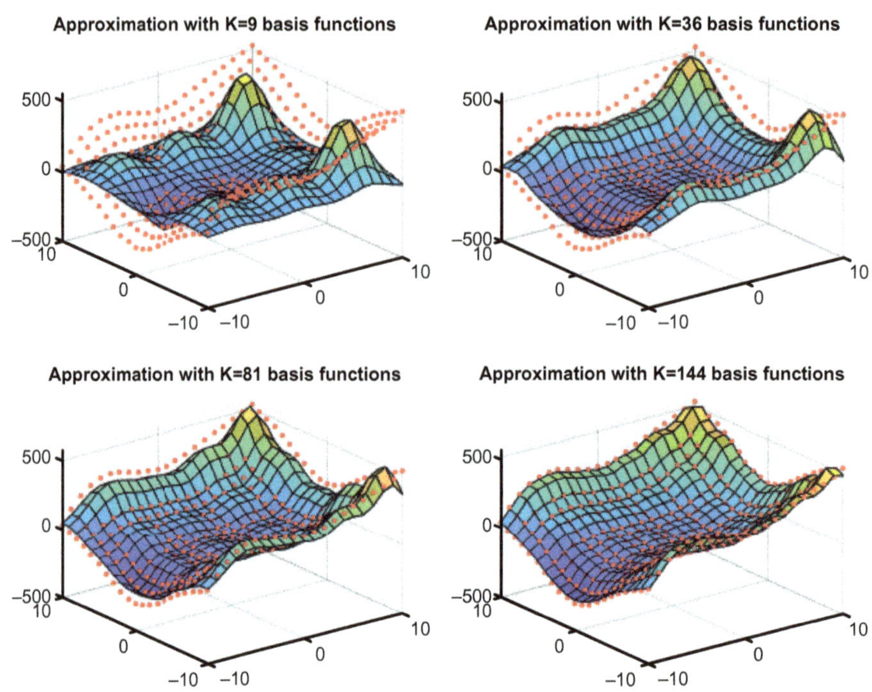

Fig. 5.18 Regression results based on different number of basis functions.

The regression results based on different numbers of basis functions are shown in Fig. 5.18. Again, we see that the approximation becomes more accurate as the value of K increases.

In addition to the different types of basis functions considered here, we further consider the following basis functions

$$\varphi_k(\mathbf{x},\mathbf{w}_k,b_k) = g\left(\mathbf{w}_k^T\mathbf{x}+b_k\right) = g\left(\sum_{i=1}^{d} w_{ki}x_i + b_k\right) = g\left(\sum_{i=0}^{d} w_{ki}x_i\right) \tag{5.45}$$

where $g(x)$, called *activation function*, can take some different forms such as a sigmoid or rectified linear function (rectified linear unit or ReLU)

$$g(x) = \begin{cases} 1/(1+e^{-x}) & \text{sigmoid function} \\ \max(0,x) = \begin{cases} 0 & x<0 \\ x & x \geq 0 \end{cases} & \text{ReLU} \end{cases} \tag{5.46}$$

Now the regression function in Eq. (5.35) can be written for multiple outputs

$$\hat{y}_j = \sum_{k=1}^{K} w_{jk}^o\,\varphi_k(\mathbf{x},\boldsymbol{\theta}_k) = \sum_{k=1}^{K} w_{jk}^o\, g\left(\sum_{i=0}^{d} w_{ki}^h x_i\right) \quad (j=1,\cdots,m) \tag{5.47}$$

which can be interpreted as the output of a three-layer *neural network* shown in Fig. 5.19 containing an input layer of d nodes, a middle or hidden layer of K nodes with weights w_{ki}^h, $(i=0,\cdots,d,\ k=1,\cdots,K)$, and an output layer of m nodes with

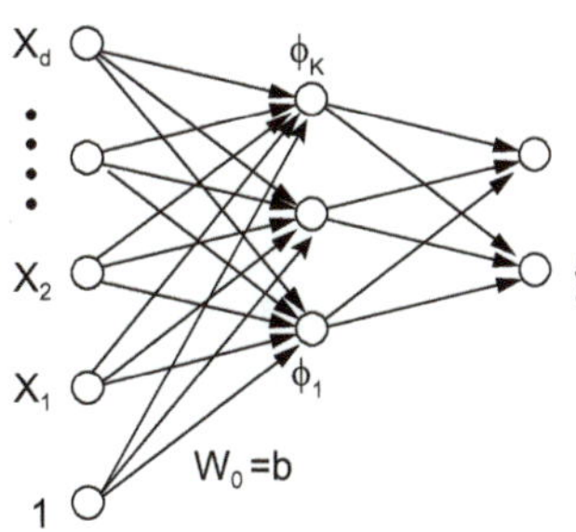

Fig. 5.19 A network model for regression.

weights w^o_{jk}, $(k = 1, \cdots, K,\ j = 1, \cdots, m)$. These nodes are also called *neurons*, as they mimic the biological neurons in the brain.

As shown in Fig. 5.19, the nodes of the input layer take the d components of a data point $\mathbf{x}$ from the dataset one at a time, while each of the K nodes of the hidden layer calculates their linear combination $\mathbf{w}_k^T\mathbf{x}$, and, together with a bias term b_k, generates the basis function $\varphi_k(\mathbf{x}) = g\left(\mathbf{w}_k^T\mathbf{x} + b_k\right)$ as its output. Finally, each node of the output layer takes these K basis function from the K hidden layer and generates as its output the regression function $\hat{y} = f(\mathbf{x}) = \sum_{k=1}^K w_k\varphi_k(\mathbf{x})$ as the model of the dataset.

This network depends on two sets of weights, those represented by w^h_{ki} for the hidden nodes, and those represented by w^o_{jk} (found in Eq. (5.43)) for the output nodes. When there are multiple nodes in the output layer, they can each approximate a different function. This three-layer network is of particular interest in the general method of artificial neural networks to be considered in Chapter 17.

Example 5.10

The method of regression based on basis functions is used to approximate the same 2-D function $f(\mathbf{x}) = f(x_1, x_2)$ in the previous example, as a linear combination of K basis functions $\varphi_k(\mathbf{x}) = \mathbf{w}_k^T\mathbf{x} + b_k$ with each parameterized by $\mathbf{w}_k$ and b_k (randomly generated).

The following MATLAB code segment shows the essential part of the program for this example, where f is a arbitrary function to be approximated.

```
a=@(x) x.*(x>0);          % ReLU activation function, or
a=@(x) 1/(1+exp(-x)); % Sigmoid activation function
g=@(x,w,b)a(w'*x+b);   % the basis function

[X,Y]=meshgrid(xmin:1:xmax, ymin:1:ymax); % define 2-D points
nx=size(X,2);             % number of samples in first dimension
ny=size(Y,1);             % number of samples in second dimension
N=nx*ny;                  % total number of data samples
Phi=zeros(N,K);          % K basis functions evaluated at N sample points
W=1-2*rand(2,K);          % random initialization of weights
b=1-2*rand(1,K);          % and biases for K basis functions
```

```
n=0;
for i=1:nx
    x(1)=X(1,i);       % first component
    for j=1:ny
        x(2)=Y(j,1);  % second component
        n=n+1;        % of the nth sample
        Y(i,j)=func(x); % function evaluated at the sample
        for k=1:K      % basis functions evaluated at the sample
            Phi(k,n)=g(x,W(:,k),b(k));
        end
    end
end
y=reshape(Y,N,1);       % convert function values to vector
w=pinv(Phi')*y;         % find weights by LS method
yhat=Phi'*w;            % reconstructed function
Yhat=reshape(yhat,m,n); % convert vector y to 2-D to compare w/ Y
```

Figures 5.20 through 5.23 show the basis functions based on both the sigmoid and ReLU activation functions, and the corresponding regression results. We note that as more basis functions are used, the more accurately the function is approximated.

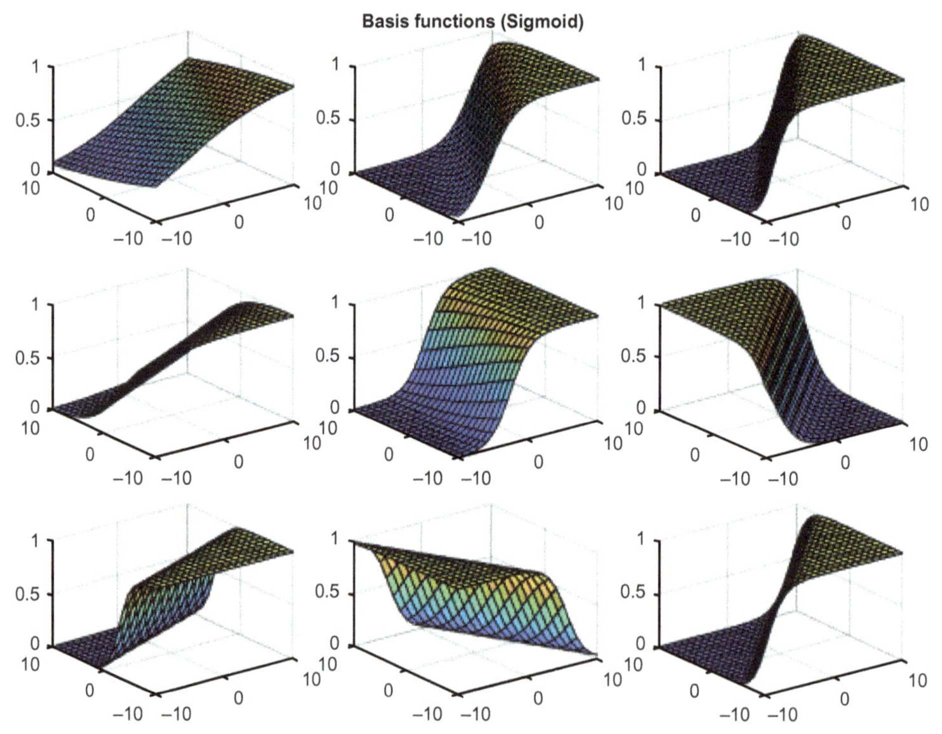

Fig. 5.20 Sigmoid basis for 2-D regression.

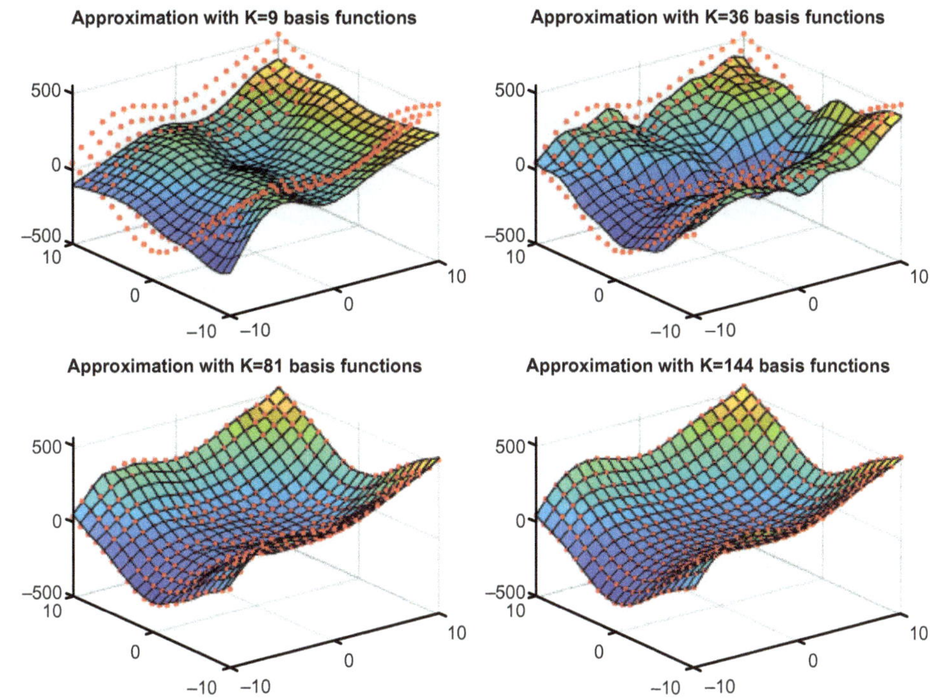

Fig. 5.21 Regression based on sigmoid basis.

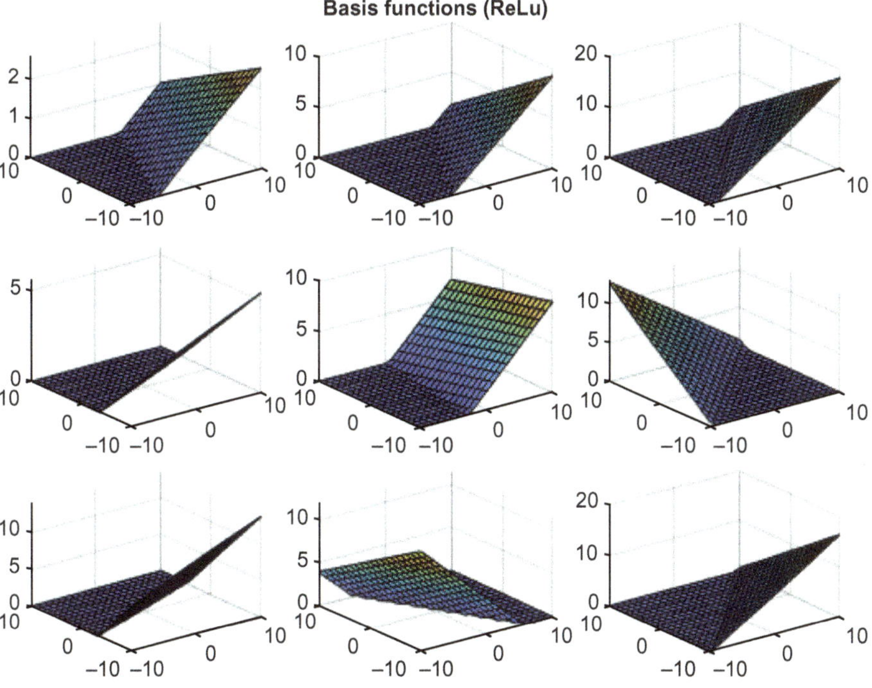

Fig. 5.22 ReLU basis for 2-D regression.

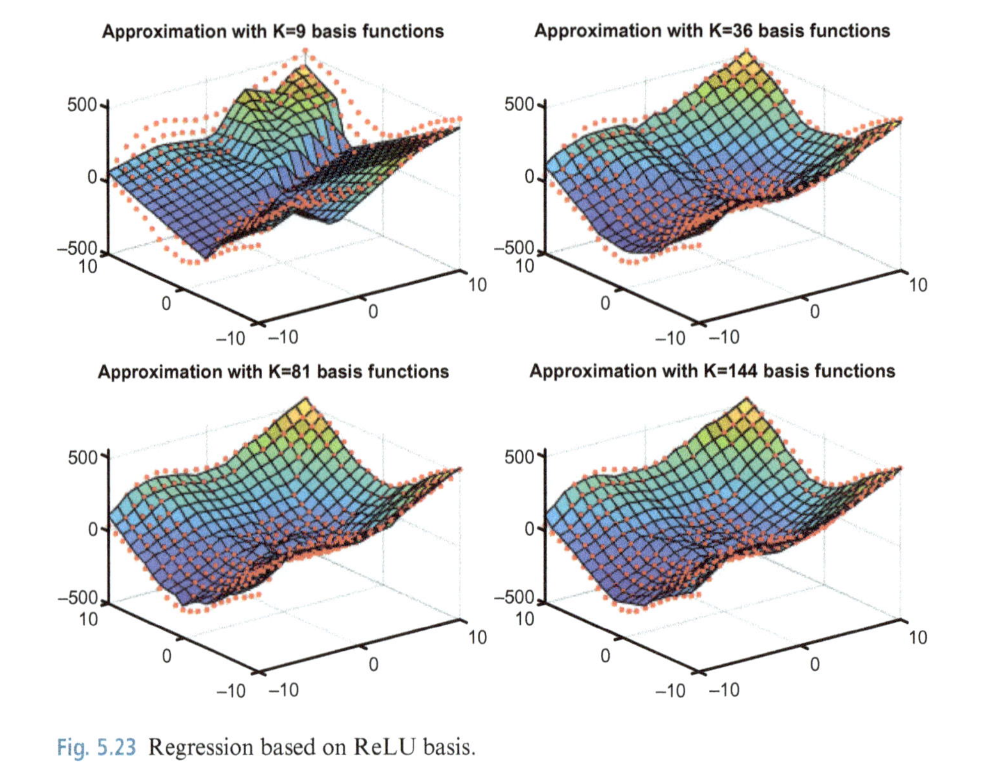

Fig. 5.23 Regression based on ReLU basis.

5.4 Bayesian Regression

The regression algorithms previously considered are frequentist methods that treat all data and parameters in a regression model as deterministic variables. However, regression can also be viewed in light of Bayesian inference, where all data and parameters in the regression model are treated as random variables. In this section, we consider two such methods, the maximum likelihood estimation (MLE) and the maximum a posterior (MAP) estimation. While the MLE method turns out to be equivalent to the least squares method, the MAP method will generate results including not only the estimated model parameter $\mathbf{w}$ but also the variance of the estimate as a confidence measurement. These methods can also be used for nonlinear regression based on basis functions.

Here we assume the model parameter $\mathbf{w}$ in the linear regression function $\hat{y} = f(\mathbf{x}, \mathbf{w}) = \mathbf{x}^T\mathbf{w}$ is a random vector. Consequently, the regression function $f(\mathbf{x}, \mathbf{w})$, as a function of $\mathbf{w}$, is also random, so is the residual $\mathbf{r} = \mathbf{y} - \hat{\mathbf{y}} = \mathbf{y} - \mathbf{X}^T\mathbf{w}$ based on the training set $\mathcal{D} = \{\mathbf{X}, \mathbf{y}\}$. Now we have

$$\mathbf{y} = \hat{\mathbf{y}} + \mathbf{r} = \mathbf{f}(\mathbf{X}) + \mathbf{r} = \mathbf{X}^T\mathbf{w} + \mathbf{r} \tag{5.48}$$

We assume $\mathbf{w}$ has a zero-mean normal prior pdf

$$p(\mathbf{w}) = \mathcal{N}(\mathbf{w}, \mathbf{0}, \mathbf{\Sigma}_w) \propto \exp\left(-\frac{1}{2}\mathbf{w}^T\mathbf{\Sigma}_w^{-1}\mathbf{w}\right), \tag{5.49}$$

based on our desire that the weights in $\mathbf{w}$ should take values around zero, instead of large values on either positive or negative, to avoid potential overfitting. Note that here the normalizing factor independent of $\mathbf{w}$ is dropped for simplicity. We further assume that all components of the residual $\mathbf{r}$ are independent of each other and they have the same variance, that is $\mathbf{\Sigma}_r = \sigma_r^2\mathbf{I}$, and $\mathbf{r}$ has a zero-mean normal distribution

$$p(\mathbf{r}) = \mathcal{N}(\mathbf{r}, \mathbf{0}, \sigma_r^2\mathbf{I}) \propto \exp\left(-\frac{1}{2\sigma_r^2}\mathbf{r}^T\mathbf{r}\right) = \exp\left(-\frac{1}{2\sigma_r^2}||\mathbf{r}||^2\right) \tag{5.50}$$

As a linear transformation of $\mathbf{w}$, the regression function $\hat{\mathbf{y}} = \mathbf{X}^T\mathbf{w}$ also has a zero-mean Gaussian distribution, and so is $\mathbf{y} = \mathbf{X}^T\mathbf{w} + \mathbf{r}$ as the sum of two Gaussians.

Given the observed data $\mathcal{D} = \{\mathbf{X}, \mathbf{y}\}$, we get its *likelihood*

$$\begin{aligned} L(\mathbf{w}|\mathcal{D}) &\propto p(\mathcal{D}|\mathbf{w}) = p(\mathbf{y}|\mathbf{X},\mathbf{w}) = p(\mathbf{X}^T\mathbf{w}+\mathbf{r}) = \mathcal{N}(\mathbf{y},\mathbf{X}^T\mathbf{w},\sigma_r^2\mathbf{I}) \\ &\propto \exp\left(-\frac{1}{2\sigma_r^2}(\mathbf{y}-\mathbf{X}^T\mathbf{w})^T(\mathbf{y}-\mathbf{X}^T\mathbf{w})\right) = \exp\left(-\frac{1}{2\sigma_r^2}||\mathbf{r}||^2\right) \end{aligned} \tag{5.51}$$

We see that the likelihood is a normal pdf of $\mathbf{y}$ (the same normal pdf but shifted by $\mathbf{X}^T\mathbf{w}$), which is the same as that of $\mathbf{r}$. Now we can used the *MLE* method to find the optimal parameter $\mathbf{w}^*$ by which the likelihood $L(\mathbf{w}|\mathcal{D})$ is maximized, or equivalently the following negative log likelihood is minimized

$$-l(\mathbf{w}|\mathbf{X},\mathbf{y}) = -\log L(\mathbf{w}|\mathbf{X},\mathbf{y}) = \frac{||\mathbf{r}||^2}{2\sigma_r^2} \propto ||\mathbf{r}||^2 \tag{5.52}$$

This is the same as the squared error $\varepsilon(\mathbf{w})$ defined in Eq. (5.8) to be minimized by the LS method previously considered. We therefore see that the maximum likelihood method and the least squares method are equivalent to each other.

We can further find the posterior $p(\mathbf{w}|\mathcal{D})$ of the model parameter $\mathbf{w}$ given the observed data $\mathcal{D}$, as the product of prior $p(\mathbf{w})$ and likelihood $L(\mathbf{w}|\mathcal{D}) \propto p(\mathbf{y}|\mathbf{X},\mathbf{w})$

$$\begin{aligned} p(\mathbf{w}|\mathcal{D}) &= p(\mathbf{w}|\mathbf{X},\mathbf{y}) = \frac{p(\mathcal{D}|\mathbf{w})\,p(\mathbf{w})}{p(\mathcal{D})} = \frac{p(\mathbf{y}|\mathbf{X},\mathbf{w})\,p(\mathbf{w})}{p(\mathbf{y}|\mathbf{X})} \\ &\propto p(\mathbf{y}|\mathbf{X},\mathbf{w})\,p(\mathbf{w}) = \mathcal{N}(\mathbf{y},\mathbf{X}^T\mathbf{w},\sigma_r^2\mathbf{I})\,\mathcal{N}(\mathbf{w},\mathbf{0},\mathbf{\Sigma}_w) \\ &\propto \exp\left[-\frac{1}{2}\left(\frac{1}{\sigma_r^2}(\mathbf{y}-\mathbf{X}^T\mathbf{w})^T(\mathbf{y}-\mathbf{X}^T\mathbf{w}) + \mathbf{w}^T\mathbf{\Sigma}_w^{-1}\mathbf{w}\right)\right] \\ &\propto \exp\left[-\frac{1}{2}\left(\frac{1}{\sigma_r^2}\mathbf{y}^T\mathbf{y} - \frac{2}{\sigma_r^2}(\mathbf{X}\mathbf{y})^T\mathbf{w} + \frac{1}{\sigma_r^2}\mathbf{w}^T\mathbf{X}\mathbf{X}^T\mathbf{w} + \mathbf{w}^T\mathbf{\Sigma}_w^{-1}\mathbf{w}\right)\right] \\ &\propto \exp\left[-\frac{1}{2}\left(\mathbf{w}^T\left(\frac{1}{\sigma_r^2}\mathbf{X}\mathbf{X}^T + \mathbf{\Sigma}_w^{-1}\right)\mathbf{w} - \frac{2}{\sigma_r^2}(\mathbf{X}\mathbf{y})^T\mathbf{w}\right)\right] \end{aligned} \tag{5.53}$$

As before, here all constant scaling factors (irrelevant to $\mathbf{w}$) are dropped for simplicity. As both the prior and the likelihood are Gaussian, this posterior, when normalized, is also Gaussian in terms of the mean $\mathbf{m}_{w/d}$ and covariance $\mathbf{\Sigma}_{w/d}$

$$\begin{aligned} p(\mathbf{w}|\mathcal{D}) = \mathcal{N}(\mathbf{w},\mathbf{m}_{w/d},\mathbf{\Sigma}_{w/d}) &\propto \exp\left[-\frac{1}{2}\left((\mathbf{w}-\mathbf{m}_{w/d})^T\mathbf{\Sigma}_{w/d}^{-1}(\mathbf{w}-\mathbf{m}_{w/d})\right)\right] \\ &\propto \exp\left[-\frac{1}{2}\left(\mathbf{w}^T\mathbf{\Sigma}_{w/d}^{-1}\mathbf{w} - 2\mathbf{m}_{w/d}^T\mathbf{\Sigma}_{w/d}^{-1}\mathbf{w}\right)\right] \end{aligned} \tag{5.54}$$

with the following covariance and mean

$$\mathbf{\Sigma}_{w/d} = \left(\frac{1}{\sigma_r^2}\mathbf{X}\mathbf{X}^T + \mathbf{\Sigma}_w^{-1}\right)^{-1}$$
$$\mathbf{m}_{w/d} = \frac{1}{\sigma_r^2}\mathbf{\Sigma}_{w/d}\mathbf{X}\mathbf{y} = \left(\mathbf{X}\mathbf{X}^T + \sigma_r^2\mathbf{\Sigma}_w^{-1}\right)^{-1}\mathbf{X}\mathbf{y} \tag{5.55}$$

Now we can find the optimal model parameter $\mathbf{w}^* = \mathbf{m}_{w/d}$ equal to the mean of the Gaussian posterior at which it is maximized. This solution is called the *MAP* estimate of the model parameter $\mathbf{w}$.

We can now predict the function value of any unlabeled test data point $\mathbf{x}_*$ based on the conditional probability distribution $p(f_*|\mathbf{x}_*, \mathcal{D})$ of the regression model $\hat{y}_* = f_* = f(\mathbf{x}_*) = \mathbf{x}_*^T\mathbf{w}$, which, as a linear transformation of the random parameters in $\mathbf{w}$, is also Gaussian (Section B.1.5).

$$p(f_*|\mathbf{x}_*, \mathcal{D}) = p(\mathbf{x}_*^T\mathbf{w}|\mathbf{x}_*, \mathbf{X}, \mathbf{y}) = \mathcal{N}(f_*, \mathbf{x}_*^T\mathbf{m}_{w/d},\ \mathbf{x}_*^T\mathbf{\Sigma}_{w/d}\mathbf{x}_*) \tag{5.56}$$

with the following mean and variance

$$\mathrm{E}[f_*|\mathbf{x}_*, \mathbf{X}, \mathbf{y}] = \mathbf{x}_*^T\mathbf{m}_{w/d} = \mathbf{x}_*^T\left(\mathbf{X}\mathbf{X}^T + \sigma_r^2\mathbf{\Sigma}_w^{-1}\right)^{-1}\mathbf{X}\mathbf{y}$$
$$\mathrm{Var}[f_*|\mathbf{x}_*, \mathbf{X}, \mathbf{y}] = \mathbf{x}_*^T\mathbf{\Sigma}_{w/d}\mathbf{x}_* = \mathbf{x}_*^T\left(\frac{1}{\sigma_r^2}\mathbf{X}\mathbf{X}^T + \mathbf{\Sigma}_w^{-1}\right)^{-1}\mathbf{x}_* \tag{5.57}$$

Furthermore, given a set of M unlabeled test data points in $\mathbf{X}_* = [\mathbf{x}_{1*}, \cdots, \mathbf{x}_{M*}]^T$, we can find

$$\hat{\mathbf{y}}_* = \mathbf{f}_* = \mathbf{f}(\mathbf{x}_*) = \begin{bmatrix} f(\mathbf{x}_{1*}) \\ \vdots \\ f(\mathbf{x}_{M*}) \end{bmatrix} = \begin{bmatrix} \mathbf{x}_{1*}^T \\ \vdots \\ \mathbf{x}_{M*}^T \end{bmatrix}\mathbf{w} = [\mathbf{x}_{1*}, \cdots, \mathbf{x}_{M*}]^T\mathbf{w} = \mathbf{X}_*^T\mathbf{w} \tag{5.58}$$

As a linear combination of Gaussian $\mathbf{w}$, $\mathbf{f}_* = \mathbf{X}_*^T\mathbf{w}$ is also Gaussian

$$p(\mathbf{f}_*|\mathbf{X}_*, \mathcal{D}) = p(\mathbf{f}_*|\mathbf{X}_*, \mathbf{X}, \mathbf{y}) = \mathcal{N}(\mathbf{f}_*, \mathbf{X}_*^T\mathbf{m}_{w/d},\ \mathbf{X}_*^T\mathbf{\Sigma}_{w/d}\mathbf{X}_*) \tag{5.59}$$

with the following mean and covariance

$$\mathrm{E}[\mathbf{f}_*|\mathbf{X}_*, \mathbf{X}, \mathbf{y}] = \mathrm{E}[\mathbf{X}_*^T\mathbf{w}] = \mathbf{X}_*^T\,\mathrm{E}[\mathbf{w}] = \mathbf{X}_*^T\mathbf{m}_{w/d} = \mathbf{X}_*^T\left(\mathbf{X}\mathbf{X}^T + \sigma_r^2\mathbf{\Sigma}_w^{-1}\right)^{-1}\mathbf{X}\mathbf{y} \tag{5.60}$$
$$\mathrm{Cov}[\mathbf{f}_*|\mathbf{X}_*, \mathbf{X}, \mathbf{y}] = \mathrm{E}[\mathbf{f}_*\mathbf{f}_*^T] = \mathrm{E}[\mathbf{X}_*^T\mathbf{w}\mathbf{w}^T\mathbf{X}_*] = \mathbf{X}_*^T\,\mathrm{E}[\mathbf{w}\mathbf{w}^T]\mathbf{X}_* = \mathbf{X}_*^T\mathbf{\Sigma}_{w/d}\mathbf{X}_* \tag{5.61}$$

The mean $\mathrm{E}[\mathbf{f}_*|\mathbf{X}_*, \mathbf{X}, \mathbf{y}]$ can be considered as the estimated function value $\hat{\mathbf{y}}_* = \mathbf{f}_* = \mathbf{X}_*^T\mathbf{w}$ at the test points in $\mathbf{X}_*$, while the variance $\mathrm{Var}[\mathbf{f}_*|\mathbf{X}_*, \mathbf{X}, \mathbf{y}]$ as the confidence or certainty of the estimation.

Such a result produced by the Bayesian method in terms of the mean and covariance of $\mathbf{f}_* = \hat{\mathbf{y}}_*$ as a random vector is quite different from the result produced by a frequentist method such as the linear least square regression in terms of an estimated value of $\hat{\mathbf{y}}_*$ as a deterministic variable. However, in the special case of $\sigma_r = 0$, that is the residual $\mathbf{r}$ is no longer treated as a random variable, then Eq. (5.60) becomes the same as Eq. (5.11) produced by linear least square regression, in other words, the linear method can be considered as a special case of the Bayesian regression when the residual is deterministic.

The method just considered is based on the linear regression model $f(\mathbf{x}) = \mathbf{x}^T\mathbf{w}$, but it can also be generalized to nonlinear functions, as in the case of regression based on basis functions discussed in Section 5.3. To do so, the regression function is assumed to be a linear combination of a set of K nonlinear mapping functions $\varphi_k(\mathbf{x})$ $(k = 1, \cdots, K)$

$$y = f(\mathbf{x}) + e = \sum_{k=1}^{K} w_k\, \varphi_k(\mathbf{x}) + r = \mathbf{w}^T \phi(\mathbf{x}) + r = \phi^T(\mathbf{x})\mathbf{w} + r \tag{5.62}$$

where

$$\mathbf{w} = [w_1, \cdots, w_K]^T, \qquad \phi(\mathbf{x}) = [\varphi_1(\mathbf{x}), \cdots, \varphi_K(\mathbf{x})]^T \tag{5.63}$$

For example, some commonly used mapping functions are listed in Eq. (5.45).

Applying this regression model to each of the N observed data points in $\mathcal{D} = \{(\mathbf{x}_n, y_n),\ (n = 1, \cdots, N)\}$, we get $\hat{y}_n = \sum_{k=1}^{K} w_k\, \varphi_k(x_n)$ with residual $r_n = y_n - \hat{y}_n$, which can be written in matrix form

$$\mathbf{y} = \begin{bmatrix} y_1 \\ \vdots \\ y_N \end{bmatrix}_{N\times 1} = \begin{bmatrix} \varphi_1(\mathbf{x}_1) & \cdots & \varphi_K(\mathbf{x}_1) \\ \vdots & \ddots & \vdots \\ \varphi_1(\mathbf{x}_N) & \cdots & \varphi_K(\mathbf{x}_N) \end{bmatrix}_{N\times K} \begin{bmatrix} w_1 \\ \vdots \\ w_K \end{bmatrix}_{K\times 1} + \begin{bmatrix} r_1 \\ \vdots \\ r_N \end{bmatrix}_{N\times 1}$$
$$= \begin{bmatrix} \phi_1^T \\ \vdots \\ \phi_N^T \end{bmatrix} \mathbf{w} + \mathbf{r} = \mathbf{\Phi}^T\mathbf{w} + \mathbf{r} \tag{5.64}$$

where, as previously defined in Eq. (5.42),

$$\phi_n = \phi(\mathbf{x}_n) = \begin{bmatrix} \varphi_1(\mathbf{x}_n) \\ \vdots \\ \varphi_K(\mathbf{x}_n) \end{bmatrix}, \qquad \mathbf{\Phi} = [\phi_1, \cdots, \phi_N] \tag{5.65}$$

As this nonlinear model $\hat{\mathbf{y}} = \mathbf{\Phi}^T\mathbf{w}$ is in the same form as the linear model $\hat{\mathbf{y}} = \mathbf{X}^T\mathbf{w}$, all previous discussion for the linear Bayesian regression is still valid for this nonlinear model, if $\mathbf{X} = [\mathbf{x}_1, \cdots, \mathbf{x}_N]_{(d+1)\times N}$ is replaced by $\mathbf{\Phi} = [\phi_1, \cdots, \phi_N]_{K\times N}$. We can obtain the covariance matrix and mean vector of the posterior $p(\mathbf{w}|\mathcal{D})$ in the K-dimensional space based on the observed data $\mathcal{D} = \{\mathbf{X}, \mathbf{y}\}$

$$\mathbf{\Sigma}_{w/d} = \left(\frac{1}{\sigma_r^2}\mathbf{\Phi}\mathbf{\Phi}^T + \mathbf{\Sigma}_w^{-1}\right)^{-1}$$
$$\mathbf{m}_{w/d} = \frac{1}{\sigma_r^2}\mathbf{\Sigma}_{w/d}\mathbf{\Phi}\mathbf{y} = \left(\mathbf{\Phi}\mathbf{\Phi}^T + \sigma_r^2\mathbf{\Sigma}_w^{-1}\right)^{-1}\mathbf{\Phi}\mathbf{y} \tag{5.66}$$

The function $\hat{\mathbf{y}}_* = \mathbf{f}(\mathbf{X}_*)$ of the multiple test points in $\mathbf{X}_* = [\mathbf{x}_{*1}, \cdots, \mathbf{x}_{*M}]^T$ is a linear transformation of $\mathbf{w}$

$$\hat{\mathbf{y}}_* = \mathbf{f}_* = \mathbf{f}(\mathbf{x}_*) = \begin{bmatrix} f_1(\mathbf{x}_*) \\ \vdots \\ f_M(\mathbf{x}_*) \end{bmatrix} = \begin{bmatrix} \phi_1^T \\ \vdots \\ \phi_M^T \end{bmatrix}_* \mathbf{w} = [\phi_1, \cdots, \phi_M]^T\mathbf{w} = \mathbf{\Phi}_*^T\mathbf{w} \tag{5.67}$$

and it also has a Gaussian joint distribution

$$p(\mathbf{f}_*|\mathbf{X}_*, \mathbf{X}, \mathbf{y}) = \mathcal{N}(\mathbf{\Phi}_*^T \mathbf{m}_{w/d},\ \mathbf{\Phi}_*^T \mathbf{\Sigma}_{w/d} \mathbf{\Phi}_*) \tag{5.68}$$

with the mean and covariance

$$\begin{aligned} \mathrm{E}[\mathbf{f}_*|\mathbf{X}_*, \mathbf{X}, \mathbf{y}] &= \mathrm{E}[\mathbf{\Phi}_*^T \mathbf{w}] = \mathbf{\Phi}_*^T\, \mathrm{E}[\mathbf{w}] = \mathbf{\Phi}_*^T \mathbf{m}_{w/d} \\ \mathrm{Cov}[\mathbf{f}_*|\mathbf{X}_*, \mathbf{X}, \mathbf{y}] &= \mathrm{E}[\mathbf{f}_* \mathbf{f}_*^T] = \mathrm{E}[\mathbf{\Phi}_*^T \mathbf{w}\mathbf{w}^T \mathbf{\Phi}_*] \\ &= \mathbf{\Phi}_*^T\, \mathrm{E}[\mathbf{w}\mathbf{w}^T] \mathbf{\Phi}_* = \mathbf{\Phi}_*^T \mathbf{\Sigma}_{w/d} \mathbf{\Phi}_* \end{aligned} \tag{5.69}$$

They can be reformulated to become

$$\begin{aligned} \mathrm{E}[\mathbf{f}_*|\mathbf{X}_*, \mathbf{X}, \mathbf{y}] = \mathbf{\Phi}_*^T \mathbf{m}_{w/d} &= \mathbf{\Phi}_*^T \left(\mathbf{\Phi}\mathbf{\Phi}^T + \sigma_r^2 \mathbf{\Sigma}_w^{-1}\right)^{-1} \mathbf{\Phi}\mathbf{y} \\ &= \mathbf{\Phi}_*^T (\mathbf{\Sigma}_w \mathbf{\Phi})(\mathbf{\Sigma}_w \mathbf{\Phi})^{-1} \left(\mathbf{\Phi}\mathbf{\Phi}^T + \sigma_r^2 \mathbf{\Sigma}_w^{-1}\right)^{-1} (\mathbf{\Phi}^{-1})^{-1} \mathbf{y} \\ &= \mathbf{\Phi}_*^T \mathbf{\Sigma}_w \mathbf{\Phi} \left[\mathbf{\Phi}^{-1} \left(\mathbf{\Phi}\mathbf{\Phi}^T + \sigma_r^2 \mathbf{\Sigma}_w^{-1}\right) \mathbf{\Sigma}_w \mathbf{\Phi}\right]^{-1} \mathbf{y} \\ &= \mathbf{\Phi}_*^T \mathbf{\Sigma}_w \mathbf{\Phi} (\mathbf{\Phi}^T \mathbf{\Sigma}_w \mathbf{\Phi} + \sigma_r^2 \mathbf{I})^{-1} \mathbf{y} \end{aligned} \tag{5.70}$$

and

$$\begin{aligned} \mathrm{Cov}[\mathbf{f}_*|\mathbf{X}_*, \mathbf{X}, \mathbf{y}] = \mathbf{\Phi}_*^T \mathbf{\Sigma}_{w/d} \mathbf{\Phi}_* &= \mathbf{\Phi}_*^T \left(\frac{1}{\sigma_r^2} \mathbf{\Phi}\mathbf{\Phi}^T + \mathbf{\Sigma}_w^{-1}\right)^{-1} \mathbf{\Phi}_* \\ &= \mathbf{\Phi}_*^T \left[\mathbf{\Sigma}_w - \mathbf{\Sigma}_w \mathbf{\Phi} (\mathbf{\Phi}^T \mathbf{\Sigma}_w \mathbf{\Phi} + \sigma_r^2 \mathbf{I})^{-1} \mathbf{\Phi}^T \mathbf{\Sigma}_w\right] \mathbf{\Phi}_* \\ &= \mathbf{\Phi}_*^T \mathbf{\Sigma}_w \mathbf{\Phi}_* - \mathbf{\Phi}_*^T \mathbf{\Sigma}_w \mathbf{\Phi} (\mathbf{\Phi}^T \mathbf{\Sigma}_w \mathbf{\Phi} + \sigma_r^2 \mathbf{I})^{-1} \mathbf{\Phi}^T \mathbf{\Sigma}_w \mathbf{\Phi}_* \end{aligned} \tag{5.71}$$

where we have used Woodbury identity in Section A.2.4.

We note that in these expressions the mapping function $\varphi(\mathbf{x})$ only shows up in the quadratic form of $\mathbf{\Phi}^T \mathbf{\Sigma}_w \mathbf{\Phi}$, $\mathbf{\Phi}^T \mathbf{\Sigma}_w \mathbf{\Phi}_*$, or $\mathbf{\Phi}_*^T \mathbf{\Sigma}_w \mathbf{\Phi}_*$. This suggests that we may consider using *kernel method* (Section 14.2, by which some significant advantage may be gained (such as in the support-vector machines). Specifically, if we replace the mn-th component $\phi^T(\mathbf{x}_m) \mathbf{\Sigma}_w \phi(\mathbf{x}_n)$ of $\mathbf{\Phi}^T \mathbf{\Sigma}_w \mathbf{\Phi}$ by a *kernel function* only in terms of $\mathbf{x}_m$ and $\mathbf{x}_n$

$$\phi^T(\mathbf{x}_m) \mathbf{\Sigma}_w \phi(\mathbf{x}_n) = k(\mathbf{x}_m, \mathbf{x}_n) = k(\mathbf{x}_n, \mathbf{x}_m) \tag{5.72}$$

then the mapping function $\varphi(\mathbf{x})$ no longer needs to be explicitly specified. This idea is called the *kernel trick*, Now we have

$$\mathbf{\Phi}^T \mathbf{\Sigma}_w \mathbf{\Phi} = \begin{bmatrix} \phi_1^T \\ \vdots \\ \phi_N \end{bmatrix} \mathbf{\Sigma}_w [\phi_1, \cdots, \phi_N] = \begin{bmatrix} k(\mathbf{x}_1, \mathbf{x}_1) & \cdots & k(\mathbf{x}_1, \mathbf{x}_N) \\ \vdots & \ddots & \vdots \\ k(\mathbf{x}_N, \mathbf{x}_1) & \cdots & k(\mathbf{x}_N, \mathbf{x}_N) \end{bmatrix} = \mathbf{K} \tag{5.73}$$

and similarly we also have $\mathbf{\Phi}^T \mathbf{\Sigma}_w \mathbf{\Phi}_* = \mathbf{K}_*$ and $\mathbf{\Phi}_*^T \mathbf{\Sigma}_w \mathbf{\Phi} = \mathbf{K}_*^T$. These kernel functions are symmetric, same as all covariance matrices. Now the mean and covariance of $\mathbf{f}_*$ can be expressed as

$$\begin{aligned} \mathrm{E}[\mathbf{f}_*] &= \mathbf{K}_*^T \left(\mathbf{K} + \sigma_r^2 \mathbf{I}\right)^{-1} \mathbf{y} \\ \mathrm{Cov}[\mathbf{f}_*] &= \mathbf{K}_{**} - \mathbf{K}_*^T \left(\mathbf{K} + \sigma_r^2 \mathbf{I}\right)^{-1} \mathbf{K}_* \end{aligned} \tag{5.74}$$

As a general summary of the method just discussed, we note that it is essentially different from the frequentist approach, in the sense that the estimated result produced is no longer a set of specific values of some unknown variables in question (function values, parameters of regression models, etc.), but the probability distribution of these variables in terms of the expectation and covariance. Consequently, we can not only treat the expectation as the estimated values, but also use the variance as the confidence or certainty of such an estimation.

A segment of MATLAB code for estimating the mean and covariance of weight vector $\mathbf{w}$ by the Bayes linear regression is given below. Here `X` and `y` are for $\mathbf{X}$ and $\mathbf{y}$ of the training data $\mathcal{D}$, and `Xstar`, `Phi` and `Phistar` are for $\mathbf{X}_*$, $\mathbf{\Phi}$ and $\mathbf{\Phi}_*$, respectively.

```
SigmaP=eye(d);    % prior variance of r in original space
SigmaQ=eye(D);    % prior variance of r in feature space
Sigmaw1=inv(X*X'/sigmaN^2+inv(SigmaP));
                  % covariance of w by Bayes regression
Sigmaw2=inv(Phi0*Phi0'/sigmaN^2+inv(SigmaQ));
                  % cov of w with basis function

w0=inv(X*X')*X*y % weight vector w by LS method
w1=Sigmaw1*X*y/sigmaN^2 % mean of w by Bayes regression
w2=Sigmaw2*Phi0*y/sigmaN^2;
                  % mean of w by Bayes regression with basis function

y0=w0'*Xstar;     % y by least squares method
y1=w1'*Xstar;     % y by Bayes regression
y2=w2'*PhistarO; % y by Bayes regression with basis function

for i=1:M
    sgm1(i)=sqrt(Xstar(:,i)'*Sigmaw1*Xstar(:,i));
                      % variance of estimated y
    sgm2(i)=sqrt(Phistar(:,i)'*Sigmaw2*Phistar(:,i));
end
```

Example 5.11

Consider a 2-D linear function

$$f(\mathbf{x}) = \mathbf{w}^T\mathbf{x} = [w_0, w_1]\begin{bmatrix} x_0 \\ x_1 \end{bmatrix} = w_0x_0 + w_1x_1 = 0.5 + 0.5\,x_1 \tag{5.75}$$

where we have assumed $x_0 = 1$, $w_0 = -1/2$ and $w_1 = 1/2$, that is the function is actually a straight line $f(x) = (-1 + x)/2$ with intercept $w_0 = -1/2$ and slope $w_1 = 1/2$. The distribution of the additive noise (residual) is $p(e) = \mathcal{N}(0, \sigma_r = 0.8)$. The prior distributions of the weights $\mathbf{w} = [w_0, w_1]^T$ are assumed to be $p(\mathbf{w}) = \mathcal{N}(\mathbf{0}, \mathbf{I})$, and posterior $p(\mathbf{w}|\mathcal{D})$ can be obtained based on the observed data $\mathcal{D} = \{\mathbf{X}, \mathbf{y}\}$, both shown in Fig. 5.24. We see that the posterior is concentrated around $\hat{w}_1 = -0.28$ and $\hat{w}_2 = 0.39$, the estimates of the true weights $w_1 = -0.5$ and $w_2 = 0.5$, based on the observed noisy data.

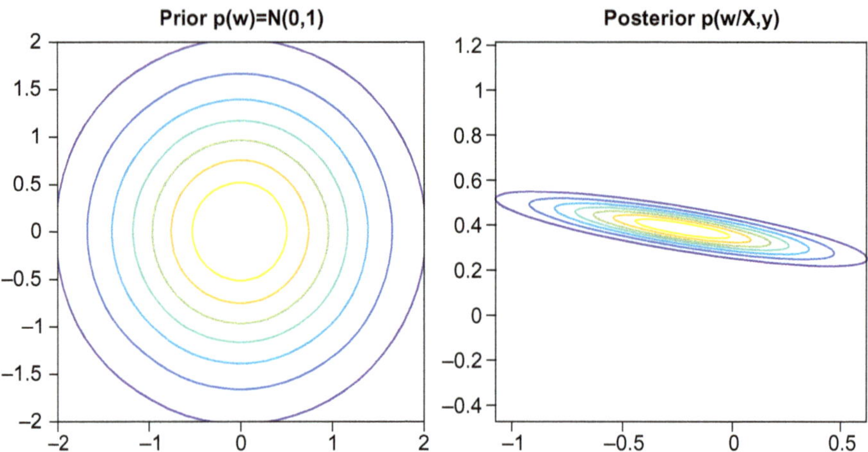

Fig. 5.24 Prior and posterior of Bayesian regression.

Various regression methods including both the least-squared method and the Bayesian linear regression methods, and in feature space spanned by a set of basis functions $\phi_i(\mathbf{x})$, $(i = 1, \cdots, D = 5)$ based on each of the three given functions are carried out with their results shown in Fig. 5.25.

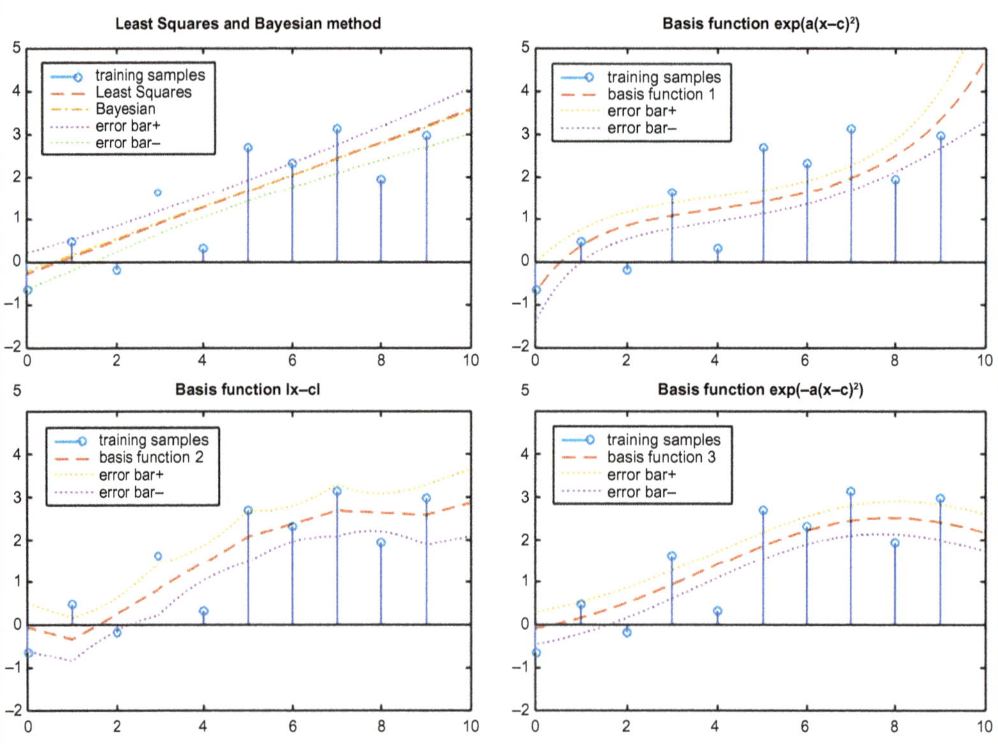

Fig. 5.25 Bayesian regression (linear and basis function based).

The LS errors for these methods are listed here

Table 5.5

Method	LS error	$\mathbf{w} = [w_1, w_2]^T$
Pseudo inverse	0.4730	$[-0.28, 0.39]^T$
Bayesian linear regression	0.4742	$[-0.22, 0.38]^T$
$\phi = \exp(a\|\|\mathbf{x} - \mathbf{c}\|\|^2)$	0.6196	
$\phi = \|\|\mathbf{x} - \mathbf{c}\|\|$	0.4429	
$\phi = \exp(-a\|\|\mathbf{x} - \mathbf{c}\|\|^2)$	0.4459	

We see that the linear methods (the pseudo inverse and the Bayesian regression) produce similar estimated weights error, but smaller error can be achieved by the same method of Bayesian linear regression if proper nonlinear mapping is used (e.g., the last two mapping functions).

PROBLEMS

1. Develop MATLAB code for both left and right pseudo inverse (Section A.6.1) of an $m \times n$ matrix $\mathbf{A}$ ($m > n$ or $m < n$).
2. Solve the over-constrained linear equation system $\mathbf{Ax} = \mathbf{b}$ and find the squared error, with the coefficient matrix

$$\mathbf{A} = \begin{bmatrix} 2 & 3 \\ 3 & -2 \\ 1 & 1 \end{bmatrix}$$

and each of the two different right-hand sides

$$\mathbf{b}_1 = \begin{bmatrix} 8 \\ -1 \\ 3 \end{bmatrix}, \qquad \mathbf{b}_1 = \begin{bmatrix} 8 \\ 1 \\ 3 \end{bmatrix}$$

3. Develop MATLAB code to solve ordinary least squares regression problems to find the $d + 1$ dimensional weight vector $\mathbf{w} = [w_0 = b, w_1, \cdots, w_d]^T$, and also other associated parameters including the sum of squares ESS, RSS, TSS, R^2, as well as the correlation coefficient ρ.
Apply your code to the following two data sets, find $\mathbf{w}$ and all parameters above, and then plot the regression curve $y = w_0 + w_1 x$ together with the data points $\{(x_i, y_i)\, i = 1, \cdots, N\}$.
   - 2-D dataset in/e176/programs/dataLS1.txt, containing two columns for (x, y) of each data point.
   - 3-D dataset in /e176/programs/dataLS2.txt, containing three columns for (x_1, x_2, y) of each data point.
4. Implement ridge regression and apply it to a set of synthetic 2-D data (generated by your own code) including an independent variable x_1 and another

closely correlated variable x_2 (a noisy version of the x_1, by, for example, `x2=x1+a*rand(1,N)`). Try different noise levels by varying a see how ill-behaved the dataset is in terms of correlation coefficient ρ between the two variables, and how ill-conditioned the problem is in terms of how close to zero one of the eigenvalues of matrix $\mathbf{XX}^T$ is. Try different λ values of the ridge regression (including $\lambda = 0$ for regular linear regression) to see how this affects the stability of the resulting coefficients w_0, w_1, w_2 and the accuracy in terms of R^2.

5. Implement the method of nonlinear regression based on each of four different basis functions considered in Section 5.3. Apply your code to solve the same 1-D regression problems given in Example 5.8. Compare results with different K values to see how the number of basis functions affects the results.
6. Develop MATLAB code to implement the method of nonlinear regression based on each of the three basis functions, Gaussian, ReLU, and sigmoid considered in Section 5.3. Apply the code to solve the same 2-D regression problems given in Examples 5.9 and 5.10.
7. Develop MATLAB code to implement the method of Bayesian regression based on different basis functions. Apply the code to the same 1-D problem in Example 5.11, and compare results with that of the LS method.
8. Apply the code developed in the previous problem to the same 2-D regression problem given in Examples 5.9 and 5.10.

6 Nonlinear Regression

6.1 Nonlinear Least Squares Regression

The goal of nonlinear least squares regression is to model the nonlinear relationship between the dependent variable y and independent variables in $\mathbf{x} = [x_1, \cdots, x_d]^T$, based on the given training dataset $\mathcal{D} = \{(\mathbf{x}_n, y_n),\ n = 1, \cdots, N\}$, by a nonlinear regression function parameterized by $\boldsymbol{\theta} = [\theta_1, \cdots, \theta_M]^T$

$$\hat{y} = f(x_1, \cdots, x_d,\ \theta_1, \cdots, \theta_M) = f(\mathbf{x}, \boldsymbol{\theta}) \tag{6.1}$$

While the specific form of the nonlinear function is typically determined based on certain prior knowledge, we need to find the parameter $\boldsymbol{\theta}$ for the model to fit the training dataset $\mathcal{D}$, so that its output $\hat{\mathbf{y}}$ matches the ground truth labeling $\mathbf{y}$ in the training set, that is, ideally the following residual is zero.

$$\mathbf{r}(\boldsymbol{\theta}) = \begin{bmatrix} r_1 \\ \vdots \\ r_N \end{bmatrix} = \begin{bmatrix} y_1 \\ \vdots \\ y_N \end{bmatrix} - \begin{bmatrix} f_1(\boldsymbol{\theta}) \\ \vdots \\ f_N(\boldsymbol{\theta}) \end{bmatrix} = \mathbf{y} - \hat{\mathbf{y}}(\boldsymbol{\theta}) = \mathbf{y} - \mathbf{f}(\mathbf{X}, \boldsymbol{\theta}) = \mathbf{0} \tag{6.2}$$

where $\hat{y}(\mathbf{x}_n, \boldsymbol{\theta}) = f_n(\boldsymbol{\theta}) = f(\mathbf{x}_n, \boldsymbol{\theta})$ is the output generated by the regression function based on the nth data point $\mathbf{x}_n$. We see that the regression problem is equivalent to solving a nonlinear equation system of N equations for the M unknown parameters in $\boldsymbol{\theta}$. As typically there are many more data samples than the unknown parameters, that is $N \gg M$, the equation $\mathbf{r} = \mathbf{0}$ is over-constrained without an exact solution. We need to find the optimal parameter $\boldsymbol{\theta}^*$ that minimizes the sum of squared error (SSE) based on the residual $\mathbf{r}$

$$\varepsilon(\boldsymbol{\theta}) = \frac{1}{2}||\mathbf{r}||^2 = \frac{1}{2}\sum_{n=1}^N r_n^2 = \frac{1}{2}\sum_{n=1}^N [y_n - f_n(\boldsymbol{\theta})]^2 \tag{6.3}$$

This optimization problem can be addressed by either the gradient descent methods or the Gauss–Newton method.

We first consider the gradient descent method based on the gradient of this error function

$$\mathbf{g}_\varepsilon(\boldsymbol{\theta}) = \frac{d}{d\boldsymbol{\theta}}\varepsilon(\boldsymbol{\theta}) = \frac{d}{d\boldsymbol{\theta}}\left(\frac{1}{2}\sum_{n=1}^N r_n^2\right) = \sum_{n=1}^N \frac{d\,r_n}{d\boldsymbol{\theta}}\, r_n$$
$$= -\sum_{n=1}^N \frac{d}{d\boldsymbol{\theta}} f_n(\boldsymbol{\theta})[y_n - f_n(\boldsymbol{\theta})] = -\mathbf{J}_f^T\,\mathbf{r} \tag{6.4}$$

where $\mathbf{J}_f$ is the $N \times M$ Jacobian matrix $\mathbf{J}_f$ of $\mathbf{f}(\mathbf{X},\boldsymbol{\theta})$. We can also obtain the mth ($m = 1,\cdots,M$) component of the gradient vector as

$$g_m(\boldsymbol{\theta}) = \frac{\partial}{\partial\theta_m}\varepsilon(\boldsymbol{\theta}) = \frac{1}{2}\frac{\partial}{\partial\theta_m}||\mathbf{r}||^2 = \frac{1}{2}\sum_{n=1}^{N}\frac{\partial}{\partial\theta_m}r_n^2 = \sum_{n=1}^{N}\frac{\partial r_n}{\partial\theta_m}r_n$$
$$= -\sum_{n=1}^{N}\frac{\partial f_n(\boldsymbol{\theta})}{\partial\theta_m}[y_n - f_n(\boldsymbol{\theta})] = -\sum_{n=1}^{N}J_{nm}\,[y_n - f_n(\boldsymbol{\theta})] = -\sum_{n=1}^{N}J_{nm}\,r_n \quad (6.5)$$

where $J_{nm} = \partial f_n(\boldsymbol{\theta})/\partial\theta_m$ is the component in the nth row and mth column of $\mathbf{J}_f$, which is simply the negative version of Jacobian of $\mathbf{r} = \mathbf{y} - \mathbf{f}(\mathbf{X},\boldsymbol{\theta})$

$$\mathbf{J}_r(\boldsymbol{\theta}) = \frac{d}{d\boldsymbol{\theta}}\mathbf{r}(\boldsymbol{\theta}) = \frac{d}{d\boldsymbol{\theta}}[\mathbf{y} - \mathbf{f}(\mathbf{X},\boldsymbol{\theta})] = -\frac{d}{d\boldsymbol{\theta}}\mathbf{f}(\mathbf{X},\boldsymbol{\theta}) = -\mathbf{J}_f(\boldsymbol{\theta}) \quad (6.6)$$

The optimal parameter $\boldsymbol{\theta}$ that minimizes $\varepsilon(\boldsymbol{\theta})$ can now be found iteratively by the gradient descent method

$$\boldsymbol{\theta}_{n+1} = \boldsymbol{\theta}_n - \delta_n\mathbf{g}_\varepsilon(\boldsymbol{\theta}_n) = \boldsymbol{\theta}_n + \delta_n\mathbf{J}_f^T\,\mathbf{r} \quad (6.7)$$

Setting the value for the step size δ_n properly may be tricky, as it depends on the specific landscape of the error hypersurface $\varepsilon(\boldsymbol{\theta})$. If δ_n is too large, a minimum may be skipped over, but if δ_n is too small the convergence of the iteration may be too slow.

This gradient method is also called *batch gradient descent (BGD)*, as the gradient in Eq. (6.4) is calculated based on the entire batch of all N data points in the dataset $\mathcal{D}$ in each iteration. When the size of the dataset is large, BGD may be slow due to the high computational complexity. In this case the method of *stochastic gradient descent (SGD)* can be used, which approximates the gradient based on only one of the N data points uniformly randomly chosen from the dataset in each iteration, that is the summation in Eq. (6.4) contains only one term corresponding to the chosen sample $\mathbf{x}_n$

$$\mathbf{g}_n(\boldsymbol{\theta}) = \frac{d}{d\boldsymbol{\theta}}\frac{1}{2}r_n^2 = r_n\frac{d\,r_n}{d\boldsymbol{\theta}} \quad (6.8)$$

The expectation of $\mathbf{g}_n$ based on one data point is the same as the true gradient (scaled by $1/N$) based on all N data points in the dataset

$$E[\mathbf{g}_n(\boldsymbol{\theta})] = \sum_{n=1}^{N}P(\mathbf{x}_n)\mathbf{g}_n(\boldsymbol{\theta}) = \sum_{n=1}^{N}\frac{1}{N}\mathbf{g}_n(\boldsymbol{\theta}) = \frac{1}{N}\sum_{n=1}^{N}\frac{dr_n}{d\boldsymbol{\theta}}r_n = \frac{1}{N}\mathbf{g}(\boldsymbol{\theta}) \quad (6.9)$$

As the data are typically noisy, the SGD gradient based on only one data point may be in a direction very different from the true gradient based on the mean of all N data points, and the resulting search path in the M–D space may be in a very jagged zig-zag shape, and the iteration may converge slowly. However, as the computational complexity in each iteration is much reduced, we can afford to carry out a large number of iterations with an overall search path generally coinciding with that of the BGD method.

Some tradeoff between BGD and SGD can be made so that the gradient is estimated based on more than one but less than all N data points in the dataset, the resulting method, called *mini-batch gradient descent*, can take advantage of both methods. The methods of stochastic and mini-batch gradient descent are widely used in machine learning, such as in the method of artificial neural networks, which can be treated as a nonlinear regression problem, considered in Part V.

We then consider the Gauss–Newton method to find the optimal parameter $\boldsymbol{\theta}^*$ of the nonlinear regression function $f(\mathbf{x}, \boldsymbol{\theta})$ by directly solving the nonlinear over-determined equation system $\mathbf{r}(\boldsymbol{\theta}) = \mathbf{y} - \mathbf{f}(\mathbf{X}, \boldsymbol{\theta}) = \mathbf{0}$ in Eq. (6.2), based on the iteration in Eq. (1.76) in Chapter 1

$$\boldsymbol{\theta}_{n+1} = \boldsymbol{\theta}_n - \mathbf{J}_r^- \, \mathbf{r} = \boldsymbol{\theta}_n - (\mathbf{J}_r^T \mathbf{J}_r)^{-1} \mathbf{J}_r^T \, \mathbf{r} = \boldsymbol{\theta}_n + (\mathbf{J}_f^T \mathbf{J}_f)^{-1} \mathbf{J}_f^T \, \mathbf{r} \tag{6.10}$$

The last equality is due to the fact that $\mathbf{J}_r = -\mathbf{J}_f$. Here $\mathbf{J}_f^T \mathbf{J}_f$ in the pseudo-inverse is actually an approximation of the Hessian $\mathbf{H}_\varepsilon$ of the error function $\varepsilon(\boldsymbol{\theta})$, as given in Eq. (2.20) in Chapter 2, and $-\mathbf{J}_f^T \mathbf{r} = \mathbf{g}_\varepsilon(\boldsymbol{\theta})$ is the gradient of the error function given in Eq. (6.4). So this iteration is essentially

$$\boldsymbol{\theta}_{n+1} = \boldsymbol{\theta}_n - \mathbf{J}_r^- \, \mathbf{r} \approx \boldsymbol{\theta}_n - \mathbf{H}_\varepsilon^{-1}(\boldsymbol{\theta}_n) \mathbf{g}_\varepsilon(\boldsymbol{\theta}_n) \tag{6.11}$$

same as the Newton–Raphson method based on the second-order Hessian as well as the first-order gradient. We therefore see that the Gauss–Newton method for minimizing the sum of squares error is essentially the same as the Newton–Raphson method, with the only difference being that the Hessian $\mathbf{H}_\varepsilon \approx \mathbf{J}_f^T \mathbf{J}_f$ as the second-order derivative of the error function $\varepsilon(\boldsymbol{\theta})$ is approximated by the Jacobian $\mathbf{J}_f \boldsymbol{\theta}$ as the first-order derivative of the regression function $\mathbf{f}(\mathbf{x}, \boldsymbol{\theta})$, making the implementation computationally easier.

In the special case of linear regression function $\mathbf{f}(\mathbf{w}) = \mathbf{x}^T \mathbf{w}$, we have

$$\hat{\mathbf{y}} = \mathbf{f}(\mathbf{X}, \mathbf{w}) = \begin{bmatrix} f(\mathbf{x}_1, \mathbf{w}) \\ \vdots \\ f(\mathbf{x}_N, \mathbf{w}) \end{bmatrix} = \begin{bmatrix} \mathbf{x}_1^T \\ \vdots \\ \mathbf{x}_N^T \end{bmatrix} \mathbf{w} = [\mathbf{x}_1, \cdots, \mathbf{x}_N]^T \mathbf{w} = \mathbf{X}^T \mathbf{w} \tag{6.12}$$

where $\mathbf{w} = [w_0, w_1, \cdots, w_d]^T$ and $\mathbf{x}_i = [1, x_{i1}, \cdots, x_{id}]^T$ are both augmented $d+1$ dimensional vectors and $\mathbf{X} = [\mathbf{x}_1, \cdots, \mathbf{x}_N]$. The Taylor series is

$$f(\mathbf{X}, \mathbf{w} + \Delta\mathbf{w}) = \mathbf{X}^T (\mathbf{w} + \Delta\mathbf{w}) = \mathbf{X}^T \mathbf{w} + \mathbf{X}^T \Delta\mathbf{w} \tag{6.13}$$

and the Jacobian is

$$\mathbf{J}_f(\mathbf{w}) = \mathbf{X}^T \tag{6.14}$$

Now Eq. (6.10) becomes

$$\begin{aligned} \mathbf{w}_{n+1} &= \mathbf{w}_n + (\mathbf{X}^T)^- (\mathbf{y} - \mathbf{X}^T \mathbf{w}_n) = \mathbf{w}_n + (\mathbf{X}\mathbf{X}^T)^{-1} \mathbf{X} (\mathbf{y} - \mathbf{X}^T \mathbf{w}_n) \\ &= \mathbf{w}_n + (\mathbf{X}\mathbf{X}^T)^{-1} \mathbf{X}\mathbf{y} - (\mathbf{X}\mathbf{X}^T)^{-1} (\mathbf{X}\mathbf{X}^T) \mathbf{w}_n \\ &= (\mathbf{X}\mathbf{X}^T)^{-1} \mathbf{X}\mathbf{y} = (\mathbf{X}^T)^- \mathbf{y} \end{aligned} \tag{6.15}$$

that is, the optimal parameter $\mathbf{w}^*$ can be found in a single step without iteration, same as that in Eq. (5.10).

In the following MATLAB code for the essential parts of the algorithm, `data(:,1)` and `data(:,2)` are the observed data containing N samples $x_1, \cdots, x_N$ in the first column and their corresponding function values $y_1, \cdots, y_N$ in the second column. In the code segment, the regression function (e.g., an exponential model) is symbolically defined, and then a function `NLregression` is called to estimate the function parameters:

```
sym x;                              % symbolic variables
a=sym('a',[3 1]);                   % symbolic parameters
f=@(x,a)a(1)*exp(a(2)*x)+a(3)|      % symbolic function
[a er]=NLregression(data(:,1),data(:,2),f,3);
                                    % call regression
```

The following code estimates the M parameters of the regression function based on Newton's method.

```
function [a er]=NLregression(x,y,f,M)   % nonlinear regression
    N=length(x);                % number of data samples
    a=sym('a',[M 1]);           % M symbolic parameters
    F=[];
    for i=1:N                   % for N given points
        F=[F; f(x(i),a)];       % evaluate function at each point
    end
    J=jacobian(F,a);            % symbolic Jacobian
    J=matlabFunction(J);        % convert to a true function
    F=matlabFunction(F);
    a1=ones(M,1);               % initial guesses for M parameters
    n=0;                        % iteration index
    da=1;
    while da>tol                % terminate if no progress made
        n=n+1;
        a0=a1;                  % update parameters
        J0=J(a0(1),a0(2));      % evaluate Jacobian based on a0
        Jinv=pinv(J0);          % pseudo inverse of Jacobian
        r=y-F(a0(1),a0(2))      % residual
        a1=a0+Jinv*r;           % Newton iteration
        er=norm(r);             % residual error
        da=norm(a0-a1);         % progress
    end
end
```

where `tol` is a small value for tolerance, such as 10^{-6}.

Example 6.1

Consider three sets of data (shown as the solid dots in the three columns in the plots in Fig. 6.1) are fitted by three different models (shown as the open circles in the plots):

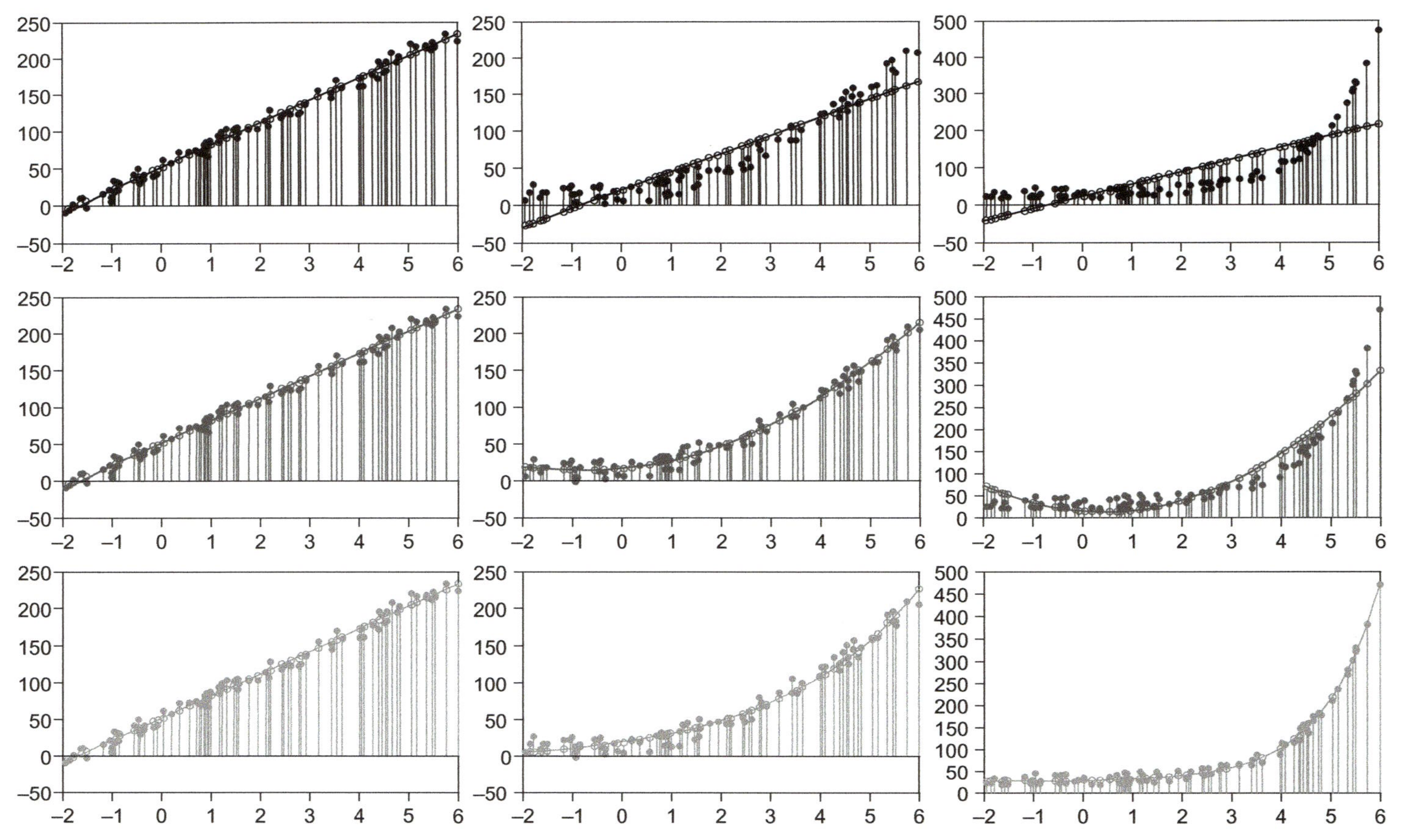

Fig. 6.1 Nonlinear regression with multiple datasets and models (1-D).

1. linear model: $y = f_1(x,a,b) = ax + b$
2. quadratic model: $y = f_2(x,a,b,c) = ax^2 + bx + c$
3. exponential model: $y = f_3(x,a,b,c) = a\,e^{bx} + c$

Table 6.1 gives the parameters for the three models to fit each of the three datasets.

Table 6.1

	Parameters	a	b	c	Error
	Model 1	30.337	51.652		7.4294
Dataset 1	Model 2	0.076	30.034	51.580	7.4223
	Model 3	5933.214	0.005	−5881.634	7.4221
	Model 1	24.519	20.089		21.0362
Dataset 2	Model 2	4.459	6.625	15.855	8.4559
	Model 3	29.791	0.346	−10.202	9.7402
	Model 1	32.608	22.008		53.0724
Dataset 3	Model 2	10.391	−9.089	12.140	28.3211
	Model 3	2.197	0.885	27.241	8.3393

We see that the first dataset can be fitted by all three models equally well with similar error (when $a \approx 0$ in model 2, it is approximately a linear functions), the second dataset can be well fitted by either the second or the third model, while the third dataset can only be fitted by the third exponential model.

Example 6.2

Based on the following three functions with four parameters $a = 1,\ b = 2,\ c = 3,\ d = 4$, three sets of synthetic 2-D data points are generated with some random noise added, shown in Fig. 6.2 as the dots in the three rows.

$$\begin{aligned} f_1(\mathbf{x}) &= ax_1 + bx_2 - c \\ f_2(\mathbf{x}) &= \exp(ax_1 + bx_2) - cx_1x_2 + d \\ f_3(\mathbf{x}) &= ax_1^2 - bx_1x_2 + cx_2^2 + d \end{aligned} \tag{6.16}$$

Then each of three datasets are fitted by the three functions as models while the optimal parameters are found by the Gauss–Newton method.

Table 6.2

	Parameters	a	b	c	d	Error
Dataset 1	Model 1	0.98	1.97	3.00		4.13
	Model 2	0.70	1.30	1.21	−4.42	6.87
	Model 3	−0.19	0.73	−0.12	−3.01	18.03
Dataset 2	Model 1	1.84	3.40	−6.09		14.70
	Model 2	0.99	2.01	2.97	3.98	3.75
	Model 3	0.37	1.13	2.66	4.87	32.45
Dataset 3	Model 1	−0.19	0.30	−5.34		16.82
	Model 2	−0.29	0.23	2.00	4.30	13.51
	Model 3	0.94	2.05	2.99	4.02	4.09

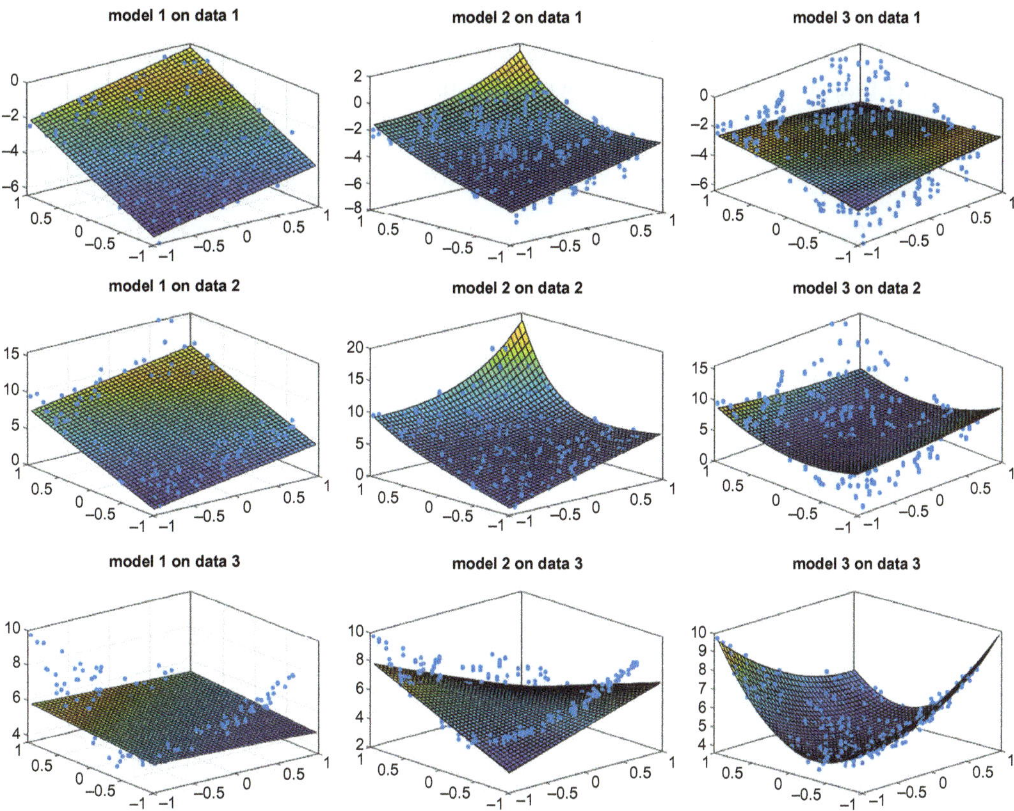

Fig. 6.2 Nonlinear regression with multiple datasets and models (2-D).

The *Levenberg–Marquardt algorithm (LMA)* or *damped least-squares method* is a method widely used to minimize a sum of squares error ε as in Eq. (6.3). The LMA is an interpolation or tradeoff between the gradient descent method based on the first-order derivative of the error function to be minimized, and the Gauss–Newton method based on both the first- and second-order derivatives of the error function. LMA can take advantage of both methods and is therefore more robust. Even if the initial guess is far from the solution corresponding to the minimum of the objective function, the iteration can still converge toward the solution.

Specifically, the LMA modifies Eq. (6.10)

$$\boldsymbol{\theta}_{n+1} = \boldsymbol{\theta}_n + \mathbf{J}_f^-(\boldsymbol{\theta}_n)[\mathbf{y} - \mathbf{f}(\mathbf{X}, \boldsymbol{\theta}_n)] = \boldsymbol{\theta}_n + (\mathbf{J}_n^T \mathbf{J}_n)^{-1} \mathbf{J}_n^T [\mathbf{y} - \mathbf{f}(\mathbf{X}, \boldsymbol{\theta}_n)] \tag{6.17}$$

to include an extra term proportional to the identity matrix $\mathbf{I}$

$$\boldsymbol{\theta}_{n+1} = \boldsymbol{\theta}_n + (\mathbf{J}_n^T \mathbf{J}_n + \lambda \mathbf{I})^{-1} \mathbf{J}_n^T [\mathbf{y} - \mathbf{f}(\mathbf{X}, \boldsymbol{\theta}_n)] \tag{6.18}$$

where $\lambda \geq 0$ is the non-negative damping factor, which is to be adjusted at each iteration. We recognize that this is actually similar to Eq. (5.34) for ridge regression.

In each iterative step, if λ is small and the error function $\varepsilon(\boldsymbol{\theta}_n)$ decreases quickly, indicating the Gauss–Newton is effective in the sense that the error function is well modeled as a quadratic function by the first- and second-order terms of the Taylor series, we will keep λ small. On the other hand, if ε decreases slowly, indicating the error function may not be well modeled as a quadratic function, due possibly to the fact that the current $\boldsymbol{\theta}_n$ is far away from the minimum, then we let λ take a large value, so that the term $\mathbf{J}_n^T\mathbf{J}_n$ becomes insignificant, and Eq. (6.18) approaches

$$\boldsymbol{\theta}_{n+1} = \boldsymbol{\theta}_n + \lambda\mathbf{J}_n^T(\mathbf{y} - \mathbf{f}(\boldsymbol{\theta})) = \boldsymbol{\theta}_n - \alpha\mathbf{g}_\varepsilon(\boldsymbol{\theta}) \tag{6.19}$$

Now $\boldsymbol{\theta}_n$ is updated along the negative direction of the gradient in Eq. (6.4)

$$\mathbf{g}_\varepsilon(\boldsymbol{\theta}) \propto \mathbf{J}^T(\mathbf{f}(\boldsymbol{\theta}) - \mathbf{y}) \tag{6.20}$$

with some step size α.

By adjusting the parameter λ, the LMA can switch between either of the two methods for the same purpose of minimizing $\varepsilon(\boldsymbol{\theta})$. When the value of λ is small, the iteration is dominated by the Gauss–Newton method that approximate the error function by a quadratic function based on the second-order derivatives. If the error reduces slowly, indicating the error function cannot be approximated by a quadratic function, a greater value of λ is used to switch to the gradient descent method.

6.2 Parameter Estimation by Natural Gradient Descent

In Section 4.1, we considered the maximum likelihood estimation (MLE) and maximum a posteriori (MAP) methods for estimating the parameter $\boldsymbol{\theta}$ of a probability model $p(\mathbf{x}|\boldsymbol{\theta})$ by maximizing the likelihood function $L(\boldsymbol{\theta}|\mathbf{x})$ or the posterior $p(\boldsymbol{\theta}|\mathbf{x})$. In the previous section we also considered how the gradient method is used to estimate the model parameter $\boldsymbol{\theta}$ by minimizing the squared error. Obviously we can also use the gradient method in MLE to maximize the likelihood. However, unlike the parameter $\boldsymbol{\theta} = [\theta_1, \cdots, \theta_m]^T$ which is treated as a vector in an m–D Euclidean space spanned by a set of orthonormal basis vectors, the family of distributions $p(\mathbf{x}|\boldsymbol{\theta})$ corresponding to all possible parameters $\boldsymbol{\theta}$ resides in a distribution space, an m-dimensional Riemannian manifold or space, in which no orthogonal basis is defined, and methods based on concepts such as inner product and Euclidean distance are no longer valid.

In this case, the difference between any two distributions $p(\mathbf{x}|\boldsymbol{\theta})$ and $p(\mathbf{x}|\boldsymbol{\theta}')$ can be measured by their Kullback–Leibler (KL) divergence $KL(p(\mathbf{x}|\boldsymbol{\theta})||p_p(\mathbf{x}|\boldsymbol{\theta}+\Delta\boldsymbol{\theta}))$, and the gradient method can still be used to find iteratively the optimal parameter that maximize the likelihood $L(\boldsymbol{\theta})$, similar to the steps that lead to Eq. (3.30). Specifically, we wanted to find the optimal $\Delta\boldsymbol{\theta}$ that maximally increases the likelihood, which is also subject to the constraint $||\Delta\boldsymbol{\theta}||^2 < C$

$$\Delta\boldsymbol{\theta}^* = \underset{||\Delta\boldsymbol{\theta}||^2<C}{\arg\max}\, L(\boldsymbol{\theta} + \Delta\boldsymbol{\theta}) \tag{6.21}$$

where $||\Delta\boldsymbol{\theta}||^2$ is defined as Eq. (B.275)

$$||\Delta\boldsymbol{\theta}||^2 = \frac{1}{2}\Delta\boldsymbol{\theta}^T\mathbf{F}(\boldsymbol{\theta})\Delta\boldsymbol{\theta} = KL(p(\mathbf{x}|\boldsymbol{\theta})||p_p(\mathbf{x}|\boldsymbol{\theta} + \Delta\boldsymbol{\theta})) < C \tag{6.22}$$

To solve this constrained optimization problem we set the derivative of the Lagrangian to zero

$$\frac{d}{d(\Delta\theta)}\left[L(\theta+\Delta\theta)+\mu\left(\frac{1}{2}\Delta\theta^T\mathbf{F}(\theta)\Delta\theta-C\right)\right]=\mathbf{g}_L(\theta)+\mu\mathbf{F}(\theta)\Delta\theta=\mathbf{0} \tag{6.23}$$

Solving this equation we get the optimal increment

$$\Delta\theta^*=-\frac{1}{\mu}\mathbf{F}^{-1}(\theta)\mathbf{g}_L(\theta) \tag{6.24}$$

and the optimal parameter can be found iteratively using this method of *natural gradient descent* (Eq. (B.3.5))

$$\theta_{n+1}=\theta_n+\Delta\theta_n=\theta_n+\lambda\mathbf{F}^{-1}\mathbf{g}_L(\theta_n)=\theta_n+\lambda\tilde{\mathbf{g}}_L(\theta_n) \tag{6.25}$$

where $\lambda = 1/\mu$ is the step size, and $\tilde{\mathbf{g}}_L(\theta_n)$ is the *natural gradient*

$$\tilde{\mathbf{g}}_L(\theta_n)=\mathbf{F}^{-1}\mathbf{g}_L(\theta_n) \tag{6.26}$$

Comparing this iteration with that in Eq. (2.15), we see that the natural gradient method is similar to the Newton–Raphson method in the sense that it is based on not only the gradient $\mathbf{g}_J(\theta)$ for the local first-order variation of the likelihood function $L(\theta)$ in the distribution space, but also the Fisher information $\mathbf{F}(\theta)$ for the more global second-order curvature in the space, by which the gradient direction is modified so that the iterative search process in the space can be carried out more efficiently.

PROBLEMS

1. Develop MATLAB code to implement the nonlinear least-squares regression algorithm considered in Chapter 6, and apply your code to resolve the 1-D nonlinear regression problem in Example 6.1 The datasets can be downloaded from here....
2. Apply your code to resolve the 2-D nonlinear regression problem such as that in Example 6.2. Three data sets can find the optimal model parameters (a, b, c, d) for your models.
   - dataset 1 in /e176/homework_e190/data3.txt
   - dataset 2 in /e176/homework_e190/data4.txt
   - dataset 3 in /e176/homework_e190/data5.txt

   You can read these data into your MATLAB code by `data=load('data3.txt')`. Each data file is organized in three columns: the first two are for the two independent variables x_1 and x_2 representing a set of data points in a 2-D space, while the third one is for the function values $f(x_1,x_2)$ corresponding to these points.

7 Logistic and Softmax Regression

7.1 Logistic Regression and Binary Classification

As mentioned previously, regression methods can be used as a binary classifier, when the regression function $f(\mathbf{x}, \boldsymbol{\theta})$ is thresholded by some constant c, which is assumed to be zero in the following without loss of generality. Once the model parameter $\boldsymbol{\theta}$ is obtained based on the training set $\mathcal{D} = \{\mathbf{X}, \mathbf{y}\}$, every point $\mathbf{x}$ in the d-dimensional feature space can be classified into one of two classes C_- and C_+

$$\text{if } f(\mathbf{x}) \begin{cases} < c \\ > c \end{cases} \quad \text{then} \quad \mathbf{x} \in \begin{cases} C_- \\ C_+ \end{cases} \tag{7.1}$$

When $c = 0$, the regression function $f(\mathbf{x})$ is thresholded by a unit step function $u(x)$ ($u(x) = 1$ if $x \geq 0$ but 0 otherwise), and the corresponding equation $f(\mathbf{x}) = 0$ represents a hypersurface in the d-dimensional space (a point if $d = 1$, a curve if $d = 2$, a surface or hypersurface if $d \geq 3$), by which the space is partitioned into two regions. In this case, $f(\mathbf{x})$ is called the *decision function* and the surface *decision boundary*. Now any point in the space is classified into either of the two classes, depending on whether it is on the positive side of the decision boundary ($f(\mathbf{x}) > 0$) or the negative side ($f(\mathbf{x}) < 0$).

This simple binary classifier suffers from the drawback that the classification result is binary and deterministic. It provides no additional information such as how reliable the result is, that is, how confident we can be regarding the result. Specifically, such a classifier simply classifies all points on one side of the boundary into the same class, without considering how far they are from the decision boundary, while one would naturally feel more confident about the classification of points far away from the boundary, than those close to it. In other words, this binary classifier may tend to overfit as it is sensitive to noisy data points close to the boundary.

This problem can be addressed by the method of *logistic regression*, which can provide the probability for any given data point to be classified to either of the two classes, as a quantitative measurement of the confidence for the classification. Based on such a probabilistic result, one can not only make a binary decision, but also gain the extra information in terms of the reliability of the decision.

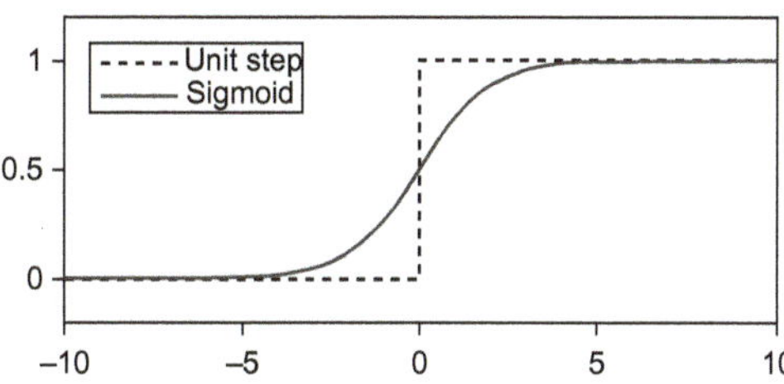

Fig. 7.1 Unit step vs. sigmoid.

The logistic regression is based on the same linear regression function $f(\mathbf{x}) = \mathbf{w}^T\mathbf{x}$. However, different from linear regression, in logistic regression, the linear function is not hard-thresholded by the unit step function to become either 0 or 1; instead, it is soft-thresholded by a sigmoid function $\sigma(f(\mathbf{x})) = \sigma(\mathbf{w}^T\mathbf{x})$ to produce a value between 0 and 1, as shown in Fig. 7.1, which can then be interpreted as the probability $p(\mathbf{x} \in C_+)$ for $\mathbf{x}$ to belong to C_+, and the probability for $\mathbf{x}$ to belong to C_- is simply $p(\mathbf{x} \in C_-) = 1 - p(\mathbf{x} \in C_+)$.

Similar to linear and nonlinear regression methods considered before, the goal of logistic regression is also to find the optimal model parameter $\mathbf{w}$ for the regression function $f(\mathbf{x}, \mathbf{w})$ to fit the training set optimally. But different from linear and nonlinear regression based on the least squares (LS) method that minimizes the squared error in Eq. (6.3), logistic regression is based on maximum likelihood (ML) estimation, which maximizes the *likelihood* of the model parameter, as discussed below.

The sigmoid function can be either the error function (cumulative Gaussian, erf)

$$\phi(x) = \int_{-\infty}^{x} \mathcal{N}(u|0,1)\, du = \frac{1}{\sqrt{2\pi}} \int_{-\infty}^{x} \exp\left(-\frac{1}{2N}u^2\right) du \tag{7.2}$$

or the logistic function

$$\sigma(x) = \frac{1}{1+e^{-x}} = \frac{e^x}{1+e^x} = \begin{cases} 1 & x \to \infty \\ 1/2 & x = 0 \\ 0 & x \to -\infty \end{cases} \tag{7.3}$$

In the following, we will only consider the logistic function for consistency and simplicity. Here is a set of properties of the logistic function

$$\begin{aligned} \sigma(-x) &= \frac{1}{1+e^x} = 1 - \frac{1}{1+e^{-x}} = 1 - \sigma(x) \\ \frac{d}{dx}\sigma(x) &= \frac{e^{-x}}{(1+e^{-x})^2} = \sigma(x)\sigma(-x) = \sigma(x)(1-\sigma(x)) \\ x &= \ln\left(\frac{\sigma(x)}{1-\sigma(x)}\right), \qquad e^x = \frac{\sigma(x)}{1-\sigma(x)} \end{aligned} \tag{7.4}$$

Moreover, a parameter w can be included in the logistic function to control the softness or smoothness of the transition between 0 and 1, as shown in Fig. 7.2.

$$\sigma(wx) = \begin{cases} 1/2 & w = 0 \\ 1/(1+e^{-wx}) & w > 0 \\ u(x) & w \to \infty \end{cases} \tag{7.5}$$

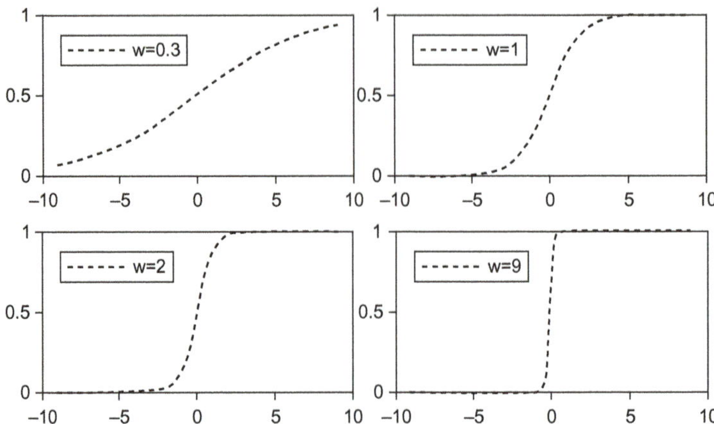

Fig. 7.2 Sigmoid with different parameter w.

We see that the argument x is thresholded by the logistic function with variable softness controlled by parameter w, between the two extreme cases of $\sigma(w\,x) = 1/2$ (no thresholding) when $w = 0$, and $\sigma(w\,x) = u(x)$ when $w \to \infty$ (hard thresholding).

Now the linear function $f(\mathbf{x}) = \mathbf{w}^T\mathbf{x} \in (-\infty, \infty)$ is mapped by the sigmoid function $\sigma(f(\mathbf{x}))$ to a real value between 0 and 1 (see Fig. 7.3), which is interpreted as the conditional probability for output $\hat{y} = 1$ given data point $\mathbf{x}$ and model parameter $\mathbf{w}$

$$p(\hat{y} = 1|\mathbf{x}, \mathbf{w}) = \sigma(\mathbf{w}^T\mathbf{x}) = \frac{1}{1 + e^{-\mathbf{w}^T\mathbf{x}}} \in (0, 1) \tag{7.6}$$

Comparing this with Eq. (7.5) for $w > 0$, we note that the weight vector $\mathbf{w}$ controls the smoothness of the thresholding.

This probability $p(\hat{y} = 1|\mathbf{x}, \mathbf{w})$ is also the probability $p(\mathbf{x} \in C_+)$ for $\mathbf{x} \in C_+$, and the probability $p(\mathbf{x} \in C_-)$ is

$$p(\hat{y} = 0|\mathbf{x}, \mathbf{w}) = 1 - p(y = 1|\mathbf{x}, \mathbf{w}) = 1 - \sigma(\mathbf{w}^T\mathbf{x}) = \sigma(-\mathbf{w}^T\mathbf{x}) = \frac{1}{1 + e^{\mathbf{w}^T\mathbf{x}}} \tag{7.7}$$

Now any given $\mathbf{x}$ can be classified into either class C_+ or class C_- depending on if $\sigma(\mathbf{w}^T\mathbf{x}) > 1/2$. Specifically, consider the following cases:

- if $f(\mathbf{x}) = \mathbf{w}^T\mathbf{x} \gg 0$, $\mathbf{x}$ is on the positive side far away from the plane $f(\mathbf{x}) = 0$, then the probability $p(\mathbf{x} \in C_+) = \sigma(f(\mathbf{x}))$ is close to 1;
- if $f(\mathbf{x}) = \mathbf{w}^T\mathbf{x} \ll 0$, $\mathbf{x}$ is on the negative side far away from the plane $f(\mathbf{x}) = 0$, then the probability $p(\mathbf{x} \in C_+) = \sigma(f(\mathbf{x}))$ is close to 0, and $p(\mathbf{x} \in C_-) = 1 - p(\mathbf{x} \in C_+)$ is close to 1;
- if $f(\mathbf{x}) \approx 0$, then both $p(\mathbf{x} \in C_+) = \sigma(f(\mathbf{x}))$ and $p(\mathbf{x} \in C_-) = 1 - \sigma(f(\mathbf{x}))$ are approximately 1/2, that is, if a point $\mathbf{x}$ is very close to the plane $f(\mathbf{x}) = 0$, then the probabilities for it to belong to either C_+ or C_- are about the same.

We note that the weight vector $\mathbf{w}$ plays the same role as w in Eq. (7.5) ($w > 0$), a larger value of $||\mathbf{w}||$ results in a sharper thresholding by the sigmoid function

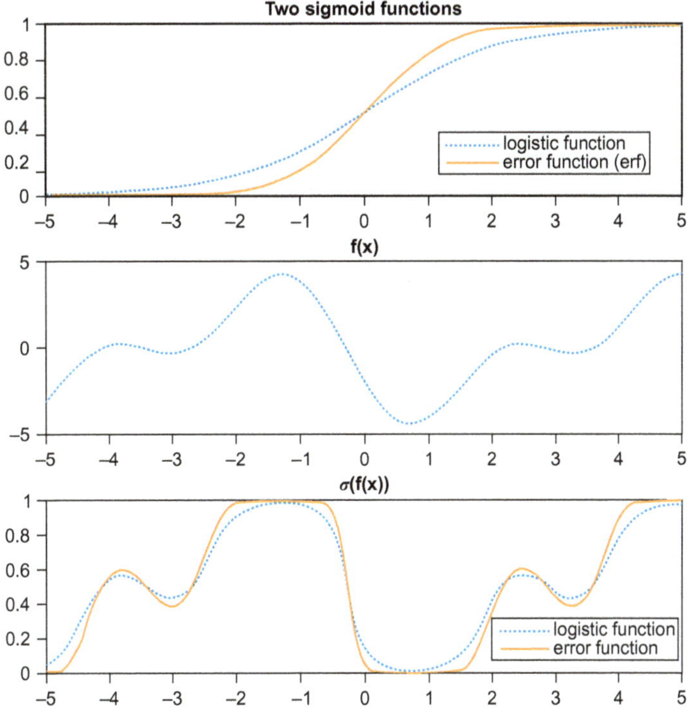

Fig. 7.3 Convert regression function to probability.

$\sigma(\mathbf{w}^T\mathbf{x})$. In other words, $\mathbf{w}$ determines not only how the classification is carried out based on the decision boundary, but also whether the classifier over and underfits, as further illustrated in the two extreme cases.

- $||\mathbf{w}|| \to \infty$: the value of the sigmoid function $\sigma(\mathbf{w}^T\mathbf{x})$ is approximately binary, close to either 0 or 1, based on which the feature space is hard-divided into two regions, each for one of the two classes. As such a binary classification is dictated mostly by a small number of data points close to the boundary between the two regions, it is prone to noise contained in these data points and tends to overfit.
- $||\mathbf{w}|| \to 0$: the sigmoid function approaches a constant $\sigma(\mathbf{w}^T\mathbf{x}) = 1/2$, totally insensitive to the locations of the data points, that is, the probability for any data point to belong to either of the two classes is always the same. As such a classifier is insensitive to the valid variations in the data, it underfits.

We therefore see that by adjusting $||\mathbf{w}||$ and thereby controlling the smoothness of the thresholding, we can make a proper tradeoff between overfitting and underfitting.

We label each sample $\mathbf{x}_n$ in the training set $\mathbf{X} = [\mathbf{x}_1, \cdots, \mathbf{x}_N]$ by either $y_n = 1$ if $\mathbf{x}_n \in C_+$, or $y_n = 0$ if $\mathbf{x}_n \in C_-$. Then the two probabilities in Eqs (7.6) and (7.7) can be combined into the following conditional probability for the model output $\hat{y}$ to match the ground truth labeling y_n

$$p(\hat{y} = y_n|\mathbf{x}_n, \mathbf{w}) = P(\hat{y} = 1|\mathbf{x}_n, \mathbf{w})^{y_n} p(\hat{y} = 0|\mathbf{x}_n, \mathbf{w})^{1-y_n} = \begin{cases} p(\hat{y} = 1|\mathbf{x}_n, \mathbf{w}) & \text{if } y_n = 1 \\ p(\hat{y} = 0|\mathbf{x}_n, \mathbf{w}) & \text{if } y_n = 0 \end{cases} \tag{7.8}$$

We can further get the likelihood function of $\mathbf{w}$ given all N samples in $\mathcal{D} = \{\mathbf{X}, \mathbf{y}\}$ (assumed i.i.d.)

$$L(\mathbf{w}|\mathcal{D}) \propto p(\mathcal{D}|\mathbf{w}) = p(\mathbf{y}|\mathbf{X}, \mathbf{w}) = \prod_{n=1}^{N} p(\hat{y} = y_n|\mathbf{x}_n, \mathbf{w})$$

$$= \prod_{n=1}^{N} p(\hat{y} = 1|\mathbf{x}_n)^{y_n} p(\hat{y} = 0|\mathbf{x}_n)^{1-y_n} = \prod_{n=1}^{N} \sigma(\mathbf{w}^T\mathbf{x}_n)^{y_n}(1 - \sigma(\mathbf{w}^T\mathbf{x}_n))^{1-y_n} \tag{7.9}$$

The optimal model parameter $\mathbf{w}$ can be found by either of the two following methods.

- *Maximum likelihood estimation (MLE)*
  The optimal model parameter $\mathbf{w}$ is defined as the one that maximizes the likelihood function $L(\mathbf{w}|\mathcal{D})$, or, equivalently, that minimizes the negative log likelihood as the objective function

$$J(\mathbf{w}) = -l(\mathbf{w}|\mathcal{D}) = -\log L(\mathbf{w}|\mathcal{D}) = -\sum_{n=1}^{N}\left[y_n \log \sigma(\mathbf{w}^T\mathbf{x}_n) + (1 - y_n)\log(1 - \sigma(\mathbf{w}^T\mathbf{x}_n))\right] \tag{7.10}$$

  This is the general MLE method that is widely used to obtain the most probable estimation of certain parameter of a model based on the given data.
- *Maximum a posteriori (MAP)*
  The optimal parameter $\mathbf{w}$ can also be defined as the one that maximizes the posterior probability

$$p(\mathbf{w}|\mathcal{D}) = p(\mathbf{w}|\mathbf{X}, \mathbf{y}) = \frac{p(\mathbf{y}|\mathbf{X}, \mathbf{w})\, p(\mathbf{w})}{p(\mathbf{y}|\mathbf{X})} \propto p(\mathbf{y}|\mathbf{X}, \mathbf{w})\, p(\mathbf{w}) \tag{7.11}$$

  where $p(\mathbf{y}|\mathbf{X}, \mathbf{w}) = L(\mathbf{w}|\mathcal{D})$ is the likelihood given in Eq. (7.9), $p(\mathbf{w})$ is the prior probability of $\mathbf{w}$ before observing the training data $\mathcal{D}$, and the denominator $p(\mathbf{y}\mathbf{X})$ is dropped as it is independent of variable $\mathbf{w}$.

  Based on the assumption that all components of $\mathbf{w}$ are independent of each other and they have zero mean and the same variance, we let the prior probability be

$$p(\mathbf{w}) = \mathcal{N}(\mathbf{0}, \sigma^2\mathbf{I}) = \frac{1}{(2\pi\sigma^2)^{d/2}} \exp\left(-\frac{||\mathbf{w}||^2}{2\sigma^2}\right) \propto \exp\left(-\frac{||\mathbf{w}||^2}{2\sigma^2}\right) \tag{7.12}$$

  where the normalizing factor $1/(2\pi\sigma^2)^{d/2}$ independent of $\mathbf{w}$ can be dropped. The assumed zero mean is attributed to the desire that the values of all weights in $\mathbf{w}$ should be around zero instead of large values on either positive or negative side, to avoid overfitting, as discussed in the following.

Now the posterior can be written as

$$p(\mathbf{w}|\mathcal{D}) \propto p(\mathbf{y}|\mathbf{X},\mathbf{w})\ p(\mathbf{w}) = \prod_{n=1}^{N} p(\hat{y}_n = y_n|\mathbf{x}_n,\mathbf{w})\ \mathcal{N}(\mathbf{0},\mathbf{\Sigma}_w)$$

$$\propto \prod_{n=1}^{N} \sigma(\mathbf{w}^T\mathbf{x}_n)^{y_n}(1-\sigma(\mathbf{w}^T\mathbf{x}_n))^{1-y_n}\ \exp\left(-\frac{||\mathbf{w}||^2}{2\sigma^2}\right) \tag{7.13}$$

The MAP method finds the optimal model parameter $\mathbf{w}$ that maximizes this posterior $p(\mathbf{w}|\mathcal{D})$ or equivalently minimizes the negative log posterior as the objective function

$$J(\mathbf{w}) = -\log p(\mathbf{w}|\mathcal{D}) = -\log p(\mathbf{y}|\mathbf{X},\mathbf{w}) - \left(-\frac{||\mathbf{w}||^2}{2\sigma^2}\right)$$

$$= -\sum_{n=1}^{N}\left[y_n \log\sigma(\mathbf{w}^T\mathbf{x}) + (1-y_n)\log(1-\sigma(\mathbf{w}^T\mathbf{x}))\right] + \frac{\lambda}{2}||\mathbf{w}||^2 \tag{7.14}$$

where λ is a hyperparameter defined as $\lambda = 1/\sigma^2$. Comparing this expression with Eq. (5.32) for ridge regression, we note that the second term can be treated as a penalty term that discourages large values of $\mathbf{w}$ to avoid overfitting. When $J(\mathbf{w})$ is minimized, a larger value of λ results in smaller values for $\mathbf{w}$, and the sigmoid function will be more smooth for a softer thresholding, and the classification is less affected by the possible noisy outliers close to the decision boundary. By adjusting λ, we can make a proper tradeoff between overfitting and underfitting.

In particular, if $\sigma \to \infty$ and $\lambda \to 0$, then $\exp(-\lambda||\mathbf{w}||^2/2) \to 1$, that is, the prior $p(\mathbf{w})$ approaches a uniform distribution (all values of $\mathbf{w}$ are equally likely), then the penalty term can be removed, and Eq. (7.14) becomes the same as Eq. (7.10), that is, the MAP solution that maximizes the posterior $p(\mathbf{w}|\mathcal{D})$ becomes the same as the MLE solution that maximizes the likelihood $L(\mathbf{w}|\mathcal{D})$.

Here we consider the MAP method which contains an extra term $||\mathbf{w}||^2//2\sigma^2$ and is therefore more general. We first find the gradient of objective function

$$\mathbf{g}_J(\mathbf{w}) = \frac{d}{d\mathbf{w}}J(\mathbf{w}) = \frac{d}{d\mathbf{w}}\left(-\log p(\mathbf{y}|\mathbf{X},\mathbf{w}) + \frac{\lambda}{2}||\mathbf{w}||^2\right)$$

$$= -\sum_{n=1}^{N}\frac{d}{d\mathbf{w}}\left[y_n\log\sigma(\mathbf{w}^T\mathbf{x}_n) + (1-y_n)\log(1-\sigma(\mathbf{w}^T\mathbf{x}_n))\right] + \lambda\mathbf{w}$$

$$= -\sum_{n=1}^{N}\left[\frac{y_n}{\sigma(\mathbf{w}^T\mathbf{x}_n)} - \frac{1-y_n}{1-\sigma(\mathbf{w}^T\mathbf{x}_n)}\right]\sigma(\mathbf{w}^T\mathbf{x}_n)(1-\sigma(\mathbf{w}^T\mathbf{x}_n))\ \mathbf{x}_n + \lambda\mathbf{w}$$

$$= \sum_{n=1}^{N}\left[\sigma(\mathbf{w}^T\mathbf{x}_n) - y_n\right]\mathbf{x}_n + \lambda\mathbf{w} = [\mathbf{x}_1,\cdots,\mathbf{x}_N]\begin{bmatrix}\sigma(\mathbf{w}^T\mathbf{x}_1) - y_1\\ \vdots \\ \sigma(\mathbf{w}^T\mathbf{x}_N) - y_N\end{bmatrix} + \lambda\mathbf{w}$$

$$= \mathbf{X}\mathbf{r} + \lambda\mathbf{w} \tag{7.15}$$

where we have used the properties of the logistic function in Eq. (7.4), and $\mathbf{r}(\mathbf{w})$ is the residual defined as

$$\mathbf{r} = \mathbf{r}(\mathbf{w}) = \begin{bmatrix} r_1(\mathbf{w}) \\ \vdots \\ r_N(\mathbf{w}) \end{bmatrix} = \begin{bmatrix} \sigma(\mathbf{w}^T\mathbf{x}_1) - y_1 \\ \vdots \\ \sigma(\mathbf{w}^T\mathbf{x}_N) - y_N \end{bmatrix} \tag{7.16}$$

containing all N residuals $r_n = \sigma(\mathbf{w}^T\mathbf{x}_n) - y_n \quad (n = 1, \cdots, N)$, the difference between the model prediction $0 < \sigma(\mathbf{w}^T\mathbf{x}_n) < 1$ and the ground truth labeling $y_n \in \{0, 1\}$.

Based on $\mathbf{g}_J(\mathbf{w})$ given in Eq. (7.15), we can find the optimal $\mathbf{w}$ that minimizes this objective function by the either the gradient descent method or Newton's method that depends on the second-order Hessian matrix $\mathbf{H}_J(\mathbf{w})$ in addition to the first-order gradient.

The following MATLAB code segment is for the gradient descent method. Here `X` contains the N samples $\mathbf{x}_1, \cdots, \mathbf{x}_N$ and `y` contains the corresponding binary labeling:

```
X=[ones(1,n); X];          % augmented X with x_0=1
w=ones(m+1,1);             % initialize weight vector
g=X*(sgm(w,X)-y)';         % gradient of log posterior
tol=10^(-6);               % tolerance
delta=0.1;                 % step size
while norm(g)>tol          % terminate when g is small enough
    w=w-delta*g;           % update weight by gradient descent
    g=X*(sgm(w,X)-y)';     % update gradient
end
```

where

```
function s=sgm(w,X)
    s=1./(1+exp(-w'*X));
end
```

We can also use Newton's method if the second-order Hessian matrix is available

$$\mathbf{H}_J(\mathbf{w}) = \frac{d}{d\mathbf{w}}\mathbf{g}_J(\mathbf{w}) = \frac{d}{d\mathbf{w}}(\mathbf{X}\mathbf{r}(\mathbf{w}) + \lambda||\mathbf{w}||) = \mathbf{X}\frac{d}{d\mathbf{w}}\mathbf{r}(\mathbf{w}) + \lambda$$
$$= \mathbf{X}\frac{d}{d\mathbf{w}}\begin{bmatrix} r_1(\mathbf{w}) \\ \vdots \\ r_N(\mathbf{w}) \end{bmatrix} + \lambda = \mathbf{X}\mathbf{J}_r(\mathbf{w}) + \lambda \tag{7.17}$$

where $\mathbf{J}_r(\mathbf{w}) = d\mathbf{r}(\mathbf{w})/d\mathbf{w}$ is the $N \times (d+1)$ Jacobian matrix of the vector function $\mathbf{r}(\mathbf{w})$, of which the nth row is

$$\frac{dr_n(\mathbf{w})}{d\mathbf{w}} = \frac{d}{d\mathbf{w}}\left[\sigma(\mathbf{w}^T\mathbf{x}_n) - y_n\right] = \sigma(\mathbf{w}^T\mathbf{x}_n)[1 - \sigma(\mathbf{w}^T\mathbf{x}_n)]\mathbf{x}_n = z_n\mathbf{x}_n \tag{7.18}$$

where we have defined $z_n = \sigma(\mathbf{w}^T\mathbf{x}_n)(1 - \sigma(\mathbf{w}^T\mathbf{x}_n))$, which is always greater than zero as $0 < \sigma(\mathbf{w}^T\mathbf{x}_n) < 1$. Now the Jacobian can be written as

$$\mathbf{J}_r(\mathbf{w}) = \begin{bmatrix} dr_1(\mathbf{w})/d\mathbf{w} \\ \vdots \\ dr_N(\mathbf{w})/d\mathbf{w} \end{bmatrix} = \begin{bmatrix} z_1\mathbf{x}_1 \\ \vdots \\ z_N\mathbf{x}_N \end{bmatrix} = \begin{bmatrix} z_1 & \cdots & 0 \\ \vdots & \ddots & \vdots \\ 0 & \cdots & z_N \end{bmatrix} \begin{bmatrix} \mathbf{x}_1^T \\ \vdots \\ \mathbf{x}_N^T \end{bmatrix} = \mathbf{Z}\mathbf{X}^T \quad (7.19)$$

where $\mathbf{Z} = \text{diag}(z_1, \cdots, z_N)$ is a positive definite diagonal matrix (as $z_n > 0$). Substituting $\mathbf{J}_r$ back into Eq. (7.17), we obtain

$$\mathbf{H}_J = \mathbf{X}\mathbf{J}_r + \lambda = \mathbf{X}\mathbf{Z}\mathbf{X}^T + \lambda \quad (7.20)$$

Since $\mathbf{Z}$ is positive definite, the following quadratic function is also positive, for any $\mathbf{y}$, we have

$$\mathbf{y}^T\mathbf{H}_J\mathbf{y} = \mathbf{y}^T\mathbf{X}\mathbf{Z}\mathbf{X}^T\mathbf{y} + \lambda\mathbf{y}^T\mathbf{y} = (\mathbf{X}^T\mathbf{y})^T\mathbf{Z}(\mathbf{X}^T\mathbf{y}) + \lambda||\mathbf{y}|| > 0 \quad (7.21)$$

indicating $\mathbf{H}_J$ is also positive definite, and the objective function $J(\mathbf{w})$ when approximated by the first three terms of its Taylor series is has a global minimum.

Having found $\mathbf{H}_J(\mathbf{w})$ as well as $\mathbf{g}_l = J(\mathbf{w})$ of the objective function $J(\mathbf{w})$, we can solve the minimization problem also by Newton's method

$$\mathbf{w}_{n+1} = \mathbf{w}_n - \mathbf{H}_J(\mathbf{w}_n)^{-1}\mathbf{g}_J(\mathbf{w}_n) = \mathbf{w}_n - (\mathbf{X}\mathbf{Z}\mathbf{X}^T)^{-1}\mathbf{X}\mathbf{r} \quad (7.22)$$

The following code segment estimates weight vector $\mathbf{w}$ by Newton method:

```
X=[ones(1,n); X];             % augmented X with x_0=1
w=ones(m+1,1);                % initialize weight vector
g=X*(sgm(w,X)-y)';            % gradient of log posterior
H=X*diag(s.*(1-s))*X';        % Hessian of log posterior
tol=10^(-6)
while norm(g)>tol             % terminate when g is small enough
    w=w-inv(H)*g;             % update weight by Newton--Raphson
    s=sgm(w,X);               % update sigma
    g=X*(s-y)';               % update gradient
    H=X*diag(s.*(1-s))*X';    % update Hessian
end
```

Once the model parameter $\mathbf{w}$ is available, an unlabeled test sample $\mathbf{x}_*$ of unknown class can be classified into either of the two classes C_1 and C_0, based on $p(y_* = 1|\mathbf{x}_*)$

$$\text{if} \quad p(y_* = 1|\mathbf{x}_*) = \sigma(\mathbf{w}^T\mathbf{x}_*) \begin{cases} > 1/2 \\ < 1/2 \end{cases}, \quad \text{then} \quad \mathbf{x}_* \in \begin{cases} C_1 \\ C_0 \end{cases} \quad (7.23)$$

In summary, we make a comparison between the linear and logistic regression methods, both can be used for binary classification. In linear regression, the function value $f(\mathbf{x}) = \mathbf{w}^T\mathbf{x}$ is hard-thresholded by the unit step function, so that any data point $\mathbf{x}$ is classified to class C_+ if $\mathbf{w}^T\mathbf{x} > 0$, or class C_- if $\mathbf{w}^T\mathbf{x} < 0$ in a deterministic fashion (with probability $p(\mathbf{x} \in C_+) \in \{0, 1\}$. Conversely, in logistic regression, the linear function $f(\mathbf{x}) = \mathbf{w}^T\mathbf{x}$ is soft-thresholded by the sigmoid function to become $p(\mathbf{x} \in C_+) = \sigma(f(\mathbf{x})) \in (0,\ 1)$, treated as the probability for $\mathbf{x}$ to belong to class C_+. Although the same linear model $f(\mathbf{x}) = \mathbf{w}^T\mathbf{x}$ is used in both cases, the two methods obtain the model parameter $\mathbf{w}$ by minimizing very different objective functions (Eqs (5.8) and (7.14)).

7.2 Softmax Regression for Multiclass Classification

In a multiclass classification problem, an unlabeled data point $\mathbf{x}$ is to be classified into one of $K > 2$ classes $\{C_1, \cdots, C_K\}$, based on the training set $\mathcal{D} = \{(\mathbf{x}_n, y_n),\ n = 1, \cdots, N\}$, where $y_n \in \{1, \cdots, K\}$ is an integer indicating $\mathbf{x}_n \in C_{y_n}$

Any binary classifier, such as the logistic regression just considered above, can be used to solve such a multiclass classification problem in either of the following two ways:

- One-vs-one (1V1): construct $K(K-1)/2$ binary classifiers for every pair of the K classes based on the training set $\{(C_i, C_j)\quad i = 2, \cdots, K,\ j = 1, \cdots, i-1\}$, then an unlabeled sample $\mathbf{x}$ is classified to one of the K classes with maximum votes out of this many binary classifications.
- One-vs-rest (1VR). Convert a K-class problem into K binary problems by regrouping the K classes into two classes $C_+ = C_i$ and $C_- = \{C_j | j = 1, \cdots, K,\ j \neq i\}$, and get the corresponding discriminant function $f(\mathbf{x})$ of the binary problem

$$\text{if } f_i(\mathbf{x}) \begin{cases} > 0 \\ < 0 \end{cases}, \text{ then } \begin{cases} \mathbf{x} \in C_+ = C_i \\ \mathbf{x} \in C_- \end{cases} \tag{7.24}$$

which represents quantitatively how much a given $\mathbf{x}$ belongs to class $C_+ = C_i$, instead of C_- containing the rest $K-1$ classes. This process is repeated K times for all $i = 1, \cdots, K$ to get

$$f_k(\mathbf{x}) = \max\{f_1(\mathbf{x}), \cdots, f_K(\mathbf{x})\}, \qquad (k = 1, \cdots, K) \tag{7.25}$$

which indicates $\mathbf{x}$ belongs to class C_k.

Alternatively, a multiclass problem with $K > 2$ can also be solved by *multinomial logistic* or *softmax regression*, which can be considered as a generalized version of the logistic regression method, based on the *softmax function* of K variables $z_1, \cdots, z_K$

$$s_k = s(z_k) = \frac{e^{z_k}}{\sum_{l=1}^K e^{z_l}}, \qquad (k = 1, \cdots, K) \tag{7.26}$$

with the following properties

$$0 < s_k < 1, \qquad \sum_{k=1}^K s_k = 1 \tag{7.27}$$

and

$$\frac{\partial s(z_i)}{\partial z_j} = \frac{\delta_{ij} e^{z_i} \sum_{l=1}^K e^{z_l} - e^{z_i} e^{z_j}}{\left(\sum_{l=1}^K e^{z_l}\right)^2}$$
$$= \delta_{ij} s(z_i) - s(z_i) s(z_j) = s(z_i)[\delta_{ij} - s(z_j)] \tag{7.28}$$

where δ_{ij} is the Dirac delta function that is 1 if $i = j$, but 0 otherwise.

Similar to the sigmoid function used to model the conditional probability $p(\hat{y} = 1|\mathbf{x}, \mathbf{w})$ for any data point $\mathbf{x}$ to belong to class C_+ based on model parameter $\mathbf{w}$ in Eq. (7.6), here the softmax function s_k is used to model the conditional probability for $\mathbf{x}$ to belong to class C_k based on some model parameter $\mathbf{W} = [\mathbf{w}_1, \cdots, \mathbf{w}_K]$

$$p(\hat{y} = k|\mathbf{x}, \mathbf{W}) = s_k = s(\mathbf{w}_k^T \mathbf{x}) = \frac{e^{\mathbf{w}_k^T \mathbf{x}}}{\sum_{l=1}^K e^{\mathbf{w}_l^T \mathbf{x}}}, \qquad (k = 1, \cdots, K) \tag{7.29}$$

where $\mathbf{W} = [\mathbf{w}_1, \cdots, \mathbf{w}_K]$ is composed of K weight vectors each associated with one of the K classes $\{C_1, \cdots, C_K\}$, to be determined in the training process based on training set $\mathcal{D}$. Same as in logistic regression, here both $\mathbf{x}$ and $\mathbf{w}$ are augmented $d + 1$ dimensional vectors.

We note that the inner product $\mathbf{w}^T \mathbf{x} = ||\mathbf{w}|| \; ||\mathbf{x}|| \cos\theta$ of vectors $\mathbf{w}$ and $\mathbf{x}$ is inversely related to the angle θ between the two vectors, that is, if a data sample $\mathbf{x}$ has a smaller angular distance to $\mathbf{w}_k$, then it has a larger inner product $\mathbf{w}_k^T \mathbf{x}$ and thereby greater probability $p(\hat{y} = k|\mathbf{x})$ to belong to class C_k. We therefore see that weight vectors $\mathbf{w}_1, \cdots, \mathbf{w}_K$ as the parameters of the softmax regression model actually represent the angular directions of the corresponding classes in the feature space.

Specially, when $K = 2$, the softmax function defined in Eq. (7.29) becomes the logistic function

$$s_1 = \frac{e^{\mathbf{w}_1^T \mathbf{x}}}{e^{\mathbf{w}_0^T \mathbf{x}} + e^{\mathbf{w}_1^T \mathbf{x}}} = \frac{1}{1 + e^{-(\mathbf{w}_1 - \mathbf{w}_0)^T \mathbf{x}}} = \sigma(\mathbf{w}^T \mathbf{x})$$
$$s_0 = \frac{e^{\mathbf{w}_0^T \mathbf{x}}}{e^{\mathbf{w}_0^T \mathbf{x}} + e^{\mathbf{w}_1^T \mathbf{x}}} = \frac{1}{1 + e^{(\mathbf{w}_1 - \mathbf{w}_0)^T \mathbf{x}}} = \sigma(-\mathbf{w}^T \mathbf{x}) = 1 - \sigma(\mathbf{w}^T \mathbf{x}) \tag{7.30}$$

where we have defined $\mathbf{w} = \mathbf{w}_1 - \mathbf{w}_0$, as the vector between the two classes of C_0 and C_1 (or C_+ and C_-), which is the normal direction of a decision plane separating the two classes, as shown in Fig. 7.4 for a 2-D case. Any point $\mathbf{x}$ of unknown class can be therefore classified into either of the two classes depending on whether its projection $\mathbf{w}^T \mathbf{x}$ onto $\mathbf{w}$ is positive or negative, that is, on which side of the decision plane it is located.

For mathematical convenience, we also label each sample $\mathbf{x}_n$ in the training set by a binary vector $\mathbf{y}_n = [y_{n1}, \cdots, y_{nK}]^T$, in addition to its integer labeling $y_n \in \{1, \cdots, K\}$. If $y_n = k$ indicating $\mathbf{x}_n \in C_k$, then the kth component of $\mathbf{y}_n$ is $y_{nk} = 1$, while other components are zero $y_{nl} = 0$ for all $l \neq k$. Note that all components of $\mathbf{y}_n$ add up to 1. The N training samples in $\mathbf{X} = [\mathbf{x}_1, \cdots, \mathbf{x}_N]$ are also labeled by their corresponding binary labelings in $\mathbf{Y} = [\mathbf{y}_1, \cdots, \mathbf{y}_N]$.

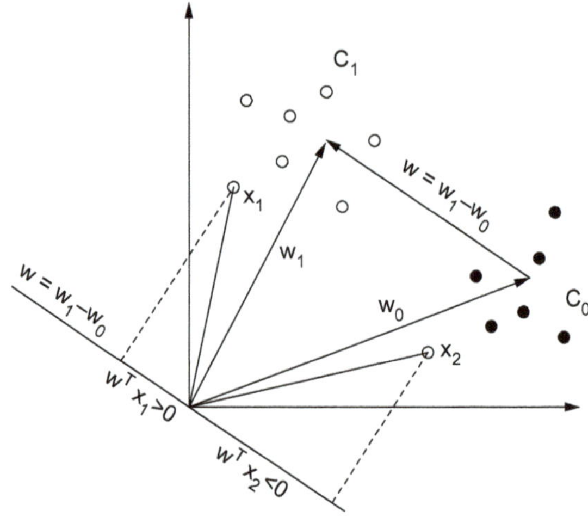

Fig. 7.4 Softmax classification.

We define the probability for $\mathbf{x}_n \in C_k$ as the softmax function based on the linear function $\mathbf{w}_k^T \mathbf{x}_n$

$$s_{nk} = p(\hat{y} = k|\mathbf{x}_n) = \frac{e^{\mathbf{w}_k^T \mathbf{x}_n}}{\sum_{l=1}^{K} e^{\mathbf{w}_l^T \mathbf{x}_n}}, \qquad (k = 1, \cdots, K, \;\; n = 1, \cdots, N) \tag{7.31}$$

and the probability for $\mathbf{x}_n$ to be correctly classified into class C_{y_n} can be written as the following product of K factors (of which $K - 1$ are equal to 1)

$$p(\hat{y} = y_n|\mathbf{x}_n, \mathbf{W}) = \prod_{k=1}^{K} p(\hat{y} = k|\mathbf{x}_n, \mathbf{W})^{y_{nk}} = \prod_{k=1}^{K} s_{nk}^{y_{nk}} \tag{7.32}$$

Our goal is to find these weight vectors in $\mathbf{W} = [\mathbf{w}_1, \cdots, \mathbf{w}_K]$ as the parameter of the softmax model for it to optimally fit the N i.i.d. data points in the training set, so that that the following likelihood is maximized

$$\begin{aligned} L(\mathbf{W}|\mathcal{D}) \propto p(\mathcal{D}|\mathbf{W}) &= \prod_{n=1}^{N} p(\hat{y} = y_n|\mathbf{x}_n, \mathbf{W}) = \prod_{n=1}^{N} \left(\prod_{k=1}^{K} s_{nk}^{y_{nk}} \right) \\ &= \prod_{n=1}^{N} \prod_{k=1}^{K} \left(\frac{e^{\mathbf{w}_k^T \mathbf{x}_n}}{\sum_{l=1}^{K} e^{\mathbf{w}_l^T \mathbf{x}_n}} \right)^{y_{nk}} \end{aligned} \tag{7.33}$$

Equivalently we can also minimize the negative log likelihood as the objective function

$$
\begin{aligned}
J(\mathbf{W}) = -l(\mathbf{W}|\mathcal{D}) &= -\log L(\mathbf{w}|\mathcal{D}) = -\sum_{n=1}^{N}\sum_{k=1}^{K} y_{nk} \log\left(\frac{e^{\mathbf{w}_k^T\mathbf{x}_n}}{\sum_{l=1}^{K} e^{\mathbf{w}_l^T\mathbf{x}_n}}\right) \\
&= -\sum_{n=1}^{N}\sum_{k=1}^{K} y_{nk}\left(\mathbf{w}_k^T\mathbf{x}_n - \log\sum_{l=1}^{K} e^{\mathbf{w}_l^T\mathbf{x}_n}\right) \\
&= -\sum_{n=1}^{N}\left[\sum_{k=1}^{K} y_{nk}\mathbf{w}_k^T\mathbf{x}_n - \sum_{k=1}^{K} y_{nk}\left(\log\sum_{l=1}^{K} e^{\mathbf{w}_l^T\mathbf{x}_n}\right)\right] \\
&= -\sum_{n=1}^{N}\left[\sum_{k=1}^{K} y_{nk}\mathbf{w}_k^T\mathbf{x}_n - \log\sum_{l=1}^{K} e^{\mathbf{w}_l^T\mathbf{x}_n}\right]
\end{aligned} \tag{7.34}
$$

The last equality is owing to the fact that the last summation of K terms is independent of k and $\sum_{k=1}^{K} y_{nk} = 1$. Also, same as in the case of $K = 2$, to avoid overfitting, we could further include in this objective function extra regularization terms $\lambda||\mathbf{w}_k||^2/2, \quad (k = 1,\cdots,K)$ to penalize large weights for smoother thresholding

$$
J(\mathbf{W}) = -l(\mathbf{W}|\mathcal{D}) + \frac{\lambda}{2}\sum_{k=1}^{K}||\mathbf{w}_k||^2 \tag{7.35}
$$

To find the optimal $\mathbf{W}$ that minimize this objective function $J(\mathbf{W})$, we find its gradient vector with respective to each of its K columns $\mathbf{w}_i$, $(i = 1,\cdots,K)$

$$
\begin{aligned}
\mathbf{g}_i(\mathbf{W}) = \frac{d}{d\mathbf{w}_i}J(\mathbf{W}) &= \frac{d}{d\mathbf{w}_i}\left[-l(\mathbf{W}|\mathcal{D}) + \frac{\lambda}{2}\sum_{k=1}^{K}||\mathbf{w}_k||^2\right] \\
&= -\sum_{n=1}^{N}\left[\sum_{k=1}^{K} y_{nk}\frac{d}{d\mathbf{w}_i}\left(\mathbf{w}_k^T\mathbf{x}_n\right) - \frac{d}{d\mathbf{w}_i}\left(\log\sum_{l=1}^{K} e^{\mathbf{w}_l^T\mathbf{x}_n}\right)\right] + \lambda\mathbf{w}_i \\
&= -\sum_{n=1}^{N}\left[\sum_{k=1}^{K} y_{nk}\delta_{ik}\mathbf{x}_n - \left(\frac{e^{\mathbf{w}_i^T\mathbf{x}_n}}{\sum_{l=1}^{K} e^{\mathbf{w}_l^T\mathbf{x}_n}}\right)\mathbf{x}_n\right] + \lambda\mathbf{w}_i \\
&= \sum_{n=1}^{N}(s_{in} - y_{in})\,\mathbf{x}_n + \lambda\mathbf{w}_i = [\mathbf{x}_1,\cdots,\mathbf{x}_N]\begin{bmatrix} s_{i1} - y_{i1} \\ \vdots \\ s_{iN} - y_{iN}\end{bmatrix} + \lambda\mathbf{w}_i \\
&= [\mathbf{x}_1,\cdots,\mathbf{x}_N]\begin{bmatrix} r_1 \\ \vdots \\ r_N\end{bmatrix} + \lambda\mathbf{w}_i = \mathbf{X}\mathbf{r}_i + \lambda\mathbf{w}_i
\end{aligned} \tag{7.36}
$$

where $r_n = s_{in} - y_{in}$ is the residual, the difference between the model output s_{in} and the binary labeling y_{in}, the ith component of $\mathbf{y}_n$ that labels $\mathbf{x}_n$ as well as y_n.

As s_{in} is a function of all K weight vectors in $\mathbf{W} = [\mathbf{w}_1,\cdots,\mathbf{w}_K]$, so is $r_{in} = s_{in} - y_{in}$, and $\mathbf{r}_i$ can be explicitly expressed as $\mathbf{r}_i(\mathbf{W})$. We note that Eq. (7.36) takes the same form as Eq. (7.15) for $K = 2$. When specially $K = 2$ with $k \in \{0,\ 1\}$, the two equations become the same. We therefore see that logistic regression is actually a special case of softmax regression.

We stack all K such $d+1$ dimensional gradient vectors $\mathbf{g}_1, \cdots, \mathbf{g}_K$ together, and get a $(d+1)K$ dimensional gradient vector of the objective function $J(\mathbf{W})$ with respect to all K parameter vectors $\mathbf{w}_1, \cdots, \mathbf{w}_K$

$$\mathbf{g}_J = \begin{bmatrix} \mathbf{g}_1 \\ \vdots \\ \mathbf{g}_K \end{bmatrix} \tag{7.37}$$

based on which the optimal parameters in $\mathbf{W}$ that minimize the objective function $J(\mathbf{W})$ can be found by the gradient descent method.

We can also find the optimal $\mathbf{W}$ by the Newton's method if we can further find the Hessian matrix $\mathbf{H}_J$ as the second derivative of $J(\mathbf{W})$, by taking the derivative of the ith gradient vector $\mathbf{g}_i(\mathbf{W})$ $(i = 1, \cdots, K)$ with respect to the jth weight vector $\mathbf{w}_j$ $(j = 1, \cdots, K)$ to obtain the second-order derivative of $J(\mathbf{W})$ with respect to both $\mathbf{w}_i$ and $\mathbf{w}_j$, $(i, j = 1, \cdots, K)$

$$\begin{aligned} \mathbf{H}_{ij} &= \frac{d^2}{d\mathbf{w}_j d\mathbf{w}_i} J(\mathbf{W}) = \frac{d}{d\mathbf{w}_j} \frac{d}{d\mathbf{w}_i} J(\mathbf{W}) = \frac{d}{d\mathbf{w}_j} \mathbf{g}_i = \frac{d}{d\mathbf{w}_j} [\mathbf{X}\mathbf{r}_i + \lambda \mathbf{w}_i] \\ &= \mathbf{X} \frac{d\mathbf{r}_i}{d\mathbf{w}_j} + \delta_{ij}\lambda = \mathbf{X}\mathbf{J}_{ij} + \delta_{ij}\lambda \end{aligned} \tag{7.38}$$

where $\mathbf{J}_{ij} = d\mathbf{r}_i / d\mathbf{w}_j$ is the $N \times (d+1)$ Jacobian matrix of $\mathbf{r}_i$ with respect to $\mathbf{w}_j$, of which the nth row is the transpose of the following vector

$$\begin{aligned} \frac{d}{d\mathbf{w}_j} r_{in} &= \frac{d}{d\mathbf{w}_j}(s_{in} - y_{in}) = \frac{d}{d\mathbf{w}_j} s_{in} = \frac{d}{d\mathbf{w}_j} s(\mathbf{w}_i^T \mathbf{x}_n) \\ &= s_{in}\left(\delta_{ij} - s_{jn}\right) \frac{d}{d\mathbf{w}_j}\left(\mathbf{w}_j^T \mathbf{x}_n\right) = z_n \mathbf{x}_n \qquad (n = 1, \cdots, N) \end{aligned} \tag{7.39}$$

where we have used Eq. (7.28) and defined $z_n = s_{in}(\delta_{ij} - s_{jn})$. Now the Jacobian can be written as

$$\mathbf{J}_{ij} = \begin{bmatrix} dr_{i1}/d\mathbf{w}_j \\ \vdots \\ dr_{iN}/d\mathbf{w}_j \end{bmatrix} = \begin{bmatrix} z_1 & \cdots & 0 \\ \vdots & \ddots & \vdots \\ 0 & \cdots & z_N \end{bmatrix} \begin{bmatrix} \mathbf{x}_1^T \\ \vdots \\ \mathbf{x}_N^T \end{bmatrix} = \mathbf{Z}\mathbf{X}^T \tag{7.40}$$

where $\mathbf{Z} = \text{diag}(z_1, \cdots, z_N)$ is an $N \times N$ diagonal matrix. Substituting $\mathbf{J}_r(\mathbf{W})$ back into the expression for $\mathbf{H}_{ij}$, we get:

$$\mathbf{H}_{ij} = \mathbf{X}\mathbf{J}_r + \delta_{ij}\lambda = \mathbf{X}\mathbf{Z}\mathbf{X}^T + \delta_{ij}\lambda = \sum_{n=1}^{N} s_{in}(\delta_{ij} - s_{jn})\mathbf{x}_n \mathbf{x}_n^T + \delta_{ij}\lambda \tag{7.41}$$

Here $\mathbf{H}_{ij}$ is a matrix of dimension $(d+1) \times (d+1)$, corresponds to $\mathbf{w}_i$ and $\mathbf{w}_j$ $(i, j = 1, \cdots, K)$. We further stack all $K \times K$ such matrices $\mathbf{H}_{ij}$ together to get the $K(d+1) \times K(d+1)$ dimensional full Hessian matrix $\mathbf{H}_J$ of $J(\mathbf{W})$ with respect to all K vectors in $\mathbf{W} = [\mathbf{w}_1, \cdots, \mathbf{w}_K]$

$$\mathbf{H}_J = \begin{bmatrix} \mathbf{H}_{11} & \cdots & \mathbf{H}_{1K} \\ \vdots & \ddots & \vdots \\ \mathbf{H}_{K1} & \cdots & \mathbf{H}_{KK} \end{bmatrix} \tag{7.42}$$

Now we can find the optimal parameter $\mathbf{W}$ iteratively by the Newton–Raphson method based on both $\mathbf{H}_J$ and $\mathbf{g}_J$

$$\begin{bmatrix} \mathbf{w}_1 \\ \vdots \\ \mathbf{w}_K \end{bmatrix}_{n+1} = \begin{bmatrix} \mathbf{w}_1 \\ \vdots \\ \mathbf{w}_K \end{bmatrix}_n - \mathbf{H}_n^{-1}\,\mathbf{g}_n = \begin{bmatrix} \mathbf{w}_1 \\ \vdots \\ \mathbf{w}_K \end{bmatrix}_n - \begin{bmatrix} \mathbf{H}_{11} & \cdots & \mathbf{H}_{1K} \\ \vdots & \ddots & \vdots \\ \mathbf{H}_{K1} & \cdots & \mathbf{H}_{KK} \end{bmatrix}_n^{-1} \begin{bmatrix} \mathbf{g}_1 \\ \vdots \\ \mathbf{g}_K \end{bmatrix}_n \tag{7.43}$$

Once $\mathbf{W} = [\mathbf{w}_1, \cdots, \mathbf{w}_K]$ is available, any unlabeled sample $\mathbf{x}$ of unknown class can be classified into one of the K classes with the maximum conditional probability given in Eq. (7.29)

$$\text{if} \quad p(\hat{y} = k|\mathbf{x}) = \max_l \; p(\hat{y} = l|\mathbf{x}), \quad \text{then} \quad \mathbf{x} \in C_k \tag{7.44}$$

Whether we should use softmax regression or K logistic regressions for a problem of K classes $C_1, \cdots, C_K$ depends on the nature of the classes. The method of softmax regression is suitable if the K classes are mutually exclusive and independent, as assumed by the method. Otherwise, K logistic regression binary classifiers are more suitable.

Here is a MATLAB function for estimating the K weight vectors in $\mathbf{W} = [\mathbf{w}_1, \cdots, \mathbf{w}_K]$ based on the training set $\mathcal{D} = \{\mathbf{X}, \mathbf{y}$ and the hyperparameter λ:

```
function W=softmaxRegression(X,y,lambda)
    [d N]=size(X);
    K=length(unique(y));       % number of classes
    X=[ones(1,N); X];          % augmented data points
    d=d+1;
    Y=zeros(K,N);
    for n=1:N
        Y(y(n),n)=1;           % generate binary labeling Y
    end
    W=zeros(d*K,1);            % initial guess of K weight vectors
    I=eye(K);
    s=zeros(K,N);
    phi=zeros(K,N);            % softmax functions
    gi=zeros(d,1);             % ith gradient
    Hij=zeros(d,d);            % ij-th Hessian
    er=9;
    tol=10^(-6);
    it=0;
    while er>tol
        it=it+1;
        W2=reshape(W,d,K);     % weight vectors in d x K 2-D array
        g=[];                  % total gradient vector
        for i=1:K
            for n=1:N
```

```
                xn=X(:,n);  % get the nth sample
                t=0;
                for k=1:K
                    wk=W2(:,k); % get the kth weight vector
                    s(k,n)=exp(wk'*xn);
                    t=t+s(k,n);
                end
                phi(i,n)=s(i,n)/t; % softmax function
            end
            gi=X*(phi(i,:)-Y(i,:))'+lambda*W2(:,i);   % ith gradient
            g=[g; gi];      % stack all gradients into a long vector
        end
        H=[];                   % total Hessian
        z=zeros(N,1);
        for i=1:K               % for the ith block row
            Hi=[];
            for j=1:K           % for the jth block in ith row
                for n=1:N
                   z(n)=phi(i,n)*((i==j)-phi(j,n));
                end
                Hij=X*diag(z)*X';
                Hi=[Hi Hij]; % append jth block
            end
            H=[H; Hi];          % append ith blow row
        end
        H=H+lambda*eye(d*K); % include regulation term
        W=W-inv(H)*g;          % update W by Newton's method
        er=norm(g);
    end
    W=reshape(W,d,K);           % reshape weight into d x K array
end
```

Here is the function for the classification of the unlabeled data samples in matrix $\mathbf{X}$ based on the weight vectors in $\mathbf{W} = [\mathbf{w}_1, \cdots, \mathbf{w}_K]$

```
function yhat=softmaxClassify(W,X)
    [d N]=size(X);          % dataset to be classified
    [d K]=size(W);          % model parameters
    X=[ones(1,N); X];       % augmented data points
    yhat=zeros(N,1);
    for n=1:N               % for each of the N samples
        xn=X(:,n);          % nth sample point
        t=0;
        for k=1:K
            wk=W(:,k);      % kth weight vector
```

```
            s(k)=exp(wk'*xn);
            t=t+s(k);
        end
        pmax=0;
        for k=1:K
            p=s(k)/t;     % probability based on softmax function
            if p>pmax
                kmax=k; pmax=p;
            end
        end
        yhat(n)=kmax;     % predicted class labeling
    end
end
```

PROBLEMS

1. Implement the method of logistic regression for binary classification, based on both (a) the gradient descent and (b) the Newton–Raphson methods. Then apply the code to these datasets:
   - dataset 1 /e176/homework/2ClassData.txt
   - dataset 2 /e176/homework/2CocentricData.txt
   - dataset 3 /e176/homework/XOR.txt

   Each data file contains three columns for x_1, x_2, y, respectively. Here y is the binary labeling (1 or −1), indicating to which of the two classes point $\mathbf{x} = [x_1, x_2]^T$ belongs.

   Compare the performances of the two methods in terms of the number of iterations and percentage error.

   Visualize your classification results by plotting all 2-D data points $\mathbf{x}_n$ in the training set and coloring them first based on their labeling y_n (0 or 1) to show their true class identity (e.g., red for class 1 and blue for class 2), and then based on your classification result y'_n. Comparing these two plots you will see how effective your classification method is.

   Show your classification results in the form of a 2×2 confusion matrix, where the entry on the ith row and jth column ($i, j = 1, 2$) is the number data points labeled by y to belong to the ith class but classified to the jth class. Ideally, if all data points are classified correctly, the confusion matrix is diagonal. Also give the percentage error, defined as the ratio of the number of misclassified data points and the total number of data points.
2. Develop MATLAB code for multiclass classification based on both (a) one-vs-one method and (b) one-vs-rest method, based on logistic regression for binary classification, then carry out classification to the following three-class datasets. Compare their results of the two methods in terms of execution time and error rates.

- dataset 1 in /e176homework/3ClassData1.txt
  The first three columns are for the components for each data point $\mathbf{x}_n = [x_{1n},\ x_{2n},\ x_{3n}]^T$, and the fourth column is the corresponding labeling y_n, indicating to which of the $K = 3$ classes $\mathbf{x}_n$ belongs.
- dataset 2 in /e176/homework/3ClassData2.txt
  The first two columns are for the components for each data point $\mathbf{x}_n = [x_{1n},\ x_{2n}]^T$, and the third column is the corresponding labeling y_n, indicating to which of the $K = 3$ classes $\mathbf{x}_n$ belongs.
- The iris dataset (https://en.wikipedia.org/wiki/Iris_flower_data_set).
  This dataset can be read into your code by MATLAB command `x=iris_dataset`, where `x` is a 4×150 matrix, each of the 150 columns is for one of the 150 data samples, containing 50 samples for each of the three species of iris (iris setosa, iris virginica, and iris versicolor).

Show your classification results in the form of a confusion matrix, of which the entry in the ith row and jth column is the number of data points in the training set belonging to the ith class according to the labeling y but classified into the jth class by your classifier. Also give the percentage error, defined as the number of data samples misclassified over the total number of samples. Compare the percentage error of the three discussed methods for each of the three datasets.

Visualize both the training dataset (the ground truth) and your classification results in 3-D space (using MATLAB function `scatter3`), color code each data point $\mathbf{x}_n$ red, green, or blue according to its ground truth class labeling y_n. Repeat the process but now with each data point $\mathbf{x}_n$ color coded according to the classification result y'_n.

3. Develop MATLAB code to implement the softmax regression method for multi-class classification, and apply it to the datasets in the previous problem. Show your results in confusion matrix and visualization. Compare your results with those obtained in the previous problem.

8 Gaussian Process Regression and Classification

8.1 Gaussian Process Regression

The *Gaussian process regression (GPR)* is yet another regression method that fits a regression function $f(\mathbf{x})$ to the data samples in the given training set. Different from all previously considered algorithms that treat the regression model as a deterministic function $f(\mathbf{x})$ that takes an explicitly specified form, the GPR treats the regression function $f(\mathbf{x})$ as a stochastic *Gaussian process (GP)*, of which the value $f(\mathbf{x})$ at any point $\mathbf{x}$ is assumed to be a random variable with a Gaussian distribution.

We can therefore assume the joint probability density distribution (pdf) of $\mathbf{f}(\mathbf{X}) = [f(\mathbf{x}_1),\dots,f(\mathbf{x}_N)]^T$ for all i.i.d. samples in the training set $\mathcal{D} = \{\mathbf{X},\mathbf{y}\}$ takes the following form

$$p(\mathbf{f}|\mathbf{X}) = \mathcal{N}(\mathbf{m}_f,\ \boldsymbol{\Sigma}_f) = \mathcal{N}(\mathbf{0},\ \boldsymbol{\Sigma}_f) \tag{8.1}$$

where the mean vector $\mathbf{m}_f = \mathbf{0}$ is assumed to be zero for convenience and without loss of generality. Based on this prior probability of the regression function $f(\mathbf{x})$, we would like to find the posterior $p(\mathbf{f}_*|\mathcal{D},\mathbf{X}_*)$ for the function of any unlabeled test data $\mathbf{X}_*$ given the training set $\mathcal{D}$.

We also assume the observed data $\mathbf{y} = \mathbf{f}(\mathbf{x}) + \mathbf{n}$ is contaminated by noise $\mathbf{n}$, with a zero-mean Gaussian pdf

$$p(\mathbf{n}) = \mathcal{N}(\mathbf{0},\ \sigma_n^2\mathbf{I}) \tag{8.2}$$

where the diagonal covariance matrix $\sigma_n^2\mathbf{I}$ is based on the assumption that all components of $\mathbf{n}$ are independent of each other. We further assume noise $\mathbf{n}$ is independent of $\mathbf{f}$, with $\text{Cov}[\mathbf{y},\mathbf{n}] = \mathbf{0}$. Now $\mathbf{y} = \mathbf{f}(\mathbf{X}) + \mathbf{n}$, as the sum of two zero-mean Gaussian variables, is also a zero mean Gaussian process.

The unknown covariance matrix $\boldsymbol{\Sigma}_f$ in Eq. (8.1) can be constructed based on the assumption that the regression function $f(\mathbf{x})$ is smooth, in the sense that any two function values $f(\mathbf{x}_m)$ and $f(\mathbf{x}_n)$ are more correlated if the distance $||\mathbf{x}_m - \mathbf{x}_n||$ between the two corresponding data points is small, but less so if the distance is large. Now the covariance of any two function values $f(\mathbf{x}_m)$ and $f(\mathbf{x}_n)$ can be modeled by the squared exponential (SE) function, also called the *kernel function*

$$\mathrm{Cov}[f(\mathbf{x}_m), f(\mathbf{x}_n)] = \exp\left(-\frac{1}{\alpha^2}||\mathbf{x}_m - \mathbf{x}_n||^2\right) = k(\mathbf{x}_m, \mathbf{x}_n) = k(\mathbf{x}_n, \mathbf{x}_m) \tag{8.3}$$

This SE function may take any value between 1 when $||\mathbf{x}_m - \mathbf{x}_n|| = 0$ and $f(\mathbf{x}_m) = f(\mathbf{x}_n)$ are maximally correlated, and 0 when $||\mathbf{x}_m - \mathbf{x}_n|| = \infty$ and $f(\mathbf{x}_m)$ and $f(\mathbf{x}_n)$ are least correlated.

Here α is a hyperparameter that controls the smoothness of $f(\mathbf{x})$. To see this, we consider two extreme cases. First, if $\alpha \to \infty$, then the SE function approaches $e^0 = 1$ and $f(\mathbf{x}_m)$ and $f(\mathbf{x}_n)$ are maximally correlated, that is, $f(\mathbf{x})$ is absolutely smooth. Second, if $\alpha \to 0$, then the SE function approaches $e^{-\infty} = 0$ and $f(\mathbf{x}_m)$ and $f(\mathbf{x}_n)$ are completely uncorrelated, that is, $f(\mathbf{x})$ is least smooth. We can set α to any value between 0 and 1, depending on the desired smoothness.

The smoothness of the function $f(\mathbf{x})$ plays an important role as it determines whether $f(\mathbf{x})$ over- or underfits the sample points in the training data. If $f(\mathbf{x})$ is not smooth enough, it may be affected by outlier samples due to noise and overfitting the data; by contrast, if $f(\mathbf{x})$ is too smooth, it may not reflect the valid signal variations and underfits the data. By adjusting α, a proper tradeoff between over and underfittings can be made.

Given $\mathrm{Cov}[f(\mathbf{x}_m), f(\mathbf{x}_n)] = k(\mathbf{x}_m,\mathbf{x}_n)$ in Eq. (8.3), the covariance matrix $\mathbf{\Sigma}_f$ in Eq. (8.1) can be expressed as

$$\mathbf{\Sigma}_f = \mathrm{Cov}[\mathbf{f},\mathbf{f}] = \begin{bmatrix} k(\mathbf{x}_1, \mathbf{x}_1) & \cdots & k(\mathbf{x}_1, \mathbf{x}_N) \\ \vdots & \ddots & \vdots \\ k(\mathbf{x}_N, \mathbf{x}_1) & \cdots & k(\mathbf{x}_N, \mathbf{x}_N) \end{bmatrix} = K(\mathbf{X},\mathbf{X}) = \mathbf{K} \tag{8.4}$$

where $\mathbf{K}$ is the kernel matrix used to model $\mathbf{\Sigma}_f$. Note that the value of all diagonal components are the same $k(\mathbf{x}_n,\mathbf{x}_n) = 1$, and the farther away an off-diagonal component $k(\mathbf{x}_m,\mathbf{x}_n)$ is from the diagonal (with greater $||\mathbf{x}_m - \mathbf{x}_n||$), the lower value it takes. In general, the off-diagonal components of $\mathbf{\Sigma}_f$ have larger values for a smooth function $f(\mathbf{x})$ than those for a less smooth function.

Now that the prior $p(\mathbf{f}|\mathbf{X}) = \mathcal{N}(\mathbf{0},\mathbf{\Sigma}_f)$ is constructed based on training set $\mathbf{X} = [\mathbf{x}_1,\cdots,\mathbf{x}_N]$, we can further find the posterior of $\mathbf{f}_* = \mathbf{f}(\mathbf{X}_*)$ for any test dataset $\mathbf{X}_* = [\mathbf{x}_{1*},\cdots,\mathbf{x}_{M*}]$ containing M unlabeled samples. In the following, we will consider first the special noise-free case with $\mathbf{n} = \mathbf{0}$ and $\mathbf{y} = \mathbf{f} + \mathbf{n} = \mathbf{f}$, and then the general case of non-zero noise case with $\mathbf{f} \neq \mathbf{y}$.

In the noise-free case, both $\mathbf{f} = \mathbf{y}$ and $\mathbf{f}_*$ are samples from the same Gaussian process, and their joint pdf conditioned on $\mathbf{X}_*$ as well as $\mathbf{X}$ can be expressed as

$$\begin{aligned} p(\mathbf{f},\mathbf{f}_*|\mathbf{X},\mathbf{X}_*) &= p\left(\begin{bmatrix} \mathbf{f} \\ \mathbf{f}_* \end{bmatrix} \middle| \mathbf{X},\mathbf{X}_*\right) = \mathcal{N}\left(\begin{bmatrix} \mathbf{0} \\ \mathbf{0} \end{bmatrix}, \begin{bmatrix} \mathbf{\Sigma}_f & \mathbf{\Sigma}_{ff_*} \\ \mathbf{\Sigma}_{f_*f} & \mathbf{\Sigma}_{f_*} \end{bmatrix}\right) \\ &= \mathcal{N}\left(\begin{bmatrix} \mathbf{0} \\ \mathbf{0} \end{bmatrix}, \begin{bmatrix} \mathbf{K} & \mathbf{K}_* \\ \mathbf{K}_*^T & \mathbf{K}_{**} \end{bmatrix}\right) \end{aligned} \tag{8.5}$$

where $\mathbf{\Sigma}_f$ is given in Eq. (8.4), and the other three covariance matrices can be similarly modeled by their corresponding kernel matrices

$$\begin{cases} \mathbf{\Sigma}_{ff_*} = \text{Cov}[\mathbf{f},\mathbf{f}_*] = \mathbf{K}_* = K(\mathbf{X},\mathbf{X}_*) \\ \mathbf{\Sigma}_{f_*f} = \text{Cov}[\mathbf{f}_*,\mathbf{f}] = \mathbf{K}_*^T = K(\mathbf{X}_*,\mathbf{X}) \\ \mathbf{\Sigma}_{f_*} = \text{Cov}[\mathbf{f}_*,\mathbf{f}_*] = \mathbf{K}_{**} = K(\mathbf{X}_*,\mathbf{X}_*) \end{cases} \tag{8.6}$$

Given this joint pdf $p(\mathbf{f},\mathbf{f}_*|\mathbf{X},\mathbf{X}_*)$ of both $\mathbf{f}$ and $\mathbf{f}_*$ in Eq. (8.5), we can further find the conditional pdf $p(\mathbf{f}_*|\mathbf{f},\mathbf{X},\mathbf{X}_*)$ of $\mathbf{f}_*$ given $\mathbf{f}$ as well as $\mathbf{X}$ and $\mathbf{X}_*$ (see properties of the Gaussian distributions in Section B.1.5)

$$p(\mathbf{f}_*|\mathbf{f},\mathbf{X},\mathbf{X}_*) = \mathcal{N}(\mathbf{m}_{f_*|f},\mathbf{\Sigma}_{f_*|f}) \tag{8.7}$$

with

$$\begin{cases} \mathbf{m}_{f_*|f} &= \text{E}[\mathbf{f}_*|\mathbf{f}] = \mathbf{K}_*^T\mathbf{K}^{-1}\mathbf{f} \\ \mathbf{\Sigma}_{f_*|f} &= \text{Cov}[\mathbf{f}_*|\mathbf{f}] = \mathbf{K}_{**} - \mathbf{K}_*^T\mathbf{K}^{-1}\mathbf{K}_* \end{cases} \tag{8.8}$$

Based on the posterior in Eq. (8.7), we can estimate $f(\mathbf{x}_*)$ at any unlabeled sample in the test set $\mathbf{X}_*$ in the noise-free case.

We next consider the noisy case with $\mathbf{n} \neq \mathbf{0}$ and $\mathbf{y} = \mathbf{f}+\mathbf{n} \neq \mathbf{f}$. Now we need to find a new posterior $p(\mathbf{f}_*|\mathbf{y},\mathbf{X},\mathbf{X}_*) = p(\mathbf{f}_*|\mathbf{X}_*,\mathcal{D})$ in terms of $\mathbf{y}$ in the training set $\mathcal{D}$ to replace $p(\mathbf{f}_*|\mathbf{f},\mathbf{X},\mathbf{X}_*)$ in Eq. (8.7) in terms of $\mathbf{f}$. To do so, we follow the same process to find first the joint pdf of $\mathbf{f}_*$ and $\mathbf{y}$ and then the conditional pdf of $\mathbf{f}_*$ given $\mathbf{y}$.

The joint pdf of $\mathbf{f}_*$ and $\mathbf{y}$ (both zero-mean Gaussian) can be expressed as

$$p\left(\begin{bmatrix} \mathbf{y} \\ \mathbf{f}_* \end{bmatrix} |\mathbf{X},\mathbf{X}_*\right) = \mathcal{N}\left(\begin{bmatrix} \mathbf{0} \\ \mathbf{0} \end{bmatrix}, \begin{bmatrix} \mathbf{\Sigma}_y & \mathbf{\Sigma}_{yf_*} \\ \mathbf{\Sigma}_{f_*y} & \mathbf{\Sigma}_{f_*} \end{bmatrix}\right) \tag{8.9}$$

where $\mathbf{\Sigma}_{f_*}$ is given in Eq. (8.6) and

$$\begin{aligned} \mathbf{\Sigma}_y &= \text{Cov}[\mathbf{y},\mathbf{y}] = \text{Cov}[\mathbf{f}+\mathbf{n},\mathbf{f}+\mathbf{n}] = \text{Cov}[\mathbf{f},\mathbf{f}] + 2\ \text{Cov}[\mathbf{f},\mathbf{n}] + \text{Cov}[\mathbf{n},\mathbf{n}] \\ &= \text{Cov}[\mathbf{f},\mathbf{f}] + \text{Cov}[\mathbf{n},\mathbf{n}] = \mathbf{\Sigma}_f + \mathbf{\Sigma}_n = \mathbf{K} + \sigma_n^2\mathbf{I} \\ \mathbf{\Sigma}_{yf_*} &= \text{Cov}[\mathbf{y},\mathbf{f}_*] = \text{Cov}[\mathbf{f}+\mathbf{n},\mathbf{f}_*] = \text{Cov}[\mathbf{f},\mathbf{f}_*] = \mathbf{\Sigma}_{ff_*} = \mathbf{K}_* \\ \mathbf{\Sigma}_{f_*y} &= \mathbf{\Sigma}_{yf_*}^T = \mathbf{K}_*^T \end{aligned} \tag{8.10}$$

where we have used the fact that $\text{Cov}[\mathbf{f},\mathbf{n}] = \text{Cov}[\mathbf{f}_*,\mathbf{n}] = \mathbf{0}$ as noise $\mathbf{n}$ is assumed to be independent of $\mathbf{f}$. Now we have

$$\text{Cov}\begin{bmatrix} \mathbf{y} \\ \mathbf{f}_* \end{bmatrix} = \begin{bmatrix} \mathbf{\Sigma}_y & \mathbf{\Sigma}_{yf_*} \\ \mathbf{\Sigma}_{f_*y} & \mathbf{\Sigma}_{f_*} \end{bmatrix} = \begin{bmatrix} \mathbf{K}+\sigma_n^2\mathbf{I} & \mathbf{K}_* \\ \mathbf{K}_*^T & \mathbf{K}_{**} \end{bmatrix} \tag{8.11}$$

We then get the desired posterior, the conditional pdf of $\mathbf{f}_*$ given $\mathbf{X}_*$ as well as the training set $\mathcal{D} = \{\mathbf{X}, \mathbf{y}\}$

$$p(\mathbf{f}_*|\mathbf{X}_*,\mathcal{D}) = p(\mathbf{f}_*|\mathbf{X}_*,\mathbf{X},\mathbf{y}) = \mathcal{N}(\mathbf{m}_{f_*},\mathbf{\Sigma}_{f_*}) \tag{8.12}$$

with the mean and covariance

$$\begin{cases} \mathbf{m}_{f_*} &= \text{E}[\mathbf{f}_*|\mathbf{y}] = \mathbf{K}_*^T\left(\mathbf{K}+\sigma_n^2\mathbf{I}\right)^{-1}\mathbf{y} \\ \mathbf{\Sigma}_{f_*} &= \text{Cov}[\mathbf{f}_*|\mathbf{y}] = \mathbf{K}_{**} - \mathbf{K}_*^T\left(\mathbf{K}+\sigma_n^2\mathbf{I}\right)^{-1}\mathbf{K}_* \end{cases} \tag{8.13}$$

We see that these expressions are the same as those in Eq. (5.74) previously obtained by Bayesian linear regression, that is, the same results can be derived by two very different methods.

In the special noise-free case when $\mathbf{n} = \mathbf{0}$ and $\sigma_n = 0$, the mean and covariance in Eq. (8.13) become the same as those in Eq. (8.8), and as $\mathbf{y} = \mathbf{f} + \mathbf{n} = \mathbf{f}$, the posterior in Eq. (8.12) becomes the same as that in Eq. (8.7)

$$p(\mathbf{f}_*|\mathbf{X}_*, \mathcal{D}) = p(\mathbf{f}_*|\mathbf{X}_*, \mathbf{X}, \mathbf{y}) = p(\mathbf{f}_*|\mathbf{X}_*, \mathbf{X}, \mathbf{f}) \tag{8.14}$$

The following MATLAB code is the essential part of the Gaussian process regression algorithm. It takes as the input the training set $\mathcal{D} = \{\mathbf{X}, \mathbf{y}\}$, the test samples in $\mathbf{X}_*$, and the noise variance σ_n^2, and it generates as the output the posterior $p(\mathbf{f}_*|\mathcal{D}, \mathbf{X}_*)$, in terms of its mean as the GPR regression function, and its covariance matrix, of which the variances on the diagonal represent the confidence or certainty of the regression.

```
xstar=xmin:0.1:xmax;              % domain of function
n=length(xstar);                  % number of test samples
Kss=Kernel(xstar,xstar);          % K(x*,x*), covariance of prior
[L p]=chol(Kss);                  % Cholesky decomposition
for i=1:k
    plot(xstar,L'*randn(n,1))     % plot sample curves from prior
    hold on
end
hold off
x=xstar(randi(n,1,N)));              % get N random samples from xstar
y=f(xstar)+randn(1,n)*sigma_n^2;     % get their noisy labeling values
Kxx=Kernel(x,x)+sigma_n^2*eye(N);    % K(x,x)
Kxs=Kernel(x,xstar);                 % K(x,x*)
Ksx=Kxs';                         % K(x*,x)
fmean=Ksx*inv(Kxx)*y;             % posterior mean, regression function
fcov=Kss-Ksx*inv(Kxx)*Kxs;        % covariance of posterior
plot(xstar,fmean,xstar,fmean+diag(fcov),xstar,fmean-diag(fcov));
[L p]=chol(fcov);                 % Cholesky decomposition
for i=1:k
    plot(xstar,L'*randn(n,1))     % plot sample curves drawn from posterior
    hold on
end
hold off
```

Here `Kernel` is a function that generates the squared exponential kernel matrix based on two sets of data samples in `X1` and `X2`:

```
function K=Kernel(X1,X2)
    a=1;                % parameter that controls smoothness of f(x)
    m=length(X1);
    n=length(X2);
    K=zeros(m,n);
```

```
    for i=1:m
        for j=1:n
            d=norm(X1(:,i)-X2(:,j));
            K(i,j)=exp(-(d/a)^2);
        end
    end
end
```

Example 8.1

As shown in Fig. 8.1, a sinusoidal function $f(x) = \cos(2\pi f x)$ (with $f = 0.2$) contaminated by Gaussian noise with $p(n) = \mathcal{N}(0, \sigma_r^2)$ ($\sigma_r = 0.2$) is sampled at $N = 7$ random positions over the duration between the limits $x_{min} = 0$ and $x_{max} = 9$.. These samples form the training dataset $\mathcal{D} = \{(x_n, y_n)\ (n = 1, \cdots, N)\}$.

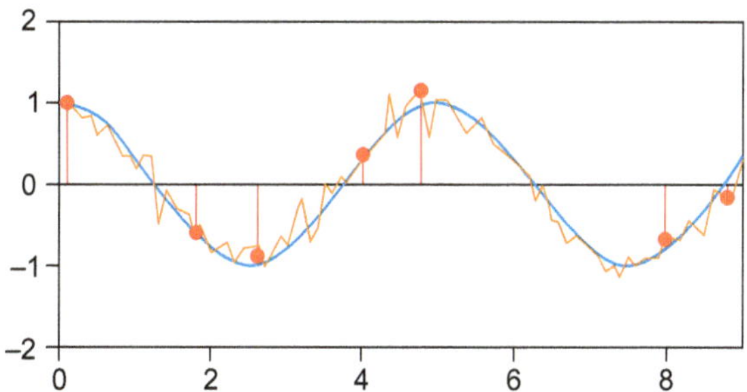

Fig. 8.1 A noisy sinusoid.

The method of GPR is used to estimate the mean and variance of $f_{i*} = f(x_{i*})$ at 91 test points x_{i*}, $(i = 1, \cdots, 91)$, evenly distributed over the entire duration. The eight panels of Fig. 8.2 show the estimated mean $\mathrm{E}[f_*]$, and the variance $\pm\mathrm{Var}[f_*]$, both above and below the mean, to form a region of uncertainty of the estimates. The estimated function values are more reliable in the region that is narrow but less so if it is wide.

The top left panel represents the prior probability $p(f_*|x_*)$ of the estimated model output $f_* = \hat{y}_*$ at point x_* with no knowledge based on the training set $\mathbf{x}$ ($N = 0$ training samples), which is assumed to have zero-mean $\mathrm{E}[f_*] = 0$ and unity variance $\mathrm{Var}[f(x_*)] = k(x_*, x_*) = 1$. The subsequent seven panels show the mean and variance of the posterior $p(f_* = \hat{y}_*|x_*, \mathcal{D})$ estimated based on $N = 1, \cdots, 7$ training samples in $\mathcal{D}$. Note that wherever an additional training sample x is available, the estimated mean $\mu_{f_*|y}$ is modified to be close to the true function value $f(x)$, and the region of uncertainty is narrowed due to the reduced variance of the posterior, that is the uncertainty of the estimate is reduced, or the confidence is increased. As shown in the last panel, when all $N = 7$ training samples are used, the general shape of the original function is reasonably reconstructed.

The eight panels in Fig. 8.3 show 12 sample curves randomly drawn from each posterior $p(f_*|\mathbf{x}_*, \mathcal{D})$. They can be treated as different interpolations for the N samples

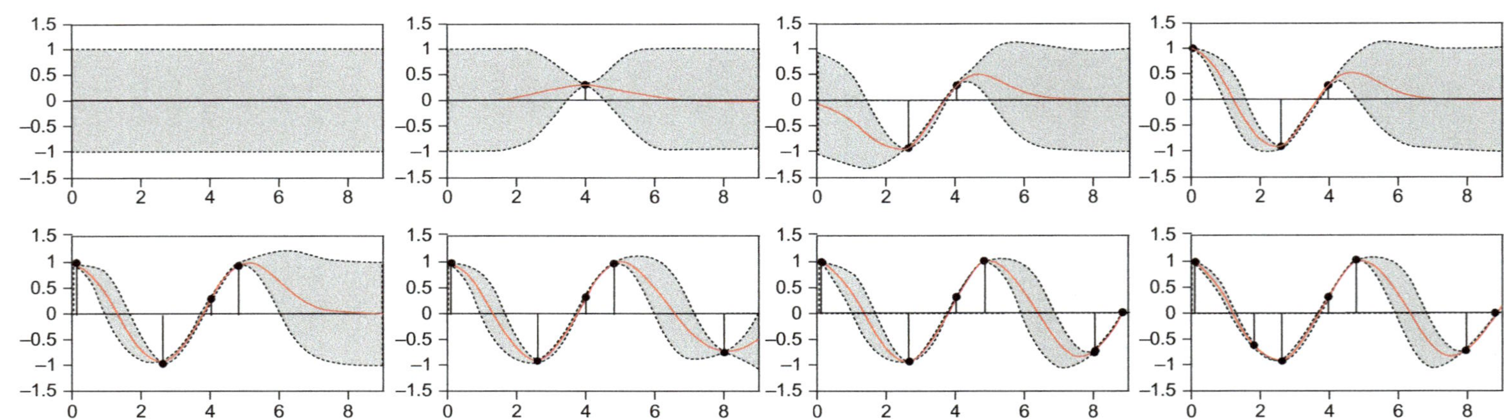

Fig. 8.2 Gaussian process regression with varying data points.

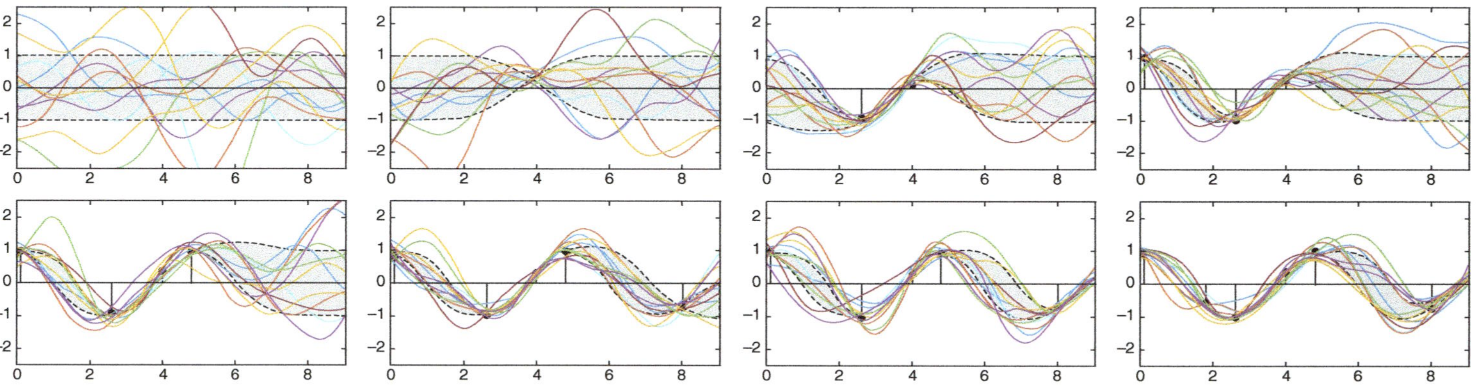

Fig. 8.3 Samples drawn from posteriors.

in $\mathcal{D}$, and any of them can be used to predict the outputs y_* at any input points x_*. The mean of these samples shown as the red curve is the same as $E[f_*]$ shown in the previous figure.

We see that wherever a training sample $f(x)$ is available at x, the variance and thereby the width of the region of uncertainty is reduced, and correspondingly the sample curves in the neighborhood of x are more tightly distributed. Specifically, we see that the sample curves drawn from the prior in the top left panel are very loosely distributed, while those samples drawn from the posterior obtained based on all $N = 7$ training points in the bottom right panel are relatively tightly distributed. In particular, note that owing to lack of training data in the interval $5 < x < 8$, the region of uncertainty is significantly wider than else where, and the sample curves are much more loosely located.

The parameter α of the kernel function plays an important role in the GPR results, as shown in the three panels in Fig. 8.4, corresponding to $\alpha = 0.5, 1.5, 3.0$. When α is

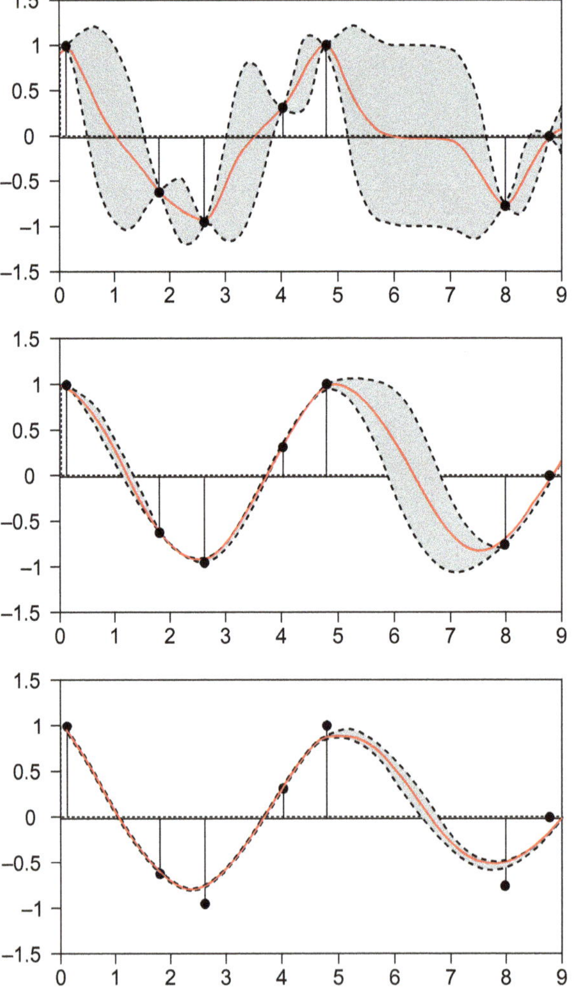

Fig. 8.4 Overfitting (top) and underfitting (bottom) by GPR.

too small (top panel), the model may overfit the data, as the regression function fits the training samples well but it not smooth and it does not interpolate regions between the training samples well (with large variance); conversely, when α is too large (bottom panel), the regression function is smooth but it may underfit the data, as it does not fit the training samples well. By adjusting the value of α carefully we can make a proper tradeoff between overfitting and underfitting as shown in the middle panel.

8.2 Gaussian Process Classifier – Binary

In both logistic regression and softmax regression considered previously, the linear regression function $\hat{y} = f(\mathbf{x}) = \mathbf{x}^T\mathbf{w}$ is converted to the probability $p(\hat{y} = k|\mathbf{x}, \mathbf{W})$ for $\mathbf{x} \in C_k$, by either the logistic function $\sigma(f(\mathbf{x}))$ for binary classification, or the softmax function $s(f(\mathbf{x}))$ for multiclass classification. Also, in GPR, the regression function $f(\mathbf{x})$ is treated as a Gaussian process. Now we consider the *Gaussian process classification (GPC)* based on the combination of both logistic/softmax regression and GPR. We will consider binary GPC based on the logistic function in this section, and then multiclass GPC based on softmax function in the following section.

Here in binary GPC, we assume the training samples are labeled by 1 or −1 (instead of 0) for mathematical convenience, that is, $\mathbf{x}_n \in C_+$ if $y_n = 1$ or $\mathbf{x}_n \in C_-$ if $y_n = -1$. Similar to logistic regression, here we also convert the regression function $f(\mathbf{x})$, treated as a Gaussian process, into the probability for $\mathbf{x} \in C_k$ by the logistic function $\sigma(f(\mathbf{x}))$, which, as a function of a random argument $f(\mathbf{x})$, is also random. We therefore define $p(y = 1|\mathbf{x}, \mathcal{D})$ as the expectation of $\sigma(f(\mathbf{x}))$ with respect to $f(\mathbf{x})$

$$p(y = 1|\mathbf{x}, \mathcal{D}) = E_f[\sigma(f(\mathbf{x}))] = \int \sigma(f(\mathbf{x}))\, p(f|\mathcal{D})\, df \tag{8.15}$$

As function $f(\mathbf{x})$ is marginalized (averaged out) by this integral, it is hidden instead of explicitly specified, and is therefore called a *latent function*. This $p(y = 1|\mathbf{x}, \mathcal{D})$ can be generalized for multiple samples in the training set $\mathbf{X} = [\mathbf{x}_1, \cdots, \mathbf{x}_N]$

$$p(\mathbf{y} = 1|\mathcal{D}) = E_f[\sigma(\mathbf{f}(\mathbf{X}))] = \int \sigma(\mathbf{f}(\mathbf{X}))\, p(\mathbf{f}|\mathcal{D})\, d\mathbf{f} \tag{8.16}$$

Here the posterior $p(\mathbf{f}|\mathcal{D})$ of $\mathbf{f}$ given $\mathcal{D}$ can be found based on Bayes' theorem in terms of the likelihood $p(\mathbf{y}|\mathbf{f}) = L(\mathbf{f}|\mathbf{y})$ and the prior $p(\mathbf{f}|\mathbf{X})$

$$p(\mathbf{f}|\mathcal{D}) = p(\mathbf{f}|\mathbf{X}, \mathbf{y}) = \frac{p(\mathbf{y}, \mathbf{f}|\mathbf{X})}{p(\mathbf{y}|\mathbf{X})} = \frac{p(\mathbf{y}|\mathbf{f})\, p(\mathbf{f}|\mathbf{X})}{p(\mathbf{y}|\mathbf{X})} \propto p(\mathbf{y}|\mathbf{f})\, p(\mathbf{f}|\mathbf{X}) \tag{8.17}$$

As always, the denominator $p(\mathbf{y}|\mathbf{X})$ independent of $\mathbf{f}$ is dropped.

We first find the likelihood $p(\mathbf{y}|\mathbf{f})$. The Gaussian process $f(\mathbf{x})$ is mapped by the logistic function into the probability for $y = 1$ representing $\mathbf{x} \in C_+$, or $y = -1$ representing $\mathbf{x} \in C_-$

$$p(y = 1|f(\mathbf{x})) = \sigma(f(\mathbf{x}))$$
$$p(y = -1|f(\mathbf{x})) = 1 - p(y = 1|f(\mathbf{x})) = 1 - \sigma(f(\mathbf{x})) = \sigma(-f(\mathbf{x})) \quad (8.18)$$

These two cases can be combined to get the conditional probability of y given $f(\mathbf{x})$

$$p(y|f) = \sigma(y\,f) = \frac{1}{1 + e^{-yf}} = \frac{e^{yf}}{1 + e^{yf}} \quad (8.19)$$

The likelihood of $\mathbf{f}$ for all N i.i.d. samples in the training set $\mathbf{X}$ is

$$p(\mathbf{y}|\mathbf{f}) = \prod_{n=1}^{N} p(y_n|f_n) = \prod_{n=1}^{N} \sigma(y_n\,f_n) \quad (8.20)$$

We then consider the prior $p(\mathbf{f}|\mathbf{X})$, which is assumed to be a zero-mean Gaussian $p(\mathbf{f}|\mathbf{X}) = \mathcal{N}(\mathbf{0}, \mathbf{\Sigma}_f)$, where the covariance matrix $\mathbf{\Sigma}_f$ is constructed based on the training set $\mathbf{X}$, the same as in GPR. Specifically, the covariance between $f(\mathbf{x}_m)$ and $f(\mathbf{x}_n)$ in the mth row and nth column of $\mathbf{\Sigma}_f$ is modeled by the SE kernel

$$\text{Cov}[f(\mathbf{x}_m), f(\mathbf{x}_n)] = k(\mathbf{x}_m, \mathbf{x}_n) = \exp\left(-\frac{1}{a^2}||\mathbf{x}_m - \mathbf{x}_n||^2\right), \qquad (m, n = 1, \cdots, N) \quad (8.21)$$

Such a covariance matrix is desired for GPC as well as GPR, so that $f(\mathbf{x}_m)$ and $f(\mathbf{x}_n)$ are more correlated if $\mathbf{x}_m$ and $\mathbf{x}_n$ are close together, but less so if they are farther apart. However the justification for such a property is different. In GPR, this property is required for the smoothness of the regression function, while here in GPC, this property is also desired so that function values $f(\mathbf{x}_m)$ and $f(\mathbf{x}_n)$ for two samples $\mathbf{x}_m$ and $\mathbf{x}_n$ close to each other in the feature space are more correlated and therefore the two samples are more likely to be classified into the same class, but less so if they are far apart.

Also, as discussed in GPR, here the parameter a in the SE controls the smoothness of the function. If $a \to \infty$, the value of the SE approaches 1, then $f(\mathbf{x}_m)$ and $f(\mathbf{x}_n)$ are highly correlated and $f(\mathbf{x})$ is very smooth; but if $a \to 0$, SE approaches 0, then $f(\mathbf{x}_m)$ and $f(\mathbf{x}_n)$ are not correlated and $f(\mathbf{x})$ is no longer smooth. By adjusting a, a proper tradeoff can be made between overfitting and underfitting.

Having found both the likelihood $p(\mathbf{y}|\mathbf{f})$ and the prior $p(\mathbf{f})$, we can obtain the posterior in Eq. (8.17)

$$\begin{aligned} p(\mathbf{f}|\mathcal{D}) &\propto p(\mathbf{y}|\mathbf{f})\,p(\mathbf{f}|\mathbf{X}) = p(\mathbf{y}|\mathbf{f})\,\mathcal{N}(\mathbf{0}, \mathbf{K}) \\ &= \prod_{n=1}^{N} \sigma(y_n\,f_n)\,\frac{1}{(2\pi)^{d/2}|\mathbf{K}|^{1/2}} \exp\left(-\frac{1}{2}\mathbf{f}^T\mathbf{K}^{-1}\mathbf{f}\right) \\ &\propto \prod_{n=1}^{N} \sigma(y_n\,f_n)\,\exp\left(-\frac{1}{2}\mathbf{f}^T\mathbf{K}^{-1}\mathbf{f}\right) \end{aligned} \quad (8.22)$$

Note that the likelihood $p(\mathbf{y}|\mathbf{f})$ is not Gaussian, as the binary labeling $\mathbf{y}$ of the training set $\mathbf{X}$ is not continuous. Consequently, the posterior $p(\mathbf{f}|\mathcal{D})$, as a product of the

Gaussian prior and non-Gaussian likelihood, is not Gaussian. However, we can carry out *Laplace approximation* and still approximate the posterior as a Gaussian

$$p(\mathbf{f}|\mathcal{D}) \approx \mathcal{N}(\mathbf{m}_{f|D}, \mathbf{\Sigma}_{f|D}) \tag{8.23}$$

of which the mean $\mathbf{m}_{f|D}$ and covariance $\mathbf{\Sigma}_{f|D}$ are to be obtained in the following. We further get the log posterior denoted by $\psi(\mathbf{f})$

$$\psi(\mathbf{f}) = \log p(\mathbf{f}|\mathcal{D}) = \sum_{n=1}^{N} \log \sigma(y_n f_n) - \frac{N}{2}\log(2\pi) - \frac{1}{2}\log|\mathbf{K}| - \frac{1}{2}\mathbf{f}^T\mathbf{K}^{-1}\mathbf{f} \tag{8.24}$$

The two middle terms independent of $\mathbf{f}$ are dropped.

Now that the posterior $p(\mathbf{f}|\mathcal{D})$ is approximated as a Gaussian, we can find the gradient vector $\mathbf{g}_\psi(\mathbf{f})$ and Hessian matrix $\mathbf{H}_\psi(\mathbf{f})$ of the log posterior

$$\mathbf{g}_\psi(\mathbf{f}) = \frac{d}{d\mathbf{f}}\psi(\mathbf{f}) = -\mathbf{\Sigma}_{f|D}^{-1}(\mathbf{f} - \mathbf{m}_{f|D}) \tag{8.25}$$

$$\mathbf{H}_\psi(\mathbf{f}) = \frac{d^2}{d\mathbf{f}^2}\psi(\mathbf{f}) = \frac{d}{d\mathbf{f}}\mathbf{g}_\psi(\mathbf{f}) = -\mathbf{\Sigma}_{f|D}^{-1} \tag{8.26}$$

By contrast, we can also find $\mathbf{g}_\psi(\mathbf{f})$ and $\mathbf{H}_\psi(\mathbf{f})$ by directly taking the first- and second-order derivatives of the log posterior

$$\mathbf{g}_\psi(\mathbf{f}) = \frac{d}{d\mathbf{f}}\psi(\mathbf{f}) = \frac{d}{d\mathbf{f}}\log p(\mathbf{y}|\mathbf{f}) - \frac{d}{d\mathbf{f}}\left(\frac{1}{2}\mathbf{f}^T\mathbf{K}^{-1}\mathbf{f}\right) = \mathbf{w} - \mathbf{K}^{-1}\mathbf{f} \tag{8.27}$$

$$\mathbf{H}_\psi(\mathbf{f}) = \frac{d^2}{d\mathbf{f}^2}\psi(\mathbf{f}) = \frac{d}{d\mathbf{f}}\mathbf{g}_\psi(\mathbf{f}) = \frac{d}{d\mathbf{f}}\left(\mathbf{w} - \mathbf{K}^{-1}\mathbf{f}\right) = \mathbf{W} - \mathbf{K}^{-1} \tag{8.28}$$

where we have defined

$$\mathbf{w} = \frac{d}{d\mathbf{f}}\log p(\mathbf{y}|\mathbf{f}), \qquad \mathbf{W} = \frac{d^2}{d\mathbf{f}^2}\log p(\mathbf{y}|\mathbf{f}) = \frac{d\mathbf{w}}{d\mathbf{f}} \tag{8.29}$$

which are, respectively, the gradient vector and Hessian matrix of the log function

$$\log p(\mathbf{y}|\mathbf{f}) = \log\prod_{n=1}^{N}\sigma(y_n f_n) = \sum_{n=1}^{N}\log\sigma(y_n f_n) = \sum_{n=1}^{N}\log\left(\frac{1}{1+\exp(-y_n f_n)}\right) \tag{8.30}$$

We first find the first- and second-order derivatives with respect to a single component f_n of $\mathbf{f} = [f_1, \cdots, f_N]^T$

$$\begin{aligned}
\frac{d}{df_n}\log p(y_n|f_n) &= \frac{d}{df_n}\log\sigma(y_n f_n) = \frac{d}{df_n}\log\left(\frac{1}{1+\exp(-y_n f_n)}\right) = \frac{y_n}{1+\exp(y_n f_n)} \\
\frac{d^2}{df_n^2}\log p(y_n|f_n) &= \frac{d^2}{df_n^n}\log\sigma(y_n f_n) = \frac{d}{df_n}\left(\frac{y_n}{1+\exp(y_n f_n)}\right) = \frac{-\exp(-y_n f_n)}{(1+\exp(-y_n f_n))^2}
\end{aligned} \tag{8.31}$$

where we have used the facts that $y_n = \pm 1$ and $y_n^2 = 1$. We then find

$$\mathbf{w} = \frac{d}{d\mathbf{f}} \log p(\mathbf{y}|\mathbf{f}) = \begin{bmatrix} d/df_1 \\ \vdots \\ d/df_N \end{bmatrix} \sum_{n=1}^{N} \log p(y_n|f_n)$$
$$= \begin{bmatrix} d \log p(y_1|f_1)/df_1 \\ \vdots \\ d \log p(y_N|f_N)/df_N \end{bmatrix} = \begin{bmatrix} \frac{y_1}{1+e^{y_1 f_1}} \\ \vdots \\ \frac{y_N}{1+e^{y_N f_N}} \end{bmatrix} \tag{8.32}$$

and

$$\mathbf{W} = \frac{d^2}{d\mathbf{f}^2} \log p(\mathbf{y}|\mathbf{f}) = \begin{bmatrix} \frac{\partial^2}{\partial f_1 \partial f_1} & \cdots & \frac{\partial^2}{\partial f_1 \partial f_N} \\ \vdots & \ddots & \vdots \\ \frac{\partial^2}{\partial f_N \partial f_1} & \cdots & \frac{\partial^2}{\partial f_N \partial f_N} \end{bmatrix} \sum_{n=1}^{N} \log p(y_n|f_n)$$
$$= \text{diag}\left[\frac{d^2}{df_1^2} \log p(y_1|f_1), \cdots, \frac{d^2}{df_N^2} \log p(y_N|f_N)\right]$$
$$= \text{diag}\left[\frac{-e^{-y_1 f_1}}{(1+e^{-y_1 f_1})^2}, \cdots, \frac{-e^{-y_N f_N}}{(1+e^{-y_N f_N})^2}\right] \tag{8.33}$$

Equating the two expressions for $\mathbf{H}_\psi(\mathbf{f})$ in Eqs (8.26) and (8.28) we get:

$$\mathbf{H}_\psi(\mathbf{f}) = \mathbf{W} - \mathbf{K}^{-1} = -\mathbf{\Sigma}_{f|D}^{-1}, \qquad \text{i.e.,} \qquad \mathbf{\Sigma}_{f|D} = (\mathbf{K}^{-1} - \mathbf{W})^{-1} \tag{8.34}$$

Equating the two expressions for $\mathbf{g}_\psi(\mathbf{f})$ in Eqs (8.25) and (8.27) we get

$$\mathbf{g}_\psi(\mathbf{f}) = -\mathbf{\Sigma}_{f|D}^{-1}(\mathbf{f} - \mathbf{m}_{f|D}) = \mathbf{w} - \mathbf{K}^{-1}\mathbf{f}$$
$$\text{i.e.,} \qquad \mathbf{m}_{f|D} = \mathbf{f} + \mathbf{\Sigma}_{f|D}\left(\mathbf{w} - \mathbf{K}^{-1}\mathbf{f}\right) = \mathbf{f} - \mathbf{H}_\psi^{-1}\mathbf{g}_\psi \tag{8.35}$$

Substituting $\mathbf{\Sigma}_{f|D}$ given in Eq. (8.34) into the equation and solving it for $\mathbf{m}_{f|D}$, we obtain

$$\mathbf{m}_{f|D} = \mathbf{f} + (\mathbf{K}^{-1} + \mathbf{W})^{-1}(\mathbf{w} - \mathbf{K}^{-1}\mathbf{f}) \tag{8.36}$$

We recognize this is actually the iteration of Newton's method solving equation $\mathbf{g}_\psi(\mathbf{f}) = \mathbf{0}$, in consistence with the fact that when $\mathbf{f} = \mathbf{m}_{f|D}$, the log posterior $\psi(\mathbf{f}) = \log p(\mathbf{f}|\mathcal{D})$, assumed to be Gaussian, reaches maximum with $\mathbf{g}_\psi(\mathbf{f}) = \mathbf{0}$. However, we note that in Eq. (8.36), $\mathbf{m}_{f|D}$ as a value of $\mathbf{f}$ is expressed in terms of both $\mathbf{w}$ and $\mathbf{W}$, which in turn are functions of $\mathbf{f}$ as given in Eqs (8.32) and (8.33), that is, $\mathbf{m}_{f|D}$ given here is not in a closed form, as $\mathbf{f}$ and parameters $\mathbf{w}$ and $\mathbf{W}$ are interdependent. We instead need to carry out an iteration during which both $\mathbf{f}$ and the parameters $\mathbf{w}$ and $\mathbf{W}$ are updated alternatively

$$\mathbf{f}_{n+1} = \mathbf{f}_n - \mathbf{H}_\psi^{-1}\,\mathbf{g}_\psi = \mathbf{f}_n + (\mathbf{K}^{-1} - \mathbf{W})^{-1}\left(\mathbf{w} - \mathbf{K}^{-1}\mathbf{f}_n\right)$$
$$= \mathbf{f}_n + (\mathbf{K}^{-1} - \mathbf{W})^{-1}\left[-(\mathbf{K}^{-1} - \mathbf{W})\mathbf{f}_n + \mathbf{w} - \mathbf{W}\mathbf{f}_n\right]$$
$$= (\mathbf{K}^{-1} - \mathbf{W})^{-1}(\mathbf{w} - \mathbf{W}\mathbf{f}_n) \tag{8.37}$$

This iterative process converges to $\mathbf{f} = \mathbf{m}_{f|D}$, at which $p(\mathbf{f}|\mathcal{D})$ is maximized, and

$$\mathbf{g}_\psi(\mathbf{f}) = \frac{d}{d\mathbf{f}}\psi(\mathbf{f}) = \mathbf{w} - \mathbf{K}^{-1}\mathbf{f} = \mathbf{0}, \qquad \text{i.e.,} \qquad \mathbf{f} = \mathbf{m}_{f|D} = \mathbf{K}\,\mathbf{w} \tag{8.38}$$

Now that $\mathbf{w}$ and $\mathbf{W}$ as well as $\mathbf{f} = \mathbf{m}_{f|D}$ are available, we can further obtain $\boldsymbol{\Sigma}_{f|D}$ in Eq. (8.34), and the posterior $p(\mathbf{f}|\mathcal{D}) \approx \mathcal{N}(\mathbf{m}_{f|D}, \boldsymbol{\Sigma}_{f|D})$.

Having found $p(\mathbf{f}|\mathcal{D})$, we can proceed to carry out classification of any test set $\mathbf{X}_*$ based on the probability (similar to Eq. (8.16))

$$p(\mathbf{y}_* = 1|\mathbf{X}_*, \mathcal{D}) = E_f[\sigma(\mathbf{f}(\mathbf{X}_*))] = \int \sigma(\mathbf{f}(\mathbf{X}_*))\; p(\mathbf{f}|\mathbf{X}_*, \mathcal{D})\, d\mathbf{f} \tag{8.39}$$

We first approximate the posterior of $\mathbf{f}_* = \mathbf{f}(\mathbf{X}_*)$ in the equation as a Gaussian

$$p(\mathbf{f}|\mathbf{X}_*, \mathcal{D}) \approx \mathcal{N}(\mathbf{m}_{f_*}, \boldsymbol{\Sigma}_{f_*}) \tag{8.40}$$

where the mean $\mathbf{m}_{f_*}$ and covariance $\boldsymbol{\Sigma}_{f_*}$ can be obtained based on the fact that both $\mathbf{f}_* = \mathbf{f}(\mathbf{X}_*)$ and $\mathbf{f} = \mathbf{f}(\mathbf{X})$ are the same Gaussian process, that is, their joint probability $p(\mathbf{f}, \mathbf{f}_*)$ is a Gaussian. The method is therefore the same as is discussed in the method of GPR. Specifically, we take the following steps:

- Find the mean $\mathbf{m}_{f_*|f} = E(\mathbf{f}_*|\mathbf{f})$ and covariance $\boldsymbol{\Sigma}_{f_*|f} = \text{Cov}[\mathbf{f}_*|\mathbf{f}]$ of $p(\mathbf{f}_*|\mathbf{X}_*, \mathbf{X}, \mathbf{f})$ conditioned on the latent function $\mathbf{f}$, same as in Eq. (8.8) for GPR

$$\begin{cases} \mathbf{m}_{f_*|f} = \mathbf{K}_*^T\mathbf{K}^{-1}\mathbf{f} \\ \boldsymbol{\Sigma}_{f_*|f} = \mathbf{K}_{**} - \mathbf{K}_*^T\mathbf{K}^{-1}\mathbf{K}_* \end{cases} \tag{8.41}$$

- Find $\mathbf{m}_{f_*} = E_f(\mathbf{f}_*)$ as the expectation of $\mathbf{m}_{f_*|f} = \mathbf{K}_*^T\mathbf{K}^{-1}\mathbf{f}$ conditioned on $\mathbf{f}$, that is, find the average of $\mathbf{m}_{f_*|f}$ over $\mathbf{f}$ based on $p(\mathbf{f}|\mathcal{D})$ (marginalize over $\mathbf{f}$)

$$\begin{aligned} \mathbf{m}_{f_*} &= E_f(\mathbf{m}_{f_*|f}) = E_f\left(\mathbf{K}_*^T\mathbf{K}^{-1}\mathbf{f}\right) = \mathbf{K}_*^T\mathbf{K}^{-1}E_f(\mathbf{f}) = \mathbf{K}_*^T\mathbf{K}^{-1}\mathbf{m}_f \\ &= \mathbf{K}_*^T\mathbf{K}^{-1}\mathbf{K}\,\mathbf{w}_{\mathbf{m}_f} = \mathbf{K}_*^T\mathbf{w}_{\mathbf{m}_f} \end{aligned} \tag{8.42}$$

  where $\mathbf{m}_f = E_f(\mathbf{f}) = \mathbf{K}\mathbf{w}_{m_f}$ given in Eq. (8.38).
- Find the covariance $\boldsymbol{\Sigma}_{m_{f_*|f}} = E_f[(\mathbf{m}_{f_*|f} - \mathbf{m}_{f_*})^2]$. Given the covariance $\boldsymbol{\Sigma}_{f|D} = (\mathbf{K}^{-1} - \mathbf{W})^{-1}$ of $\mathbf{f}$ in Eq. (8.34), we can further find the covariance of $\mathbf{m}_{f_*|f} = \mathbf{K}_*^T\mathbf{K}^{-1}\mathbf{f}$ as a linear combination of $\mathbf{f}$ (recall if $\mathbf{y} = \mathbf{A}\mathbf{x}$, then $\boldsymbol{\Sigma}_y = \mathbf{A}\Sigma_x\mathbf{A}^T$)

$$\boldsymbol{\Sigma}_{m_{f_*|f}} = \mathbf{K}_*^T\mathbf{K}^{-1}\,\boldsymbol{\Sigma}_{f|D}\,\mathbf{K}^{-1}\mathbf{K}_* = \mathbf{K}_*^T\mathbf{K}^{-1}(\mathbf{K}^{-1} - \mathbf{W})^{-1}\mathbf{K}^{-1}\mathbf{K}_* \tag{8.43}$$

- Find the covariance $\boldsymbol{\Sigma}_{f_*}$ of $\mathbf{f}_*$ as the sum of $\boldsymbol{\Sigma}_{f_*|f} = E[(\mathbf{f}_* - \mathbf{m}_{f_*|f})(\mathbf{f}_* - \mathbf{m}_{f_*|f})^T]$ for the variation of $\mathbf{f}_*$ with respect to $\mathbf{m}_{f_*|f}$, and $\boldsymbol{\Sigma}_{m_{f_*|f}} = E_f[(\mathbf{m}_{f_*|f} - \mathbf{m}_{f_*})^2]$ for the variation of $\mathbf{m}_{f_*|f}$ with respect to $\mathbf{m}_{f_*}$

$$\begin{aligned} \boldsymbol{\Sigma}_{f_*} &= \boldsymbol{\Sigma}_{f_*|f} + \boldsymbol{\Sigma}_{m_{f_*|f}} \\ &= (\mathbf{K}_{**} - \mathbf{K}_*^T\mathbf{K}^{-1}\mathbf{K}_*) + (\mathbf{K}_*^T\mathbf{K}^{-1}(\mathbf{K}^{-1} - \mathbf{W})^{-1}\mathbf{K}^{-1}\mathbf{K}_*) \\ &= \mathbf{K}_{**} - \mathbf{K}_*^T\mathbf{K}^{-1}\mathbf{K}_* + \mathbf{K}_*^T\mathbf{K}^{-1}[\mathbf{K} - \mathbf{K}(\mathbf{K} - \mathbf{W}^{-1})^{-1}\mathbf{K}]\mathbf{K}^{-1}\mathbf{K}_* \\ &= \mathbf{K}_{**} - \mathbf{K}_*^T(\mathbf{K} - \mathbf{W}^{-1})^{-1}\mathbf{K}_* \end{aligned} \tag{8.44}$$

  Here we have used the identity for $(\mathbf{A} + \mathbf{B})^{-1}$ given in Section A.2.4.

Now that the posterior $p(\mathbf{f}|\mathbf{X}_*, \mathcal{D}) = \mathcal{N}(\mathbf{m}_{f_*}, \mathbf{\Sigma}_{f_*})$ is approximated as a Gaussian with mean and covariance given in Eqs (8.42) and (8.44), we can finally carry out Eq. (8.16) to find the probability for the test points in $\mathbf{X}_*$ to belong to class C_1

$$p(\mathbf{y}_* = 1|\mathbf{X}_*, \mathcal{D}) = \int \sigma(\mathbf{f}_*)\, p(\mathbf{f}_*|\mathbf{X}_*, \mathcal{D})\, d\mathbf{f}_* \\ = E_f[\sigma(\mathbf{f}(\mathbf{X}_*))] = \sigma(E_f(\mathbf{f}(\mathbf{X}_*))) = \sigma(\mathbf{m}_{f_*}) \qquad (8.45)$$

As for GPR, the certainty or confidence of this classification result is represented by the variances on the diagonal of the covariance $\mathbf{\Sigma}_{f_*}$.

The MATLAB code for the essential parts of this algorithm follows. Here `X` and `y` are for the training data $\mathcal{D} = \{\mathbf{X}, \mathbf{y}\}$, and `Xs` is an array composed of test vectors. First, the essential segment of the main program takes in the training and test data, generates the covariance matrices $\mathbf{K}$, $\mathbf{K}_*$, and $\mathbf{K}_{**}$ represented by `K`, `Ks`, and `Kss`, respectively. The function `Kernel` is exactly the same as the one used for Gaussian process regression. This code segment then further calls a function `findPosteriorMean` which finds the mean and covariance of $p(\mathbf{f}|\mathbf{X}, \mathbf{y})$ based on covariance of training data ($\mathbf{K} = \text{Cov}[\mathbf{X}]$ and $\mathbf{y}$, and computes the mean $\mathbf{m}_{f_*|y}$ and covariance $\mathbf{\Sigma}_{f_*|y}$ based on Eqs (8.42) and (8.44), respectively. The sign function of $\mathbf{m}_{f_*|y}$ indicates the classification of the test data points in $\mathbf{X}_*$.

```
K=Kernel(X,X);                % cov(f,f), covariance of prior p(f|X)
Ks=Kernel(X,Xs);              % cov(f_*,f)
Kss=Kernel(Xs,Xs);            % cov(f_*,f_*)
[Sigmaf W w]=findPosteriorMean(K,y);
                              % get mean/covariance of p(f|D), W, w
meanfD=Ks'*w;                 % mean of p(f_*|X_*,D)
SigmafD=Kss-Ks'*inv(K-inv(W))*Ks;  % covariance of p(f_*|X_*,D)
ys=sign(meanfD);              % binary classification of test data
p=1./(1+exp(-meanfD));        % p(y_*=1|X_*,D) as logistic function
```

The function `findPosteriorMean` listed in the following code uses Newton's method to find the mean and covariance of the posterior $p(\mathbf{f}|\mathcal{D})$ of the latent function $\mathbf{f}$ based on the training data, and returns them in `meanf` and `covf`, respectively, together with `w` and `W` for the gradient and Hessian of the likelihood $p(\mathbf{y}|\mathbf{f})$, to be used for computing $\mathbf{m}_{f_*|D}$ and $\mathbf{\Sigma}_{f_*|D}$.

```
function [meanf covf W w]=findPosteriorMean(K,y)
    % K: covariance of prior of p(f|X)
    % y: labeling of training data X
    % w: gradient vector of log p(y|f)
    % W: Hessian matrix of log p(y|f)
    % meanf: mean of p(f|X,y)
    % covf: covariance of p(f|X,y)
    n=length(y);          % number of training samples
    f0=zeros(n,1);        % initial value of latent function
    f=f0;
    er=1;
```

```
    while er > 10^(-9)  % Newton method to get f that maximizes p(f|X,y)
        e=exp(-y.*f);
        w=y.*e./(1+e);              % update w
        W=diag(-e./(1+e).^2);       % update W
        f=inv(inv(K)-W)*(w-W*f0); % iteration to get f from previous f0
        er=norm(f-f0);              % difference between 2 consecutive f's
        f0=f;                       % update f
    end
    meanf=f;                        % mean of f
    covf=inv(inv(K)-W);             % coviance of f
end
```

Example 8.2

The GPC method is trained by the two classes shown in Fig. 8.5, represented by 100 red points and 80 blue points drawn from two Gaussian distributions $\mathcal{N}(\mathbf{m}_0, \mathbf{\Sigma}_0)$ and $\mathcal{N}(\mathbf{m}_1, \mathbf{\Sigma}_1)$, where

$$\mathbf{m}_0 = \begin{bmatrix} 1 \\ 0 \end{bmatrix}, \quad \mathbf{m}_1 = \begin{bmatrix} -1 \\ 0 \end{bmatrix}, \quad \mathbf{\Sigma}_0 = \begin{bmatrix} 1 & 0 \\ 0 & 1 \end{bmatrix}, \quad \mathbf{\Sigma}_1 = \begin{bmatrix} 1 & 0.9 \\ 0.9 & 1 \end{bmatrix} \tag{8.46}$$

The 3-D plot of these two normalized Gaussian distributions is also shown in the figure.

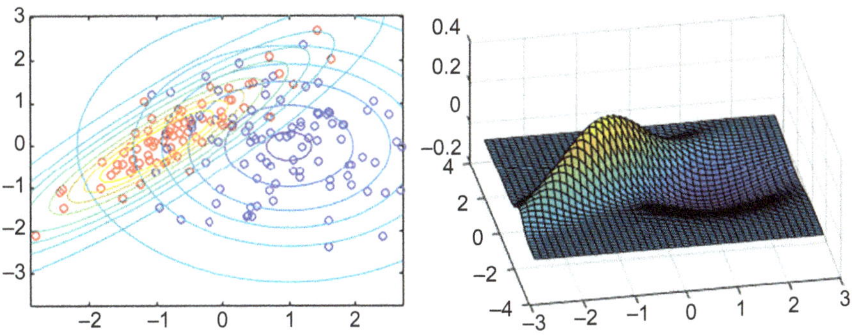

Fig. 8.5 Gaussian process for classification (Binary).

The three classification results shown in the three panels in Fig. 8.6 correspond to three different values $\alpha = 0.4,\ 0.2,\ 0.03$ for the kernel function. On the left, the 2-D space is partitioned into red and blue regions corresponding to the two classes based on the sign function of $\mathbf{m}_{f_*|y}$; on the right, the 3-D distributions of $\sigma(\mathbf{m}_{f_*|y})$ represent the estimated probability for $\mathbf{x}_*$ to belong to either class (not normalized), to be compared with the original Gaussian distributions in Fig. 8.5, from which the training samples were drawn.

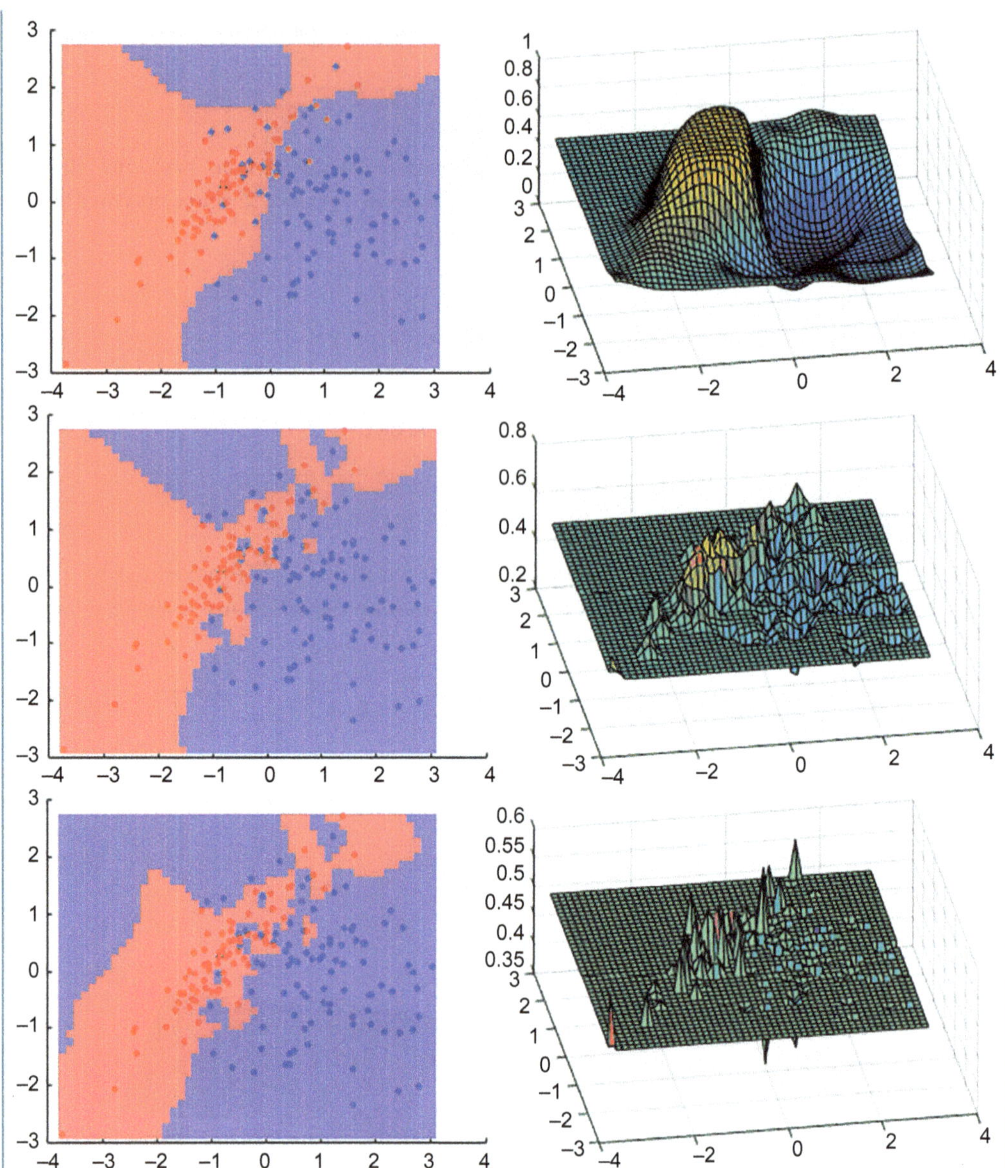

Fig. 8.6 Classification results with different α values.

We see that wherever there is evidence represented by the training samples of either class in red or blue, there are high probabilities for the neighboring points to belong to either C_1 or C_0 represented by the positive or negative peaks in the 3-D plots. Data points far away from any evidence will have low probability to belong to either class.

We make the following observations for three different values of the parameter α in SE.

- $\alpha = 0.4$ (top row), the space is partitioned into three regions, with 20 out of 180 training points misclassified. The estimated distribution is smooth.
- $\alpha = 0.2$ (middle row), the space is fragmented into several more pieces for the two classes (blue islands inside the red region and vice versa), with five out of the 180 training points misclassified. The estimated distribution is jagged.
- $\alpha = 0.03$ (bottom row), the space is partitioned into still more pieces, with all 180 training points correctly classified. The estimated distribution is spiky.

Although the error rate is lowest when α is small, the classification result is not necessarily the best as it may well be an overfitting of the noisy data. We conclude that by adjusting parameter α, we can make a proper tradeoff between error rate and overfitting.

8.3 Gaussian Process Classifier – Multiclass

The binary GPC considered previously can be generalized to multiclass GPC based on softmax function, similar to the how binary classification based on logistic function is generalized to multiclass classification. First of all, we define the following variables for each class C_k, $(k = 1, \cdots, K)$ of the K classes $\{C_1, \cdots, C_K\}$:

- **Class labeling**
  Binary labeling $\mathbf{y}_k = [y_{1k}, \cdots, y_{Nk}]^T$ indicating whether each of the N samples in the training set $\mathbf{X} = [\mathbf{x}_1, \cdots, \mathbf{x}_N]$ belongs to C_k

$$y_{nk} = \begin{cases} 1 & \text{if } \mathbf{x}_n \in C_k \\ 0 & \text{if } \mathbf{x}_n \notin C_k \end{cases} \tag{8.47}$$

  In other words, each training sample $\mathbf{x}_n \in C_k$ is labeled by K binary variables $y_{nk} = 1$ and $y_{nl} = 0$ for all $l = 1, \cdots, K$, $l \neq k$ (same as in softmax regression), as well as an integer labeling $y_n = k$.
- **Latent functions**
  Latent function $\mathbf{f}_k = [f_{1k}, \cdots, f_{Nk}]^T$, of which the nth component $f_{nk} = f_k(\mathbf{x}_n)$ $(n = 1, \cdots, N)$ is kth Gaussian process associated with C_k evaluated at the nth training sample $\mathbf{x}_n$ in the training set $\mathbf{X} = [\mathbf{x}_1, \cdots, \mathbf{x}_N]$.
- **Probability modeling**
  Probability $\mathbf{p}_k = [p_{1k}, \cdots, p_{Nk}]^T$, of which the nth component p_{nk} $(n = 1, \cdots, N)$ is the probability for $\mathbf{x}_n \in C_k$, modeled by the softmax function based on f_{nk} (same as in Eq. (7.31) in softmax regression)

$$p_{nk} = p(\hat{y} = k|\mathbf{x}_n) = \frac{e^{f_{nk}}}{\sum_{l=1}^{K} e^{f_{nl}}}, \qquad (k = 1, \cdots, K, \;\; n = 1, \cdots, N) \tag{8.48}$$

  Note that if $y_{nk} = 1$ then necessarily $y_{nl} = 0$ for all $l = 1, \cdots, K$ but $l \neq k$, that is, the K variables $y_{n1}, \cdots, y_{nK}$ are not independent. Given $p(y_{nk} = 1)$, we do not need to consider $p(y_{nl} = 0)$ for any $l \neq k$.

The probability for $\mathbf{x}_n$ to be correctly classified into class C_{y_n} can be written as the following product of K factors (of which $K-1$ are equal to 1, same as Eq. (7.32) in softmax regression)

$$p(\hat{y} = y_n|\mathbf{x}_n,\mathbf{f}) = \prod_{k=1}^{K} p(y_{nk} = 1|\mathbf{x}_n,\mathbf{f})^{y_{nk}} = \prod_{k=1}^{K} p_{nk}^{y_{nk}} \tag{8.49}$$

Based on $\mathbf{y}_k$, $\mathbf{f}_k$, and $\mathbf{p}_k$ for all $k = 1,\cdots,K$, we further define the following KN dimensional vectors

$$\mathbf{y} = \begin{bmatrix} \mathbf{y}_1 \\ \vdots \\ \mathbf{y}_K \end{bmatrix} = \begin{bmatrix} y_{11} \\ \vdots \\ y_{N1} \\ \vdots \\ y_{1K} \\ \vdots \\ y_{NK} \end{bmatrix}, \quad \mathbf{f} = \begin{bmatrix} \mathbf{f}_1 \\ \vdots \\ \mathbf{f}_K \end{bmatrix} = \begin{bmatrix} f_{11} \\ \vdots \\ f_{N1} \\ \vdots \\ f_{1K} \\ \vdots \\ f_{NK} \end{bmatrix}, \quad \mathbf{p} = \begin{bmatrix} \mathbf{p}_1 \\ \vdots \\ \mathbf{p}_K \end{bmatrix} = \begin{bmatrix} p_{11} \\ \vdots \\ p_{N1} \\ \vdots \\ p_{1K} \\ \vdots \\ p_{NK} \end{bmatrix} \tag{8.50}$$

The posterior of $\mathbf{f}$ given the training set $\mathcal{D}$ can be found based on the Bayesian theorem as (same as in Eq. (8.17) in the binary case)

$$p(\mathbf{f}|\mathcal{D}) = p(\mathbf{f}|\mathbf{X},\mathbf{y}) = \frac{p(\mathbf{y}|\mathbf{f})\; p(\mathbf{f}|\mathbf{X})}{p(\mathbf{y}|\mathbf{X})} \propto p(\mathbf{y}|\mathbf{f})\; p(\mathbf{f}|\mathbf{X}) \tag{8.51}$$

We first find the likelihood based on all p_{nk} (same as Eq. (7.33) in softmax regression)

$$L(\mathbf{f}|\mathbf{y}) \propto p(\mathbf{y}|\mathbf{f}) = \prod_{n=1}^{N} p(\hat{y} = y_n|\mathbf{x}_n,\mathbf{f}) = \prod_{n=1}^{N}\left(\prod_{k=1}^{K}(p_{nk})^{y_{nk}}\right) = \prod_{n=1}^{N}\prod_{k=1}^{K}\left(\frac{e^{f_{nk}}}{\sum_{h=1}^{K} e^{f_{nh}}}\right)^{y_{nk}} \tag{8.52}$$

As each sample $\mathbf{x}_n$ belongs to only one of the K classes, only one of $\{y_{n1},\cdots,y_{nK}\}$ can be 1 while all others are 0, consequently, the product just mentioned contains only N probabilities raised to the power of $y_{nk} = 1$, each for one of the N samples belonging to a certain class, while all other probabilities raised to the power of $y_{nk} = 0$ become 1 and not considered.

We next assume the prior probability of the latent function $\mathbf{f}_k$ for each class C_k to be a zero-mean Gaussian process $p(\mathbf{f}_k|\mathbf{X}) = \mathcal{N}(\mathbf{0},\boldsymbol{\Sigma}_k)$, where the covariance matrix $\boldsymbol{\Sigma}_k$ is constructed based on the SE for its mnth component

$$\begin{aligned} \text{Cov}[f_k(\mathbf{x}_m), f_k(\mathbf{x}_n)] &= \text{Cov}[f_{mk}, f_{nk}] = k(\mathbf{x}_m,\mathbf{x}_n) \\ &= \exp\left(-\frac{1}{a^2}||\mathbf{x}_m - \mathbf{x}_n||^2\right), \qquad (m,n = 1,\cdots,N) \end{aligned} \tag{8.53}$$

Then the prior of $\mathbf{f}$ containing all K such latent functions is also a Gaussian $p(\mathbf{f}|\mathbf{X}) = \mathcal{N}(\mathbf{0},\mathbf{K})$, where $\mathbf{0}$ is a KN-dimensional zero vector and $\mathbf{K}$ is a $KN \times KN$ block diagonal matrix

$$\mathbf{K} = \begin{bmatrix} \mathbf{K}_1 & \mathbf{0} & \cdots & \mathbf{0} \\ \mathbf{0} & \mathbf{K}_2 & \cdots & \mathbf{0} \\ \vdots & \vdots & \ddots & \vdots \\ \mathbf{0} & \cdots & \mathbf{0} & \mathbf{K}_K \end{bmatrix} \tag{8.54}$$

with $\mathbf{K}_1 = \cdots = \mathbf{K}_K$ as defined in Eq. (8.4) on the diagonal. All off-diagonal blocks are zero as the latent functions of different classes are uncorrelated.

Having found both the likelihood $p(\mathbf{y}|\mathbf{f})$ and prior $p(\mathbf{f}|\mathbf{X})$, we can write the posterior as

$$p(\mathbf{f}|\mathcal{D}) \propto p(\mathbf{f}|\mathbf{X}, \mathbf{y}) \propto p(\mathbf{y}|\mathbf{f})\ p(\mathbf{f}|\mathbf{X}) = \prod_{n=1}^{N}\prod_{k=1}^{K}\left(\frac{e^{f_{nk}}}{\sum_{h=1}^{K} e^{f_{nh}}}\right)^{y_{nk}} \mathcal{N}(\mathbf{0}, \mathbf{K}) \tag{8.55}$$

We note that the likelihood $p(\mathbf{y}|\mathbf{f})$ is not Gaussian, as $y_{nk} \in \{0, 1\}$ is a binary labeling, instead of a continuous variable. As a product of the Gaussian prior and non-Gaussian likelihood, the posterior $p(\mathbf{f}|\mathcal{D})$ is not a Gaussian process either. However, for convenience, we still assume it is approximately Gaussian $p(\mathbf{f}|\mathcal{D}) \approx \mathcal{N}(\mathbf{f}, \mathbf{m}_{f|D}, \boldsymbol{\Sigma}_{f|D})$, of which the mean $\mathbf{m}_{f|D}$ and covariance $\boldsymbol{\Sigma}_{f|D}$ are to be found.

Taking the log of the posterior, we get

$$\begin{aligned}\psi(\mathbf{f}) &= \log p(\mathbf{f}|\mathcal{D}) = \log p(\mathbf{y}|\mathbf{f}) + \log p(\mathbf{f}|\mathbf{X}) \\ &= \sum_{n=1}^{N}\sum_{k=1}^{K} y_{nk}\left(f_{nk} - \log\sum_{h=1}^{K} e^{f_{nh}}\right) + \log \mathcal{N}(\mathbf{0}, \mathbf{K}) \\ &= \mathbf{y}^T\mathbf{f} - \sum_{n=1}^{N}\log\sum_{h=1}^{K} e^{f_{nh}} - \frac{NK}{2}\log(2\pi) - \frac{1}{2}\log|\mathbf{K}| - \frac{1}{2}\mathbf{f}^T\mathbf{K}^{-1}\mathbf{f}\end{aligned} \tag{8.56}$$

where we have used the fact that $\sum_{k=1}^{K} y_{nk} = 1$. The gradient of $\psi(\mathbf{f})$ is

$$\begin{aligned}\mathbf{g}_\psi &= \frac{d}{d\mathbf{f}}\psi(\mathbf{f}) = \frac{d}{d\mathbf{f}}\left(\mathbf{y}^T\mathbf{f} - \sum_{n=1}^{N}\log\sum_{h=1}^{K} e^{f_{nh}} - \frac{NK}{2}\log(2\pi) - \frac{1}{2}\log|\mathbf{K}| - \frac{1}{2}\mathbf{f}^T\mathbf{K}^{-1}\mathbf{f}\right) \\ &= \mathbf{y} - \mathbf{p} - \mathbf{K}^{-1}\mathbf{f}\end{aligned} \tag{8.57}$$

where the second term $\mathbf{p}$ comes from

$$\sum_{n=1}^{N}\frac{d}{d\mathbf{f}}\log\sum_{h=1}^{K} e^{f_{nh}} = \sum_{n=1}^{N}\begin{bmatrix} 0 \\ \vdots \\ 0 \\ \frac{e^{f_{nk}}}{\sum_{h=1}^{K} e^{f_{nh}}} \\ 0 \\ \vdots \\ 0 \end{bmatrix} = \sum_{n=1}^{N}\begin{bmatrix} 0 \\ \vdots \\ 0 \\ p_{nk} \\ 0 \\ \vdots \\ 0 \end{bmatrix} = \mathbf{p} \tag{8.58}$$

Note that $\mathbf{p} = \mathbf{p}(\mathbf{f})$ is a function of $\mathbf{f}$.

As the posterior $p(\mathbf{f}|\mathcal{D}) \approx \mathcal{N}(\mathbf{m}_{f|D},\mathbf{\Sigma}_{f|D})$ is approximated as a Gaussian, which reaches maximum at its mean $\mathbf{m}_{f|D}$, and so does the log prior $\psi(\mathbf{f}) = \log p(\mathbf{f}|\mathcal{D})$, and the gradient of $psi(\mathbf{f})$ is zero at $\mathbf{f} = \mathbf{m}_{f|D}$

$$\mathbf{g}_\psi(\mathbf{f})\Big|_{\mathbf{f}=\mathbf{m}_{f|D}} = \mathbf{y} - \mathbf{p} - \mathbf{K}^{-1}\mathbf{f}\Big|_{\mathbf{f}=\mathbf{m}_{f|D}} = \mathbf{y} - \mathbf{p}_{m_f} - \mathbf{K}^{-1}\mathbf{m}_{f|D} = \mathbf{0}, \tag{8.59}$$

i.e.,

$$\mathbf{m}_{f|D} = \mathbf{K}(\mathbf{y} - \mathbf{p}_{m_f}) \tag{8.60}$$

where $\mathbf{p}_{m_f} = \mathbf{p}(\mathbf{m}_{f|D})$ is the defined vector evaluated at $\mathbf{f} = \mathbf{m}_{f|D}$.

We further get the Hessian matrix of $\psi(\mathbf{f})$

$$\mathbf{H}_\psi(\mathbf{f}) = \frac{d^2}{d\mathbf{f}^2}\psi(\mathbf{f}) = \frac{d}{d\mathbf{f}}\mathbf{g}_\psi = \frac{d}{d\mathbf{f}}\left(\mathbf{y} - \mathbf{K}^{-1}\mathbf{f} - \mathbf{p}\right) = -\mathbf{K}^{-1} - \mathbf{W} = -\mathbf{\Sigma}_{f|D}^{-1} \tag{8.61}$$

The last equality is due to the property of the Gaussian distribution given in Section B.1.5, from which we get the KN by KN covariance matrix of $p(\mathbf{f}|\mathcal{D})$

$$\mathbf{\Sigma}_{f|D} = -\mathbf{H}_\psi^{-1}(\mathbf{f}) = (\mathbf{K}^{-1} + \mathbf{W})^{-1} \tag{8.62}$$

Here $\mathbf{W} = d\mathbf{p}/d\mathbf{f}$ is a $KN \times KN$ Jacobian matrix of $\mathbf{p}$, of which the $ijkl$th component $(i,j = 1,\cdots,N,\;\; k,l = 1,\cdots,K)$ is

$$\begin{aligned}\frac{\partial}{\partial f_{jl}} p_{ik} &= \frac{\partial}{\partial f_j^l}\left[\frac{e^{f_{ik}}}{\sum_{h=1}^K e^{f_{ih}}}\right] = \frac{e^{f_{ik}}\left(\sum_{h=1}^K e^{f_{ih}}\right)\delta_{kl} - e^{f_{ik}}e^{f_{jl}}}{\left(\sum_{h=1}^K e^{f_{ih}}\right)^2}\delta_{ij} \\ &= \left[\frac{e^{f_{ik}}}{\sum_{h=1}^K e^{f_{ih}}}\delta_{kl} - \frac{e^{f_{ik}}e^{f_{jl}}}{\left(\sum_{h=1}^K e^{f_{ih}}\right)^2}\right]\delta_{ij} = (p_{ik}\delta_{kl} - p_{ik}p_{jl})\delta_{ij},\end{aligned} \tag{8.63}$$

and $\mathbf{W}$ can be written also in two terms

$$\mathbf{W} = \text{diag}(\mathbf{p}) - \mathbf{P}\mathbf{P}^T \tag{8.64}$$

where the first term $\text{diag}(\mathbf{p})$ is a diagonal matrix containing all KN components of $\mathbf{p}$ along the diagonal ($k = l$ and $i = j$), and $\mathbf{P}$ in the second term is a KN by N matrix composed of K $N \times N$ diagonal matrices $\text{diag}(\mathbf{p}_k)$

$$\mathbf{P} = \begin{bmatrix}\text{diag}(\mathbf{p}_1)\\ \vdots \\ \text{diag}(\mathbf{p}_K)\end{bmatrix}, \quad \text{diag}(\mathbf{p}_K) = \begin{bmatrix} p_{1k} & 0 & \cdots & 0\\ 0 & p_{2k} & \cdots & 0 \\ \vdots & \vdots & \ddots & \vdots \\ 0 & \cdots & 0 & p_{Nk}\end{bmatrix}, \quad (k = 1,\cdots,K) \tag{8.65}$$

so that

$$\mathbf{P}\mathbf{P}^T = \begin{bmatrix} \text{diag}(\mathbf{p}_1) \\ \vdots \\ \text{diag}(\mathbf{p}_K) \end{bmatrix} [\text{diag}(\mathbf{p}_1), \cdots, \text{diag}(\mathbf{p}_K)]$$
$$= \begin{bmatrix} \text{diag}(\mathbf{p}_1^2) & \cdots & \text{diag}(\mathbf{p}_1)\,\text{diag}(\mathbf{p}_K) \\ \vdots & \ddots & \vdots \\ \text{diag}(\mathbf{p}_K)\,\text{diag}(\mathbf{p}_1) & \cdots & \text{diag}(\mathbf{p}_K^2) \end{bmatrix} \tag{8.66}$$

with

$$\text{diag}(\mathbf{p}_K)\,\text{diag}(\mathbf{p}_l) = \begin{bmatrix} p_1^k p_1^l & 0 & \cdots & 0 \\ 0 & p_1^k p_1^l & \cdots & 0 \\ \vdots & \vdots & \ddots & \vdots \\ 0 & \cdots & 0 & p_{Nk} p_{Nl} \end{bmatrix} \tag{8.67}$$

Having found both the gradient $\mathbf{g}_\psi = \mathbf{y} - \mathbf{p} - \mathbf{K}^{-1}\mathbf{f}$ and Hessian $\mathbf{H}_\psi = -(\mathbf{K}^{-1} + \mathbf{W})$, we can further find the mean $\mathbf{m}_{f|D}$ at which $\psi(\mathbf{f})$ achieves maximum by the following iteration of Newton's method

$$\begin{aligned} \mathbf{f}_{n+1} &= \mathbf{f}_n - \mathbf{H}_\psi^{-1}\mathbf{g}_\psi = \mathbf{f}_n + (\mathbf{K}^{-1} + \mathbf{W})^{-1}(\mathbf{y} - \mathbf{K}^{-1}\mathbf{f}_n - \mathbf{p}) \\ &= \mathbf{f}_n + (\mathbf{K}^{-1} + \mathbf{W})^{-1}(-(\mathbf{K}^{-1} + \mathbf{W})\mathbf{f}_n + \mathbf{W}\mathbf{f}_n + \mathbf{y} - \mathbf{p}) \\ &= (\mathbf{K}^{-1} + \mathbf{W})^{-1}(\mathbf{W}\mathbf{f}_n + \mathbf{y} - \mathbf{p}) \end{aligned} \tag{8.68}$$

We note that during the iteration, we need to update not only $\mathbf{f}$, but also $\mathbf{p}$ as a functions of $\mathbf{f}$ given in Eq. (8.58).

Now that we have got both $\mathbf{m}_{f|D}$ and $\mathbf{\Sigma}_{f|D}$ of $\mathbf{f}$, we can further get $\mathbf{m}_{f_*}$ and $\mathbf{\Sigma}_{f_*}$ of $\mathbf{f}_*$.

- Obtain mean $\mathbf{m}_{f_*}$
  Similar to Eq. (8.42) in the binary case, we have the following based on $\mathbf{f} = \mathbf{K}(\mathbf{y} - \mathbf{p})$ in Eq. (8.60)

  $$\begin{aligned} \mathbf{m}_{f_*} &= E_f(\mathbf{m}_{f_*|f}) = E_f(\mathbf{K}_*^T\mathbf{K}^{-1}\mathbf{f}) = \mathbf{K}_*^T\mathbf{K}^{-1}E_f(\mathbf{f}) \\ &= \mathbf{K}_*^T\mathbf{K}^{-1}\,[\mathbf{K}(\mathbf{y} - \mathbf{p}_{m_f})] = \mathbf{K}_*^T(\mathbf{y} - \mathbf{p}_{m_f}) \end{aligned} \tag{8.69}$$

  As the K classes are uncorrelated, that is, $\mathbf{K}$ is block-diagonal, this expression can be separated into K equations each for one of the classes

  $$\mathbf{m}_{f_*^k} = (\mathbf{K}_*^k)^T(\mathbf{y}^k - \mathbf{p}^k), \qquad (k = 1, \cdots, K) \tag{8.70}$$

- Obtain covariance $\mathbf{\Sigma}_{f_*}$
  Similar to Eq. (8.44) in the binary case, we have the following based on $\mathbf{\Sigma}_{f|D} = (\mathbf{K}^{-1} + \mathbf{W})^{-1}$ in Eq. (8.62):

  $$\mathbf{\Sigma}_{m_{f_*|f}} = \mathbf{K}_*^T\mathbf{K}^{-1}\,\mathbf{\Sigma}_{f|D}\,\mathbf{K}^{-1}\mathbf{K}_* = \mathbf{K}_*^T\mathbf{K}^{-1}(\mathbf{K}^{-1} + \mathbf{W})^{-1}\mathbf{K}^{-1}\mathbf{K}_* \tag{8.71}$$

where $\mathbf{W} = \text{diag}(\mathbf{p}) - \mathbf{P}\mathbf{P}^T$ is given in Eq. (8.64) with $\mathbf{p}$ evaluated at $\mathbf{m}_{f|D}$. Then, similar to Eq. (8.44), we get

$$\boldsymbol{\Sigma}_{f_*} = \mathbf{K}_{**} - \mathbf{K}_*^T(\mathbf{K} + \mathbf{W}^{-1})^{-1}\mathbf{K}_* \tag{8.72}$$

Now we can further obtain the probability for $\mathbf{x}_*$ to belong to C_k based on the softmax function in Eq. (8.48)

$$p_*^k = \frac{e^{m_{f_*}^k}}{\sum_{l=1}^K e^{m_{f_*}^l}}, \qquad (k = 1, \cdots, K) \tag{8.73}$$

and classify $\mathbf{x}_*$ to class C_k if $p_*^k = \max\{p_*^1, \cdots, p_*^K\}$.

Moreover, the certainty or confidence of this classification result can be found from $\boldsymbol{\Sigma}_{f_*}$.

The MATLAB code for the essential parts of the algorithm is listed in the following. First, the following code segment carries out the classification of n given test samples based on the N training set $\mathcal{D} = \{\mathbf{X}, \mathbf{y}\}$.

```
K=Kernel(X,X);      % covariance of prior of p(f|X)
Ks=Kernel(X,Xs);
Kss=Kernel(Xs,Xs);
[meanf Sigmaf p W]=findPosteriorMeanMC(K,y,C);
                    % find mean and covariance of p(f|D), and W, p
Sigmafy=Kss-Ks'*inv(K+inv(W))*Ks; % covariance of p(f|D)
p=reshape(p,N,C);
y=reshape(y,N,C);
for k=1:C
    meanfD(:,k)=Ks'*(y(:,k)-p(:,k));
                    % mean of p(f_*|X_*,D) for kth class
end
for i=1:n           % for each of n test samples
    d=sum(exp(meanfD(i,:)));
                    % denominator of softmax function
    [pmax k]=max(exp(meanfD(i,:))/d);
                    % find class with max probability
    ys(i)=k;        % label ith sample as member of kth class
    pr(i,k)=pmax;   % probability of ith sample belonging to kth class
end
```

This code segment calls the following function, which computes the mean and covariance of $p(\mathbf{f}|\mathbf{X}, \mathbf{y})$ by Newton's method:

```
function [meanf Sigmaf p W]=findPosteriorMeanMC(K,y,C)
    % get mean and covariance of p(f|X,y) by Newton's method, given {X,y}
    n=length(y);         % number of training samples
    f0=zeros(n,1);       % initial value of latent function
    f=zeros(n,1);
    er=1;
```

```
        k=0;
        while er > 10^(-9)     % find f that maximizes p(f|X,y)
            k=k+1;
            [p W]=findp(f,N,C);
                               % call a function to find vector p and matrix W
            f=inv(inv(K)+W)*(W*f0+y-p);
                               % iteratively update value of f
            er=norm(f-f0);     % difference between consecutive iterations
            f0=f;              % update f
        end
        meanf=f;
        Sigmaf=inv(inv(K)+W);
    end
```

The following function called by the previous function finds vector $\mathbf{p}$ and matrix $\mathbf{W} = \text{diag}(\mathbf{p}) - \mathbf{P}\mathbf{P}^T$:

```
function [p W]=findp(f,N,C)  % find vector p and matrix W=diag(p)-P*P'
    F=reshape(f,N,C)';       % kth row contains N samples of class k
    p=zeros(C,N);            % initialize p
    for n=1:N                % for each of N training samples
        d=sum(exp(F(:,n)));  % sum of all C terms in denominator
        for k=1:C            % for all C classes
            p(k,n)=exp(F(k,n))/d;
        end
    end
    P=[];
    for k=1:C                % generate P
        P=[P; diag(p(k,:))]; % stack C diagonal matrices
    end
    p=reshape(p',N*C,1);     % convert p into a column vector
    W=diag(p)-P*P';          % generate W matrix
end
```

Example 8.3

This example illustrates the classification of the same dataset of three classes used before. The top two panels of Fig. 8.7 show the distributions of the three classes in the training data set, while the bottom two panels show the classification results in terms of the partitioning of the feature space (bottom left) and the posterior distribution $p(\mathbf{f}_*|\mathbf{X}, \mathbf{y}, \mathbf{X}_*)$ (bottom right), which can be compared with the distribution of the training set (top right). The confusion matrix of the classification result is shown here, with the error rate $30/600 = 0.045$

$$\begin{bmatrix} 197 & 3 & 0 \\ 0 & 195 & 5 \\ 1 & 21 & 178 \end{bmatrix}$$

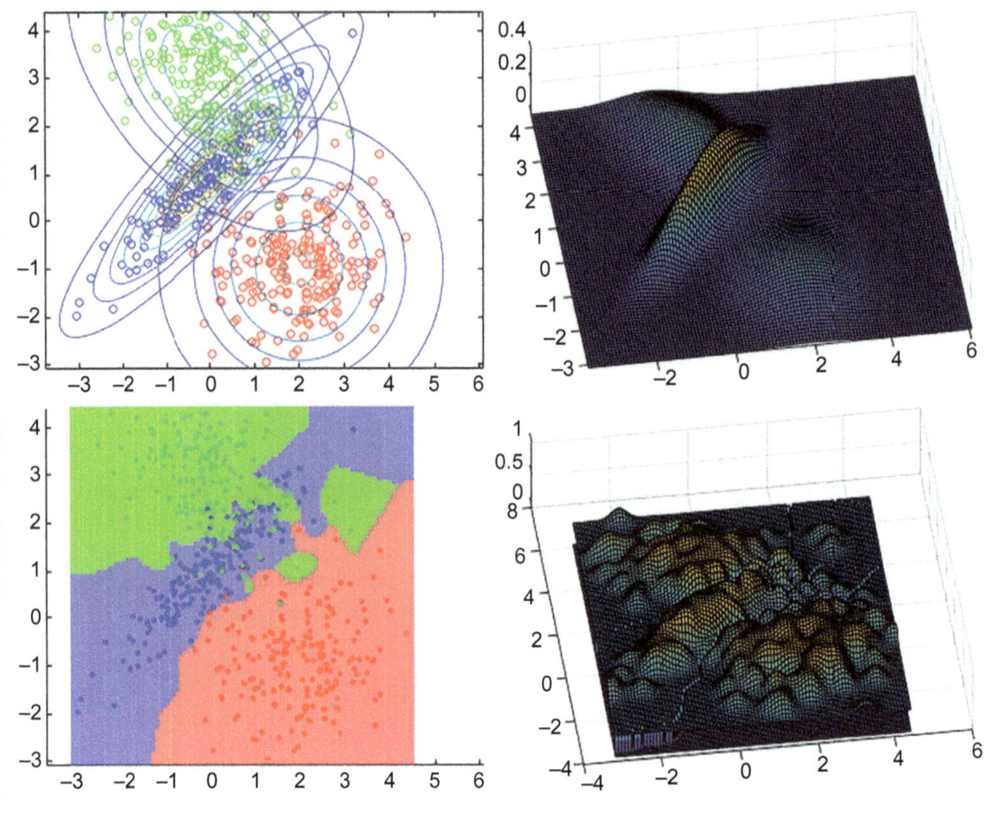

Fig. 8.7 Gaussian process for classification (multiclass).

Example 8.4

This example illustrates the classification of the XOR data set as shown in Fig. 8.8. The confusion matrix of the classification result is shown, with the error rate 13/400 = 0.0325

$$\begin{bmatrix} 193 & 7 \\ 6 & 194 \end{bmatrix}$$

The results in Examples 8.3 and 8.4 can be compared with those in Examples 13.6 and 13.7 obtained using the naive Bayes classification method (Section 13.3), and Examples 18.4 and 18.5, obtained using the backpropagation classification method (Section 18.2). All of these results are based on the same datasets. While the GPC method tends to achieve lower error rates, it may also be more susceptible to overfitting. To mitigate this, the parameter of the squared exponential kernel should be adjusted to strike an appropriate balance between overfitting and classification error.

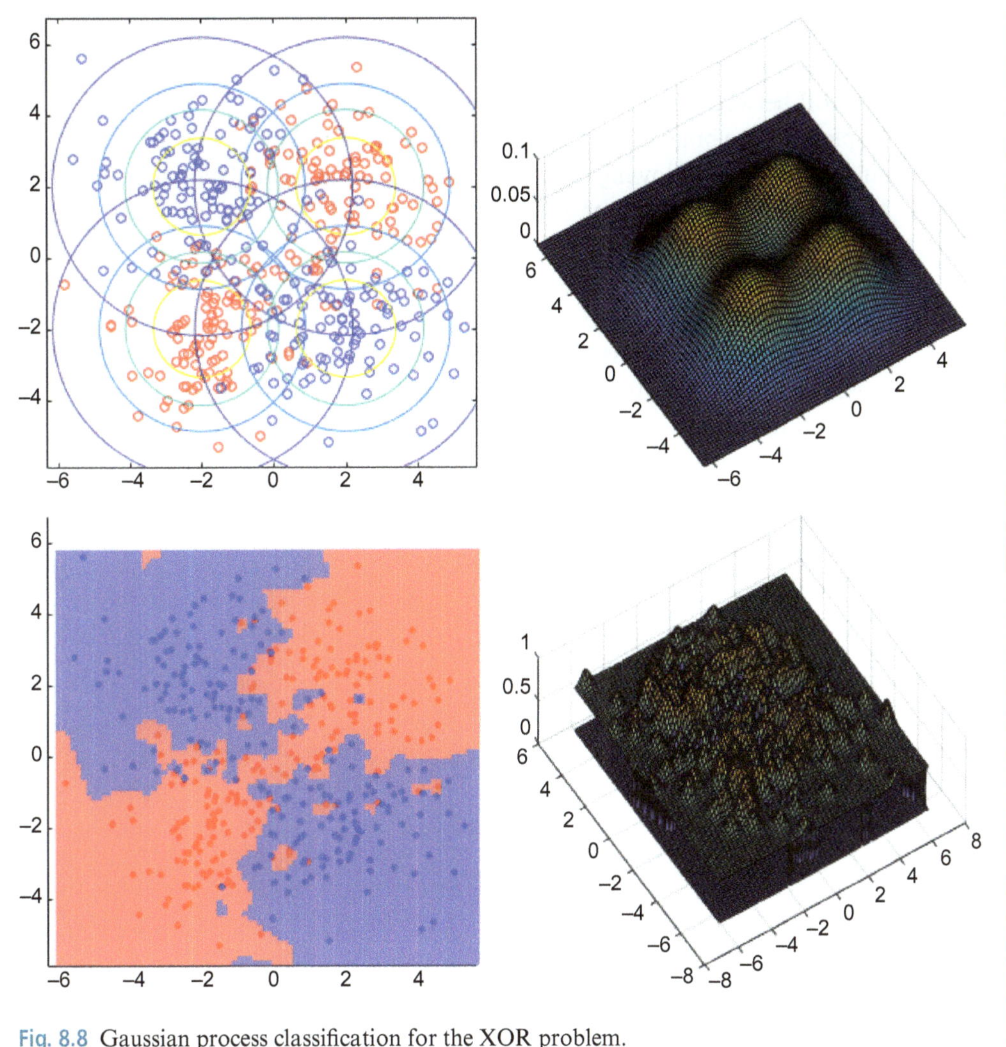

Fig. 8.8 Gaussian process classification for the XOR problem.

PROBLEMS

1. Implement the method of Gaussian process regression and carry out regression based on the dataset in problem as in Example 8.1 with $N = 8$ training samples randomly drawn from a noisy sinusoidal function $y = f(x) + n = \cos(2\pi f x) + n$ in the domain `x=0:0.1:9`. Here $f = 0.2$ is the frequency and the noise has a Gaussian distribution with $(\mu_n = 0,\ \sigma_n^2)$. Try different values of σ (e.g., $\sigma = 0.2, 0.5, 1, \cdots$).

   Plot your results in terms of the mean $E[f_*]$ for the regression function and variance $\pm\,\mathrm{Var}[f_*]$ for the range of certainty as three curves, the same as in Figs 8.2 and 8.3.

Try different values for the hyperparameter α (variable a in Eq. (8.3)) to see how the smoothness of the regression function is affected, and to make a proper tradeoff between underfitting and overfitting.

2. Implement the binary Gaussian process classifier and then carry out binary classification of the following dataset containing data points belonging to either of two classes (labeled by 1 or -1).

   /e176/programs/GPBCdata.txt

   Try three different α values for the squared exponential function to show underfitting, overfitting, and proper fitting (similar Example 8.2).

3. Implement the multiclass Gaussian process classifier and then carry out multiclass classification of the following dataset containing data points belonging to any of three classes (labeled by 1, 2, or 3).

   /e176/programs/GPMCdata.txt

   Try three different α values for the squared exponential function to show underfitting, overfitting, and proper fitting.

Part III

Feature Extraction

In Part III, we will discuss various methods for *feature selection* or *extraction*, which can be considered as a preprocessing stage for the task of *pattern classification*, one of the main tasks in machine learning (ML) to be discussed in Part IV, for the goal of classifying a set of *patterns* into one of some K classes or clusters. Here the generic term *pattern* means any objects of interest, each described by a set of d *features* denoted as a column vector $\mathbf{x} = [x_1, \cdots, x_d]^T$ in a d-dimensional *feature space*, of which each component is some value that quantitatively measures the corresponding feature used to describe the pattern. The task of classification can therefore be considered to partition the feature space into K regions each corresponding to one of the K classes or clusters.

Obviously the computational complexity of an operation on the dataset is a function (e.g., linear, polynomial, exponential, etc.) of the dimensionality d of the feature space. When d is large, it is often desirable to reduce it to some significantly smaller value $d' \ll d$, while still maintaining most of the information in the dataset relevant to the task at hand, such as the variation of the features that describe the patterns or distinguish different classes of the patterns. For example, in ML and data science, it is often desirable to visualize the dataset in a high ($d > 3$) dimensional feature space to gain some intuitive sense of the dataset. It is therefore necessary to reduce the dimensionality to three or even two while maintaining the signal variation of interest in the data as much as possible. As another example in classification, the computational complexity of an algorithm may be too high if the dimensionality of the feature space is high. To reduce the computational complexity, we may consider reducing the dimensionality while still preserving the *separability* of the data points of different classes.

Feature selection or feature extraction can be used to reduce the dimensionality of a given dataset, by either selecting d' features directly from the d original ones (with $C_d^{d'} = d!/d'!(d - d')!$ ways to do so), or generating d' new features, each as a function, such as the linear combinations, of the d original ones. In either case, the essential information of interest in the data needs to be maximally maintained in the resulting lower dimensional space spanned by the d' new features.

In the following chapters we will first consider how information of interest, such as class separability, can be measured quantitatively, and then discuss specific methods for information extraction and dimensionality reduction, such as the methods of *principal component analysis (PCA)*, *independent component analysis (ICA)* and their variations, widely used in signal processing, data science, and ML.

9 Feature Selection

9.1 Distances and Separability Measurements

Before considering various feature selection methods we first define some typical distances used to quantitatively measure the difference between two entities, such as data points and groups of data points. In the following, we assume there are N samples in the dataset, of which N_k samples belong to class C_k, that is, $N_1 + \cdots + N_K = N$, and we define $P_k = N_k/N$ as the prior probability for any sample $\mathbf{x}$ randomly selected from the dataset to belong to C_k.

- **Distance between two data points**
  The distance between two points $\mathbf{x}$ and $\mathbf{y}$ in the d-dimensional feature space can be measured by the p-norm of the difference $\mathbf{x} - \mathbf{y}$ (Eq. (A.203) in Section A.4.1)

$$d_p(\mathbf{x}, \mathbf{y}) = ||\mathbf{x} - \mathbf{y}||_p = \left(\sum_{i=1}^{d} |x_i - y_i|^p \right)^{1/p} \qquad 1 \leq p \leq \infty \tag{9.1}$$

  Specially, consider the following three cases:
  - $p = 1$, *city block or Manhattan distance*

$$d_1(\mathbf{x}, \mathbf{y}) = ||\mathbf{x} - \mathbf{y}||_1 = \sum_{i=1}^{d} |x_i - y_i| \tag{9.2}$$

  - $p = 2$, *Euclidean distance*

$$d_2(\mathbf{x}, \mathbf{y}) = ||\mathbf{x} - \mathbf{y}||_2 = \sqrt{\sum_{i=1}^{d} |x_i - y_i|^2} \tag{9.3}$$

  - $p = \infty$, the *Chebyshev distance*

$$d_\infty(\mathbf{x}, \mathbf{y}) = ||\mathbf{x} - \mathbf{y}||_\infty = \max\{|x_1 - y_1|, \cdots, |x_d - y_d|\} \tag{9.4}$$

- **Intra-class distance**
  An intra-class distance measures how widely (or narrowly) all samples in each class C_k are distributed in the space, that is, how tightly all data points in the class are clustered. It needs to be small for good separability.

- The *max diameter*

$$d_{max}(C_k) = \max_{\mathbf{x},\mathbf{y}\in C_k} d(\mathbf{x},\mathbf{y}) \tag{9.5}$$

- The *average diameter*

$$d_{average}(C_k) = \frac{1}{N_k(N_k-1)} \sum_{\mathbf{x},\mathbf{y}\in C_k, \mathbf{x}\neq\mathbf{y}} d(\mathbf{x},\mathbf{y}) \tag{9.6}$$

- The average distance from each sample to the mean vector of the cluster:

$$\frac{1}{N_k} \sum_{\mathbf{x}\in C_k} d(\mathbf{x},\mathbf{m}_k), \quad \text{where} \quad \mathbf{m}_k = \frac{1}{N_k} \sum_{\mathbf{x}\in C_k} \mathbf{x} \tag{9.7}$$

- The covariance of all samples in the class represents the tightness of the samples in the class

$$\Sigma_k = \frac{1}{N_k} \sum_{\mathbf{x}\in C_k} (\mathbf{x}-\mathbf{m}_k)(\mathbf{x}-\mathbf{m}_k)^T \tag{9.8}$$

and its determinant or trace can be used as such a scalar measurement of the tightness

$$\det \Sigma_k = \prod_{i=1}^{d} \lambda_i, \quad \operatorname{tr} \Sigma_k = \sum_{i=1}^{d} \lambda_i \tag{9.9}$$

As $\mathbf{\Sigma}_k$ is positive semi-definite, all of its eigenvalues λ_i are non-negative and so are its determinant and trace.

- **Distance between a point and a cluster/class of points**
The distance between a single point $\mathbf{x}$ and a cluster C_k can be defined differently depending on how the cluster is characterized.
  - The *max and min distances:* This distance assumes no additional knowledge of the cluster except its member data points.

$$d_{max}(\mathbf{x},C_k) = \max_{\mathbf{y}\in C_k} d(\mathbf{x},\mathbf{y}), \quad d_{min}(\mathbf{x},C_k) = \min_{\mathbf{y}\in C_k} d(\mathbf{x},\mathbf{y}) \tag{9.10}$$

  - The *centroid distance:* This distance assumes each cluster C_k of points is characterized by their mean $\mathbf{m}_k$ representing their central location in the feature space.

$$d_{centroid}(\mathbf{x},C_k) = d_2(\mathbf{x},\mathbf{m}_k) = ||\mathbf{x}-\mathbf{m}_k||^2 = (\mathbf{x}-\mathbf{m}_k)^T(\mathbf{x}-\mathbf{m}_k) \tag{9.11}$$

  - The *Mahalanobis distance:* This distance assumes each cluster C_k of points is characterized by their covariances in $\mathbf{\Sigma}_k$ representing how they are distributed (the tightness-looseness, the shape and orientation of their distribution) as well as their mean $\mathbf{m}_k$ for the central position.

$$d_M(\mathbf{x},C_k) = (\mathbf{x}-\mathbf{m}_k)^T \Sigma_k^{-1} (\mathbf{x}-\mathbf{m}_k) \tag{9.12}$$

Note that this distance is positively related to the distance between $\mathbf{x}$ and $\mathbf{m}_k$, but negatively related to $\mathbf{\Sigma}_k$, that is, the more tightly they are clustered, the greater the distance $d_M(\mathbf{x},C_k)$.

- **Inter-class distance**
  An intra-class distance measures the difference between two classes. It needs to be large for good separability.
  - The *max and min distances*

$$d_{max}(C_i, C_j) = \max_{\mathbf{x} \in C_i, \mathbf{y} \in C_j} d(\mathbf{x}, \mathbf{y}), \quad d_{min}(C_i, C_j) = \min_{\mathbf{x} \in C_i, \mathbf{y} \in C_j} d(\mathbf{x}, \mathbf{y}) \quad (9.13)$$

  - The *average distance* between two clusters is the average of all pair-wise distances (e.g., Euclidean) between members of the two classes

$$d_{average}(C_i, C_j) = \frac{1}{N_i N_j} \sum_{\mathbf{x} \in C_i, \mathbf{y} \in C_j} d(\mathbf{x}, \mathbf{y}) \quad (9.14)$$

  - The *centroid distance* is the distance (e.g., Euclidean) between the centroids (mean vectors) of the two classes

$$d_{centroid}(C_i, C_j) = d(\mathbf{m}_i, \mathbf{m}_j) \quad (9.15)$$

  - The *Bhattacharyya distance*

$$\begin{aligned} d_B(C_i, C_j) &= \frac{1}{4} (\mathbf{m}_i - \mathbf{m}_j)^T \left(\frac{\Sigma_i + \Sigma_j}{2} \right)^{-1} (\mathbf{m}_i - \mathbf{m}_j) \\ &+ \log \left[\frac{|(\Sigma_i + \Sigma_j)/2|}{(|\Sigma_i| \, |\Sigma_j|)^{1/2}} \right] \end{aligned} \quad (9.16)$$

  Similar to the Mahalanobis distance, the first term is for the difference between the mean vectors of the two clusters but negatively related to the average of their covariances. The second term reflects the difference between the covariances of the two clusters, which is always positive owing to the *AM–GM inequality* (algebraic mean vs. geometric mean) of the arithmetic and geometric means

$$\frac{1}{n} \sum_{i=1}^{n} x_i \geq \left(\prod_{i=1}^{n} x_i \right)^{1/n} \quad (9.17)$$

  Note that even when $\mathbf{m}_i = \mathbf{m}_j$ and therefore the first term is zero, the Bhattacharyya distance is still greater than zero if $\mathbf{\Sigma}_i \neq \mathbf{\Sigma}_j$.

The inter and intra-class distances can also be defined based on a set of scatter matrices:

- Total scatter matrix
  The total scatter matrix is the same as the covariance matrix of the entire dataset

$$\mathbf{S}_T = \frac{1}{N}\sum_{\text{all }\mathbf{x}}(\mathbf{x}-\mathbf{m})(\mathbf{x}-\mathbf{m})^T = \frac{1}{N}\sum_{k=1}^{K}\sum_{\mathbf{x}\in C_k}(\mathbf{x}-\mathbf{m})(\mathbf{x}-\mathbf{m})^T \tag{9.18}$$

Owing to the total mean $\sum_{\mathbf{X}}\mathbf{x} = \mathbf{m}$, rank $\mathbf{S}_T) \le N-1$. The equality holds if all N samples are independent.

- Within-class (intra-class) scatter matrix

$$\mathbf{S}_W = \sum_{k=1}^{K} P_k\boldsymbol{\Sigma}_k = \sum_{k=1}^{K}\frac{N_k}{N}\boldsymbol{\Sigma}_k = \frac{1}{N}\sum_{k=1}^{K}\sum_{\mathbf{x}\in C_k}(\mathbf{x}-\mathbf{m}_k)(\mathbf{x}-\mathbf{m}_k)^T \tag{9.19}$$

Owing to the K means $\mathbf{m}_k = \sum_{\mathbf{x}\in C_k}\mathbf{x}/N_k$, $(k = 1,\cdots,K)$ as K constraints, rank$(\mathbf{S}_W) = \sum_{k=1}^{K}(N_k-1) \le N-K$. Again, the equality holds if all N samples are independent.

- Between-class (inter-class) scatter matrix

$$\mathbf{S}_B = \sum_{k=1}^{K} P_k(\mathbf{m}_k-\mathbf{m})(\mathbf{m}_k-\mathbf{m})^T = \sum_{k=1}^{K}\frac{N_k}{N}(\mathbf{m}_k-\mathbf{m})(\mathbf{m}_k-\mathbf{m})^T \tag{9.20}$$

Owing to total mean $\mathbf{m} = \sum_{k=1}^{K} N_k\mathbf{m}_k/N$ as an additional constraint, rank$(\mathbf{S}_B) \le K-1$.

We can show that the total scatteredness of the dataset is contributed by the within-class and between-class scatteredness, that is, $\mathbf{S}_T = \mathbf{S}_W + \mathbf{S}_B$

$$\begin{aligned}
\mathbf{S}_T &= \frac{1}{N}\sum_{k=1}^{K}\sum_{\mathbf{x}\in C_k}(\mathbf{x}-\mathbf{m})(\mathbf{x}-\mathbf{m})^T = \frac{1}{N}\sum_{k=1}^{K}\sum_{\mathbf{x}\in C_k}(\mathbf{x}-\mathbf{m}_k+\mathbf{m}_k-\mathbf{m})(\mathbf{x}-\mathbf{m}_k+\mathbf{m}_k-\mathbf{m})^T\\
&= \frac{1}{N}\sum_{k=1}^{K}\sum_{\mathbf{x}\in C_k}[(\mathbf{x}-\mathbf{m}_k)(\mathbf{x}-\mathbf{m}_k)^T + (\mathbf{x}-\mathbf{m}_k)(\mathbf{m}_k-\mathbf{m})^T\\
&\quad +(\mathbf{m}_k-\mathbf{m})(\mathbf{x}-\mathbf{m}_k)^T + (\mathbf{m}_k-\mathbf{m})(\mathbf{m}_k-\mathbf{m})^T]\\
&= \frac{1}{N}\sum_{k=1}^{K}\sum_{\mathbf{x}\in C_k}(\mathbf{x}-\mathbf{m}_k)(\mathbf{x}-\mathbf{m}_k)^T + \frac{1}{K}\sum_{k=1}^{K}\sum_{\mathbf{x}\in C_k}(\mathbf{m}_k-\mathbf{m})(\mathbf{m}_k-\mathbf{m})^T\\
&= \sum_{k=1}^{K}\frac{N_k}{N}\boldsymbol{\Sigma}_k + \sum_{k=1}^{K}\frac{N_k}{N}(\mathbf{m}_k-\mathbf{m})(\mathbf{m}_k-\mathbf{m})^T = \mathbf{S}_W + \mathbf{S}_B
\end{aligned} \tag{9.21}$$

where the two middle terms are both zero as $\sum_{\mathbf{x}\in C_k}(\mathbf{x}-\mathbf{m}_k) = \mathbf{0}$.

For better separability, we want the within-class scatteredness to be small but the between-class scatteredness to be large. However, as these scatter matrices cannot be directly compared in terms of their sizes, we instead consider their traces (or determinants) as the scalar measurements for the separability

$$\begin{aligned}
J_B &= \text{tr }\mathbf{S}_B = \text{tr}\left[\sum_{k=1}^{C} P_k(\mathbf{m}_k-\mathbf{m})(\mathbf{m}_k-\mathbf{m})^T\right]\\
&= \sum_{i=1}^{C} P_k \text{ tr}\left[(\mathbf{m}_k-\mathbf{m})(\mathbf{m}_k-\mathbf{m})^T\right] = \sum_{i=1}^{C} P_k\ ||\mathbf{m}_k-\mathbf{m}||^2
\end{aligned} \tag{9.22}$$

and

$$J_W = \text{tr } \mathbf{S}_W = \text{tr} \left[\sum_{k=1}^{K} P_k \sum_{\mathbf{x} \in C_k} (\mathbf{x} - \mathbf{m}_k)(\mathbf{x} - \mathbf{m}_k)^T \right]$$
$$= \sum_{k=1}^{K} P_k \sum_{\mathbf{x} \in C_k} \text{tr} \left[(\mathbf{x} - \mathbf{m}_k)(\mathbf{x} - \mathbf{m}_k)^T \right] = \sum_{k=1}^{K} P_k \sum_{\mathbf{x} \in C_k} ||\mathbf{x} - \mathbf{m}_k||^2 \quad (9.23)$$

We see that J_B is the weighted average of the Euclidean distances between $\mathbf{m}_k$ and total $\mathbf{m}$ for the entire dataset containing all K classes, and J_W is the weighted average of the Euclidean distances between all $\mathbf{x} \in C_k$ to $\mathbf{m}_k$ for all K classes $C_1, \cdots, C_K$. We also recall that the trace or determinant of a matrix is the sum or product of all of its eigenvalues (Eq. (A.105) in Section A.3.1).

To achieve a high separability, we would like to to maximize $\text{tr}\,\mathbf{S}_B$ and minimize $\text{tr}\,\mathbf{S}_W$. In the same space with a constant $\mathbf{S}_T = \mathbf{S}_B + \mathbf{S}_W$, maximizing $\text{tr}\,\mathbf{S}_B$ is equivalent to minimizing $\text{tr}\,\mathbf{S}_W$. However, to measure the separability across different spaces with different $\mathbf{S}_T$, we need to normalize $\mathbf{S}_B$ by either $\mathbf{S}_T$ or equivalently $\mathbf{S}_W$, and redefine the objective function as either of the following

$$J_{B/W} = \text{tr}(\mathbf{S}_W^{-1}\mathbf{S}_B) = \text{tr}(\mathbf{S}_B\mathbf{S}_W^{-1}), \qquad J_{B/T} = \text{tr}(\mathbf{S}_T^{-1}\mathbf{S}_B) = \text{tr}(\mathbf{S}_B\mathbf{S}_T^{-1}) \quad (9.24)$$

To reduce the dimensionality of the feature space while maximally conserving the separability in the original d–D space, we can select $d' < d$ of the d original features that maximize either $J_{B/W}$ of $J_{B/T}$ as an objective function.

PROBLEMS

In many of the problems in this homework and subsequent ones it is necessary to develop code to carry out feature extraction, dimensionality reduction, and classification/clustering, to be tested on both real and synthetic datasets. The following sample code can be used to generate a synthetic dataset containing multiple classes, each composed of a set of normally distributed data points, based on user specified parameters including dimensionality d, number of classes K, and number of samples in each class (adding up to the total number of samples N), as well as the means and covariances (for the positions and shape of the normal distribution, respectively). The code generates a $d \times N$ array $\mathbf{X}$ for all data points and an N–D vector $\mathbf{y}$ for the labelings, to be used as the ground truth for any supervised learning algorithm. Note that the repeatability of the algorithm is desired, a fixed seed for the random number generator needs to be specified.

```
function [X y]=sampleDataset
    K=4;                    % number of classes
    d=3;                    % dimensionality of dataset
    rng(1);                 % seed for random number generator
    Nk=[150 200 250];       % number of samples in each class
    N=sum(Nk);              % total number of samples
```

```
    mu=zeros(d,K);          % mean vectors for all classes
    Sigma=zeros(d,d,K);     % covariance matrices for all classes
    Sigma(:,:,1)=eye(3);    % covariance matrix for each class
    Sigma(:,:,2)=eye(3);    % feel free to specify any other form
    Sigma(:,:,3)=eye(3);
    mu(:,1)=[0; 0; 0];      % specify mean vector for each class
    mu(:,2)=[0; 0; 1];
    mu(:,3)=[0; 1; 0];
    mu=mu*4;                % scaling factor (for varying separability)
    X=[];                   % for all data samples (d by N matrix)
    y=[];                   % for their labelings (1 by N vector)
    for k=1:K               % for each of K classes
        Xk=mvnrnd(mu(:,k),Sigma(:,:,k), Nk(k))';
                            % generate normally distributed dataset
        X=[X Xk];           % concatenate all samples to X
        y=[y k*ones(1,Nk(k))];  % concatenate their labeling to y
        switch k
            case 1
                color='r';
            case 2
                color='g';
            case 3
                color='b';
        end
        scatter3(Xk(1,:),Xk(2,:),Xk(3,:),[],color,'.') % visualize
        hold on
    end
    hold off
    data=[X; y];
    save('data34.txt','data','-ASCII') % save dataset as ASCII file
end
```

The following code can be used to read in the data file generated by the code above, and any dataset composed of data array **X** and the labeling vector **y** with data points in arbitrary order.

```
data=load('data34.txt','data');  % read disk file into data
d=size(data,1)-1          % dimensionality of data
N=size(data,2);           % total number of samples
X=data(1:d,:);            % all N d-dimensional sample vectors
y=data(d+1,:);            % and their labelings
K=length(unique(y));      % number of classes
for k=1:K                 % for each of the K classes
    idx=find(y==k);       % indices of samples in class k
    Xk=X(:,idx)           % collect all samples in class k
```

```
        mean(Xk')               % mean vector of class k
        cov(Xk')                % covariance matrix of class k
    end
```

In addition to synthetic data, some real datasets are also used in both homework problems and examples in the current and future chapters, such as the iris flower dataset (https://en.wikipedia.org/wiki/Iris_flower_data_set), which was first introduced in 1936 by British statistician and biologist Ronald Fisher, and has been widely used for benchmarking various classification algorithms in recent decades. This dataset contains 150 samples of iris flowers of three different species (setosa, virginica, and versicolor) each containing 50 samples characterized by four features: the length and width of sepals and petals of the flower. This dataset can be read into your MATLAB code by a built-in Metlab function:

```
X=iris_dataset;             % read in iris data
[d, N]=size(data);          % dimensionality and number of samples
y=[ones(1,50) 2*ones(1,50) 3*ones(1,50)];  % data labelings
```

1. Generate and visualize multiclass synthetic data generated by this code. Try different parameter values (e.g., $d = 2,3$, and $K = (3, 4, 5,\cdots))$, and experiment with different mean vectors and covariance matrices to see how the locations and shapes of distributions of the data points are affected.
2. Generate a dataset (e.g., $d = 3$ and $K = 5$) and find all $K(K-1)/2$ pair-wise distances $\{d_p(\mathbf{m}_i,\mathbf{m}_j),\ i = 2,\cdots,K,\ j = 1,\cdots,i-1\}$ between the mean vectors of any two of different classes for each value of $p = 1,\ 2,\ \infty$. Display your results as the lower triangular matrix (all components below the main diagonal of a $K \times K$ matrix).
3. Find all pair-wise Bhattacharyya distances between any two classes, and show them as the lower triangular matrix, and compare them with those obtained in the previous problem.
4. Find the between-class, within-class, and total scatter matrices and verify that they do satisfy $\mathbf{S}_B + \mathbf{S}_W = \mathbf{S}_T$.
5. Repeat all of these problems for the iris dataset.

10 Principal Component Analysis

The *principal component analysis (PCA)* is a method widely used for visualizing, processing and analyzing large datasets of high dimensionalities. The PCA is capable of significantly reducing the dimensionality of the dataset, while still maximally keeping the signal variation or information of interest in the dataset measured in a certain way, so that the subsequent processing and analysis of the dataset can be carried out much more effectively and efficiently.

As always, we assume the given dataset is in the form of a $d \times N$ matrix $\mathbf{X} = [\mathbf{x}_1, \cdots, \mathbf{x}_N]$, of which each column vector $\mathbf{x} = [x_1, \cdots, x_d]^T$ is a data sample, a point in the d-dimensional feature space. In the context of pattern classification, each component x_i of $\mathbf{x}$ is some measurement of one of the d features of a pattern. Given $\mathbf{X}$, the PCA carries out a special orthogonal transformation $\mathbf{y} = \mathbf{A}^T\mathbf{x}$, called *Karhunen–Loève transform (KLT)*, in such a way that the components in the resulting vector $\mathbf{y}$ are completely decorrelated and the energy or information contained in the data is redistributed optimally so that most of the energy is concentrated in a small number of components called *principal components*. Based on these two highly desirable properties, the KLT can be used for data compression in data processing and feature extraction in pattern recognition.

10.1 Covariance and Correlation

In PCA, each pattern $\mathbf{x} = [x_1, \cdots, x_d]^T$ is treated as a random vector of which each component x_i is a random variable with mean and variance

$$\begin{aligned}
\mu_i &= \mathrm{E}[\mathbf{x}_i] = \int x_i\, p(x_i)\, dx_i \\
\sigma_i^2 &= \mathrm{E}[(x_i - \mu_i)^2] = \mathrm{E}[x_i^2] - \mu_i^2 = \int x_i^2\, p(x_i)\, dx_i - \mu_{x_i}^2 \\
&(i = 1, \cdots, d)
\end{aligned} \tag{10.1}$$

and the *covariance* of x_i and x_j is

$$\begin{aligned}
\sigma_{ij}^2 &= \mathrm{E}[(x_i - \mu_i)(x_j - \mu_j)] = \mathrm{E}[x_i x_j] - \mu_i \mu_j \\
&= \int\int x_i x_j\, p(x_i, x_j)\, dx_i\, dx_j - \mu_i \mu_j \qquad (i, j = 1, \cdots, d)
\end{aligned} \tag{10.2}$$

The mean vector and covariance matrix of $\mathbf{x}$ are

$$\mathbf{m}_x = \mathrm{E}[\mathbf{x}] = \begin{bmatrix} \mu_1 \\ \vdots \\ \mu_d \end{bmatrix}$$
$$\boldsymbol{\Sigma}_x = \mathrm{E}[(\mathbf{x}-\mathbf{m}_x)(\mathbf{x}-\mathbf{m}_x)^T] = \mathrm{E}[\mathbf{x}\mathbf{x}^T] - \mathbf{m}_x\mathbf{m}_x^T$$
$$= \begin{bmatrix} \sigma_{11}^2 & \cdots & \sigma_{1d}^2 \\ \vdots & \ddots & \vdots \\ \sigma_{d1}^2 & \cdots & \sigma_{dd}^2 \end{bmatrix} \tag{10.3}$$

Usually the joint probability density function (pdf) $p(\mathbf{x}) = p(x_1, \cdots, x_d)$ of the random vector $\mathbf{x}$ is unknown, and the mean vector $\mathbf{m}_x$ and covariance matrix $\boldsymbol{\Sigma}_x$ of $\mathbf{x}$ can only be estimated by the method of *maximum likelihood estimation (MLE)* (Section B.3.3) based on a set of observed data samples in data set $\mathbf{X}$

$$\hat{\mathbf{m}}_x = \frac{1}{N}\sum_{n=1}^{N} \mathbf{x}_n, \qquad \hat{\boldsymbol{\Sigma}} = \frac{1}{N-1}\sum_{n=1}^{N}(\mathbf{x}_n - \hat{\mathbf{m}})(\mathbf{x}_n - \hat{\mathbf{m}})^T = \frac{1}{N}\sum_{n=1}^{N}\mathbf{x}_n\mathbf{x}_n^T - \hat{\mathbf{m}}\hat{\mathbf{m}}^T \tag{10.4}$$

Note that the rank of the $d \times d$ estimated covariance matrix $\hat{\boldsymbol{\Sigma}}$ is at most $N-1$, due to the N samples in the dataset $\mathbf{X}$, assumed to be are independent, and the additional constraint

$$\sum_{n=1}^{N}(\mathbf{x}_n - \hat{\mathbf{m}}) = \sum_{n=1}^{N}\mathbf{x}_n - N\hat{\mathbf{m}} = \mathbf{0} \tag{10.5}$$

The variance $\sigma_i^2 = \mathrm{E}[(x_i - \mu_{x_i})^2]$ can be treated as the dynamic energy contained in x_i, or the amount of information carried in x_i, while the trace $\mathrm{tr}\,\boldsymbol{\Sigma}_x = \sum_{i=1}^{d}\sigma_i^2$ can be considered as the total amount of dynamic energy contained in $\mathbf{x}$. Also, the covariance $\sigma_{ij}^2 = \mathrm{E}[(x_i - \mu_{x_i})(x_j - \mu_{x_j})]$ can be considered as the *mutual energy*, a measure of the correlation between x_i and x_j. By normalizing the covariance σ_{ij}^2, we get the *correlation coefficient* between x_i and x_j

$$\rho_{ij} = \frac{\sigma_{ij}^2}{\sqrt{\sigma_i^2\,\sigma_j^2}} = \frac{\sigma_{ij}^2}{\sigma_i\,\sigma_j} \tag{10.6}$$

that measures how the two random variables x_i and x_j are correlated.

- $\rho_{ij} = \pm 1$ or $|\rho_{ij}| = 1$: x_i and x_j are maximally correlated. The information contained in the two variables is completely redundant, that is, given the value of one of them, the value of the other is known.
- $-1 < \rho_{ij} < 1$ or $|\rho_{ij}| < 1$: x_i and x_j are correlated to different extents. The information they each carry has certain redundancy.
- $\rho_{ij} = 0$: x_i and x_j are uncorrelated. They each carry their own independent information with no redundancy.

We see that the correlation coefficient ρ_{ij} that measures how much two random variables are correlated also measures the redundancy of the information they

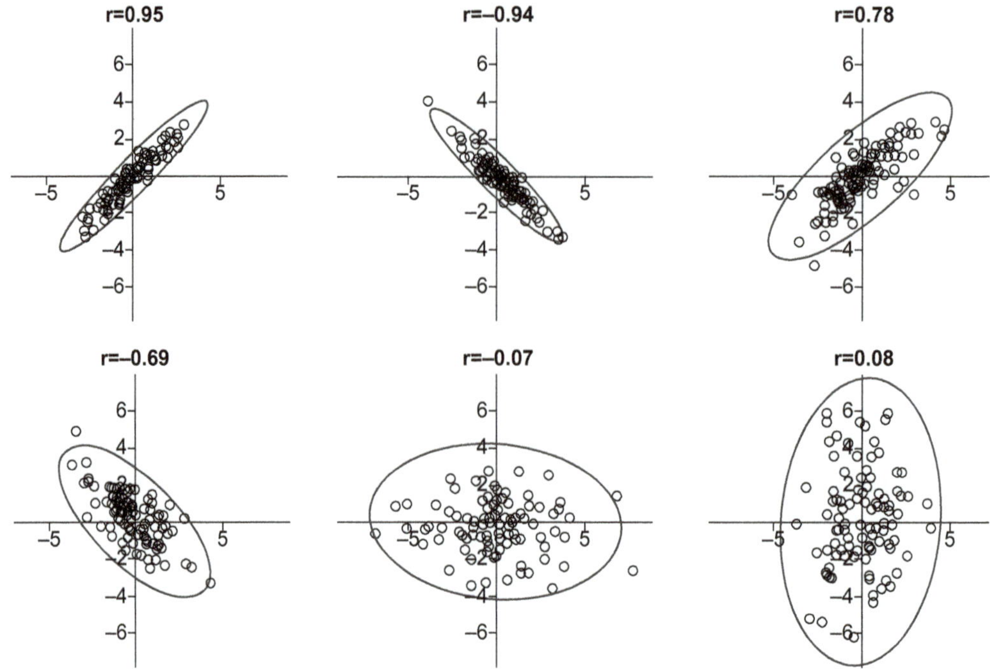

Fig. 10.1 Normally distributed datasets with different covariance matrices.

carry. When there exists some data redundancy in the data, it is possible to carry out data compression by some method such as the PCA based on the covariance $\mathbf{\Sigma}$ to reduce the data redundancy, so that the data size can be significantly reduced while the information (dynamic energy) contained in the data is still mostly conserved.

For example, six normally distributed 2-D datasets are generated with zero mean and the following covariance matrices

$$\mathbf{\Sigma}_1 = \begin{bmatrix} 1.0 & 0.95 \\ 0.95 & 1.0 \end{bmatrix}, \quad \mathbf{\Sigma}_2 = \begin{bmatrix} 1.0 & -0.95 \\ -0.95 & 1.0 \end{bmatrix}, \quad \mathbf{\Sigma}_2 = \begin{bmatrix} 1.0 & 0.7 \\ 0.7 & 1.0 \end{bmatrix}$$

$$\mathbf{\Sigma}_4 = \begin{bmatrix} 1.0 & -0.7 \\ -0.7 & 1.0 \end{bmatrix}, \quad \mathbf{\Sigma}_5 = \begin{bmatrix} 3 & 0 \\ 0 & 1 \end{bmatrix}, \quad \mathbf{\Sigma}_6 = \begin{bmatrix} 1 & 0 \\ 0 & 3 \end{bmatrix}$$

Figure 10.1 shows the 2-D data points together with the contour line (isoline) of the corresponding normal distribution based on each of the six covariance matrices. The correlation coefficient is also shown on top of each dataset.

10.2 Karhunen–Loève Transformation

A data point in a d-dimensional space is represented by a vector $\mathbf{x} = [x_1, \cdots, x_d]^T$, of which the components are the coordinates along the d standard orthonormal basis vectors $\{\mathbf{e}_1, \cdots, \mathbf{e}_d\}$ that span the d–D space

$$\mathbf{x} = \begin{bmatrix} x_1 \\ x_2 \\ \vdots \\ x_d \end{bmatrix} = x_1 \begin{bmatrix} 1 \\ 0 \\ \vdots \\ 0 \end{bmatrix} + x_2 \begin{bmatrix} 0 \\ 1 \\ \vdots \\ 0 \end{bmatrix} + \cdots + x_d \begin{bmatrix} 0 \\ \vdots \\ 0 \\ 1 \end{bmatrix} = \sum_{i=1}^{d} x_i \mathbf{e}_i$$
$$= \begin{bmatrix} \mathbf{e}_1 & \cdots & \mathbf{e}_d \end{bmatrix} \begin{bmatrix} x_1 \\ \vdots \\ x_d \end{bmatrix} = \mathbf{I}\mathbf{x} \tag{10.7}$$

where $\mathbf{I} = [\mathbf{e}_1, \cdots, \mathbf{e}_d]$ is the identity matrix, and the ith component $x_i = \mathbf{e}_i^T \mathbf{x}$ is the projection of $\mathbf{x}$ onto the ith standard basis vector $\mathbf{e}_i$.

However, as shown in Fig. 10.2, the space can also be spanned by another orthonormal basis as the column vectors in any orthogonal matrix $\mathbf{A} = [\mathbf{a}_1, \cdots, \mathbf{a}_d]$ satisfying

$$\mathbf{a}_i^T \mathbf{a}_j = \delta_{ij} = \begin{cases} 1 & i = j \\ 0 & i \neq j \end{cases} \tag{10.8}$$

and the vector can also be equivalently represented as a weighted sum of these basis vectors

$$\mathbf{x} = \sum_{i=1}^{d} x_i \mathbf{e}_i = \sum_{i=1}^{d} y_i \mathbf{a}_i = \begin{bmatrix} \mathbf{a}_1 & \cdots & \mathbf{a}_d \end{bmatrix} \begin{bmatrix} y_1 \\ \vdots \\ y_d \end{bmatrix} = \mathbf{A}\mathbf{y} \tag{10.9}$$

where $\mathbf{y} = [y_1, \cdots, y_d]^T$ is a vector containing all d coordinates in the directions of the basis vector $\{\mathbf{a}_i, \cdots, \mathbf{a}_d\}$.

The orthogonal matrix $\mathbf{A}$ satisfies $\mathbf{A}^T = \mathbf{A}^{-1}$, that is, $\mathbf{A}\mathbf{A}^T = \mathbf{A}^T\mathbf{A} = \mathbf{I}$. Premultiplying $\mathbf{A}^T$ on both sides of the equation, we obtain $\mathbf{A}\mathbf{x} = \mathbf{A}^T\mathbf{A}\mathbf{y} = \mathbf{A}^{-1}\mathbf{A}\mathbf{y} = \mathbf{y}$, that is,

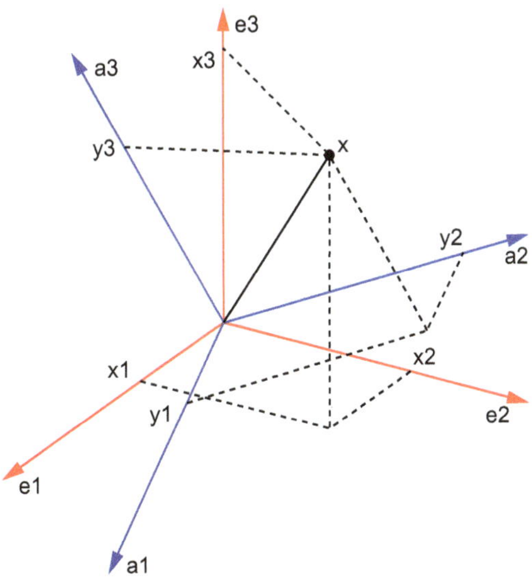

Fig. 10.2 Orthogonal transform as a rotation of the coordinate system.

$$\mathbf{y} = \begin{bmatrix} y_1 \\ \vdots \\ y_d \end{bmatrix} = \mathbf{A}^T\mathbf{x} = \begin{bmatrix} \mathbf{a}_1^T \\ \vdots \\ \mathbf{a}_d^T \end{bmatrix} \mathbf{x} = \begin{bmatrix} \mathbf{a}_1^T\mathbf{x} \\ \vdots \\ \mathbf{a}_d^T\mathbf{x} \end{bmatrix} \tag{10.10}$$

where the ith coordinate $y_i = \mathbf{a}_i^T\mathbf{x}$ is the projection of $\mathbf{x}$ onto the ith basis vector $\mathbf{a}_i$. These two equations form an orthogonal transform pair

$$\begin{cases} \mathbf{y} = \mathbf{A}^T\mathbf{x} & \text{forward transform} \\ \mathbf{x} = \mathbf{A}\mathbf{y} & \text{inverse transform} \end{cases} \tag{10.11}$$

The basis vectors in $\mathbf{A} = [\mathbf{a}_1, \cdots, \mathbf{a}_d]$ can be considered as a rotated version of the standard basis in $\mathbf{I} = [\mathbf{e}_1, \cdots, \mathbf{e}_d]$, and the norm or length of the vector before and after the transform remain the same

$$||\mathbf{x}||^2 = \mathbf{x}^T\mathbf{x} = \left(\sum_{i=1}^d y_i\mathbf{a}_i\right)^T \left(\sum_{j=1}^d y_j\mathbf{a}_j\right) = \sum_{i=1}^d\sum_{j=1}^d y_iy_j\, \mathbf{a}_i^T\mathbf{a}_j = \sum_{i=1}^d y_i^2 = ||\mathbf{y}||^2 \tag{10.12}$$

If $\mathbf{x}$ is treated as a random vector, then its linear transform $\mathbf{y} = \mathbf{A}^T\mathbf{x}$ is also a random vector, with the following mean and covariance

$$\begin{aligned} \mathbf{m}_y &= E[\mathbf{y}] = E[\mathbf{A}^T\mathbf{x}] = \mathbf{A}^T\, E[\mathbf{x}] = \mathbf{A}^T\mathbf{m}_x \\ \mathbf{\Sigma}_y &= E[\mathbf{y}\mathbf{y}^T] - \mathbf{m}_y\mathbf{m}_y^T = E[\mathbf{A}^T\mathbf{x}\mathbf{x}^T\mathbf{A}] - \mathbf{A}^T\mathbf{m}_x\mathbf{m}_x^T\mathbf{A} \\ &= \mathbf{A}^T[E[\mathbf{x}\mathbf{x}^T] - \mathbf{m}_x\mathbf{m}^T]\mathbf{A} = \mathbf{A}^T\mathbf{\Sigma}_x\mathbf{A} \end{aligned} \tag{10.13}$$

The *KLT* is just a special case of the orthogonal transforms considered here, in the same form of $\mathbf{y} = \mathbf{V}^T\mathbf{x}$, where the orthogonal transform matrix $\mathbf{V} = [\mathbf{v}_1, \cdots, \mathbf{v}_d]$ is the eigenvector matrix of the covariance matrix $\mathbf{\Sigma}_x$ of $\mathbf{x}$, composed of the d normalized eigenvectors of $\mathbf{\Sigma}_x$. As in general the covariance matrix $\mathbf{\Sigma}_x$ is symmetric and positive definite, its eigenvalues $\{\lambda_1, \cdots, \lambda_d\}$ are real and positive, and its eigenvectors are orthogonal

$$\lambda_d \geq \cdots \geq \lambda_1 \geq 0, \qquad \mathbf{v}_i^T\mathbf{v}_j = \delta_{ij} = \begin{cases} 0 & i \geq j \\ 1 & i = j \end{cases} \qquad (i, j = 1, \cdots, d) \tag{10.14}$$

that is $\mathbf{V}$ is an orthogonal matrix satisfying $\mathbf{V}^T = \mathbf{V}^{-1}$ or $\mathbf{V}^T\mathbf{V} = \mathbf{I}$.

The d eigenvalues $\{\lambda_1, \cdots, \lambda_d\}$ and the corresponding eigenvectors $\{\mathbf{v}_1, \cdots, \mathbf{v}_d\}$ can then be found by solving the eigenequations

$$\Sigma_x\mathbf{v}_i = \lambda_i\mathbf{v}_i \qquad (i = 1, \cdots, d) \tag{10.15}$$

Note that the eigenvector $\mathbf{v}_i$ is not unique, as the eigenequation still holds when $\mathbf{v}_i$ is arbitrarily scaled by any constant. These d eigenequations can be combined to be expressed in matrix form

$$\mathbf{\Sigma}_x\mathbf{V} = \mathbf{\Sigma}_x[\mathbf{v}_1, \cdots, \mathbf{v}_d] = [\mathbf{v}_1, \cdots, \mathbf{v}_d] \begin{bmatrix} \lambda_1 & \cdots & 0 \\ \vdots & \ddots & \vdots \\ 0 & \cdots & \lambda_d \end{bmatrix} = \mathbf{V}\mathbf{\Lambda} \tag{10.16}$$

where $\mathbf{\Lambda} = \text{diag}(\lambda_1, \cdots, \lambda_d)$ is a diagonal matrix.

Pre-multiplying $\mathbf{v}_i^T$ on both sides of the eigenequation in Eq. (10.15), we get:

$$\mathbf{v}_i^T \Sigma_x \mathbf{v}_i = \lambda_i \mathbf{v}_i^T \mathbf{v}_i = \lambda_i \qquad (i = 1, \cdots, d) \tag{10.17}$$

or in matrix form, pre-multiplying $\mathbf{V}^T = \mathbf{V}^{-1}$ on both sides of Eq. (10.16), we have

$$\mathbf{V}^T \Sigma_x \mathbf{V} = \mathbf{V}^{-1}\mathbf{V}\mathbf{\Lambda} = \mathbf{\Lambda} \tag{10.18}$$

Taking inverse on both sides, we also get

$$(\mathbf{V}^T \Sigma_x \mathbf{V})^{-1} = \mathbf{V}^{-1}\Sigma_x^{-1}(\mathbf{V}^T)^{-1} = \mathbf{V}^T \Sigma_x^{-1} \mathbf{V} = \mathbf{\Lambda}^{-1} \tag{10.19}$$

We see that $\mathbf{\Sigma}_x^{-1}$ and Σ_x share the same eigenvectors, their eigenvalues are reciprocal of each other, and they are both diagonalized by the orthogonal eigenvector matrix $\mathbf{V}$ to become $\mathbf{\Lambda}$ and $\mathbf{\Lambda}^{-1}$, respectively.

Based on the orthogonal eigenvector matrix $\mathbf{V}$, the KLT is defined as

$$\mathbf{y} = \begin{bmatrix} y_1 \\ \vdots \\ y_d \end{bmatrix} = \mathbf{V}^T\mathbf{x} = \begin{bmatrix} \mathbf{v}_1^T \\ \vdots \\ \mathbf{v}_d^T \end{bmatrix} \mathbf{x} = \begin{bmatrix} \mathbf{v}_1^T\mathbf{x} \\ \vdots \\ \mathbf{v}_d^T\mathbf{x} \end{bmatrix} \tag{10.20}$$

of which the ith component $y_i = \mathbf{v}_i^T\mathbf{x}$ is the projection of vector $\mathbf{x}$ onto the ith eigenvector $\mathbf{v}_i$. Pre-multiplying $\mathbf{V} = (\mathbf{V}^T)^{-1}$ on both sides, we obtain

$$\mathbf{V}\mathbf{y} = \mathbf{V}\mathbf{V}^T\mathbf{x} = \mathbf{V}\mathbf{V}^{-1}\mathbf{x} = \mathbf{x} \tag{10.21}$$

These two equations form the KLT pair:

$$\begin{cases} \mathbf{y} = \mathbf{V}^T\mathbf{x} & \text{forward transform} \\ \mathbf{x} = \mathbf{V}\mathbf{y} & \text{inverse transform} \end{cases} \tag{10.22}$$

10.3 Optimality of the KLT

The KLT is the optimal orthogonal transform among all possible orthogonal transforms $\mathbf{y} = \mathbf{A}^T\mathbf{x}$ based on any orthogonal matrix $\mathbf{A}$ satisfying $\mathbf{A}^T\mathbf{A} = \mathbf{I}$, in the sense that the KLT completely decorrelates the signal, and it maximally compacts the energy (information) contained in the signal, as shown in the following.

- **KLT completely decorrelates signal components**
  The mean vector and covariance matrix of the random vector $\mathbf{y} = \mathbf{V}^T\mathbf{x}$ after the KLT can be found as

$$\mathbf{m}_y = \mathrm{E}[\mathbf{y}] = \mathrm{E}[\mathbf{V}^T\mathbf{x}] = \mathbf{V}^T\,\mathrm{E}[\mathbf{x}] = \mathbf{V}^T\mathbf{m}_x \tag{10.23}$$

$$\begin{aligned} \Sigma_y &= \mathrm{E}[(\mathbf{y}-\mathbf{m}_y)(\mathbf{y}-\mathbf{m}_y)^T] = \mathrm{E}[\mathbf{V}^T(\mathbf{x}-\mathbf{m}_x)][\mathbf{V}^T(\mathbf{x}-\mathbf{m}_x)]^T \\ &= \mathbf{V}^T E\left[(\mathbf{x}-\mathbf{m}_x)(\mathbf{x}-\mathbf{m}_x)^T\right]\mathbf{V} = \mathbf{V}^T\Sigma_x\mathbf{V} = \mathbf{\Lambda} \\ &= \begin{bmatrix} \lambda_1 & \cdots & 0 \\ \vdots & \ddots & \vdots \\ 0 & \cdots & \lambda_d \end{bmatrix} = \begin{bmatrix} \sigma_{y_1}^2 & \cdots & 0 \\ \vdots & \ddots & \vdots \\ 0 & \cdots & \sigma_{y_d}^2 \end{bmatrix} \end{aligned} \tag{10.24}$$

Based on the fact that the covariance matrix $\mathbf{\Sigma}_y$ is a diagonal matrix, we have the following two observations.

- The zero off-diagonal component $\sigma^2_{ij} = 0$ $(i \neq j)$ for the covariance between any two components y_i and y_j indicates they are completely decorrelated after the KLT, that is, each component carries its own independent information.
- The nonzero diagonal component for the variance $\lambda_i = \sigma^2_i = \mathrm{E}[(y_i - \mu_{y_i})^2]$ $(i = 1, \cdots, c)$ of the ith component y_i represents the dynamic energy or amount of information contained in y_i. As $\mathbf{\Sigma}_x$ and $\mathbf{\Sigma}_y = \mathbf{\Lambda}$ share the same eigenvalues $\{\lambda_1, \cdots, \lambda_d\}$, and therefore have the same trace

$$\mathcal{E}_x = \operatorname{tr} \mathbf{\Sigma}_x = \sum_{i=1}^{d} \lambda_i = \operatorname{tr} \mathbf{\Sigma}_y = \mathcal{E}_y \tag{10.25}$$

  We see that the total amount of energy or information contained in the signal $\mathbf{x}$ before the KLT and $\mathbf{y}$ after the KLT remains the same, that is, the energy or information contained in the signal is conserved by the KLT. This is generally true for all orthogonal transforms $\mathbf{y} = \mathbf{A}^T\mathbf{x}$ (such as the Fourier transform), due to the fact that the eigenvalues remain the same under any orthogonal transform $\mathbf{B} = \mathbf{R}^T\mathbf{A}\mathbf{R}$, that is, matrices $\mathbf{A}$ and $\mathbf{B}$ share the same eigenvalues, and thereby the same trace and determinant.

- **KLT optimally compacts signal energy**

  While the total energy contained in the signal is conserved by any orthogonal transform, the distribution of this energy among the d components before and after the transform may be very different. We can show that the KLT is optimal in the sense that it redistributes the total energy in such a way that it is maximally compacted into a subset of components of $\mathbf{y} = \mathbf{V}^T\mathbf{x}$.

  Consider a generic orthogonal transform $\mathbf{y} = \mathbf{A}^T\mathbf{x}$ based on an arbitrary orthogonal matrix $\mathbf{A} = [\mathbf{a}_1, \cdots, \mathbf{a}_d]$ satisfying $\mathbf{A}^T = \mathbf{A}^{-1}$

$$\mathbf{y} = \begin{bmatrix} y_1 \\ \vdots \\ y_d \end{bmatrix} = \mathbf{A}^T\mathbf{x} = \begin{bmatrix} \mathbf{a}_1^T \\ \vdots \\ \mathbf{a}_d^T \end{bmatrix} \mathbf{x} \tag{10.26}$$

  We also have

$$y_i = \mathbf{a}_i^T\,\mathbf{x}, \qquad \mu_{y_i} = E(y_i) = \mathbf{a}_i^T\,E(\mathbf{x}) = \mathbf{a}_i^T\mathbf{m}_x, \qquad (i = 1, \cdots, d) \tag{10.27}$$

  We then consider the energy contained in $d' < d$ out of the d components of $\mathbf{y} = \mathbf{A}^T\mathbf{x}$ expressed as a function of the transform matrix $\mathbf{A} = [\mathbf{a}_1, \cdots, \mathbf{a}_d]$

$$\begin{aligned} \mathcal{E}'(\mathbf{A}) &= \sum_{i=1}^{d'} \sigma^2_{y_i} = \sum_{i=1}^{d'} E\left[(y_i - \mu_{y_i})^2\right] = \sum_{i=1}^{d'} E\left[\mathbf{a}_i^T(\mathbf{x} - \mathbf{m}_{x_i})\,(\mathbf{x} - \mathbf{m}_{x_i})^T\mathbf{a}_i\right] \\ &= \sum_{i=1}^{d'} \mathbf{a}_i^T E\left[(\mathbf{x} - \mathbf{m}_{x_i})(\mathbf{x} - \mathbf{m}_{x_i})^T\right]\mathbf{a}_i^T = \sum_{i=1}^{d'} \mathbf{a}_i^T \mathbf{\Sigma}_x \mathbf{a}_i \end{aligned} \tag{10.28}$$

  We can find the optimal orthogonal transform matrix $\mathbf{A}$ that maximizes this partial energy $\mathcal{E}'$ by solving the following optimization problem with the constraint that the column vectors of $\mathbf{A}$ are normalized with $\mathbf{a}_j^T\mathbf{a}_j = 1$

$$\begin{cases} \text{maximize: } \mathcal{E}'(\mathbf{A}) = \mathcal{E}'(\mathbf{a}_1, \cdots, \mathbf{a}_{d'}) \\ \text{subject to: } \ \mathbf{a}_j^T \mathbf{a}_j = 1 \quad (j = 1, \cdots, d') \end{cases} \tag{10.29}$$

The Lagrangian function of this constrained optimization problem is

$$\mathcal{L}(\mathbf{A}) = \mathcal{E}'(\mathbf{A}) - \sum_{j=1}^{d'} \lambda_j (\mathbf{a}_j^T \mathbf{a}_j - 1) \tag{10.30}$$

To find the optimal $\mathbf{A}$, we set the partial derivative with respect to each $\mathbf{a}_i$ $(i = 1, \cdots, d')$ to zero

$$\begin{aligned} \frac{\partial}{\partial \mathbf{a}_i} \mathcal{L}(\mathbf{A}) &= \frac{\partial}{\partial \mathbf{a}_i} \left[\sum_{j=1}^{d'} (\mathbf{a}_j^T \Sigma_x \mathbf{a}_j - \lambda_j \mathbf{a}_j^T \mathbf{a}_j + \lambda_j) \right] \\ &= \frac{\partial}{\partial \mathbf{a}_i} \left(\mathbf{a}_i^T \Sigma_x \mathbf{a}_i - \lambda_i \mathbf{a}_i^T \mathbf{a}_i \right) = 2\Sigma_x \mathbf{a}_i - 2\lambda_i \mathbf{a}_i = \mathbf{0} \end{aligned} \tag{10.31}$$

The resulting equation $\Sigma_x \mathbf{a}_i = \lambda_i \mathbf{a}_i$ happens to be the eigenequation of the covariance matrix $\mathbf{\Sigma}_x$, and the column vectors of $\mathbf{A}$ must be the eigenvectors of Σ_x, that is, the optimal transform is the KLT $\mathbf{A} = \mathbf{V}$.

$$\mathcal{E}'(\mathbf{V}) = \sum_{i=1}^{d'} \mathbf{v}_i^T \Sigma_x \mathbf{v}_i = \sum_{i=1}^{d'} \lambda_i \le \sum_{i=1}^{d} \lambda_i = \mathcal{E}_y = \mathcal{E}_x \tag{10.32}$$

To maximize the partial energy $\mathcal{E}'(\mathbf{V})$ contained in the d' components we carry out the KLT based on the eigenvectors $\mathbf{v}_i$ corresponding to the d' largest eigenvalues of Σ_x: $\lambda_1 \ge \lambda_2 \ge \cdots \lambda_{d'} \cdots \ge \lambda_d$.

10.4 Geometric Interpretation of the KLT

The optimality of the KLT can be demonstrated geometrically, based on the assumption that the random vector $\mathbf{x} = [x_1, \cdots, x_d]^T$ has a normal pdf

$$p(\mathbf{x}) = \mathcal{N}(\mathbf{m}_x, \mathbf{\Sigma}_x) = \frac{1}{(2\pi)^{d/2} |\mathbf{\Sigma}_x|^{1/2}} \exp\left[-\frac{1}{2} (\mathbf{x} - \mathbf{m}_x)^T \mathbf{\Sigma}_x^{-1} (\mathbf{x} - \mathbf{m}_x) \right] \tag{10.33}$$

with mean vector $\mathbf{m}_x$ and covariance matrix $\mathbf{\Sigma}_x$. The shape of this normal distribution in the d-dimensional space can be represented by an iso-hypersurface in the space determined by

$$\mathcal{N}(\mathbf{m}_x, \mathbf{\Sigma}_x) = c_0 \tag{10.34}$$

where c_0 is some constant, or equivalently

$$(\mathbf{x} - \mathbf{m}_x)^T \mathbf{\Sigma}_x^{-1} (\mathbf{x} - \mathbf{m}_x) = c_1 \tag{10.35}$$

where c_1 is another constant related to c_0. As $\mathbf{\Sigma}^{-1}$ is positive definite as well as $\mathbf{\Sigma}$, this equation represents an hyper ellipsoid in the d-dimensional space. In particular, when $d = 2$, $\mathbf{x} = [x_1, x_2]^T$, with positive definite $\mathbf{\Sigma}_x^{-1}$

$$\mathbf{\Sigma}_x^{-1} = \begin{bmatrix} A & B/2 \\ B/2 & C \end{bmatrix} \quad \text{and} \quad \left| \mathbf{\Sigma}_x^{-1} \right| = AC - B^2/4 > 0 \tag{10.36}$$

then this quadratic equation becomes

$$(\mathbf{x}-\mathbf{m}_x)^T\Sigma_x^{-1}(\mathbf{x}-\mathbf{m}_x) = [x_1-\mu_{x_1}, x_2-\mu_{x_2}]\begin{bmatrix} A & B/2 \\ B/2 & C \end{bmatrix}\begin{bmatrix} x_1-\mu_{x_1} \\ x_2-\mu_{x_2} \end{bmatrix}$$
$$= A(x_1-\mu_{x_1})^2 + B(x_1-\mu_{x_1})(x_2-\mu_{x_2}) + C(x_2-\mu_{x_2})^2 = c_1$$

representing an ellipse (instead of other quadratic curves such as hyperbola and parabola) centered at $\mathbf{m}_x = [\mu_1, \mu_2]^T$. When $d = 3$, the quadratic equation represents an ellipsoid. In general when $d > 3$, the equation $\mathcal{N}(\mathbf{x}, \mathbf{m}_x, \Sigma_x) = c_0$ represents a hyper-ellipsoid in the d-dimensional space.

Substituting $\mathbf{x} = \mathbf{V}\mathbf{y}$ and $\mathbf{m}_x = \mathbf{V}\mathbf{m}_y$ into the iso-surface equation Eq. (10.35), we get the equation for the hyper-ellipsoid after the KLT

$$(\mathbf{x}-\mathbf{m}_x)^T\mathbf{\Sigma}_x^{-1}(\mathbf{x}-\mathbf{m}_x) = [\mathbf{V}(\mathbf{y}-\mathbf{m}_y)]^T\mathbf{\Sigma}_x^{-1}\mathbf{V}\mathbf{y}-\mathbf{m}_y)$$
$$= (\mathbf{y}-\mathbf{m}_y)^T\mathbf{V}^T\mathbf{\Sigma}_x^{-1}\mathbf{V}(\mathbf{y}-\mathbf{m}_y) = (\mathbf{y}-\mathbf{m}_y)^T\mathbf{\Lambda}^{-1}(\mathbf{y}-\mathbf{m}_y)$$
$$= \sum_{i=1}^{d}\frac{(y_i-\mu_{y_i})^2}{\lambda_i} = \sum_{i=1}^{d}\frac{(y_i-\mu_{y_i})^2}{\sigma_{y_i}^2} = c_1 \tag{10.37}$$

This a standardized hyper-ellipsoid. We therefore see that the KLT $\mathbf{y} = \mathbf{V}^T\mathbf{x}$ is actually a rotation of the coordinate system of the d-dimensional space, which is spanned by the standard basis $\{\mathbf{e}_1, \cdots, \mathbf{e}_d\}$ before the KLT and the eigenvectors $\{\mathbf{v}_1, \cdots, \mathbf{v}_d\}$ as the basis vectors after the KLT.

As a result, the principal semi-axes of the ellipsoid representing the Gaussian distribution of the dataset become in parallel with the axes of the new coordinate system, that is, the ellipsoid becomes standardized. Moreover, the length of the ith principal semi-axis is proportional to the standard deviation $\sigma_{y_i} = \sqrt{\lambda_i}$ of the ith variable y_i. This is the reason why KLT possesses the two desirable properties: (a) the decorrelation of the signal components and (b) redistribution and compaction of the energy or information contained in the signal, as illustrated in Fig. 10.3.

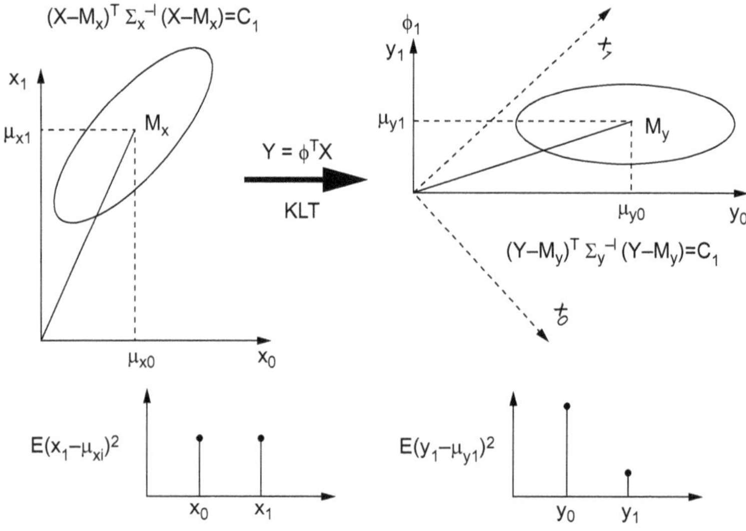

Fig. 10.3 Standardization of ellipsoids by the KLT.

Example 10.1

The top row of Fig. 10.4 shows three given datasets with the following parameters

$$\mathbf{\Sigma}_1 = \begin{bmatrix} 2.964 & 0.544 \\ 0.544 & 4.687 \end{bmatrix}, \quad \mathbf{\Sigma}_2 = \begin{bmatrix} 5.036 & 3.083 \\ 3.083 & 2.830 \end{bmatrix}, \quad \mathbf{\Sigma}_3 = \begin{bmatrix} 3.043 & -3.076 \\ -3.076 & 3.930 \end{bmatrix} \tag{10.38}$$

$$\rho_1 = 0.146, \qquad \rho_2 = 0.817, \qquad \rho_3 = -0.889 \tag{10.39}$$

Then the KLT is carried out to rotate the coordinate system from the original standard basis to the eigenvectors of their covariance matrices, with the corresponding eigenvalues

$$\lambda_{1,2} = (4.845,\ 2.807), \qquad \lambda_{1,2} = (7.207,\ 0.659), \qquad \lambda_{1,2} = (6.594,\ 0.379) \tag{10.40}$$

The datasets after the KLT are shown in the bottom row of the figure. We see that in each of the three cases, the ellipse representing the distribution of the data points becomes standardized, with its semi-axes proportional to the eigenvalues, which, when squared, are the variances of the principal components.

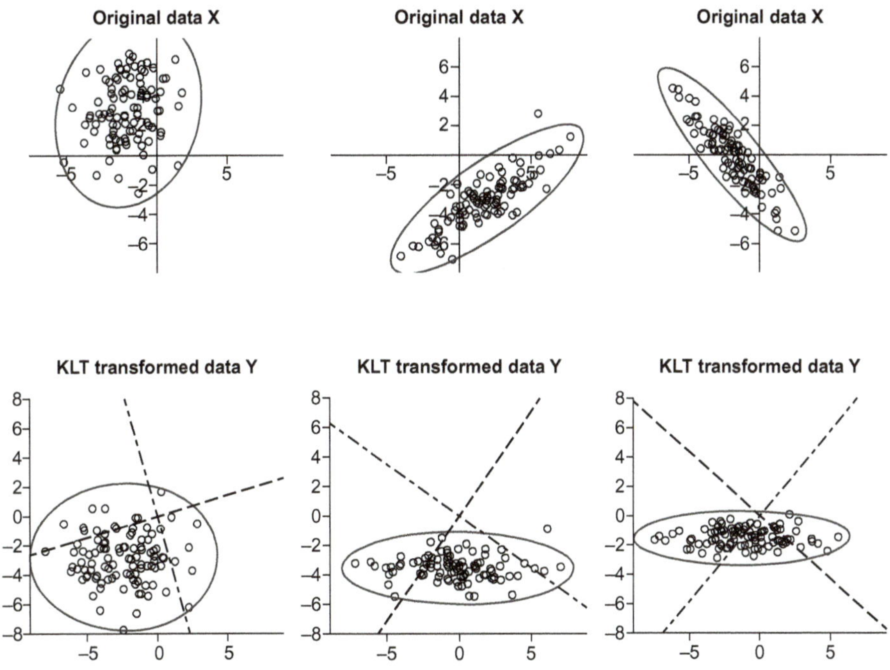

Fig. 10.4 Three datasets before and after the KLT.

10.5 Computation of the KLT

Owing to KLT's properties of signal decorrelation and energy compaction, it can be used for data compression by reducing the dimensionality of the dataset. Specifically, we carry out the following five steps.

1. Find the mean vector $\mathbf{m}_x$ and the covariance matrix Σ_x of the random vectors $\mathbf{x}$ based on the given dataset $\mathbf{X} = [\mathbf{x}_1, \cdots, \mathbf{x}_N]$ containing N samples of $\mathbf{x}$. Note that the rank of $\mathbf{\Sigma}$ is no greater than $N - 1$.
2. Find the eigenvalues $\lambda_1 \geq \lambda_2 \geq \cdots \geq \lambda_d$ of Σ_x, sorted in descending order, and their corresponding eigenvectors $\{\mathbf{v}_1, \mathbf{v}_2, \cdots, \mathbf{v}_d\}$.
3. Determine the reduced dimensionality $d' < d$ so that the percentage of energy contained after the transform is no less than a given threshold (e.g., 95%)
$$\frac{\sum_{i=1}^{d'} \lambda_i}{\sum_{i=1}^{d} \lambda_i} = \frac{\sum_{i=1}^{d'} \sigma_{y_i}^2}{\sum_{i=1}^{d} \sigma_{y_i}^2} \geq 0.95 \tag{10.41}$$
4. Construct a $d \times d'$ transform matrix composed of the d eigenvectors corresponding to the d largest eigenvalues of Σ_x
$$\mathbf{V}' = [\mathbf{v}_1, \cdots, \mathbf{v}_{d'}]_{d \times d'} \tag{10.42}$$
and carry out the KLT based on $\mathbf{V}'$
$$\mathbf{y} = \begin{bmatrix} y_1 \\ \vdots \\ y_{d'} \end{bmatrix}_{d' \times 1} = \mathbf{V}'^T \mathbf{x} = \begin{bmatrix} & \mathbf{v}_1^T & \\ & \vdots & \\ & \mathbf{v}_{d'}^T & \end{bmatrix}_{d' \times d} \begin{bmatrix} x_1 \\ \vdots \\ \vdots \\ x_d \end{bmatrix}_{d \times 1} \tag{10.43}$$
This is a lossy compression by which the dimensionality is reduced from d to $d' < d$, with the following error representing the percentage of information lost
$$\sum_{i=d'+1}^{d} \lambda_i / \sum_{i=1}^{d} \lambda_i \tag{10.44}$$
But as $\lambda_{d'+1}, \cdots, \lambda_d$ are the smallest $d - d'$ eigenvalues, the error is minimum (e.g., 5%).
5. Reconstruct $\mathbf{x}$ by inverse KLT
$$\mathbf{x} = \begin{bmatrix} x_1 \\ \vdots \\ \vdots \\ x_d \end{bmatrix}_{d \times 1} = \mathbf{V}' \mathbf{y} = \begin{bmatrix} \mathbf{v}_1 & \cdots & \mathbf{v}_{d'} \end{bmatrix}_{d \times d'} \begin{bmatrix} y_1 \\ \vdots \\ y_{d'} \end{bmatrix}_{d' \times 1} = \sum_{i=1}^{d'} y_i \mathbf{v}_i \tag{10.45}$$

The rank of the $d \times d$ estimated covariance matrix $\hat{\mathbf{\Sigma}}_x$ in Eq. (10.4) based on the dataset $\mathbf{X} = [\mathbf{x}_1, \cdots, \mathbf{x}_N]$ is no greater than $N - 1$. When $N \leq d$, then the rank of $\hat{\mathbf{\Sigma}}$ is at most $N - 1 < d$, that is, there are no more than $N - 1$ non-zero eigenvalues while all remaining $d - (N-1) = d - N + 1$ eigenvalues are zero. In this case, this estimated covariance matrix $\hat{\mathbf{\Sigma}}_x$ is not invertible.

Furthermore, if the dimensionality $d \gg N$ is much higher than the number of data samples, the computational complexity $O(d^3)$ for solving this eigenvalue problem of $\hat{\mathbf{\Sigma}}_x$ may be expensive. In this case, we can find the $N-1$ non-zero eigenvalues and their corresponding eigenvectors of $\hat{\mathbf{\Sigma}}_x$ by solving the eigenvalue problem of an $N \times N$ matrix with much reduced complexity of $O(N^3)$.

Specifically, for convenience and without loss of generality, we assume that the data have zero mean $\mathbf{m}_x = \mathbf{0}$, so that the covariance matrix of the given dataset $\mathbf{X}$ can be estimated as

$$\hat{\mathbf{\Sigma}}_x = \frac{1}{N}\sum_{n=1}^{N}(\mathbf{x}_n - \mathbf{m}_x)(\mathbf{x}_n - \mathbf{m}_x)^T = \frac{1}{N}\sum_{n=1}^{N}\mathbf{x}_n\mathbf{x}_n^T$$
$$= \frac{1}{N}[\mathbf{x}_1, \cdots, \mathbf{x}_N]\begin{bmatrix} \mathbf{x}_1^T \\ \vdots \\ \mathbf{x}_N^T \end{bmatrix} = \frac{1}{N}\mathbf{X}\mathbf{X}^T \tag{10.46}$$

and the eigenequation of this $d \times d$ matrix can be written as

$$\hat{\mathbf{\Sigma}}_x\mathbf{v}_i = \frac{1}{N}\mathbf{X}\mathbf{X}^T\mathbf{v}_i = \lambda_i\mathbf{v}_i \tag{10.47}$$

Premultiplying $\mathbf{X}^T$ on both sides, we get the eigenequation of an $N \times N$ matrix $\mathbf{X}^T\mathbf{X}/N$

$$\frac{1}{N}\mathbf{X}^T\mathbf{X}\left(\mathbf{X}^T\mathbf{v}_i\right) = \lambda_i\left(\mathbf{X}^T\mathbf{v}_i\right) \quad \text{i.e.,} \quad \frac{1}{N}\mathbf{X}^T\mathbf{X}\mathbf{u}_i = \lambda_i\mathbf{u}_i \tag{10.48}$$

where $\mathbf{u}_i = \mathbf{X}^T\mathbf{v}_i$ is ith eigenvector of $\mathbf{X}^T\mathbf{X}/N$. We see that $\mathbf{X}\mathbf{X}^T/N$ and $\mathbf{X}^T\mathbf{X}/N$ share the same non-zero eigenvalues. When $N < d$, we can solve the eigenvalue problem in Eq. (10.48) with complexity $O(N^3)$, instead of that of the original problem in Eq. (10.47) with $O(d^3)$, to obtain the $N-1$ non-zero eigenvalues λ_i, $(i = 1, \cdots, N-1)$, and the corresponding eigenvectors $\mathbf{u}_i = \mathbf{X}^T\mathbf{v}_i$. To find the eigenvectors $\mathbf{v}_i$ of the original matrix $\hat{\mathbf{\Sigma}}_x = \mathbf{X}\mathbf{X}^T/N$, we pre-multiply $\mathbf{X}$ on both sides of the eigenequation in Eq. (10.48) and obtain

$$\frac{1}{N}\mathbf{X}\mathbf{X}^T\left(\mathbf{X}\mathbf{u}_i\right) = \lambda_i\left(\mathbf{X}\mathbf{u}_i\right) \tag{10.49}$$

Comparing this with Eq. (10.47), we see that $\mathbf{X}\mathbf{u}_i = \mathbf{v}_i$, and they become the same when normalized.

10.6 Comparison with Other Orthogonal Transforms

Here we compare the KLT, in terms of signal decorrelation and energy compaction, with other orthogonal transforms such as discrete cosine transform (DCT) and Fourier transform (DFT), as well as identity transform (IT) (no transform), based on two images of cloud and sand as shown in the following.

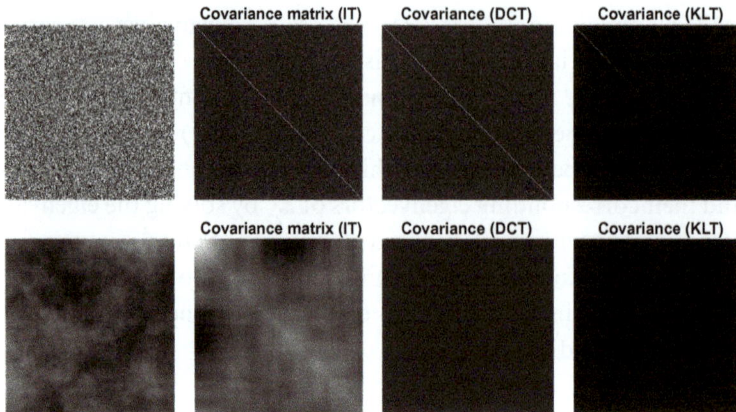

Fig. 10.5 Images of clouds and sands.

We treat each column of the image as an observation (instantiation) of a 1-D random vector $\mathbf{x}$, and apply an orthogonal transforms $\mathbf{y} = \mathbf{A}^T\mathbf{x}$ to $\mathbf{x}$ based on the covariance matrix $\mathbf{\Sigma}_x$, and compare the corresponding covariance matrices Σ_y after the transform to see how well each transform decorrelates the signal and compacts its energy.

Figure 10.5 shows two images (left) and the corresponding covariance matrices in image form (pixel values proportional to the values in the covariance matrix) after three orthogonal transforms of IT (no transform), DCT, and KLT.

- Sand image
  In the second panel showing the covariance after IT, the off-diagonal pixels are dark, indicating the pixels in the original image are not highly correlated at all. In the third panel, the covariance after DCT looks similar to that before the transform, indicating that the DCT has little effect in terms of signal decorrelation and energy compaction. Finally, in the right-most panel showing the covariance after KLT, all off-diagonal elements are zero, that is, the signal components are completely decorrelated.
- Cloud image
  In the second panel for the covariance after IT, there exist some bright areas off the diagonal, indicating that many signal components close to each other are highly correlated. In the third panel for the covariance after DCT, the values of the off-diagonal elements are significantly reduced, indicating that the signal components are significantly decorrelated. Finally, in the right-most panel showing the covariance after KLT, all off-diagonal elements are zero, that is, the signal components are completely decorrelated.

The effect of energy redistribution and compaction of these different transforms are also shown in Fig. 10.6, where the variances of signal components are sorted and plotted.

We see that, owing to the physical nature of the clouds and sand grains, the textures in the corresponding images are very different. In the cloud image, the neighboring pixels are highly correlated, while in the sand image, they are not much

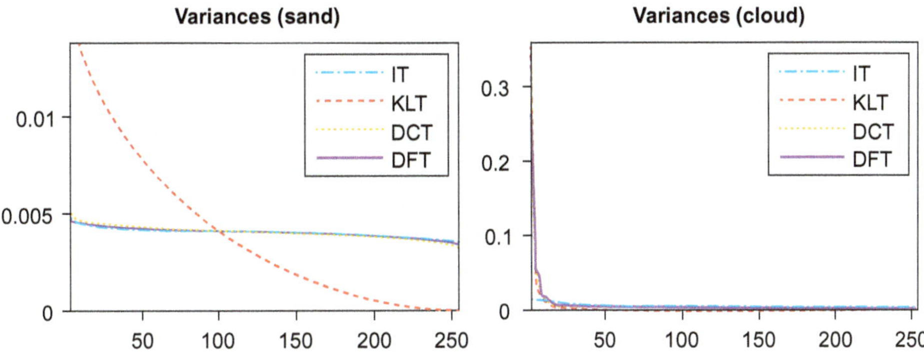

Fig. 10.6 Energy compaction and redistribution by the KLT.

correlated at all. Correspondingly, the DCT can effectively decorrelate the cloud image, but much less so in the sand image. But in either case, whether the pixels in the image are highly correlated or not, the KLT can always completely decorrelate the signal and optimally compact the energy.

The following table further demonstrates the effect of energy compaction quantitatively in terms of the number of components out of the total of 256 components needed after the transform in order to keep a certain percentage of the total dynamic energy (information). For example, if it is desired to keep 95% of the total energy contained in the original signal, 230 components are needed with no transform, 22 are needed after DCT, and only 13 are needed after KLT. Note that DCT's performance is reasonably close to that of the optimal KLT in this case.

Table 10.1

Percentage	90%	95%	99%	100%
IT	209	230	250	256
DCT	10	22	97	256
KLT	7	13	55	256

Although KLT is optimal in terms of signal decorrelation and energy compaction, it is not as convenient as other transforms for two reasons. First, the KLT is data-specific, that is, the transform matrix is the eigenvector matrix of the covariance matrix of the dataset, which needs to be estimated based on sufficient amount of data samples, while all other orthogonal transforms are data independent. Second, the computational cost of the KLT is significantly higher than that of other transforms, due to the $O(d^3)$ complexity for solving the eigenvalue problem of the $d \times d$ matrix covariance matrix $\mathbf{\Sigma}_x$, while for most other popular orthogonal transforms fast algorithm exists with $O(d \log_2 d)$ complexity instead of $O(d^2)$ required by the KLT.

10.7 Application to Image Data

The KLT can be applied to a stack of N images each containing K pixels, as shown on the left of Fig. 10.7. Depending on the specific need (such as data compression and feature extraction), the KLT can be carried out in either of two alternative manners, which define the random signal vector $\mathbf{x}$ differently, as shown in the following:

- A K-dimensional vector $\mathbf{x}_n \quad (n = 1, \cdots, N)$ can be formed by all K pixels of each of the N images by concatenating the rows (or columns) of the image array (same as what we did for the sand and cloud images). All N such vectors are treated as the column vectors of a $K \times N$ data array $\mathbf{X} = [\mathbf{x}_1, \cdots, \mathbf{x}_N]$. The $K \times K$ covariance matrix of these K–D column vectors can be estimated (assuming with zero mean) as

$$\Sigma = \frac{1}{N} \sum_{n=1}^{N} \mathbf{x}_n \mathbf{x}_n^T = \frac{1}{N} [\mathbf{x}_1, \cdots, \mathbf{x}_N] \begin{bmatrix} \mathbf{x}_1^T \\ \vdots \\ \mathbf{x}_N^T \end{bmatrix} = \frac{1}{N} \left(\mathbf{X}\mathbf{X}^T \right)_{K \times K} \tag{10.50}$$

- An N-dimensional vector can be formed by the pixels at the same position (e.g., in the ith row and jth column of the image array) from all N images. There are K such vectors each for one pixel in the images, and they are actually the rows of matrix $\mathbf{X}$ just defined, or the columns of $\mathbf{X}^T$. The $N \times N$ covariance matrix of these N–D row vectors can be estimated as

$$\Sigma' = \frac{1}{K} \left(\mathbf{X}^T \mathbf{X} \right)_{N \times N}. \tag{10.51}$$

As discussed previously, these two different covariance matrices share the same eigenvalues. The eigenequations for $\mathbf{X}^T\mathbf{X}$ and $\mathbf{X}^T\mathbf{X}$ (with the constant coefficients $1/K$ and $1/N$ neglected) are

$$\mathbf{X}^T \mathbf{X} \mathbf{v} = \lambda \mathbf{v}, \qquad \mathbf{X}\mathbf{X}^T \mathbf{u} = \mu \mathbf{u}. \tag{10.52}$$

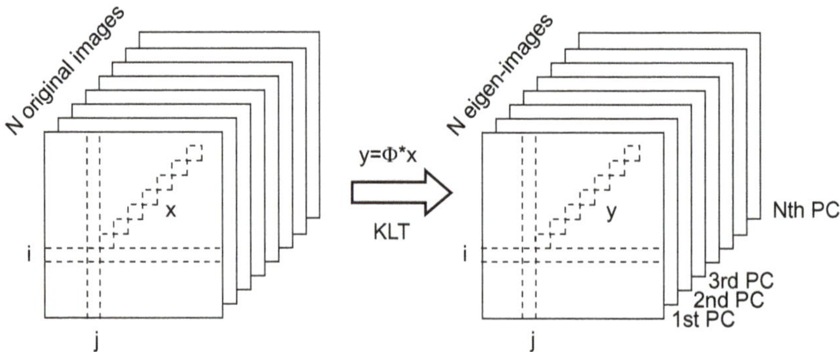

Fig. 10.7 The KLT of a set of images.

Premultiplying $\mathbf{X}^T$ on both sides of the second equation we get

$$\mathbf{X}^T\mathbf{X}[\mathbf{X}^T\mathbf{u}] = \mu[\mathbf{X}^T\mathbf{u}]. \tag{10.53}$$

which is actually the first eigenequation with the same eigenvalue $\mu = \lambda$ and eigenvector $\mathbf{v} = \mathbf{X}^T\mathbf{u}$ when normalized. The two covariance matrices Σ and Σ' have the same rank $R = \min(N, K)$ (if $\mathbf{X}$ is not degenerate) and therefore the same number of non-zero eigenvalues. Consequently, the KLT can be carried out based on either matrix with the same effects in terms of the signal decorrelation and energy compaction. As the number of pixels in the image is typically much greater than the number of images, $K > N$, we will take the second approach to treat the pixels in the same position in all N images as a sample of an N-dimensional random signal vector and carry out the KLT based on the $N \times N$ covariance matrix $\hat{\Sigma}$.

We can now carry out the KLT to each of the K N-dimensional vectors $\mathbf{x}$ corresponding to each pixel of the N images to obtain another N-dimensional vector $\mathbf{y} = \mathbf{v}^*\mathbf{x}$ for the same pixel of a set of N *eigenimages*, as shown on the right of Fig. 10.7. After the KLT, most of the energy/information contained in the N images, representing the signal variations among all N images, is concentrated in the first few eigenimages corresponding to the greatest eigenvalues, while the remaining eigenimages can be omitted without losing much energy/information. This is the foundation for various KLT-based image compression and feature extraction algorithms. The subsequent operations such as image recognition and classification can all be carried out in a much lower dimensional space.

We now consider some of such applications.

- **Video image compression**
  A sequence of $N = 8$ frames of a video of a moving escalator and their eigenimages are shown, respectively, in the upper and lower parts of Fig. 10.8.

  It is interesting to observe that the first eigenimage corresponding to the greatest eigenvalue (left panel of the third row of the figure) represents mostly the static scene of the image frames representing the main variations in the image (carrying most of the energy), while the subsequent eigenimages represent mostly the motion in the video, the variation between the frames. For example, the motion of the people riding on the escalator is mostly reflected by the first few eigenimages following the first one, while the motion of the escalator stairs is mostly reflected in the subsequent eigenimages.

  The 8×8 covariance matrix and the energy distribution among the eight components plot before and after the KLT are shown in Fig. 10.9.

  We see that, owing to the spatial correlation between nearby pixels, the covariance matrix before the KLT (left) can be modeled by the squared exponential function, while the covariance matrix after the KLT (middle) is completely decorrelated and the energy is highly compacted into a small number of principal components (here the first component), as also clearly shown in the comparison of the energy distribution before and after the KLT (right).

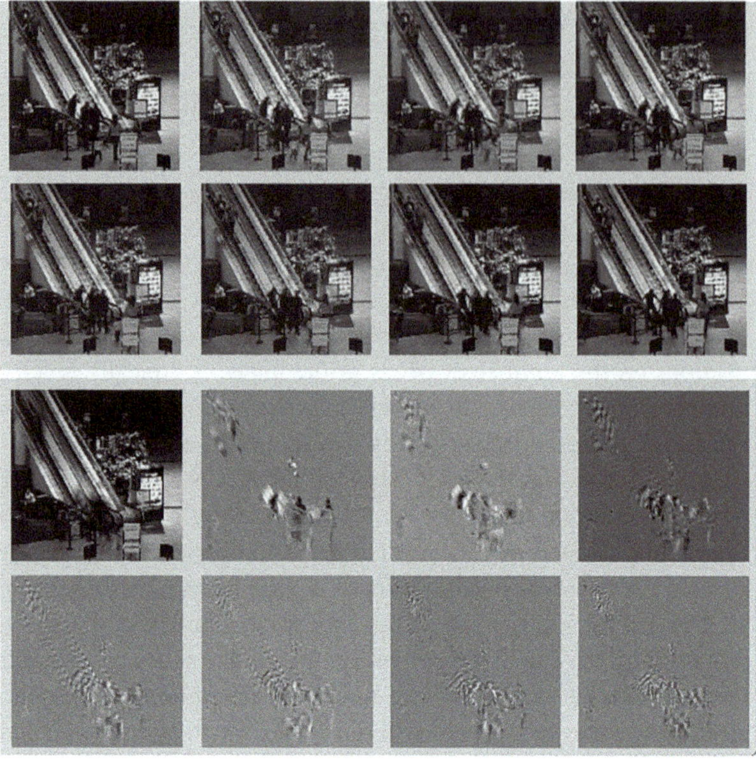

Fig. 10.8 Eight frames of a video before (top) and after (bottom) the KLT.

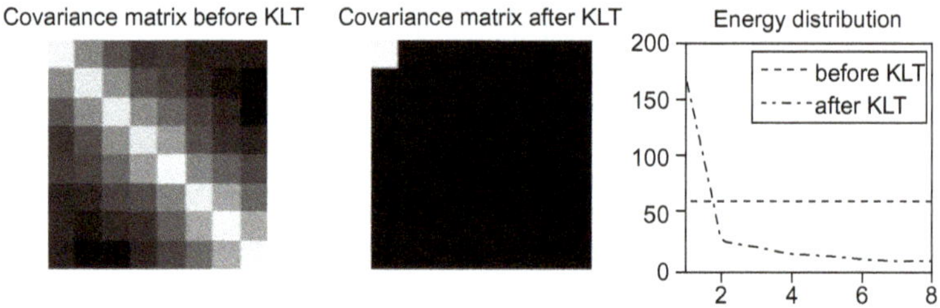

Fig. 10.9 Covariance matrix before (left) and after (middle) the KLT, and energy redistribution (right).

Fig. 10.10 Face images.

- **Eigen-face and face recognition**
  The KLT is applied to a stack of 20 face images ($N = 20$) shown in Fig. 10.10 (credit to AT&T Laboratories Cambridge, downloadable from https://cam-orl.co.uk/facedatabase.html), to generate the *eigen-faces* in Fig. 10.11.

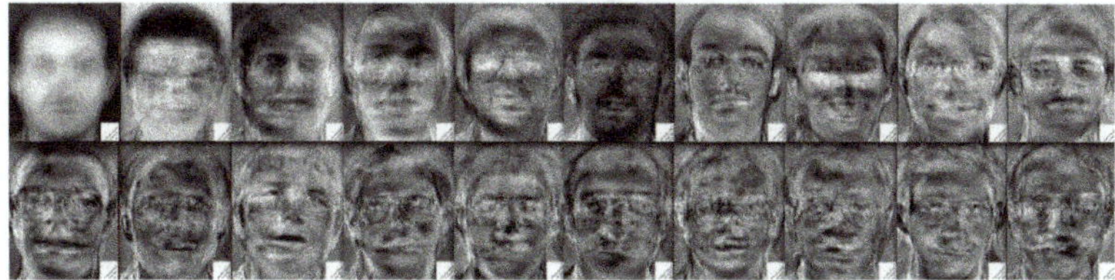

Fig. 10.11 The eigen-faces.

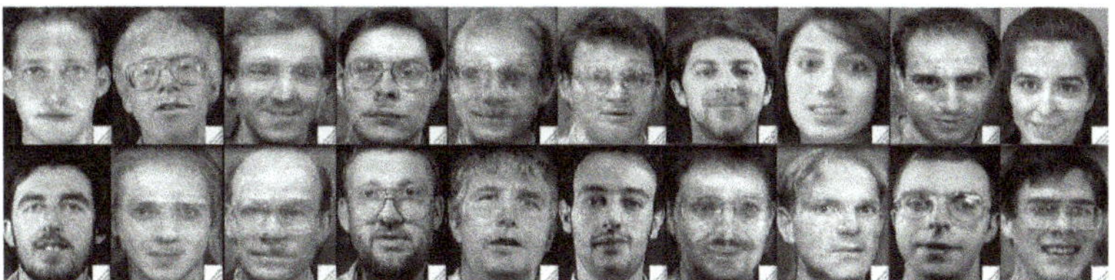

Fig. 10.12 Reconstructed faces.

Table 10.2 shows the percentage of energy contained in each of the components

Table 10.2

Components	1	2	3	4	5	6	7	8	9
Percentage energy	48.5	11.6	6.1	4.6	3.8	3.7	2.6	2.5	1.9
Accumulative energy	48.5	60.1	66.2	70.8	74.6	78.3	81.0	83.5	85.4

10	11	12	13	14	15	16	17	18	19	20
1.9	1.8	1.6	1.5	1.4	1.3	1.2	1.1	1.1	0.9	0.8
87.3	89.	90.7	92.2	93.6	94.9	96.1	97.2	98.2	99.2	100.0

The faces are then reconstructed based on 95% of the total information (15 out of 20 components) as shown in Fig. 10.12.

- **Handwritten digit recognition**

  The goal here is to recognize handwritten digits from 0 to 9 given in image form of 16×16 pixels, in a dataset containing $N = 224 \times 10$ samples for the $K = 10$ digits. A small subset of the samples is shown in Fig. 10.13. All $16 \times 16 = 256$ pixels in the image of each sample can be converted into a $d = 256$ vector by concatenating all columns (or rows) of the image. Then the KLT can be carried out to reduce the dimensionality from $d = 256$ to some $d' \ll d$, based on either the covariance matrix $\Sigma_x = \mathbf{S}_T$ of all $N = 2240$ sample vectors representing the overall distribution of the dataset, or the between-class scatter matrix $\mathbf{S}_B$ previously considered representing the separability of the ten classes.

  Specifically, we use the eigenvectors corresponding to the d' greatest eigenvalues of the covariance matrix or the between-class scatter matrix to form a d' by d transform matrix. After the KLT by the data, certain classification algorithms can be carried out in the much reduced d'-dimensional space.

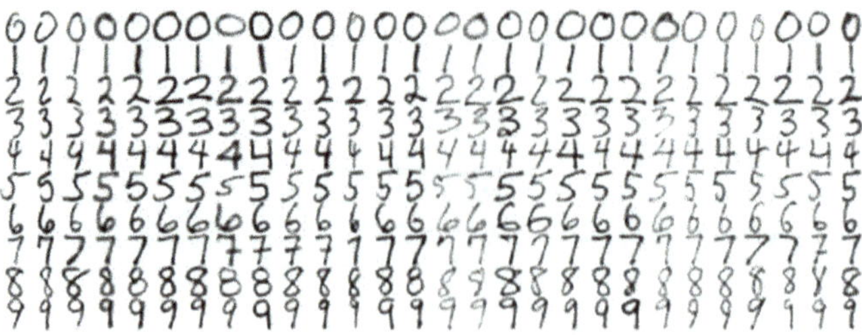

Fig. 10.13 Ten handwritten digits.

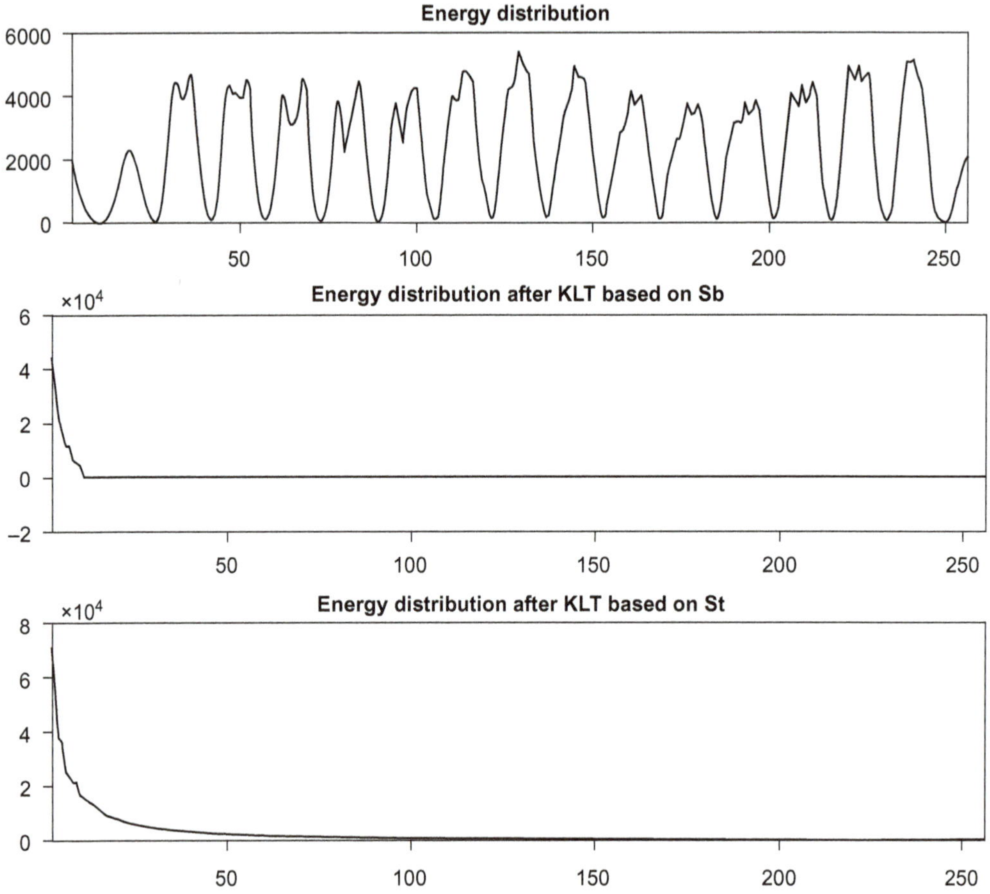

Energy distribution

Energy distribution after KLT based on Sb

Energy distribution after KLT based on St

Fig. 10.14 Energy distribution before (top) and after the KLT based on $\mathbf{S}_B$ (middle), and $\mathbf{S}_T$ (bottom).

The energy distribution over all $d = 256$ signal components is plotted in Fig. 10.14 for both the original signal before the KLT (top) and after the KLT based on both $\mathbf{S}_B$ (middle) and $\mathbf{S}_T$ (bottom).

For the KLT based on $\mathbf{S}_T$ with rank $N - 1$, $d' = 79$ components are needed to keep 95.1% of the total energy, in comparison to the KTL based

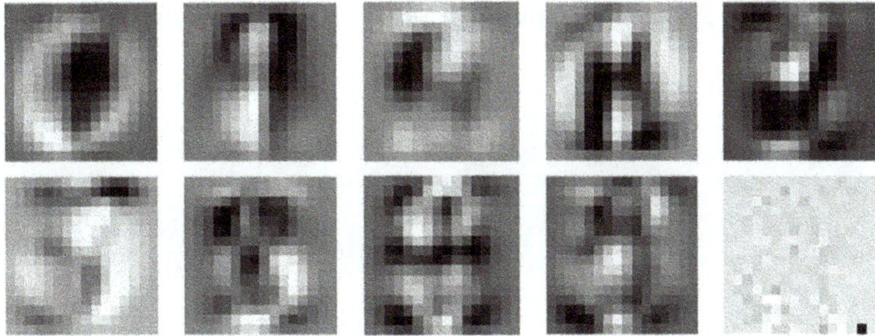

Fig. 10.15 The first ten eigen-digits.

on $\mathbf{S}_B$ with rank $\text{rank}(\mathbf{S}_B) = K - 1 = 9$, requiring are only $d' = 9$ principal components corresponding to the same number of non-zero eigenvalues of $\mathbf{S}_B$ to keep 100% of the total energy representing the separability information. The percentage energy contained in these non-zero eigenvalues are: 28.6%, 22.7%, 14.4%, 10.9%, 7.1%, 7.0%, 3.8%, 3.2%, 2.4%.

The $d = 256$-dimensional eigenvectors can be visualized when converted into 16×16 eigen-digits, representing the basis by which any original face images can be represented as a linear combination of such eigen-digits as shown in Fig. 10.15. We see that the 10th eigen-digit corresponding to a zero eigenvalue contains no meaningful information but random noise.

If we only keep the first two or three principal components (corresponding to the greatest eigenvalues) after the KLT, the dataset can be visualized as shown in Fig. 10.16. The sample points in each of the ten different classes are color coded. It can be seen that even when the dimensions are much reduced from $N = 256$ to $d = 3$ or even $d = 2$, it is still possible to separate the ten classes reasonably well.

10.8 PCA for Feature Extraction

In the PCA method considered in previous sections, the transform matrix is the eigenvector matrix of the covariance matrix of the dataset, of which the diagonal components are the variances of the signal components that measure the dynamic energy or information contained in all data points in the entire dataset.

In the context of classification, we can also use the PCA method to reduce the data dimensionality while maximally preserving the information of interest, which in this case should be pertinent to the separability of data points into different classes, instead of just the generic variation in all data points independent of their classes. Specifically, the transform matrix $\mathbf{A} = [\mathbf{a}_1, \cdots, \mathbf{a}_{d'}]$ in the linear transformation

$$\mathbf{y} = \mathbf{A}^T \mathbf{x} \tag{10.54}$$

needs to be constructed based on certain measurement of the class separability, such as the within and between scatter matrices considered in Chapter 9, instead of the covariance matrix of the entire dataset.

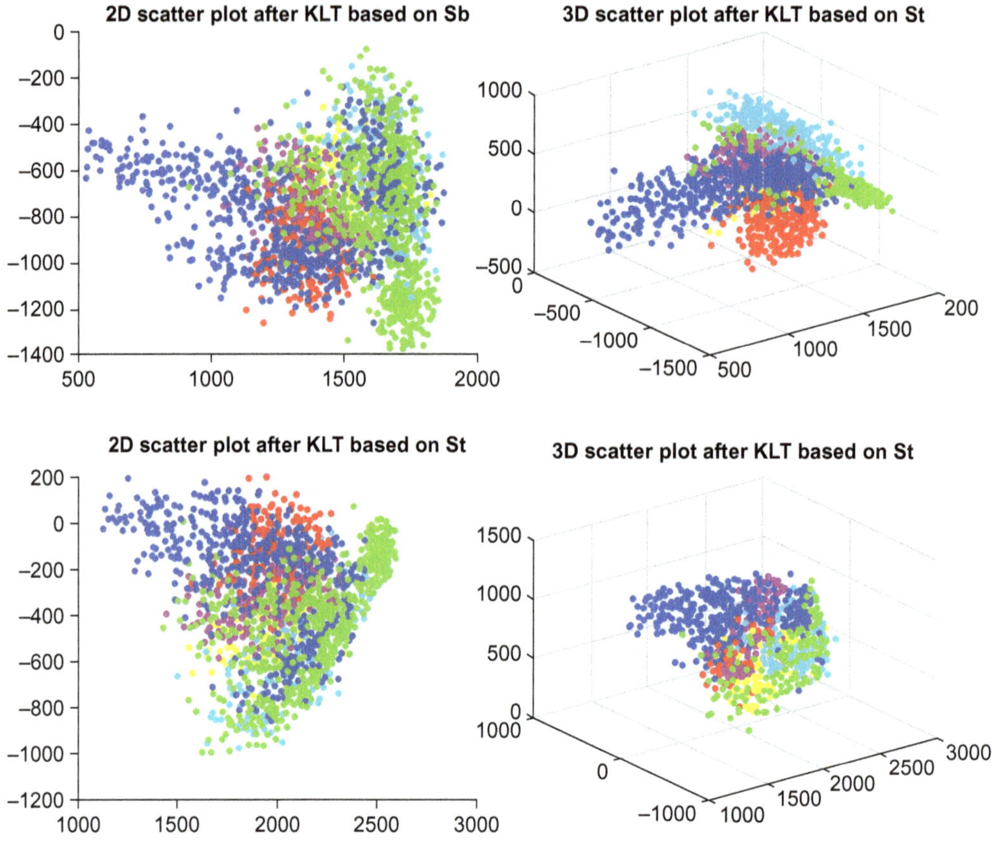

Fig. 10.16 Visualization of ten classes of digits.

- **Optimal transformation for maximizing** $\text{tr}(\mathbf{S}_B)$
  We first consider the objective function $J_B(\mathbf{A})$ in the new d'–D space as a function of the transform matrix $\mathbf{A}$

$$J(\mathbf{A}) = \text{tr}\,(\mathbf{S}_B^y) = \text{tr}\,(\mathbf{A}^T\,\mathbf{S}_B^x\,\mathbf{A}) = \text{tr}\begin{bmatrix}\mathbf{a}_1^T\\ \vdots \\ \mathbf{a}_{d'}^T\end{bmatrix}\mathbf{S}_B^x[\mathbf{a}_1,\cdots,\mathbf{a}_{d'}]$$

$$= \text{tr}\begin{bmatrix}\mathbf{a}_1^T\\ \vdots \\ \mathbf{a}_{d'}^T\end{bmatrix}[\mathbf{S}_B^x\mathbf{a}_1,\cdots,\mathbf{S}_B^x\mathbf{a}_{d'}] = \sum_{i=1}^{d'}\mathbf{a}_i^T\mathbf{S}_B^x\mathbf{a}_i \tag{10.55}$$

To find optimal $\mathbf{A}$ that maximizes $\text{tr}(\mathbf{S}_B^y)$, we need to solve the following constrained maximization problem

$$\begin{cases}\text{maximize} & J(\mathbf{A}) = \text{tr}(\mathbf{A}^T\mathbf{S}_B^x\mathbf{A}) = \sum_{i=1}^{d'}\mathbf{a}_i^T\mathbf{S}_B^x\mathbf{a}_i \\ \text{subject to} & ||\mathbf{a}_j||^2 = \mathbf{a}_j^T\mathbf{a}_j = 1 \quad (j=1,\cdots,d')\end{cases} \tag{10.56}$$

where the constraint equations are to guarantee that the column vectors of $\mathbf{A}$ are normalized and therefore finite. This constrained optimization problem can be solved by Lagrange multiplier method. We first construct the Lagrangian function

$$L(\mathbf{A},\boldsymbol{\lambda}) = J(\mathbf{A}) - \sum_{j=1}^{d'} \lambda_j(\mathbf{a}_j^T\mathbf{a}_j - 1) \tag{10.57}$$

and then set its gradient with respect to each column vector $\mathbf{a}_i$ of $\mathbf{A}$ to zero

$$\begin{aligned}\frac{\partial}{\partial \mathbf{a}_i} L(\mathbf{A},\boldsymbol{\lambda}) &= \frac{\partial}{\partial \mathbf{a}_i}\left[\sum_{j=1}^{d'}(\mathbf{a}_j^T\mathbf{S}_B^x\mathbf{a}_j - \lambda_j\mathbf{a}_j^T\mathbf{a}_j + \lambda_j)\right]\\ &= \frac{\partial}{\partial \mathbf{a}_i}\left[\mathbf{a}_i^T\mathbf{S}_B^x\mathbf{a}_i - \lambda_i\mathbf{a}_i^T\mathbf{a}_i\right] = 2\mathbf{S}_B^x\mathbf{a}_i - 2\lambda_i\mathbf{a}_i = 0, \quad \text{i.e.,} \quad \mathbf{S}_B\mathbf{a}_i = \lambda_i\mathbf{a}_i\\ &\quad (i = 1,\cdots,n)\end{aligned} \tag{10.58}$$

This happens to be the ith eigenequation of $\mathbf{S}_B$, indicating the column vectors of the optimal $\mathbf{A}$ have to be the orthogonal eigenvectors $\{\mathbf{v}_1,\cdots,\mathbf{v}_{d'}\}$ of the symmetric matrix $\mathbf{S}_B^x$, that is, $\mathbf{A} = [\mathbf{a}_1,\cdots,\mathbf{a}_{d'}] = \mathbf{V} = [\mathbf{v}_1,\cdots,\mathbf{v}_{d'}]$. Now the d' new features can be obtained as

$$\mathbf{y} = \mathbf{A}^T\mathbf{x} = \begin{bmatrix}\mathbf{v}_1^T\\ \vdots \\ \mathbf{v}_{d'}^T\end{bmatrix}_{d'\times d} \mathbf{x}_{d\times 1}, \quad \text{or} \quad y_i = \mathbf{v}_i^T\mathbf{x}, \;\; (i = 1,\cdots,d') \tag{10.59}$$

and

$$J(\mathbf{A}) = \sum_{i=1}^{d'} \mathbf{v}_i^T\mathbf{S}_B\mathbf{v}_i = \sum_{i=1}^{d'} \mathbf{v}_i^T\mathbf{v}_i\lambda_i = \sum_{i=1}^{d'} \lambda_i > 0 \tag{10.60}$$

We note that as $\mathbf{S}_B^x$ is symmetric and positive definite, its eigenvalues are positive and its normalized eigenvectors are orthonormal: $\mathbf{v}_i^T\mathbf{v}_j = \delta_{ij}$. To maximize $J(\mathbf{A})$, the eigenvectors in $\mathbf{A}$ need to correspond to the d' greatest eigenvalues of $\mathbf{S}_B$.

- **Optimal transformation for maximizing** $\text{tr}(\mathbf{S}_{B/T})$

The previous method only maximizes $\mathbf{S}_B$ without taking $\mathbf{S}_W$ or $\mathbf{S}_T = \mathbf{S}_B + \mathbf{S}_W$ into consideration. As $\mathbf{S}_W^y$ in the d'–D space may have also changed as well as $\mathbf{S}_B^y$. To maximize the separability, we need to maximize $\mathbf{S}_B$ while at the same time also minimize $\mathbf{S}_W$. Or, equivalently, we need to maximize $\mathbf{S}_B^y$ normalized by $\mathbf{S}_T^y$

$$\begin{aligned}\mathbf{S}_{B/T}^y &= (\mathbf{S}_T^y)^{-1}\,\mathbf{S}_B^y = (\mathbf{A}^T\mathbf{S}_T^x\mathbf{A})^{-1}(\mathbf{A}^T\mathbf{S}_B^x\mathbf{A})\\ &= \mathbf{A}^{-1}(\mathbf{S}_T^x)^{-1}(\mathbf{A}^T)^{-1}\mathbf{A}^T\mathbf{S}_B^x\mathbf{A}\\ &= \mathbf{A}^{-1}(\mathbf{S}_T^x)^{-1}\mathbf{S}_B^x\mathbf{A} = \mathbf{A}^{-1}\mathbf{S}_{B/T}^x\mathbf{A}\end{aligned} \tag{10.61}$$

We note that different from the symmetric matrix $\mathbf{S}_B$, matrix $\mathbf{S}_{B/T}^y$ as a product of two symmetric matrices is no longer symmetric, that is, its eigenvectors are not orthogonal in general, and this method cannot be used, we will find the optimal matrix $\mathbf{A}$ in some other way.

We first consider the special case of $d' = 1$, that is, $\mathbf{A} = \mathbf{a}$ is an $d \times 1$ vector and $y = \mathbf{a}^T\mathbf{x}$ is a scalar. We want to maximize the following objective function in a 1-D space

$$(\mathbf{S}_T^y)^{-1}\,\mathbf{S}_B^y = (\mathbf{a}^T\mathbf{S}_T^x\mathbf{a})^{-1}(\mathbf{a}^T\mathbf{S}_B^x\mathbf{a}) = \frac{s_B^y}{s_T^y} = \frac{\mathbf{a}^T\mathbf{S}_B^x\mathbf{a}}{\mathbf{a}^T\mathbf{S}_T^x\mathbf{a}} = R(\mathbf{a}) \tag{10.62}$$

This function $R(\mathbf{a})$ is the *Rayleigh quotient* (Section A.3.3) of the two symmetric matrices $\mathbf{S}_B^x$ and $\mathbf{S}_T^x$. The optimal transform vector $\mathbf{a}$ that maximizes $R(\mathbf{a})$ can be found by solving the corresponding *generalized eigenvalue problem* (Section A.3.2)

$$\mathbf{S}_B^x\mathbf{v}_i = \lambda_i\mathbf{S}_T^x\mathbf{v}_i, \qquad (i = 1,\cdots,d) \tag{10.63}$$

where $\lambda_i = R(\mathbf{a})$ is an eigenvalue and $\mathbf{v}_i$ the corresponding eigenvector of $\mathbf{S}_{B/T} = (\mathbf{S}_T^x)^{-1}\mathbf{S}_B^x$. Obviously, the transform vector $\mathbf{a}$ that maximizes $R(\mathbf{a})$ is the eigenvector corresponding to the greatest eigenvalue $\lambda_{\max} \geq \lambda_i$ $(i = 1,\cdots,d)$.

We next generalize this method to the case of $d' > 1$. The matrix form of the eigenequation is

$$\mathbf{S}_B^x\mathbf{V} = \mathbf{S}_T^x\mathbf{V}\mathbf{\Lambda} \tag{10.64}$$

where $\mathbf{\Lambda} = \text{diag}(\lambda_1,\cdots,\lambda_d)$ is the diagonal eigenvalue matrix of $(\mathbf{S}_T^x)^{-1}\mathbf{S}_B^x$, and $\mathbf{V} = [\mathbf{v}_1,\cdots,\mathbf{v}_d]$ the eigenvector matrix which is no longer orthogonal in general. This generalized eigenvalue problem can be solved by finding the matrix $\mathbf{V}$ that diagonalizes both $\mathbf{S}_T^x$ and $\mathbf{S}_B^x$ at the same time (same as how to get Eq. (A.137))

$$\begin{cases} \mathbf{S}_B^y = \mathbf{V}^T\mathbf{S}_B^x\mathbf{V} = \mathbf{\Lambda} \\ \mathbf{S}_T^y = \mathbf{V}^T\mathbf{S}_T^x\mathbf{V} = \mathbf{I} \end{cases} \tag{10.65}$$

Left multiplying $(\mathbf{S}_T^y)^{-1} = \mathbf{I}$ to the first, we get

$$\mathbf{S}_{B/T}^y = (\mathbf{S}_T^y)^{-1}\mathbf{S}_B^y = \mathbf{\Lambda} \tag{10.66}$$

The optimal transform matrix is composed of the eigenvectors $\mathbf{A} = [\mathbf{v}_1,\cdots,\mathbf{v}_{d'}]$ corresponding to the d' greatest of all d eigenvalues $\lambda_1 \geq \cdots \geq \lambda_{d'} \geq \cdots \geq \lambda_d$, so that

- the signal components in $\mathbf{y} = [y_1,\cdots,y_{d'}]^T$ are completely decorrelated, each component $y_i = \mathbf{v}_i^T\mathbf{x}$ carries certain separability information $s_B^{(y_i)}/s_T^{(y_i)} = \lambda_i$ $(i = 1,\cdots,d')$, independent of others;
- the total separability contained in the d'–D space, as the sum of the d' greatest eigenvalues $\sum_{i=1}^{d'}\lambda_i$, is maximized.

PROBLEMS

In this problem set and some later ones as well as many examples in future chapters, we will again use the handwritten digit dataset considered in Section 10.7. A small subset of the dataset is shown in Fig. 10.13, while the complete dataset containing

2240 handwritten digits can be downloaded from the Cambridge University Press website at www.cambridge.org/wang-iml.

Also, an ASCII file containing a 256×2240 array $\mathbf{X}$ for all 2240 data samples in vector form and a 256-D vector $\mathbf{y}$ for their class labelings can be downloaded here: /e176/homework/data0to9.txt

This ASCII file can be read by the following code

```
data=load('data0to9.txt','data');  $ read ASCII file
d=size(data,1)-1          % dimensionality of data
N=size(data,2);           % total number of samples
X=data(1:d,:);            % all N d-dimensional sample vectors
y=data(d+1,:);            % and their labelings
K=length(unique(y));      % number of classes
for k=1:K                 % for each of the K classes
    idx=find(y==k);       % indices of points in class k
    Xk=X(:,idx)           % collect all points in class k
end
```

1. Carry out KLT for the 4-D iris dataset. Display the 4×4 covariance matrices before and after the transform, and verify (1) the trace (sum of all diagonal components of the variance matrix) is conserved, (2) all off-diagonal components become zero after the KLT, that is, all signal components are completely decorrelated.
2. Visualize the 4-D iris dataset in both 2-D and 3-D spaces spanned, respectively, by the first two and three principal components. Color-code all 150 data points as red, green, or blue according to their class identities.
3. Carry out KLT for the handwritten digit dataset and observe the effects of signal decorrelation and energy redistribution of the transform. Display the 256×256 covariance matrices both before and after the transform in image form, same as in Figs 10.5 and 10.9, and plot their diagonal components (representing the dynamic energy or information contained in each data component), same as in Fig. 10.14.
4. Visualize the handwritten digit dataset in both 2-D and 3-D spaces spanned, respectively, by the first two and three principal components. Color code all data points according to their class identities to appreciate how difficult it is to separate them into different classes. (That is what you need to do in the future.)
5. Find all pair-wise Bhattacharyya distances $d_B(C_i, C_j)$, $i, i = 2, \cdots, 10$, $j = 1, \cdots, i-1\}$ between any two of the ten classes in the handwritten digits dataset, and display them in a lower triangular distance matrix.

   Note:

   - As the 256-D covariance matrix $\mathbf{\Sigma}_k$ is estimated based on only 224 samples of class C_k, the rank of this estimated covariance matrix is no more than $224 - 1 = 223$.
   - As all 224 samples are for the same digit, they are similar to each other and highly correlated, the rank of $\mathbf{\Sigma}_k$ is likely to be much lower than 223.

We therefore see that the estimated $\mathbf{\Sigma}_k$ is non-invertible as its rank is much smaller than its dimension, and so is the sum of two such covariance matrices (needed in the Bhattacharyya distance). Consequently, the dimensionality of the dataset must be reduced significantly ($d' \ll d = 256$) to become smaller than the rank of $\mathbf{\Sigma}_k$ by the KLT, before the Bhattacharyya distances can be calculated.

6. Typically the KLT matrix is calculated as the eigenvector matrix of the covariance matrix $\mathbf{\Sigma}$ (same as the total scatter matrix $\mathbf{S}_T$) of the entire dataset, and the resulting principal components reflect the signal variations of all data points in the dataset. Alternatively, the KLT can also be carried out based on the between-class scatter matrix $\mathbf{S}_B$ or $\mathbf{S}_{B/T}$ to maximize the trace (or equivalently based on the within-class scatter matrix $\mathbf{S}_W$ to minimize its trace), if the signal variation of interest is the separability of different classes in the context of classification.

   Now carry out such a KLT to reduce the dimensionality of both the iris and handwritten digit datasets based on $\mathbf{S}_B$. Visualize the datasets in both the 2-D and 3-D spaces spanned by the greatest principal components, with all data points color coded according to their class identities.
7. As discussed in Section 10.7, the method of PCA can be applied to image data based on the covariance matrices given in Eq. (10.7). Now carry out PCA to the face images, downloadable from https://cam-orl.co.uk/facedatabase.html). Observe the resulting eigen-faces to see how different types of variations among all faces are extracted by the eigen-faces corresponding to the first few greatest eigenvalues.
8. Repeat the previous problem based on the images of handwritten numbers. Observe the resulting eigen-digits.

11 Variations of PCA

In this chapter we consider a set of methods that are either some variations of the principal component analysis (PCA) method in the previous chapter or closely related to it to achieve similar effects for feature selection and extraction.

11.1 Kernel Methods

In the kernel method, all data points as a vector $\mathbf{x} = [x_1, \cdots, x_d]^T$ in the original d-dimensional feature space are mapped by a *kernel function* $\mathbf{z} = \phi(\mathbf{x})$ into a higher, possibly infinite, dimensional space

$$\mathbf{x} \Longrightarrow \mathbf{z} = \phi(\mathbf{x}) \tag{11.1}$$

The presumption of the method is that the data points only appear in the form of inner product $\mathbf{x}_m^T \mathbf{x}_n$ in the algorithm, then based on the *kernel trick*, the kernel function $\mathbf{z} = \phi(\mathbf{x})$ never needs to be actually carried out. In fact, the form of the kernel function $\phi(\mathbf{x})$ and the dimensionality of the higher dimensional space do not need to be explicitly specified or known.

The motivation for such a kernel mapping is that the relevant operations such as classification and clustering may be carried out much more effectively once the dataset is mapped to the higher dimensional space. For example, classes not linearly separable in the original d-dimensional feature space may be trivially separable in a higher dimensional space, as illustrated by the following examples.

Example 11.1

In 1-D space, two classes $C_- = \{x | (a \le x \le b)\}$ and $C_+ = \{x | (x \le a) \text{ or } (x \ge b)\}$ shown in the first row of Fig. 11.1 are not linearly separable. However, by the following mapping from 1-D space to 2-D space

$$\mathbf{z} = \phi(x) = \begin{bmatrix} z_1 \\ z_2 \end{bmatrix} = \begin{bmatrix} x \\ (x - (a+b)/2)^2 \end{bmatrix} \tag{11.2}$$

the two classes can be separated by thresholding in the second dimension of the 2-D space.

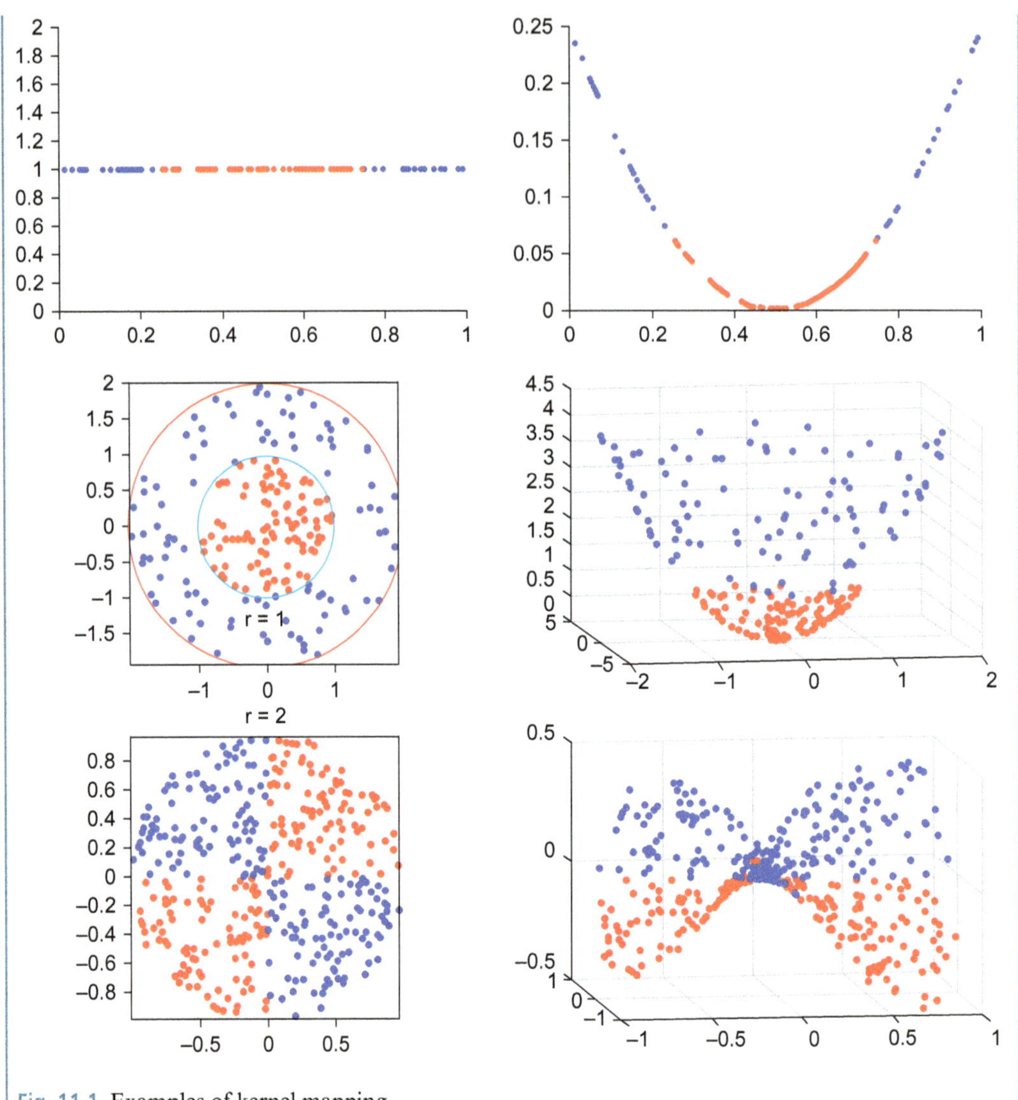

Fig. 11.1 Examples of kernel mapping.

Example 11.2

This method can be generalized to spaces of higher dimensions. For example, the two classes in 2-D space shown in the middle row of Fig. 11.1 are not linearly separable: $C_- = \{\mathbf{x},\ ||\mathbf{x}|| < D\}$ and $C_+ = \{\mathbf{x},\ ||\mathbf{x}|| > D\}$. However, by the following mapping from 2-D space to 3-D space

$$\mathbf{z} = \phi(\mathbf{x}) = \begin{bmatrix} z_1 \\ z_2 \\ z_3 \end{bmatrix} = \begin{bmatrix} x_1 \\ x_2 \\ x_1^2 + x_2^2 \end{bmatrix} \tag{11.3}$$

the two classes can be trivially separated linearly by thresholding in the third dimension of the 3-D space.

Example 11.3

In 2-D space, in the exclusive OR dataset shown in the bottom row of Fig. 11.1 the two classes of C_- containing points in quadrants I and III, and C_+ containing points in quadrants II and IV are not linearly separable. However, by mapping the data points to a 3-D space

$$\mathbf{z} = \phi(\mathbf{x}) = \begin{bmatrix} z_1 \\ z_2 \\ z_3 \end{bmatrix} = \begin{bmatrix} x_1 \\ x_2 \\ x_1 x_2 \end{bmatrix} \tag{11.4}$$

the two classes can be separated by simply thresholding in the third dimension of the 3-D space.

Definition: A kernel is a function that takes two vectors $\mathbf{x}_m$ and $\mathbf{x}_n$ as arguments and returns the inner product of their images $\mathbf{z}_m = \phi(\mathbf{x}_m)$ and $\mathbf{z}_n = \phi(\mathbf{x}_n)$

$$K(\mathbf{x}_m, \mathbf{x}_n) = \phi(\mathbf{x}_m)^T \phi(\mathbf{x}_n) = \mathbf{z}_m^T \mathbf{z}_n \tag{11.5}$$

The kernel function takes as input some two vectors $\mathbf{x}_m$ and $\mathbf{x}_n$ in the original space, and returns a scalar value as the inner product $\mathbf{z}_m$ and $\mathbf{z}_n$ in some higher dimensional space. If the data points in the original space only appear in the form of inner product, then the kernel function $\mathbf{z} = \phi(\mathbf{x})$, called the kernel-induced *implicit* mapping, never needs to be explicitly specified, and the dimension of the new does not even need to be known.

The following is a set of commonly used kernel functions $K(\mathbf{x}, \mathbf{x}')$, which can be represented as an inner product of two vectors $\mathbf{z} = \phi(\mathbf{x})$ and $\mathbf{z}' = \phi(\mathbf{x}')$ in a higher dimensional space.

- **Linear kernel**
  Given two vectors $\mathbf{x}$ and $\mathbf{x}'$, the linear kernel is defined as the true inner product, without any kernel mapping

$$K(\mathbf{x}, \mathbf{x}') = \mathbf{x}^T \mathbf{x}' = \sum_{i=1}^{d} x_i x_i' \tag{11.6}$$

- **Polynomial kernels**
  The binomial theorem states

$$(x + y)^n = \sum_{k=0}^{n} \binom{n}{k} x^{n-k} y^k = \sum_{k=0}^{n} \frac{n!}{k!(n-k)!} x^{n-k} y^k \tag{11.7}$$

where the binomial coefficient

$$\binom{n}{k} = \frac{n!}{k!(n-k)!} \tag{11.8}$$

is the number of ways to distribute n items into two bins (k in one and $n-k$ in the other). This result can be generalized to the multinomial case

$$\begin{aligned}(x_1 + \cdots + x_d)^n &= \sum_{\sum_{i=1}^d k_i = n} \binom{n}{k_1, \cdots, k_d} x_1^{k_1} \cdots x_d^{k_d} \\ &= \sum_{\sum_{i=1}^d k_i = n} \frac{n!}{k_1! \cdots k_d!} x_1^{k_1} \cdots x_d^{k_d}\end{aligned} \tag{11.9}$$

where the multinomial coefficient

$$\binom{n}{k_1, \cdots, k_d} = \frac{n!}{k_1! \cdots k_d!} \tag{11.10}$$

is the number of ways to distribute n balls into d bins with k_i balls in the ith bin, and the summation is over all possible ways to obtain d non-negative integers $k_1, \cdots, k_d$ that add up to n.

Now consider the homogeneous polynomial kernel for d-dimensional vectors $\mathbf{x} = [x_1, \cdots, x_d]^T$ defined as

$$\begin{aligned}K(\mathbf{x}, \mathbf{x}') &= (\mathbf{x}^T \mathbf{x}')^n = (x_1 x_1' + \cdots + x_d x_d')^n \\ &= \sum_{\sum_{i=1}^d k_i = n} \frac{n!}{k_1! \cdots k_d!} \left((x_1 x_1')^{k_1} \cdots (x_d x_d')^{k_d} \right) \\ &= \phi(\mathbf{x})^T \phi(\mathbf{x}') = \mathbf{z}^T \mathbf{z}'\end{aligned} \tag{11.11}$$

where

$$\mathbf{z} = \phi(\mathbf{x}) = \left[\sqrt{\frac{n!}{k_1! \cdots k_d!}} \left(x_1^{k_1} \cdots x_d^{k_d} \right), \left(k_i \geq 0, \sum_{i=1}^d k_i = n \right) \right]^T \tag{11.12}$$

In particular, when $d = 2$ and $n = 2$, the polynomial kernel defined over 2-D vectors $\mathbf{x} = [x_1, x_2]^T$ is

$$\begin{aligned}K(\mathbf{x}, \mathbf{x}') &= (\mathbf{x}^T \mathbf{x}')^2 = (x_1 x_1' + x_2 x_2')^2 \\ &= (x_1 x_1')^2 + 2 x_1 x_1' x_2 x_2' + (x_2 x_2')^2 = \phi(\mathbf{x})^T \phi(\mathbf{x}') = \mathbf{z}^T \mathbf{z}'\end{aligned} \tag{11.13}$$

where $\mathbf{z} = \phi(\mathbf{x}) = [x_1^2, \sqrt{2} x_1 x_2, x_2^2]$ is a mapping from $\mathbf{x}$ in 2-D space to $\mathbf{z}$ in 3-D space.

A non-homogeneous polynomial kernel is defined as

$$K(\mathbf{x}, \mathbf{x}') = (1 + \mathbf{x}^T \mathbf{x}')^n \tag{11.14}$$

- **The radial basis function (RBF) kernel**
  The RBF kernel is defined as

$$K(\mathbf{x}, \mathbf{x}') = e^{-||\mathbf{x} - \mathbf{x}'||^2 / 2\sigma^2} = e^{-\gamma ||\mathbf{x} - \mathbf{x}'||^2} \tag{11.15}$$

where $\gamma = 1/2\sigma^2$ is a parameter that can be adjusted to fit each specific dataset. This kernel can be written as the inner product of two infinite dimensional vectors (for simplicity, we assume $\sigma = 1$)

$$\begin{aligned}
K(\mathbf{x}, \mathbf{x}') &= e^{-||\mathbf{x}-\mathbf{x}'||^2/2} = e^{-||\mathbf{x}||^2/2}\, e^{-||\mathbf{x}'||^2/2}\, e^{\mathbf{x}^T\mathbf{x}'} = e^{-||\mathbf{x}||^2/2}\, e^{-||\mathbf{x}'||^2/2} \sum_{n=0}^{\infty} \frac{(\mathbf{x}^T\mathbf{x}')^n}{n!} \\
&= e^{-||\mathbf{x}||^2/2}\, e^{-||\mathbf{x}'||^2/2} \sum_{n=0}^{\infty} \left[\frac{1}{n!} \sum_{\sum_{i=1}^d k_i = n} \frac{n!}{k_1! \cdots k_d!} \left((x_1 x'_1)^{k_1} \cdots (x_d x'_d)^{k_d} \right) \right] \\
&= \sum_{n=0}^{\infty} \sum_{\sum_{i=1}^d k_i = n} \left(e^{-||\mathbf{x}||^2/2} \frac{x_1^{k_1} \cdots x_d^{k_d}}{\sqrt{k_1! \cdots k_d!}} \right) \left(e^{-||\mathbf{x}'||^2/2} \frac{x_1'^{k_1} \cdots x_d'^{k_d}}{\sqrt{k_1! \cdots k_d!}} \right) \\
&= \phi(\mathbf{x})^T \phi(\mathbf{x}') = \mathbf{z}^T \mathbf{z}'
\end{aligned} \tag{11.16}$$

where

$$\mathbf{z} = \phi(\mathbf{x}) = \left[e^{-||\mathbf{x}||^2/2} \frac{x_1^{k_1} \cdots x_d^{k_d}}{\sqrt{k_1! \cdots k_d!}}, \left(n = 0, \cdots, \infty, \sum_{k=1}^{n} k_i = n \right) \right]^T \tag{11.17}$$

is a vector in an infinite dimensional space. In particular, when $d = 1$ we have

$$\begin{aligned}
K(x, x') &= e^{-(x-x')^2/2} = e^{-x^2/2}\, e^{-x'^2/2}\, e^{xx'} = e^{-x^2/2}\, e^{-x'^2/2} \sum_{n=0}^{\infty} \frac{(xx')^n}{n!} \\
&= \sum_{n=0}^{\infty} (e^{-x^2/2}\, x^n / \sqrt{n!})\, (e^{-x'^2/2}\, x'^n / \sqrt{n!})
\end{aligned} \tag{11.18}$$

where $\mathbf{z} = \phi(x) = \left[e^{-x^2/2}\, x^n / \sqrt{n!},\ (n = 0, \cdots, \infty) \right]^T$ is a kernel function that maps a 1-D space into an infinite dimensional space.

11.2 Kernel PCA

The PCA method maps the dataset to a lower dimensional space, and the kernel method maps the dataset to a higher dimensional space. These two methods are combined in kernel PCA to take advantage of separability in the high dimensional space and computational efficiency in the low dimensional space.

Specifically, we assume the given dataset $\mathbf{X} = [\mathbf{x}_1, \cdots, \mathbf{x}_N]$ in the original d-dimensional space is mapped into $\mathbf{Z} = [\mathbf{z}_1, \cdots, \mathbf{z}_N]$ in a high dimensional space. Correspondingly, the inner product $\mathbf{x}_m^T \mathbf{x}_n$ in the original space is mapped into an inner product in the high dimensional space by a kernel function

$$\mathbf{z}_m^T \mathbf{z}_n = K(\mathbf{x}_m, \mathbf{x}_n) = k_{mn} \qquad (m, n = 1, \cdots, N) \tag{11.19}$$

The N by N kernel matrix can be found as

$$\mathbf{K} = \mathbf{Z}^T\mathbf{Z} = \begin{bmatrix} \mathbf{z}_1^T \\ \vdots \\ \mathbf{z}_N^T \end{bmatrix} [\mathbf{z}_1, \cdots, \mathbf{z}_N] = \begin{bmatrix} \mathbf{z}_1^T\mathbf{z}_1 & \cdots & \mathbf{z}_1^T\mathbf{z}_N \\ \vdots & \ddots & \vdots \\ \mathbf{z}_N^T\mathbf{z}_1 & \cdots & \mathbf{z}_N^T\mathbf{z}_N \end{bmatrix} = \begin{bmatrix} k_{11} & \cdots & k_{1N} \\ \vdots & \ddots & \vdots \\ k_{N1} & \cdots & k_{NN} \end{bmatrix} \tag{11.20}$$

To carry out Karhunen–Loève transform (KLT) in the high dimensional space, we need to estimate the covariance matrix of $\mathbf{z} = \phi(\mathbf{x})$

$$\mathbf{\Sigma}_z = \frac{1}{N}\sum_{n=1}^{N}(\mathbf{z}_n - \mathbf{m}_z)(\mathbf{z}_n - \mathbf{m}_z)^T = \frac{1}{N}\sum_{n=1}^{N}\mathbf{z}_n\mathbf{z}_n^T - \mathbf{m}_z\mathbf{m}_z^T \tag{11.21}$$

If we assume the mean vector of all N samples in $\mathbf{Z} = [\mathbf{z}_1, \cdots, \mathbf{z}_N]$ is zero $\mathbf{m}_z = \mathbf{0}$ (to be justified later), $\mathbf{\Sigma}_z$ can be written as (same as Eq. (10.46))

$$\mathbf{\Sigma}_z = \frac{1}{N}\sum_{n=1}^{N}(\mathbf{z}_n - \mathbf{m}_z)(\mathbf{z}_n - \mathbf{m}_z)^T = \frac{1}{N}\sum_{n=1}^{N}\mathbf{z}_n\mathbf{z}_n^T = \frac{1}{N}[\mathbf{z}_1, \cdots, \mathbf{z}_N]\begin{bmatrix} \mathbf{z}_1^T \\ \vdots \\ \mathbf{z}_N^T \end{bmatrix} = \frac{1}{N}\mathbf{Z}\mathbf{Z}^T \tag{11.22}$$

We further get its eigenequations

$$\mathbf{\Sigma}_z\mathbf{v}_n = \left(\frac{1}{N}\mathbf{Z}\mathbf{Z}^T\right)\mathbf{v}_n = \lambda_n\mathbf{v}_n \qquad (n = 1, 2, \cdots) \tag{11.23}$$

To express this equations in terms of the kernel matrix $\mathbf{K} = \mathbf{K}^T\mathbf{K}$, we pre-multiplying $\mathbf{Z}^T$ on both sides of the eigenequation to obtain (same as Eq. (10.48))

$$\frac{1}{N}\mathbf{Z}^T\mathbf{Z}\left(\mathbf{Z}^T\mathbf{v}_n\right) = \frac{1}{N}\mathbf{K}\mathbf{u}_n = \lambda_n\left(\mathbf{Z}^T\mathbf{v}_n\right) = \lambda_n\mathbf{u}_n \tag{11.24}$$

where we have defined $\mathbf{u}_n = \mathbf{Z}^T\mathbf{v}_n$. Solving this eigenequation of the $N \times N$ matrix $\mathbf{K}/N$ we get its eigenvectors $\mathbf{u}_n$ corresponding to the eigenvalues λ_n, the same as the non-zero eigenvalues of $\mathbf{\Sigma}_z = \mathbf{Z}\mathbf{Z}^T/N$. The eigenvectors $\mathbf{v}_n$ of $\mathbf{\Sigma}_z$ in Eq. (11.23) can now be written as

$$\mathbf{v}_n = \frac{1}{\lambda_n N}\mathbf{Z}\left(\mathbf{Z}^T\mathbf{v}_n\right) = \frac{1}{\lambda_n N}\mathbf{Z}\mathbf{u}_n \tag{11.25}$$

which can be further normalized by imposing the condition $||\mathbf{v}_n||^2 = 1$

$$\begin{aligned} ||\mathbf{v}_n||^2 &= \mathbf{v}_n^T\mathbf{v}_n = \frac{1}{(\lambda_n N)^2}(\mathbf{Z}\mathbf{u}_n)^T(\mathbf{Z}\mathbf{u}_n) = \frac{1}{(\lambda_n N)^2}\mathbf{u}_n^T\mathbf{Z}^T\mathbf{Z}\mathbf{u}_n \\ &= \frac{1}{(\lambda_n N)^2}\mathbf{u}_n^T\mathbf{K}\mathbf{u}_n = \frac{1}{\lambda_n N}\mathbf{u}_n^T\mathbf{u}_n = \frac{1}{\lambda_n N}||\mathbf{u}_n||^2 = 1 \end{aligned} \tag{11.26}$$

where we have used the fact $\mathbf{K}\mathbf{u}_n = \lambda_n N\mathbf{u}_n$ in Eq. (11.24). We can get the normalized eigenvectors of $\mathbf{\Sigma}_z$ in Eq. (11.23)

$$\mathbf{v}_n = \frac{1}{\sqrt{\lambda_n N}}\mathbf{u}_n \tag{11.27}$$

Now we can carry out KLT for any data point $\mathbf{z} = \phi(\mathbf{x})$ in the high dimensional space by

$$\mathbf{y} = \mathbf{V}^T\mathbf{z} = [\mathbf{v}_1, \cdots, \mathbf{v}_N]^T\mathbf{z} \tag{11.28}$$

of which each component $y_n = \mathbf{v}_n^T\mathbf{z}$ is the projection of $\mathbf{z} = \phi(\mathbf{x})$ onto one of the eigenvectors $\mathbf{v}_n$ of $\mathbf{\Sigma}_z$ $(n = 1, \cdots, N)$

$$\begin{aligned} y_n &= \mathbf{z}^T\mathbf{v}_n = \mathbf{z}^T\left(\frac{1}{\lambda_n N}\mathbf{Z}\mathbf{u}_n\right) = \frac{1}{\lambda_n N}\mathbf{z}^T[\mathbf{z}_1, \cdots, \mathbf{z}_N]\mathbf{u}_n \\ &= \frac{1}{\lambda_n N}[\mathbf{z}^T\mathbf{z}_1, \cdots, \mathbf{z}^T\mathbf{z}_N]\,\mathbf{u}_n = \frac{1}{\lambda_n N}[k_1, \cdots, k_N]\,\mathbf{u}_n = \frac{1}{\lambda_n N}\mathbf{k}^T\mathbf{u}_n \end{aligned} \tag{11.29}$$

where we have defined $\mathbf{k} = [k_1, \cdots, k_N]^T$ with $k_n = \mathbf{z}^T\mathbf{z}_n$. Although neither $\mathbf{z}$ nor $\mathbf{v}_n = \mathbf{Z}\mathbf{u}_n/\lambda_n N$ is explicitly available, their inner product $y_n = \mathbf{z}^T\mathbf{v}_n$ can still be obtained based on the kernel function $\mathbf{k}$ and $\mathbf{u}$. As all data points in the high dimensional space appear only in the form of an inner product in the equation above, as well as in the eigenequation Eq. (11.24), the kernel mapping $\mathbf{z} = \phi(\mathbf{x})$ never needs to be explicitly carried out.

This discussion is based on the assumption that in the high dimensional space the data have zero mean $\mathbf{m}_z = \mathbf{0}$. This cannot be achieved by subtracting the mean from the data to get $\tilde{\mathbf{z}} = \mathbf{z} - \mathbf{m}_z$, as neither $\mathbf{z}$ nor $\mathbf{m}_z$ is available without actually carrying out the kernel mapping $\mathbf{z} = \phi(\mathbf{x})$. However, we can find the inner product of any two zero-mean data points $\tilde{\mathbf{z}}_m$, $\tilde{\mathbf{z}}_n$, $(m, n = 1, \cdots, N)$ in the high dimensional space

$$\begin{aligned} \tilde{k}_{mn} &= \tilde{\mathbf{z}}_m^T\tilde{\mathbf{z}}_n = (\mathbf{z}_m - \mathbf{m}_z)^T(\mathbf{z}_n - \mathbf{m}_z) = \left(\mathbf{z}_m - \frac{1}{N}\sum_{k=1}^N \mathbf{z}_k\right)^T\left(\mathbf{z}_n - \frac{1}{N}\sum_{l=1}^N \mathbf{z}_l\right) \\ &= \mathbf{z}_m^T\mathbf{z}_n - \mathbf{z}_m^T\left(\frac{1}{N}\sum_{l=1}^N \mathbf{z}_l\right) - \left(\frac{1}{N}\sum_{k=1}^N \mathbf{z}_k^T\right)\mathbf{z}_n + \frac{1}{N^2}\sum_{k=1}^N\sum_{l=1}^N \mathbf{z}_k^T\mathbf{z}_l \\ &= k_{mn} - \frac{1}{N}\sum_{l=1}^N k_{ml} - \frac{1}{N}\sum_{k=1}^N k_{kn} + \frac{1}{N^2}\sum_{k=1}^K\sum_{l=1}^N k_{kl} \end{aligned} \tag{11.30}$$

The kernel matrix composed of all $N \times N$ such inner products can be written as

$$\tilde{\mathbf{K}} = \mathbf{K} - \mathbf{1}_N\mathbf{K} - \mathbf{K}\mathbf{1}_N + \mathbf{1}_N\mathbf{K}\mathbf{1}_N \tag{11.31}$$

where

$$\mathbf{1}_N = \frac{1}{N}\begin{bmatrix} 1 & \cdots & 1 \\ \vdots & \ddots & \vdots \\ 1 & \cdots & 1 \end{bmatrix}_{N\times N} \tag{11.32}$$

In the considered algorithm, if we replace the kernel matrix $\mathbf{K}$ for $\mathbf{z}$ with non-zero mean by $\tilde{\mathbf{K}}$ for $\tilde{\mathbf{z}}$ with zero mean, then the assumption that $\mathbf{z}$ has zero mean $\mathbf{m}_z$ becomes valid. The steps of the kernel PCA algorithm are summarized below.

- Given the dataset $\mathbf{X} = \{\mathbf{x}_1, \cdots, \mathbf{x}_N\}$, obtain $\mathbf{K}$ based on kernel mapping $k_{mn} = k(\mathbf{x}_m, \mathbf{x}_n)$ for all $m, n = 1, \cdots, N$, and then find $\tilde{\mathbf{K}}$, as in Eq. (11.31).

- Solve eigenvalue problem of $\tilde{\mathbf{K}}$ to get the eigenvalue λ_n and the corresponding eigenvector $\mathbf{u}_n$ for all $i = 1, \cdots, N$.
- Project each data point $\mathbf{z} = \phi(\mathbf{x})$ (implicit) onto the nth eigenvector $\mathbf{v}_n$ to obtain the principal component: $y_n = \mathbf{z}^T\mathbf{v}_n = \mathbf{k}^T\mathbf{u}_n/\lambda_n N$ for each $n = 1, \cdots, N$, as in Eq. (11.29).
- Carry out whatever operation (e.g., feature selection, classification) in the subspace spanned by the first $M < N$ principal components $\{y_1, \cdots, y_M\}$ in the high dimensional feature space, containing most of the information.

Note that in kernel PCA methods, the complexity for solving the eigenvalue problem of the $N \times N$ kernel matrix $\mathbf{K}$ is $O(N^3)$, instead of $O(d^3)$ for the linear PCA based on the $d \times d$ covariance matrix $\mathbf{\Sigma}_x$.

The following shows a MATLAB function for the kernel PCA algorithm which takes the dataset $\mathbf{X} = [\mathbf{x}_1, \cdots, \mathbf{x}_N]$ as input and generates the eigenvalues in $\mathbf{d}$ and the dataset in the transform space $\mathbf{Y}$. The function `Kernel(X,X)` calculates the kernel matrix $\mathbf{K}$.

```
function [Y d]=KPCA(X)
    [D N]=size(X);
    K=Kernel(X,X);                 % kernel matrix of data X
    one=ones(N)/N;
    Kt=K-one*K-K*one+one*K*one;    % Kernel matrix of data with zero mean
    [U D]=eig(Kt);                 % solve eigenvalue problem for K
    [d idx]=sort(diag(D),'descend');
                                   % sort eigenvalues in descending order
    U=U(:,idx);
    D=D(:,idx);
    for n=1:N
        U(:,n)=U(:,n)/d(n)/N;
    end
    Y=(K*U)';
end
```

The linear and kernel PCAs (based on RBF kernel) are applied to three datasets and compared as shown in Figs 11.2 through 11.4.

11.3 Factor Analysis and Expectation Maximization

The method of *factor analysis (FA)* models a set of d observed *manifest variables* in $\mathbf{x} = [x_1, \cdots, x_d]^T$ as a linear combination of a set of $d' < d$ unobserved hidden *latent variables* or *common factors* in $\mathbf{z} = [z_1, \cdots, z_{d'}]^T$, to explain and reveal the variability and dependency among the d observed variables, typically correlated, in terms of the latent variables, assumed to be independent and therefore uncorrelated. The method of FA is therefore similar to the method of PCA in

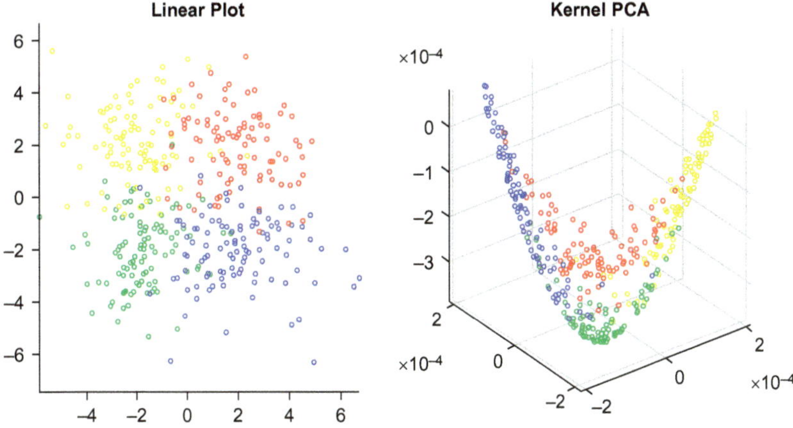

Fig. 11.2 Linear and kernel PCAs for a 2-D dataset of four classes.

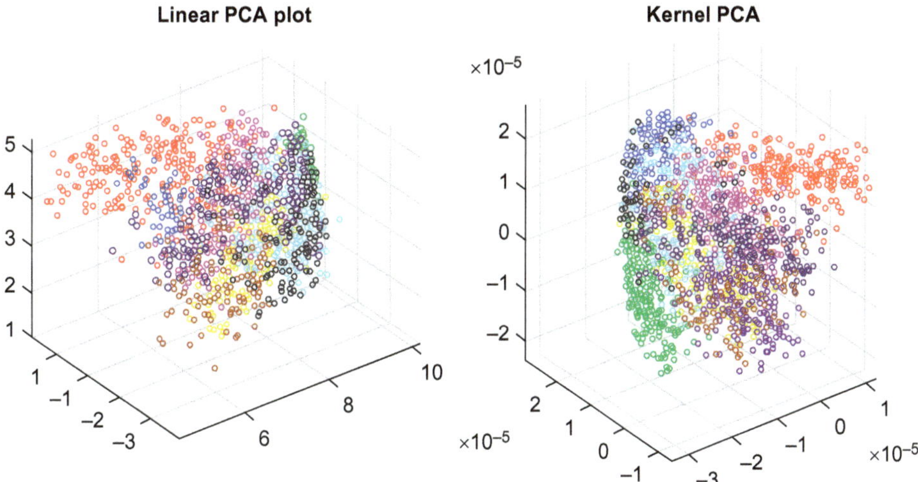

Fig. 11.3 Linear and kernel PCAs for the 256-D handwritten digits.

the sense that they both aim at extracting the essential information contained in a multivariate dataset as some linear combination of all signal components and thereby reducing the dimensionality of the dataset while preserving most of the information contained in it.

Specifically, we assume each of the observed variables in $\mathbf{x}$ is a linear combination of the d' factors in $\mathbf{z}$

$$x_i = \sum_{j=1}^{d'} w_{ij} z_j + e_i = [w_{i1}, \cdots, w_{id'}] \begin{bmatrix} z_1 \\ \vdots \\ z_{d'} \end{bmatrix} + e_i, \qquad (i = 1, \cdots, d) \tag{11.33}$$

or in matrix form

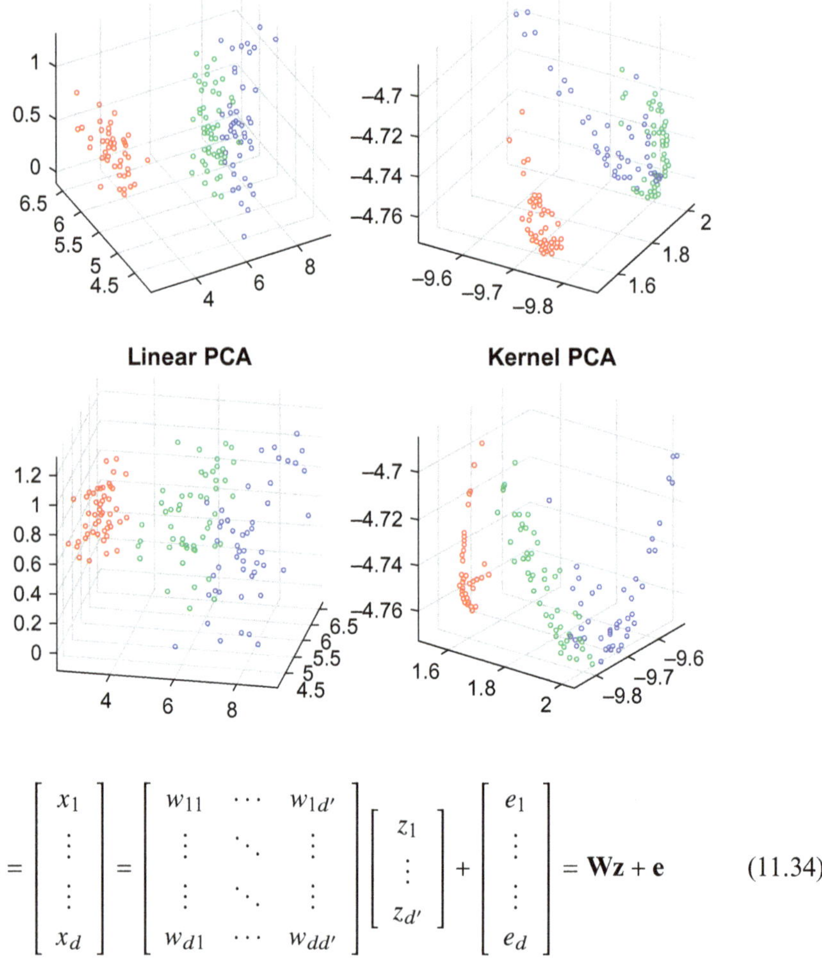

Fig. 11.4 Linear and kernel PCAs for the iris dataset.

$$\mathbf{x} = \begin{bmatrix} x_1 \\ \vdots \\ \vdots \\ x_d \end{bmatrix} = \begin{bmatrix} w_{11} & \cdots & w_{1d'} \\ \vdots & \ddots & \vdots \\ \vdots & \ddots & \vdots \\ w_{d1} & \cdots & w_{dd'} \end{bmatrix} \begin{bmatrix} z_1 \\ \vdots \\ z_{d'} \end{bmatrix} + \begin{bmatrix} e_1 \\ \vdots \\ \vdots \\ e_d \end{bmatrix} = \mathbf{W}\mathbf{z} + \mathbf{e} \tag{11.34}$$

where $\mathbf{W}$ is a $d \times d'$ *factor loading matrix*, and e_i is the noise associated with x_i. Also, for simplicity and without loss of generality, we assume the dataset has a zero mean. If $\mathbf{W}$ were available, the d' factors in $\mathbf{z}$ can be found by solving this over-determined linear equation system of d equations but $d' < d$ unknowns by the least-squares method (with minimum squared error $||\mathbf{e}||^2$)

$$\hat{\mathbf{z}} = \mathbf{W}^{-}\mathbf{x} = (\mathbf{W}^T\mathbf{W})^{-1}\mathbf{W}^T\mathbf{x} \tag{11.35}$$

where $\mathbf{W}^{-} = (\mathbf{W}^T\mathbf{W})^{-1}\mathbf{W}^T$ is the left pseudo-inverse of $\mathbf{W}$.

However, as $\mathbf{W}$ is unavailable, it needs to be estimated together with $\mathbf{z}$ at the same time, based on the given dataset $\mathbf{X} = [\mathbf{x}_1, \cdots, \mathbf{x}_N]$, typically $N \gg d$. This can be done by the general method of *expectation-maximization (EM)*.

Specifically, we treat both $\mathbf{z}$ and $\mathbf{e}$ as random vectors, and make the following assumptions:

- The latent variables in $\mathbf{z}$ are of zero mean, independent of each other, and with unity variance

$$\mathbf{m}_z = E[\mathbf{z}] = \mathbf{0}, \qquad \boldsymbol{\Sigma}_z = \text{Cov}[\mathbf{z}] = E[\mathbf{z}\mathbf{z}^T] = \mathbf{I} \tag{11.36}$$

and they are normally distributed

$$p(\mathbf{z}) = \mathcal{N}(\mathbf{0}, \mathbf{I}) \tag{11.37}$$

- The noise components in $\mathbf{e}$ have zero mean, and they are independent of each other with a diagonal covariance matrix

$$\mathbf{m}_e = \mathrm{E}[\mathbf{e}] = \mathbf{0}, \qquad \boldsymbol{\Sigma}_e = \mathrm{Cov}[\mathbf{e}] = \mathrm{E}[\mathbf{e}\mathbf{e}^T] = \boldsymbol{\Psi} = \begin{bmatrix} \psi_1 & \cdots & 0 \\ \vdots & \ddots & \vdots \\ 0 & \cdots & \psi_d \end{bmatrix} \tag{11.38}$$

and they are also normally distributed

$$p(\mathbf{e}) = \mathcal{N}(\mathbf{0}, \boldsymbol{\Psi}) \tag{11.39}$$

- The latent variables and the noise are independent of each other:

$$\boldsymbol{\Sigma}_{ze} = \mathrm{Cov}[\mathbf{z}, \mathbf{e}] = \mathrm{E}[(\mathbf{z} - \mathbf{m}_z)(\mathbf{e} - \mathbf{m}_e^T)] = \mathrm{E}[\mathbf{z}\mathbf{e}^T] = \mathbf{0} \tag{11.40}$$

The two matrices $\mathbf{W}$ and $\boldsymbol{\Psi}$ defined as the parameters of the FA model are denoted by $\theta = \{\mathbf{W}, \boldsymbol{\Psi}\}$.

Based on the given assumptions, we would like to find the conditional pdf $p(\mathbf{z}|\mathbf{x})$ of the latent variable $\mathbf{z}$ given the observed variable $\mathbf{x}$. To do so, we first find the pdf of $\mathbf{x} = \mathbf{W}\mathbf{z} + \mathbf{e}$, which, as a linear combination of the two normally distributed random vectors $\mathbf{z}$ and $\mathbf{e}$, is also normally distributed with $p(\mathbf{x}) = \mathcal{N}(\mathbf{m}_x, \boldsymbol{\Sigma}_x)$, where

$$\mathbf{m}_x = \mathrm{E}[\mathbf{x}] = \mathrm{E}[\mathbf{W}\mathbf{z} + \mathbf{e}] = \mathbf{W}\,\mathrm{E}[\mathbf{z}) + \mathrm{E}(\mathbf{e}] = \mathbf{0} \tag{11.41}$$

$$\begin{aligned} \boldsymbol{\Sigma}_x &= \mathrm{Cov}[\mathbf{x}] = \mathrm{E}[(\mathbf{x} - \mathbf{m}_x)(\mathbf{x} - \mathbf{m}_x^T)] = \mathrm{E}[\mathbf{x}\mathbf{x}^T] = E\left[(\mathbf{W}\mathbf{z} + \mathbf{e})(\mathbf{W}\mathbf{z} + \mathbf{e})^T\right] \\ &= \mathbf{W}\,\mathrm{E}[\mathbf{z}\mathbf{z}^T]\mathbf{W}^T - \mathrm{E}[\mathbf{e}\mathbf{z}^T]\mathbf{W}^T - \mathbf{W}\,\mathrm{E}[\mathbf{z}\mathbf{e}^T] + \mathrm{E}[\mathbf{e}\mathbf{e}^T] \\ &= \mathbf{W}\mathbf{W}^T + \boldsymbol{\Psi} \end{aligned} \tag{11.42}$$

As both $\mathbf{m}_x$ and $\boldsymbol{\Sigma}_x$ are based on the model parameter $\boldsymbol{\theta} = \{\mathbf{W}, \boldsymbol{\Psi}\}$, the pdf of $\mathbf{x}$ is conditional on $\boldsymbol{\theta}$

$$p(\mathbf{x}|\theta) = \mathcal{N}(\mathbf{m}_x, \boldsymbol{\Sigma}_x) = \mathcal{N}(\mathbf{0}, \mathbf{W}\mathbf{W}^T + \boldsymbol{\Psi}) \tag{11.43}$$

The joint distribution $p(\mathbf{z}, \mathbf{x})$ is also Gaussian with a zero mean

$$\mathbf{m} = E\left(\begin{bmatrix} \mathbf{z} \\ \mathbf{x} \end{bmatrix}\right) = \begin{bmatrix} \mathbf{m}_z \\ \mathbf{m}_x \end{bmatrix} = \begin{bmatrix} \mathbf{0} \\ \mathbf{0} \end{bmatrix} \tag{11.44}$$

and a covariance $\boldsymbol{\Sigma}$ composed of four submatrices

$$\boldsymbol{\Sigma} = \begin{bmatrix} \boldsymbol{\Sigma}_{zz} & \boldsymbol{\Sigma}_{zx} \\ \boldsymbol{\Sigma}_{xz} & \boldsymbol{\Sigma}_{xx} \end{bmatrix} \tag{11.45}$$

where

$$\begin{aligned} \boldsymbol{\Sigma}_{zz} &= \mathbf{I} \\ \boldsymbol{\Sigma}_{xx} &= \mathbf{W}\mathbf{W}^T + \boldsymbol{\Psi} \\ \boldsymbol{\Sigma}_{zx} &= \boldsymbol{\Sigma}_{zx}^T = \mathrm{E}[(\mathbf{z} - \mathbf{m}_z)(\mathbf{x} - \mathbf{m}_x)^T] = \mathrm{E}[\mathbf{z}\mathbf{x}^T] = \mathrm{E}[\mathbf{z}(\mathbf{W}\mathbf{z} + \mathbf{e})^T] \\ &= \mathrm{E}[\mathbf{z}\mathbf{z}^T]\mathbf{W}^T + \mathrm{E}[\mathbf{z}\mathbf{e}^T] = \mathbf{I}\mathbf{W}^T + \mathbf{0} = \mathbf{W}^T \end{aligned} \tag{11.46}$$

The normal distribution $p(\mathbf{z}, \mathbf{x}|\boldsymbol{\theta})$ can now be expressed as

$$p(\mathbf{x}, \mathbf{z}|\boldsymbol{\theta}) = p\left(\begin{bmatrix} \mathbf{z} \\ \mathbf{x} \end{bmatrix}\right) = \mathcal{N}(\mathbf{m}, \boldsymbol{\Sigma}) = \mathcal{N}\left(\begin{bmatrix} \mathbf{0} \\ \mathbf{0} \end{bmatrix}, \begin{bmatrix} \boldsymbol{\Sigma}_{zz} & \boldsymbol{\Sigma}_{zx} \\ \boldsymbol{\Sigma}_{xz} & \boldsymbol{\Sigma}_{xx} \end{bmatrix}\right)$$
$$= \mathcal{N}\left(\begin{bmatrix} \mathbf{0} \\ \mathbf{0} \end{bmatrix}, \begin{bmatrix} \mathbf{I} & \mathbf{W}^T \\ \mathbf{W} & \mathbf{W}\mathbf{W}^T + \boldsymbol{\Psi} \end{bmatrix}\right) \tag{11.47}$$

Based on this joint pdf of both $\mathbf{x}$ and $\mathbf{z}$, we can further find the desired conditional pdfs of both $p(\mathbf{x}|\mathbf{z})$ and $p(\mathbf{z}|\mathbf{x})$ (Section B.1.5)

- $p(\mathbf{x}|\mathbf{z}, \theta) = \mathcal{N}(\mathbf{m}_{x|z}, \boldsymbol{\Sigma}_{x|z})$, with

$$\begin{cases} \mathbf{m}_{x|z} &= \mathbf{m}_x + \boldsymbol{\Sigma}_{xz}\boldsymbol{\Sigma}_{zz}^{-1}(\mathbf{z} - \mathbf{m}_z) = \mathbf{W}\mathbf{z} \\ \boldsymbol{\Sigma}_{x|z} &= \boldsymbol{\Sigma}_{xx} - \boldsymbol{\Sigma}_{xz}\boldsymbol{\Sigma}_{zz}^{-1}\boldsymbol{\Sigma}_{zx} = \mathbf{W}\mathbf{W}^T + \boldsymbol{\Psi} - \mathbf{W}\mathbf{W}^T = \boldsymbol{\Psi} \end{cases} \tag{11.48}$$

- $p(\mathbf{z}|\mathbf{x}, \theta) = \mathcal{N}(\mathbf{m}_{z|x}, \boldsymbol{\Sigma}_{z|x})$, with

$$\begin{cases} \mathbf{m}_{z|x} &= \mathbf{m}_z + \boldsymbol{\Sigma}_{zx}\boldsymbol{\Sigma}_{xx}^{-1}(\mathbf{x} - \mathbf{m}_x) = \mathbf{W}^T(\mathbf{W}\mathbf{W}^T + \boldsymbol{\Psi})^{-1}\mathbf{x} = \mathbf{B}\mathbf{x} \\ \boldsymbol{\Sigma}_{z|x} &= \boldsymbol{\Sigma}_{zz} - \boldsymbol{\Sigma}_{zx}\boldsymbol{\Sigma}_{xx}^{-1}\boldsymbol{\Sigma}_{xz} = \mathbf{I} - \mathbf{W}^T(\mathbf{W}\mathbf{W}^T + \boldsymbol{\Psi})^{-1}\mathbf{W} = \mathbf{I} - \mathbf{B}\mathbf{W} \end{cases} \tag{11.49}$$

where we have defined

$$\mathbf{B} = \mathbf{W}^T(\mathbf{W}\mathbf{W}^T + \boldsymbol{\Psi})^{-1} \tag{11.50}$$

Note that while $p(\mathbf{z}) = \mathcal{N}(\mathbf{0}, \mathbf{I})$ has zero mean and diagonal covariance, the conditional distribution $p(\mathbf{z}|\mathbf{x}, \boldsymbol{\theta}) = \mathcal{N}(\mathbf{m}_{z|x}, \boldsymbol{\Sigma}_{z|x})$ has non-zero mean $\mathbf{m}_{z|x}$ and non-diagonal covariance $\boldsymbol{\Sigma}_{z|x}$.

The computational complexity for the inversion of the $N \times N$ matrix $\mathbf{W}\mathbf{W}^T + \boldsymbol{\Psi}$ is $O(N^3)$. However, by applying the Woodbury matrix identity (Section A.2.4)

$$(\boldsymbol{\Psi} + \mathbf{W}\mathbf{W}^T)^{-1} = \boldsymbol{\Psi}^{-1} - \boldsymbol{\Psi}^{-1}\mathbf{W}(\mathbf{I} + \mathbf{W}^T\boldsymbol{\Psi}^{-1}\mathbf{W})^{-1}\mathbf{W}^T\boldsymbol{\Psi}^{-1} \tag{11.51}$$

where $\boldsymbol{\Psi}$ as a diagonal matrix can be easily inverted, and $\mathbf{I} + \mathbf{W}^T\boldsymbol{\Psi}^{-1}\mathbf{W}$ is an $d' \times d'$ matrix that can be inverted with complexity $O(d'^3) \ll O(N^3)$.

The model parameter $\boldsymbol{\theta} = \{\mathbf{W}, \boldsymbol{\Psi}\}$ can now be estimated based on the given dataset $\mathbf{X}$ by the EM algorithm, an iterative process of the following two steps.

- **The E-step**

  Find the expectation of the log-likelihood function of the model parameter $\boldsymbol{\theta}$, to be maximized in the following M-step.

  Find the likelihood function of $\boldsymbol{\theta}$ based on the observed dataset $\mathbf{X} = [\mathbf{x}_1, \cdots, \mathbf{x}_N]$ (all samples assumed to be i.i.d.)

$$L(\theta|\mathbf{X}, \mathbf{z}) = p(\mathbf{X}, \mathbf{z}|\theta) = \prod_{n=1}^N p(\mathbf{x}_n, \mathbf{z}|\theta) = \prod_{n=1}^N p(\mathbf{x}_n|\mathbf{z}, \theta)\; p(\mathbf{z}|\theta) \tag{11.52}$$

  and the log-likelihood

$$\log L(\theta|\mathbf{X}, \mathbf{z}) = \sum_{n=1}^N (\log p(\mathbf{x}_n|\mathbf{z}, \theta) + \log p(\mathbf{z})) \tag{11.53}$$

The second term can be dropped as $p(\mathbf{z}) = \mathcal{N}(\mathbf{0}, \mathbf{I})$ is independent of the model parameter $\boldsymbol{\theta}$ and therefore irrelevant to the maximization of the log-likelihood with respect to $\boldsymbol{\theta}$.

We then find the expectation of the log-likelihood function, denoted by Q, with respect to the latent variable $\mathbf{z}$ based on Eq. (11.48)

$$\begin{aligned}
Q &= E_{z|x}\left[\log L(\theta|\mathbf{X},\mathbf{z})\right] = \sum_{n=1}^{N} E_{z|x_n}\left[\log p(\mathbf{x}_n|\mathbf{z},\theta)\right] \\
&= \sum_{n=1}^{N} E_{z|x_n} \log\left[\frac{1}{(2\pi)^{d/2}|\boldsymbol{\Sigma}_{x|z}|^{1/2}} \exp\left(-\frac{1}{2}(\mathbf{x}_n - \mathbf{m}_{x|z})^T \boldsymbol{\Sigma}_{x|z}^{-1}(\mathbf{x}_n - \mathbf{m}_{x|z})\right)\right] \\
&= \sum_{n=1}^{N} E_{z|x_n}\left[-\frac{d}{2}\log(2\pi) - \frac{1}{2}\log|\boldsymbol{\Psi}| - \frac{1}{2}(\mathbf{x}_n - \mathbf{Wz})^T\boldsymbol{\Psi}^{-1}(\mathbf{x}_n - \mathbf{Wz})\right] \\
&= -\frac{dN}{2}\log(2\pi) - \frac{N}{2}\log|\boldsymbol{\Psi}| - \frac{1}{2}\sum_{n=1}^{N} E_{z|x_n}\left[(\mathbf{x}_n - \mathbf{Wz})^T\boldsymbol{\Psi}^{-1}(\mathbf{x}_n - \mathbf{Wz})\right] \\
&= C - \frac{N}{2}\log|\boldsymbol{\Psi}| - \frac{1}{2}\sum_{n=1}^{N} E_{z|x_n}\left[\mathbf{x}_n^T\boldsymbol{\Psi}^{-1}\mathbf{x}_n - 2\mathbf{x}_n^T\boldsymbol{\Psi}^{-1}\mathbf{Wz} + \mathbf{z}^T\mathbf{W}^T\boldsymbol{\Psi}^{-1}\mathbf{Wz}\right]
\end{aligned} \tag{11.54}$$

where we have used the fact that

$$\mathbf{x}_n^T\boldsymbol{\Psi}^{-1}(\mathbf{Wz}) = (\mathbf{Wz})^T\boldsymbol{\Psi}^{-1}\mathbf{x}_n \tag{11.55}$$

as both $\boldsymbol{\Psi}^{-1}$ and $\boldsymbol{\Psi}$ are diagonal and therefore also symmetric. Here the constant $C = -dN\log(2\pi)/2$ can be dropped.

- **The M-step**

  Find the optimal model parameter $\theta = \{\mathbf{W}, \boldsymbol{\Psi}\}$ that maximizes the expectation of the log-likelihood Q obtained in the E-step.

  We set to zero the derivative of Q with respective each of the two parameters in $\theta = \{\mathbf{W}, \boldsymbol{\Psi}\}$ and solve the resulting equations.

  - Find $\mathbf{W}$

$$\begin{aligned}
\frac{\partial Q}{\partial \mathbf{W}} &= \frac{\partial}{\partial \mathbf{W}}\sum_{n=1}^{N} E_{z|x_n}\left[-\mathbf{x}_n^T\boldsymbol{\Psi}^{-1}\mathbf{x}_n + 2\mathbf{x}_n^T\boldsymbol{\Psi}^{-1}\mathbf{Wz} - \mathbf{z}^T\mathbf{W}^T\boldsymbol{\Psi}^{-1}\mathbf{Wz}\right] \\
&= \sum_{n=1}^{N} E_{z|x_n}\left[\frac{\partial}{\partial \mathbf{W}}\left(2\mathbf{x}_n^T\boldsymbol{\Psi}^{-1}\mathbf{Wz} - \mathbf{z}^T\mathbf{W}^T\boldsymbol{\Psi}^{-1}\mathbf{Wz}\right)\right] \\
&= \sum_{n=1}^{N} E_{z|x_n}\left[2\boldsymbol{\Psi}^{-1}\mathbf{x}_n\mathbf{z}^T - 2\boldsymbol{\Psi}^{-1}\mathbf{Wzz}^T\right] \\
&= 2\boldsymbol{\Psi}^{-1}\sum_{n=1}^{N}\left(\mathbf{x}_n E_{z|x_n}[\mathbf{z}^T] - \mathbf{W}E_{z|x_n}[\mathbf{zz}^T]\right) = \mathbf{0}
\end{aligned} \tag{11.56}$$

Solving for $\mathbf{W}$ we obtain

$$\mathbf{W} = \left(\sum_{n=1}^{N} \mathbf{x}_n E_{z|x_n}[\mathbf{z}^T] \right) \left(\sum_{n=1}^{N} E_{z|x_n}[\mathbf{z}\mathbf{z}^T] \right)^{-1} \tag{11.57}$$

where $E_{z|x_n}[\mathbf{z}]$ and $E_{z|x_n}[\mathbf{z}\mathbf{z}^T]$ can be found in Eq. (11.49)

$$\begin{cases} E_{z|x_n}[\mathbf{z}] & = & \mathbf{m}_{z|x_n} = \mathbf{B}\mathbf{x}_n \\ E_{z|x_n}[\mathbf{z}\mathbf{z}^T] & = & \mathbf{\Sigma}_{z|x_n} + \mathbf{m}_{z|x_n}\mathbf{m}_{z|x_n}^T = \mathbf{I} - \mathbf{B}\mathbf{W} + \mathbf{B}\mathbf{x}_n\mathbf{x}_n^T\mathbf{B}^T \end{cases} \tag{11.58}$$

The second equation is due to the fact that $\mathbf{\Sigma}_z = \mathrm{E}[\mathbf{z}\mathbf{z}^T] - \mathbf{m}_z\mathbf{m}_z^T$. Here $E_{z|x_n}[\mathbf{z}]$ can be considered as the estimation of the $\mathbf{z}$, while $E_{z|x_n}[\mathbf{z}\mathbf{z}^T]$ the uncertainty of the estimation.

- Find $\mathbf{\Psi}$

$$\begin{aligned} \frac{\partial Q}{\partial \mathbf{\Psi}^{-1}} &= \frac{\partial}{\partial \mathbf{\Psi}^{-1}} \left[N \log |\mathbf{\Psi}| + \sum_{n=1}^{N} E_{z|x_n} \left[\mathbf{x}_n^T \mathbf{\Psi}^{-1} \mathbf{x}_n - 2\mathbf{x}_n^T \mathbf{\Psi}^{-1} \mathbf{W}\mathbf{z} + \mathbf{z}^T \mathbf{W}^T \mathbf{\Psi}^{-1} \mathbf{W}\mathbf{z} \right] \right] \\ &= -N\mathbf{\Psi} + \sum_{n=1}^{N} E_{z|x_n} \left[\mathbf{x}_n\mathbf{x}_n^T - 2\mathbf{x}_n\mathbf{z}^T\mathbf{W}^T + \mathbf{W}(\mathbf{z}\mathbf{z}^T)\mathbf{W}^T \right] \\ &= -N\mathbf{\Psi} + \sum_{n=1}^{N} \mathbf{x}_n\mathbf{x}_n^T - 2\sum_{n=1}^{N} \mathbf{x}_n E_{z|x_n}[\mathbf{z}]^T\mathbf{W}^T + \mathbf{W}\sum_{n=1}^{N} E_{z|x}[\mathbf{z}\mathbf{z}^T]\mathbf{W}^T = \mathbf{0} \end{aligned} \tag{11.59}$$

Solving for $\mathbf{\Psi}$ we have

$$\mathbf{\Psi} = \frac{1}{N} \mathrm{diag}\left(\sum_{n=1}^{N} \mathbf{x}_n\mathbf{x}_n^T - 2\left(\sum_{n=1}^{N} \mathbf{x}_n E_{z|x_n}[\mathbf{z}]^T \right) \mathbf{W}^T + \mathbf{W}\left(\sum_{n=1}^{N} E_{z|x_n}[\mathbf{z}\mathbf{z}^T] \right) \mathbf{W}^T \right) \tag{11.60}$$

where diag($\mathbf{A}$) is the operation that sets all off-diagonal elements of matrix $\mathbf{A}$ to zero, so that the resulting matrix is indeed diagonal as what $\mathbf{\Psi}$ should be. We further replace $\mathbf{W}$ in front of the last term by that in Eq. (11.57) and get

$$\mathbf{\Psi} = \frac{1}{N} \mathrm{diag}\left(\sum_{n=1}^{N} \mathbf{x}_n\mathbf{x}_n^T - \sum_{n=1}^{N} \mathbf{x}_n E_{z|x_n}[\mathbf{z}]^T\mathbf{W}^T \right) \tag{11.61}$$

In summary, here are the steps of the EM algorithm.

1. Initialize parameters $\theta_{old} = \{\mathbf{W}_{old}, \mathbf{\Psi}_{old}\}$.
2. E-step:
   Find $E_{z|x}(\mathbf{z})$ and $E_{z|x}(\mathbf{z}\mathbf{z}^T)$ in Eq (11.58) based on $\mathbf{m}_{z|x_n}$ and $\mathbf{\Sigma}_{z|x_n}$ in Eq. (11.49), which in turn is based on $\boldsymbol{\theta}_{old} = \{\mathbf{W}_{old}, \mathbf{\Psi}_{old}\}$.
3. M-step:
   Find $\boldsymbol{\theta}_{new} = \{\mathbf{W}_{new}, \mathbf{\Psi}_{new}\}$ in Eqs (11.57) and (11.61), based on $E_{z|x}(\mathbf{z})$ and $E_{z|x}(\mathbf{z}\mathbf{z}^T)$.
4. Terminate if convergence criterion is satisfied, otherwise replace θ_{old} by θ_{new} and return to the E-step.

As the E-step and M-step of are interdependent on each other, they need to be carried out iteratively based on certain initial guess of the parameters in $\boldsymbol{\theta}$. In general, the EM algorithm is an iterative method for obtaining the ML or MAP estimates of the parameters in a statistical model, based on some unobserved latent variables. The EM algorithm is widely used in ML and data science, such as in the method of probabilistic PCA considered in the following section, and the Gaussian mixture model for clustering analysis to be considered in Section 15.2.

11.4 Probabilistic PCA

The method of *probabilistic PCA (PPCA)* can be considered as a special case of factor analysis based on the model $\mathbf{x} = \mathbf{W}\mathbf{z} + \mathbf{e}$, when the random noise $\mathbf{e}$ is assumed to have an isotropic normal distributions with covariance matrix $\boldsymbol{\Psi} = \varepsilon\mathbf{I}$. More specially, if we further let $\varepsilon \to 0$ approach zero, the iteration of the EM algorithm for the PPCA becomes extremely simple, as we will see in the following.

Same as in Eqs (11.37) and (11.39) for FA, here we also have the normal distribution of both $\mathbf{z}$ and $\mathbf{e}$

$$p(\mathbf{z}) = \mathcal{N}(\mathbf{0}, \mathbf{I}), \qquad p(\mathbf{e}) = \mathcal{N}(\mathbf{0}, \varepsilon\mathbf{I}) \tag{11.62}$$

and the normal distribution of the observed data $\mathbf{x} = \mathbf{W}\mathbf{z} + \mathbf{e}$, same as in Eq. (11.43)

$$p(\mathbf{x}) = \mathcal{N}(\mathbf{0}, \mathbf{W}\mathbf{W}^T + \varepsilon\mathbf{I}) \tag{11.63}$$

The goal here is to estimate ε and $\mathbf{W}$ as the model parameter, based on the observed data dataset $\mathbf{X} = [\mathbf{x}_1, \cdots, \mathbf{x}_N]$.

Now Eqs (11.49) through (11.51) are rewritten as

$$\begin{cases} \mathbf{m}_{z|x_n} = E_{z|x_n}(\mathbf{z}) = \mathbf{B}\mathbf{x}_n \\ \boldsymbol{\Sigma}_{z|x_n} = Cov_{z|x_n}(\mathbf{z}) = \mathbf{I} - \mathbf{B}\mathbf{W} \end{cases} \tag{11.64}$$

$$\mathbf{B} = \mathbf{W}^T(\mathbf{W}\mathbf{W}^T + \varepsilon\mathbf{I})^{-1} \tag{11.65}$$

$$(\mathbf{W}\mathbf{W}^T + \varepsilon\mathbf{I})^{-1} = \varepsilon^{-1}\mathbf{I} - \varepsilon^{-1}\mathbf{W}(\varepsilon\mathbf{I} + \mathbf{W}^T\mathbf{W})^{-1}\mathbf{W}^T \tag{11.66}$$

The two steps of the EM algorithm become:

- E-step: Get the expectation of the log-likelihood by replacing $\boldsymbol{\Psi}$ in Eq. (11.54) by $\varepsilon\mathbf{I}$

$$Q = -\frac{dN}{2}\log\varepsilon - \frac{1}{2\varepsilon}\sum_{n=1}^{N} E_{z|x_n}\left[\mathbf{x}_n^T\mathbf{x}_n - 2\mathbf{x}_n^T\mathbf{W}\mathbf{z} + \mathbf{z}^T\mathbf{W}^T\mathbf{W}\mathbf{z}\right] \tag{11.67}$$

- M-step: Find the optimal model parameters $\mathbf{W}$ and ε that maximize the expectation of the log-likelihood Q, by setting the derivative of Q with respect to the parameters to zero and solve the resulting equations. For $\mathbf{W}$, the result is the same as Eq. (11.57)

$$\mathbf{W}_{new} = \left(\sum_{n=1}^{N} \mathbf{x}_n E_{z|x_n}[\mathbf{z}^T]\right) \left(\sum_{n=1}^{N} E_{z|x_n}[\mathbf{z}\mathbf{z}^T]\right)^{-1} \tag{11.68}$$

where

$$\begin{cases} E_{z|x_n}[\mathbf{z}] & = & \mathbf{m}_{z|x_n} = \mathbf{B}\mathbf{x}_n \\ E_{z|x_n}[\mathbf{z}\mathbf{z}^T] & = & \mathbf{\Sigma}_{z|x_n} + \mathbf{m}_{z|x_n}\mathbf{m}_{z|x_n}^T \end{cases} \tag{11.69}$$

To find ε, we solve the equation

$$\frac{\partial Q}{\partial \varepsilon} = -\frac{dN}{2\varepsilon} + \frac{1}{2\varepsilon^2}\sum_{n=1}^{N} E_{z|x_n}\left[\mathbf{x}_n^T\mathbf{x}_n - 2\mathbf{x}_n^T\mathbf{W}\mathbf{z} + \mathbf{z}^T\mathbf{W}^T\mathbf{W}\mathbf{z}\right] = 0 \tag{11.70}$$

and get

$$\begin{aligned} \varepsilon &= \frac{1}{dN}\sum_{n=1}^{N}\left[\mathbf{x}_n^T\mathbf{x}_n - 2\mathbf{x}_n^T\mathbf{W}E_{z|x_n}(\mathbf{z}) + tr(\mathbf{W}^T E_{z|x_n}(\mathbf{z}\mathbf{z}^T)\mathbf{W})\right] \\ &= \frac{1}{dN}\sum_{n=1}^{N}\left[||\mathbf{x}_n||^2 - 2E_{z|x_n}(\mathbf{z}^T)\mathbf{W}^T\mathbf{x}_n + tr(E_{z|x_n}(\mathbf{z}\mathbf{z}^T)\mathbf{W}^T\mathbf{W})\right] \end{aligned} \tag{11.71}$$

where we have used the fact that $\mathbf{a}^T\mathbf{b} = tr(\mathbf{a}\mathbf{b}^T)$.

In summary, here are the steps in the EM algorithm for PPCA.

1. Initialize parameters $\boldsymbol{\theta}_{old} = \{\mathbf{W}_{old}, \varepsilon_{old}\}$.
2. E-step: Based on $\boldsymbol{\theta}_{old}$, find $\mathbf{m}_{z|x_n}$ and $\mathbf{\Sigma}_{z|x_n}$ in Eq. (11.64), and $E_{z|x_n}(\mathbf{z})$ and $E_{z|x_n}(\mathbf{z}\mathbf{z}^T)$ in Eq. (11.69).
3. M-step: Find $\boldsymbol{\theta}_{new} = \{\mathbf{W}_{new}, \varepsilon_{new}\}$ by evaluating Eqs (11.68) and (11.71).
4. Terminate if convergence criterion is satisfied, otherwise replace θ_{old} by θ_{new} and return to step 2.

Specially, if we further assume $\varepsilon \to 0$ and the latent variables in $\mathbf{z}$ are deterministic, that is, $E_{z|x}(\mathbf{z}) = \mathbf{z}$. At the limit where $\varepsilon = 0$, $\mathbf{x}$ is simply a linear combination of the latent variables in $\mathbf{z}$

$$\mathbf{x} = \mathbf{W}\mathbf{z} + \varepsilon\mathbf{I} \overset{\varepsilon\to 0}{\Longrightarrow} \mathbf{W}\mathbf{z} = [\mathbf{w}_1, \cdots, \mathbf{w}_{d'}]\begin{bmatrix} z_1 \\ \vdots \\ z_{d'} \end{bmatrix} = \sum_{i=1}^{d'} z_i\mathbf{w}_i \tag{11.72}$$

Comparing this expression of $\mathbf{x}$ with Eq. (10.45)

$$\mathbf{x} = \mathbf{V}'\mathbf{y} = \sum_{i=1}^{d'} y_i\mathbf{v}_i \tag{11.73}$$

we see that the column vectors in $\mathbf{W} = [\mathbf{w}_1, \cdots, \mathbf{w}_{d'}]$ for the PPCA are similar to the column vectors of $\mathbf{V}$, the eigenvector matrix of $\mathbf{\Sigma}_x$, in the linear PCA, as both of them can be viewed as the principal directions, an alternative set of basis vectors that also span the same d-dimensional feature space in which $\mathbf{x}$ resides. When $\mathbf{x}$ is expressed as a linear combination of these basis vectors, it can be approximated by only $d' < d$ of the d such basis vectors weighted by the d' principal components in $\mathbf{y} = [y_1, \cdots, y_{d'}]^T$ in linear PCA, or the d' factors in $\mathbf{z} = [z_1, \cdots, z_{d'}]^T$ in PPCA.

Here are the two EM steps for the PPCA.

- E-step: If $\varepsilon \to 0$, Eq. (11.64) can be written as

$$\begin{aligned} \mathbf{m}_{z|x_n} &= E_{z|x_n}(\mathbf{z}) = \mathbf{z}_n = \mathbf{B}\mathbf{x}_n = \mathbf{W}^T(\mathbf{W}\mathbf{W}^T + \varepsilon\mathbf{I})^{-1}\mathbf{x}_n \\ &\stackrel{\varepsilon \to 0}{\Longrightarrow} \mathbf{W}^T(\mathbf{W}\mathbf{W}^T)^{-1}\mathbf{x}_n \qquad (n = 1, \cdots, N) \end{aligned} \tag{11.74}$$

All N such equations can be written in matrix form as

$$\mathbf{Z} = [\mathbf{z}_1, \cdots, \mathbf{z}_n] = \mathbf{W}^T(\mathbf{W}\mathbf{W}^T)^{-1}[\mathbf{x}_1, \cdots, \mathbf{x}_N] = \mathbf{W}^T(\mathbf{W}\mathbf{W}^T)^{-1}\mathbf{X} \tag{11.75}$$

As $\mathbf{W}$ is a $d \times d'$ matrix ($d' < d$), the rank of the $d \times d$ matrix $\mathbf{W}\mathbf{W}^T$ is at most d', that is, the inverse of $\mathbf{W}\mathbf{W}^T$ does not exist and we cannot find $\mathbf{z}_n = \mathbf{W}^T(\mathbf{W}\mathbf{W}^T)^{-1}\mathbf{x}_n$. In fact the transformation matrix $\mathbf{W}^T(\mathbf{W}\mathbf{W}^T)^{-1}$ is the *right pseudo inverse* (Section A.6.1) of the $d \times d'$ matrix $\mathbf{W}$, which exists only if $d' \geq d$.

However, we note that the model $\mathbf{x} = \mathbf{W}\mathbf{z}$ is an over-constrained linear system of d equations but only $d' < d$ variables in $\mathbf{z} = [z_1, \cdots, z_{d'}]^T$, and the least-squares solution that minimizes the squared error $||\mathbf{x} - \mathbf{W}\mathbf{z}||^2$ can be found by the *left pseudo inverse* (Section A.6.1) of $\mathbf{W}$

$$\mathbf{z}_n = (\mathbf{W}^T\mathbf{W})^{-1}\mathbf{W}^T\mathbf{x}_n = \mathbf{W}^{-}\mathbf{x}_n \qquad (n = 1, \cdots, N) \tag{11.76}$$

and we can get the latent variables in $\mathbf{Z} = [\mathbf{z}_1, \cdots, \mathbf{z}_N]$ given all N data points in $\mathbf{X} = [\mathbf{x}_1, \cdots, \mathbf{x}_N]$

$$\mathbf{Z} = (\mathbf{W}^T\mathbf{W})^{-1}\mathbf{W}^T\mathbf{X} = \mathbf{W}^{-}\mathbf{X} \tag{11.77}$$

which is based on the old value of $\mathbf{W}$ previously estimated.

- M-step: Now Eq. (11.68) can be written as

$$\begin{aligned} \mathbf{W} &= \left(\sum_{n=1}^{N} \mathbf{x}_n E_{z|x_n}(\mathbf{z}^T)\right)\left(\sum_{n=1}^{N} E_{z|x_n}(\mathbf{z}\mathbf{z}^T)\right)^{-1} \\ &= [\mathbf{x}_1, \cdots, \mathbf{x}_N]\begin{bmatrix} \mathbf{z}_1^T \\ \vdots \\ \mathbf{z}_N^T \end{bmatrix}\left([\mathbf{z}_1, \cdots, \mathbf{z}_N]\begin{bmatrix} \mathbf{z}_1^T \\ \vdots \\ \mathbf{z}_N^T \end{bmatrix}\right)^{-1} \\ &= \mathbf{X}\mathbf{Z}^T(\mathbf{Z}\mathbf{Z}^T)^{-1} = \mathbf{X}\mathbf{Z}^{-} \end{aligned} \tag{11.78}$$

by which $\mathbf{W}$ is updated based on $\mathbf{Z}$ found previously in the E-step. We note that $\mathbf{Z}^T(\mathbf{Z}\mathbf{Z}^T)^{-1}$ is the right pseudo inverse of the $d \times N$ matrix $\mathbf{Z}$, which exists if $d \leq N$.

Again, as the E-step and M-steps are interdependent on each other, they need to be carried out iteratively

$$\begin{aligned} \text{E-step:} \qquad & \mathbf{Z} = (\mathbf{W}^T\mathbf{W})^{-1}\mathbf{W}^T\mathbf{X} = \mathbf{W}^-\mathbf{X} \\ \text{M-step:} \qquad & \mathbf{W} = \mathbf{X}\mathbf{Z}^T(\mathbf{Z}\mathbf{Z}^T)^{-1} = \mathbf{X}\mathbf{Z}^- \end{aligned} \tag{11.79}$$

We note that given the dataset $\mathbf{X}$, the FA model $\mathbf{x} = \mathbf{W}\mathbf{z}$ can be expressed as an over-determined matrix equation

$$\mathbf{X}_{d\times N} = [\mathbf{x}_1, \cdots, \mathbf{x}_N] = \mathbf{W}_{d\times d'}[\mathbf{z}_1, \cdots, \mathbf{z}_N] = \mathbf{W}_{d\times d'}\mathbf{Z}_{d'\times N} \tag{11.80}$$

or, taking transpose of the equation, we get

$$\mathbf{X}^T_{N\times d} = \mathbf{Z}^T_{N\times d'}\mathbf{W}^T_{d'\times d} \tag{11.81}$$

Given $\mathbf{X}$, this equation can be interpreted in two ways.

- If $\mathbf{W}$ is available, this is a system of d equations but $d' < d$ unknowns in each of the N columns of $\mathbf{Z}$, which is solved as in the E-step.
- If $\mathbf{Z}$ is available, this is a system of N equations but $d' < N$ unknowns in each of the d rows of $\mathbf{W}$, which is solved as in the M-step.

In either step, the pseudo inverse method can be used to minimize the squared error $||\mathbf{X} - \mathbf{W}\mathbf{Z}||^2$.

When compared to the regular PCA method that finds all d eigenvalues and the corresponding eigenvectors all at once by solving the eigenequation $\mathbf{\Sigma}_x\mathbf{w} = \lambda\mathbf{w}$, the PPCA method only finds the d' column vectors of matrix $\mathbf{W}$ as the basis vectors, as the basis vectors that span a subspace with much reduced dimensionality but containing the most essential information in the data. However, unlike the PCA, the basis vectors in PPCA are not necessarily orthogonal to each other.

The implementation of the PPCA is extremely simple, as shown in the following MATLAB code below. Based on some initialized $\mathbf{W}$, the EM iteration is composed of the E-step and M-step:

```
while er>tol
    Z=inv(W'*W)*W'*X;          % find Z given W in E-step
    Wnew=X*Z'*inv(Z*Z');       % find W given Z in M-step
    er=norm(Wnew-W);           % the LS error
    W=Wnew;
end
```

Figures 11.5 and 11.6 show the first few iterations of the PPCA based on the EM method presented, for a set of data points in $\mathbf{X} = [\mathbf{x}_1, \cdots, \mathbf{x}_N]$ in 2-D and 3-D. The straight lines in the plots represents the column vectors of $\mathbf{W}$, which is randomly initialized, but quickly approach the directions corresponding to the principal components obtained by the PCA method, along which the data points are most widely spread.

Figure 11.7 shows the dataset of 10 handwritten numbers from 0 to 9, containing 2240 data points in a 256-D space, but now linearly mapped to a 3-D space spanned by $d' = 3$ factors found by the PPCA method.

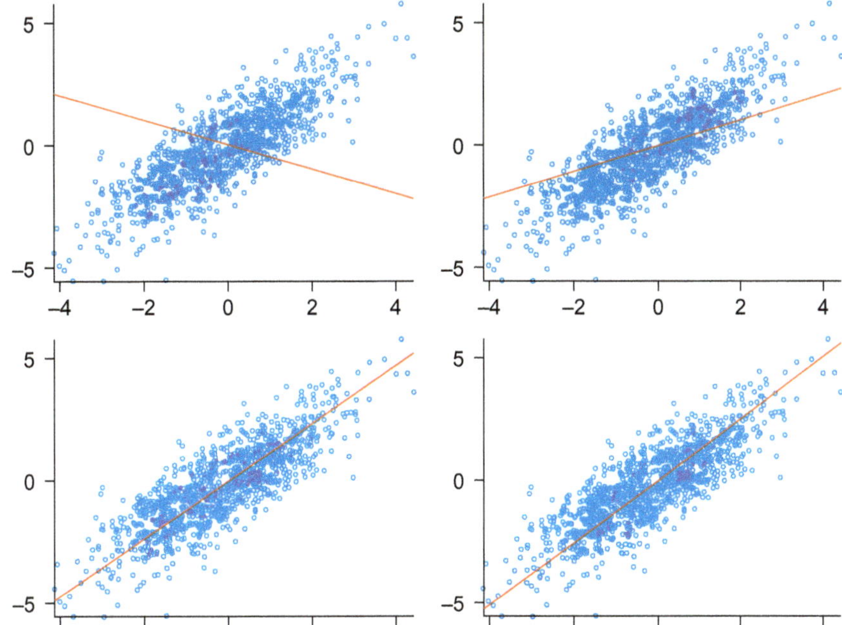

Fig. 11.5 Column vectors in **W** iteratively approach the principal directions of dataset (2-D).

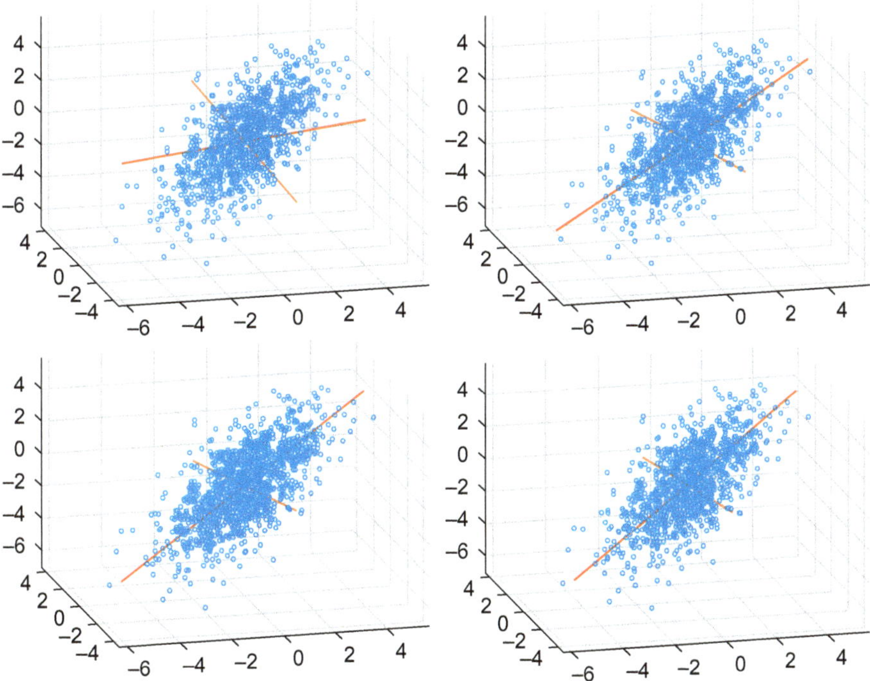

Fig. 11.6 Column vectors in **W** iteratively approach the principal directions of dataset (3-D).

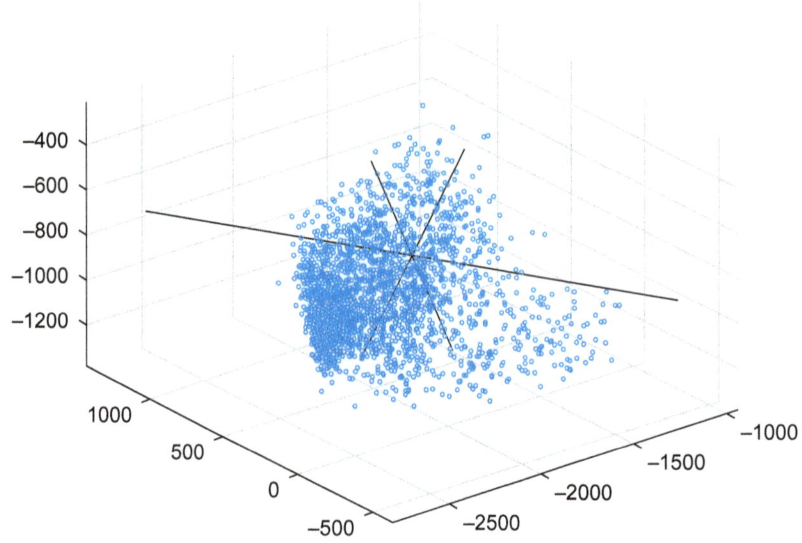

Fig. 11.7 First three components of the handwritten digits dataset by the PPCA.

11.5 Classical Multidimensional Scaling

The goal of *multidimensional scaling (MDS)* is to reduce the dimensionality of the dataset representing a set of N objects of interest, each described by a set of d features and represented as a vector $\mathbf{x} = [x_1, \cdots, x_d]^T$ in a d-dimensional space, while the pairwise similarity relationship between any two of these objects are preserved. MDS can be used to visualize datasets in a high dimensional space. Specially, here we consider the *classical MDS (cMDS)*, also known as *principal coordinates analysis (PCoA)*, as one of the methods in MDS.

Specifically, we represent the pairwise similarity of a set of N objects $\mathbf{X} = [\mathbf{x}_1, \cdots, \mathbf{x}_N]$ by the $N \times N$ similarity matrix $\mathbf{D}_x = [\ d_{ij}^x\ ]$, of which the ijth component d_{ij}^x is certain measurement of the similarity between $\mathbf{x}_i$ and $\mathbf{x}_j$. The goal is to map the dataset $\mathbf{X}$ in the d-dimensional space into $\mathbf{Y} = [\mathbf{y}_1, \cdots, \mathbf{y}_N]$ in a lower d'-dimensional space, in which the similarity matrix $\mathbf{D}_x$ is optimally approximated by $\mathbf{D}_y = [\ d_{ij}^y]$. When $d' \le 3$, these data points in $\mathbf{Y}$ can be visualized. MDS can be considered as a method for dimension reduction. In some cases only the pairwise similarities d_{ij}^x $(i, j = 1, \cdots, N)$ are given, while the N objects in $\mathbf{X}$ are not explicitly given, and the dimensionality d may be unknown.

We assume the given pairwise similarity is the Euclidean distance $d_{ij} = ||\mathbf{x}_i - \mathbf{x}_j||_2$ between $\mathbf{x}_i$ and $\mathbf{x}_j$

$$d_{ij}^2 = ||\mathbf{x}_i - \mathbf{x}_j||^2 = (\mathbf{x}_i - \mathbf{x}_j)^T(\mathbf{x}_i - \mathbf{x}_j) = \mathbf{x}_i^T\mathbf{x}_i - 2\mathbf{x}_i^T\mathbf{x}_j + \mathbf{x}_j^T\mathbf{x}_j \tag{11.82}$$

and obtain

$$\mathbf{D}_x^2 = \begin{bmatrix} & & \\ & d_{ij}^2 & \\ & & \end{bmatrix}_{N\times N} = \begin{bmatrix} & & \\ & \mathbf{x}_i^T\mathbf{x}_i - 2\mathbf{x}_i^T\mathbf{x}_j + \mathbf{x}_j^T\mathbf{x}_j & \\ & & \end{bmatrix}_{N\times N} = \mathbf{X}_r - 2\mathbf{X}^T\mathbf{X} + \mathbf{X}_c \tag{11.83}$$

where

$$\mathbf{X}_r = \begin{bmatrix} ||\mathbf{x}_1||^2 & \cdots & ||\mathbf{x}_1||^2 \\ \vdots & \ddots & \vdots \\ ||\mathbf{x}_N||^2 & \cdots & ||\mathbf{x}_N||^2 \end{bmatrix} = \mathbf{X}_c^T$$

$$\mathbf{X}^T\mathbf{X} = \begin{bmatrix} \mathbf{x}_1{}^T \\ \vdots \\ \mathbf{x}_N^T \end{bmatrix} [\mathbf{x}_1, \cdots, \mathbf{x}_N] = \begin{bmatrix} \mathbf{x}_1^T\mathbf{x}_1 & \cdots & \mathbf{x}_1^T\mathbf{x}_N \\ \vdots & \ddots & \vdots \\ \mathbf{x}_N^T\mathbf{x}_1 & \cdots & \mathbf{x}_N^T\mathbf{x}_N \end{bmatrix} \tag{11.84}$$

Our goal is to find $\mathbf{Y} = [\mathbf{y}_1, \cdots, \mathbf{y}_N]$ in a lower dimensional space so that its pairwise similarity matrix $\mathbf{D}_y$ matches $\mathbf{D}_x$ optimally, in the sense that the following objective function is minimized

$$J(\mathbf{Y}) = ||\mathbf{D}_y^2 - \mathbf{D}_x^2|| \tag{11.85}$$

We note that such a solution $\mathbf{Y}$ is not unique, as any translated version of $\mathbf{Y}$ has the same pairwise similarity matrix and is also a solution. We therefore seek to find a unique solution $\mathbf{Y}$ in which all data points are centralized, i.e, the mean of all points in $\mathbf{Y}$ is zero

$$\bar{\mathbf{y}} = \frac{1}{N}\sum_{n=1}^{N} \mathbf{y}_n = \mathbf{0} \tag{11.86}$$

To do so, we introduce an $N \times N$ symmetric centering matrix $\mathbf{C} = \mathbf{I} - \mathbf{1}/N = \mathbf{C}^T$ (Section A.8.1), where $\mathbf{1} = \mathbf{e}\mathbf{e}^T$ is an $N \times N$ matrix with all components equal 1. Pre-multiplying $\mathbf{C}$ to a column vector $\mathbf{a} = [a_1, \cdots, a_N]^T$ removes the mean $\bar{a} = \sum_{i=1}^N a_i/N$ of $\mathbf{a}$ from each of its component

$$\mathbf{Ca} = \mathbf{Ia} - \frac{1}{N}\mathbf{1a} = \begin{bmatrix} 1 & \cdots & 0 \\ \vdots & \ddots & \vdots \\ 0 & \cdots & 1 \end{bmatrix} \begin{bmatrix} a_1 \\ \vdots \\ a_N \end{bmatrix} - \frac{1}{N} \begin{bmatrix} 1 & \cdots & 1 \\ \vdots & \ddots & \vdots \\ 1 & \cdots & 1 \end{bmatrix} \begin{bmatrix} a_1 \\ \vdots \\ a_N \end{bmatrix} = \begin{bmatrix} a_1 - \bar{a} \\ \vdots \\ a_N - \bar{a} \end{bmatrix} \tag{11.87}$$

Taking transpose of Eq. (11.87) we get

$$(\mathbf{Ca})^T = \mathbf{a}^T\mathbf{C} = [a_1 - \bar{a}, \cdots, a_N - \bar{a}] \tag{11.88}$$

that is post-multiplying $\mathbf{C}$ to a row vector $\mathbf{a}^T = [a_1, \cdots, a_N]$ removes the mean $\bar{a}$ from each of its component.

We can now post-multiply $\mathbf{C}$ to $\mathbf{X}$ to remove the mean of each row of $\mathbf{X}$

$$\mathbf{XC} = [\mathbf{x}_1, \cdots, \mathbf{x}_N]\mathbf{C} = [\bar{\mathbf{x}}_1, \cdots, \bar{\mathbf{x}}_N] = \bar{\mathbf{X}} \tag{11.89}$$

that is the nth column becomes

$$\bar{\mathbf{x}}_n = \mathbf{x}_n - \bar{\mathbf{x}}, \quad \text{where} \quad \bar{\mathbf{x}} = \frac{1}{N}\sum_{n=1}^{N} \mathbf{x}_n \tag{11.90}$$

and pre-multiply $\mathbf{C}$ to $\mathbf{X}^T$ to remove the mean of each column of $\mathbf{X}^T$

$$\mathbf{CX}^T = \bar{\mathbf{X}}^T \tag{11.91}$$

We further apply double centering to $\mathbf{D}_x^2$ by pre- and post-multiplying $\mathbf{C}$ and obtain

$$\begin{aligned}\mathbf{CD}_x^2\mathbf{C} &= \mathbf{C}\left(\mathbf{X}_r - 2\mathbf{X}^T\mathbf{X} + \mathbf{X}_c\right)\mathbf{C} = \mathbf{CX}_r\mathbf{C} - 2\mathbf{C}\left(\mathbf{X}^T\mathbf{X}\right)\mathbf{C} + \mathbf{CX}_c\mathbf{C} \\ &= -2\mathbf{C}\left(\mathbf{X}^T\mathbf{X}\right)\mathbf{C} = -2\,(\mathbf{XC})^T\,(\mathbf{XC}) = -2\bar{\mathbf{X}}^T\bar{\mathbf{X}}\end{aligned}$$

Note that the first term $\mathbf{CX}_r\mathbf{C} = \mathbf{C}(\mathbf{X}_r\mathbf{C}) = \mathbf{C0} = \mathbf{0}$, as all components of each row of $\mathbf{X}_r$ are the same as their mean, and removing the mean results in a zero row vector, $\mathbf{X}_r\mathbf{C} = \mathbf{0}$; and the last term $\mathbf{CX}_c\mathbf{C} = (\mathbf{CX}_c)\mathbf{C} = \mathbf{0C} = \mathbf{0}$, as all components of each column of $\mathbf{X}_c$ are the same as their mean, and removing the mean results in a zero column vector.

Now the similarity matrix $\mathbf{D}_x^2$ of all centralized data points in $\bar{\mathbf{X}}$ with zero mean can be represented by the *Gram matrix* of $\bar{\mathbf{X}}$

$$\mathbf{G}_{\bar{x}} = \bar{\mathbf{X}}^T\bar{\mathbf{X}} = -\frac{1}{2}\mathbf{CD}_x^2\mathbf{C} \tag{11.92}$$

We further assume all data points in $\mathbf{Y}$ are also centered with zero mean and their corresponding similarity matrix is also represented by $\mathbf{G}_{\bar{y}}$, then the objective function can be expressed as

$$J(\mathbf{Y}) = ||\mathbf{G}_{\bar{y}} - \mathbf{G}_{\bar{x}}|| = ||\bar{\mathbf{Y}}^T\bar{\mathbf{Y}} - \mathbf{G}_{\bar{x}}|| \tag{11.93}$$

To find $\bar{\mathbf{Y}}$ that minimizes $J(\mathbf{Y})$, we first consider the equation $\bar{\mathbf{Y}}^T\bar{\mathbf{Y}} - \mathbf{G}_{\bar{x}} = \mathbf{0}$ so that $J(\mathbf{Y}) = 0$. In other words, we want to express $\mathbf{G}_{\bar{x}}$ as the inner product of some matrix with itself. To do so, we carry out eigenvalue decomposition of $\mathbf{G}_x$ to find its eigenvalue matrix $\mathbf{\Lambda}$ and the orthogonal eigenvector matrix $\mathbf{V}$ satisfying $\mathbf{G}_x\mathbf{V} = \mathbf{V\Lambda}$, that is,

$$\begin{aligned}\mathbf{G}_{\bar{x}} &= \mathbf{V\Lambda V}^{-1} = \mathbf{V\Lambda V}^T = [\mathbf{v}_1, \cdots, \mathbf{v}_N]\begin{bmatrix}\lambda_1 & \cdots & 0 \\ \vdots & \ddots & \vdots \\ 0 & \cdots & \lambda_N\end{bmatrix}\begin{bmatrix}\mathbf{v}_1^T \\ \vdots \\ \mathbf{v}_N^T\end{bmatrix} \\ &= \left(\mathbf{V\Lambda}^{1/2}\right)\left(\mathbf{\Lambda}^{1/2}\mathbf{V}^T\right) = \left(\mathbf{\Lambda}^{1/2}\mathbf{V}^T\right)^T\left(\mathbf{\Lambda}^{1/2}\mathbf{V}^T\right)\end{aligned}$$

Note that as $\mathbf{G}_{\bar{x}}$ is symmetric, the eigenvalues are real and the eigenvectors are orthogonal, that is, $\mathbf{V}^T = \mathbf{V}^{-1}$ is the orthogonal matrix. Now we can find the desired $\mathbf{Y}$ as

$$\mathbf{Y} = \mathbf{\Lambda}^{1/2}\mathbf{V}^T = \begin{bmatrix}\sqrt{\lambda_1} & \cdots & 0 \\ \vdots & \ddots & \vdots \\ 0 & \cdots & \sqrt{\lambda_N}\end{bmatrix}\begin{bmatrix}\mathbf{v}_1^T \\ \vdots \\ \mathbf{v}_N^T\end{bmatrix} = \begin{bmatrix}\sqrt{\lambda_1}\mathbf{v}_1^T \\ \vdots \\ \sqrt{\lambda_N}\mathbf{v}_N^T\end{bmatrix} \tag{11.94}$$

Each column $\mathbf{y}_i$ in $\mathbf{Y} = [\mathbf{y}_1, \cdots, \mathbf{y}_N]$ is an N–D vector. If all N dimensions of these $\mathbf{y}_n$ are used, $\mathbf{Y}^T\mathbf{Y} = \mathbf{G}_x$ and $J(\mathbf{Y} = 0$. To reduce the dimensionality to $d' < d$, we construct $\mathbf{Y}'$ by the first d' rows of $\mathbf{Y}$ corresponding to the d' greatest eigenvalues of $\lambda_1 \geq \cdots \geq \lambda_{d'} \geq \cdots \geq \lambda_N$ and their corresponding eigenvectors $\mathbf{v}_1, \cdots, \mathbf{v}_{d'}$. The error is

$$J(\mathbf{Y}') = ||\mathbf{Y}'^T\mathbf{Y}' - \mathbf{Y}^T\mathbf{Y}|| = ||\sum_{i=1}^{d'} \lambda_i \mathbf{v}_i \mathbf{v}^T - \sum_{i=1}^{N} \lambda_i \mathbf{v}_i \mathbf{v}^T|| = ||\sum_{i=d'}^{N} \lambda_i \mathbf{v}_i \mathbf{v}^T|| \tag{11.95}$$

In summary, here is the PCoA algorithm:

- given the pairwise similarity, an $N \times N$ matrix $\mathbf{D}_x = [\ d_{ij}\ ]$, construct the squared proximity matrix $\mathbf{D}_x^2 = [\ d_{ij}^2\ ]$;
- apply double centering to get $\mathbf{G}_x = -\mathbf{C}\mathbf{D}_x^2\mathbf{C}/2$;
- find the d' greatest eigenvalues $\lambda_1, \cdots, \lambda_{d'}$ of $\mathbf{G}_x$, and the corresponding eigenvectors $\mathbf{v}_1, \cdots, \mathbf{v}_{d'}$;
- obtain the map

$$\mathbf{Y}_{d' \times N} = \mathbf{\Lambda}_{d'}^{1/2}\mathbf{V}_{d'} = \begin{bmatrix} \sqrt{\lambda_1}\mathbf{v}_1^T \\ \vdots \\ \sqrt{\lambda_{d'}}\mathbf{v}_{d'}^T \end{bmatrix} \tag{11.96}$$

The MATLAB functions for the PCA and PCoA are listed in the following

```
function Y=PCA(X)
    [V D]=eig(cov(X'));            % solve eigenequation of covariance
    [d idx]=sort(diag(D),'descend');
                                   % sort eigenvalues in descend order
    V=V(:,idx);                    % reorder corresponding eigenvectors
    Y=V'*X;                        % carry out KLT
end
```

```
function Y=PCoA(X)
    N=size(X,2);
    D=zeros(N);                    % pairwise similarity matrix
    C=eye(N)-ones(N)/N;            % centering matrix
    for i=1:N
        for j=1:N
            D(i,j)=(norm(X(:,i)-X(:,j)))^2;
        end
    end
    B=-C*D*C/2;
    [V D]=eig(B);                  % solve eigenequation for matrix B
    [d idx]=sort(diag(D),'descend');
                                   % sort eigenvalues in descend order
    V=V(:,idx);                    % reorder eigenvectors
    Y=sqrt(D)*V';                  % carry out PCoA
end
```

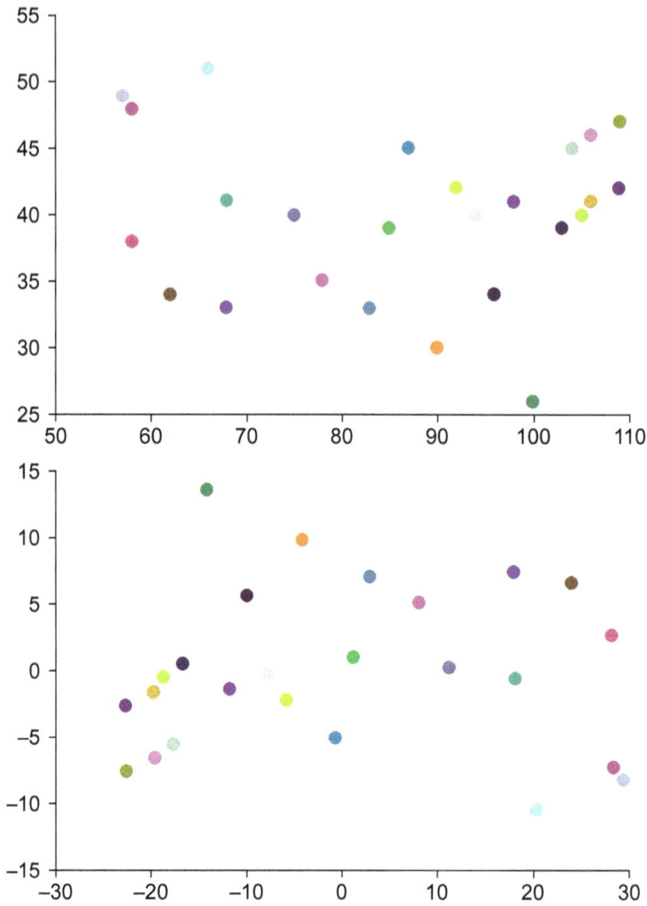

Fig. 11.8 Reconstruction of the locations of north America cities by the MDS.

In Fig. 11.8, the top panel shows the locations of some major cities in North America, and the bottom panel shows the reconstructed city locations by the MDS method based on the pairwise distance of these cities. Note that the reconstructed map is a rotated and centralized version of the original map.

The methods of PCA and PCoA are compared when applied to the 4-D iris dataset and the 256-D handwritten digits dataset, with their corresponding visualizations in both 2-D and 3-D shown in Figs 11.9 and 11.10.

11.6 t-Distributed Stochastic Neighbor Embedding

The method of PCA considered previously is often used to reduce the dimensionality of a given dataset $\mathbf{X} = [\mathbf{x}_1, \cdots, \mathbf{x}_N]$ by an orthogonal mapping from the high dimensional space to a low dimensional space. However, sometimes this method based on the linear KLT may not be able to properly interpret complex nonlinear relationships between the variables that spanned the feature space, as it may not be able to maintain the relative pairwise distances between the data points in

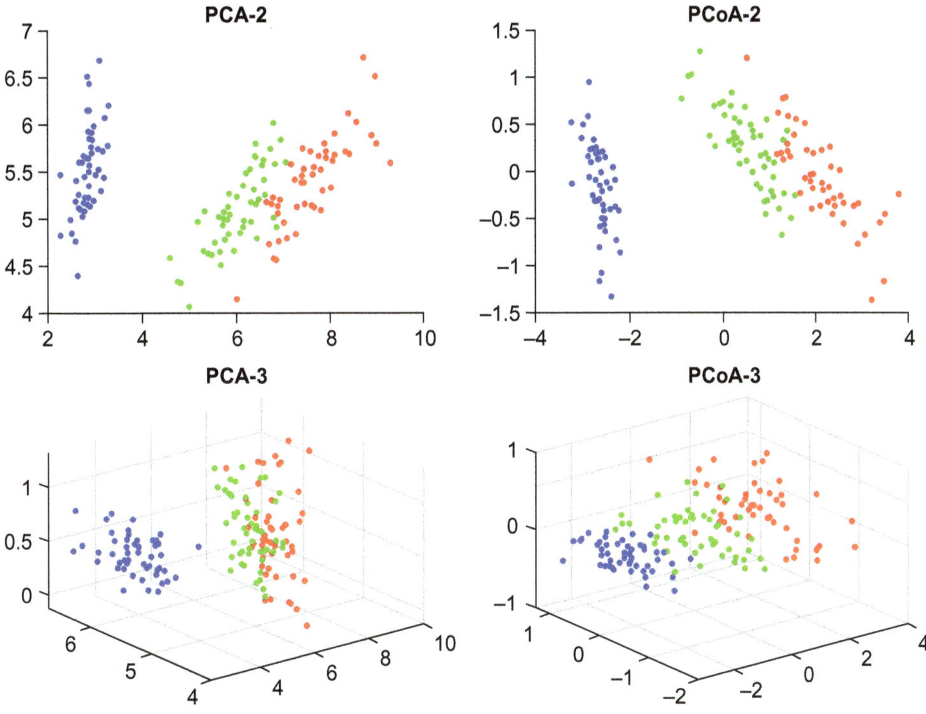

Fig. 11.9 Visualization of the iris dataset based on both PCA and PCoA.

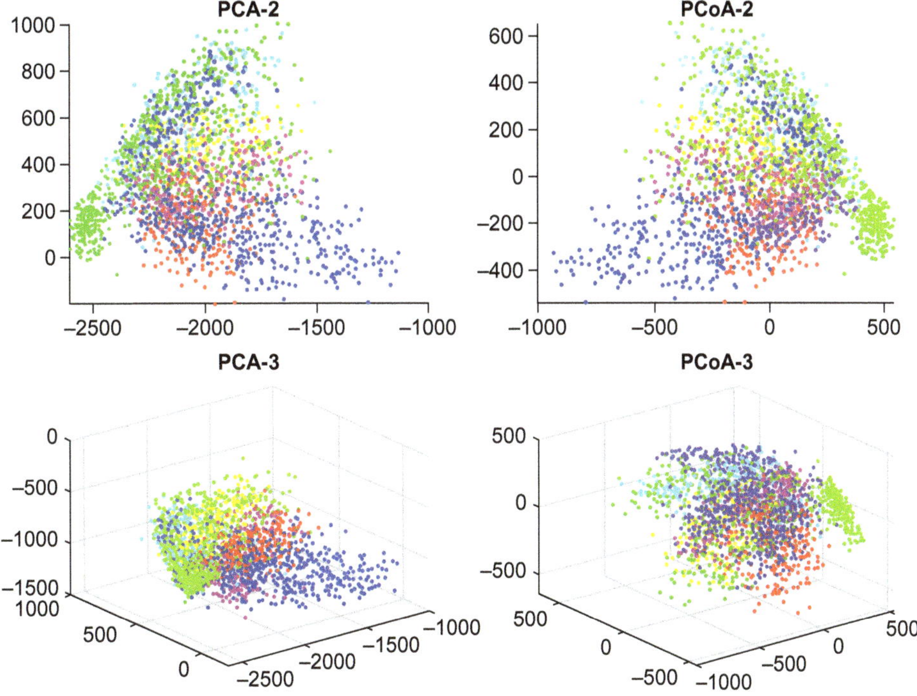

Fig. 11.10 Visualization of the handwritten digits dataset based on both PCA and PCoA.

the resulting lower dimensional space. For example, in the low dimensional space after the transform, those dissimilar data points far apart from each other in a large scale in the original high dimensional space may be well represented in the low dimensional space by the major principal components along the axes with large variances. However, the differences between those points dissimilar with each other in some smaller scales may not be properly represented, or even completely neglected, if their dissimilarity is mostly represented in the dimensions with small variances. In other words, the PCA method tends to emphasize large-scale global structures of the dataset, while ignoring the small-scale local structures, which may be equally important or even of greater interest, and therefore also need to be properly represented.

Also, when the dimensionality is much reduced from a high-D space to a low-D space, the crowding problem occurs. In a high-D space, a data point may have a large number of similar neighbors of roughly equal distance; however, in a low-D space, the number of equal-distance neighbors is significantly reduced (e.g., 8 immediate neighbors in a 3-D space, 4 in a 2-D space, 2 in a 1-D space). Thus the space available for the neighbors of a point is reduced while the dimensionality is reduced. As a result, when a large number of neighbors in the high-D are mapped to a low-D space, they are forced to spread out. Some then may become closer to those that are originally farther away in the high-D space, thereby reducing the gaps between potential clusters of similar points.

In contrast, the t-distributed Stochastic neighbor embedding (tSNE) method is a nonlinear method that is based on probability distributions of the data points being neighbors, and it attempts to preserve the structure at all scales, but emphasizing more on the small-scale structures, by mapping nearby points in high-D space to nearby points in low-D space.

In the high-D space, the similarity between any data point $\mathbf{x}_i$ and each of other points $\mathbf{x}_j$ in the dataset $\mathbf{X}$ is defined based on the Gaussian kernel as a conditional probability

$$p_{j/i} = \frac{\exp(-||\mathbf{x}_i - \mathbf{x}_j||^2/2\sigma_i^2)}{\sum_{k=1}^N \sum_{l=1}^N \exp(-||\mathbf{x}_k - \mathbf{x}_l||^2/2\sigma_i^2)} \tag{11.97}$$

where σ_i^2 is the variance of the data points in the local neighborhood of $\mathbf{x}_i$.

We see that if any $\mathbf{x}_j$ is close to $\mathbf{x}_i$, $||\mathbf{x}_i - \mathbf{x}_j||$ is small and $p_{j/i}$ is large, indicating $\mathbf{x}_j$ is similar to $\mathbf{x}_i$; but if $\mathbf{x}_j$ is far away from $\mathbf{x}_j$, $||\mathbf{x}_i - \mathbf{x}_j||$ is large and p_{ij} is small, indicating $\mathbf{x}_j$ is dissimilar to $\mathbf{x}_i$. Also, as $\sum_i \sum_j p_{ij} = 1$, that is, p_{ij} is normalized and can therefore be considered as a probability that any two points $\mathbf{x}_j$ and $\mathbf{x}_i$ are similar to each other and are therefore neighbors in the feature space.

As a similarity measurement between any two points $\mathbf{x}_i$ and $\mathbf{x}_j$, the conditional probability $p_{j/i}$ defined here needs to be modified for two reasons.

First, the Gaussian kernel p_{ij} should be independent of the local density of the data points, which may vary drastically in different local regions of the feature space, so that similarities of different scales can be equally well represented. To achieve this goal, we adjust the variance σ_i^2 in the Gaussian kernel $p_{j/i}$ for each data point $\mathbf{x}_i$ in such a way that the entropy $H(\mathbf{x}_i) = -\sum_j p_{j/i} \log_2 p_{j/i}$ for all data points are the same as some prespecified value. As shown in Section B.2.1, the entropy of the Gaussian distribution $\mathcal{N}(x, \mu, \sigma)$ is $\log(2\pi e\sigma^2)/2$, a monotonic function of the variance σ^2. The similarity measurement based on $p_{j/i}$ of the same entropy for all data points is essentially the same as the Mahalanobis distance (Eq. (9.12) and the Bhattacharyya distance (Eq. (9.16)), in the sense that these similarity measurements are all normalized by the variance σ^2, and they are therefore independent with respect to the local data density and scale.

Second, as $p_{j/i} \neq p_{i/j}$ as a conditional probability defined here is asymmetrical, it needs to be modified so that it becomes symmetric as a similarity measurement should be. We therefore redefine the probability measurement as the average of two conditional probabilities

$$p_{ij} = \frac{p_{i/j} + p_{j/i}}{2} \tag{11.98}$$

Once the dataset $\mathbf{X}$ in the high dimensional space is mapped to $\mathbf{Y} = [\mathbf{y}_1, \cdots, \mathbf{y}_N]$ in the low dimensional space, we also need to measure the similarity between any two data points. To do so, we further define such a similarity based on Student's t-distribution of degree 1 (Eq. (B.140))

$$q_{ij} = \frac{(1 + ||\mathbf{y}_i - \mathbf{y}_j||^2)^{-1}}{\sum_{k=1}^N \sum_{l=1}^N (1 + ||\mathbf{y}_k - \mathbf{y}_l||^2)^{-1}} = \frac{(1 + d_{ij}^2)^{-1}}{\sum_{k=1}^N \sum_{l=1}^N (1 + d_{ij}^2)^{-1}} = \frac{(1 + d_{ij}^2)^{-1}}{D} \tag{11.99}$$

where D denotes the denominator

$$D = \sum_{k=1}^N \sum_{l=1}^N (1 + d_{kl}^2)^{-1} \tag{11.100}$$

and $d_{kl} = ||\mathbf{y}_k - \mathbf{y}_l|| = \left[(\mathbf{y}_k - \mathbf{y}_l)^T (\mathbf{y}_k - \mathbf{y}_l)\right]^{1/2}$ is the Euclidean distance between any two data points $\mathbf{y}_k$ and $\mathbf{y}_l$. Again as q_{ij} is normalized, it can be interpreted as a probability measurement of the similarity between any two points in the low dimensional space.

The t-distribution with one degree of freedom is also a bell-shaped function, similar to the Gaussian distribution, but it has a lower peak in the center but higher tails on the sides, as shown in Fig. 11.11, which also illustrates why the similarity is measure based on the Gaussian distribution in the high-D space but the t-distribution in the low-D space. Specifically, consider any point x_i and two of its neighbors x_j and x'_j in the high-D space also illustrated in the figure.

- Point $x_j = -0.9$ (red on the left) similar to $x_i = 0$ as measured by the Gaussian in the high-D space is mapped to $y_j = -0.6$ in the low-D space measured by the t-distribution, even closer to $y_i = x_i$.
- Point $x'_j = 1.3$ (red on the right) not very similar to $x_i = 0$ as measured by the Gaussian in the high-D space is mapped to $y'_j = 1.9$ in the low-D space measured by the t-distribution, even farther away from $y_i = x_i$.

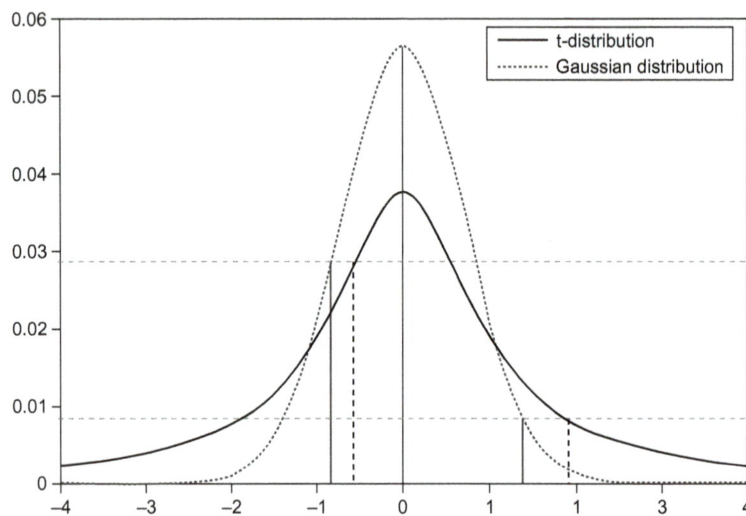

Fig. 11.11 The t-distribution compared to the Gaussian distribution.

In other words, when all data points x_n, $(n = 1, \cdots, N)$ in the high-D space are nonlinearly mapped to y_n, $(n = 1, \cdots, N)$ in the low-D space in such a way that their similarities measured by the Gaussian before the mapping match those measured by the t-distribution after the mapping, similar points become even closer while dissimilar points become farther apart from each other. We note that whether two points are considered to be similar or not depends on the variance of the Gaussian, a hyperparameter that needs to be properly determined based on the specific dataset in question.

The goal of the tSNE method is to find the optimal mapping from all points in dataset $\mathbf{X}$ in the high-D space to $\mathbf{Y}$ in the low-D space that minimizes the discrepancy between the two sets of similarity measurements p_{ij} before the mapping and q_{ij} after the mapping, which can be quantitatively measured by the Kullback–Leibler (KL) divergence (Section B.2.2)

$$J = D_{KL}(P||Q) = \sum_{i=1}^{N}\sum_{j=1}^{N} p_{ij}\log\frac{p_{ij}}{q_{ij}} = \sum_{i=1}^{N}\sum_{j=1}^{N}\left(p_{ij}\log p_{ij} - p_{ij}\log q_{ij}\right) \quad (11.101)$$

The KL divergence is not symmetric as $D_{KL}(P||Q) \neq D_{KL}(Q||P)$, that is, it can not be viewed as a distance between the two sets of probabilities. However, this asymmetry is actually a desired feature. Consider two cases.

- If p_{ij} is large but q_{ij} is small, then D_{KL} is large, and such a mapping is heavily penalized. As the result, similar points that are close to each other in the high-D space tend to be mapped to points also close to each other in the low-D space.
- If p_{ij} is small but q_{ij} is large, then D_{KL} is small, and such a mapping is not heavily penalized. As the result, dissimilar points that are far apart in the high-D space are allowed to be mapped to points closer to each other in the low-D space.

We therefore see that minimizing the KL divergence has the effect of preserving the local structure of the dataset, but not necessarily the global structure.

The optimal points in $\mathbf{Y}$ in the low-D space that minimize the objective function $J = D_{KL}(P||Q)$ can be found by the gradient descent method. The ith component $\partial J/\partial \mathbf{y}_i$ of the gradient can be found by the chain rule, based on $d_{ij} = d_{ji}$ treated as an intermediate variable

$$\frac{\partial J}{\partial \mathbf{y}_i} = \sum_{j=1}^{N} \left(\frac{\partial J}{\partial d_{ij}} \frac{\partial d_{ij}}{\partial \mathbf{y}_i} + \frac{\partial J}{\partial d_{ij}} \frac{\partial d_{ij}}{\partial \mathbf{y}_i} \right) = 2 \sum_{j=1}^{N} \frac{\partial J}{\partial d_{ij}} \frac{\partial d_{ij}}{\partial \mathbf{y}_i} \tag{11.102}$$

Note that $J = D_{KL}(P||Q)$ is a function of both p_{ij} as a constant, and $q_{ij} = (1 + d_{ij}^{-1})/D$, which in turn is a function of d_{ij}. The first derivative is

$$\begin{aligned} \frac{\partial J}{\partial d_{ij}} &= \frac{\partial}{\partial d_{ij}} \left[\sum_{k=1}^{N} \sum_{l=1}^{N} (p_{kl} \log p_{kl} - p_{kl} \log q_{kl}) \right] \\ &= -\frac{\partial}{\partial d_{ij}} \sum_{k=1}^{N} \sum_{l=1}^{N} p_{kl} \log q_{kl} = -\frac{\partial}{\partial d_{ij}} \sum_{k=1}^{N} \sum_{l=1}^{N} p_{kl} (\log(q_{kl} D) - \log D) \\ &= -\sum_{k=1}^{N} \sum_{l=1}^{N} p_{kl} \frac{\partial}{\partial d_{ij}} \log(q_{kl} D) + \sum_{k=1}^{N} \sum_{l=1}^{N} p_{kl} \frac{1}{D} \frac{\partial D}{\partial d_{ij}} \end{aligned} \tag{11.103}$$

To proceed, we consider the two terms separately. There is only one non-zero term in the summation of the first term

$$p_{ij} \frac{1}{q_{ij} D} \frac{\partial}{\partial d_{ij}} (1 + d_{ij}^2)^{-1} = \frac{p_{ij}}{q_{ij}} \frac{2 d_{ij} (1 + d_{ij}^2)^{-2}}{D} \tag{11.104}$$

As $\sum_{k=1}^{N} \sum_{l=1}^{N} p_{kl} = 1$, the second term becomes

$$\frac{1}{D} \frac{\partial D}{\partial d_{ij}} = \frac{1}{D} \frac{\partial}{\partial d_{ij}} \left(\sum_{k=1}^{N} \sum_{l=1}^{N} (1 + d_{ij}^2)^{-1} \right) = \frac{2 d_{ij} (1 + d_{ij}^2)^{-2}}{D} \tag{11.105}$$

Now Eq. (11.103) can be written as

$$\frac{\partial J}{\partial d_{ij}} = 2 d_{ij} \left(\frac{p_{ij}}{q_{ij}} - 1 \right) \frac{(1 + d_{ij}^2)^{-2}}{D} = 2 d_{ij} (p_{ij} - q_{ij}) (1 + d_{ij}^2)^{-1} \tag{11.106}$$

The last equality is due to Eq. (11.99).

The second derivative in Eq (11.102) is

$$\frac{\partial d_{ij}}{\partial \mathbf{y}_i} = \frac{\partial}{\partial \mathbf{y}_i}\left[(\mathbf{y}_i - \mathbf{y}_j)^T(\mathbf{y}_i - \mathbf{y}_j)\right]^{1/2}$$
$$= \frac{1}{2}\left[(\mathbf{y}_i - \mathbf{y}_j)^T(\mathbf{y}_i - \mathbf{y}_j)\right]^{-1/2} 2(\mathbf{y}_i - \mathbf{y}_j) = \frac{\mathbf{y}_i - \mathbf{y}_j}{d_{ij}} \tag{11.107}$$

Finally, substituting Eqs (11.106) and (11.107) into Eq. (11.102) we get the gradient of the objective function J with respect to each of the map points $\mathbf{y}_i$

$$\frac{\partial J}{\partial \mathbf{y}_i} = 2\sum_{j=1}^{N} \frac{\partial J}{\partial d_{ij}} \frac{\partial d_{ij}}{\partial \mathbf{y}_i} = 4\sum_{j=1}^{N}(p_{ij} - q_{ij})(1 + d_{ij}^2)^{-1}(\mathbf{y}_i - \mathbf{y}_j)$$
$$= 4\sum_{j=1}^{N} \frac{p_{ij} - q_{ij}}{1 + ||\mathbf{y}_i - \mathbf{y}_j||^2}(\mathbf{y}_i - \mathbf{y}_j) = 4\sum_{j=1}^{N} w_{ij}(\mathbf{y}_i - \mathbf{y}_j)$$
$$= \left(\sum_{j=1}^{N} w_{ij}\right)\mathbf{y}_i - \sum_{j=1}^{N} w_{ij}\mathbf{y}_j \tag{11.108}$$

where $w_{ij} = (p_{ij} - q_{ij})/(1 + d_{ij}^2)$ for all $i, j = 1, \cdots, N$. Having found the gradient vector, we can iteratively approach the optimal data points in the low-D space that minimize the discrepancy $J = D_{KL}(P||Q)$

$$\mathbf{y}_{n+1} = \mathbf{y}_n - \delta\frac{\partial J}{\delta \mathbf{y}} + \alpha_n(\mathbf{y}_n - \mathbf{y}_{n-1}) \tag{11.109}$$

where δ is the step size (learning rate) and α is the coefficient for the momentum term in the direction of the previous iteration step (Eq. (2.55)).

The following MATLAB function takes as input the data set $\mathbf{X}$ in the original high dimensional space, and generates as output the $N \times N$ matrix $\mathbf{P}$ of which the component p_{ij} in the ith row and jth column is defined in Eq. (11.98)

```
function P=data2prob(X)          % Given dataset X, find P matrix
    x2=sum(X.^2);                % squared distances between points
    D=x2+x2'-2*(X'*X);           % pair-wise squared distances
    N=size(D,1);                 % number of data points
    P=zeros(N);                  % initializing matrix P
    H0=3;                        % desired entropy value
    tol=1e-4;                    % tolerance
    for i=1:N                    % for each of the N data points
        d2=D(i,[1:i-1,i+1:N]);   % squared distance between xi and others
        sigma=1;                 % default value of variance
        sigmamax=9*sigma;        % initial upper limit
        sigmamin=0;              % initial lower limit
        H=0;
        while abs(H-H0)>tol      % adjust sigma for desired entropy
            p=exp(-d2/sigma);    % Gaussian kernel
            p=p/sum(p);          % normalization
            H=-sum(p.*log(p));   % actual entropy based on current sigma
```

```
            if H<H0              % H too small
               sigmaamin=sigma; % update upper limit
               sigma=(sigma+sigmamax)/2; % increase sigma
            else                 % H too small
               sigmamax=sigma; % update lower limit
               sigma=(sigma+sigmamin)/2; % reduce sigma
            end
        end
        P(i,[1:i-1,i+1:N])=p;  % Save resulting p for xi
    end
    P=(P+P')/2;                % make P symmetric
    P=P/sum(P(:));             % normalization
end
```

The following MATLAB function takes as inputs matrix **P** generated by the previous function, *dim* as the dimension of the low dimensional space, and δ is the step size, and generates as output **Y** for the data points in the low dimensional space

```
function Y=tSNE(P,dim,delta)
    N=size(P,1);
    Y=0.0001*randn(dim,N);    % random initialization of data
    dY=zeros(dim,N);          % increment for gradient descent
    alpha=0.5;                % coefficient for the momentum term
    it=0;
    while it<200
        it=it+1;
        Y0=Y;
        s=sum(Y.^2);          % squared distances between points
        T=1./(1+s+s'-2*(Y'*Y)); % t-distribution
        Q=T./sum(T(:));       % normalization
        W=(P-Q).*T;
        g=Y*(diag(sum(W,1))-W); % gradient of KL-divergence
        Y=Y-delta*g+alpha*dY % gradient descent with momentum
        Y=Y-mean(Y,2);        % remove mean from dataset
        dY=Y-Y0;              % increment as momentum for next step
    end
end
```

The tSNE method is applied to the dataset of handwritten digits of 2240 256-D data points to visualize the dataset in 2-D and 3-D spaces as shown in Fig. 11.12. Comparing these results with those shown previously in Fig. 10.16, we see that the tSNE method is obviously more effective than the PCA method in two or three dimensions out of the original 256-D space.

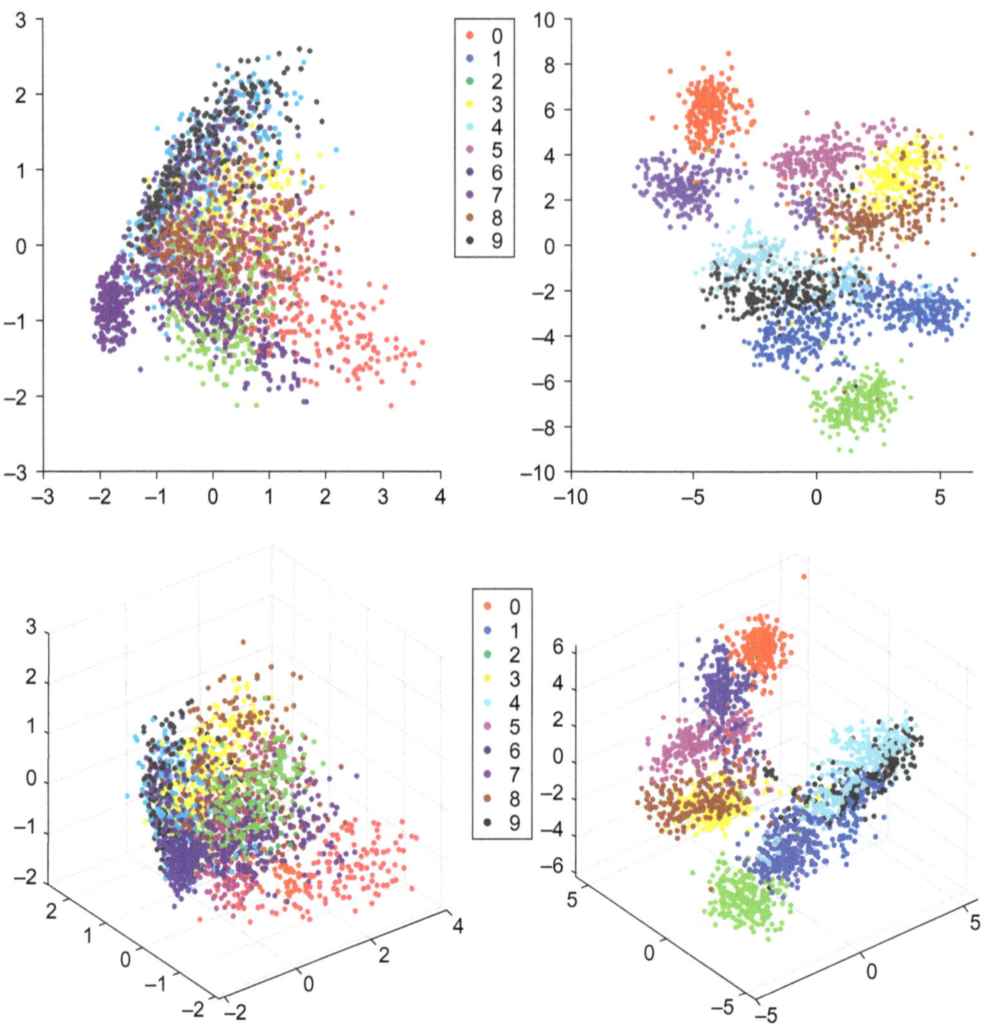

Fig. 11.12 Visualization of 256-D dataset by the PCA (left) and tSNE (right) methods.

PROBLEMS

1. Carry out kernel PCA to visualize the 4-D iris dataset in both 2-D and 3-D space, with all data points color coded according to their class identities. Compare your results with those obtained in the previous homework.
2. Carry out kernel PCA to visualize the 256-D handwritten digit dataset in both 2-D and 3-D space, with all data points color coded according to their class identities. Compare your results with those obtained in the previous homework.
3. Carry out probabilistic PCA to visualize the 256-D handwritten digit dataset in both 2-D and 3-D space, with all data points color coded according to their class identities. Compare your results with those obtained in the previous problem.
4. Use the PCoA method to reconstruct the locations of some cities in the US based on their pairwise distances, which can be obtained from their longitude–latitude

coordinates available here: https://www.latlong.net/category/cities-236-15.html or here: https://www.geodatos.net/en/coordinates/united-states. Compare your reconstructed map with the ground truth.

5. Apply PCoA method to visualize the iris dataset in both 2-D and 3-D spaces and compare the results with those obtained by the PCA method.
6. Apply PCoA method to visualize the handwritten digit dataset in both 2-D and 3-D spaces and compare the results with those obtained by the PCA method.
7. Apply the tSNE method to visualize the handwritten digit dataset in both 2-D and 3-D spaces and compare the results with those obtained by the PCA method.

12 Independent Component Analysis

The *independent component analysis (ICA)* is a computational method widely used in signal processing and data mining to reveal the independent signal components hidden in an observed dataset. The methods of ICA and principal component analysis (PCA) considered in the previous chapters are similar in the sense that they both intend to find some essential signal components embedded in the data. However, the ICA is different from the PCA as it is based on a different assumption that each of the variables in the observed multivariate data is some linear combination of a set of independent signal components. These independent components are to be separated based on a fundamental statistic property that the distribution of a linear combination of a set of random variables tends to be more Gaussian than those of the individual variables.

A typical problem that can be addressed by the ICA is the *blind source separation (BSS)* problem, for the purpose of separating and reconstructing a set of independent source signals from their linear mixtures. For example, in the "cocktail party problem," the goal is to reconstruct the individual voices of multiple speakers from a set of microphones recording all the conversations. The word "blind" means that no prior knowledge about either the sources or the mixing process is assumed to be available, except that the source signals are statistically independent. Although this BSS problem seems severely underconstrained, the method of ICA can find nearly unique solutions satisfying certain properties.

In the following, we will first discuss the theoretical background of the ICA methods and then specifically consider two different ICA algorithms.

12.1 Independence and Non-Gaussianity

The goal of independent component analysis is to reconstruct a set of m independent source signals, called *independent components* and denoted by $\mathbf{s} = [s_1, \cdots, s_m]^T$, from another set of n observed variables denoted by $\mathbf{x} = [x_1, \cdots, x_n]^T$, each assumed to be a linear combination of the source signals

$$\mathbf{x} = \mathbf{A}\mathbf{s} = \sum_{i=1}^{m} \mathbf{a}_i s_i \tag{12.1}$$

where $\mathbf{A} = [\mathbf{a}_1, \cdots, \mathbf{a}_m]$ is an $n \times m$ *mixing matrix* of unknown values. We assume $n \geq m$ to avoid underdetermination. Given a dataset $\mathbf{X} = [\mathbf{x}_1, \cdots, \mathbf{x}_N]$ containing N observed data samples $\mathbf{x}_n = \mathbf{A}\mathbf{s}_n$ $(n = 1, \cdots, N)$, the ICA is to recover the independent source signals $\mathbf{s}_1, \cdots, \mathbf{s}_N$, based on a linear reconstruction model

$$\mathbf{y} = \mathbf{W}\mathbf{x} = \mathbf{W}\mathbf{A}\mathbf{s} \tag{12.2}$$

where $\mathbf{W}$ is a $m \times n$ reconstruction matrix, and $\mathbf{y} = [y_1, \cdots, y_m]^T$ contains m reconstructed signals $y_i = \mathbf{w}_i^T\mathbf{x}$ to match the corresponding independent signals s_i in $\mathbf{s}$, and $\mathbf{w}_i^T$ is the ith row vector of $\mathbf{W}$. It is obvious that for the reconstructed signal $\mathbf{y}$ to match exactly the source $\mathbf{s}$, ideally we would like to find $\mathbf{W} = \mathbf{A}^{-1}$. But as $\mathbf{A}$ is unknown, we need to find $\mathbf{W}$ in some different way, as the main task of ICA algorithms.

As ICA is an unsupervised method without labeled training data for the ground truth regarding either $\mathbf{A}$ or $\mathbf{s}$, we cannot find the reconstruction matrix $\mathbf{W}$ in closed form by the least squares method that minimizes the squared error $||\mathbf{y} - \mathbf{s}||^2$. Instead we have to find the optimal $\mathbf{W}$ as well as $\mathbf{s}$ based only on the requirement that the components in $\mathbf{y} = \mathbf{W}\mathbf{x}$ need to be as independent of each other as possible.

The method of ICA can be compared with the PCA. Both methods aim to find a matrix $\mathbf{W}$ so that a given data vector $\mathbf{x}$ is mapped to another vector $\mathbf{y} = \mathbf{W}\mathbf{x}$. In PCA, the transform matrix $\mathbf{W} = \mathbf{V}^T$ (Eq. (10.20)) is the eigenvector matrix of the covariance matrix $\mathbf{\Sigma}_x$ of $\mathbf{x}$, that is, the data point $\mathbf{x}$ is projected onto the eigenvectors $\mathbf{v}_i$ as the orthogonal basis of the space, and components $y_i = \mathbf{v}_i^T\mathbf{x}$ of $\mathbf{y}$ are completely decorrelated. (Only in the special case when $\mathbf{y} = [y_1, \cdots, y_n]$ is Gaussian, are its components also independent.) While in ICA, $\mathbf{W}$ is in general not orthogonal, and the components are independent and necessarily also uncorrelated. In this sense, the ICA is a more powerful method as it satisfies a stronger requirement for $\mathbf{W}$.

Here are some additional assumptions about the source signal $\mathbf{s}$ and the observed data $\mathbf{x} = \mathbf{A}\mathbf{s}$:

- The number n for the observed variables in $\mathbf{x}$ is not less than the number m for the independent source signals in $\mathbf{s}$. We assume $m = n$ in the following.
- The source signal $\mathbf{s}$ has a zero mean $\text{E}[\mathbf{s}] = \mathbf{0}$ without loss of generality, and therefore so is the observed data: $\text{E}[\mathbf{x}] = \text{E}[\mathbf{A}\mathbf{s}] = \mathbf{A}\,\text{E}[\mathbf{s}] = \mathbf{0}$.
- The source components are stochastically independent and therefore uncorrelated, and they have to be non-Gaussian for reasons to be discussed in this section.
- The estimation of $\mathbf{A}$ and $\mathbf{s}$ is up to a scaling factor. Let

$$\mathbf{C} = \text{diag}(c_1, \cdots, c_n), \qquad \mathbf{C}^{-1} = \text{diag}(1/c_1, \cdots, 1/c_n) \tag{12.3}$$

and $\mathbf{A}' = \mathbf{A}\mathbf{C}^{-1}$ and $\mathbf{s}' = \mathbf{C}\mathbf{s}$, then we have

$$\mathbf{x} = \mathbf{A}\mathbf{s} = [\mathbf{A}\mathbf{C}^{-1}][\mathbf{C}\mathbf{s}] = \mathbf{A}'\mathbf{s}' \tag{12.4}$$

- Owing to the arbitrary scaling factor, all independent components in $\mathbf{s}$ can be assumed to have unit variance $\sigma_i^2 = \mathrm{E}[s_i^2] = 1$. Also, as they are independent and necessarily uncorrelated, that is

$$\sigma_{ij}^2 = \mathrm{E}[s_i s_j] = \delta_{ij}, \quad \text{i.e.,} \quad E(\mathbf{s}\mathbf{s}^T) = \mathbf{I} \tag{12.5}$$

- The estimated independent components are not in any particular order. When the order of the corresponding elements in both $\mathbf{s}$ and $\mathbf{A}$ is rearranged, $\mathbf{x} = \mathbf{A}\mathbf{s}$ still holds.

The theoretical foundation of ICA algorithms is the *central limit theorem*, which states that the distribution of the linear combination of N independent random variables approaches Gaussian as $N \to \infty$. Specifically in a BSS problem, the observed variables in $\mathbf{x} = \mathbf{A}\mathbf{s}$ as a linear combination of the independent components $\{s_1, \cdots, s_m\}$ in $\mathbf{s}$ are necessarily more Gaussian than $\mathbf{s}$. To reconstruct $\mathbf{s}$ from $\mathbf{x}$, we need to reverse the process to reduce the Gaussianity. Based on this idea that *non-Gaussianity* is independence, the ICA method can be seen as the reverse process of the central limit theorem, with the goal of finding an optimal matrix $\mathbf{W}$ so that the components of the reconstructed $\mathbf{y} = \mathbf{W}\mathbf{x}$ are least Gaussian and therefore most independent.

The ICA can therefore be considered as an optimization problem for finding $\mathbf{W}$ that maximizes a certain objective function that measures either the non-Gaussianity or the independence of $\mathbf{y} = \mathbf{W}\mathbf{x}$. We consider two main ICA algorithms. The first one, called FastICA, maximizes the non-Gaussianity of $\mathbf{y}$, while the second one as a maximum likelihood estimate (MLE) method estimates matrix $\mathbf{A}$ by maximizing its likelihood, or equivalently the independence of the source signal $\mathbf{s}$ or the reconstructed signal $\mathbf{y}$. Due to their different criteria, these two algorithms may perform differently.

12.2 Preprocessing and Whitening

Before discussing the specific algorithms, we first consider the following operations as the preprocessing of the observed data so that all assumptions previously made here about the data $\mathbf{x}$ made and required by the algorithm are satisfied.

- **Centering**
  Subtract $\mathrm{E}[\mathbf{x}]$ from $\mathbf{x}$ so that it becomes centered with zero mean. By doing so, we also have $\mathrm{E}[\mathbf{x}] = \mathrm{E}[\mathbf{A}\mathbf{s}] = \mathbf{A}\ \mathrm{E}[\mathbf{s}] = \mathbf{0}$, that is, $E[\mathbf{s}] = \mathbf{A}^{-1}E[\mathbf{x}] = \mathbf{0}$.
- **Whitening**
  Find the eigenvalue matrix $\mathbf{\Lambda}$ and eigenvector matrix $\mathbf{V}$ of the covariance matrix $E[\mathbf{x}\mathbf{x}^T]$ so that

$$\mathbf{V}^T E[\mathbf{x}\mathbf{x}^T]\mathbf{V} = \mathbf{\Lambda}, \quad \text{i.e.,} \quad \mathbf{\Lambda}^{-1/2}\mathbf{V}^T E[\mathbf{x}\mathbf{x}^T]\mathbf{V}\mathbf{\Lambda}^{-1/2} = \mathbf{I} \tag{12.6}$$

As $E[\mathbf{x}\mathbf{x}^T]$ is symmetric, $\mathbf{V}^T = \mathbf{V}^{-1}$ is orthogonal. Then carry out a linear transform

$$\mathbf{x}' = (\mathbf{\Lambda}^{-1/2}\mathbf{V}^T)\mathbf{x}, \qquad \text{i.e.,} \qquad \mathbf{x} = (\mathbf{\Lambda}^{-1/2}\mathbf{V}^T)^{-1}\mathbf{x}' = (\mathbf{V}\Lambda^{1/2})\mathbf{x}' \tag{12.7}$$

so that the components of the data are not only decorrelated (same as the Karhunen–Loève transform (KLT)) but also normalized with unity variance:

$$\begin{aligned} \mathrm{E}[\mathbf{x}'\mathbf{x}'^T] &= \mathrm{E}[(\mathbf{\Lambda}^{-1/2}\mathbf{V}^T\mathbf{x})(\mathbf{\Lambda}^{-1/2}\mathbf{V}^T\mathbf{x})^T] = \mathrm{E}[\mathbf{\Lambda}^{-1/2}\mathbf{V}^T\mathbf{x}\mathbf{x}^T\mathbf{V}\Lambda^{-1/2}] \\ &= \mathbf{\Lambda}^{-1/2}\mathbf{V}^T\,\mathrm{E}[\mathbf{x}\mathbf{x}^T]\mathbf{V}\Lambda^{-1/2} = \mathbf{I} \end{aligned} \tag{12.8}$$

that is $\mathbf{x}'$ is whitened with covariance $\sigma_{ij}^2 = \delta_{ij}$.

Once $\mathbf{x}$ is centered and whitened to become $\mathbf{x}' = (\mathbf{\Lambda}^{-1/2}\mathbf{V}^T)\mathbf{x}$, the reconstruction becomes

$$\mathbf{y} = \mathbf{W}\mathbf{x} = \mathbf{W}(\mathbf{V}\Lambda^{1/2})\mathbf{x}' = \mathbf{W}'\mathbf{x}' \tag{12.9}$$

where $\mathbf{W}' = \mathbf{W}\mathbf{V}\Lambda^{1/2}$ is the reconstruction matrix for whitened $\mathbf{x}'$. We also have

$$\mathrm{E}[\mathbf{x}'\mathbf{x}'^T] = \mathrm{E}[(\mathbf{A}\mathbf{s})(\mathbf{A}\mathbf{s})^T] = \mathbf{A}\,\mathrm{E}[\mathbf{s}\mathbf{s}^T]\mathbf{A}^T = \mathbf{A}\mathbf{A}^T = \mathbf{I} \tag{12.10}$$

As $\mathbf{y} = \mathbf{W}'\mathbf{x}'$ is to match $\mathbf{s}$, we assume they have the same covariance $\mathrm{E}[\mathbf{y}\mathbf{y}^T] = \mathrm{E}[\mathbf{s}\mathbf{s}^T] = \mathbf{I}$ (Eq. (12.5)), and further obtain

$$\mathbf{I} = \mathrm{E}[\mathbf{y}\mathbf{y}^T] = \mathrm{E}[(\mathbf{W}'\mathbf{x}')\,(\mathbf{W}'\mathbf{x}')^T] = \mathbf{W}'\,\mathrm{E}[\mathbf{x}'\mathbf{x}'^T]\mathbf{W}'^T = \mathbf{W}'\mathbf{W}'^T \tag{12.11}$$

We see that after centering and whitening, both the mixing matrix $\mathbf{A}$ and the reconstruction matrix $\mathbf{W}'$ becomes orthogonal and can therefore be interpreted as rotation matrices. For convenience, in the following discussion we will still denote the whitened data samples by $\mathbf{x}$ and the reconstruction matrix by $\mathbf{W}$.

12.3 Non-Gaussianity and FastICA

In ICA, we want to find the optimal $\mathbf{W}$ that maximizes the non-Gaussianity of the reconstructed $\mathbf{y} = \mathbf{W}\mathbf{x}$. In general, the non-Gaussianity of a random variable y with a density function $p(y)$ can be measured by

- **Kurtosis**
  The excess kurtosis (Section B.2.4) of the centered and whitened random variable x (with $\mu_x = 0$ and $\sigma_x = 1$) is defined as:

$$\mathrm{Kurt}(x) = E\{x^4\} - 3 \tag{12.12}$$

  which is zero only when x has a Gaussian distribution. A non-Gaussian distribution is either supergaussian if its kurtosis is positive or subgaussian if its kurtosis is negative. Therefore the absolute value of the excess kurtosis or kurtosis squared can be used to measure the non-Gaussianity of a distribution. However, as kurtosis is very sensitive to noisy outliers, it is not a robust measurement of non-Gaussianity.

- **Negentropy**
  The negentropy of a random variable y is defined as the difference between its entropy $H(y)$ (Section B.2.1) and the entropy of a Gaussian variable g with the same variance σ^2:

$$J(y) = H(g) - H(y) = C - H(y) \geq 0 \tag{12.13}$$

  where $C = H(g) = \log(2\pi e\sigma^2)/2$ ((Eq. (B.183)) is a constant independent of y. Here $J(y)$ is non-negative as a Gaussian distribution has the maximum entropy among all distributions with the same variance. Negentropy $J(y)$ measures non-Gaussianity as it takes a larger value when y is less Gaussian. The negentropy defined for a scalar random variable y can be generalized to a random vector $\mathbf{y} = [y_1, \cdots, y_m]^T$

$$J(\mathbf{y}) = H(\mathbf{y}_G) - H(\mathbf{y}) = C - H(\mathbf{y}) \geq 0 \tag{12.14}$$

While kurtosis and negentropy of a random variable y are good measurements for the non-Gaussianity of y, they both depend on its distribution $p(y)$, which is typically not readily available, and the negentropy $J(\mathbf{y})$ as previously theoretically defined can only be approximated based on the maximum-entropy principle

$$J(y) \approx \sum_i c_i(\mathrm{E}[G_i(y)] - \mathrm{E}[G_i(g)])^2 \tag{12.15}$$

where g is again a Gaussian variable and both y and g are assumed to have zero mean and unit variance, and G_i is some non-quadratic function such as the following also shown in Fig. 12.1

$$G_1(y) = \frac{1}{a} \log \cosh\,(a\, y), \qquad G_2(y) = -\exp(-y^2/2) \tag{12.16}$$

where $1 \leq a \leq 2$ is a constant. This approximated negentropy is always greater than zero except when x is Gaussian. In particular, when only a single term is used in Eq. (12.15), we have

$$J(y) \approx (\mathrm{E}[G(y)] - \mathrm{E}[G(g)])^2 \tag{12.17}$$

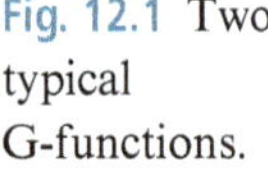

Fig. 12.1 Two typical G-functions.

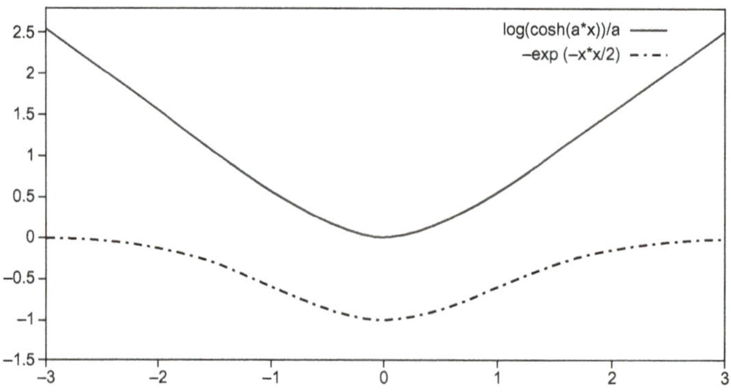

Since the second term is a constant, we want to maximize $E\{G(y)\}$ to maximize $J(y)$.

Based on this approximated negentropy, the FastICA algorithm finds the reconstruction matrix $\mathbf{W}$ that maximizes the sum of the approximated negentropies of all m components of the reconstructed signal $\mathbf{y} = \mathbf{W}\mathbf{x}$

$$\sum_{i=1}^{m} E[G(y_i)] = \sum_{i=1}^{m} E[G(\mathbf{w}_i^T \mathbf{x})] \tag{12.18}$$

where $y_i = \mathbf{w}_i^T \mathbf{x}$ is the ith component of $\mathbf{y}$ and $\mathbf{w_i}^T$ is the ith row vector of $\mathbf{W}$.

We first consider computing $\mathbf{w}_i$ for y_i as one of the m components (with the subscript i dropped for now). Specifically, we need to find a vector $\mathbf{w}$ that maximizes $E[G(y)]$ as the objective function subject to the constraint that $\mathbf{w}$ has to be normalized, that is, $\mathbf{w}^T\mathbf{w} = 1$. To solve this constrained optimization problem we first construct the Lagrangian function:

$$L(\mathbf{w}) = E[G(\mathbf{w}^T \mathbf{x})] - \frac{\lambda}{2}(\mathbf{w}^T\mathbf{w} - 1) \tag{12.19}$$

and then set to zero its derivative with respect to $\mathbf{w}$:

$$\mathbf{g}_L(\mathbf{w}) = \frac{d}{d\mathbf{w}} L(\mathbf{w}) = E[\mathbf{x}\frac{d}{d\mathbf{w}} G(\mathbf{w}^T\mathbf{x})] - \lambda\mathbf{w} = E[\mathbf{x}\, G'(\mathbf{w}^T\mathbf{x})] - \lambda\mathbf{w} = 0 \tag{12.20}$$

where $G'(z) = dG(z)/dz$ is the derivative of function $G(z)$. This equation system can be solved for $\mathbf{w}$ iteratively by the Newton–Raphson method

$$\mathbf{w} \Leftarrow \mathbf{w} - \mathbf{J}_g^{-1}(\mathbf{w})\mathbf{g}(\mathbf{w}) \tag{12.21}$$

Here $\mathbf{J}_f(\mathbf{w})$ is the Jacobian matrix of $\mathbf{g}(\mathbf{w}) = dL(\mathbf{w})/d$:

$$\mathbf{J}_g(\mathbf{w}) = \frac{d}{d\mathbf{w}}\mathbf{g}(\mathbf{w}) = E[\mathbf{x}\mathbf{x}^T G''(\mathbf{w}^T\mathbf{x})] - \lambda\mathbf{I} \tag{12.22}$$

where $G''(z) = d^2G(z)/d^2z$ is the second-order derivative of $G(z)$. We further approximate the first term on the right to get

$$E[\mathbf{x}\mathbf{x}^T G''(\mathbf{w}^T\mathbf{x})] \approx E[\mathbf{x}\mathbf{x}^T]\ E[G''(\mathbf{w}^T\mathbf{x})] = E[G''(\mathbf{w}^T\mathbf{x})]\mathbf{I} \tag{12.23}$$

where $E[\mathbf{x}\mathbf{x}^T] = \mathbf{I}$ as the dataset is assumed to be whitened, then the Jacobian becomes diagonal

$$J_f(\mathbf{w}) = \left(E[G''(\mathbf{w}^T\mathbf{x})] - \lambda\right)\mathbf{I} \tag{12.24}$$

and the Newton–Raphson iteration becomes

$$\mathbf{w}_{n+1} = \mathbf{w}_n - \frac{1}{E[G''(\mathbf{w}_n^T\mathbf{x})] - \lambda}\left(E[\mathbf{x}\, G'(\mathbf{w}_n^T\mathbf{x})] - \lambda\mathbf{w}\right) \tag{12.25}$$

Multiplying both sides by $c = \lambda - E[G''(\mathbf{w}_n^T\mathbf{x})]$, we have

$$c\mathbf{w}_{n+1} = E[\mathbf{x}G'(\mathbf{w}_n^T\mathbf{x})] - E[G''(\mathbf{w}_n^T\mathbf{x})]\mathbf{w} \tag{12.26}$$

where the scaling factor c can be removed as the updated $\mathbf{w}$ will be normalized.

This algorithm can then be repeated to get all row vectors in $\mathbf{W}$ each for one of the m row vectors $\mathbf{w}_i$. However, if this process is carried out independently, the resulting vectors may not be orthogonal to each other as required. To address this issue, the Gram–Schmidt method in Section A.1.2 can be used. Specifically, during each iteration of the computation for $\mathbf{w}_i$ for $i > 1$, before it is normalized, we also subtract its projection onto each of the previously found $\mathbf{w}_1, \cdots, \mathbf{w}_{i-1}$ (Eq. (A.39)), so that it is guaranteed to be orthogonal to all of them. All vectors computed in this will guarantee that the final matrix $\mathbf{W}$ is orthogonal satisfying $\mathbf{W}^T = \mathbf{W}^{-1}$ as required.

Now we have the following FastICA algorithm that is carried out m times each for one of the vectors $\mathbf{w}_i$, $(i = 1, \cdots, m)$.

1. Choose an initial random guess for $\mathbf{w}_i$
2. Iterate:

$$\mathbf{w} = \mathrm{E}[\mathbf{x}G'(\mathbf{w}^T\mathbf{x})] - \mathrm{E}[G''(\mathbf{w}^T\mathbf{x})]\mathbf{w} \tag{12.27}$$

   where the expectation is estimated as the average of all N samples in the dataset.
3.

$$\mathbf{w} \longleftarrow \mathbf{w} - \sum_{j=1}^{i-1} (\mathbf{w}^T\mathbf{w}_j)\mathbf{w}_j \tag{12.28}$$

4. Normalize:

$$\mathbf{w} \longleftarrow \mathbf{w}/||\mathbf{w}|| \tag{12.29}$$

5. If not converged, go back to step 2.

The following code segment shows the essential part of the FastICA algorithm, in which the reconstruction matrix $\mathbf{W}$ is calculated based on dataset $\mathbf{X}$ containing N observed data samples. Here`Imax` is the maximum number of iterations and `tol` is the tolerance, for example, 10^{-6}.

```
syms u
G=-exp(-u^2/2);
G=log(cosh(u));               % symbolic function G
g=matlabFunction(diff(G));    % G'
dg=matlabFunction(diff(diff(G))); % G''
[d N]=size(X);                % size of dataset
[X T]=whiten(X);              % function that whitens dataset
W=[];                         % initialize weight matrix W
for i=1:d                     % for each weight vector w
    w=rand(d,1);              % random initialization
    for i=1:Imax
        wold=w;               % old weight vector
        wx=w'*X;
        w=X*g(wx)'-sum(dg(wx))*w; % new weight vector
```

```
            if i>1
                w=w-W*W'*w;      % Gram--Schmidt process
            end
            w=w/norm(w);         % renormalization
            if norm(w-wold)<tol % convergence
                break
            end
        end
        W=[W w];
    end
    Y=W'*X;                      % signal Reconstruction by ICA
    W=W*T;
```

Also, the following code lists the whitening function `whiten` used in the code that takes the dataset `X` as input and generates the centered and whitened data and the whitening transform matrix `T` as the output

```
function [X T]=whitening(X)
    [d N]=size(X);
    X=X-repmat(mean(X,2),1,N);
    [U S V]=svd(X);
    T=sqrt(N)*diag(1./(diag(S(1:d,1:d))))*U';
    X=T*X;
end
```

12.4 Likelihood and Independence Maximization

The ideal reconstruction matrix $\mathbf{W}$ can also be estimated by solving an optimization problem based on the following different but equivalent criteria all related to the assumed independence of the source signal $\mathbf{s}$, or, equivalently, the desired independence of the reconstructed signal $\mathbf{y} = \mathbf{W}\mathbf{x} = \mathbf{W}\mathbf{A}\mathbf{s}$.

- The optimal $\mathbf{W}$ can be found so that the reconstructed signal $\mathbf{y} = \mathbf{W}\mathbf{x}$ is maximally independent measured by the mutual information (Eq. (B.201)) shared by its components $\{y_1, \cdots, y_m\}$:

$$
\begin{aligned}
I(\mathbf{y}) &= \sum_{i=1}^{n} H(y_i) - H(\mathbf{y}) = -\sum_{i=1}^{n} \mathrm{E}\left[\log p(y_i)\right] - H(\mathbf{W}\mathbf{x}) \\
&= -\mathrm{E}\left[\sum_{i=1}^{n} \log p(y_i)\right] - \log\, |det\, \mathbf{W}| - H(\mathbf{x})
\end{aligned} \tag{12.30}
$$

  where $H(\mathbf{y}) = H(\mathbf{W}\mathbf{x}) = H(\mathbf{x}) + \log\, |\det \mathbf{W}|$ (Eq. (B.188)), and the last term $H(\mathbf{x})$ is unrelated to $\mathbf{W}$ and can therefore be dropped.
- Matrix $\mathbf{A}$ can be estimated so that $p(\mathbf{x})$ of the observed data $\mathbf{x}$ most closely matches $p(\mathbf{x}|\mathbf{A})$ based on the assumed independence of the components in the source $\mathbf{s}$ (Eq. (B.85))

$$p(\mathbf{x}|\mathbf{A}) = p(\mathbf{As}) = |\det \mathbf{A}^{-1}|\, p(\mathbf{s}) = |\det \mathbf{A}^{-1}| \prod_{i=1}^{m} p_i(s_i) \tag{12.31}$$

Now the optimal **A** can be found as the one that minimizes the Kullback–Leibler (KL) divergence between $p(\mathbf{x}$ and $p(\mathbf{x}|\mathbf{A})$

$$\begin{aligned} D_{KL}(p(\mathbf{x})||p(\mathbf{x}|\mathbf{A}) &= \int p(\mathbf{x}) \log \frac{p(\mathbf{x})}{p(\mathbf{x}|\mathbf{A})} d\mathbf{x} \\ &= \int p(\mathbf{x}) \log p(\mathbf{x}) d\mathbf{x} - \int p(\mathbf{x}) \log\left(|\det \mathbf{A}^{-1}| \prod_{i=1}^{m} p_i(s_i) \right) d\mathbf{x} \\ &= H(\mathbf{x}) - \log |\det \mathbf{A}^{-1}| \int p(\mathbf{x}) d\mathbf{x} - \int p(\mathbf{x}) \left(\sum_{i=1}^{m} \log p_i(s_i) \right) d\mathbf{x} \\ &= H(\mathbf{x}) - \log |\det \mathbf{A}^{-1}| - E_x \left[\sum_{i=1}^{m} \log p_i(s_i) \right] \end{aligned} \tag{12.32}$$

Here the first term $H(\mathbf{x})$ independent of **A** can be dropped, and the expectation with respect to the observed data **x** in the last term can be estimated as the average of the all samples in the dataset $\mathbf{X} = [\mathbf{x}_1, \cdots, \mathbf{x}_N]$.

- Matrix **A** can also be estimated by maximizing its likelihood given the observed dataset **X** containing N independent and identically distributed (i.i.d.) samples

$$\begin{aligned} L(\mathbf{A}|\mathbf{X}) = p(\mathbf{X}|\mathbf{A}) &= \prod_{n=1}^{N} p(\mathbf{x}_n|\mathbf{A}) = \prod_{n=1}^{N} |\det \mathbf{A}^{-1}| \prod_{i=1}^{m} p_i(s_{in}) \\ &= |\det \mathbf{A}^{-1}|^N \prod_{n=1}^{N} \prod_{i=1}^{m} p_i(s_{in}) \end{aligned} \tag{12.33}$$

where $p(\mathbf{x}_n|\mathbf{A})$ is given in Eq. (12.31) and s_{in} is the ith component of $\mathbf{s}_n$. We further have the log-likelihood

$$\begin{aligned} l(\mathbf{A}|\mathbf{X}) = \log L(\mathbf{A}|\mathbf{X}) &= \log \left[|\det \mathbf{A}^{-1}|^N \prod_{n=1}^{N} \prod_{i=1}^{m} p_i(s_{in}) \right] \\ &= N \left[\log |\det \mathbf{A}^{-1}| + \frac{1}{N} \sum_{n=1}^{N} \sum_{i=1}^{m} p_i(s_{in}) \right] \\ &\propto \log |\det \mathbf{A}^{-1}| + E_x \left[\sum_{i=1}^{m} \log p_i(s_i) \right] \end{aligned} \tag{12.34}$$

Here the average over all N data samples in the second term is treated as the estimated expectation with respect to **x**.

All three of these cases are equivalent, as maximizing the log-likelihood in Eq. (12.34) is equivalent to minimizing mutual information of either the reconstructed signal **y** or the source signal **s** in Eq. (12.30) or (12.32), thereby maximizing their independence.

Also, we see that for the reconstructed signal $\mathbf{y} = \mathbf{Wx} = \mathbf{WAs}$ to match perfectly the source signal **s**, the reconstruction matrix simply needs to be $\mathbf{W} = \mathbf{A}^{-1}$ so that

$y_i = \mathbf{w}_i^T \mathbf{x} = s_i$ for all $i = 1, \cdots, m$, and the three criteria become exactly the same. We can now find the optimal $\mathbf{W}$ that maximizes Eq. (12.34) as the objective function

$$J(\mathbf{W}) = \log |\det \mathbf{W}| + E\left[\sum_{i=1}^{m} \log p_i(\mathbf{w}_i^T \mathbf{x})\right] \tag{12.35}$$

The gradient of this objective function is

$$\mathbf{g}_J(\mathbf{W}) = \frac{d}{d\mathbf{W}} J(\mathbf{W}) = \frac{d}{d\mathbf{W}} \log |\det \mathbf{W}| + E\left[\frac{d}{d\mathbf{W}} \sum_{i=1}^{m} \log p_i(\mathbf{w}_i^T \mathbf{x})\right] \tag{12.36}$$

The first term is (Eq. (B.245))

$$\frac{d}{d\mathbf{W}} \log |\det \mathbf{W}| = (\mathbf{W}^{-1})^T = (\mathbf{W}^T)^{-1} \tag{12.37}$$

For each term in the summation of the second term, $\mathbf{w}_i^T \mathbf{x} = y_i$ is the ith component of $\mathbf{y}$, and the derivative with respect to the kth component w_{ki} of $\mathbf{w}_i$ is

$$\frac{d}{d\, w_{ki}} \log p_i(\mathbf{w}_i^T \mathbf{x}) = \frac{d}{d\, w_{ki}} \log p_i(y_i) = \frac{p_i'(y_i)}{p_i(y_i)}\, x_k = \phi_i x_k \tag{12.38}$$

where we have defined

$$\phi_i = \phi(y_i) = \frac{p'(y_i)}{p(y_i)} \tag{12.39}$$

If we further assume specifically $p(y) = 1/\cosh(y)$ as a non-Gaussian distribution, and obtain

$$\phi(y) = \frac{(1/\cosh(y))'}{1/\cosh(y)} = -\frac{\sinh(y)}{\cosh(y)} = -\tanh(y) \tag{12.40}$$

then the second term of Eq. (12.36) can be written as $\mathrm{E}[\boldsymbol{\phi}(\mathbf{y})\mathbf{x}^T]$, where $\boldsymbol{\phi}(\mathbf{y}) = [\phi_1, \cdots, \phi_m]^T$, and the gradient becomes

$$\mathbf{g}_J(W) = (\mathbf{W}^{-1})^T + \mathrm{E}[\boldsymbol{\phi}\mathbf{x}^T] \tag{12.41}$$

Now the optimal $\mathbf{W}$ that maximizes $J(\mathbf{W})$ can be found iteratively based on the gradient ascent method

$$\mathbf{W}_{n+1} = \mathbf{W}_n + \delta \mathbf{g}_J(\mathbf{W}_n) = \mathbf{W}_n + \delta\left[(\mathbf{W}_n^{-1})^T + \mathrm{E}[\boldsymbol{\phi}(\mathbf{y})\mathbf{x}^T]\right] \tag{12.42}$$

where δ is the step size, and the expectation can be estimated as the average of all N data samples.

The gradient $\mathbf{g}_J(\mathbf{W})$ can be modified by post-multiplying $\mathbf{W}^T\mathbf{W}$ to obtain the *natural gradient* (Section B.3.4)

$$\left[(\mathbf{W}^T)^{-1} + \mathrm{E}[\boldsymbol{\phi}\mathbf{x}^T]\right] \mathbf{W}^T\mathbf{W} = \mathbf{W} + \mathrm{E}[\boldsymbol{\phi}\, \mathbf{x}^T\mathbf{W}^T\, \mathbf{W}] = \mathbf{W} + \mathrm{E}[\boldsymbol{\phi}\mathbf{y}^T\mathbf{W}] \tag{12.43}$$

The last equality is due to $\mathbf{W}\mathbf{x} = \mathbf{y}$. Now the iteration becomes

$$\mathbf{W}_{n+1} = \mathbf{W}_n + \delta \mathbf{g}_J(\mathbf{W}_n) = \mathbf{W}_n + \delta\left(\mathbf{W}_n + \mathrm{E}[\boldsymbol{\phi}\mathbf{y}^T\mathbf{W}_n]\right) \tag{12.44}$$

Based this natural gradient along a modified search direction in each step of the iteration, it will converge more quickly.

The following MATLAB code segment shows the essential part of the MLE algorithm, where `Imax` is the maximum number of iterations, `\lambda` is the step size, and `tol` is the tolerance, for example 10^{-6}. In the algorithm, data preprocessing (centering and whitening) is not required and is therefore optional. Here matrix $\mathbf{W}$ can be updated based on either the gradient or the natural gradient which converges much more quickly.

```
    [d N]=size(X);          % size of dataset
    T=eye(d);               % default whitening transform matrix
    if preprocessing        % optional preprocessing
        [X T]=whiten(X);
    end
    W=eye(d);               % initialize W
    lambda=0.5;             % step size
    tol=10^(-6);            % tolerance
    for i=1:Imax
        Y=W*X;              % reconstruction of y
%        Wnew=W+lambda*(inv(B)'-tanh(Y)*X'/N);
                                                % update W by gradient
        Wnew=W+lambda*(W-tanh(Y)*(Y'*W)/N);  % or natural gradient
        if mean(abs(Wnew(:)-W(:))) < tol     % check convergence
            break
        end
        W=Wnew;
    end
    W=W*T;
```

Example 12.1

This example demonstrates the ICA algorithm based on the maximum likelihood method for the reconstruction of two independent source signals in $\mathbf{s} = [s_1, s_2]^T$, shown in the first panel in Fig. 12.2, based on the observed data $\mathbf{x} = [x_1, x_2]^T = \mathbf{A}_0\mathbf{s}$, shown in the second panel in the figure.

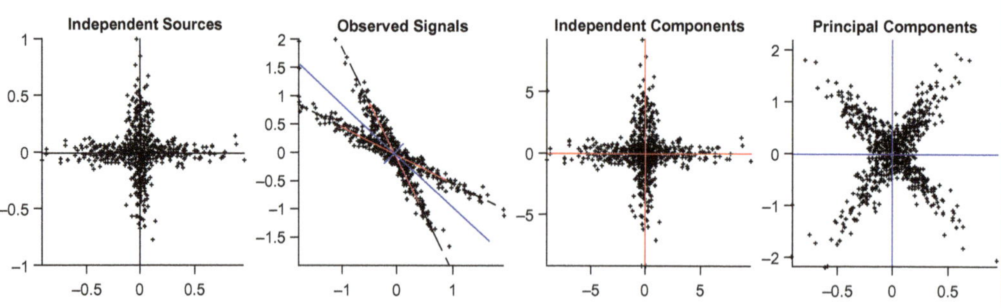

Fig. 12.2 Reconstruction of independent components by ICA based on machine learning (ML) (2D).

The two reconstructed independent components by the ICA algorithm are shown in the third panel, which match the ground truth source signals in the first panel very well.

Also the estimated mixing matrix $\mathbf{A}$ is listed here, together with the ground truth $\mathbf{A}_0$ (not used in the algorithm)

$$\mathbf{A} = \begin{bmatrix} 0.243 & -0.111 \\ -0.120 & 0.218 \end{bmatrix}, \qquad \mathbf{A}_0 = \begin{bmatrix} 2 & -1 \\ -1 & 2 \end{bmatrix} \tag{12.45}$$

We see that the ground truth $\mathbf{A}_0$ is matched by the estimated $\mathbf{A}$ very closely, when scaled up by a factor of 9.

In the second panel, we also plotted the column vectors of the ground truth mixing matrix $\mathbf{A}_0$ (black dashed lines), together with those of the estimated mixing matrix $\mathbf{A}$ (red lines). These two sets of lines match each other very closely, both representing the two independent sources.

For comparison, the PCA method is also used to find the principal components of the dataset, as shown by the blue lines in the fourth panel. They are also plotted in the second panel, as the eigenvectors of the covariance matrix of the dataset, weighted by the corresponding eigenvalues. As expected, these eigenvectors are along the major and minor axes of an ellipse approximating the overall distribution of observed data points. We see that the principal components are very different from the independent components.

The same algorithm is also applied to a 3-D dataset and the results are shown in Fig. 12.3.

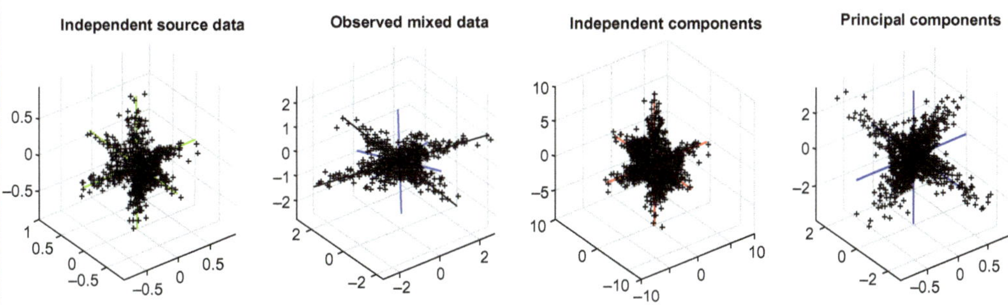

Fig. 12.3 Reconstruction of independent components by ICA based on ML (3-D).

Example 12.2

This example illustrates the results of the FastICA algorithm when applied to reconstruct the independent components of three observed time signals shown in the second panel of Fig. 12.4, each as a linear combination of some three independent source signals (noisy square wave, triangular wave, and sinusoidal wave) as shown in the first panel.

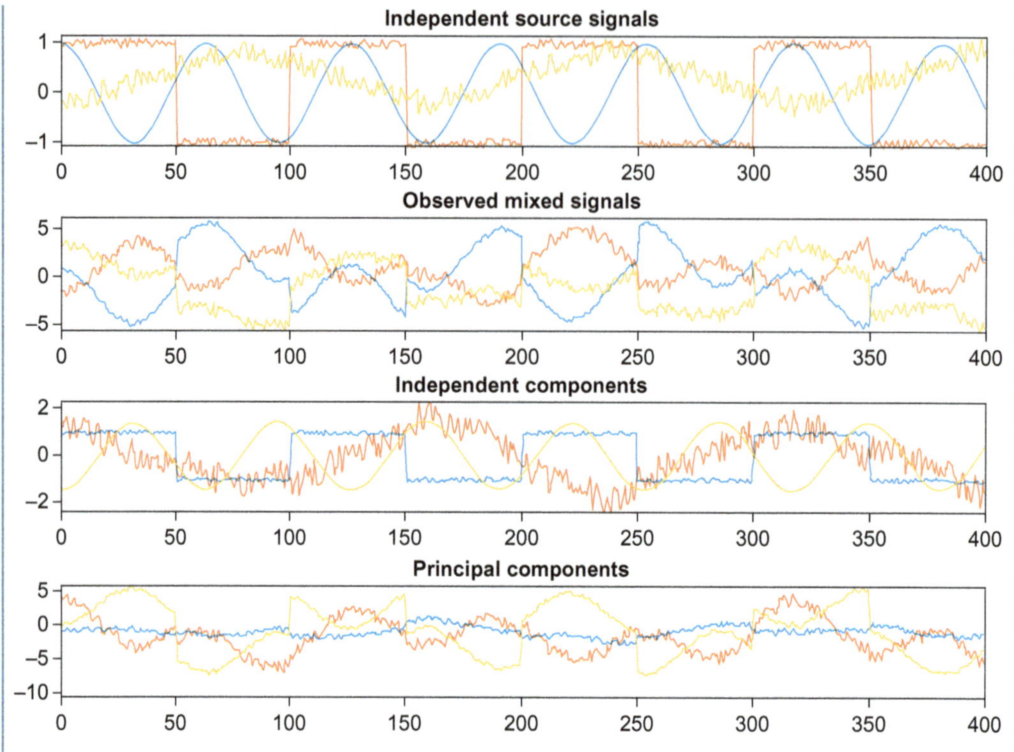

Fig. 12.4 Reconstruction of independent components in time signals by FastICA.

The reconstructed independent components, shown in the third panel, closely match the ground truth source signals (first panel), up to some scaling factors (could be negative). For comparison, the principal components obtained by the PCA method are also shown in the fourth panel, which are very different from the desired independent components.

PROBLEMS

1. Implement the ICA algorithm based on maximum likelihood method and apply your code to both the 2-D and 3-D datasets used in Example 12.1.
   The independent source signals can be generated by the following code

```
function S=data(d,N)
    c=30;                    % variance of independent signals
    if d==2                  % 2D signals
        m=[0 0]';            % zero mean
        C1=[1 0; 0 c];
        C2=[c 0; 0 1];       % covariance of two independent sets
        S=mvnrnd(m,C1,N)';   % generate 2 sets of N data points
        S=[S mvnrnd(m,C2,N)']; % with covariances C1 and C2
```

```
    elseif d==3                 % 3D signals
        m=[0 0 0]';             % zero mean
        C1=eye(3);  C2=C1;  C3=C1;
        C1(1,1)=c;  C2(2,2)=c;  C3(3,3)=c;
                                % covariance of three independent sets
        S=mvnrnd(m,C1,N)';      % generate 3 sets of N data points
        S=[S mvnrnd(m,C2,N)'];  % with covariances C1, C2 and C3
        S=[S mvnrnd(m,C3,N)'];
    end
    S = S./max(abs(S(:)));      % normalization
end
```

The observed data can then be obtained by linearly combining the independent source signals:

```
if d==2
    A=[2 -1; -1 2];                     % 2D mixing matrix
elseif d==3
    A=[3 -2 1; -2 1 3; 1 2 -3];         % 3D mixing matrix
end
X=A*S;      % generate observed data by mixing independent sources
```

2. Implement the algorithm of FastICA and apply your code to some observed time signals each as a linear combination of some independent components (such as square, sinusoidal, and triangular waves of different frequencies) contaminated by some random noise, similar to those used in Example 12.4.

Part IV

Classification

In Part IV, we consider *pattern classification*. As one of the most important topics in machine learning, classification can be considered as a learning process that finds out the underneath structure of the given dataset. Many real-world problems can be formulated as to classify all data points in the given dataset into different categories. For example, in image recognition, the given data may be a set of images of various objects of interest (e.g., different animals or handwritten alphanumeric symbols), and the goal is to recognize these objects by classifying them into different categories each for one type of the objects. Such objects of interest are generally referred to as *patterns*, and the categories are the *classes*. In such cases, pattern classification is also called *pattern (or object) recognition*.

A classification problem can be generally formulated so as to classify a set of N data samples, each represented as a vector $\mathbf{x} = [x_1, \cdots, x_d]^T$, a point in the d-dimensional feature space, into a set of K classes denoted by $\{C_1, \cdots, C_K\}$, in such a way that all similar points are classified into the same class.

In the context of probabilistic methods of classification, these data points are assumed to be independent samples taken from a population with the same probability distribution, that is, they can be treated as a set of *independent and identically distributed (i.i.d.)* samples of the population. Geometrically, the task of classifying similar data samples into the same classes can be considered as to partition the feature space into a set of regions each corresponding to one of the K classes. The boundary between any two such regions is called the *decision boundary*, and the goal of a specific classification algorithm, called a classifier, is to determine these decision boundaries based on the given dataset.

Classification is a *supervised* learning process if each sample $\mathbf{x}_n$ in the given dataset is associated with some labeling y_n, representing some prior knowledge of the data, such as the class identities of the samples. For example, we can use $y_n \in \{1, \cdots, N\}$ to indicate sample $\mathbf{x}_n$ is known to belong to class C_{y_n}, or in the case of binary classification, we can use either $y_n \in \{0, 1\}$ or $y_n \in \{-1, 1\}$ to indicate $\mathbf{x}$ belongs to either of the two classes. How the samples are labeled depends on the specific classification algorithms.

Such a set of labeled data samples is called the *training set* and is represented by $\mathcal{D} = \{(\mathbf{x}_n, y_n)\ (n = 1, \cdots, N)\}$, which can also be denoted by $\mathcal{D} = \{\mathbf{X}, \mathbf{y}\}$, where $\mathbf{X} = [\mathbf{x}_1, \cdots, \mathbf{x}_N]$ is $d \times N$ array containing all N samples, and $\mathbf{y} = [y_1, \cdots, y_N]^T$ is an N–D vector containing their corresponding labelings. We can further denote by N_k the number of samples all labeled to belong to class C_k, so that $N = \sum_{k=1}^{K} N_k$.

Supervised classification is a two-stage process, in which a classifier is first constructed by learning the training set in the training stage, and then tested in the testing stage where any unlabeled data samples can be classified into one of the classes. We will consider different types of classification algorithms in Chapters 13 and 14.

If the samples in $\mathbf{X}$ are not labeled, that is, $\mathbf{y}$ is unavailable in the given dataset $\mathcal{D}$, then classification becomes an *unsupervised* learning process, called *clustering*, and the task becomes to group or cluster all data points into a set of clusters in such a way that similar data points (close to each other in the feature space) are in the same cluster while dissimilar ones (far away from each other) are in different clusters. Unsupervised clustering is similar to supervised classification in the sense that they both partition the feature space into different regions each for one of the classes/clusters containing similar data samples, but clustering is very different from classification in the sense that the partitioning is totally based on how the data points are distributed in the feature space, without any prior knowledge, such as the labeling. We will consider various unsupervised clustering algorithms in Chapter 15.

A wide variety of supervised and unsupervised learning algorithms will be discussed in the chapters of this Part, as well as those in Part V (for neural network methods), all for the same purpose of classification or clustering. It is natural to ask then, why do we need so many different algorithms? which of these algorithms is the best of them all? The answer is it depends. Every learning algorithm may be better suited for a certain type of learning problems than others, depending on many specifics such as the computational cost, the size of the dataset, how noisy or reliable the dataset is, how separable the classes are, etc. Moreover, even for the same learning algorithm, there are typically many *hyperparameters* to be adjusted in order to achieve the optimal performance on a specific dataset. Therefore the reader is encouraged to experiment with different settings of the hyperparameters of each learning algorithm, and compare multiple algorithms for classification or clustering when applied to solve the same problem.

13 Statistic Classification

13.1 Discriminative vs. Generative Methods for Classification

Many supervised learning algorithms in both regression and classification are based on some assumed learning model with a set of parameters generally represented by a vector $\boldsymbol{\theta} = [\theta_1, \cdots, \theta_m]^T$. Once a specific model to use is determined, the main learning task is to obtain $\boldsymbol{\theta}$ based on available training dataset.

All such learning methods can be characterized based on one of two types of models, as illustrated in Fig. 13.1.

These two different models are illustrated in Fig. 13.1.

- *Discriminative models*

  A discriminative algorithm is typically a deterministic method aiming at establishing a model that fits the training set in some optimal way, such as a set of decision boundaries that partitions the feature space into different regions each including all data points with the same labeling. In this case, only those data samples close to the boundaries play an important role, while all other samples farther away from the boundaries may be mostly ignored.

  The discriminative methods find the decision boundary in the feature space directly based on the data points in the training set, in general they

  - are simpler than the generative approach requiring problem-specific knowledge for building models of the data,
  - are effective in producing accurate result when the dataset is large
  - provide no insight or interpretation regarding the data and no uncertainty estimate

  Typical discriminative methods include:

  - k-nearest neighbors method
  - logistic regressions
  - support vector machines
  - perceptrons and neural networks
  - decision trees and random forests

- *Generative models*

  The generative methods formulate the learning problem in a probabilistic framework to establish a model for the underlying structure of the data as an effort to explain how the data were generated, and then finds specifically the decision

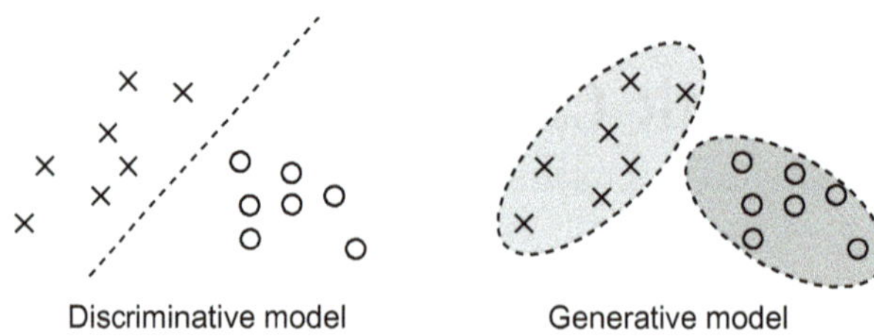

Fig. 13.1 Discriminative vs. generative models.

boundaries based the model. Typically such a method finds certain probabilities of interest, such as the joint probability $p(\mathbf{x}, y|\boldsymbol{\theta})$ of both $\mathbf{x}$ and the corresponding y in the training set, or the conditional probability $p(y|\mathbf{x}, \boldsymbol{\theta})$ of the class symbol y, given the observed pattern $\mathbf{x}$. In either case, the probability is parameterized by $\boldsymbol{\theta}$, to be estimated in the learning process based on the training set. Once $\boldsymbol{\theta}$ is obtained, the probabilistic model becomes available, based on which the class identity of any unlabeled sample can be determined. For example, once the conditional probability $p(y|\mathbf{x})$ is available, we can classify a data point $\mathbf{x}$ to a class labeled by y with the greatest $p(y|\mathbf{x})$, and we can further get the confidence for such a decision measured by the probability. For a probabilistic method, all data samples in the training set are needed for the estimation their overall probability density distribution.

Here are the properties of the generative methods:

- allow the use of problem-specific knowledge for building the model,
- can provide explanation and interpretation of the data,
- may be less prone to overfitting than a discriminative method,
- can provide uncertainty estimate
- may not be as accurate as the discriminative methods if the model does not fit the dataset well.

Typical generative methods include:

- minimum distance classifier
- naive Bayes classifier
- Gaussian mixture classifier
- hidden Markov model

13.2 k-nearest Neighbors and Minimum Distance Classifiers

Here we first consider a set of simple supervised classification algorithms that assign an unlabeled sample $\mathbf{x}$ to one of the K known classes based on set N of *training samples* $\mathbf{X} = [\mathbf{x}_1, \cdots, \mathbf{x}_N]$, where each sample $\mathbf{x}_n$ is labeled by $y_n = k \in \{1, \cdots, K\}$, indicating it belongs to class C_k.

- **k-nearest neighbors (k-NN) classifier**
  Given an unlabeled pattern $\mathbf{x}$, we first find its *k-nearest neighbors* in the training dataset, and then assign $\mathbf{x}$ to one of the K classes by a majority vote of the

k neighbors based on their class labelings. The voting can be weighted so that closer neighbors are more heavily weighted than those that are farther away. In particular, when $k = 1$, $\mathbf{x}$ is assigned to the class of its closest neighbor.

While the k-NN method is simple and straightforward, its computational cost is high as classifying any unlabeled pattern $\mathbf{x}$ requires computing distances to all data points in the training set.

- **Minimum distance classifier**

  The minimum distance classifier is based on the most straightforward idea that an unlabeled data point $\mathbf{x}$ should be classified into the closest class in the feature space, among all K such classes $\{C_1, \cdots, C_K\}$ in the training dataset

$$\text{if } d(\mathbf{x}, C_k) = min\{d(\mathbf{x}, C_i) \;\; i = 1, \cdots, C\} \quad \text{then } \mathbf{x} \in C_k \tag{13.1}$$

where $d(\mathbf{x}, C_k)$ can be any of the point-to-class distances considered in Section 9.1, such as the centroid distance $d(\mathbf{x}, \mathbf{m}_k) = ||\mathbf{x} - \mathbf{m}||$, and the Mahalanobis distance

$$d_M(\mathbf{x}, C_k) = (\mathbf{x} - \mathbf{m}_k)^T \Sigma_k^{-1} (\mathbf{x} - \mathbf{m}_k) \tag{13.2}$$

based on the estimated mean and covariance matrix of all data points belonging to the kth class C_k

$$\mathbf{m}_k = \frac{1}{N_k} \sum_{n=1}^{N_k} \mathbf{x}_n, \qquad \mathbf{\Sigma}_k = \frac{1}{N_k} \sum_{n=1}^{N_k} (\mathbf{x}_n - \mathbf{m})^T (\mathbf{x}_n - \mathbf{m}) \tag{13.3}$$

We need to compare different point-to-class distances to see which of them measures the similarity between $\mathbf{x}$ and each class C_k most accurately, as shown in the following examples.

Example 13.1

Here we consider a simple binary classification problem in a 1-D space as illustrated in the left panel in Fig. 13.2, based on the centroid distance, the Euclidean distance $d_2(\mathbf{x}, \mathbf{m}_k)$ between the data point and the mean of all samples in each of the two classes represented by their corresponding Gaussian probability density distribution (pdf):

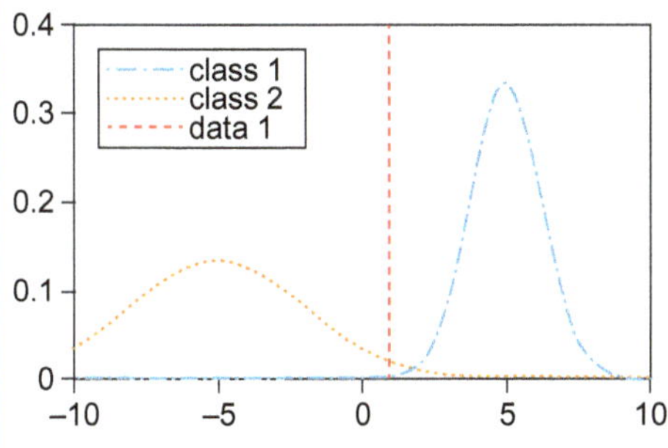

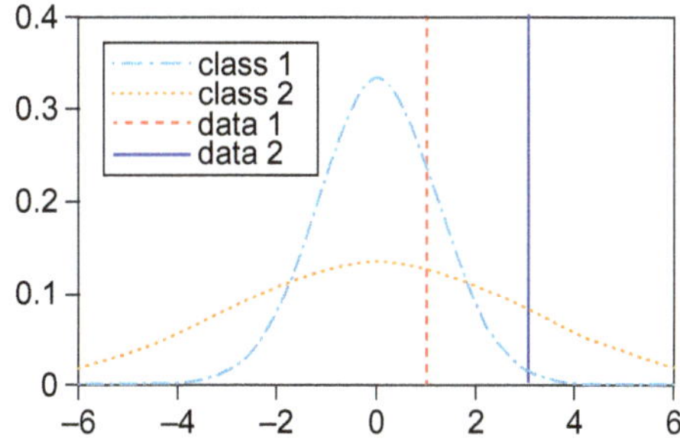

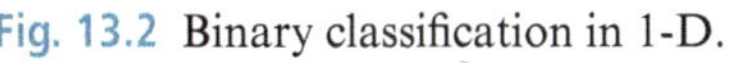

Fig. 13.2 Binary classification in 1-D.

$$C_1 \sim \mathcal{N}(m_1, \sigma_1^2) = \mathcal{N}(5, 1.2^2), \qquad C_2 \sim \mathcal{N}(m_2, \sigma_2^2) = \mathcal{N}(-5, 3^2) \tag{13.4}$$

Using the centroid distance based on only the two means $m_1 = 5$ and $m_2 = -5$, a data sample $x = 1$ is classified to class C_1 with a closer distance to its mean: $d_2(x, m_1) = 4 < d_2(x, m_2) = 6$. However, as shown in the figure, x seems to be more similar to class C_2, as it is more widely distribution than C_1 owing to its greater variance $\sigma_2 = 3 > \sigma_1 = 1.2$, which is not taken into consideration in this case.

Example 13.2

Here we use the Mahalanobis distances between x and each of the two classes, so that the variances of the two classes as well as their means are taken into consideration. In the 1-D case this distance is simply $d_M(x, C_k) = (x - m_k)^2/\sigma_k^2$, which is proportional to $(x - m_k)^2$ but inversely proportional to σ_k^2. As $d_M(x_1, m_1) = 11.11 > d_M(x_1, m_2) = 4.0$, $x = 1$ is classified to C_2.

Example 13.3

Now consider classifying the two samples $x_1 = 1$ and $x_2 = 3$ shown in the right panel in Fig. 13.2 into either of two classes both with a normal distribution:

$$C_1 \sim \mathcal{N}(0, 1.2^2), \qquad C_2 \sim \mathcal{N}(0, 3^2) \tag{13.5}$$

The Mahalanobis distances $d_M(x_i, C_k) = (x_i - m_k)^2/\sigma_k^2$, $(i = 1, 2,\ k = 1, 2)$ between each of the two samples and each of the two classes are listed as

Table 13.1

	Class 1	Class 2
x_1	$1/1.2^2 = 0.69$	$1/3^2 = 0.11$
x_2	$3^3/1.2^2 = 6.25$	$3^2/3^2 = 1$

As $m_1 = m_2 = 0$, $|x_i - m_1|^2 = |x_i - m_2|^2$ for both x_1 and x_2, their Mahalanobis distances are determined by the variances σ_2^2, σ_2^2 alone, and they are both classified into C_2, with a greater variance $\sigma_2^2 > \sigma_2^2$ and therefore smaller distance. However, as shown in the figure, it seems $x_1 = 1$ should be classified to C_1. We therefore see that sometimes minimum distance classification based on the Mahalanobis distance may not be reliable, and some better method needs to be considered, to be discussed in the following section.

13.3 Naive Bayes Classification

The method of naive Bayes (NB) classification is a classical supervised classification algorithm, which is first trained by a training set of N samples $\mathbf{X} = [\mathbf{x}_1, \cdots, \mathbf{x}_N]$ and their corresponding labelings $\mathbf{y} = [y_1, \cdots, y_N]^T$, and then used to classify any unlabeled sample $\mathbf{x}T$ into class C_k with the maximum *posterior probability*. As indicated by the name, naive Bayes classification is based on Bayes' theorem

$$P(C_k|\mathbf{x}) = \frac{p(\mathbf{x}, C_k)}{p(\mathbf{x})} = \frac{p(\mathbf{x}|C_k)P(C_k)}{p(\mathbf{x})} \tag{13.6}$$

where

- $P(C_k)$, or briefly P_k, is the *prior probability* that any randomly selected sample $\mathbf{x}$ belongs to class C_k without any prior knowledge of the sample. If the training set is a fair representation of all patterns of different classes in the entire population, in the sense that the data points in training set are uniformly sampled from the distribution they represent, then we can assume

$$P_k = \frac{N_k}{N} = \frac{N_k}{\sum_{l=1}^K N_l}, \qquad (k = 1, \cdots, K) \qquad \text{satisfying} \quad \sum_{k=1}^K P_k = 1 \tag{13.7}$$

  where N_k is the number of data samples in the training set labeled to belong to class C_k. This estimation is based on the assumption that the training samples are evenly drawn from the entire population, and are therefore a fair representation of all K classes.
- $p(\mathbf{x}|C_k) = L(C_k|\mathbf{x})$ is the *likelihood* for any observed $\mathbf{x}$ to belong to C_k, which is proportional to the conditional probability of $\mathbf{x}$ given $\mathbf{x} \in C_k$, assumed to be a Gaussian in terms of the mean vector $\mathbf{m}_k$ and covariance matrix $\mathbf{\Sigma}_k$

$$p(\mathbf{x}|C_k) = \mathcal{N}(\mathbf{x}, \mathbf{m}_k, \Sigma_k) = \frac{1}{(2\pi)^{d/2}\, |\Sigma_k|^{1/2}} \exp\left[-\frac{1}{2}(\mathbf{x} - \mathbf{m}_k)^T \Sigma_k^{-1} (\mathbf{x} - \mathbf{m}_k)\right] \tag{13.8}$$

  This assumption is based on the fact that the Gaussian distribution has the maximum entropy (uncertainty) among all probability density functions with the same covariance, that is, it imposes the least amount of unsupported constraint on the model for the dataset.
- $p(\mathbf{x}, C_k)$ is the joint probability of $\mathbf{x}$ and C_k

$$p(\mathbf{x}, C_k) = p(\mathbf{x}|C_k)P(C_k) = P(C_k|\mathbf{x})p(\mathbf{x}) \tag{13.9}$$

- $p(\mathbf{x})$ is the distribution of any data sample in the dataset, independent of its class identity, the weighted sum of all likelihood $p(\mathbf{x}|C_k)$ for $k = 1, \cdots, K$, that is the joint probability $p(\mathbf{x}, C_k)$ marginalized over all K classes

$$p(\mathbf{x}) = \sum_{k=1}^{K} p(\mathbf{x}, C_k) = \sum_{k=1}^{K} P_k \, p(\mathbf{x}|C_k)$$
$$= \sum_{k=1}^{K} P_k \left[\frac{1}{(2\pi)^{d/2} \, |\Sigma_k|^{1/2}} \exp\left(-\frac{1}{2} (\mathbf{x} - \mathbf{m}_k)^T \Sigma_k^{-1} (\mathbf{x} - \mathbf{m}_k) \right) \right] \quad (13.10)$$

- $P(C_k|\mathbf{x})$ is the posterior probability that a data sample $\mathbf{x}$ belongs to class C_k based on the observed values of $\mathbf{x}$.

Based on this Bayes' theorem, the naive Bayes method classifies an unlabeled data sample $\mathbf{x}$ to class C_k with the maximum posterior probability (*maximum a posteriori (MAP)*)

$$\text{if } P(C_k|\mathbf{x}) = \max_l \{P(C_l|\mathbf{x}), \ (l = 1, \cdots, K)\}, \quad \text{then } \mathbf{x} \in C_k \quad (13.11)$$

As $p(\mathbf{x})$ is common to all K classes (independent of k), it plays no role in the relative comparison among the K classes, and can therefore be dropped, that is, $\mathbf{x}$ is classified to C_k with maximal $p(\mathbf{x}|C_k) \, P_k$. In the case where the P_k is not available due to the lack of any prior knowledge, it has to be assumed to be the same for all K classes (equally likely), then it is dropped and the method becomes maximum likelihood estimation (MLE).

The naive Bayes classifier is an optimal classifier in the sense that the classification error is minimum. To see this, consider an arbitrary boundary between two classes C_1 and C_2 that partitions the 1-D feature space into two regions R_1 and R_2, as shown in Fig. 13.3. The probability of a misclassification is the joint probability $P(\mathbf{x} \in C_i \cap \mathbf{x} \in R_j)$ for a sample $\mathbf{x} \in C_i$ but falling in R_j. The total probability of error owing to misclassification can be expressed as

$$P(error) = P((\mathbf{x} \in R_2) \cap (\mathbf{x} \in C_1)) + P((\mathbf{x} \in R_1) \cap (\mathbf{x} \in C_2))$$
$$= P(\mathbf{x} \in R_2/C_1) \, P_1 + P(\mathbf{x} \in R_1/C_2) \, P_2$$
$$= P_1 \int_{R_2} p(\mathbf{x}/C_1) d\mathbf{x} + P_2 \int_{R_1} p(\mathbf{x}/C_2) d\mathbf{x} \quad (13.12)$$

As shown in Fig. 13.3, it is obvious that the Bayes classifier is indeed optimal, because the decision boundary $p(\mathbf{x}|C_1)P_1 = p(\mathbf{x}|C_2)P_2$ between two classes C_1 and C_2 (the bottom plot) guarantees the classification error (shaded area) to be minimum.

To find the likelihood function $p(\mathbf{x}|C_k) = \mathcal{N}(\mathbf{x}, \mathbf{m}_k, \mathbf{\Sigma}_k)$, we need to first estimate $\mathbf{m}_k$ and $\mathbf{\Sigma}_k$ based on the N_k training samples in class C_k in the training set

$$\mathbf{m}_k = \frac{1}{N_k} \sum_{i=1}^{N_k} \mathbf{x}_i, \quad \mathbf{\Sigma}_k = \frac{1}{N_k} \sum_{i=1}^{N_k} (\mathbf{x}_i - \mathbf{m}_k)(\mathbf{x}_i - \mathbf{m}_k)^T, \quad (\mathbf{x}_i \in C_k) \quad (13.13)$$

Once both $P_k = N_k/N$ and $p(\mathbf{x}|C_k) = \mathcal{N}(\mathbf{x}, \mathbf{m}_k, \mathbf{\Sigma}_k)$ are available, the classifier is trained, and the posterior $P(C_k|\mathbf{x})$ can be calculated for the classification in Eq. (13.11).

As the classification is based on the relative comparison of the posterior $P(C_k|\mathbf{x})$, any monotonic mapping of the posterior, such as the logarithmic function, can be equivalently used to simplify the computation

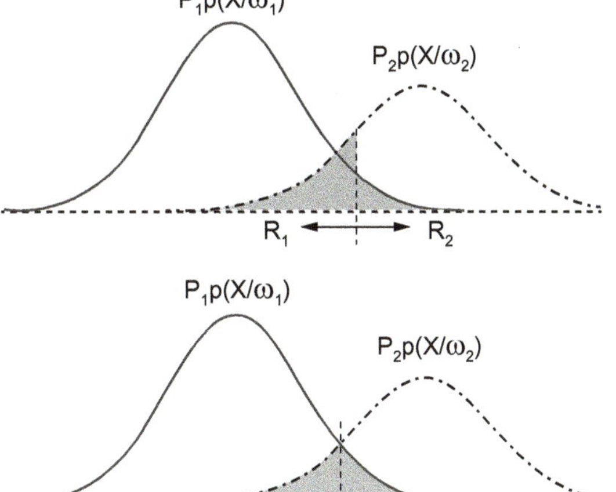

Fig. 13.3 Minimum error of Naive Bayes classifier.

$$\begin{aligned}\log P(C_k|\mathbf{x}) &= \log[p(\mathbf{x}|C_k)P_k/p(\mathbf{x})] = \log p(\mathbf{x}|C_k) + \log P_k - \log p(\mathbf{x}) \\ &= -\frac{1}{2}(\mathbf{x}-\mathbf{m}_k)^T\Sigma_k^{-1}(\mathbf{x}-\mathbf{m}_k) \\ &\quad -\frac{N}{2}\log(2\pi) - \frac{1}{2}\log|\mathbf{\Sigma}_k| + \log P_k - \log p(\mathbf{x})\end{aligned} \tag{13.14}$$

We can further drop the constant terms $\log p(\mathbf{x})$ and $N\log(2\pi)/2$ shared by all classes and obtain the *quadratic discriminant function*

$$D_k(\mathbf{x}) = -\frac{1}{2}(\mathbf{x}-\mathbf{m}_k)^T\Sigma_k^{-1}(\mathbf{x}-\mathbf{m}_k) - \frac{1}{2}\log|\mathbf{\Sigma}_k| + \log P_k \tag{13.15}$$

Now the classification can be represented by

$$\text{if } D_l(\mathbf{x}) = \max\{D_k(\mathbf{x}),\ (k = 1,\cdots,K)\}, \quad \text{then } \mathbf{x} \in C_l \tag{13.16}$$

Equivalently, we can also treat the negative of this discriminant function as a distance between point $\mathbf{x}$ and class C_k with $\mathbf{m}_k$ and $\mathbf{\Sigma}_k$

$$d_k(\mathbf{x}, C_k) = (\mathbf{x}-\mathbf{m}_k)^T\Sigma_k^{-1}(\mathbf{x}-\mathbf{m}_k) + \log|\mathbf{\Sigma}_k| - 2\log P_k \tag{13.17}$$

for minimum distance classification. We note that the first term is the Mohalanobis distance between $\mathbf{x}$ and C_k. However, the additional terms in the expression contribute to better classification performance, as illustrated in the following example.

Example 13.4

Reconsider Example 13.3 shown in the left panel of Fig. 13.4, but now using the negative of the discriminant function treated in Eq. (13.17) as the distance. In this case, the last term $logP_k$ of the distance are dropped as they are assumed to be the same for both classes, and the distances $d(x_i, C_k) = (x_i - m_k)^2/\sigma_k^2 - \log|\sigma_k^2|$, $(i = 1, 2,\ k = 1, 2)$ between each of the two samples and each of the two classes are listed as

Table 13.2

	Class 1	Class 2
x_1	$0.69 + \log(1.2^2) = 1.06$	$0.11 + \log(3^2) = 2.31$
x_2	$6.25 + \log(1.2^2) = 6.61$	$1 + \log(3^2) = 3.20$

As $d(x_1,C_1) = 1.06 < d(x_1,C_2) = 2.31$, x_1 is classified into class C_1 as desired, and as $d(x_1,C_1) = 1.06 < d(x_1,C_2) = 2.31$, x_2 is still classified into C_2, the same as before. We see that the previous misclassification of x_1 into C_2 based on the Mahalanobis distance is now corrected, due to the additional term $\log(\sigma^2)$. Moreover, the right panel of Fig. 13.4 shows how the 1-D space is partitioned by the decision boundaries (the two vertical lines) into regions for C_1 (inside the two vertical lines) and C_2 (outside the vertical lines).

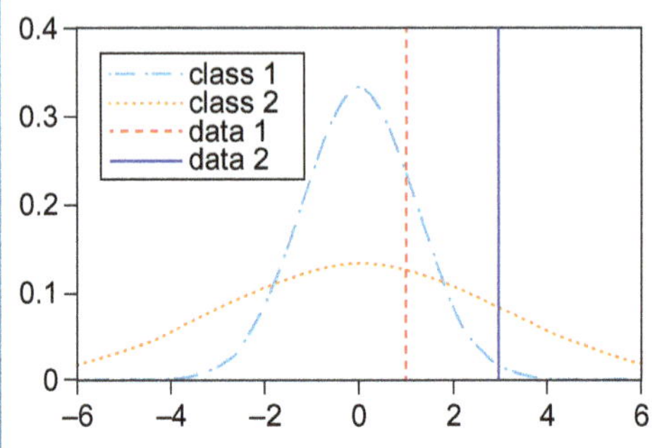

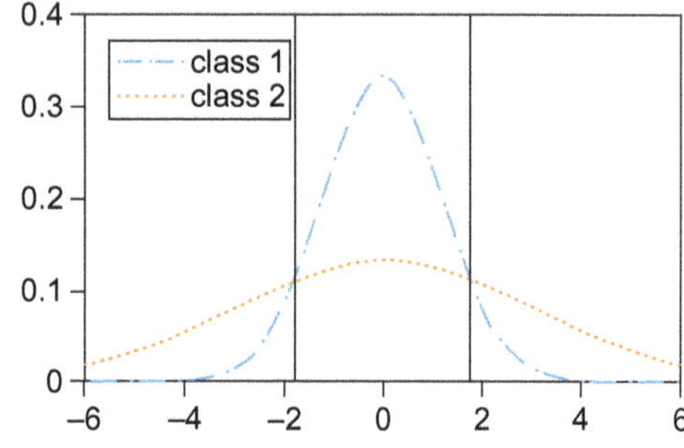

Fig. 13.4 Binary Classification in 1-D revisited.

In the more general case of higher dimensional multiple classes, the feature space is partitioned into regions corresponding to the K classes by the decision boundaries between every pair of classes C_i and C_j, represented by the equation $D_i(\mathbf{x}) = D_j(\mathbf{x})$

$$-\frac{1}{2}(\mathbf{x}-\mathbf{m}_i)^T\Sigma_i^{-1}(\mathbf{x}-\mathbf{m}_i) - \frac{1}{2}\log|\Sigma_i| + \log P_i$$
$$= -\frac{1}{2}(\mathbf{x}-\mathbf{m}_j)^T\Sigma_j^{-1}(\mathbf{x}-\mathbf{m}_j) - \frac{1}{2}\log|\Sigma_j| + \log P_j \tag{13.18}$$

which can be written as

$$\mathbf{x}^T\mathbf{W}\mathbf{x} + \mathbf{w}^T\mathbf{x} + w = 0 \tag{13.19}$$

where

$$\mathbf{W} = -\frac{1}{2}(\Sigma_i^{-1} - \Sigma_j^{-1})$$
$$\mathbf{w} = \Sigma_i^{-1}\mathbf{m}_i - \Sigma_j^{-1}\mathbf{m}_j$$
$$w = -\frac{1}{2}(\mathbf{m}_i^T\Sigma_i^{-1}\mathbf{m}_i - \mathbf{m}_j^T\Sigma_j^{-1}\mathbf{m}_j) - \frac{1}{2}\log\frac{|\Sigma_i|}{|\Sigma_j|} + \log\frac{P_i}{P_j} \tag{13.20}$$

In general, this decision boundary is a quadratic hypersurface (hypersphere, hyperellipsoid, hyperparabola, or hyperhyperbola), by which an unlabeled data sample

$\mathbf{x}$ is classified into either of the two classes C_i and C_j based on whether it is on the positive or negative side of the surface

$$\text{if } \mathbf{x}^T\mathbf{W}\mathbf{x} + \mathbf{w}^T\mathbf{x} + w \begin{cases} > 0 \\ < 0 \end{cases}, \quad \text{then } \mathbf{x} \in \begin{cases} C_i \\ C_j \end{cases} \tag{13.21}$$

We further consider several special cases:

- All classes have the same prior

$$P_i = P_j \quad (i, j = 1, \cdots, K) \tag{13.22}$$

then the last term of $D_i(\mathbf{x})$ is zero, and the naive Bayes classification as a MAP method becomes an MLE method, we have

$$D_k(\mathbf{x}) = -\frac{1}{2}(\mathbf{x} - \mathbf{m}_k)^T \Sigma_k^{-1} (\mathbf{x} - \mathbf{m}_k) - \frac{1}{2} \log |\Sigma_k| \tag{13.23}$$

- All classes have the same covariance matrix $\mathbf{\Sigma}_i = \mathbf{\Sigma}$, the discriminant function becomes

$$D_i(\mathbf{x}) = -\frac{1}{2}(\mathbf{x} - \mathbf{m}_i)^T \Sigma^{-1} (\mathbf{x} - \mathbf{m}_i) + \log P_i \tag{13.24}$$

Moreover, if all P_i are the same and the second term is dropped, $D_i(\mathbf{x})$ becomes the negative *Mahalanobis distance*.

The boundary equation $D_i(\mathbf{x}) = D_j(\mathbf{x})$ between C_i and C_j becomes

$$-\frac{1}{2}(\mathbf{x} - \mathbf{m}_i)^T \mathbf{\Sigma}^{-1} (\mathbf{x} - \mathbf{m}_i) + \log P_i = -\frac{1}{2}(\mathbf{x} - \mathbf{m}_j)^T \mathbf{\Sigma}^{-1} (\mathbf{x} - \mathbf{m}_j) + \log P_j \tag{13.25}$$

As the quadratic terms $\mathbf{x}^T \mathbf{\Sigma}^{-1} \mathbf{x}$ on both sides of the equation are the same, it becomes a linear equation

$$\mathbf{w}^T\mathbf{x} + w = 0 \tag{13.26}$$

where

$$\mathbf{w} = \Sigma^{-1}(\mathbf{m}_i - \mathbf{m}_j) \tag{13.27}$$

and

$$w = -\frac{1}{2}(\mathbf{m}_i^T \Sigma^{-1} \mathbf{m}_i - \mathbf{m}_j^T \Sigma^{-1} \mathbf{m}_j) + \log \frac{P_i}{P_j} \tag{13.28}$$

This linear equation represents a hyperplane between the two points $\mathbf{m}_i$ and $\mathbf{m}_j$ and perpendicular to the straight line $\Sigma^{-1}(\mathbf{m}_i - \mathbf{m}_j)$ (the straight line $\mathbf{m}_i - \mathbf{m}_j$ rotated by matrix Σ^{-1}).

- All classes have the same isotropic distribution

$$\Sigma_i = \sigma^2 \mathbf{I} = \text{diag}[\sigma^2, \cdots, \sigma^2] \tag{13.29}$$

then $|\Sigma_i| = \sigma^{2d}$ and $D_i(\mathbf{x})$ becomes

$$D_i(\mathbf{x}) = -\frac{||\mathbf{x} - \mathbf{m}_i||^2}{2\sigma^2} + \log P_i \tag{13.30}$$

Note that the term log $|\Sigma_i|$ has been dropped from the original expression of $D_i(\mathbf{x})$ as it is now the same for all classes.

The boundary equation $D_i(\mathbf{x}) = D_j(\mathbf{x})$ becomes

$$\frac{||\mathbf{x} - \mathbf{m}_i||^2}{2\sigma^2} - \log P_i = \frac{||\mathbf{x} - \mathbf{m}_j||^2}{2\sigma^2} - \log P_j \tag{13.31}$$

which can be simplified to a linear equation

$$\mathbf{w}^T\mathbf{x} + w = 0 \tag{13.32}$$

where

$$\mathbf{w} = \mathbf{m}_i - \mathbf{m}_j, \qquad w = -(\mathbf{m}_i^T\mathbf{m}_i - \mathbf{m}_j^T\mathbf{m}_j) + 2\sigma^2 \log \frac{P_i}{P_j} \tag{13.33}$$

This linear equation represents a hyperplane between the two points $\mathbf{m}_i$ and $\mathbf{m}_j$ and perpendicular to the straight line passing through these points.

- Further, if all classes have the same P_i, $D_i(\mathbf{x})$ becomes

$$D_i(\mathbf{x}) = -(\mathbf{x} - \mathbf{m}_i)^T(\mathbf{x} - \mathbf{m}_i) = -||\mathbf{x} - \mathbf{m}_i||^2 \tag{13.34}$$

and the Bayes classifier becomes minimum distance classifier based on the centroid distance, the Euclidean distance between $\mathbf{x}$ and the mean $\mathbf{m}_k$ of class C_k (maximizing $D_i(\mathbf{x})$ is equivalent to minimizing $||\mathbf{x} - \mathbf{m}_i||^2$).

The MATLAB functions for training and testing are listed below

```
function [M S P]=NBtraining(X,y)    % naive Bayes training
    [d N]=size(X);                  % dimensions of dataset
    K=length(unique(y));            % number of classes
    M=zeros(d,K);                   % mean vectors
    S=zeros(d,d,K);                 % covariance matrices
    P=zeros(1,K);                   % prior probabilities
    for k=1:K                       % for each of K classes
        idx=find(y==k);             % indices of samples
        Xk=X(:,idx);                % get all samples in class
        P(k)=length(idx)/N;         % prior probability
        M(:,k)=mean(Xk');           % mean vector
        S(:,:,k)=cov(Xk');          % covariance matrix
    end
end

function yhat=NBtesting(X,M,S,P)    % naive Bayes testing
    [d N]=size(X);                  % dimensions of dataset
    K=length(P);                    % number of classes
    for k=1:K
        InvS(:,:,k)=inv(S(:,:,k));  % inverse of covariance
        Det(k)=det(S(:,:,k));       % determinant of covariance
    end
```

```
    for n=1:N                           % for each of N samples
        x=X(:,n);
        dmax=-inf;
        for k=1:K
            xm=x-M(:,k);
            d=-(xm'*InvS(:,:,k)*xm)/2;
            d=d-log(Det(k))/2+log(P(k)); % discriminant function
            if d>dmax
                yhat(n)=k;              % assign nth sample to kth class
                dmax=d;
            end
        end
    end
end
```

The classification result in terms of the estimated class identity $\hat{\mathbf{y}}$ of the training samples compared with the ground truth class labeling $\mathbf{y}$ can be represented in the form of a $K \times K$ confusion matrix, of which the element in the ith row and jth column is the number of training samples labeled as a member of the ith class but actually classified into the jth class. For an ideal classifier that classifies all data samples correctly, this confusion matrix should be a diagonal matrix. But for a nonideal classifier, its error rate can be found as the percentage of the number of misclassified samples (sum of all off-diagonal elements) out of the total number of samples. The following code below shows the function that generates such a confusion matrix based on the estimated class identity $\hat{\mathbf{y}}$ and the ground truth labeling $\mathbf{y}$

```
function [Cm er]=ConfusionMatrix(yhat,ytrain)
    N=length(ytrain);               % number of test samples
    K=length(unique(ytrain));       % number of classes
    Cm=zeros(K);                    % the confusion matrices
    for n=1:N
        i=ytrain(n);
        j=yhat(n);
        Cm(i,j)=Cm(i,j)+1;
    end
    er=1-sum(diag(Cm))/N;           % error percentage
    end
```

Example 13.5

This example shows the classification of two concentric classes with one surrounding the other in 2-D space, as shown in Fig. 13.5. The feature space is partitioned into two regions by an ellipse and 9 out of 200 samples misclassified, that is, with an error rate of $9/200 = 4.5\%$. The confusion matrix is

$$\begin{bmatrix} 92 & 8 \\ 1 & 99 \end{bmatrix}$$

indicating 8 of the 100 samples belonging to class 1 are misclassified to class 2, and 1 of the 100 samples belonging to class 2 is misclassified to class 1.

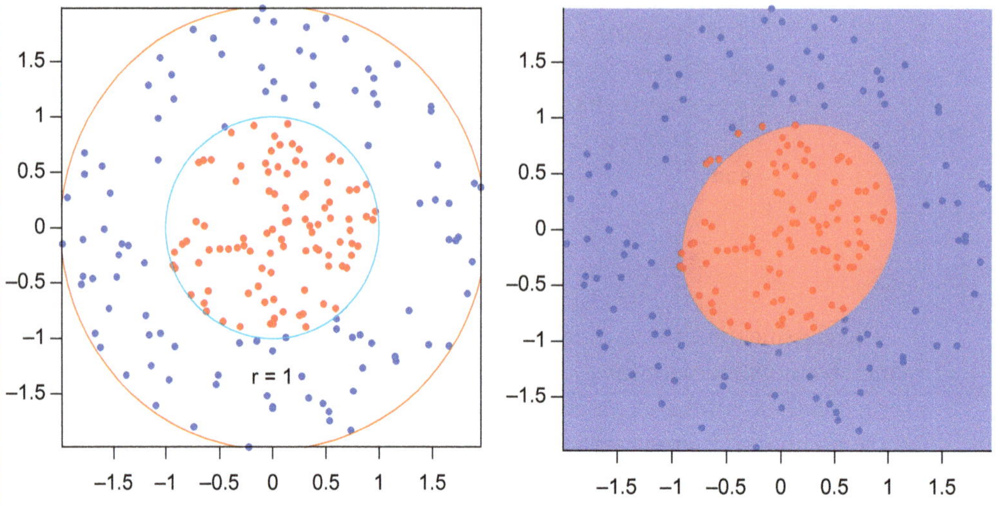

Fig. 13.5 Naive Bayes classification of concentric binary classes.

Example 13.6

This example shows the classification of a two-class exclusive-or (XOR) data set with significant overlap, as shown in Fig. 13.6, in terms of the confusion matrix and the error rate of $61/400 = 15.25\%$. Note that the decision boundaries are a pair of hyperbolas. The confusion matrix is

$$\begin{bmatrix} 175 & 25 \\ 36 & 164 \end{bmatrix}$$

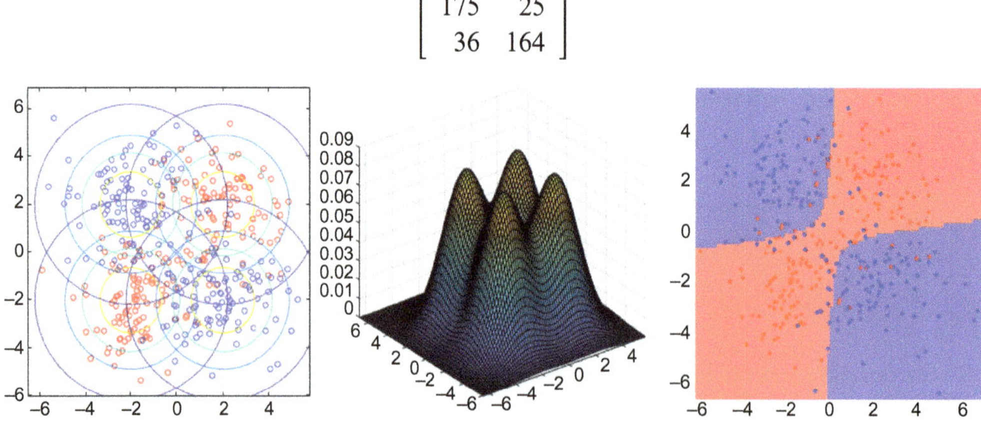

Fig. 13.6 Naive Bayes classification of XOR binary classes.

Example 13.7

The left panel of Fig. 13.7 shows data points in 2-D space drawn from three distributions shown in the middle panel, while the right one shows the partitioning of the space by quadratic decision boundaries obtained by the naive Bayes classifier. The confusion matrix of the classification result is shown with the error rate $49/600 = 8.17\%$.

$$\begin{bmatrix} 196 & 3 & 1 \\ 0 & 191 & 9 \\ 3 & 33 & 164 \end{bmatrix}$$

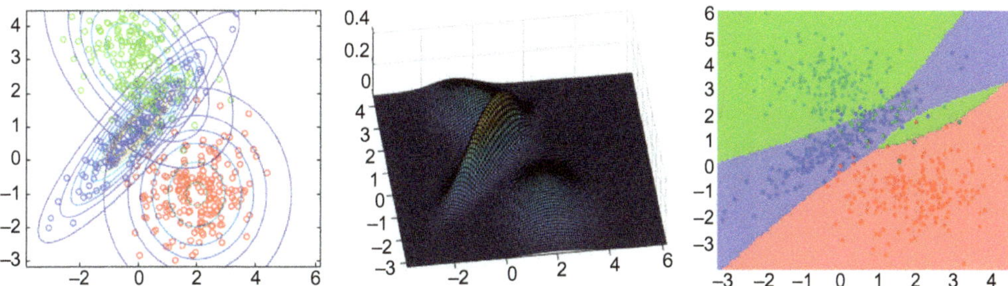

Fig. 13.7 Naive Bayes classification of three classes.

Example 13.8

Figure 13.8 shows a 3-D feature space containing $K = 3$ groups of data points (in red, green and blue), which is partitioned by three quadratic decision boundaries $D_i(\mathbf{x}) = D_j(\mathbf{x})$ (Eq. (13.19)) into three regions for each C_l of the three classes with maximum discriminant function $D_l(\mathbf{x})$ (Eq. (13.16).

The same classification result can also be reached by carrying out three binary classifications each based on one of the three decision boundaries followed by a majority vote. For example, a data point in the red group will win at least two out of the three binary classifications and it is therefore classified into the red class. This method can be used to carry out multiclass classification based on any binary classifier, such as the support vector machine, to be considered in Chapter 14.

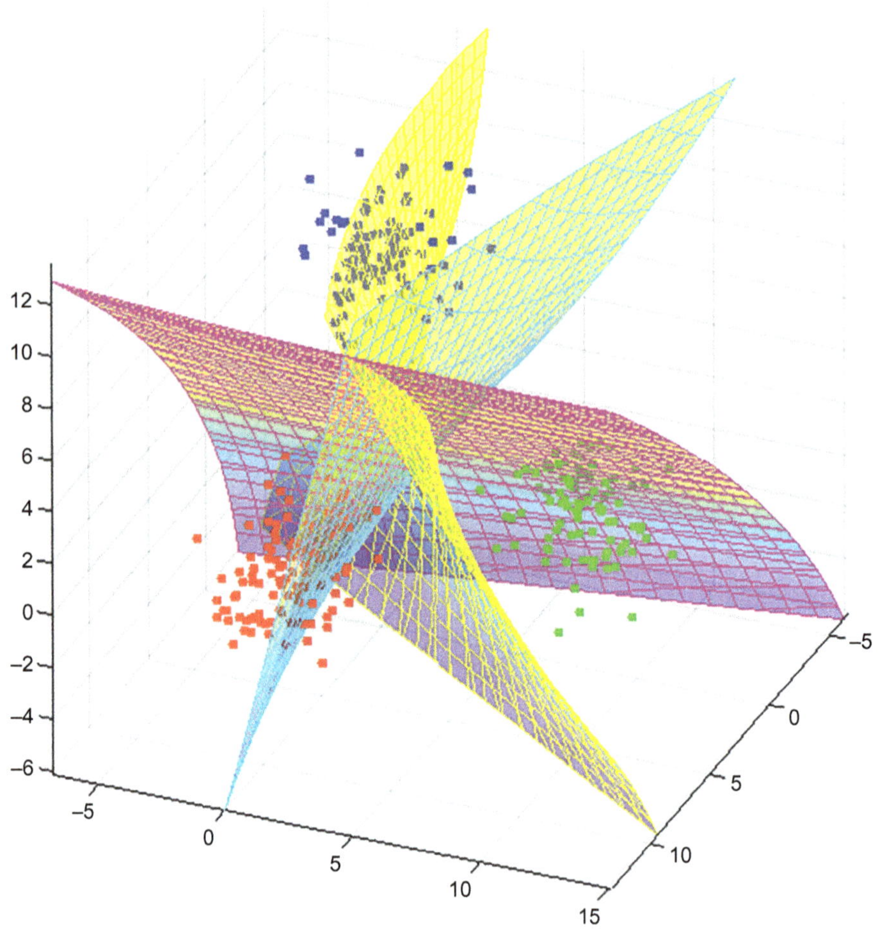

Fig. 13.8 Partitioning of a 3-D space by quadratic decision boundaries.

Example 13.9

Figure 13.9 shows a subset of the dataset of handwritten digits each in a 16×16 image form treated as a 256-dimensional vector.

The dataset is visualized by mapping the data points from the original 256-D space to 3-D or even 2-D space based on either the linear Karhunen–Loève transform (KLT) (Section 10.2), or the nonlinear t-distributed stochastic neighbor embedding (tSNE) (Section 11.6), as shown respectively in the left and right panels of Figs 13.10 and 13.11.

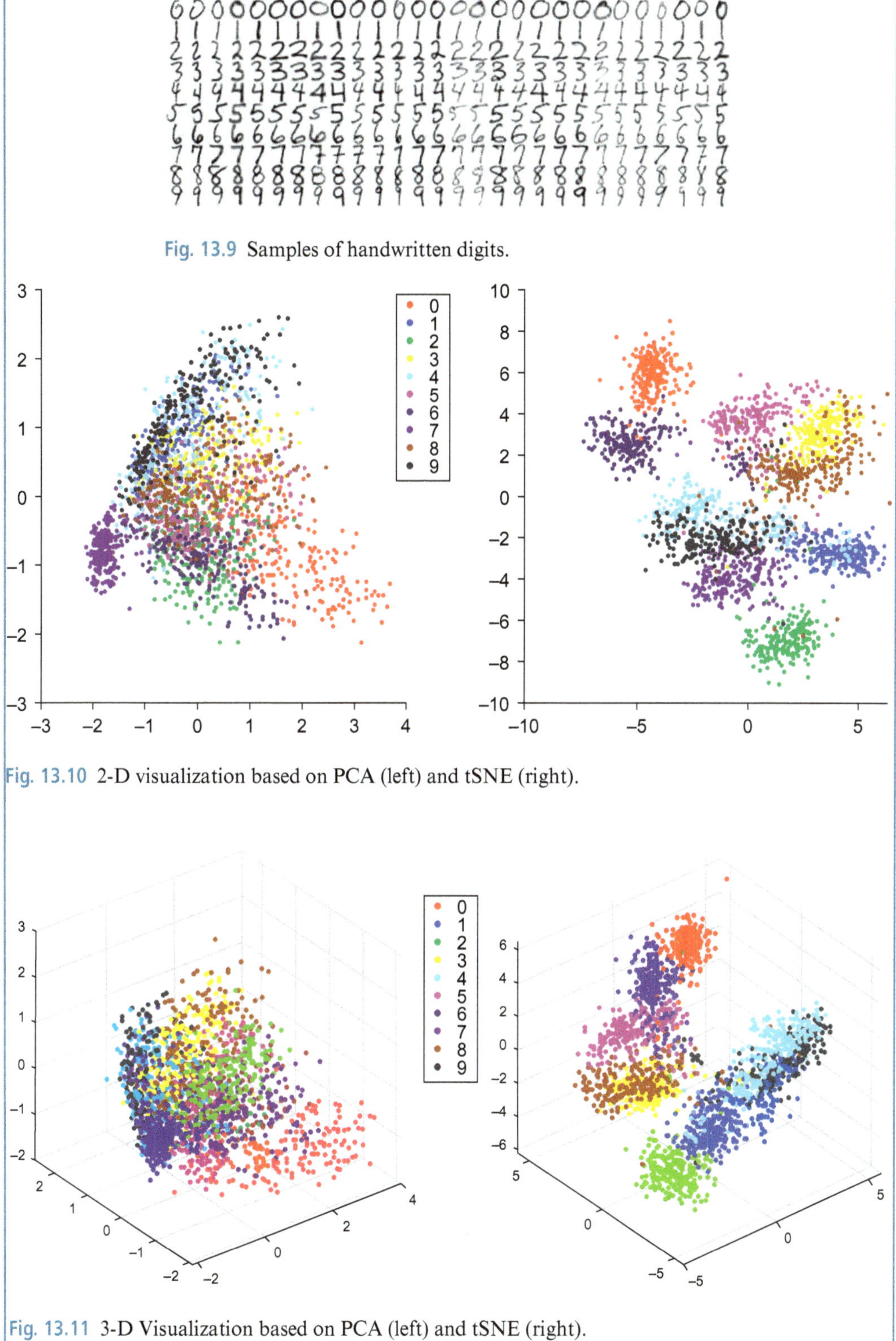

Fig. 13.9 Samples of handwritten digits.

Fig. 13.10 2-D visualization based on PCA (left) and tSNE (right).

Fig. 13.11 3-D Visualization based on PCA (left) and tSNE (right).

To use the naive-Bayes classifier for the classification of this dataset, we need to estimate the covariance matrix $\boldsymbol{\Sigma}_k$ based on the 224 samples in each of the $K = 10$ class. As the rank of the matrix is $224 - 1 = 223 < 256$, it is not invertible and $\boldsymbol{\Sigma}_k^{-1}$ needed in the algorithm does not exist. We therefore have to reduce the dimensionality by, either KLT or tSNE, from 256 to some value (e.g., 100) smaller than the number of samples used for estimating the covariance matrix.

The classifier is first trained on half of the data samples randomly selected samples and then tested on the other half. The classification results are given in the following in terms of the confusion matrices of the training samples (first) and the testing samples (second). All 1,120 training samples are correctly classified with zero error rate, while out of the remaining 1,120 test samples, 198 are misclassified with an error rate of 17.7%.

$$\begin{bmatrix} 107 & 0 & 0 & 0 & 0 & 0 & 0 & 0 & 0 & 0 \\ 0 & 108 & 0 & 0 & 0 & 0 & 0 & 0 & 0 & 0 \\ 0 & 0 & 105 & 0 & 0 & 0 & 0 & 0 & 0 & 0 \\ 0 & 0 & 0 & 118 & 0 & 0 & 0 & 0 & 0 & 0 \\ 0 & 0 & 0 & 0 & 110 & 0 & 0 & 0 & 0 & 0 \\ 0 & 0 & 0 & 0 & 0 & 120 & 0 & 0 & 0 & 0 \\ 0 & 0 & 0 & 0 & 0 & 0 & 128 & 0 & 0 & 0 \\ 0 & 0 & 0 & 0 & 0 & 0 & 0 & 104 & 0 & 0 \\ 0 & 0 & 0 & 0 & 0 & 0 & 0 & 0 & 109 & 0 \\ 0 & 0 & 0 & 0 & 0 & 0 & 0 & 0 & 0 & 111 \end{bmatrix}$$

$$\begin{bmatrix} 101 & 0 & 0 & 0 & 0 & 0 & 0 & 2 & 0 & 0 \\ 0 & 58 & 0 & 0 & 0 & 0 & 0 & 0 & 0 & 0 \\ 0 & 1 & 105 & 0 & 0 & 0 & 0 & 1 & 0 & 0 \\ 5 & 2 & 6 & 101 & 2 & 4 & 0 & 28 & 4 & 4 \\ 0 & 15 & 0 & 0 & 102 & 0 & 0 & 4 & 0 & 9 \\ 5 & 4 & 1 & 1 & 0 & 99 & 1 & 1 & 2 & 1 \\ 0 & 17 & 1 & 0 & 2 & 0 & 95 & 2 & 0 & 0 \\ 0 & 1 & 0 & 0 & 0 & 0 & 0 & 62 & 0 & 0 \\ 6 & 15 & 6 & 4 & 6 & 1 & 0 & 18 & 109 & 9 \\ 0 & 3 & 0 & 0 & 2 & 0 & 0 & 2 & 0 & 90 \end{bmatrix}$$

The dimensionality of the dataset can be reduced further down to some value much lower such as a single digits, by either the principal component analysis (PCA) or tSNE methods, and then the naive Bayes classification is carried out in the low dimensional space. The error rates of the two classification methods when applied to this dataset (with all 2,240 samples used in both training and testing) are compared in the following table and plotted in Fig. 13.12. It is interesting to note that when the dimensionality is reduced to lower than 7, the error rates based on tSNE are much lower than those based on PCA, but for higher dimensionalities, PCA slightly outperforms tSNE.

Table 13.3

Dimensions	1	2	3	4	5	6	7	8	9	10
PCA	0.69	0.48	0.25	0.18	0.15	0.11	0.08	0.06	0.06	0.05
tSNE	0.50	0.15	0.10	0.08	0.08	0.08	0.08	0.08	0.08	0.07

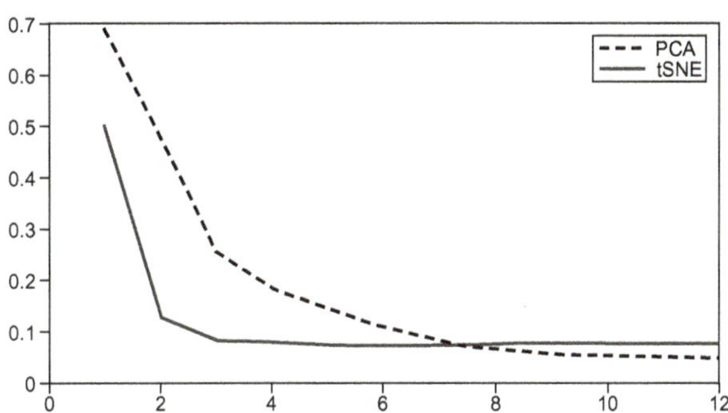

Fig. 13.12 Comparison of KLT and tSNE for feature selection.

13.4 Adaptive Boosting

Adaptive boosting (AdaBoost) is a supervised classification algorithm that belongs to the general method of ensemble learning which is first discussed in Section 4.4. AdaBoost is a binary classifier based on the training set containing N labeled data samples $\{(\mathbf{x}_n, y_n) \mid n = 1, \cdots, N\}$, where each sample $\mathbf{x}_n$ is labeled by $y_n \in \{-1, +1\}$, indicating to which of the two classes C_- and C_+ it belongs.

AdaBoost learning is an iterative process. In the tth iteration, each of the N training samples is classified into one of the two classes by a *weak classifier* $h_t(\mathbf{x}_n) \in \{-1, +1\}$, as a hypothesis. Here the classifier is said to be weak in the sense that its performance only needs to be better than chance. If $\mathbf{x}_n$ is correctly classified, then $h_t(\mathbf{x}_n) = y_n = \pm 1$, that is, $y_n\, h_t(\mathbf{x}_n) = 1$, otherwise it is misclassified with $h_t(\mathbf{x}_n) = -y_n = \pm 1$, that is, $y_n\, h_t(\mathbf{x}_n) = -1$.

The weighted overall error rate is defined as

$$\varepsilon_t = \frac{\sum_{y_n\, h_t(\mathbf{x}_n)=-1} w_t(n)}{\sum_{n=1}^{N} w_t(n)} \tag{13.35}$$

where $w_t(n)$ is the weight for the nth training sample $\mathbf{x}_n$ at the tth iteration. For this error rate to be lower than chance, it is required that $\varepsilon_t < 1/2$. We need to find the best weak classifier that minimizes this error ε_t. We also find the correct rate as

$$1 - \varepsilon_t = 1 - \frac{\sum_{y_n\, h_t(\mathbf{x}_n)=-1} w_t(n)}{\sum_{n=1}^{N} w_t(n)} = \frac{\sum_{y_n\, h_t(\mathbf{x}_n)=1} w_t(n)}{\sum_{n=1}^{N} w_t(n)} \tag{13.36}$$

At the beginning of the iteration with $t = 0$, all N weights are initialized to 1, $w_0(1) = \cdots = w_0(N) = 1$, and the weighted error

$$\varepsilon_0 = \frac{1}{N} \sum_{y_n h_0(\mathbf{x}_n)=-1} w_0(n) = \frac{\text{number of misclassified samples}}{\text{total number of samples}} \tag{13.37}$$

is the probability for any training sample $\mathbf{x}$ to be misclassified by the weak classifier h_0. During the iteration when $t > 0$, this weight $w_t(n)$ will be modified depending on whether $\mathbf{x}_n$ is classified correctly in the subsequent iterations.

At each iteration, we also construct a *strong* or *boosted classifier* $H_t(\mathbf{x}_n)$ as a linear combination (boosting) of all previous weak classifiers $h_i(\mathbf{x}_n)$, $(i \le t)$

$$\begin{aligned} F_t(\mathbf{x}_n) &= \alpha_t h_t(\mathbf{x}_n) + F_{t-1}(\mathbf{x}_n) = \alpha_t h_t(\mathbf{x}_n) + \alpha_{t-1} h_{t-1}(\mathbf{x}_n) + F_{t-2}(\mathbf{x}_n) \\ &= \cdots = \sum_{i=1}^{t} \alpha_i h_i(\mathbf{x}_n) \qquad (n = 1, \cdots, N) \end{aligned} \tag{13.38}$$

where α_i is some coefficient for $h_i(\mathbf{x}_n)$ in the ith iteration, to be determined as shown in the following. The strong classifier is then defined as the sign function of $F_t(\mathbf{x})$

$$H_t(\mathbf{x}_n) = sign[F_t(\mathbf{x}_n)] = \begin{cases} +1 & F_t(\mathbf{x}_n) > 0 \\ -1 & F_t(\mathbf{x}_n) < 0 \end{cases} \tag{13.39}$$

by which a training sample $\mathbf{x}_n$ is classified into one of the two classes. The magnitude $|F_t(\mathbf{x}_n)|$ represents the confidence of this decision.

The weight $w_t(n)$ of $\mathbf{x}_n$ in the tth iteration $(t > 0)$ is defined as

$$w_t(n) = e^{-y_n F_{t-1}(\mathbf{x}_n)} \begin{cases} > 1 & \text{if } y_n\, F_{t-1}(\mathbf{x}_n) < 0 \text{ (misclassification)} \\ < 1 & \text{if } y_n\, F_{t-1}(\mathbf{x}_n) > 0 \text{ (correct classification)} \end{cases} \tag{13.40}$$

which can be obtained from $w_{t-1}(n)$ of the previous iteration

$$\begin{aligned} w_t(n) &= e^{-y_n\, F_{t-1}(\mathbf{x}_n)} = e^{-y_n\, [\alpha_{t-1} h_{t-1}(\mathbf{x}_n) + F_{t-2}(\mathbf{x}_n)]} \\ &= e^{-\alpha_{t-1}\, y_n h_{t-1}(\mathbf{x}_n)}\, e^{-y_n\, F_{t-2}(\mathbf{x}_n)} = e^{-\alpha_{t-1}\, y_n h_{t-1}(\mathbf{x}_n)}\, w_{t-1}(n) \end{aligned} \tag{13.41}$$

and this can be carried out recursively all the way back to the first iteration with $w_1(k) = e^{-\alpha_0\, y_n h_0(\mathbf{x}_n)} w_0(n)$.

The performance of the strong classifier can be measured by the *exponential loss*, defined as the sum of all N weights

$$E_{t+1} = \sum_{n=1}^{N} w_{t+1}(n) = \sum_{n=1}^{N} e^{-y_n\, F_t(\mathbf{x}_n)} = \sum_{n=1}^{N} e^{-\alpha_t\, y_n\, h_t(\mathbf{x}_n)}\, w_t(n) \qquad (t > 1) \tag{13.42}$$

Now we can find coefficient α_t in Eq. (13.38) as the one that minimizes the exponential loss E_{t+1}. To do so, we first separate each of the N terms into two parts, corresponding to whether $\mathbf{x}_n$ is classified by h_t correctly or incorrectly

$$E_{t+1} = \sum_{n=1}^{N} w_t(n) e^{-\alpha_t\, y_n\, h_t(\mathbf{x}_n)} = \sum_{y_n h_t(\mathbf{x}_n)=-1} w_t(n) e^{\alpha_t} + \sum_{y_n h_t(\mathbf{x}_n)=1} w_t(n) e^{-\alpha_t} \tag{13.43}$$

and then set its derivative with respect to α_t to zero:

$$\frac{dE_{t+1}}{d\alpha_t} = \sum_{y_n h_t(\mathbf{x}_n)=-1} w_t(n) e^{\alpha_t} - \sum_{y_n h_t(\mathbf{x}_n)=1} w_t(n) e^{-\alpha_t} = 0 \tag{13.44}$$

Solving this equation we get α_t as the optimal coefficient that minimizes E_{t+1}

$$\alpha_t = \frac{1}{2}\ln\left(\frac{\sum_{y_n h_t(\mathbf{x}_n)=1} w_t(n)}{\sum_{y_n h_t(\mathbf{x}_n)=-1} w_t(n)}\right) = \ln\sqrt{\frac{1-\varepsilon_t}{\varepsilon_t}} > 0 \tag{13.45}$$

which is greater than zero because $\varepsilon_t < 1/2$, $1 - \varepsilon_t > 1/2$, and

$$e^{\alpha_t} = \sqrt{\frac{1-\varepsilon_t}{\varepsilon_t}} > 1, \quad e^{-\alpha_t} = \sqrt{\frac{\varepsilon_t}{1-\varepsilon_t}} < 1 \tag{13.46}$$

We see that if a weak classifier h_t has a small error ε_t, it will be weighted by a large α_t and therefore contributes more to the strong classifier H_t; conversely, if h_t has a large error ε_t, it will be weighted by a small α_t and contributes less to H_t. As the strong classifier H_t takes advantage of all previous weak classifiers $h_1, \cdots, h_t$, each of which may be more effective in a certain region of the N–D feature space than others, H_t can be expected to be a much more accurate classifier than any of the weak classifiers.

Replacing t by $t + 1$ in Eq. (13.41), we now have

$$w_{t+1}(n) = w_t(n)\, e^{-\alpha_t y_n h_t(\mathbf{x}_n)} = \begin{cases} w_t(n)\sqrt{\frac{\varepsilon_t}{1-\varepsilon_t}} < w_t(n) & \text{if } y_n h_t(\mathbf{x}_n) = 1 \\ w_t(n)\sqrt{\frac{1-\varepsilon_t}{\varepsilon_t}} > w_t(n) & \text{if } y_n h_t(\mathbf{x}_n) = -1 \end{cases} \tag{13.47}$$

We see that if $y_n h_t(\mathbf{x}_n) = 1$, that is, $\mathbf{x}_n$ is classified correctly by h_t, it will be weighted more lightly by $w_{t+1}(n) < w_t(n)$ in the next iteration, but if $y_n h_t(\mathbf{x}_n) = -1$, that is, $\mathbf{x}_n$ is classified incorrectly by h_t, it will be weighted more heavily by $w_{t+1}(k) > w_t(k)$ in the next iteration, thereby it will be emphasized and have a better chance to be corrected to have more accurately classified by h_{t+1} in the next iteration.

We further consider the ratio of the exponential losses of two consecutive iterations

$$\begin{aligned} \frac{E_{t+1}}{E_t} &= \frac{\sum_{n=1}^N w_{t+1}(n)}{\sum_{n=1}^N w_t(n)} = \frac{\sum_{n=1}^N w_t(n) e^{-\alpha_t y_n h_t(\mathbf{x}_n)}}{\sum_{n=1}^N w_t(k)} \\ &= \frac{\sum_{y_n h_t(\mathbf{x}_n)=-1} w_t(n)}{\sum_{n=1}^N w_t(n)} e^{\alpha_t} + \frac{\sum_{y_n h_t(\mathbf{x}_n)=1} w_t(n)}{\sum_{n=1}^N w_t(n)} e^{-\alpha_t} \\ &= \varepsilon_t \sqrt{\frac{1-\varepsilon_t}{\varepsilon_t}} + (1-\varepsilon_t)\sqrt{\frac{\varepsilon_t}{1-\varepsilon_t}} = 2\sqrt{\varepsilon_t(1-\varepsilon_t)} \le 1 \end{aligned} \tag{13.48}$$

This ratio is twice the geometric average $\sqrt{\varepsilon_t(1-\varepsilon_t)} \le 1$ of ε_t and $1 - \varepsilon_t$, which reaches its maximum when $\varepsilon_t = 1 - \varepsilon_t = 1/2$. However, as $\varepsilon_t < 1/2$, the ratio is always smaller than 1. We see that the exponential cost can be approximated as an exponentially decaying function from its initial value $E_0 = \sum_{n=1}^N w_0(n) = N$ for some $\varepsilon < 1/2$

$$\lim_{t\to\infty} E_t \approx \lim_{t\to\infty} \left(2\sqrt{\varepsilon(1-\varepsilon)}\right)^t E_0 = 0 \tag{13.49}$$

We can therefore conclude that the exponential loss will always decrease, that is, the error of AdaBoost will always converge to zero.

The weak classifier used in each iteration is typically implemented as a *decision stump*, a binary classifier that partitions the N–D feature space into two regions. Specifically, as a simple example, a coordinate descent method can be used by which all training samples are projected onto the ith dimension of the feature space $\mathbf{d}_i$

$$x_n = P_{\mathbf{d}_i}(\mathbf{x}_n) = \frac{\mathbf{x}_n^T \mathbf{d}_i}{||\mathbf{d}_i||}, \qquad (n = 1, \cdots, N) \tag{13.50}$$

and then classified into two classes by a threshold T along the direction of $\mathbf{d}_i$

$$h(x_n) = \begin{cases} +1 & \text{if } x_n < T \\ -1 & \text{if } x_n > T \end{cases} \tag{13.51}$$

The optimal decision stump is obtained when the following weighted error is minimized with respect to the threshold T along that direction of $\mathbf{d}_i$

$$\varepsilon_t = \sum_{n=1}^{N} w_t(n)\ \delta(h_t(x_n) - y_n) = \sum_{h_t(x_n) \neq y_n} w_t(n) \tag{13.52}$$

Alternatively, the weak classifier can also be obtained based on the *PCA* by partitioning the N–D feature space along the directions of the eigenvectors of the between-class scatter matrix

$$\mathbf{S}_b = \frac{1}{N}\left[N_-(\mathbf{m}_{-1} - \mathbf{m})(\mathbf{m}_{-1} - \mathbf{m})^T + N_+(\mathbf{m}_{+1} - \mathbf{m})(\mathbf{m}_{+1} - \mathbf{m})^T\right] \tag{13.53}$$

where N_- and N_+ are the numbers of samples in the two classes ($N_- + N_+ = N$), and

$$\mathbf{m}_\pm = \frac{1}{N_\pm} \sum_{y_n = \pm 1} w(n)\, \mathbf{x}_n, \qquad \mathbf{m} = \frac{1}{N} \sum_{n=1}^{N} w(n)\, \mathbf{x}_n \tag{13.54}$$

are the weighted mean vectors of the two classes and the total weighted mean vector of all N samples in the training set. The training samples are likely to be better separated along these directions of the eigenvectors of $\mathbf{S}_b$, which measures the separability of the two classes. The eigenequations of $\mathbf{S}_b$ are

$$\mathbf{S}_b = \mathbf{V}\Lambda\mathbf{V}^{-1} = \mathbf{V}\Lambda\mathbf{V}^T \tag{13.55}$$

As $\mathbf{S}_b$ is symmetric, the eigenvector matrix $\mathbf{V}$ is orthonormal and its columns can be used as an orthogonal basis that spans the feature space as well as the standard basis. The same binary threshold classification considered for the weak classifiers can be readily applied along these PCA bases.

The following lists the MATLAB code of the main iteration loop of the algorithm, followed by the essential functions called by the main loop. Here `T` is the maximum number of iteration, and `N` is the total number of training samples.

The algorithm can be carried out in the feature space based on either the PCA basis (columns of the eigenvector matrix $\mathbf{V}$), or the original standard basis (columns of the identity matrix $\mathbf{I}$).

The *decision stump* is implemented by a binary classifier, which partitions all training samples projected onto a 1-D space into two groups by a threshold value, which needs to be optimal in terms of the number of misclassification. A parameter `plt` is used to indicate the polarity of the binary classification, that is, which of the two class C_1 and C_0 is on the lower (or higher) side of the threshold.

```
Plt=[];                % polarity of binary classification
Tr=[];                 % transformation of basis vectors
Th=[];                 % threshold of binary classification
Alpha=[];              % alpha values
h=zeros(T,N);          % h functions
F=zeros(T,N);          % F functions
w=ones(T,N);           % weights for N training samples
Er=N;                  % error initialized to N
t=0;                   % iteration index
while t<T  & Er>0      % the main iteration
    t=t+1;
    [tr th plt er]=WeakClassifier(X,y,w(t,:),PCA);
                                  % weak classifier
    alpha=log(sqrt((1-er)/er)); % update alpha
    Alpha=cat(1,Alpha,alpha);    % record alpha
    Tr=cat(2,Tr,tr');            % record transform vector
    Th=cat(1,Th,th);             % record threshold
    Plt=cat(1,Plt,plt);          % record polarity
    x=tr*X;                      % carry out transform
    c=sqrt(er/(1-er));
    for n=1:N                    % update weights
        h(t,n)=h_function(x(n),th,plt);  % find h function
      if h(t,n)*y(n)<0
          w(t+1,n)=w(t,n)/c;     % update weights
      else
          w(t+1,n)=w(t,n)*c;
      end
    end
    F(t,:)=Alpha(t)*h(t,:);      % get F functions
    if t>1
        F(t,:)=F(t,:)+F(t-1,:);
    end
    Er=sum(sign(F(t,:).*y)==-1);  % error of strong classifier,
                                  % number of misclassifications
    fprintf('%d: %d/%d=%f\n',t,Er,N,Er/N)
end
```

Here are the functions called by the main iteration loop

```
function [Tr,Th Plt Er]=WeakClassifier(X,y,w,pca)
    % X: N columns each for one of the N training samples
    % y: labeling of X
```

```
% w: weights for N training samples
% pca: use PCA dimension if pca~=0
% Er: minimum error among all D dimensions
% Tr: transform vector (standard or PCA basis)
% Th: threshold value
% Plt: polarity
[D N]=size(X);
n0=sum(y>0);                % number of samples in class C+
n1=sum(y<0);                % number of samples in class C-
if pca                      % find PCA basis
    for i=1:D
        Y(i,:)=w.*X(i,:);
    end
    X0=Y(:,find(y>0));  % all samples in class C+
    X1=Y(:,find(y<0));  % all samples in class C-
    m0=mean(X0')';          % mean of C+
    m1=mean(X1')';          % mean of C-
    mu=(n0*m0+n1*m1)/N; % over all mean
    Sb=(n0*(m0-mu)*(m0-mu)'+n1*(m1-mu)*(m1-mu)')/N;
                            % between-class scatter matrix
    [v d]=eig(Sb);          % eigenvector and eigenvalue of Sb
else
    v=eye(N);               % standard basis
end
Er=9e9;                     % initialize min error
for i=1:D                   % for each of D dimensions
    tr=v(:,i)';             % get an eigenvector of Sb
    x=tr*X;                 % rotate the vector
    [th plt er]=BinaryClassifier(x,y,w);
                            % binary classify N samples in 1-D
    er=0;
    for n=1:N
        h(n)=h_function(x(n),th,plt);
                            % h-function of nth sample
        if h(n)*y(n)<0  % if misclassified
            er=er+w(n); % add error
        end
    end
    er=er/sum(w);           % total error of ith dimension
    if er<Er                % a smaller error
        Er=er;              % update min error
        Plt=plt;            % polarity
        Th=th;              % threshold
        Tr=tr;              % transform vector
```

```
            end
        end
    end
    function h=h_function(x, th, plt)
        if xor(x>th, plt)
            h=1;
        else
            h=-1;
        end
    end
    function [Th Plt Er]=BinaryClassifier(x,y,w)
        N=length(x);
        [x1 i]=sort(x);          % sort 1-D data x
        y1=y(i);                 % reorder targets
        w1=w(i);                 % reorder weights
        Er=9e9;                  & initialize min error
        for n=1:N-1              % for N-1 ways of binary classification
            e0=sum(w1(find(y1(1:n)==1)))+sum(w1(n+find(y1(n+1:N)==-1)));
            e1=sum(w1(find(y1(1:n)~=1)))+sum(w1(n+find(y1(n+1:N)~=-1)));
            if e1 > e0           % polarity: left -1, right +1
                plt=0; er=e0;
            else                 % polarity: left +1, right -1
                plt=1; er=e1;
            end
            if er<Er             % a smaller error
                Er=er;           % update min error
                Plt=plt;         % polarity
                Th=(x1(k)+x1(k+1))/2;   % threshold
            end
        end
    end
```

Example 13.10

This example shows the classification of the two-class XOR dataset (used previously for testing the naive Bayes method), containing two classes of $N = 200$ samples each in 2-D space, represented by the red and blue dots in Fig. 13.13. The intermediate results after 4, 8, 16 and 32 iterations shows the progress of the iteration, when the error rate continuously reduces from 134/400, 85/400, 81/400, 58/100. Figure 13.14 shows a sequence of 350 binary participations of the 2-D space (top left) and the final classification result (top right). The error rate eventually reduced to zero as guaranteed by the AdaBoost method. Although all training samples are correctly classified eventually, the result suffers the problem of overfitting.

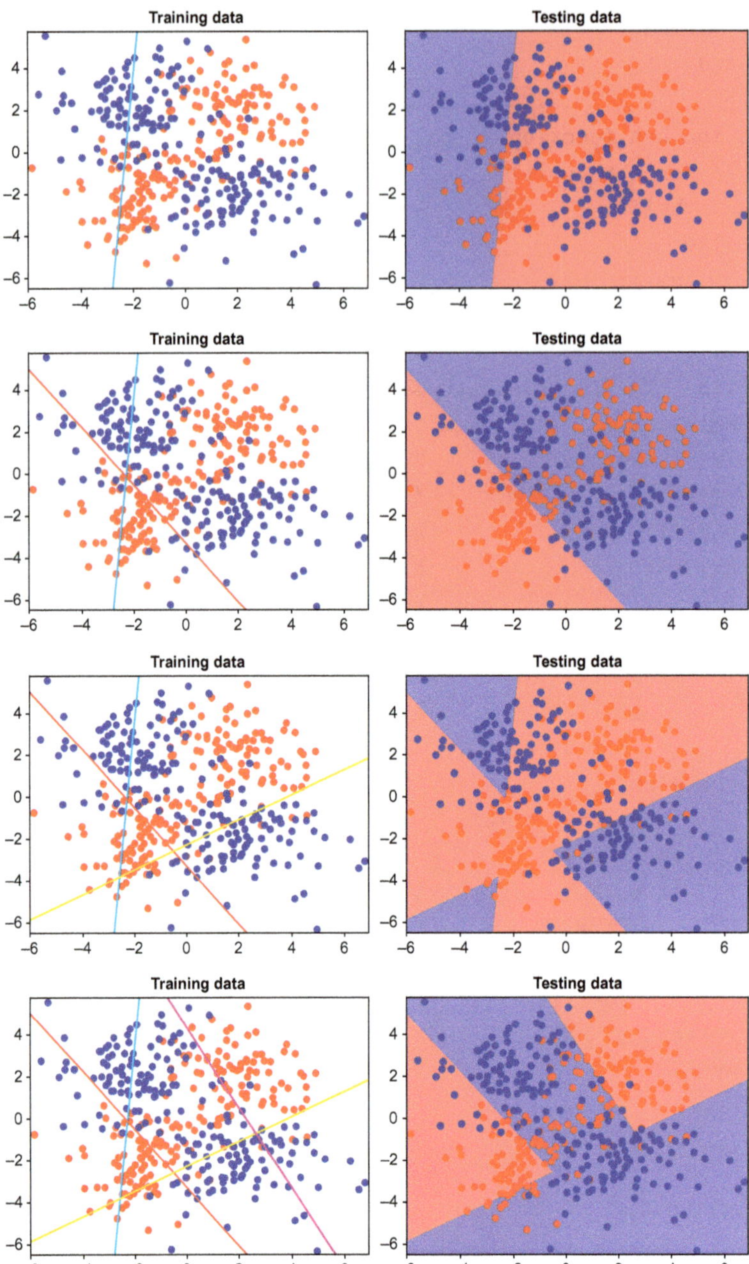

Fig. 13.13 Classification by AdaBoost on XOR dataset (early iterations).

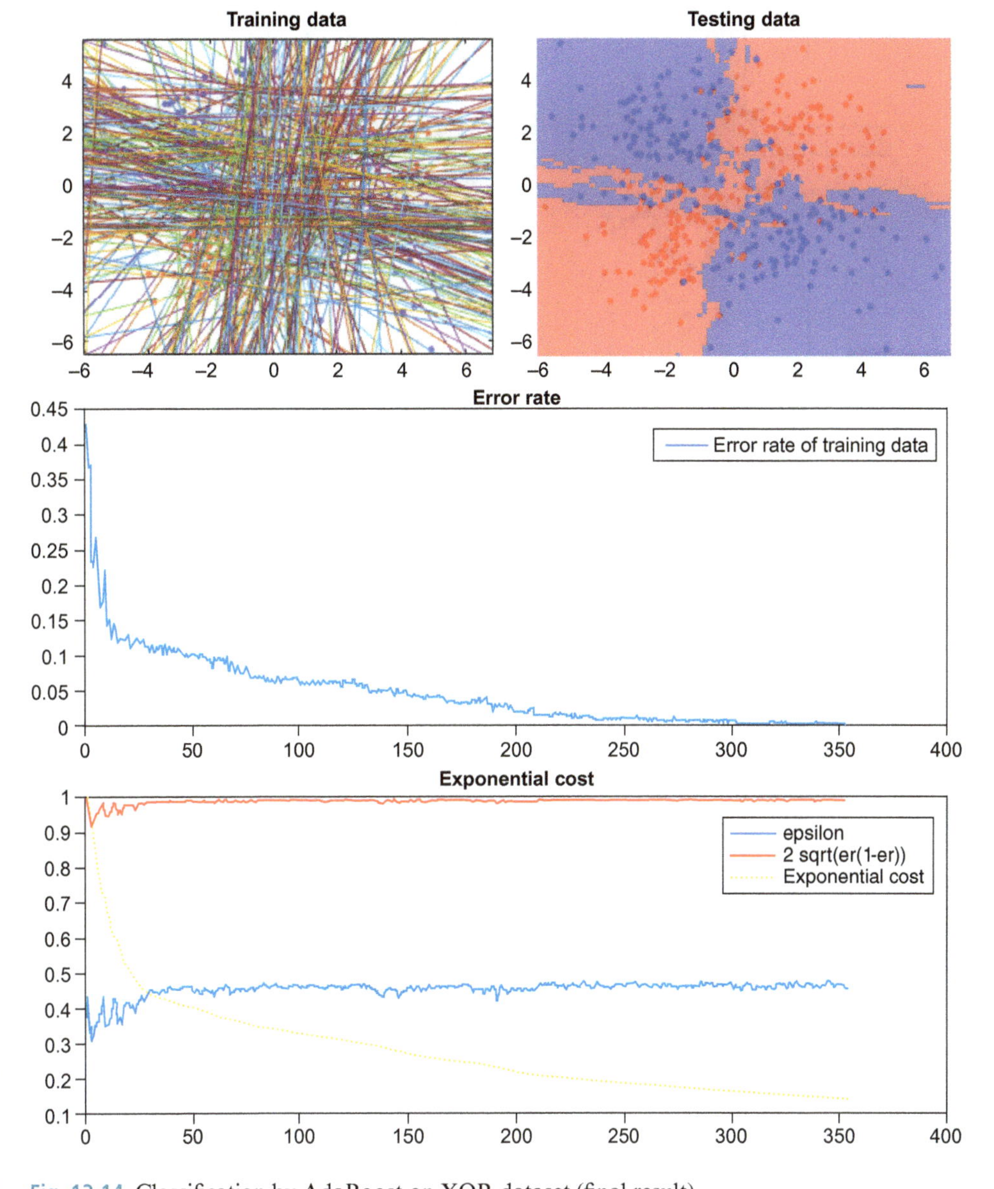

Fig. 13.14 Classification by AdaBoost on XOR dataset (final result).

Example 13.11

Figure 13.15 shows the partitioning of the 2-D space by the AdaBoost algorithm trained on data samples from two concentric classes, along both the standard basis vectors (left) and the PCA directions (right). The PCA method performs significantly better in this example as its error reduces faster and it converges to zero after 61 iterations, while the error of the method based on the standard axes (in vertical and horizontal directions) does not go down to zero even after 120 iterations. This faster reduction of error of the PCA method can be easily understood as many different directions are used to partition the space into two regions, while in the coordinate descent method the partitioning of the space is limited to either of the two directions.

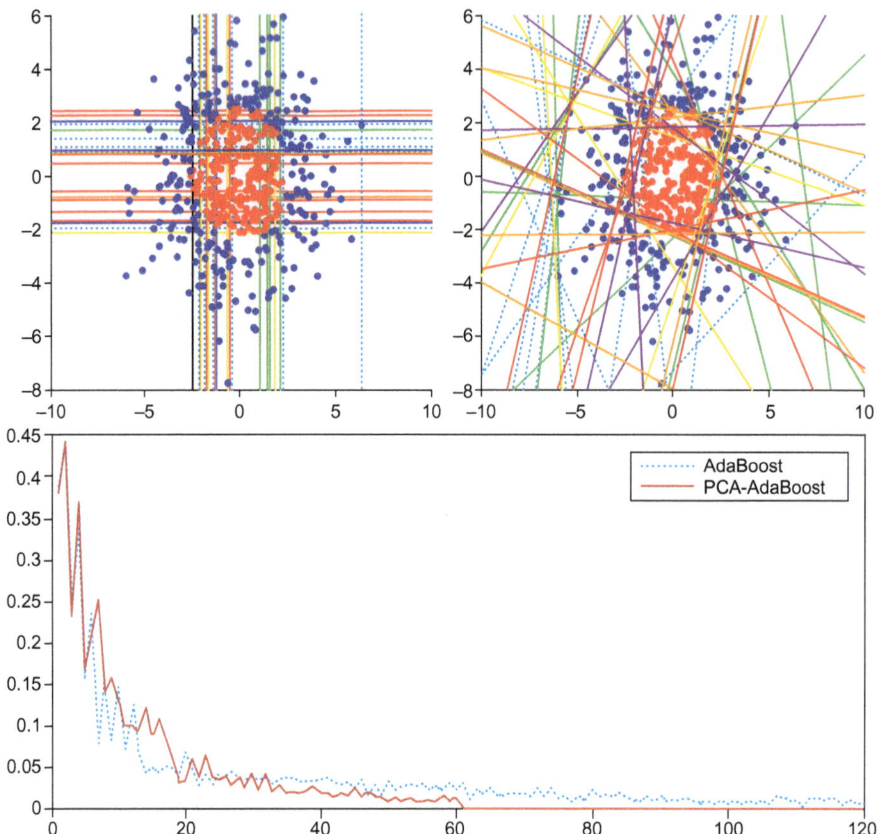

Fig. 13.15 AdaBoost classification along basis vectors and PCA directions.

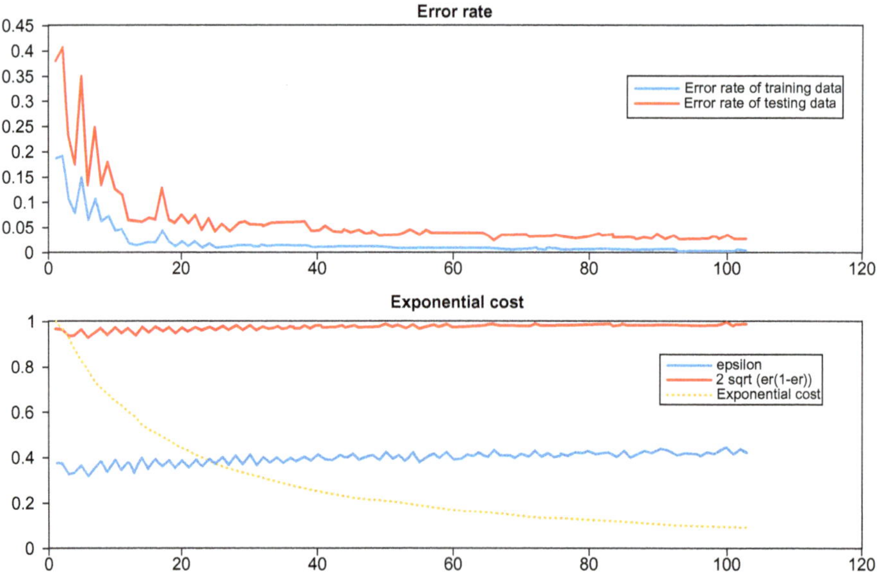

Fig. 13.16 Progress of iterative training of AdaBoost.

Figure 13.16 shows how the AdaBoost algorithm is iteratively trained. The error rate of both the training and testing samples are plotted as the iteration progresses (top). Also, the weighted error ε_t, the exponential costs E_t, and the ratio $E_{t+1}/E_t = 2\sqrt{\varepsilon_t(1-\varepsilon_t)}$ are all plotted (bottom). We see that $\varepsilon_t < 0.5$ and $E_{t+1}/E_t < 1$ for all steps, and the exponential cost E_t attenuates exponentially toward zero.

Example 13.12

This example shows the classification of a dataset containing 200 samples of two classes (100 each) in an XOR pattern in the 2-D space, by the AdaBoost method based on both coordinate descent and PCA methods, as shown in Figs 13.17 and 13.18, respectively. In both cases, half of the dataset (top-left) is used for training, while the other half for testing (top-right).

We see that the coordinate descent method performs poorly in both the slow convergence (more than 400 iterations) during training and high classification error rate (more than 1/3) during testing. The partitioning of the 2-D space is an obvious over fitting of the training set instead of reflecting the actual distribution of the two classes (four clusters). By contrast, the PCA method converges quickly (10 iterations) and classifies the testing samples with very low error rate. The space is clearly partitioned into two regions corresponding to the two classes.

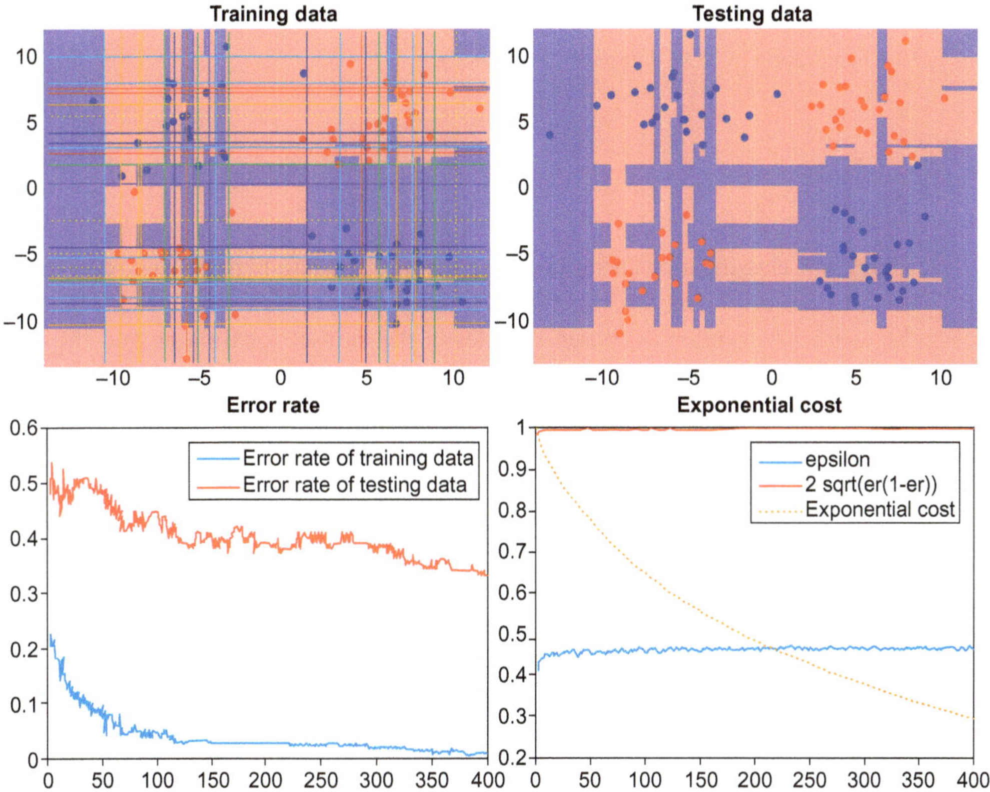

Fig. 13.17 Classification of XOR dataset (coordinate descent).

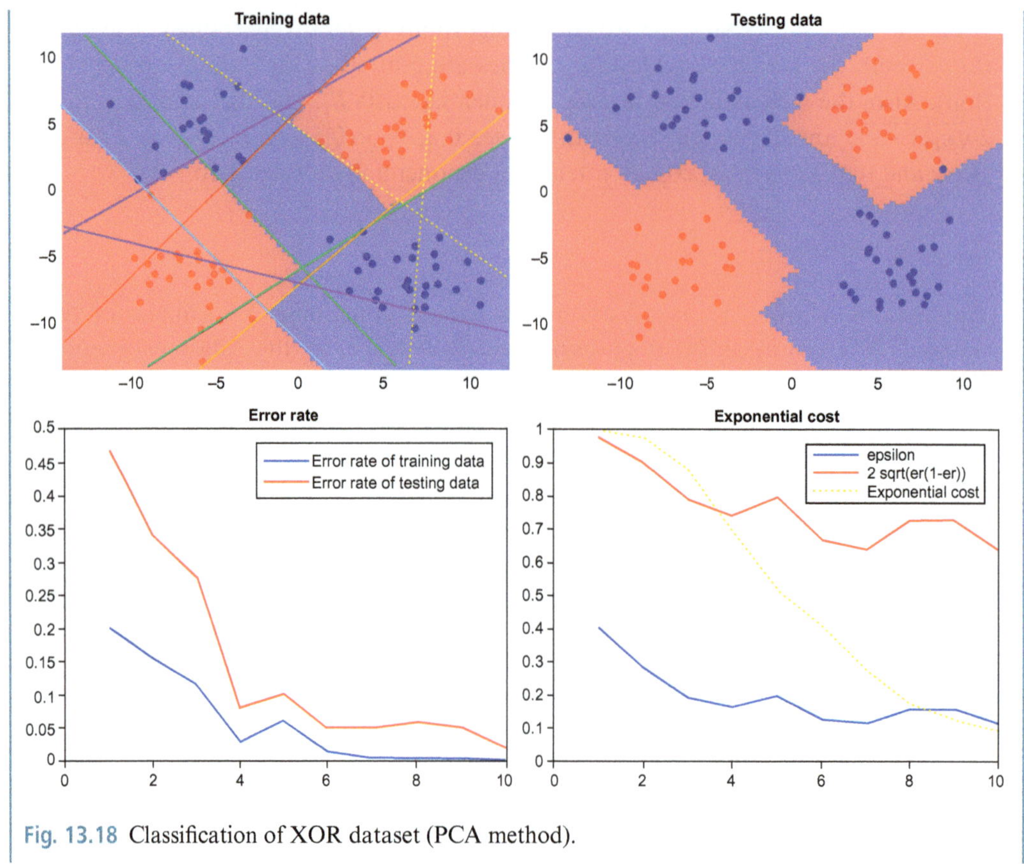

Fig. 13.18 Classification of XOR dataset (PCA method).

PROBLEMS

Carry out classification using both the iris and the handwritten digit datasets by each of the following three methods. In each case, cross validate your algorithm by using 50% randomly chosen samples in the dataset for training and the other 50% for testing. Show your classification results in the confusion matrix together with the error rate, the sum of all off-diagonal components of the confusion matrix divided by the total number of samples.

1. the k-nearest neighbor method;
2. the minimum distance method based on Mahananobis distance;
3. the naive Bayes method;

14 Support Vector Machine

14.1 Maximum Margin and Support Vectors

The *support vector machine (SVM)* is a supervised binary classifier trained by a data set $\{(\mathbf{x}_n, y_n),\quad n = 1, \cdots, N\}$, from which sample $\mathbf{x}_n$ belongs to either class C_+ if labeled by $y_n = 1$ or class C_- if labeled by $y_n = -1$. The goal of the training process is to partition the feature space into two halves by a *decision plane* H_0, described by the following linear *decision equation*

$$f(\mathbf{x}) = \mathbf{w}^T\mathbf{x} + b = \sum_{i=1}^{d} x_i w_i + b = 0 \tag{14.1}$$

in terms of its normal vector $\mathbf{w}$ and intercept b. Once these two parameters are determined in the training process, any unlabeled sample $\mathbf{x}$ can be classified into either of the two classes

$$\text{if} \quad f(\mathbf{x}) = \mathbf{w}^T\mathbf{x} + b \begin{cases} > 0 \\ = 0 \\ < 0 \end{cases}, \quad \text{then} \quad \begin{cases} \mathbf{x} \in C_+ \\ \mathbf{x} \in H_0 \\ \mathbf{x} \in C_- \end{cases} \tag{14.2}$$

Obviously the underlying assumption of this simple idea is the linear separability of the two classes. However, as we will see later, the SVM can also be applied to nonlinear and multiclass classifications.

We note that Eq. (14.1) is actually the same as Eq. (5.1) for linear regression, when the linear regression function $f(\mathbf{x})$ is thresholded by the zero plane to become an equation $f(\mathbf{x}) = 0$. This is illustrated in Fig. 14.1, which is essentially the same as Fig. 5.1. Although both SVM and linear regression methods need to obtain the optimal parameters $\mathbf{w}$ and b, they are very different in that the SVM, as a classification method, needs to determine $\mathbf{w}$ and b so that the resulting decision plane separates the training samples optimally, in the sense that its distance to the closest sample on either side, called *support vectors*, is maximized.

By moving the constant term in Eq. (14.1) to the right-hand side and dividing both sides by $||\mathbf{x}||$, this equation can be rewritten as

$$p_{\mathbf{w}}(\mathbf{x}) = \frac{\mathbf{w}^T\mathbf{x}}{||\mathbf{w}||} = -\frac{b}{||\mathbf{w}||} \tag{14.3}$$

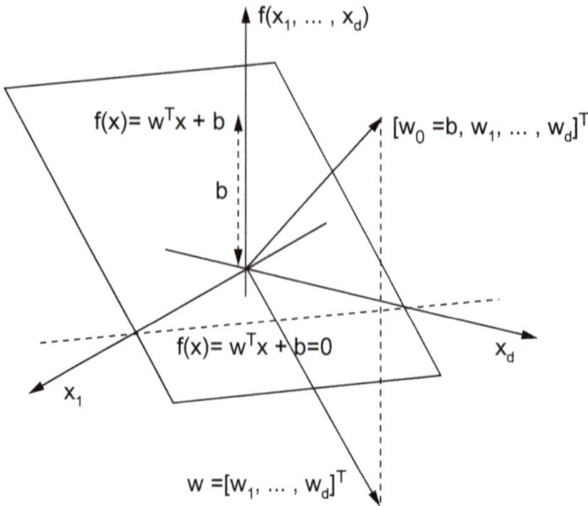

Fig. 14.1 Binary partition by thresholding a linear function.

where $p_{\mathbf{w}}(\mathbf{x}) = \mathbf{w}^T\mathbf{x}/||\mathbf{w}||$ is the projection of any point $\mathbf{x} \in H_0$ on the decision plane H_0 onto its normal direction $\mathbf{w}$. The absolute value of this projection is the distance between H_0 and the origin

$$d(H_0, \mathbf{0}) = |p_{\mathbf{w}}(\mathbf{x})| = \frac{|b|}{||\mathbf{w}||} \tag{14.4}$$

We further find the projection of any point $\mathbf{x}' \notin H_0$ off the decision plane H_0 onto $\mathbf{w}$ as $p_{\mathbf{w}}(\mathbf{x}') = \mathbf{w}^T\mathbf{x}'/||\mathbf{w}||$, and its distance to H_0 is

$$d(H_0, \mathbf{x}') = \left|p_{\mathbf{w}}(\mathbf{x}') - p_{\mathbf{w}}(\mathbf{x})\right| = \left|\frac{\mathbf{w}^T\mathbf{x}'}{||\mathbf{w}||} - \left(-\frac{b}{||\mathbf{w}||}\right)\right| = \frac{|\mathbf{w}^T\mathbf{x}' + b|}{||\mathbf{w}||} = \frac{|f(\mathbf{x}')|}{||\mathbf{w}||} \tag{14.5}$$

Figure 14.2 shows the projections of 2-D points, $\mathbf{x} \in H_0$, together with $\mathbf{x}'$ and $\mathbf{x}''$ on opposite sides of H_0, onto its normal direction $\mathbf{w}$.

We would like to find the optimal decision plane H_0 that separates the training samples belonging to the two different classes, assumed to be linearly separable, in such a way that its distance to the support vectors, denoted by $\mathbf{x}_{sv}$, is maximized

$$d(H_0, \mathbf{x}_{sv} \in C_+) = d(H_0, \mathbf{x}_{sv} \in C_-) = d(H_0, \mathbf{x}_{sv}) = \frac{|f(\mathbf{x}_{sv})|}{||\mathbf{w}||} \tag{14.6}$$

We further consider two additional planes H_+ and H_- that are in parallel with H_0 and pass through the support vectors $\mathbf{x}_{sv}$ on either side of H_0

$$f_{\pm}(\mathbf{x}) = f(\mathbf{x}) \pm c = \mathbf{w}^T\mathbf{x} + b \pm c = \mathbf{w}^T\mathbf{x} + b \pm 1 = 0 \tag{14.7}$$

Here for convenience we have set $c = 1$, as these two equations can be arbitrarily scaled. Based on Eq. (14.4), the distances from H_+ or H_- to the origin can be written as

$$d(H_{\pm}, \mathbf{0}) = \frac{|b \pm 1|}{||\mathbf{w}||} \tag{14.8}$$

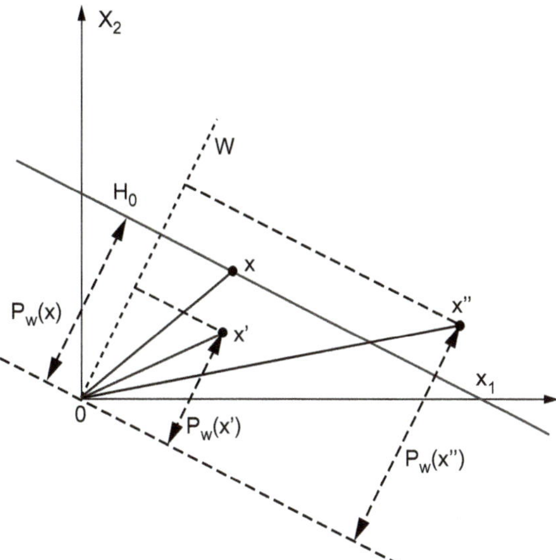

Fig. 14.2 Decision plane and decision margin.

and their distances to the decision plane H_0, called the decision *margin*, can be found as

$$d(H_\pm, H_0) = \left| d(H_\pm, \mathbf{0}) - d(H_0, \mathbf{0}) \right| = \left| \frac{|b \pm 1|}{||\mathbf{w}||} - \frac{|b|}{||\mathbf{w}||} \right| = \frac{1}{||\mathbf{w}||} \tag{14.9}$$

To maximize this margin, we need to minimize $||\mathbf{w}||$.

For these planes to correctly separate all samples in the training set $\{(\mathbf{x}_n, y_n)\ \ (n = 1, \cdots, N)\}$, they need to satisfy the following two conditions

$$\begin{cases} f_-(\mathbf{x}_n) = \mathbf{w}^T\mathbf{x}_n + b - 1 \geq 0 & \text{if } y_n = 1 \\ f_+(\mathbf{x}_n) = \mathbf{w}^T\mathbf{x}_n + b + 1 \leq 0 & \text{if } y_n = -1 \end{cases} \qquad (n = 1, \cdots, N) \tag{14.10}$$

where the equality is satisfied by the support vectors on H_+ or H_-, while the inequalities are satisfied by all other samples farther away from H_0 behind H_+ or H_-. These two conditions can be combined to become

$$y_n(\mathbf{w}^T\mathbf{x}_n + b) \geq 1, \qquad (n = 1, \cdots, N) \tag{14.11}$$

Now the task of finding the optimal decision plane H_0 can be formulated as a constrained minimization problem

$$\begin{aligned} \text{minimize:} \quad & \frac{1}{2}||\mathbf{w}||^2 = \frac{1}{2}\mathbf{w}^T\mathbf{w} \\ \text{subject to:} \quad & y_n(\mathbf{x}_n^T\mathbf{w} + b) \geq 1, \quad (n = 1, \cdots, N) \end{aligned} \tag{14.12}$$

Figure 14.3 illustrates the three planes in the 2-D case, including the decision plane H_0, together with H_+ and H_-, as well as the decision margin $1/||\mathbf{w}||$.

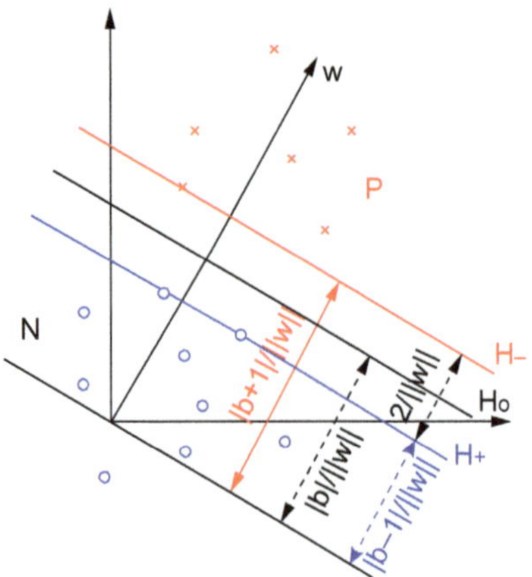

Fig. 14.3 Decision plane and margin.

Example 14.1

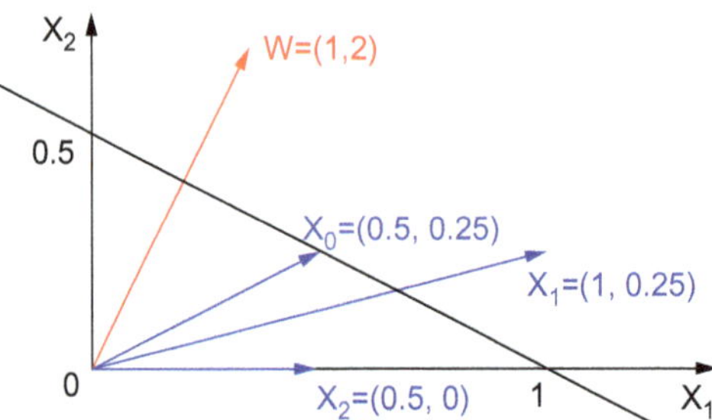

Fig. 14.4 Decision plane and support vectors.

In Fig. 14.4, the straight line in 2-D space, denoted by H_0, is described by the following linear equation

$$f(\mathbf{x}) = \mathbf{w}^T\mathbf{x} + b = [w_1, w_2]\begin{bmatrix} x_1 \\ x_2 \end{bmatrix} + b = [1, 2]\begin{bmatrix} x_1 \\ x_2 \end{bmatrix} - 1 = x_1 + 2x_2 - 1 = 0 \quad (14.13)$$

The distance from H_0 to the origin is

$$d(H_0, \mathbf{0}) = \frac{|b|}{||\mathbf{w}||} = \frac{1}{\sqrt{w_1^2 + w_2^2}} = \frac{1}{\sqrt{1^2 + 2^2}} = \frac{1}{\sqrt{5}} = 0.447 \quad (14.14)$$

which is invariant with respect to any scaling of the equation. Consider three points in the space.

- $\mathbf{x}_0 = [0.5,\ 0.25]^T$, $f(\mathbf{x}_0) = 0.5 + 2 \times 0.25 - 1 = 0$, that is, $\mathbf{x}_0$ is on the plane. Its distance to the plane is $d(\mathbf{x}_0, H) = f(\mathbf{x}_0)/||\mathbf{w}|| = 0$.
- $\mathbf{x}_1 = [1,\ 0.25]^T$, $f(\mathbf{x}_1) = 1 + 2 \times 0.25 - 1 = 0.5 > 0$, that is, $\mathbf{x}_1$ is above the straight line. Its distance to the plane is $d(\mathbf{x}_1, H) = f(\mathbf{x}_1)/||\mathbf{w}|| = |0.5|/\sqrt{5} = 0.2235$.

- $\mathbf{x}_2 = [0.5,\ 0]^T$, $f(\mathbf{x}_2) = 0.5 + 2 \times 0 - 1 = -0.5 < 0$, that is, $\mathbf{x}_2$ is below the straight line. Its distance to the plane is $d(\mathbf{x}_2, H) = f(\mathbf{x}_2)/||\mathbf{w}|| = |-0.5|/\sqrt{5} = 0.2235$.

As the two points $\mathbf{x}_1 = [1,\ 0.25]^T$ and $\mathbf{x}_2 = [0.5,\ 0]^T$ have equal distance to H_0, the straight line $f(\mathbf{x}) = x_1 + 2x_2 - 1 = 0$, we assume they are two support vectors $\mathbf{x}_{sv}$ on either side of H_0

$$|f(\mathbf{x}_{sv})| = |\mathbf{w}^T\mathbf{x}_1 + b| = |\mathbf{w}^T\mathbf{x}_2 + b| = 0.5 \tag{14.15}$$

Based on the decision equation scaled by 2

$$f(\mathbf{x}) = \mathbf{w}^T\mathbf{x} + b = 2x_1 + 4x_2 - 2 = 0 \tag{14.16}$$

we further obtain the equations for the two planes H_+ and H_- both parallel o H_0:

$$H_+ :\ 2x_1 + 4x_2 - 2 - 1 = 2x_1 + 4x_2 - 3 = 0$$
$$H_- :\ 2x_1 + 4x_2 - 2 + 1 = 2x_1 + 4x_2 - 1 = 0$$

Their distances to the origin are

$$d(H_+, \mathbf{0}) = \frac{|-3|}{||\mathbf{w}||} = 1.341, \quad d(H_-, \mathbf{0}) = \frac{|-1|}{||\mathbf{w}||} = 0.447 \tag{14.17}$$

and the distance between them is

$$d(H_-, H_+) = d(H_+, \mathbf{0}) - d(H_-, \mathbf{0})1.341 - 0.447 = 0.894 = \frac{2}{||\mathbf{w}||} \tag{14.18}$$

which is indeed twice the distance $d(H_\pm, H_0) = 1/||\mathbf{w}|| = 0.447$.

For reasons to be discussed later, instead of directly solving the constrained minimization problem in Eq. (14.12), now called the *primal problem*, we actually solve the *dual problem* (Section 3.3). We first construct the Lagrangian function of the primal problem, called the *primal function*

$$L_p(\mathbf{w}, b, \boldsymbol{\alpha}) = \frac{1}{2}\mathbf{w}^T\mathbf{w} + \sum_{n=1}^{N} \alpha_n(1 - y_n(\mathbf{w}^T\mathbf{x}_n + b)) \tag{14.19}$$

where $\alpha_1, \cdots, \alpha_N$ are the *Lagrange multipliers*. For this minimization problem with non-positive constraints, the Lagrangian multipliers are required to be negative, α_n, according to Table 3.2 in Section 3.2, if a minus sign is used for the second term. However, to be consistent with all SVM literature, we use the positive sign for the second term and require $\alpha_n \geq 0$. Note that if $\mathbf{x}_n$ is a support vector on either H_+ or H_-, that is, the equality constraint $y_n(\mathbf{w}^T\mathbf{x}_n + b) = 1$ holds, then $\alpha_n > 0$; but if $\mathbf{x}_n$ is not a support vector, the equality constraint does not hold, and $\alpha_n = 0$.

As discussed in subsection 3.3, we can solve this constrained optimization problem by solving an equivalent dual problem, that is, we first find the lower bound of $L_p(\mathbf{w}, b, \boldsymbol{\alpha})$ with respect to $\mathbf{w}$ and b, and then find the tightest bound with respect to $\boldsymbol{\alpha}$). Such a solution has to satisfy the Karush–Kuhn–Tucker (KKT) conditions (Eqs (3.43) through (3.46)).

- Stationarity:

$$\frac{\partial}{\partial b}L_p(\mathbf{w},b) = \frac{\partial}{\partial b}\left[\frac{1}{2}\mathbf{w}^T\mathbf{w} + \sum_{n=1}^N \alpha_n(1 - y_n(\mathbf{w}^T\mathbf{x}_n + b))\right]$$
$$= \sum_{n=1}^N \alpha_n y_n = 0 \tag{14.20}$$

and

$$\frac{\partial}{\partial \mathbf{w}}L_p(\mathbf{w},b) = \frac{\partial}{\partial \mathbf{w}}\left[\frac{1}{2}\mathbf{w}^T\mathbf{w} + \sum_{n=1}^N \alpha_n(1 - y_n(\mathbf{w}^T\mathbf{x}_n + b))\right]$$
$$= \mathbf{w} - \sum_{n=1}^N \alpha_n y_n \mathbf{x}_n = \mathbf{0}, \tag{14.21}$$

that is

$$\mathbf{w} = \sum_{n=1}^N \alpha_n y_n \mathbf{x}_n \tag{14.22}$$

- Primal feasibility

$$y_n(\mathbf{w}^T\mathbf{x} + b) - 1 \geq 0, \qquad (n = 1, \cdots, N) \tag{14.23}$$

- Dual feasibility

$$\alpha_n \geq 0, \qquad (n = 1, \cdots, N) \tag{14.24}$$

- Complementarity

$$\alpha_n[y_n(\mathbf{w}^T\mathbf{x} + b) - 1] = 0 \tag{14.25}$$

To find the minimum as the lower bound of the primal function $L_p(\mathbf{w}, b)$ in Eq. (14.19), we substitute Eqs (14.20) and (14.22) back into the primal function, and get the *dual function*

$$L_d(\boldsymbol{\alpha}) = \min_{\mathbf{w},b} L_p(\mathbf{w},b,\boldsymbol{\alpha}) = \min_{\mathbf{w},b}\left[\frac{1}{2}\mathbf{w}^T\mathbf{w} + \sum_{n=1}^N \alpha_n(1 - y_n(\mathbf{w}^T\mathbf{x}_n + b))\right]$$
$$= \min_{\mathbf{w},b}\left[\frac{1}{2}\mathbf{w}^T\mathbf{w} + \sum_{n=1}^N \alpha_n - \mathbf{w}^T\sum_{n=1}^N \alpha_n y_n \mathbf{x}_n - b\sum_{n=1}^N \alpha_n y_n\right]$$
$$= \frac{1}{2}\left(\sum_{m=1}^N \alpha_m y_m \mathbf{x}_m\right)^T\left(\sum_{n=1}^N \alpha_n y_n \mathbf{x}_n\right)$$
$$+ \sum_{n=1}^N \alpha_n - \left(\sum_{m=1}^N \alpha_m y_m \mathbf{x}_m\right)^T\left(\sum_{n=1}^N \alpha_n y_n \mathbf{x}_n\right)$$
$$= \sum_{n=1}^N \alpha_n - \frac{1}{2}\sum_{n=1}^N\sum_{m=1}^N \alpha_n\alpha_m y_n y_m \left(\mathbf{x}_n^T\mathbf{x}_m\right) \tag{14.26}$$

For this dual function $L_d(\boldsymbol{\alpha})$ to be the tightest lower bound of the primal function, it needs to be maximized with respect to $\boldsymbol{\alpha}$, subject to the constraint imposed by Eq. (14.20)

$$\text{maximize:} \quad L_d(\boldsymbol{\alpha}) = \sum_{n=1}^{N} \alpha_n - \frac{1}{2} \sum_{n=1}^{N} \sum_{m=1}^{N} \alpha_n \alpha_m y_n y_m \left(\mathbf{x}_n^T \mathbf{x}_m \right) = \mathbf{1}^T \boldsymbol{\alpha} - \frac{1}{2} \boldsymbol{\alpha}^T \mathbf{Q} \boldsymbol{\alpha}$$

$$\text{subject to:} \quad \sum_{n=1}^{N} \alpha_n y_n = \mathbf{y}^T \boldsymbol{\alpha} = 0, \quad \alpha_n \geq 0 \quad (n = 1, \cdots, N) \tag{14.27}$$

where $\mathbf{Q}$ is an N by N symmetric matrix of which the component in the mth row and nth column is $Q(m,n) = y_m y_n \mathbf{x}_m^T \mathbf{x}_n$ $(m,n = 1, \cdots, N)$. Now we have converted the primal problem of constrained minimization of the primal function $L_p(\mathbf{w}, b, \boldsymbol{\alpha})$ with respect to $\mathbf{w}$ and b to its dual problem of linearly constrained maximization of the dual function $L_d(\boldsymbol{\alpha})$ with respect to $\{\alpha_1, \cdots, \alpha_N\}$. Solving this quadratic programming (QP) problem (Section 3.6) we obtain all Lagrange multipliers $\alpha_n \geq 0$, $(n = 1, \cdots, N)$. All training samples $\mathbf{x}_n$ corresponding to positive Lagrange multipliers $\alpha_n > 0$ are support vectors, while others corresponding to $\alpha_n = 0$ are not support vectors.

We can now find the solution $\mathbf{w}^*$ based on Eq. (14.22)

$$\mathbf{w}^* = \sum_{n=1}^{N} \alpha_n y_n \mathbf{x}_n = \sum_{n \in sv} \alpha_n y_n \mathbf{x}_n = \sum_{x_n \in C_+,\, n \in sv} \alpha_n \mathbf{x}_n - \sum_{x_n \in C_-,\, n \in sv} \alpha_n \mathbf{x}_n \tag{14.28}$$

Note that the normal vector is the difference between the weighted means of the support vectors in class C_+ (first term) and class C_- (second term), and it is determined only by the support vectors.

We can further find b^* based on Eq. (14.25) for complementarity of the KKT conditions. As the Lagrangian multiplier α_n corresponding to any support vector $\mathbf{x}_n$ is non-zero, we must have $y_n(\mathbf{w}^T \mathbf{x}_n + b) - 1 = 0$, which is actually the equality of Eq. (14.11) satisfied by all support vectors

$$y_n \left(\mathbf{w}^T \mathbf{x}_n + b \right) = 1 \quad \text{or} \quad \mathbf{w}^T \mathbf{x}_n + b = y_n, \qquad (n \in sv) \tag{14.29}$$

(as $y_n^2 = 1$). Solving for b we get the optimal b^*

$$b^* = y_n - \mathbf{w}^T \mathbf{x}_n = y_n - \sum_{m \in sv} \alpha_m y_m (\mathbf{x}_m^T \mathbf{x}_n), \qquad (n \in sv) \tag{14.30}$$

All support vectors should yield the same result. Computationally we simply obtain b^* as the average of the above for all support vectors.

We note that both $\mathbf{w}$ and b depend only on the support vectors on planes H_+ and H_-, corresponding to a positive $\alpha_n > 0$, while all remaining samples corresponding to $\alpha_n = 0$ are farther away from the decision plane H_0, behind either H_+ or H_-. We see that the SVM is solely based on those support vectors in the training set. Once they are identified during the training process, all other non-support vectors are irrelevant.

Once $\mathbf{w}$ and b are obtained in the training process, any unlabeled sample $\mathbf{x}$ can be classified into either of the two classes depending on whether the following decision function is greater or smaller than zero

$$
\begin{aligned}
f(\mathbf{x}) = \mathbf{w}^T\mathbf{x} + b &= \left(\sum_{n\in sv} \alpha_n y_n \mathbf{x}_n \right)^T \mathbf{x} + b \\
&= \sum_{n\in sv} \alpha_n y_n \, (\mathbf{x}_n^T \mathbf{x}) + b \quad \begin{cases} > 0 & \mathbf{x} \in C_+ \\ < 0 & \mathbf{x} \in C_- \end{cases}
\end{aligned} \tag{14.31}
$$

We also get

$$
\begin{aligned}
||\mathbf{w}||^2 = \mathbf{w}^T\mathbf{w} &= \left(\sum_{n\in sv} \alpha_n y_n \mathbf{x}_n^T \right) \left(\sum_{m\in sv} \alpha_m y_m \mathbf{x}_m \right) \\
&= \sum_{n\in sv} \alpha_n y_n \sum_{m\in sv} \alpha_m y_m (\mathbf{x}_n^T \mathbf{x}_m) \\
&= \sum_{n\in sv} \alpha_n y_n (y_n - b) = \sum_{n\in sv} \alpha_n (1 - y_n b) \\
&= \sum_{n\in sv} \alpha_n - b \sum_{n\in sv} \alpha_n y_n = \sum_{n\in sv} \alpha_n
\end{aligned} \tag{14.32}
$$

The last equality is due to equality constraints of the dual problem $\sum_{n=1}^N \alpha_n y_n = 0$. Now the decision margin can be written as

$$
\frac{1}{||\mathbf{w}||} = \left(\sum_{n\in sv} \alpha_n \right)^{-1/2} \tag{14.33}
$$

The SVM algorithm for binary classification can now by summarized as the following steps

- find the Lagrange multipliers $\{\alpha_1, \cdots, \alpha_N\}$ by solving the QP problem in Eq. (14.27);
- find the normal vector $\mathbf{w}$ by Eq. (14.28);
- find the bias b by Eq. (14.30);
- classify unlabeled $\mathbf{x}$ by Eq. (14.31).

As the last expression of the decision function in Eq. (14.31) depends only on α_n as well as $\mathbf{x}_n$ and y_n in the training set, the normal vector $\mathbf{w}$ is never needed, and the second step for calculating $\mathbf{w}$ by Eq. (14.28) can be dropped.

We prefer to solve this dual problem not only because it is easier than the original primal problem, but also, more importantly, for the reason that all data points appear only in the form of an inner product $\mathbf{x}_n^T\mathbf{x}_m$ in the dual problem, allowing the *kernel method* to be used, as discussed later.

In the following MATLAB code for the essential part of the algorithm, `X` and `y` are, respectively, the data array composed of N training vectors $[\mathbf{x}_1, \cdots, \mathbf{x}_N]$ and their corresponding labelings $y_1, \cdots, y_N$, and `IPmethod` is a function that implements the *interior point method* (Section 3.7) for solving a general QP problem of minimizing $\mathbf{x}^T\mathbf{Q}\mathbf{x}/2 + \mathbf{c}^T\mathbf{x}$ subject to the linear equality constraints $\mathbf{A}\mathbf{x} = \mathbf{b}$ and $\mathbf{x} \geq \mathbf{0}$.

```
[X y]=getData;          % get the training data
[m n]=size(X);
Q=(y'*y).*(X'*X);       % compute the quadratic matrix
c=-ones(n,1);           % coefficients of linear term
A=y;                    % coefficient matrix
b=0;                    % constants of linear equality constraints
alpha=0.1*ones(n,1);    % initial guess of solution

[alpha mu lambda]=IPmethod(Q,A,c,b,alpha);
                        % solve QP for alpha (interior point method)
I=find(abs(alpha)>10^(-3));  % indices of non-zero alphas
asv=alpha(I);           % non-zero alphas
Xsv=X(:,I);             % support vectors
ysv=y(:,I);             % their labels
w=sum(repmat(ysv.*asv,m,1).*Xsv,2);  % normal vector (not needed)
bias=mean(ysv-w'*Xsv);               % bias
```

Example 14.2

The training set contains two classes of 2-D points with Gaussian distributions generated based on the following mean vectors and covariance matrices

$$\mathbf{m}_- = \begin{bmatrix} 3.5 \\ 0 \end{bmatrix}, \quad \mathbf{S}_- = \begin{bmatrix} 1 & 0.5 \\ 0.5 & 1 \end{bmatrix}, \quad \mathbf{m}_+ = \begin{bmatrix} 0 \\ 3.5 \end{bmatrix}, \quad \mathbf{S}_+ = \begin{bmatrix} 1 & 0 \\ 0 & 1 \end{bmatrix}$$

The SVM algorithm finds the weight vector $\mathbf{w}$ and constant b of the optimal boundary between the two classes based on three support vectors, all listed in the following. Note that decision boundary is completely dictated by the three support vectors and the margin distance between them is maximized.

Table 14.1

n	α_n	$\mathbf{x}_n = [x_1, x_2]$	y_n
40	0.52	$x = [4.43, 2.44]$	−1
52	3.11	$x = [1.17, 0.74]$	−1
103	3.64	$x = [1.30, 1.64]$	1

$$\mathbf{w} = \begin{bmatrix} -1.25 \\ 2.39 \end{bmatrix}, \qquad b = -1.31$$

The binary classification result is shown in Fig. 14.5.

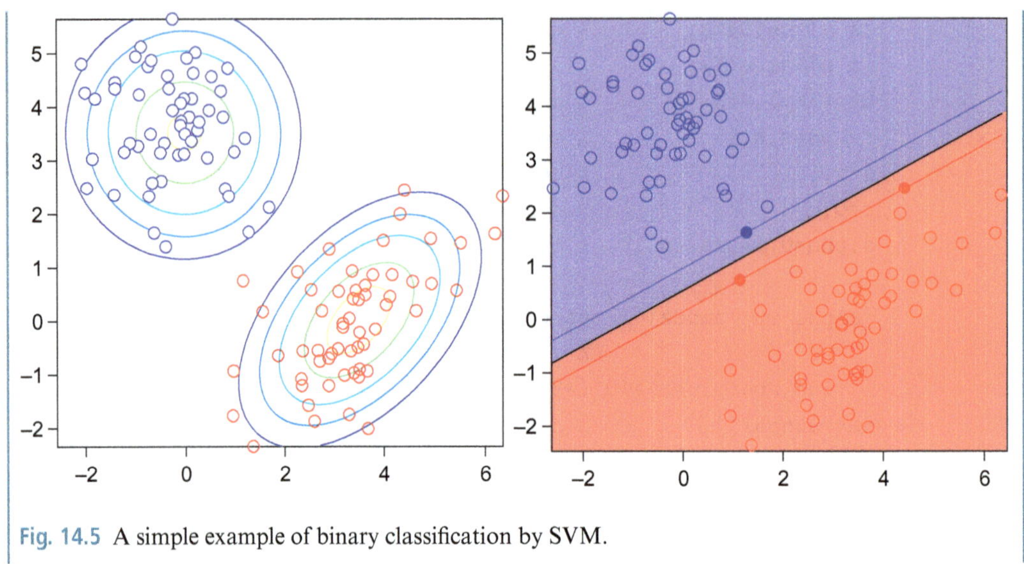

Fig. 14.5 A simple example of binary classification by SVM.

14.2 Kernel SVM

The SVM algorithm presented in the previous section converges only if the data points of the two classes in the training set are linearly separable. If this is not the case, we can consider the *kernel method* first introduced in Section 11.1.

Kernel mapping can be applied to the SVM algorithm as all data points appear in the algorithm are in the form of an inner product. Specifically, during the training process, we replace the inner product $\mathbf{x}_m^T\mathbf{x}_n$ in both Eq. (14.27) and Eq. (14.30) by the kernel function $K(\mathbf{x}_m, \mathbf{x}_n)$ to obtain

$$\begin{aligned} \text{maximize:} \quad & L_d(\boldsymbol{\alpha}) = \sum_{n=1}^N \alpha_n - \frac{1}{2}\sum_{n=1}^N \sum_{m=1}^N \alpha_n\alpha_m y_n y_m K(\mathbf{x}_n, \mathbf{x}_m) \\ \text{subject to:} \quad & \sum_{n=1}^N \alpha_n y_n = \mathbf{y}^T\boldsymbol{\alpha} = 0, \quad \alpha_n \geq 0 \quad (n = 1, \cdots, N) \end{aligned} \tag{14.34}$$

and

$$b = y_n - \sum_{m\in sv} \alpha_m y_m K(\mathbf{x}_m, \mathbf{x}_n), \qquad (n \in sv) \tag{14.35}$$

We also replace the inner product $\mathbf{x}_n^T\mathbf{x}$ in Eq. (14.31) by $K(\mathbf{x}_n, \mathbf{x})$ for the classification of any unlabeled point $\mathbf{z} = \phi(\mathbf{x})$

$$f(\mathbf{z}) = \mathbf{w}^T\mathbf{z} + b = \sum_{n\in sv} \alpha_n y_n\, (\mathbf{z}_n^T\mathbf{z}) + b = \sum_{n\in sv} \alpha_n y_n\, K(\mathbf{x}_n, \mathbf{x}) + b \quad \begin{cases} > 0 & \mathbf{x} \in C_+ \\ < 0 & \mathbf{x} \in C_- \end{cases} \tag{14.36}$$

Again, we note that the normal vector $\mathbf{w}$ in Eq. (14.28) never needs to be explicitly calculated. As now both the training and classification are carried out in some higher dimensional space $\mathbf{z} = \phi(\mathbf{x})$, in which the classes are more likely linearly separable, the classification can be more effectively. More generally, the kernel method can be applied to any algorithm, so long as the data always appear in the form of an inner product.

14.3 Soft Margin SVM

When the two classes C_- and C_+ are not linearly separable, the condition for the optimal hyperplane in Eq. (14.11) can be relaxed by including a slack variable $\xi_n \geq 0$ representing an extra error term

$$y_n(\mathbf{x}_n^T\mathbf{w} + b) \geq 1 - \xi_n, \quad (n = 1,\cdots,N) \tag{14.37}$$

For better separate data points belonging to different classes, this error ξ_n needs to be minimized as well as $||\mathbf{w}||$. Now the primal problem in Eq. (14.12) can be modified to the following minimization problem with an objective function and two sets of inequality (non-negative) constraints

$$\begin{aligned} \text{minimize:} \quad & \frac{1}{2}||\mathbf{w}||^2 + C\sum_{n=1}^{N}\xi_n \\ \text{subject to:} \quad & y_n(\mathbf{x}_n^T\mathbf{w} + b) + \xi_n - 1 \geq 0, \qquad \xi_n \geq 0; \quad (n = 1,\cdots,N) \end{aligned} \tag{14.38}$$

The value of the parameter C can be adjusted to make a proper tradeoff between underfitting emphasizing the global distribution of the training samples, and overfitting emphasizing the local samples close to the decision boundary H_0. A large C value emphasizes the minimization of the error term in the objective function, and the resulting H_0 is mostly dictated by a small number of local support vectors in the region between the class, including possibly some noisy outliers, thereby risking the problem of overfitting. By contrast, a small C value de-emphasizes the error term and the classifier can tolerate greater errors, thereby allowing more data samples farther away from H_0 to become support vectors and play a role in determining H_0. Consequently the classification trends to better reflect the overall distribution of the data samples, but with the risk of overfitting.

This constrained minimization problem can be converted to the minimization of the corresponding Lagrangian

$$L_p(\mathbf{w}, b, \xi, \alpha, \mu) = \frac{1}{2}||\mathbf{w}||^2 + C\sum_{n=1}^{N}\xi_n - \sum_{n=1}^{N}\alpha_n[y_n(\mathbf{w}^T\mathbf{x} + b) + \xi_n - 1] - \sum_{n=1}^{N}\beta_n\xi_n \tag{14.39}$$

where β_n as well as α_n are the Lagrange multipliers for the two sets of constraints. As they are non-negative (instead of non-positive as in the previous section), minus sign is used in front of the last two terms for the constraints, so that the Lagrange multipliers α_n and β_n are still required to be positive (consistent with Table 3.2). The KKT conditions (Eqs (3.43) through (3.46)) of this minimization problem are listed here (for all $n = 1,\cdots,N$)

- Stationarity

$$\frac{\partial L_p}{\partial \mathbf{w}} = \mathbf{w} - \sum_{n=1}^{N} y_n \alpha_n \mathbf{x}_n = 0, \quad \text{i.e.,} \quad \mathbf{w} = \sum_{n=1}^{N} y_n \alpha_n \mathbf{x}_n; \tag{14.40}$$

$$\frac{\partial L_p}{\partial b} = \sum_{n=1}^{N} y_n \alpha_n = 0; \tag{14.41}$$

$$\frac{\partial L_p}{\partial \xi_n} = C - \alpha_n - \beta_n = 0; \tag{14.42}$$

- Primal feasibility

$$y_n(\mathbf{w}^T \mathbf{x}_n + b) + \xi_n - 1 \geq 0, \quad \xi_n \geq 0, \tag{14.43}$$

- Complementarity

$$\alpha_n[y_n(\mathbf{w}^T \mathbf{x}_n + b) + \xi_n - 1] = 0, \quad \beta_n \xi_n = 0, \tag{14.44}$$

- Dual feasibility

$$\alpha_n \geq 0, \quad \beta_n \geq 0 \tag{14.45}$$

Depending on the value of α_n, there exist three possible cases, because of the complementarity (Eq. (14.44)):

- If $\alpha_n = 0$, then $\beta_n = C - \alpha_n = C > 0$, $\xi_n = 0$, and we have

$$y_n(\mathbf{w}^T \mathbf{x}_n + b) - 1 \geq 0 \tag{14.46}$$

- If $0 < \alpha_n < C$, that is, $\alpha_n \neq 0$, then we have

$$y_n(\mathbf{w}^T \mathbf{x}_n + b) + \xi_n - 1 = 0 \tag{14.47}$$

  But as $\beta_n = C - \alpha_n > 0$, $\xi_n = 0$ and this equation becomes

$$y_n(\mathbf{w}^T \mathbf{x}_n + b) - 1 = 0 \tag{14.48}$$

- If $\alpha_n = C \neq 0$, then we have

$$y_n(\mathbf{w}^T \mathbf{x}_n + b) + \xi_n - 1 = 0 \tag{14.49}$$

  But as $\beta_n = C - \alpha_n = 0$, $\xi_n \geq 0$ and this equation becomes

$$y_n(\mathbf{w}^T \mathbf{x}_n + b) - 1 \leq 0 \tag{14.50}$$

In all three cases the expression $y_n(\mathbf{w}^T \mathbf{x}_n + b) - 1$ can be rewritten as

$$y_n(\mathbf{w}^T \mathbf{x}_n + b) - 1 = y_n(\mathbf{w}^T \mathbf{x}_n + b) - y_n^2 = y_n[(\mathbf{w}^T \mathbf{x}_n + b) - y_n] = y_n E_n \tag{14.51}$$

where $E_n = (\mathbf{w}^T \mathbf{x}_n + b) - y_n$ is the error, or the difference between the actual and desired output.

We summarize all three of these cases to get

$$y_n E_n \begin{cases} \geq 0 & \text{if } \alpha_n = 0 \\ = 0 & \text{if } 0 < \alpha_n < C \\ \leq 0 & \text{if } \alpha_n = C \end{cases} \tag{14.52}$$

These results derived from the KKT conditions in Eqs (14.40) through (14.45) are an alternative form of the KKT conditions, which will be used in SMO algorithm to check if the KKT conditions are violated by any α_n.

The dual of the primal problem in Eq. (14.39) can be obtained by substituting the first set of equations (stationarity) into the primal Lagrangian

$$\begin{aligned}\text{maximize} \quad & L_d(\alpha) = \sum_{n=1}^{N} \alpha_n - \frac{1}{2}\sum_{n=1}^{N}\sum_{m=1}^{N} \alpha_n\alpha_m y_n y_m \mathbf{x}_m^T \mathbf{x}_n \\ & + C\sum_{n=1}^{N} \xi_n - \sum_{n=1}^{N} \alpha_n \xi_n - \sum_{n=1}^{N} \beta_n \xi_n \\ & = \mathbf{1}^T\alpha - \frac{1}{2}\alpha^T\mathbf{Q}\alpha + C\sum_{n=1}^{N} \xi_n - \sum_{n=1}^{N} \alpha_n \xi_n - \sum_{n=1}^{N} \beta_n \xi_n \\ \text{subject to} \quad & 0 \le \alpha_n \le C, \quad \sum_{n=1}^{N} \alpha_n y_n = 0 \end{aligned} \tag{14.53}$$

Note that owing to the condition $C = \alpha_n + \beta_n$, the Lagrangian multipliers ξ_n and β_n no longer appear in the Lagrangian L_d of the dual problem, which happens to be identical to that of the linearly separable problem discussed in previous section. Also, since $\alpha_n \ge 0$, $\beta_n \ge 0$ and $C = \alpha_n + \beta_n$, we have $0 \le \alpha_n = C - \beta_n \le C$. Having solved this QP problem for $\alpha_1, \cdots, \alpha_N$, we get the optimal decision plane in terms of the normal vector of the decision plane based on Eq. (14.28)

$$\mathbf{w} = \sum_{n=1}^{N} y_n\alpha_n\mathbf{x}_n = \sum_{n\in sv} y_n\alpha_n\mathbf{x}_n \tag{14.54}$$

and the bias term b based on Eq. (14.30)

$$b = y_n - \sum_{i\in sv} y_i\alpha_i\mathbf{x}_i^T\mathbf{x}_n = y_n - \sum_{i\in sv} y_i\alpha_i K(\mathbf{x}_i, \mathbf{x}_n), \qquad (n \in sv) \tag{14.55}$$

where the inner product $\mathbf{x}_k^T\mathbf{x}$ is replaced by the kernel function $K(\mathbf{x}_k, \mathbf{x})$ if it is nonlinear. Computationally, we take the average of such b values of all support vectors as the offset term.

Once $\mathbf{w}$ and b are obtained, any unlabeled point $\mathbf{x}$ can be classified into either of the two classes depending on whether it is on the positive or negative side of the decision plane

$$\mathbf{w}^T\mathbf{x} + b = \sum_{i\in sv} y_i\alpha_i\mathbf{x}_i^T\mathbf{x} + b = \sum_{i\in sv} y_i\alpha_i K(\mathbf{x}_i, \mathbf{x}) + b \begin{cases} > 0, & \mathbf{x} \in C_+ \\ < 0, & \mathbf{x} \in C_- \end{cases} \tag{14.56}$$

where the inner product is replaced by a kernel function if it is nonlinear.

14.4 Sequential Minimal Optimization Algorithm

The *sequential minimal optimization (SMO)* is a more efficient algorithm for solving the SVM problem, compared with the generic QP algorithms such as the interior-point method. The SMO algorithm can be considered as a method of decomposition, by which an optimization problem of multiple variables is decomposed into a series of sub-problems each optimizing an objective function of a small number of variables, typically only one, while all other variables are treated as constants that remain unchanged in the sub-problem. For example, the coordinate descent algorithm is just such a decomposition method, which solves a problem in a multidimensional space by converting it into a sequence of sub-problems each in a 1-D space.

When applying the decomposition method to the soft margin SVM problem, we could consider optimizing only one variable α_i at a time, while treating all remaining variables $\alpha_1,\cdots,\alpha_{i-1},\alpha_{i+1},\cdots,\alpha_N$ as constants. However, due to the constraint $\sum_{n=1}^N \alpha_n y_n = 0$, α_i is a linear combination of $N-1$ constants and is therefore also a constant. We therefore need to consider two variables at a time, denoted by α_i and α_j, while treating the remaining $N-2$ variables as constants. The objective function of such a sub-problem can be obtained by dropping all constant terms independent of the two selected variables α_i and α_j in the original objective function in Eq. (14.53)

$$\begin{aligned}
\text{maximize} \quad & L(\alpha_i,\alpha_j) = \alpha_i + \alpha_j - \frac{1}{2}\left(\alpha_i^2 \mathbf{x}_i^T\mathbf{x}_i + \alpha_j^2 \mathbf{x}_j^T\mathbf{x}_j + 2\alpha_i\alpha_j y_i y_j \mathbf{x}_i^T\mathbf{x}_j\right) \\
& -\alpha_i y_i \left(\sum_{n\neq i} \alpha_n y_n \mathbf{x}_n^T\right)\mathbf{x}_i - \alpha_j y_j \left(\sum_{n\neq j} \alpha_n y_n \mathbf{x}_n^T\right)\mathbf{x}_j \\
& = \alpha_i + \alpha_j - \frac{1}{2}\left(\alpha_1^2 K_{ii} + \alpha_2^2 K_{jj} + 2\alpha_i\alpha_j y_i y_j K_{ij}\right) \\
& -\alpha_i y_i \sum_{n\neq i,j} \alpha_n y_n K_{ni} - \alpha_j y_j \sum_{n\neq i,,j} \alpha_n y_n K_{nj} \\
\text{subject to} \quad & 0 \le \alpha_i,\alpha_j \le C, \quad \sum_{n=1}^N \alpha_n y_n = 0
\end{aligned} \tag{14.57}$$

Here all inner products $\mathbf{x}_m^T\mathbf{x}_n$ are replaced by a kernel the corresponding kernel function $K_{mn} = K(\mathbf{x}_m,\mathbf{x}_n)$.

This maximization problem can be solved iteratively. Out of all previous values α_n^{old} ($n = 1,\cdots,N$), the two selected variables are updated to become α_i^{new} and α_j^{new} by which $L(\alpha_i,\alpha_j)$ is maximized subject to the two constraints.

We rewrite the second constraint as

$$\alpha_i y_i + \alpha_j y_j = -\sum_{n\neq i,j} \alpha_n y_n \tag{14.58}$$

and multiply both sides by y_i to get

$$y_i^2\alpha_i + y_i y_j \alpha_j = \alpha_i + s\alpha_j = \left(-\sum_{n\neq i,j} \alpha_n y_n\right) y_i = \delta, \quad \text{i.e.,} \quad \alpha_i = \delta - s\alpha_j \tag{14.59}$$

where $y_i^2 = 1$, and we have defined

$$s = y_i y_j, \qquad \delta = -\sum_{n \neq i,j} \alpha_n y_n y_i \tag{14.60}$$

As δ is independent of α_i and α_j, it remains the same before and after they are updated, and we have

$$\alpha_i^{new} + s\alpha_j^{new} = \alpha_i^{old} + s\alpha_j^{old} = \delta \tag{14.61}$$

that is

$$\Delta\alpha_i = \alpha_i^{new} - \alpha_i^{old} = -s(\alpha_j^{new} - \alpha_j^{old}) = -s\Delta\alpha_j \tag{14.62}$$

We now consider a closed-form solution for updating α_i and α_j in each iteration of this two-variable optimization problem. We first rewrite Eq. (14.28) as

$$\sum_{n \neq i,j} \alpha_n y_n \mathbf{x}_n = \mathbf{w} - \alpha_i y_i \mathbf{x}_i - \alpha_j y_j \mathbf{x}_j \tag{14.63}$$

so that the two summations in $L(\alpha_i,\alpha_j)$, now denoted by v_k $(k = i,j)$, can be written as

$$\begin{aligned} v_k &= \sum_{n \neq i,j} \alpha_n y_n K_{nk} = \sum_{n=1}^{N} \alpha_n y_n K_{nk} - \alpha_i y_i K_{ik} - \alpha_j y_j K_{jk} \\ &= \left(\sum_{n=1}^{N} \alpha_n y_n K_{nk} + b \right) - b - \alpha_i y_i K_{ik} - \alpha_j y_j K_{jk} \\ &= u_k - b - \alpha_i y_i K_{ik} - \alpha_j y_j K_{jk} \end{aligned} \tag{14.64}$$

where

$$u_k = f(\mathbf{z}_k) = \sum_{n=1}^{N} \alpha_n y_n K_{nk} + b \tag{14.65}$$

is the output in Eq. (14.36) corresponding to input $\mathbf{x}_k$. Now the objective function can be written as

$$L(\alpha_i,\alpha_j) = \alpha_i + \alpha_j - \frac{1}{2}(\alpha_i^2 K_{ii} + \alpha_j^2 K_{jj} + 2s\alpha_i\alpha_j K_{ij}) - \alpha_i y_i v_i - \alpha_j y_j v_j \tag{14.66}$$

Substituting $\alpha_i = \delta - s\alpha_j$ (Eq. (14.62)) into $L(\alpha_i,\alpha_j)$, we can rewrite it as a function of a single variable α_j alone

$$\begin{aligned} L(\alpha_j) &= \delta + (1-s)\alpha_j - \frac{1}{2}\left[(\delta - s\alpha_j)^2 K_{ii} + \alpha_j^2 K_{jj} + 2s(\delta - s\alpha_j)\alpha_j K_{ij}\right] \\ &\quad -(\delta - s\alpha_j) y_i v_i - \alpha_j y_j v_j \\ &= \frac{1}{2}(2K_{ij} - K_{ii} - K_{jj})\alpha_j^2 \\ &\quad +[1 - s + s\delta(K_{ii} - K_{ij}) + y_j(v_i - v_j)]\alpha_j + \text{Const} \end{aligned} \tag{14.67}$$

where $s^2 = (y_i y_j)^2 = 1$, and scalar Const represents all constant terms independent of α_i and α_j and can therefore be dropped. Now the objective function becomes a quadratic function of the single variable α_j. Consider the coefficients of this quadratic function:

- The coefficient of α_j^2

$$\frac{1}{2}(2K_{ij} - K_{ii} - K_{jj}) = \frac{1}{2}\eta \tag{14.68}$$

where η is defined as

$$\begin{aligned}\eta &= 2K_{ij} - K_{ii} - K_{jj} = 2\mathbf{z}_i^T\mathbf{z}_j - \mathbf{z}_i^T\mathbf{z}_i - \mathbf{z}_j^T\mathbf{z}_j = -(\mathbf{z}_i - \mathbf{z}_j)^T(\mathbf{z}_i - \mathbf{z}_j) \\ &= -||\mathbf{z}_i - \mathbf{z}_j||^2 \le 0\end{aligned} \tag{14.69}$$

- The coefficient of α_j

$$\begin{aligned}&1 - s + s\delta(K_{ii} - K_{ij}) + y_j(v_i - v_j) \\ &= 1 - s + (s\alpha_i + \alpha_j)(K_{ii} - K_{ij}) \\ &\quad + y_j[(u_i - b - \alpha_i y_i K_{ii} - \alpha_j y_j K_{ij}) - (u_j - b - \alpha_i y_i K_{ij} - \alpha_j y_j K_{jj})] \\ &= 1 - s + s\alpha_1(K_{ii} - K_{ij}) + \alpha_2(K_{ii} - K_{ij}) \\ &\quad + y_j(u_i - u_j) - \alpha_i s K_{ii} - \alpha_j K_{ij} + \alpha_i s K_{ij} + \alpha_j K_{jj} \\ &= y_j^2 - s - \alpha_j(2K_{ij} - K_{ii} - 2K_{jj}) + y_j(u_i - u_j) \\ &= y_j(u_i - y_j - u_j + y_j) - \alpha_j(2K_{ij} - K_{ii} - K_{jj}) \\ &= y_j(E_i - E_j) - \alpha_j\eta\end{aligned} \tag{14.70}$$

where we have defined

$$E_k = u_k - y_k = \sum_{n=1}^{N} \alpha_n^{old} y_n K_{nk} + b - y_k \qquad (k = 1, 2) \tag{14.71}$$

as the difference between the desired output y_k and the actual output based on the previous values of the variables α_n^{old} $(n = 1, \cdots, N)$.

Now the quadratic objective function can be rewritten as

$$L(\alpha_j) = \frac{1}{2}\eta\alpha_j^2 + \left[y_j(E_i - E_j) - \alpha_j^{old}\eta\right]\alpha_j \tag{14.72}$$

Given the previous values α_n^{old} $(n = 1, \cdots, N)$ in the coefficients, we will find α_j^{new} that maximizes $L(\alpha_j)$. Consider its first- and second-order derivatives

$$\begin{aligned}\frac{d}{d\alpha_j}L(\alpha_j) &= \eta\alpha_j + y_j(E_i - E_j) - \alpha_j^{old}\eta \\ \frac{d^2}{d\alpha_j^2}L(\alpha_j) &= \eta \le 0\end{aligned} \tag{14.73}$$

As the second-order derivative of $L(\alpha_j)$ is $\eta \le 0$, it has a maximum. Solving the equation $d\,L(\alpha_j)/d\alpha_j = 0$, we obtain α_j^{new} that maximizes $L(\alpha_j)$ based on the old value α^{old}

$$\alpha_j^{new} = \alpha_j^{old} + \frac{y_j(E_j - E_i)}{\eta} = \alpha_j^{old} + \Delta\alpha_j, \qquad \Delta\alpha_j = \alpha_j^{new} - \alpha_j^{old} = \frac{y_j(E_j - E_i)}{\eta}, \tag{14.74}$$

Note that this is an unconstrained solution, which needs to be modified if the constraints in Eq. (14.57) are violated. As shown in Fig. 14.6, α_i and α_j are inside the square of size C, owing to the first constraint $0 \le \alpha_i,\ \alpha_j \le C$; also, due to the

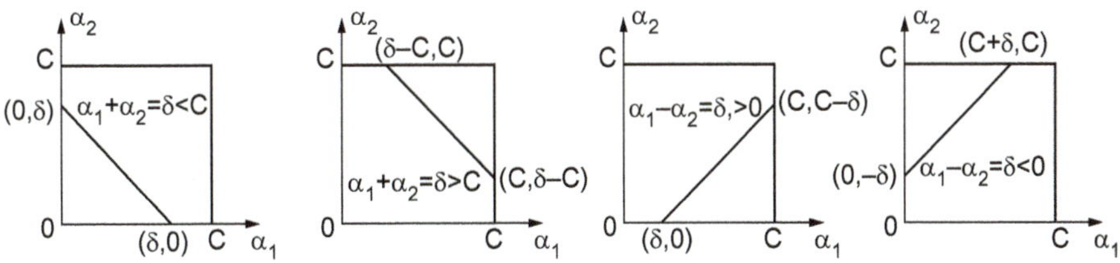

Fig. 14.6 Four cases chile updating α_i and α_i.

second constraint $\alpha_i + s\alpha_j = \delta$ (Eq. (14.59)), they have to be on the diagonal line segment inside the square.

The panels in Fig. 14.6 illustrate the four possible cases depending on whether $s = 1$ ($y_i = y_j = \pm 1$, panels 1 and 2), or $s = -1$ ($y_i = -y_j = \pm 1$, panels 3 and 4), and whether $\delta < C$ (panels 1 and 3) or $\delta > C$ (panels 2 and 4), which can be further summarized in terms of the lower bound L and upper bound H of α_2:

- If $s = 1$, then $\alpha_i + s\alpha_j = \alpha_i + \alpha_j = \delta$,
  - if $\delta < C$, then $\min(\alpha_j) = 0$, $\max(\alpha_j) = \delta$ (first panel)
  - if $\delta > C$, then $\min(\alpha_j) = \delta - C$, $\max(\alpha_j) = C$ (second panel)

  Combining these two cases, we have

$$\begin{cases} L = \max(0,\ \delta - C) = \max(0,\ \alpha_i + \alpha_j - C) \\ H = \min(\delta,\ C) = \min(\alpha_i + \alpha_j,\ C) \end{cases} \tag{14.75}$$

- If $s = -1$, then $\alpha_i + s\alpha_j = \alpha_i - \alpha_j = \delta$,
  - if $\delta > 0$, then $\min(\alpha_j) = 0$, $\max(\alpha_j) = C - \delta$ (third panel)
  - if $\delta < 0$, then $\min(\alpha_j) = -\delta$, $\max(\alpha_j) = C$ (fourth panel)

  Combining these two cases, we have

$$\begin{cases} L = \max(0,\ -\delta) = \max(0,\ \alpha_j - \alpha_i) \\ H = \min(C - \delta,\ C) = \min(C + \alpha_j - \alpha_i,\ C) \end{cases} \tag{14.76}$$

Now we apply the constraints in terms of L and H to t optimal solution α_j^{new} obtained previously to get α_j that is feasible as well as optimal

$$\alpha_j^{new} \Longleftarrow \begin{cases} H & \text{if } \alpha_j^{new} \geq H \\ \alpha_j^{new} & \text{if } L < \alpha_2^{new} < H \\ L & \text{if } \alpha_j^{new} \leq L \end{cases} \tag{14.77}$$

We can further find α_i based on Eq. (14.62)

$$\alpha_i^{new} = \alpha_i^{old} - s\Delta\alpha_j = \alpha_i^{old} - s(\alpha_j^{new} - \alpha_j^{old}) \tag{14.78}$$

and update the weight vector based on Eq. (14.40)

$$\begin{aligned} \mathbf{w}^{new} &= \sum_{n \neq i,j} y_n \alpha_n^{old} \mathbf{x}_n + y_i \alpha_i^{new} \mathbf{x}_i + y_j \alpha_j^{new} \mathbf{x}_j \\ &= \mathbf{w}^{old} - y_i \alpha_i^{old} \mathbf{x}_i - y_j \alpha_j^{old} \mathbf{x}_j + y_i \alpha_i^{new} \mathbf{x}_i + y_j \alpha_j^{new} \mathbf{x}_j \\ &= \mathbf{w}^{old} + y_i \Delta\alpha_i \mathbf{x}_i + y_j \Delta\alpha_j \mathbf{x}_j = \mathbf{w}^{old} + \Delta\mathbf{w} \end{aligned} \tag{14.79}$$

where $\Delta\mathbf{w} = \mathbf{w}^{new} - \mathbf{w}^{old} = y_i\Delta\alpha_i\mathbf{x}_i + y_j\Delta\alpha_j\mathbf{x}_j$. To update b, consider:

$$\begin{aligned}\Delta E_k &= E_k^{new} - E_k^{old} = u_k^{new} - u_k^{old} \\ &= \sum_{n=1}^{N} \alpha_n^{new} y_n K_{nk} + b^{new} - \sum_{n=1}^{N} \alpha_n^{old} y_n K_{nk} - b^{old} \\ &= y_i\Delta\alpha_i K_{ik} + y_j\Delta\alpha_j K_{jk} + b^{new} - b^{old} \qquad (k = 1,2)\end{aligned} \tag{14.80}$$

Consider the following cases:

- If $0 < \alpha_k < C$ for either $k = i$ or $k = j$, then $y_k E_k = 0$ according to Eq. (14.52). We let $E_k^{new} = 0$ and solve to obtain

$$b_k^{new} = b_k^{old} - (E_k^{old} + y_i\Delta\alpha_i K_{ik} + y_j\Delta\alpha_j K_{jk}) \qquad (k = 1 \text{ or } 2) \tag{14.81}$$

- If $0 < \alpha_k < C$ for both $k = i$ and $k = j$, then $b_i^{new} = b_j^{new}$.
- If $\alpha_k = 0$ or $\alpha_k = C$ for both $k = i$ and $k = 2$, then $E_k \neq 0$, and $b_i^{new} \neq b_j^{new}$, their average $b^{new} = (b_i^{new} + b_j^{new})/2$ can be used.

This computation for maximizing $L(\alpha_i,\alpha_j)$ is carried out iteratively each time when a pair of variables α_i and α_j is selected to be updated. Specifically, we select a variable α_i that violates any of the KKT conditions given in Eq. (14.52) in the outer loop of the SMO algorithm, and then pick a second variable α_j in the inner loop of the algorithm, both to be optimized. The selection of these variables can be either random (e.g., by following their order), or guided by some heuristics, such as always choosing the variable with maximum step size $\Delta\alpha_n = \alpha_n^{new} - \alpha_n^{old}$. This iterative process is repeated until convergence when the KKT conditions are satisfied by all $\alpha_1,\cdots,\alpha_N$ variables, that is, no more variables need to be updated. All data points corresponding to $\alpha_n \neq 0$ are the support vectors, based on which we can find the offset b by Eq. (14.55) and classify any unlabeled point $\mathbf{x}$ by Eq. (14.56).

The MATLAB code for the essential part of the SMO algorithm is listed in the following, based mostly on a simplified SMO algorithm with some modifications. In particular, the if-statement

```
if (alpha(i)<C & yE<-tol) | (alpha(i)>0 & yE>tol)
```

checks the three KKT conditions in Eq. (14.52), where $yE = y_n E_n$ and tol is a small constant. The first half checks if the first two of the three conditions are violated, while the second half checks if the second two of the three conditions are violated.

```
[X y]=getData;            % get training data
[m n]=size(X);            % size of data
alpha=zeros(1,n);         % alpha variables
bias=0;                   % initial bias
it=0;                     % iteration index
while (it<maxit)          % number of iterations less than maximum
    it=it+1;
    changed_alphas=0;     % number of changed alphas
    N=length(y);          % number of support vectors
```

```
for i=1:N            % for each alpha_i
    Ei=sum(alpha.*y.*K(X,X(:,i)))+bias-y(i);
    yE=Ei*y(i);
    if (alpha(i)<C & yE<-tol) | (alpha(i)>0 & yE>tol)
                     % KKT violation
        for j=[1:i-1,i+1:N]   % for each alpha_j unequal alpha_i
            Ej=sum(alpha.*y.*K(X,X(:,j)))+bias-y(j);
            ai=alpha(i);      % alpha_i old
            aj=alpha(j);      % alpha_j old
            if y(i)==y(j)     % s=y_i y_j=1
                L=max(0,alpha(i)+alpha(j)-C);
                H=min(C,alpha(i)+alpha(j));
            else              % s=y_i y_j=-1
                L=max(0,alpha(j)-alpha(i));
                H=min(C,C+alpha(j)-alpha(i));
            end
            if L==H           % skip when L==H
                continue
            end
            eta=2*K(X(:,j),X(:,i))-K(X(:,i),X(:,i))-K(X(:,j),X(:,j));
            alpha(j)=alpha(j)+y(j)*(Ej-Ei)/eta;   % update alpha_j
            if alpha(j) > H
                alpha(j) = H;
            elseif alpha(j) < L
                alpha(j) = L;
            end
            if norm(alpha(j)-aj) < tol   % skip if no change
                continue
            end
            alpha(i)=alpha(i)-y(i)*y(j)*(alpha(j)-aj);
                                                  % find alpha_i
            bi = bias - Ei - y(i)*(alpha(i)-ai)*K(X(:,i),X(:,i))...
                -y(j)*(alpha(j)-aj)*K(X(:,j),X(:,i));
            bj = bias - Ej - y(i)*(alpha(i)-ai)*K(X(:,i),X(:,j))...
                -y(j)*(alpha(j)-aj)*K(X(:,j),X(:,j));
            if 0<alpha(i) & alpha(i)<C
                bias=bi;
             elseif 0<alpha(j) & alpha(j)<C
                bias=bj;
            else
                bias=(bi+bj)/2;
            end
            changed_alphas=changed_alphas+1;  % one more alpha changed
        end
    end
end
```

```
    if changed_alphas==0       % no more changed alpha, quit
        break
    end
    I=find(alpha~=0);          % indices of non-zero alphas
    alpha=alpha(I);            % find non-zero alphas
    Xsv=X(:,I);                % find support vectors
    ysv=y(I);                  % their corresponding indices
end                            % end of iteration
nsv=length(ysv);               % number of support vectors
```

Here K(X,x) is a function that takes an $m \times n$ matrix $\mathbf{X} = [\mathbf{x}_1, \cdots, \mathbf{x}_n]$ and an n–D vector $\mathbf{x}$ and produces the kernel functions $K(\mathbf{x}_i, \mathbf{x})$, $(i = 1, \cdots, n)$ as output.

```
function ker=K(X,x)
    global kfunction
    global gm                        % parameter for Gaussian kernel
    [m n]=size(X);
    ker=zeros(1,n);
    if kfunction=='l'
        for i=1:n
            ker(i)=X(:,i).'*x; % linear kernel
        end
    elseif kfunction=='g'
        for i=1:n
            ker(i)=exp(-gm*norm(X(:,i)-x)); % Gaussian kernel
        end
    elseif kfunction=='p'
        for i=1:n
            ker(i)=(X(:,i).'*x).^3; % polynomial kernel
        end
    end
end
```

Having found all non-zero $\alpha_n \neq 0$ and the corresponding support vectors, we further find the offset or bias term:

```
    bias=0;
    for i=1:nsv
        bias=bias+(ysv(i)-sum(ysv.*alpha.*K(Xsv,Xsv(:,i))));
    end
    bias=bias/nsv;
```

Any unlabeled data point $\mathbf{x}$ is classified into class C_+ or C_-, depending on whether the following expression is greater or smaller than zero:

```
    y=sum(alpha.*ysv.*K(Xsv,x))+bias;
```

Example 14.3

The same training set used previously to test the SVM with a hard margin is used again to test the SVM with a soft margin. Two results are shown in Fig. 14.7, corresponding to two different values $C = 1.2$ and $C = 0.05$ for the weight of the error term $\sum_{n=1}^{N} \xi_n$ in the objective function. We see that when $C = 1.2$ is large, the number of support vectors (the four solid dots) is small owing to the small errors ξ_n allowed, and the decision boundary determined by these support vectors independent of the rest of the data set (circles) is mostly dictated by the local points close to the boundary, not necessarily a good reflection of the global distribution of the data points. This result is similar to the previous one generated by the hard-margin SVM.

Conversely, when $C = 0.1$ is small, a greater number of data points become support vectors due to the larger (the 13 solid dots) error ξ_n allowed, and the resulting linear decision line determined by these support vectors reflects global distribution of the data points, and it better separates the two classes in terms of their mean positions.

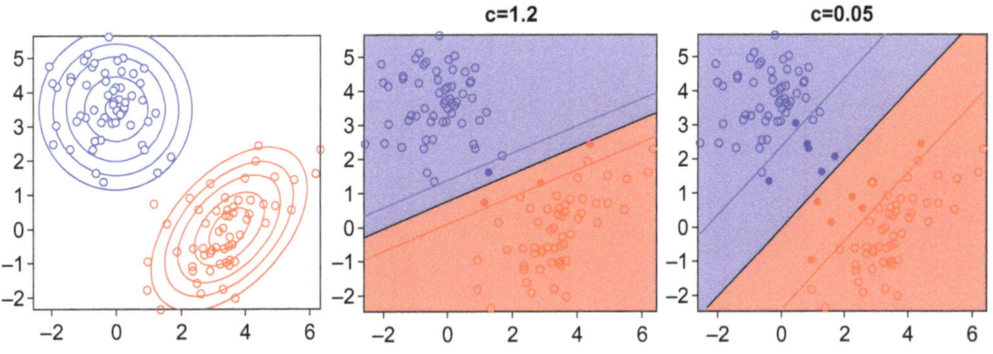

Fig. 14.7 Classification by SVM based on SMO.

Example 14.4

The classification results for the XOR data set are shown in Fig. 14.8 with $C = 1.2$ and $C = 0.05$. As the two classes in the XOR data set are not linearly separable, a Gaussian or radial basis function (RBF) kernel is used to map the data from 2-D space to an infinite dimensional space, which is partitioned into two regions for the two classes. When $C = 1.2$ is large, the decision boundary is dictated by a relatively small number of support vectors (solid dots, with small error ξ_n) and all data points are classified correctly, but it is more prone to the overfitting problem, if some of the small number of support vectors are outliers owing to noise.

By contrast, when $C = 0.05$ is small, the decision boundary is determined by a greater number of support vectors, better reflecting the global distribution of the data points. However, as the local points close to the boundary are not relatively de-emphasized some misclassification occurs, some nine red dots are classified into the blue region incorrectly.

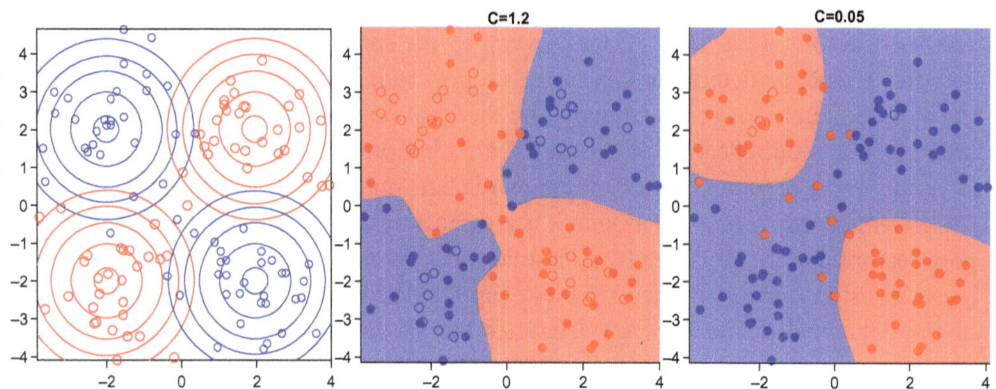

Fig. 14.8 Classification of XOR dataset by SVM based on SMO.

Two additional examples of binary classification based on SMO are shown in Figs 14.9 and 14.10.

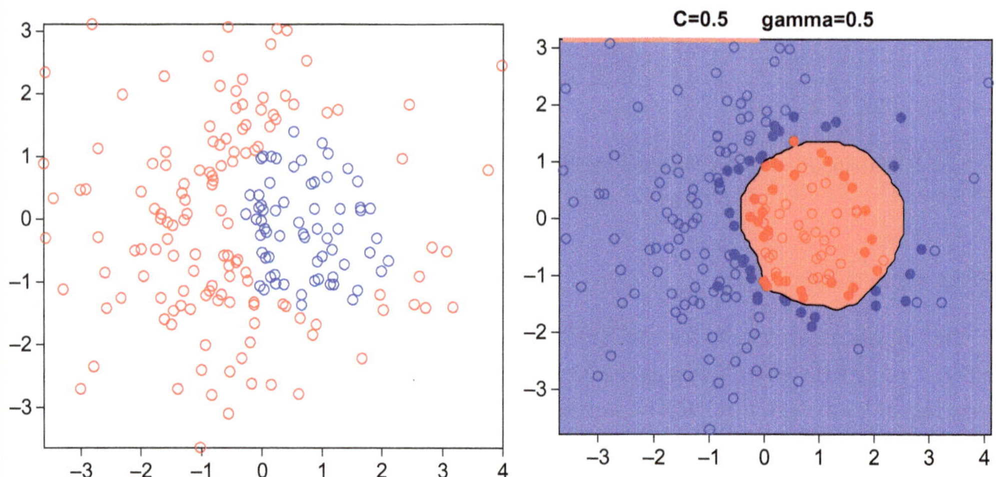

Fig. 14.9 Example of classification by SMO.

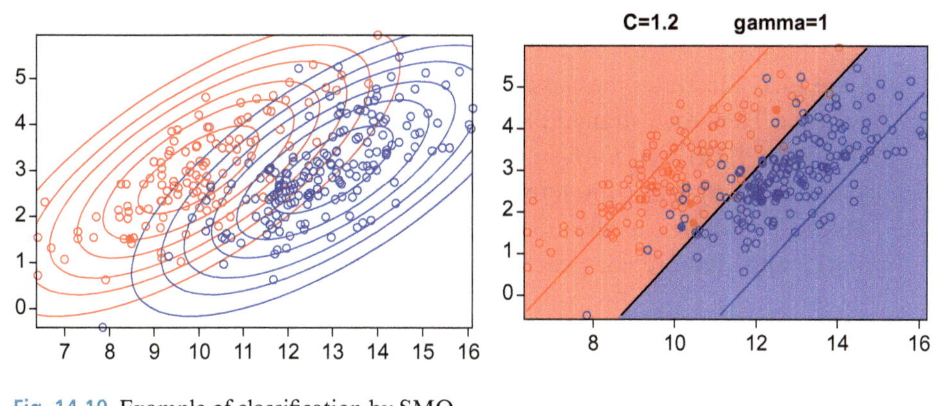

Fig. 14.10 Example of classification by SMO.

14.5 Multiclass Classification

The SVM algorithm is inherently a binary classifier, but it can be generalized to multiclass problems as well. In this section, we consider a general K-class classifier based on a training set $\mathbf{X} = [\mathbf{x}_1, \cdots, \mathbf{x}_N]$, of which each sample $\mathbf{x}_n$ is labeled by $y_n \in \{1, \cdots, K\}$ to indicate $\mathbf{x}_n \in C_{y_n}$.

We first consider two straightforward and imperial methods for multiclass classification based directly on binary SVM.

- One-Vs-One (1V1)
  Any unlabeled $\mathbf{x}$ is classified to a class which receives the maximum votes out of the $K(K-1)/2$ binary classifications between every pair of the K classes.
- One-Vs-Rest (1VR)
  This method converts a K-class problem ($K > 2$) into K binary problems. First, we regroup the K classes $C_1, \cdots, C_K$ into two classes $C_+ = C_k$ and $C_- = \{C_l | l = 1, \cdots, K,\ l \neq k\}$ and find the corresponding decision function $f_k(\mathbf{x})$ representing quantitatively to what extent a given $\mathbf{x}$ belongs to class $C_+ = C_k$ (if $f_k(\mathbf{x}) > 0$), instead of C_- containing all remaining $K-1$ classes (if $f_k(\mathbf{x}) < 0$). After this process has been carried out for all $k = 1, \cdots, K$, the $\mathbf{x}$ can be classified as

$$\text{if} \quad f_l(\mathbf{x}) = \max\{f_1(\mathbf{x}), \cdots, f_K(\mathbf{x})\}, \quad \text{then} \quad \mathbf{x} \in C_l \tag{14.82}$$

We now consider another method for multiclass classification by directly generalizing the binary SVM so that all K classes can be separated by the decision boundaries in such a way that the margin between any two of the K classes is maximized. First, as in the binary case, we define a linear function for each of the K classes

$$f_k(\mathbf{x}) = \mathbf{w}_k^T \mathbf{x} + b_k \qquad (k = 1, \cdots, K) \tag{14.83}$$

Once the parameters $\mathbf{w}_k$ and b_k are determined during training, any unlabeled point $\mathbf{x}$ can be classified as

$$\text{if } f_l(\mathbf{x}) = \max_k\{f_k(\mathbf{x}) \ \ (k = 1, \cdots, K)\}, \quad \text{then } \mathbf{x} \in C_l \tag{14.84}$$

Figure 14.11 illustrates the classification of $K = 3$ classes in $d = 2$ dimensional feature space. Here each straight line H_k (plane or hyperplane if $d \geq 3$) is determined by equation $f_k(\mathbf{x}) = \mathbf{w}_k^T\mathbf{x} + b_k = 0$, $(k = 1, \cdots, K)$, with normal direction in $\mathbf{w}_k$ and distance $d(H_k, \mathbf{0}) = |b_k|/||\mathbf{w}_k||$ to the origin (the same as in Eq. (14.4)).

According to the classification rule in Eq. (14.84), the 2-D feature space is partitioned into three regions each for one of the $K = 3$ classes by the decision boundaries B_{12}, B_{23}, B_{31}, where B_{ij} between classes C_i and C_j is composed of all points $\mathbf{x} \in B_{ij}$ satisfying $f_i(\mathbf{x}) = f_j(\mathbf{x})$. Like in binary SVM case, we further define for each of the K classes two additional straight lines (planes or hyperplanes if $d \geq 3$) in parallel with and on either side of H_k, denoted by $H_{k\pm}$ and represented by equations $f_k(\mathbf{x}) = \mathbf{w}_k^T\mathbf{x} + b_k \pm c_k$. This distance between these straight lines and H_k is (the same as in Eq. (14.9))

$$d(H_{k\pm}, H_k) = \frac{|c_k|}{||\mathbf{w}_k||} \tag{14.85}$$

Also as shown in the figure, the intersections of these parallel lines (two for each of class) further determine two straight lines parallel to the decision boundary B_{12}, defining the decision margin between the support vectors of C_1 and C_2. It can be seen that this decision margin is related to both $|c_1|/||\mathbf{w}_1||$ and $|c_2|/||\mathbf{w}_2||$, which are maximized if $||\mathbf{w}_1||$ and $||\mathbf{w}_2||$ are minimized.

Now the multiclass classification problem can be formulated as to find the parameters $\mathbf{w}_k$ and b_k for each of the K classes, so that $\sum_{k=1}^K ||\mathbf{w}_k||^2$ is minimized and thereby the decision margins for all boundaries B_{ij} are maximized, under the constraint that every training sample $\mathbf{x}_n$ labeled by y_n is correctly classified with a decision margin of 1 (the distance between the support vectors on opposite sides of B_{ij} is 2)

$$\begin{aligned} &\text{minimize} && \frac{1}{2}\sum_{k=1}^K ||\mathbf{w}_k||^2 \\ &\text{subject to} && \mathbf{w}_{y_n}^T\mathbf{x}_n + b_{y_n} \geq \mathbf{w}_k^T\mathbf{x}_n + b_k + 2 \quad (k = 1, \cdots, K, \ k \neq y_n) \end{aligned} \tag{14.86}$$

Similar to Eq. (14.37) in soft margin SVM, the constraints of this optimization problem can be relaxed by including in each constraint an extra error term $\xi_{nk} \geq 0$, which also needs to be minimized. Now the problem can be reformulated as

$$\begin{aligned} &\text{minimize} && \frac{1}{2}\sum_{k=1}^K ||\mathbf{w}_k||^2 + C\sum_{k=1}^K\sum_{n=1}^N \xi_{nk} \\ &\text{subject to} && \mathbf{w}_{y_n}^T\mathbf{x}_n + b_{y_n} \geq \mathbf{w}_k^T\mathbf{x}_n + b_k + 2 - \xi_{nk} \\ &&& \xi_{nk} \geq 0 \\ &&& (n = 1, \cdots, N, \qquad k = 1, \cdots, K, \ k \neq y_n) \end{aligned} \tag{14.87}$$

Once this constrained optimization problem is solved we get $\mathbf{w}_k$ and b_k for all $k = 1, \cdots, K$, and we can find $f_k(\mathbf{x}) = \mathbf{w}_k^T\mathbf{x} + b_k$ for any unlabeled sample $\mathbf{x}$ and classify it based on Eq. (14.84).

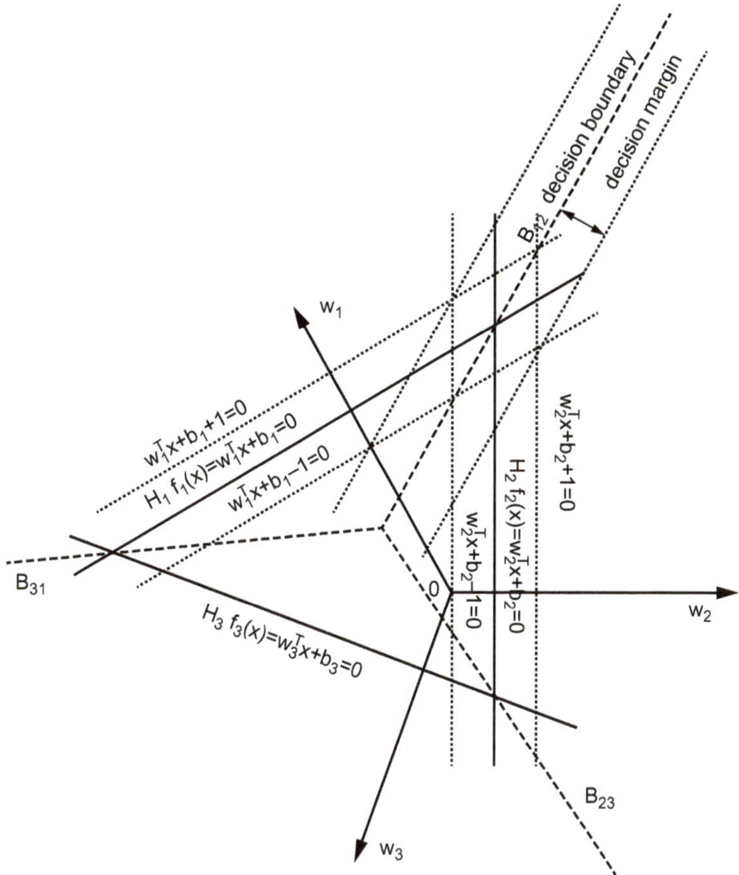

Fig. 14.11 Multiclass SMV.

Similar to the binary SVM, we approach this optimization problem by considering the dual problem. We first construct the Lagrangian, the primal function, of this constrained optimization problem

$$
\begin{aligned}
L_p(\mathbf{W},\mathbf{b},\boldsymbol{\xi},\boldsymbol{\alpha},\boldsymbol{\beta}) &= \frac{1}{2}\sum_k ||\mathbf{w}_k||^2 + C\sum_k\sum_n \xi_{nk} \\
&- \sum_k\sum_n \alpha_{nk}\left[(\mathbf{w}_{y_n}-\mathbf{w}_k)^T\mathbf{x}_n + b_{y_n} - b_k - 2 + \xi_{nk}\right] \\
&- \sum_k\sum_n \beta_{nk}\xi_{nk}
\end{aligned}
\tag{14.88}
$$

where $\mathbf{W}$, $\mathbf{b}$, $\boldsymbol{\xi}$, $\boldsymbol{\alpha}$, and $\boldsymbol{\beta}$ are matrices and vectors containing the corresponding parameters $\mathbf{w}_k$, b_k, ξ_{nk}, α_{nk}, and β_{nk} for all $k = 1,\cdots,K$ and $n = 1,\cdots,N$. For the special case $k = y_n$ not included in the constraints, $\alpha_{ny_n} = \beta_{ny_n} = 0$, and the constraint becomes an equality $0 = 2 - \xi_{ny_n}$, that is, $\xi_{ny_n} = 2$.

The KKT conditions for the Lagrangian in Eq. (14.88) are listed here (similar to Eqs (14.40) through (14.45) for the Lagrangian in Eq. (14.39)):

- Stationarity

$$\frac{\partial}{\partial \mathbf{w}_k} L_p(\mathbf{w}, b, \xi, \alpha, \beta) = \mathbf{w}_k + \sum_n \alpha_{nk} \mathbf{x}_n - \sum_n c_{nk} A_n \mathbf{x}_n = 0 \tag{14.89}$$

that is

$$\mathbf{w}_k = \sum_n (c_{nk} A_n - \alpha_{nk}) \mathbf{x}_n \tag{14.90}$$

where we have defined

$$c_{nk} = \begin{cases} 1 & \text{if } k = y_n \\ 0 & \text{else} \end{cases}, \quad \text{and} \quad A_n = \sum_k \alpha_{nk} \tag{14.91}$$

$$\frac{\partial}{\partial b_k} L_p(\mathbf{w}, b, \xi, \alpha, \beta) = -\sum_n c_{nk} A_n + \sum_n \alpha_{nk} = 0 \tag{14.92}$$

$$\frac{\partial}{\partial \xi_{nk}} L_p(\mathbf{w}, b, \xi, \alpha, \beta) = C - \alpha_{nk} - \beta_{nk} = 0 \tag{14.93}$$

- Primal and dual feasibilities

$$\begin{aligned} &(\mathbf{w}_{y_n} - \mathbf{w}_k)^T \mathbf{x}_n + b_{y_n} - b_k - 2 + \xi_{nk} \geq 0 \\ &\alpha_{nk} \geq 0, \quad \beta_{nk} \geq 0, \quad \xi_{nk} \geq 0 \\ &\qquad (n = 1, \cdots, N, \quad k = 1, \cdots, K, \ k \neq y_n) \end{aligned} \tag{14.94}$$

- Complementarity

$$\alpha_{nk}[(\mathbf{w}_{y_n} - \mathbf{w}_k)^T \mathbf{x}_n + b_{y_n} - b_k - 2 + \xi_{nk}] = 0 \tag{14.95}$$

These KKT conditions need to be satisfied by the solution of the optimization problem, including $\mathbf{w}_k^*$, b_k^*, and ξ_{nk}^* that need to satisfy the stationarity of the KKT conditions. We can therefore substitute Eq. (14.90) into Eq. (14.88) and obtain

$$\begin{aligned} L_p(\mathbf{W}, \mathbf{b}, \xi, \alpha, \beta) &= \frac{1}{2} \sum_k \left(\sum_m (c_{mk} A_m - \alpha_{mk}) \right) \left(\sum_n (c_{nk} A_n - \alpha_{nk}) \right) \mathbf{x}_m^T \mathbf{x}_n \\ &- \sum_k \sum_n \alpha_{nk} \left[\sum_m (c_{my_n} A_m - \alpha_{my_n}) \mathbf{x}_m^T \mathbf{x}_n - \sum_m (c_{mk} A_m - \alpha_{mk}) \mathbf{x}_m^T \mathbf{x}_n + b_{y_n} - b_k - 2 \right] \\ &- \sum_k \sum_n \alpha_{nk} \xi_{nk} + C \sum_k \sum_n \xi_{nk} - \sum_k \sum_n \beta_{nk} \xi_{nk} \end{aligned} \tag{14.96}$$

This expression can be significantly simplified. First, we note that the last three terms cancel out due to Eq. (14.93). Second, the two terms containing b_{y_n} and b_k can be written as

$$\sum_k \sum_n \alpha_{nk} b_{y_n} = \sum_k \sum_n c_{nk} \alpha_{nk} b_k = \sum_k b_k \sum_n A_n c_{nk}$$
$$\sum_k \sum_n \alpha_{nk} b_k = \sum_k b_k \sum_n \alpha_{nk} \tag{14.97}$$

and they also cancel out due to Eq. (14.92).

Now we can rewrite L_p as a function of only the variable $\boldsymbol{\alpha}$

$$\begin{aligned}
L_d(\boldsymbol{\alpha}) &\overset{1}{=} 2\sum_k \sum_n \alpha_{nk} + \sum_k \sum_m \sum_n \left[\frac{1}{2} c_{mk} c_{nk} A_m A_n - c_{mk} \alpha_{nk} A_m + \frac{1}{2} \alpha_{mk} \alpha_{nk} \right. \\
&\quad \left. - c_{my_n} \alpha_{nk} A_m + \alpha_{nk} \alpha_{my_n} + c_{mk} \alpha_{nk} A_m - \alpha_{nk} \alpha_{mk} \right] \mathbf{x}_m^T \mathbf{x}_n \\
&= 2\sum_k \sum_n \alpha_{nk} + \sum_k \sum_m \sum_n \left[\frac{1}{2} c_{mk} c_{nk} A_m A_n - \frac{1}{2} \alpha_{mk} \alpha_{nk} \right. \\
&\quad \left. - c_{my_n} \alpha_{nk} A_m + \alpha_{nk} \alpha_{my_n} \right] \mathbf{x}_m^T \mathbf{x}_n \\
&\overset{2}{=} 2\sum_k \sum_n \alpha_{nk} + \sum_k \sum_m \sum_n \left[-\frac{1}{2} c_{my_n} A_m A_n - \frac{1}{2} \alpha_{mk} \alpha_{nk} + \alpha_{nk} \alpha_{my_n} \right] \mathbf{x}_m^T \mathbf{x}_n
\end{aligned} \tag{14.98}$$

where the labeled equal signs are due to the following:

1. The two cross terms of the second part can be combined as

$$\sum_m \sum_n \sum_k c_{mk} \alpha_{nk} A_m = \sum_m \sum_n \sum_k c_{nk} \alpha_{mk} A_n \tag{14.99}$$

2. As $c_{my_n} = 1$ if $y_m = y_n$ but zero otherwise, we have

$$\sum_k c_{mk} c_{nk} = c_{my_n} = \begin{cases} 1 & y_m = y_n \\ 0 & \text{otherwise} \end{cases} \tag{14.100}$$

   and the first term of the second part can be written as $A_m A_n c_{my_n}/2$. Also, due to Eq. (14.92), the third term can be written as $-c_{my_n} c_{nk} A_n A_m = -c_{my_n} A_n A_m$ (when summed over k) and combined with the first one.

Equation (14.98) is the quadratic dual function that needs to be maximized with respect to $\boldsymbol{\alpha}$ for it to be the tightest lower bound, subject to the linear constraint imposed by Eq. (14.92). We therefore have the following dual problem

$$\begin{aligned}
\text{maximize} \quad & L_d(\boldsymbol{\alpha}) \\
\text{subject to} \quad & \sum_n A_n c_{nk} = \sum_n \alpha_{nk} \\
& 0 \le \alpha_{nk} \le C, \quad \alpha_{ny_n} = 0 \\
& (n = 1, \cdots, N, \quad k = 1, \cdots, K, \ k \ne y_n)
\end{aligned} \tag{14.101}$$

This is a QP problem containing NK variables α_{nk}, $(n = 1, \cdots,\ k = 1, \cdots, K)$, solving which we get the optimal $\boldsymbol{\alpha}^*$.

We can further obtain all K optimal weight vectors $\mathbf{w}_k^*$ by Eq. (14.90), and $\mathbf{b}^*$ based on the complementarity of the KKT conditions, by solving the simultaneous equations in Eq. (14.95) (similar to how we obtain b^* in binary case in Eq. (14.30)). Having found both optimal parameters $\mathbf{w}_k^*$ and b_k^*, we can classify any unlabeled pattern $\mathbf{x}$ by Eq. (14.84).

We note that ass the data points in Eq. (14.98) appear only in the form of an inner product, kernel method can be used to map the problem into a high dimensional space where the data points are better separable.

The algorithm discussed here can be considered the most classic multiclass SVM algorithm as it is the first SVM algorithm to address multiclass problems by directly generalizing the binary SVM. Owing to some of its drawback such as the size of variable α in the dual problem (NK versus N in Eq. (14.27) in the binary case), various algorithms are developed later for improvement in computational complexity.

14.6 Kernelized Bayes Classifier

The naive Bayes classification is based on a quadratic decision function and is therefore unable to classify data that are not quadratically separable. However, as shown in the following, the Bayes method can be reformulated so that all data points appear in the form of an inner product, and the kernel method can be used to map the original space into a higher dimensional space in which all groups can be separated even though they are not quadratically separable in the original space.

Consider a binary Bayes classifier by which any sample $\mathbf{x}$ is classified into either of the two classes C_+ and C_- depending on whether $\mathbf{x}$ is on the positive or negative side of the quadratic decision surface (Eq. (13.21))

$$f(\mathbf{x}) = \mathbf{x}^T\mathbf{W}\mathbf{x} + \mathbf{w}^T\mathbf{x} + w \begin{cases} > 0, & \mathbf{x} \in C_+ \\ < 0, & \mathbf{x} \in C_- \end{cases} \tag{14.102}$$

As show in Eq. (13.20), here

$$\begin{aligned} \mathbf{W} &= -\frac{1}{2}(\Sigma_+^{-1} - \Sigma_-^{-1}) \\ \mathbf{w} &= \Sigma_+^{-1}\mathbf{m}_+ - \Sigma_-^{-1}\mathbf{m}_- \\ w &= -\frac{1}{2}(\mathbf{m}_+^T\Sigma_+^{-1}\mathbf{m}_+ - \mathbf{m}_-^T\Sigma_-^{-1}\mathbf{m}_-) - \frac{1}{2}\log\frac{|\Sigma_+|}{|\Sigma_-|} + \log\frac{P(C_+)}{P(C_-)} \end{aligned} \tag{14.103}$$

where $\mathbf{\Sigma}_+$ and $\mathbf{\Sigma}_-$ are, respectively, the covariance matrices of the samples in C_+ and C_-,

$$\mathbf{m}_+ = \frac{1}{N_+}\sum_{x \in C_+} \mathbf{x}, \quad \mathbf{m}_- = \frac{1}{N_-}\sum_{x \in C_-} \mathbf{x} \tag{14.104}$$

are their mean vectors, and $P_+ = N_+/N$ and $P_- = N_-/N$ are their prior probabilities. Specially if $\mathbf{\Sigma}_+ = \mathbf{\Sigma}_- = \mathbf{\Sigma}$ and therefore $\mathbf{W} = \mathbf{0}$, the quadratic decision surface becomes a linear decision plane described by

$$f(\mathbf{x}) = \mathbf{w}^T\mathbf{x} + w = \left[\boldsymbol{\Sigma}^{-1}(\mathbf{m}_+ - \mathbf{m}_-)\right]^T \mathbf{x} + w = 0 \tag{14.105}$$

in the same form as the decision equation for the support vector machine. An unlabeled sample $\mathbf{x}$ is classified into either of the two classes depending on which side of a threshold $-w$ its projection onto the normal direction $\mathbf{w}$ of the decision plane lies.

As matrices $\mathbf{W}$, $\boldsymbol{\Sigma}_+^{-1}$, and $\boldsymbol{\Sigma}_-^{-1}$ are all symmetric, they can be written in the following eigen-decomposition form

$$\mathbf{W} = \mathbf{V}\boldsymbol{\Lambda}\mathbf{V}^T = \mathbf{U}\mathbf{U}^T, \quad \boldsymbol{\Sigma}_+^{-1} = \mathbf{V}_+\boldsymbol{\Lambda}_+\mathbf{V}_+^T = \mathbf{U}_+\mathbf{U}_+^T, \quad \boldsymbol{\Sigma}_-^{-1} = \mathbf{V}_-\boldsymbol{\Lambda}_-\mathbf{V}_-^T = \mathbf{U}_-\mathbf{U}_-^T \tag{14.106}$$

where $\mathbf{U} = \mathbf{V}\boldsymbol{\Lambda}^{1/2}$, $\mathbf{U}_+ = \mathbf{V}_+\boldsymbol{\Lambda}_+^{1/2}$ and $\mathbf{U}_- = \mathbf{V}_-\boldsymbol{\Lambda}_-^{1/2}$. We can write vector $\mathbf{w}$ as

$$\mathbf{w} = \Sigma_+^{-1}\mathbf{m}_+ - \Sigma_-^{-1}\mathbf{m}_- = \mathbf{U}_+\mathbf{U}_+^T \frac{1}{N_+} \sum_{\mathbf{x}_+ \in C_+} \mathbf{x}_+ - \mathbf{U}_-\mathbf{U}_-^T \frac{1}{N_-} \sum_{\mathbf{x}_- \in C_-} \mathbf{x}_- \tag{14.107}$$

Any unlabeled sample $\mathbf{x}$ can now be classified into either of the two classes based on its decision function

$$\begin{aligned} f(\mathbf{x}) &= \mathbf{x}^T\mathbf{W}\mathbf{x} + \mathbf{w}^T\mathbf{x} + w \\ &= \mathbf{x}^T\mathbf{U}\mathbf{U}^T\mathbf{x} + \left(\frac{1}{N_+}\sum_{\mathbf{x}_+} \mathbf{x}_+^T\mathbf{U}_+\mathbf{U}_+^T\right)\mathbf{x} - \left(\frac{1}{N_-}\sum_{\mathbf{x}_-} \mathbf{x}_-^T\mathbf{U}_-\mathbf{U}_-^T\right)\mathbf{x} + w \\ &= \mathbf{z}^T\mathbf{z} + \frac{1}{N_+}\sum_{\mathbf{z}_{++}} \left(\mathbf{z}_{++}^T\mathbf{z}_+\right) - \frac{1}{N_-}\sum_{\mathbf{z}_{--}} \left(\mathbf{z}_{--}^T\mathbf{z}_-\right) + w \end{aligned} \tag{14.108}$$

where

$$\begin{cases} \mathbf{z}_{++} = \mathbf{U}_+^T\mathbf{x}_+ & (\mathbf{x}_+ \in C_+) \\ \mathbf{z}_{--} = \mathbf{U}_-^T\mathbf{x}_- & (\mathbf{x}_- \in C_-) \end{cases}, \qquad \begin{cases} \mathbf{z} = \mathbf{U}^T\mathbf{x} \\ \mathbf{z}_+ = \mathbf{U}_+^T\mathbf{x} \\ \mathbf{z}_- = \mathbf{U}_-^T\mathbf{x} \end{cases} \tag{14.109}$$

As all data points appear in the form of an inner product, the kernel method can be used by replacing all inner products in the decision function by the corresponding kernel functions

$$f(\mathbf{x}) = K(\mathbf{z},\mathbf{z}) + \frac{1}{N_+}\sum_{\mathbf{z}_{++}} K(\mathbf{z}_{++},\mathbf{z}_+) - \frac{1}{N_-}\sum_{\mathbf{z}_{--}} K(\mathbf{z}_{--},\mathbf{z}_-) + b = p(\mathbf{x}) + b \tag{14.110}$$

This is the decision function in the new higher dimensional space, where $p(\mathbf{x})$ is defined as a function composed of all terms in $f(\mathbf{x})$ except the last offset term b, which is to replace w in the original space. To find this b, we first map all training samples into a 1-D space $x_n = p(\mathbf{x}_n)$, $(n = 1,\cdots,N)$ and sort them together with their corresponding labelings $y_1,\cdots,y_N$, and then search through all $N-1$ possible ways to partition them into two groups indexed, respectively, by $1,\cdots,k$ and $k+1,\cdots,N$ to find the optimal k corresponding to the maximum labeling consistency measured by

$$\left|\sum_{n=1}^{k} y_n\right| + \left|\sum_{n=k+1}^{N} y_n\right|, \qquad (k = 1,\cdots,N-1) \tag{14.111}$$

The middle point $(x_k + x_{k+1})/2$ between x_k and x_{k+1} is used as the optimal threshold to separate the training samples into two classes in the 1-D space, that is, the offset is $b = -(x_k + x_{k+1})/2$. Now the unlabeled point $\mathbf{x}$ can be classified into either of the two classes C_+ and C_-

$$p(\mathbf{x}) + b \begin{cases} > 0, & \mathbf{x} \in C_+ \\ < 0, & \mathbf{x} \in C_- \end{cases} \tag{14.112}$$

In general, when all data points are kernel mapped to a higher dimensional space, the two classes can be more easily separated, even by a hyperplane based on the linear part of the decision function without the second-order term. This allows the assumption that the two classes have the same covariance matrix so that $\mathbf{W} = -(\boldsymbol{\Sigma}_+ - \boldsymbol{\Sigma}_-)/2 = \mathbf{0}$ and the second-order term is dropped. This is the justification for the following two special cases.

- If we approximate both $\boldsymbol{\Sigma}_+$ and $\boldsymbol{\Sigma}_-$ by their average $\boldsymbol{\Sigma} = (\boldsymbol{\Sigma}_+ + \boldsymbol{\Sigma}_-)/2$, then $\mathbf{W} = \mathbf{0}$ and the decision function of any $\mathbf{x}$ becomes

$$\begin{aligned} f(\mathbf{x}) &= \mathbf{w}^T\mathbf{x} + w = \left[\boldsymbol{\Sigma}^{-1}(\mathbf{m}_+ - \mathbf{m}_-)\right]^T \mathbf{x} + w \\ &= \left[\mathbf{U}\mathbf{U}^T\left(\frac{1}{N_+}\sum_{\mathbf{x}_+}\mathbf{x}_+ - \frac{1}{N_-}\sum_{\mathbf{x}_-}\mathbf{x}_-\right)\right]^T \mathbf{x} + w \\ &= \frac{1}{N_+}\sum_{\mathbf{x}_+}\mathbf{x}_+^T\mathbf{U}\mathbf{U}^T\mathbf{x} - \frac{1}{N_-}\sum_{\mathbf{x}_-}\mathbf{x}_-^T\mathbf{U}\mathbf{U}^T\mathbf{x} + w \\ &= \frac{1}{N_+}\sum_{\mathbf{z}_+}\mathbf{z}_+^T\mathbf{z} - \frac{1}{N_-}\sum_{\mathbf{z}_-}\mathbf{z}_-^T\mathbf{z} + w \end{aligned} \tag{14.113}$$

where $\mathbf{U}\mathbf{U}^T = \boldsymbol{\Sigma}^{-1}$, $\mathbf{z}_+ = \mathbf{U}^T\mathbf{x}_+$, $\mathbf{z}_- = \mathbf{U}^T\mathbf{x}_-$, and $\mathbf{z} = \mathbf{U}^T\mathbf{x}$. This decision function can be converted to the following if the kernel method is used

$$f(\mathbf{x}) = \frac{1}{N_+}\sum_{\mathbf{z}_+}K(\mathbf{z}_+,\mathbf{z}) - \frac{1}{N_-}\sum_{\mathbf{z}_-}K(\mathbf{z}_-,\mathbf{z}) + b \tag{14.114}$$

- More specially, if $\boldsymbol{\Sigma}_+ = \boldsymbol{\Sigma}_- = \mathbf{I}$, then this decision function becomes

$$\begin{aligned} f(\mathbf{x}) &= \mathbf{w}^T\mathbf{x} + w = (\mathbf{m}_+ - \mathbf{m}_-)^T \mathbf{x} + w \\ &= \frac{1}{N_+}\sum_{\mathbf{x}_+}\mathbf{x}_+^T\mathbf{x} - \frac{1}{N_-}\sum_{\mathbf{x}_-}\mathbf{x}_-^T\mathbf{x} + w \end{aligned} \tag{14.115}$$

which can be converted to the following if the kernel method is used

$$f(\mathbf{x}) = \frac{1}{N_+}\sum_{\mathbf{x}_+}K(\mathbf{x}_+,\mathbf{x}) - \frac{1}{N_-}\sum_{\mathbf{x}_-}K(\mathbf{x}_-,\mathbf{x}) + b \tag{14.116}$$

Note that if the kernel method is used to replace an inner product by a kernel function $K(\mathbf{x},\mathbf{x}') = \phi(\mathbf{x})^T\phi(\mathbf{x}')$, we need to map all data points to $\phi(\mathbf{x}_n)$, $(n = 1,\cdots,N)$ in the higher dimensional space, instead of only mapping their means $\phi(\mathbf{m}_+)$ and $\phi(\mathbf{m}_-)$, because the mapping of a sum is not equal to the sum of the mapped points if the kernel is not linear

$$\phi(\mathbf{m}_+ - \mathbf{m}_-) \neq \phi(\mathbf{m}_+) - \phi(\mathbf{m}_-), \qquad \phi(\mathbf{m}_\pm) = \phi\left(\frac{1}{n_\pm}\sum_{\mathbf{x}\in C_\pm}\mathbf{x}\right) \neq \frac{1}{n_\pm}\sum_{\mathbf{x}\in C_\pm}\phi(\mathbf{x}) \tag{14.117}$$

The three cases are summarized as

- Type I, Eq. (14.116)

$$f(\mathbf{x}) = \frac{1}{N_+}\sum_{\mathbf{x}_+} K(\mathbf{x}_+, \mathbf{x}) - \frac{1}{N_-}\sum_{\mathbf{x}_-} K(\mathbf{x}_-, \mathbf{x}) + b \tag{14.118}$$

- Type II, Eq. (14.113)

$$f(\mathbf{x}) = \frac{1}{N_+}\sum_{\mathbf{z}_+} K(\mathbf{z}_+, \mathbf{z}) - \frac{1}{N_-}\sum_{\mathbf{z}_-} K(\mathbf{z}_-, \mathbf{z}) + b \tag{14.119}$$

- Type III, Eq. (14.110)

$$f(\mathbf{x}) = \mathbf{z}^T\mathbf{z} + \frac{1}{N_+}\sum_{\mathbf{z}_{++}} K(\mathbf{z}_{++}, \mathbf{z}_+) - \frac{1}{N_-}\sum_{\mathbf{z}_{--}} K(\mathbf{z}_{--}, \mathbf{z}_-) + b \tag{14.120}$$

The MATLAB code for the essential part of the algorithm is listed here. Given `X` and `y` for the data array composed of N training vectors $\mathbf{X} = [\mathbf{x}_1, \cdots, \mathbf{x}_N]$ each labeled by the corresponding component $\mathbf{y} = [y_1, \cdots, y_N]$, the code carries out the training and then classifies any unlabeled data point into either of the two classes. Parameter `type` selects any one of the three different versions of the algorithm, and the function `K(X,x)` returns a 1-D array containing all N kernel functions $K(\mathbf{x}_i, \mathbf{x})$, $(i = 1, \cdots, N)$ of the column vectors in $\mathbf{X}$ and vector $\mathbf{x}$.

```
X=getData;
[m n]=size(X);
X0=X(:,find(y>0));   n0=size(X0,2); % separate C+ and C-
X1=X(:,find(y<0));   n1=size(X1,2);
m0=mean(X0,2);     C0=cov(X0');       % mean and covariance
m1=mean(X1,2);     C1=cov(X1');
if type==1
    for i=1:n
        x(i)=sum(K(X0,X(:,i)))/n0-sum(K(X1,X(:,i)))/n1;
    end
elseif type==2
    C=inv(C0+C1);    [V D]=eig(C);    U=(V*D^(1/2))';
    Z=U*X;  Z0=U*X0;    Z1=U*X1;
    for i=1:n
        x(i)=sum(K(Z0,Z(:,i)))/n0-sum(K(Z1,Z(:,i)))/n1;
    end
elseif type==3
    C0=inv(C0);    C1=inv(C1);    W=-(C0-C1)/2;
    [V D]=eig(W);       U=(V*D^(1/2)).';      Z=U*X;
    [V0 D0]=eig(C0);  U0=(V0*D0^(1/2))';  Z0=U0*X;  Z00=U0*X0;
    [V1 D1]=eig(C1);  U1=(V1*D1^(1/2))';  Z1=U1*X;  Z11=U1*X1;
    for i=1:n
        x(i)=K(Z(:,i),Z(:,i))+sum(K(Z00,Z0(:,i)))/n0-sum(K(Z11,Z1(:,i)))/n1;
    end
```

```
end
[x I]=sort(x);    y=y(I);     % sort 1-D data and their labelings
smax=0;
for i=1:n-1                   % find optimal threshold value b
    s=abs(sum(y(1:i)))+abs(sum(y(i+1:n)));
    if s>smax
        smax=s;  b=-(x(i)+x(i+1))/2;
    end
end
```

Note that $\mathbf{W} = -(\boldsymbol{\Sigma}_0^{-1} - \boldsymbol{\Sigma}_1^{-1})/2$ may be either positive or negative definite, and its eigenvalue matrix $\mathbf{D}$ may contain negative values and $\mathbf{D}^{1/2}$ may contain complex values. Given any unlabeled data point $\mathbf{x}$, the following code is carried out

```
if type==1
    y(i)=sum(K(X0,x))/n0-sum(K(X1,x))/n1+b;
elseif type==2
    Z=U*x;
    y(i)=sum(K(Z0,z))/n0-sum(K(Z1,z))/n1+b;
elseif type==3
    z=U*x;     z0=U0*x;    z1=U1*x;
    y(i)=K(z,z)+sum(K(Z00,z0))/n0-sum(K(Z11,z1))/n1+b;
end
```

to classify $\mathbf{x}$ into either C_+ if $y > 0$, or C_- if $y < 0$.

In comparison with the SVM method, which requires solving a QP problem by certain iterative algorithm (either the interior point method or the SMO method), the kernel-based Bayes method is closed form and therefore extremely efficient computationally. Moreover, as shown in the following examples, this method is also highly effective as its classification results are comparable and often more accurate than those of the SVM method.

We now show a few examples to test all three types of the kernel-based Bayes method based on a set of simulated 2-D data. Both linear kernel and RBF kernel $K(\mathbf{x},\mathbf{y}) = \exp(-\gamma||\mathbf{x} - \mathbf{y}||)$. The value of the parameter γ used in the examples is 5, but it can be fine tuned in a wide range (e.g., $1 < \gamma < 9$) to make a proper tradeoff between accuracy and avoiding overfitting. The performances of these method are also compared with the SVM method implemented by the MATLAB function

```
fitcsvm(X',y,'Standardize',true,'KernelFunction','linear','KernelScale','auto'))
```

Example 14.5

Four different binary classifiers are applied to two 2-D datasets, with their results compared in terms of the correct rates listed, and their partitionings of the 2-D space shown in Figs 14.12 through 14.14.

- Test 1: Two sets of data are generated based on the following mean vectors and covariance matrices

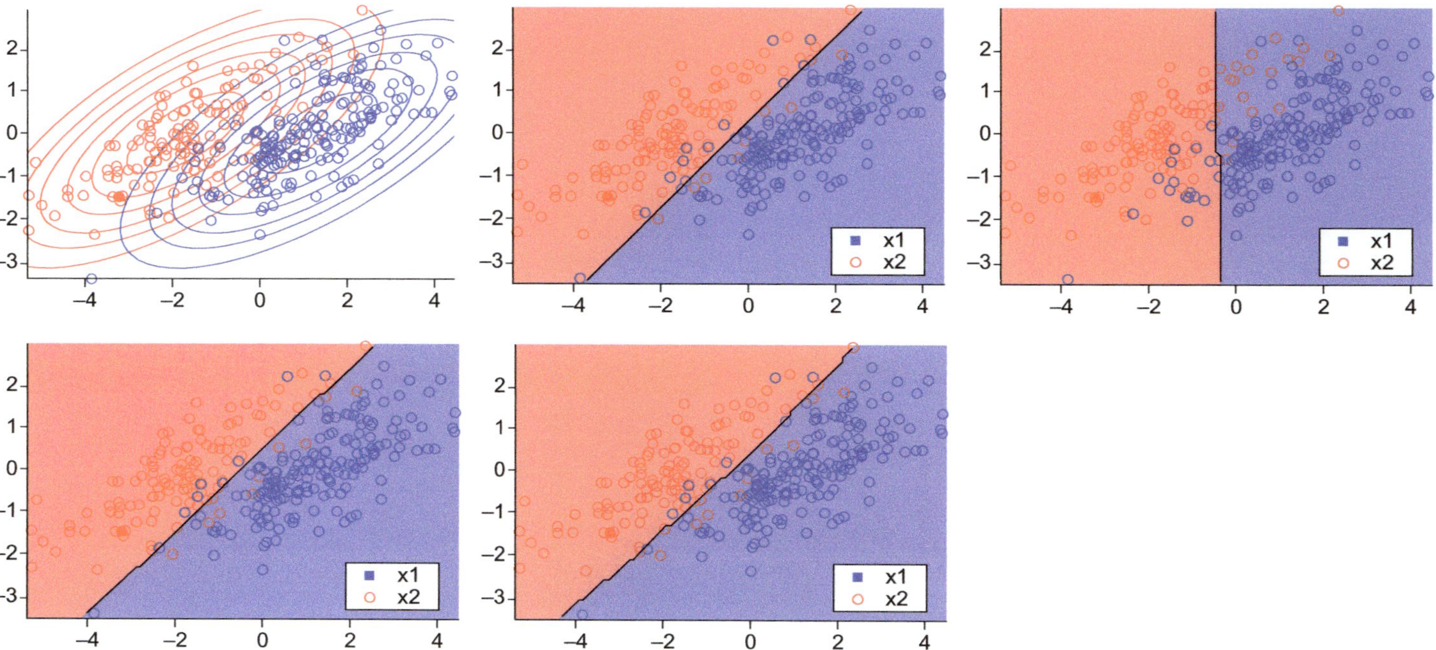

Fig. 14.12 Test 1 of kernelized Bayes classifier.

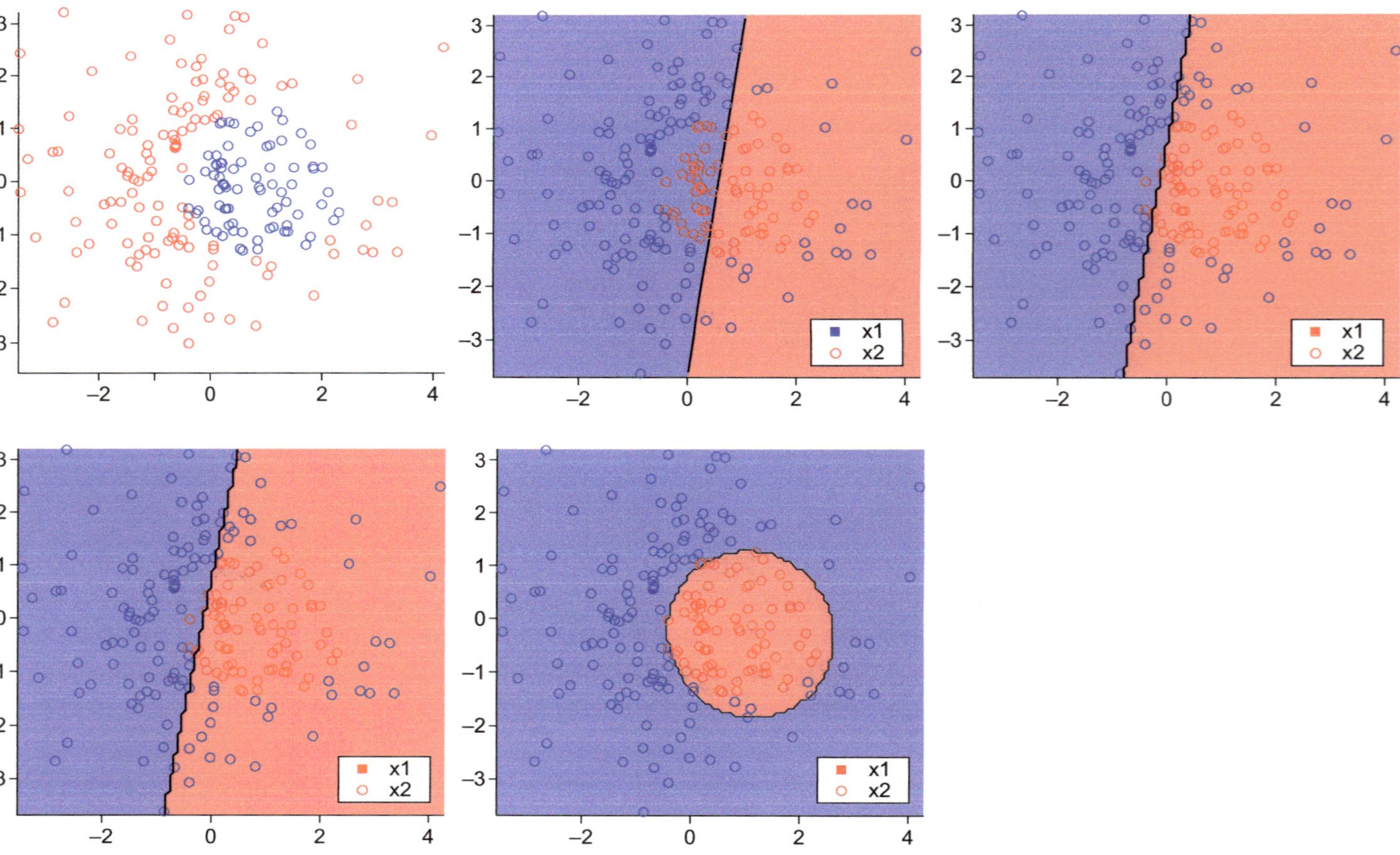

Fig. 14.13 Test 2 of kernelized Bayes classifier.

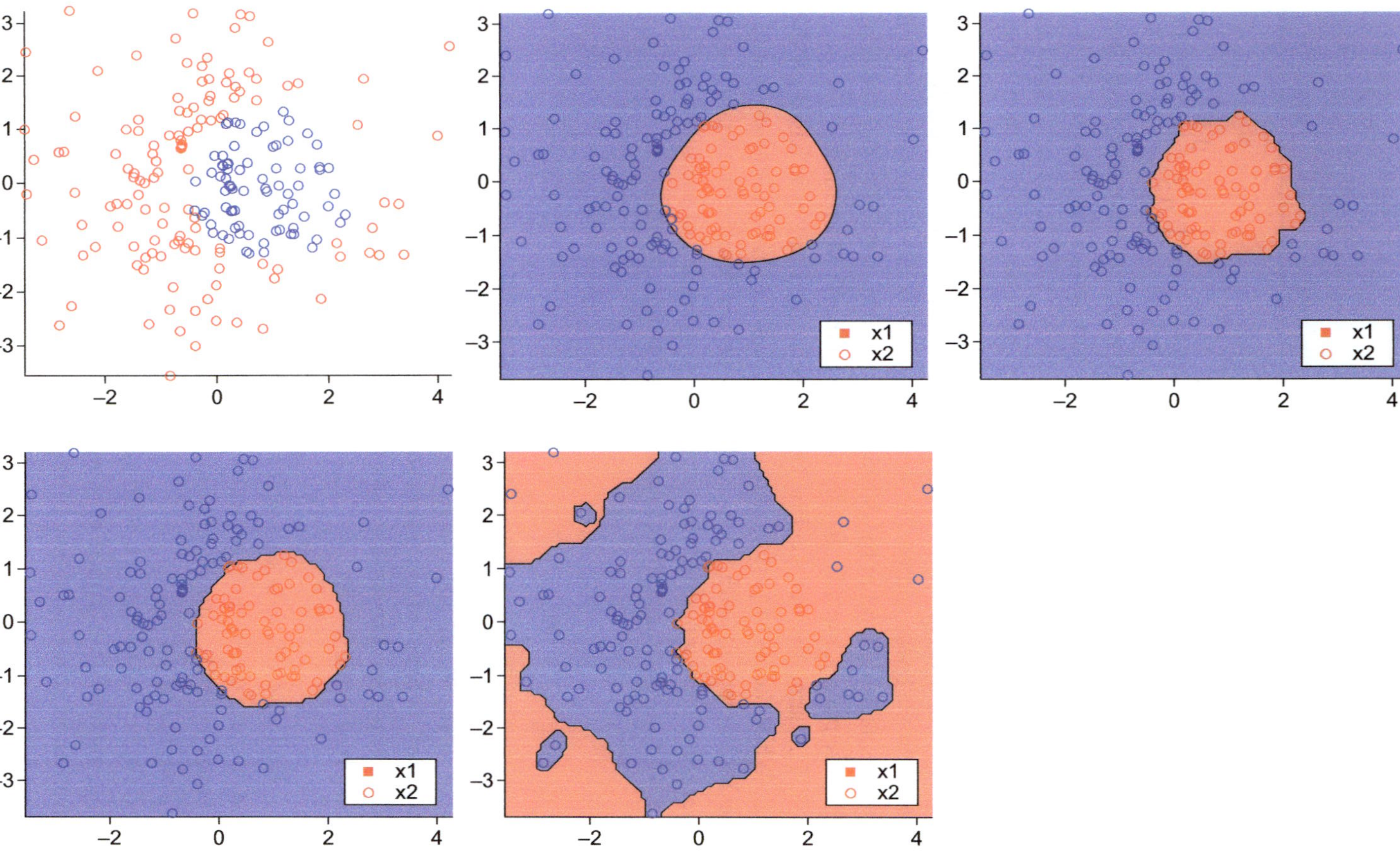

Fig. 14.14 Test 3 of kernelized Bayes classifier.

Table 14.2

	Kernel	MATLAB SVM	Kernel Bayes I	Kernel Bayes II	Kernel Bayes III
Test 1	Linear	93.0%	88.0%	94.0%	94.0%
Test 2	Linear	73.0%	80.0%	79.5%	96.5%
Test 3	RBF	98.5%	100%	100%	100%

$$\mathbf{m}_+ = \begin{bmatrix} -1 \\ 0 \end{bmatrix}, \quad \mathbf{m}_- = \begin{bmatrix} +1 \\ 0 \end{bmatrix}, \quad \boldsymbol{\Sigma}_+ = \boldsymbol{\Sigma}_- = 3 \times \begin{bmatrix} 1 & 0.5 \\ 0.5 & 0.5 \end{bmatrix} \tag{14.121}$$

The linear kernel is used for all methods. The kernel Bayes methods II and III achieve the best result (94%), slightly better than that of the standard SVM (93%). But the kernel Bayes method I performs poorly (84.7%), as it is based only on the means of the two classes without taking into consideration their covariances representing the distribution of the data points.

Test 2: Two sets of data not linearly separable are classified by the four methods all based on the linear kernel. The kernel Bayes III achieves the best result (96.5%), due to its quadratic term in the decision function, the kernel Bayes I and II perform much more poorly, but still slightly better than the SVM method.

- Test 3: The same dataset used in Test 2 is classified but now based on the RBF kernel. While the SVM performs reasonably well (98.5%), all three versions of the kernel Bayes method (with $\gamma = 5$) achieve the perfect result with all points of the two classes completely separated. However, the partitioning of the space by the kernel Bayes III is highly fragmented, indicating an obvious overfitting.

Example 14.6

Three datasets (XOR, concentric rings, multiclusters) are generated using the MATLAB code at: `https://www.mathworks.com/help/stats/support-vector-machines-for-binary-classification.html`.

The following four tests are carried out for the three versions of the kernel Bayes all based on a parameter $\gamma = 5$.

- Test 1: Exclusive OR dataset, linear kernel
  As the data are not linearly separable, all methods performed poorly except kernelized Bayes Type III with a second-order term in the decision function. The results are shown in Fig. 14.15.
- Test 2: Exclusive OR dataset, RBF kernel
  All four methods performed very well, but all three variations of the kernelized Bayes method achieved higher correct rates than the SVM. The results are shown in Fig. 14.16.
- Test 3: Concentric ring dataset, RBF kernel. The results are shown in Fig. 14.17.
- Test 4: Multicluster dataset, RBF kernel. The results are shown in Fig. 14.18.

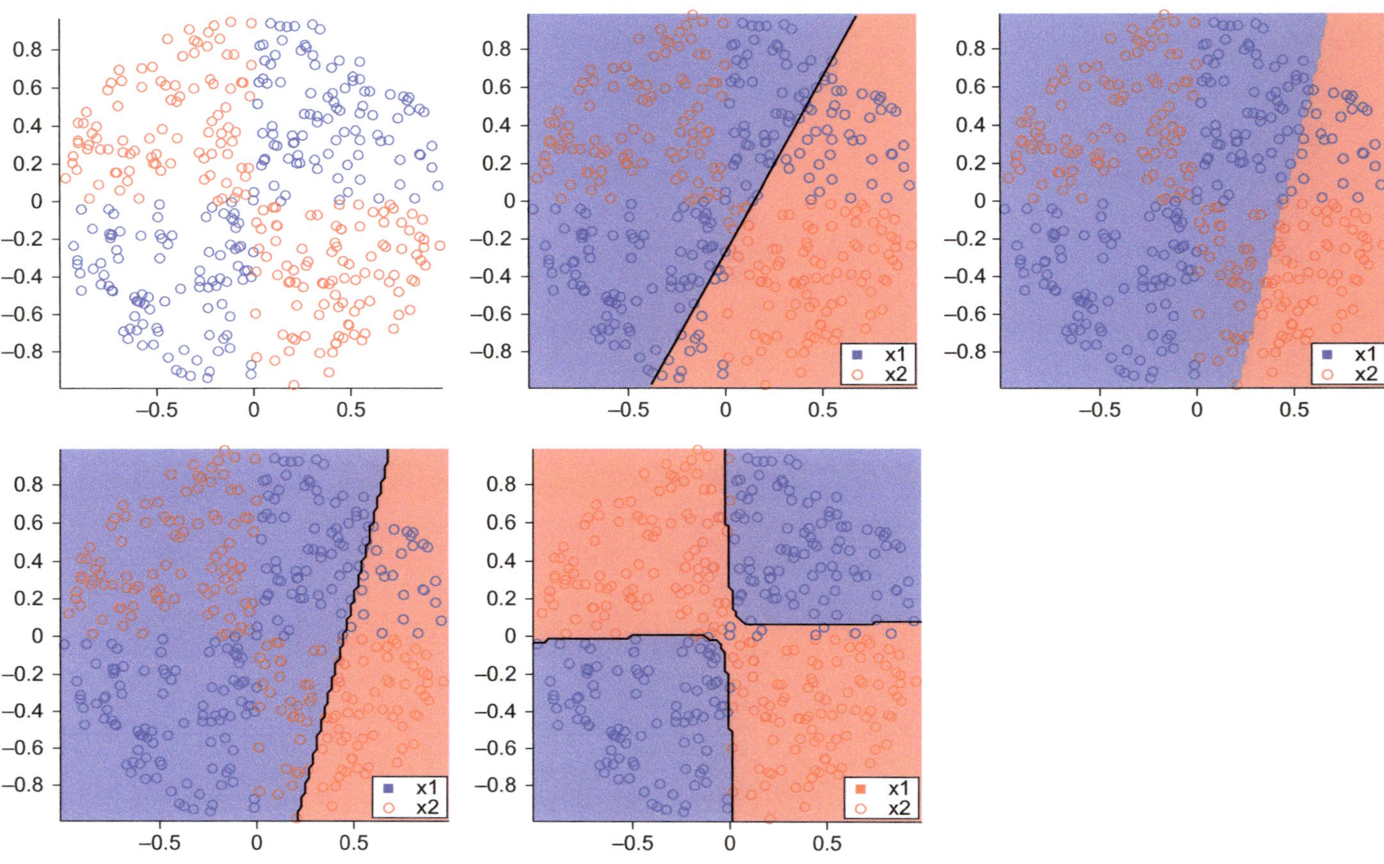

Fig. 14.15 Example 14.6 test 1.

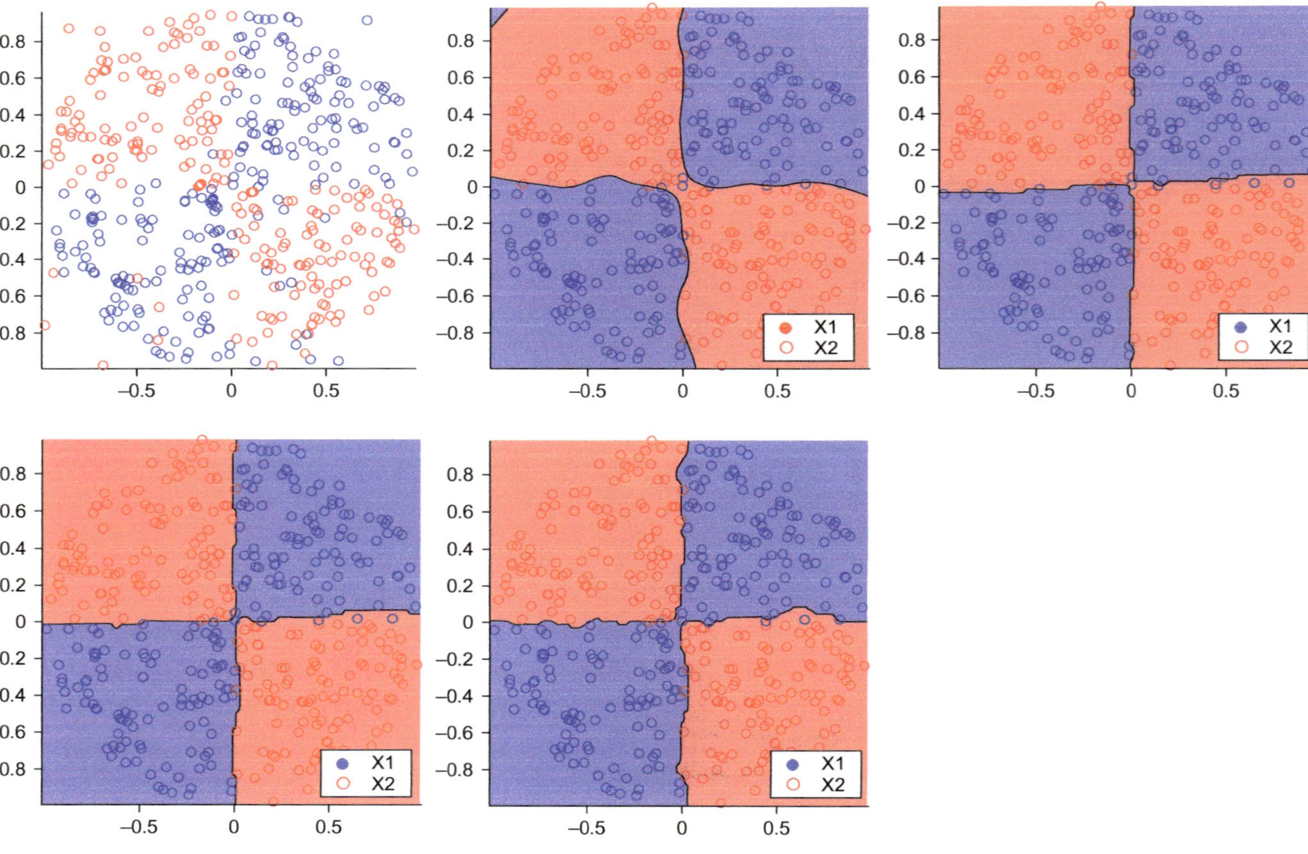

Fig. 14.16 Example 14.6 test 2.

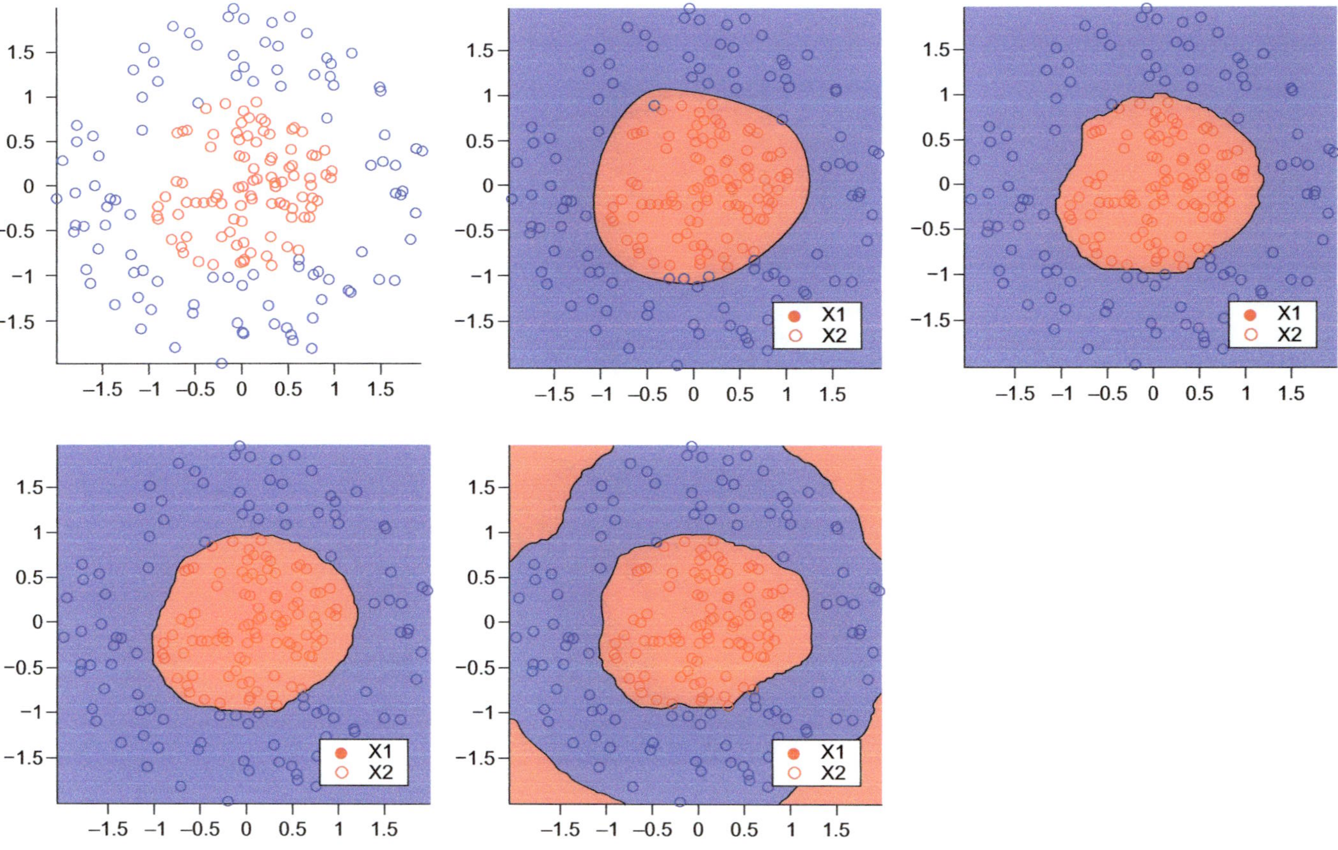

Fig. 14.17 Example 14.6 test 3.

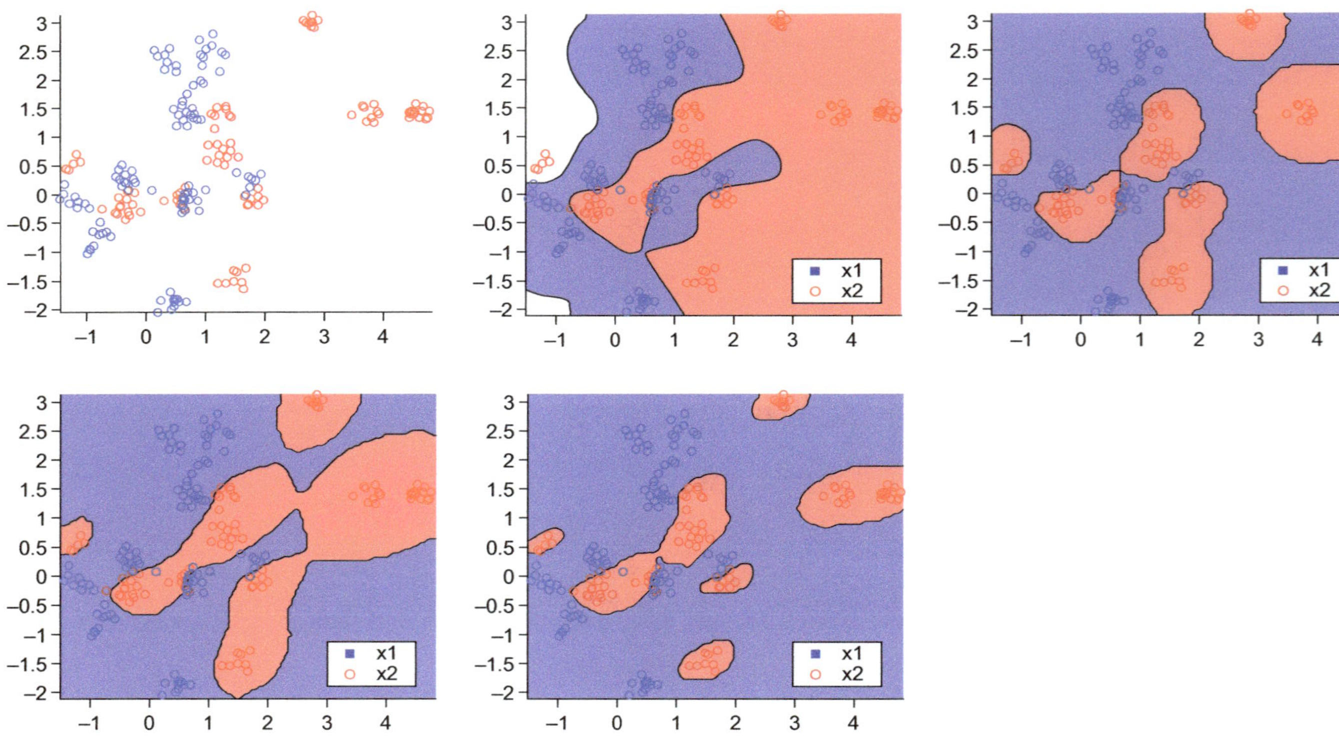

Fig. 14.18 Example 14.6 test 4.

The classification correct rates of the four tests are listed in Table 14.3.

Table 14.3

	Kernel	MATLAB SVM	Kernel Bayes I	Kernel Bayes II	Kernel Bayes III
Test 1	Linear	58.75%	60.0%	60.25%	97.0%
Test 2	RBF	97.75%	98.50%	98.50%	99.50%
Test 3	RBF	98.0%	100%	100%	100%
Test 4	RBF	96.0%	98.5%	95.5%	97.0%

Example 14.7

In this example both the kernel Bayes and SVM methods (by MATLAB function `fitcsvm`, see `https://www.mathworks.com/help/stats/fitcsvm.html`) are applied to the iris dataset. Their results are compared in Fig. 14.19. Here only two (third and fourth) of the four features are used, with half of the samples used for training while the other half for testing. Note that the second class (setosa in green) and third class (versicolor in blue) not linearly separable can be better separated by the kernel Bayes method.

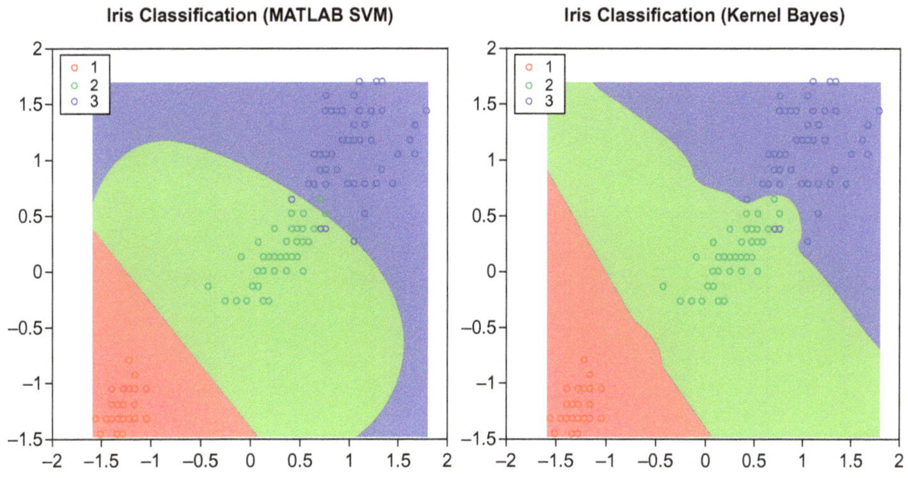

Fig. 14.19 Comparison of kernel Bayes and SVM applied to iris data.

PROBLEMS

In this homework, you are to develop your own code to implement the binary and multiclass SVM algorithms. You are also encouraged to compare your results with those obtained by the built-in SVM functions in MATLAB.

1. Develop code to implement the SVM algorithm using your own function for QP by the interior point algorithm (Section 3.7), and built-in QP function *quadprog* in MATLAB, based on the RBF kernel $K(\mathbf{x}_i, \mathbf{x}_j) = c\exp(-\gamma||\mathbf{x}_i - \mathbf{x}_j||^2)$.

   Apply your code to carry out binary classification for each of the following 2-D datasets. Experiment with different values for the two hyper-parameters c and γ in the kernel function and check to see how they affect the classification result.

   Visualize the partitioning of the 2-D feature space by classifying all points in the space (with proper resolution) into either of the two classes, as with the examples in
   - /e176/programs/data60.txt
   - /e176/programs/data61.txt
   - /e176/programs/data62.txt
   - /e176/programs/data63.txt
   - /e176/programs/data64.txt
2. Carry out multiclass classification for both the iris and handwritten datasets, using your binary SVM code and the method of 1V1 or 1VR. Show your classification results in the form of confusion matrix as well as error rates.
3. Develop code for multiclass SVM and carry out classification for the iris and handwritten datasets. Show the classification results in the form of confusion matrix as well as error rates.

15 Clustering Analysis

All supervised classification algorithms previously considered are based on the assumed availability of some training data in which each of the data patterns $\mathbf{x}_n \in \mathbf{X} = [\mathbf{x}_1, \cdots, \mathbf{x}_N\}]$ is labeled by $y_n \in \mathbf{y} = [y_1, \cdots, y_n]^T$, that is, the class they each belong is known. However, when such a training set is not available, there is still the need to explore some potential structure of the data in the form of a set of unknown number K of clusters $\{C_1, \cdots, C_K\}$, each composed of a set of similar data points close to each other in the feature space. This can be accomplished by the unsupervised method of *clustering analysis*, based only on the given dataset $\mathbf{X} = [\mathbf{x}_1, \cdots, \mathbf{x}_N]$, without any additional prior knowledge, such as the labeling of the data samples. We will now consider a few algorithms for clustering analysis.

15.1 k-Means Clustering

As suggested by the name of k-means clustering, in this method a set of K mean vectors $\mathbf{m}_1, \cdots, \mathbf{m}_K$ are used to represent the clusters in the feature space, based on the assumption that there exist a set of K clusters in the dataset.

The k-means clustering algorithm can be formulated as an optimization problem to minimize an objective function

$$J = \sum_{n=1}^{N} \sum_{k=1}^{K} P_{nk} ||\mathbf{x}_n - \mathbf{m}_k||^2 \tag{15.1}$$

where

$$P_{nk} = \begin{cases} 1 & \text{if } k = \arg\min_l ||\mathbf{x}_n - \mathbf{m}_l|| \\ 0 & \text{otherwise} \end{cases} \tag{15.2}$$

indicates that $\mathbf{x}_n$ is assigned to the kth cluster C_k if its distance to the mean $\mathbf{m}_k$ is minimum among all K such distances. To minimize the objective function J with respective to $\mathbf{m}_k$, we set its derivative with respect to each $\mathbf{m}_k$ to zero

$$\frac{d}{d\mathbf{m}_k} J = 2 \sum_{n=1}^{N} P_{nk} ||\mathbf{x}_n - \mathbf{m}_k|| = \mathbf{0} \qquad k = 1, \cdots, K \tag{15.3}$$

and solve the resulting equation for $\mathbf{m}_k$

$$\mathbf{m}_k = \frac{\sum_{n=1}^{N} P_{nk} \mathbf{x}_n}{\sum_{n=1}^{N} P_{nk}} = \frac{1}{N_k} \sum_{\mathbf{x} \in C_k} \mathbf{x} \quad (k = 1, \cdots, K) \tag{15.4}$$

where $N_k = \sum_{n=1}^{N} P_{nk}$ is the number of all samples assigned to cluster C_k, and the summation is over all samples assigned to C_k.

As P_{nk} depends on $\mathbf{m}_k$ which in turn also depends on P_{nk}, these steps can be carried out iteratively, starting with some K randomly initialized mean vectors, which are then modified iteratively until convergence, as shown in the following steps.

1. Set $t = 0$, and initialize randomly the K mean vectors, such as any K samples randomly chosen from the dataset: $\mathbf{m}_1^{(0)}, \mathbf{m}_2^{(0)}, \ldots, \mathbf{m}_K^{(0)}$, set iteration index to zero $t = 1$.
2. Assign each sample $\mathbf{x}$ in the dataset to one of the K clusters according to its distance to the corresponding mean vector

$$\text{if} \quad ||\mathbf{x} - \mathbf{m}_k^{(t)}||^2 = \min_{1 \le l \le K} ||\mathbf{x} - \mathbf{m}_l^{(t)}||^2, \quad \text{then} \quad \mathbf{x} \in C_k^{(t)} \tag{15.5}$$

3. Update the mean vectors to get the new mean $\mathbf{m}_k^{(t+1)}$ so that the objective function J is minimized

$$\mathbf{m}_k^{(t+1)} = \frac{1}{N_k} \sum_{\mathbf{x} \in C_k} \mathbf{x}, \quad (k = 1, \cdots, K) \tag{15.6}$$

4. Terminate if the algorithm has converged:

$$\mathbf{m}_k^{(l+1)} = \mathbf{m}_k^{(t)} \quad (k = 1, \cdots, K) \tag{15.7}$$

   Otherwise, $t \leftarrow t + 1$, go back to step 2.

The k-means clustering is simple and effective, but it has the main drawback that the ground truth number of clusters in the dataset is typically unknown, and it may be difficult to determine its value without any prior knowledge. One way to address this issue is to carry out the clustering process multiple times with different K values, and then determine the optimal K value by comparing the clustering results to see which best reflects the intrinsic structure of the dataset in terms of how the data points are distributed. How well the clustering result fits the data can be measured quantitatively by the separability of the clusters, such as the ratio between the intra and inter-cluster distances, or the ration between-cluster scatteredness and the total scatteredness of the entire dataset, that is, $tr(\mathbf{S}_T^{-1}\mathbf{S}_B)$ in Eq. (9.24). The K value corresponding to the highest separability among all resulting clusters can be adopted as the optimal one. Such a quantitative measurement can also be used as a criterion for other clustering algorithms, including the method of competitive learning network to be considered in Section 19.1.

The issue of unknown parameter K can also be addressed by modifying its value during the iteration. For example, if the inter-cluster distance between two small clusters is smaller than a predetermined threshold, they can be merged into one, also, if the intra-cluster distance of a large cluster is greater than a predetermined threshold, it can be split into two. This method of modifying the clustering results by merging and splitting leads to the algorithm of *iterative self-organizing data analysis technique (ISODATA)*. However, this algorithm is highly empirical and it suffers from the drawback that it may be difficult to predetermine those thresholds without any prior knowledge or experience. It may therefore become a highly interactive process during which the threshold values themselves may need to be modified to better fit the data.

The MATLAB code for the iteration loop of the k-means algorithm is listed in the following, where `Mold` and `Mnew` are, respectively, the K mean vectors before and after each modification. The iteration terminates when the mean vectors no longer change.

```
Mnew=X(:,randi(N,1,K));       % pick any K samples as initial means
er=inf;
while er>0                    % main iteration
    Mold=Mnew;
    Mnew=zeros(d,K);          % initialize new means
    Nk=zeros(1,K);
    for n=1:N                 % for all N samples
        x=X(:,n);
        dmin=inf;
        for k=1:K             % for all K clusters
            d=norm(x-Mold(:,k));
            if d<dmin
                dmin=d;  j=k;
            end
        end
        Nk(j)=Nk(j)+1;
        Mnew(:,j)=Mnew(:,j)+x;
    end
    for k=1:K
        if Nk(k)>0
            Mnew(:,k)=Mnew(:,k)/Nk(k);
        end
    end
    er=norm(Mnew-Mold);       % terminate if means no longer change
end
```

Example 15.1

The k-means algorithm is applied to a simulated dataset in 3-D space with $C = 4$ clusters. The three panels in Fig. 15.1 show the results corresponding to $K = C - 1$ (left), $K = C$ (middle), and $K = C + 1$ (right). The initial positions of the K means are marked by the black squares, while their subsequent positions throughout the iteration are marked by smaller dots. The iteration terminates once the means have moved to the centers of the clusters and no longer change positions.

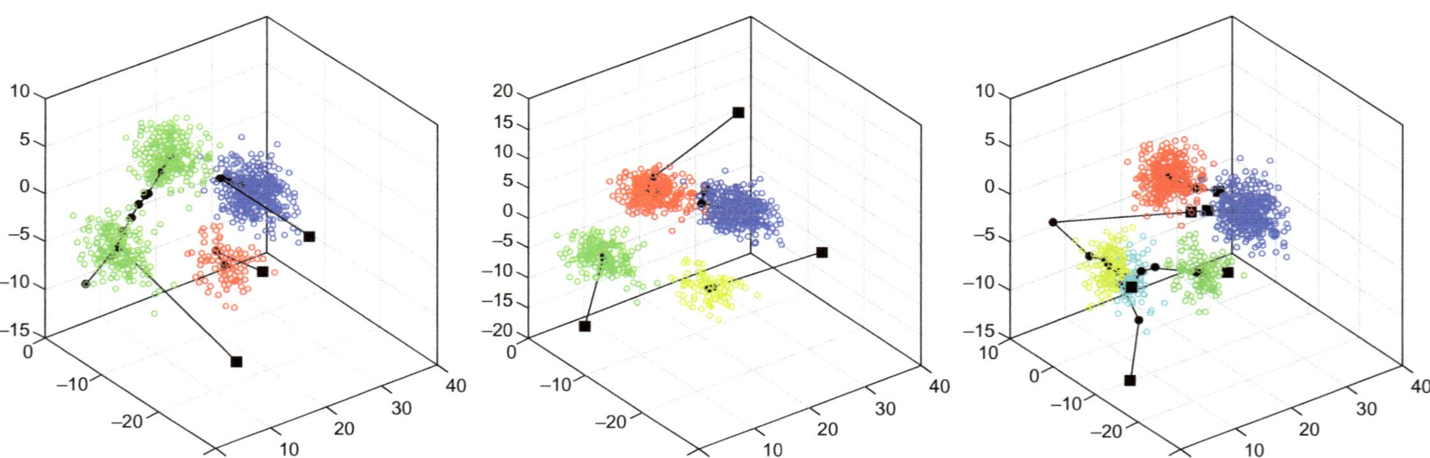

Fig. 15.1 k-means clustering with $K = 3$ (left), $K = 4$ (middle), and $K = 5$ (right).

The clustering results corresponding to $K = 3, 4, 5$ can be evaluated by the separability $tr(\mathbf{S}_T^{-1}\mathbf{S}_B)$, the K intra-cluster distance $tr(\boldsymbol{\Sigma}_k)$ $(k = 1, \cdots, K)$ of the resulting clusters

Table 15.1

	$K = C - 1 = 3$			$K = C = 4$				$K = C + 1 = 5$				
Separability	1.76			2.56				2.58				
Intra-cluster distance	9.1	44.3	11.8	10.8	12.7	11.1	9.1	10.8	9.1	11.1	8.9	9.4

and the $K(K-1)/2$ inter-cluster (Bhattacharyya) distances (between any two of the K clusters)

$K = C - 1 = 3$

	1	2
2	10.9	
3	21.9	184.7

$K = C = 4$

	1	2	3
2	4.0		
3	5.1	1.6	
4	2.4	2.3	4.4

$K = C + 1 = 5$

	1	2	3	4
2	4.0			
3	5.1	1.6		
4	7.0	4.3	0.3	
5	17.3	15.5	8.9	11.1

(15.8)

We see that when $K = 3 < C = 4$, the intra-cluster distance of the second cluster is significantly greater than the other two, indicating the cluster may contain two smaller clusters. Also, when $K = 5 > C = 4$, the inter-cluster distance between clusters 3 and 4 is significantly smaller than others, indicating the two clusters are too close and can therefore be merged.

While the k-means method is simple and effective, it has the main shortcoming that the number of clusters K needs to be predetermined while it is typically unknown ahead of time without any prior knowledge. In this case, one could try different K values and compare the corresponding results in terms of the intra and inter cluster distances, as well as the separabilities of the resulting clusters, as shown in the example. Moreover, if the intra-cluster distance of a cluster is too large indicating it may contain more than one cluster (e.g., $K = 3$ in the example), it can be split; conversely, if the inter-cluster distance between two clusters is too small (e.g., $K = 5$ in the example), the two clusters may belong to the same cluster and need to be merged. Following such merging and/or splitting, a few more iterations can be carried out to make sure the final clustering results are optimal. Figure 15.2 shows the clustering results of the same dataset but modified by merging and splitting. The left panel is for $K = 3$, but the second cluster (green) is split, while the right panel is for $K = 5$, when the fourth (yellow) and fifth (cyan) clusters are merged.

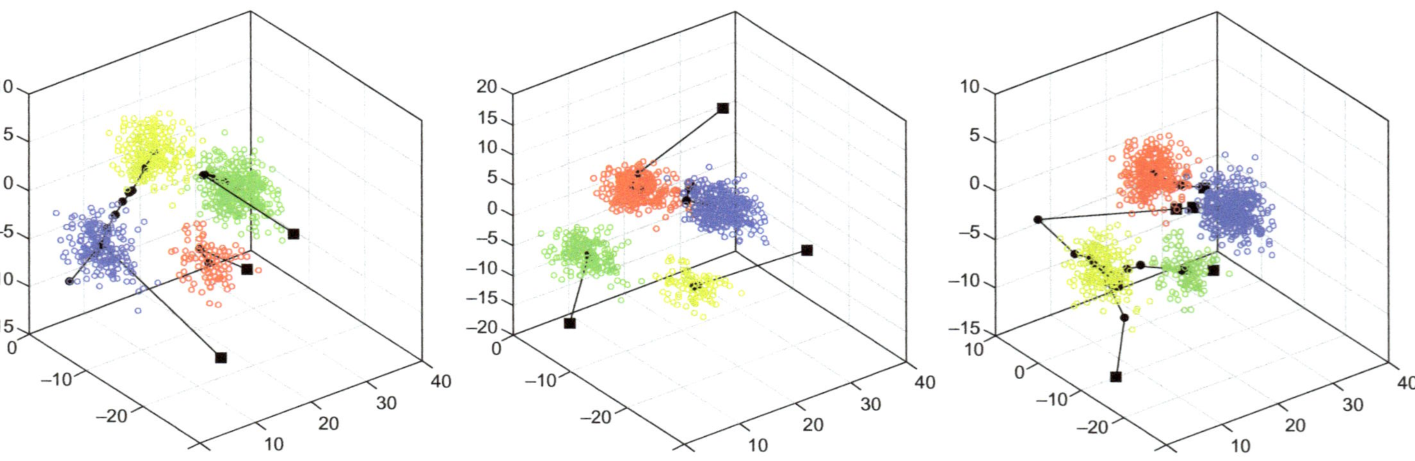

Fig. 15.2 k-means clustering with merge and split.

Example 15.2

The k-means clustering method is applied to the dataset of handwritten digits from 0 to 9 used previously. The clustering result is visualized in Fig 15.3, based on the Karhunen–Loève transform (KLT) that maps the data samples in the original 256-D space into the 3-D space spanned by the first three eigenvectors with the greatest eigenvalues of the covariance matrix of the dataset. The ground truth labelings are color coded as shown on the left, while the clustering result is shown on the right. We see that the clustering results match the original data reasonably well.

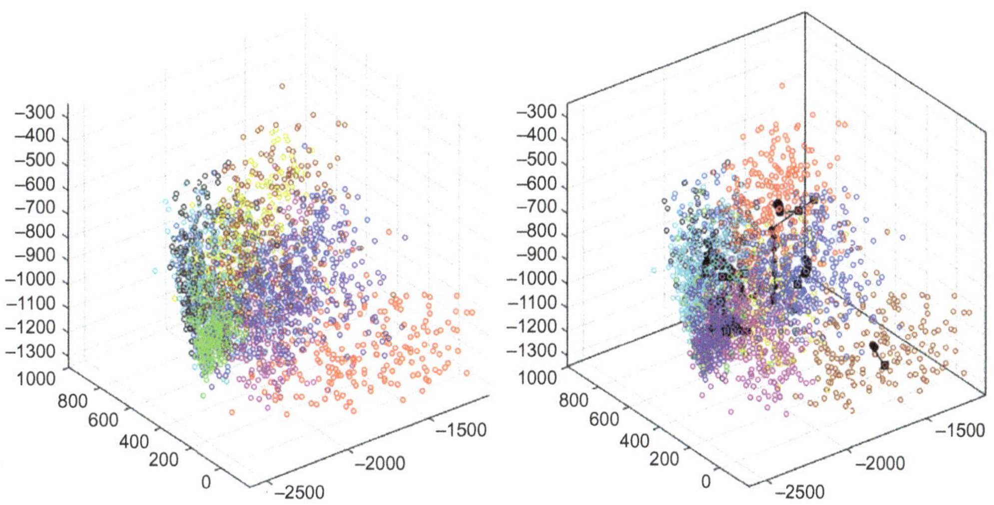

Fig. 15.3 k-means clustering of handwritten digits, ground truth classes (left) vs. clusters (right).

The clustering result can also be shown in the confusion matrix previously used for supervised classification. Here the columns and rows represent the class labelings (the ground truth not used clustering) and the columns represent the clusters based on the clustering result, that is, the element in the ith row and jth column is the number of samples labeled to belong to the ith class but assigned to the jth cluster. Ideally, there should be only one nonzero value in each row indicating all samples in a class are assigned to the same cluster, and one nonzero value in each column indicating all samples assigned to a cluster are from the same class. Based on this confusion matrix, we can also define the error rate as the ratio between the number of miss-clustered samples, the sum of all components in the confusion matrix excluding the greatest one in each row, and the total number of samples, sum of all components in the matrix.

$$\begin{bmatrix} 1 & 0 & 25 & 165 & 27 & 2 & 3 & 0 & 1 & 0 \\ 0 & 209 & 0 & 0 & 2 & 0 & 13 & 0 & 0 & 0 \\ 2 & 3 & 2 & 0 & 8 & 2 & 14 & 151 & 0 & 42 \\ 103 & 0 & 0 & 1 & 5 & 3 & 83 & 0 & 0 & 29 \\ 0 & 28 & 1 & 0 & 15 & 2 & 62 & 0 & 115 & 1 \\ 44 & 0 & 1 & 1 & 167 & 0 & 8 & 0 & 3 & 0 \\ 0 & 6 & 160 & 14 & 44 & 0 & 0 & 0 & 0 & 0 \\ 1 & 6 & 0 & 0 & 3 & 122 & 74 & 2 & 1 & 15 \\ 82 & 1 & 1 & 1 & 33 & 3 & 47 & 2 & 0 & 54 \\ 4 & 2 & 0 & 1 & 0 & 28 & 119 & 1 & 64 & 5 \end{bmatrix} \tag{15.9}$$

We note that as the resulting clusters are in random order unrelated to the ordering of the class identities in the dataset (ground truth not used in clustering), it is in general not the case (as in supervised classification) that the diagonal components of the confusion matrix always take larger values while all off-diagonal components take small values (ideally zero). In clustering, the largest value, assumed to be in the jth column, in each row, assumed to be the ith one, indicates that most of the samples known to belong to the ith class are assigned to the jth cluster. For a good clustering result, there should be only one large value in each row indicating most of the samples in a class are assigned to the same cluster, and there should be only one large value in each column indicating most samples assigned to a cluster are from the same class.

To further visualize the clustering result obtained by the k-means method, we convert all data points from 256-D vectors back into 16×16 image form, and then display the averages of all samples in each of the ten clusters as shown in Fig. 15.4. We see that each of the ten clusters clearly represents one distinct digit.

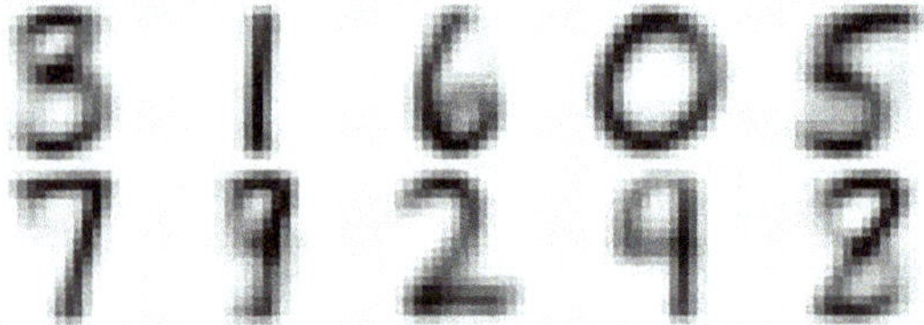

Fig. 15.4 Visualization of the clusters found by k-means.

15.2 Gaussian Mixture Model

The k-means clustering assigns each data point $\mathbf{x}$ to one of the K clusters based on the Euclidean distances between the point and the means of all clusters, without considering the overall distribution of the data points in each cluster represented by the covariance as well as the mean.

This issue can be addressed by the generative method of *Gaussian mixture model (GMM)* which describes the clusters in a global manner, based on the assumption that each of the K clusters $\{C_1,\cdots,C_K\}$ in the given dataset can be modeled by a Gaussian distribution $\mathcal{N}(\mathbf{m}_k,\mathbf{\Sigma}_k)$, $(k=1,\cdots,K)$ in terms of the mean vector $\mathbf{m}_k$ and covariance matrix $\mathbf{\Sigma}_k$. The overall distribution of the entire dataset $\mathbf{X}=[\mathbf{x}_1,\cdots,\mathbf{x}_N]$ is the weighted sum of the K Gaussian distributions

$$\begin{aligned} p(\mathbf{x}) &= \sum_{k=1}^{K} P_k \mathcal{N}(\mathbf{x},\mathbf{m}_k,\mathbf{\Sigma}_k) \\ &= \sum_{k=1}^{K} P_k \left[\frac{1}{(2\pi)^{d/2}|\mathbf{\Sigma}_k|^{1/2}} \exp\left(-\frac{1}{2}(\mathbf{x}-\mathbf{m}_k)^T\mathbf{\Sigma}_k^{-1}(\mathbf{x}-\mathbf{m}_k)\right)\right] \end{aligned} \tag{15.10}$$

where P_k is the weight for the kth Gaussian $\mathcal{N}(\mathbf{m}_k,\mathbf{\Sigma}_k)$, satisfying

$$\int_{-\infty}^{\infty} p(\mathbf{x})\,d\mathbf{x} = \sum_{k=1}^{K} P_k \int_{-\infty}^{\infty} \mathcal{N}(\mathbf{x},\mathbf{m}_k,\mathbf{\Sigma}_k)\,d\mathbf{x} = \sum_{k=1}^{K} P_k = 1 \tag{15.11}$$

The GMM clustering is a method of *expectation maximization (EM)*, by which all model parameters denoted by $\boldsymbol{\theta}=\{P_k,\mathbf{m}_k,\mathbf{\Sigma}_k,\ (k=1,\cdots,K)\}$ is first estimated based on the given dataset, and then all data samples in the dataset are assigned to one of the clusters described by these parameters.

We note that the GMM model in Eq. (15.10) is actually the same as Eq. (13.10) in the naive Bayes classification. These two methods are similar in the sense that each cluster or class C_k is modeled by a Gaussian $\mathcal{N}(\mathbf{m}_k,\mathbf{\Sigma}_k)$, weighted by P_k, and the model parameters $\mathbf{m}_k$ and $\mathbf{\Sigma}_k$, as well as P_k, are to be estimated based on the given dataset.

However, the two methods are different in that the dataset $\mathbf{X}$ in the supervised naive Bayes method is labeled by $\mathbf{y}$, which is not available in unsupervised GMM. Instead, here we introduce a latent or hidden variable $\mathbf{z}=[z_1,\cdots,z_K]^T$ as the cluster labeling of each sample $\mathbf{x}$ in the dataset $\mathbf{X}$. All components of $\mathbf{z}_n$ are binary random variables $z_k\in\{0,1\}$, of which only one is 1 (e.g., $z_l=1$) indicating $\mathbf{x}$ belongs to the lth cluster C_l, while all others are 0. As $\mathbf{x}$ must belong to one of the K clusters, these K binary variables add up to 1, that is, $\sum_{k=1}^{K} z_k=1$.

We consider the following probabilities defined for each of the K clusters C_k $(k=1,\cdots,K)$:

- The *prior probability* P_k, $(k=1,\cdots,K)$ for any unobserved data sample in the dataset to belong to C_k, they satisfy

$$\sum_{k=1}^{K} P_k = 1 \tag{15.12}$$

  These K prior probabilities add up to 1, that is, the K events $z_k=1$ $(k=1,\cdots,K)$ are mutually exclusive and complementary, as any sample $\mathbf{x}$ belongs to one and only one of the K clusters.
- The probability distribution of all samples in cluster C_k, assumed to be a Gaussian

$$p(\mathbf{x}|z_k=1,\boldsymbol{\theta}) = \mathcal{N}(\mathbf{x},\mathbf{m}_k,\mathbf{\Sigma}_k) \tag{15.13}$$

- The joint distribution of $\mathbf{x}$ and $z_k = 1$

$$p(\mathbf{x}, z_k = 1|\boldsymbol{\theta}) = p(\mathbf{x}|z_k = 1, \boldsymbol{\theta})\ P(z_k = 1) = P_k \mathcal{N}(\mathbf{x}, \mathbf{m}_k, \boldsymbol{\Sigma}_k) \tag{15.14}$$

  where $P(z_{nk} = 1) = P_k$ is the probability for any $\mathbf{x}_n$ to belong to C_k.

Marginalizing this joint probability over the latent variable z_k, we obtain the GMM, the distribution $p(\mathbf{x})$ of any sample $\mathbf{x}$ regardless to which cluster it belongs

$$p(\mathbf{x}|\boldsymbol{\theta}) = \sum_{k=1}^{K} p(\mathbf{x}, z_k = 1 fe|\boldsymbol{\theta}) = \sum_{k=1}^{K} p(\mathbf{x}|z_k = 1, \boldsymbol{\theta})\ P(z_k = 1) = \sum_{k=1}^{K} P_k \mathcal{N}(\mathbf{x}, \mathbf{m}_k, \boldsymbol{\Sigma}_k) \tag{15.15}$$

Note that Eqs (15.12) through (15.15) are the same as Eqs (13.7) through (13.10) in the naive Bayes classifier, respectively.

All such probabilities defined for $z_k = 1$ can be generalized to $\mathbf{z} = [z_1, \cdots, z_K]^T$ for all K

$$p(\mathbf{z}|\boldsymbol{\theta}) = \prod_{k=1}^{K} P_k^{z_k} \tag{15.16}$$

$$p(\mathbf{x}|\mathbf{z}, \boldsymbol{\theta}) = \prod_{k=1}^{K} \mathcal{N}(\mathbf{x}, \mathbf{m}_k, \boldsymbol{\Sigma}_k)^{z_k} \tag{15.17}$$

$$p(\mathbf{x}, \mathbf{z}|\boldsymbol{\theta}) = p(\mathbf{z})\ p(\mathbf{x}|\mathbf{z}, \boldsymbol{\theta}) = \prod_{k=1}^{K} (P_k \mathcal{N}(\mathbf{m}_k, \boldsymbol{\Sigma}_k))^{z_k} \tag{15.18}$$

We further define $\mathbf{Z} = [\mathbf{z}_1, \cdots, \mathbf{z}_N]$ of which the nth column $\mathbf{z}_n = [z_{n1}, \cdots, z_{nK}]^T$ is the labeling of the nth sample $\mathbf{x}_n$ in the dataset $\mathbf{X}$ containing N independent and identically distributed (i.i.d.) samples. Note that here $\mathbf{Z}$ is defined in the same way as $\mathbf{Y} = [\mathbf{y}_1, \cdots, \mathbf{y}_N]$ in softmax regression (Section 7.2), both treated as the labeling of $\mathbf{X}$, with the only difference that $\mathbf{Y}$ is provided in the training data available for a supervised method, but here $\mathbf{Z}$ is a latent variable, not available in the unlabeled dataset in unsupervised clustering.

For a specific sample $\mathbf{x}_n$ in the dataset, the joint probability in Eq. (15.18) can be written as

$$p(\mathbf{x}_n, \mathbf{z}_n|\boldsymbol{\theta}) = \prod_{k=1}^{K} (P_k \mathcal{N}(\mathbf{x}_n, \mathbf{m}_k, \boldsymbol{\Sigma}_k))^{z_{nk}}, \qquad (n = 1, \cdots, N) \tag{15.19}$$

based on which the likelihood function of the GMM model parameter $\boldsymbol{\theta}$ can be expressed as

$$\begin{aligned} L(\boldsymbol{\theta}|\mathbf{X}, \mathbf{Z}) &= p(\mathbf{X}, \mathbf{Z}|\boldsymbol{\theta}) = p\left([\mathbf{x}_1, \cdots, \mathbf{x}_N], [\mathbf{z}_1, \cdots, \mathbf{z}_N] \middle| \mathbf{m}_k, \boldsymbol{\Sigma}_k, P_k \ \ (k = 1, \cdots, K)\right) \\ &= \prod_{n=1}^{N} p(\mathbf{x}_n, \mathbf{z}_n|\boldsymbol{\theta}) = \prod_{n=1}^{N} \prod_{k=1}^{K} (P_k \mathcal{N}(\mathbf{x}_n, \mathbf{m}_k, \boldsymbol{\Sigma}_k))^{z_{nk}} \end{aligned} \tag{15.20}$$

and the log likelihood is

$$
\begin{aligned}
l(\boldsymbol{\theta}|\mathbf{X},\mathbf{Z}) &= \log L(\boldsymbol{\theta}|\mathbf{X},\mathbf{Z}) = \log p(\mathbf{X},\mathbf{Z}|\boldsymbol{\theta}) \\
&= \log\left[\prod_{n=1}^{N}\prod_{k=1}^{K}(P_k\mathcal{N}(\mathbf{x}_n,\mathbf{m}_k,\boldsymbol{\Sigma}_k))^{z_{nk}}\right] \\
&= \sum_{n=1}^{N}\sum_{k=1}^{K} z_{nk}\,[\log P_k + \log\mathcal{N}(\mathbf{x}_n,\mathbf{m}_k,\boldsymbol{\Sigma}_k)]
\end{aligned}
\tag{15.21}
$$

Similar to the general method of maximum likelihood estimation (MLE) that finds the optimal model parameter $\boldsymbol{\theta}$ that maximizes the likelihood $L(\boldsymbol{\theta})$, here the GMM method finds the model parameter $\boldsymbol{\theta} = \{P_k,\mathbf{m}_k,\boldsymbol{\Sigma}_k,\ (k = 1,\cdots,K)\}$ that maximizes the expectation of the likelihood function $l(\boldsymbol{\theta}|\mathbf{X},\mathbf{Z})$. However, as GMM is an unsupervised method, the ground truth labeling is not available in the given dataset, it has to be carried out in two separate steps. First, in the E-step the expectation of the log likelihood function is estimated based on the previously estimated model parameters; second, in the M-step the estimated parameters that maximize the expectation of the log likelihood are obtained and the previous estimation is updated. As these two steps are interdependent of each other, they need to be carried out alternatively in an iteration from some initial guess of the parameters. This process is called an EM method due to these E and M steps.

Now we consider the specific computations taking place in each of the two steps.

- **E-step:** Find the expectation of the log likelihood function.
  We first find the posterior probability for an observed $\mathbf{x}_n$ to belong to cluster C_k, indicated by $z_{nk} = 1$ (and $z_{nl} = 0$ for all $l \neq k$)

$$
P_{nk} = P(z_{nk} = 1|\mathbf{x}_n,\boldsymbol{\theta}) = \frac{p(\mathbf{x}_n, z_{nk} = 1|\boldsymbol{\theta})}{p(\mathbf{x}_n|\boldsymbol{\theta})} = \frac{P_k\,\mathcal{N}(\mathbf{x}_n;\mathbf{m}_k\boldsymbol{\Sigma}_k)}{\sum_{l=1}^{K} P_l\,\mathcal{N}(\mathbf{x}_n;\mathbf{m}_l\boldsymbol{\Sigma}_l)}
$$
$$
(n = 1,\cdots,N;\ k = 1,\cdots,K) \tag{15.22}
$$

  As $\mathbf{x}_n$ has to belong to one of the K clusters, these K posterior probabilities add up to 1, same as the prior probabilities $P(z_k = 1) = P_k$

$$
\sum_{k=1}^{K} P(z_k = 1) = \sum_{k=1}^{K} P_k = 1, \qquad \sum_{k=1}^{K} P(z_{nk} = 1|\mathbf{x}_n,\boldsymbol{\theta}) = \sum_{k=1}^{K} P_{nk} = 1 \tag{15.23}
$$

  Once this posterior $P_{nk} = p(z_{nk} = 1|\mathbf{x}_n,\boldsymbol{\theta})$ is calculated based on Eq. (15.22) for all $n = 1,\cdots,N)$, they are used to assign $\mathbf{x}_n$ to cluster C_k if $P_{nk} = \max_l P_{nl}$.

  We note that the definition of P_{nk} is similar to the softmax function s_{nk} with the only difference that here the weighted Gaussian function are used instead of the softmax function used in softmax regression.

  We also note that the posterior probability P_{nk} just defined represents a *soft* decision, in the sense that it is possible for any $\mathbf{x}_n$ to belong to any C_k with probability $0 < P_{nk} < 1$ for all $k = 1,\cdots,K$, instead of a *hard* decision, in the sense that $\mathbf{x}_n$ belongs to only one specific C_k with $P_{nk} = 1$, while $P_{nl} = 0$ for all $l \neq k$, as in k-means clustering.

We can now find the expectation of the log likelihood with respect to the latent variables in $\mathbf{Z}$

$$
\begin{aligned}
E_Z\left(\log L(\boldsymbol{\theta}|\mathbf{X},\mathbf{Z})\right) &= E_Z\left[\sum_{n=1}^{N}\sum_{k=1}^{K} z_{nk}\left[\log P_k + \log \mathcal{N}(\mathbf{x}_n,\mathbf{m}_k,\boldsymbol{\Sigma}_k)\right]\right] \\
&= \sum_{n=1}^{N}\sum_{k=1}^{K} E_Z(z_{nk})\left[\log P_k + \log \mathcal{N}(\mathbf{x}_n,\mathbf{m}_k,\boldsymbol{\Sigma}_k)\right] \\
&= \sum_{n=1}^{N}\sum_{k=1}^{K} P_{nk}\left[\log P_k + \log \mathcal{N}(\mathbf{x}_n,\mathbf{m}_k,\boldsymbol{\Sigma}_k)\right] \qquad (15.24)
\end{aligned}
$$

The last equality is due to the following

$$
E_z(z_{nk}) = 1\ P(z_{nk}=1|\mathbf{x}_n) + 0\ P(z_{nk}=0|\mathbf{x}_n) = P(z_{nk}=1|\mathbf{x}_n) = P_{nk} \qquad (15.25)
$$

- **M-step:** Find the optimal model parameters that maximize the expectation of the log likelihood in Eq. (15.24).

  We first set to zero the derivatives of the expectation of the log likelihood with respect to each of the parameters in $\boldsymbol{\theta} = \{P_k,\ \mathbf{m}_k,\ \boldsymbol{\Sigma}_k,\ (k=1,\cdots,K)\}$, and then solve the resulting equations to get the optimal parameters.

  - Find P_k:

    As all K of P_ks need to satisfy the constraint $\sum_{k=1}^{K} P_k = 1$, we first construct the Lagrangian function composed of an extra Lagrange multiplier term for the constraint as well as the log likelihood as the objective function

$$
L(\boldsymbol{\theta},\lambda) = \sum_{n=1}^{N}\sum_{k=1}^{K} P_{nk}\left[\log P_k + \log \mathcal{N}(\mathbf{x}_n,\mathbf{m}_k,\boldsymbol{\Sigma}_k)\right] + \lambda\left(\sum_{k=1}^{K} P_k - 1\right) \qquad (15.26)
$$

and set it derivative with respect to P_k to zero

$$
\begin{aligned}
\frac{\partial}{\partial P_k} L(\boldsymbol{\theta},\lambda) &= \frac{\partial}{\partial P_k}\left[\sum_{n=1}^{N}\sum_{k=1}^{K} P_{nk}\left[\log P_k + \log \mathcal{N}(\mathbf{x}_n,\mathbf{m}_k,\boldsymbol{\Sigma}_k)\right] + \lambda\left(\sum_{k=1}^{K} P_k - 1\right)\right] \\
&= \sum_{n=1}^{N} P_{nk}\frac{1}{P_k} + \lambda = 0 \qquad (15.27)
\end{aligned}
$$

Multiplying both sides by P_k, we obtain

$$
\sum_{n=1}^{N} P_{nk} + P_k\lambda = N_k + P_k\lambda = 0 \qquad (15.28)
$$

where we have defined

$$
N_k = \sum_{n=1}^{N} P_{nk} \qquad (15.29)
$$

that satisfies

$$
\sum_{k=1}^{K} N_k = \sum_{k=1}^{K}\left(\sum_{n=1}^{N} P_{nk}\right) = \sum_{n=1}^{N}\left(\sum_{k=1}^{K} P_{nk}\right) = \sum_{n=1}^{N} 1 = N \qquad (15.30)
$$

Summing Eq. (15.28) over all $k = 1, \cdots, K$, we get

$$\sum_{k=1}^{K} (N_k + P_k \lambda) = \sum_{k=1}^{K} N_k + \lambda \left(\sum_{k=1}^{K} P_k \right) = N + \lambda = 0 \tag{15.31}$$

Substituting $\lambda = -N$ back into Eq. (15.28), we obtain the expression for the prior

$$p(z_k = 1) = P_k = \frac{N_k}{N} = \frac{1}{N} \sum_{n=1}^{N} P_{nk} \tag{15.32}$$

This is the same as the prior P_k in Eq. (13.9) in naive Bayes classification, but here N_k defined in Eq. (15.29) is the sum of the probabilities for all N data samples to belong to C_k, instead of the number of data samples in C_k (unknown in this unsupervised case).

- Find $\mathbf{m}_k$

$$\begin{aligned}
\frac{\partial}{\partial \mathbf{m}_k} E_Z(\log L(\boldsymbol{\theta}|\mathbf{X}, \mathbf{Z})) &= \frac{\partial}{\partial \mathbf{m}_k} \left[\sum_{n=1}^{N} \sum_{l=1}^{K} P_{nl} \left[\log P_l + \log \mathcal{N}(\mathbf{x}_n, \mathbf{m}_l, \boldsymbol{\Sigma}_l) \right] \right] \\
&= \sum_{n=1}^{N} P_{nk} \frac{\partial}{\partial \mathbf{m}_k} \log \mathcal{N}(\mathbf{x}_n, \mathbf{m}_k, \boldsymbol{\Sigma}_k) \\
&\overset{*}{=} \sum_{n=1}^{N} P_{nk} \frac{\partial}{\partial \mathbf{m}_k} \left[-\frac{1}{2} (\mathbf{x}_n - \mathbf{m}_k)^T \boldsymbol{\Sigma}_k^{-1} (\mathbf{x}_n - \mathbf{m}_k) \right] \\
&= \frac{1}{2} \sum_{n=1}^{N} P_{nk} \boldsymbol{\Sigma}_k^{-1} (\mathbf{x}_n - \mathbf{m}_k) = \mathbf{0}
\end{aligned} \tag{15.33}$$

Note that here $\overset{*}{=}$ indicates that we have neglected the scaling coefficient $(2\pi)^{-d/2} |\boldsymbol{\Sigma}_k|^{-1/2}$ of the Gaussian, which is independent of $\mathbf{m}_k$. Multiplying $2\boldsymbol{\Sigma}_k$ on both sides, we have

$$\sum_{n=1}^{N} P_{nk} (\mathbf{x}_n - \mathbf{m}_k) = \sum_{n=1}^{N} P_{nk} \mathbf{x}_n - \sum_{n=1}^{N} P_{nk} \mathbf{m}_k = \sum_{n=1}^{N} P_{nk} \mathbf{x}_n - N_k \mathbf{m}_k = \mathbf{0} \tag{15.34}$$

Solving for $\mathbf{m}_k$ we get

$$\mathbf{m}_k = \frac{1}{N_k} \sum_{n=1}^{N} P_{nk} \mathbf{x}_n \tag{15.35}$$

- Find $\boldsymbol{\Sigma}_k$

$$\frac{\partial}{\partial \boldsymbol{\Sigma}_k} E_Z(\log L(\boldsymbol{\theta}|\mathbf{X}, \mathbf{Z})) = \frac{\partial}{\partial \boldsymbol{\Sigma}_k} \left[\sum_{n=1}^{N} \sum_{l=1}^{K} P_{nl} \left[\log P_l + \log \mathcal{N}(\mathbf{x}_n, \mathbf{m}_l, \boldsymbol{\Sigma}_l) \right] \right]$$

$$= \sum_{n=1}^{N} P_{nk} \frac{\partial}{\partial \boldsymbol{\Sigma}_k} \log \mathcal{N}(\mathbf{x}_n, \mathbf{m}_k, \boldsymbol{\Sigma}_k)$$

$$= -\frac{1}{2}\sum_{n=1}^{N} P_{nk}\left[\frac{\partial}{\partial \mathbf{\Sigma}_k}\log|\mathbf{\Sigma}_k| + \frac{\partial}{\partial \mathbf{\Sigma}_k}(\mathbf{x}_n - \mathbf{m}_k)^T\mathbf{\Sigma}_k^{-1}(\mathbf{x}_n - \mathbf{m}_k)\right]$$

$$= -\frac{1}{2}\sum_{n=1}^{N} P_{nk}\left[\mathbf{\Sigma}_k^{-1} - \mathbf{\Sigma}_k^{-1}(\mathbf{x}_n - \mathbf{m}_k)(\mathbf{x}_n - \mathbf{m}_k)^T\mathbf{\Sigma}_k^{-1}\right] = \mathbf{0} \qquad (15.36)$$

Here we have used the following facts (for more details see *Matrix Cookbook* `https://www.math.uwaterloo.ca/~hwolkowi/matrixcookbook.pdf`)

$$\frac{d}{d\mathbf{A}}\log|\mathbf{A}| = (\mathbf{A}^{-1})^T, \qquad \frac{d}{d\mathbf{A}}\left(\mathbf{a}^T\mathbf{A}^{-1}\mathbf{b}\right) = -(\mathbf{A}^{-1})^T\mathbf{a}\mathbf{b}^T(\mathbf{A}^{-1})^T \qquad (15.37)$$

Pre and post multiplying $\mathbf{\Sigma}_k$ on both sides of this equation we have

$$\sum_{n=1}^{N} P_{nk}\left(\mathbf{\Sigma}_k - (\mathbf{x}_n - \mathbf{m}_k)(\mathbf{x}_n - \mathbf{m}_k)^T\right) = \mathbf{0} \qquad (15.38)$$

Solving for $\mathbf{\Sigma}_k$, we get

$$\mathbf{\Sigma}_k = \frac{1}{N_k}\sum_{n=1}^{N} P_{nk}(\mathbf{x}_n - \mathbf{m}_k)(\mathbf{x}_n - \mathbf{m}_k)^T \qquad (15.39)$$

We note that all model parameters in $\boldsymbol{\theta} = \{P_k, \mathbf{m}_k, \mathbf{\Sigma}_k, (k = 1, \cdots, K)\}$ found, respectively, in Eqs (15.32), (15.35), and (15.39) in the M-step depend on P_{nk} in Eq. (15.22) in the E-step, which in turn is a function of these parameters, that is, the E-step and M-step need to be carried out in an alternative and iterative fashion from some initial values of the parameters until convergence.

In summary, here are the steps of GMM clustering algorithm:

1. Initialize means $\mathbf{m}_k$, covariance $\mathbf{\Sigma}_k$ and coefficient P_k.
2. The E-step:
   Find P_{nk} for all N data points and all K clusters and then N_k

$$P_{nk} = P(r_k = 1|\mathbf{x}_n) = \frac{P_k\,\mathcal{N}(\mathbf{x}_n; \mathbf{m}_k, \mathbf{\Sigma}_k)}{\sum_{l=1}^{K} P_l\,\mathcal{N}(\mathbf{x}_n; \mathbf{m}_l, \mathbf{\Sigma}_l)}, \qquad N_k = \sum_{n=1}^{N} P_{nk} \qquad (15.40)$$

3. The M-step:
   Recalculate the parameters that maximize the likelihood function

$$P_k = \frac{N_k}{N}$$
$$\mathbf{m}_k = \frac{1}{N_k}\sum_{n=1}^{N} P_{nk}\mathbf{x}_n$$
$$\mathbf{\Sigma}_k = \frac{1}{N_k}\sum_{n=1}^{N} P_{nk}(\mathbf{x}_n - \mathbf{m}_k)(\mathbf{x}_n - \mathbf{m}_k)^T \qquad (15.41)$$

4. Terminate the iteration if the parameters have converged, that is, the difference between their values of two consecutive iterations is sufficiently small. Otherwise go back to step 2. The probability for each sample $\mathbf{x}_n$ to belong to cluster C_l is $p_{nl} = P(z_{nl} = 1|\mathbf{x}_n, \boldsymbol{\theta})$, and it is therefore assigned to C_k if $P_{nk} = \max_l p_{nl}$.

The following MATLAB code is the essential part of the algorithm including the initialization of parameters and the iteration loop. The function `mvnpdf(X,mk(:,k),Ck(:,:,k))` evaluate the Gaussian pdf of the kth cluster $\mathcal{N}(\mathbf{x}_n;\mathbf{m}_k,\mathbf{\Sigma}_k)$ at all N samples in dataset X.

```
[N d]=size(X);                  % N samples of d dimensions
mk=zeros(d,K);                  % old mean vectors
Mk=zeros(d,K);                  % new mean vectors
Ck=zeros(d,d,K);                % covariance matrices
p=zeros(N,K);                   % posterior
Pnk=zeros(N,K);                 % Pnk
Pk=zeros(K,1);                  % prior Pk=Nk/N
Nk=zeros(1,K);                  % number of samples in kth cluster
for k=1:K                       % initial parameters for kth cluster
    sd=randi(N,1,2*d);          % random indices of subsamples
    mk(:,k)=mean(X(sd,:))';     % mean of these samples
    Ck(:,:,k)=cov(X(sd,:));     % covariance of these samples
    Pk(k)=1/K;                  % prior probability
end
er=1;
while er>10^(-4)
    for k=1:K                   % for each of K clusters
        p(:,k)=Pk(k)*mvnpdf(X,mk(:,k),Ck(:,:,k)); % weighted posterior
    end
    D=sum(p,2);                 % sum over all K terms
    er=0;
    for k=1:K                   % update parameters for kth cluster
        Pnk(:,k)=p(:,k)./D;     %
        Nk(k)=sum(Pnk(:,k));    % Nk
        Pk(k)=Nk(k)/N;          % Pk
        if Nk(k)>0
            Mk(:,k)=sum(X.*repmat(Pnk(:,k),[1,d]),1)/Nk(k);     % mean
            Xm=X-repmat(Mk(:,k)',[N,1]);
            Ck(:,:,k)=Xm'*(Xm.*repmat(Pnk(:,k),[1,d]))/Nk(k); % covariance
        end
        er=er+norm(mk(:,k)-Mk(:,k));  % difference between two iterations
    end
    mk=Mk;                         % keep iterating
end
```

We can show that the k-means algorithm is actually a special case of the EM algorithm, when all covariance matrices are the same $\mathbf{\Sigma}_k = \sigma^2\mathbf{I}$, and σ^2 approaches to zero. In this case we have

$$p(\mathbf{x}|z_k = 1,\boldsymbol{\theta}) = \mathcal{N}(\mathbf{x}|\mathbf{m}_k,\sigma^2\mathbf{I}) = \frac{1}{(2\pi)^{d/2}\sigma} \exp\left(-\frac{1}{2\sigma^2}||\mathbf{x}-\mathbf{m}_k||^2\right) \tag{15.42}$$

and the probability for any $\mathbf{x}_n \in \mathbf{X}$ to belong to cluster C_k is

$$P_{nk} = P(z_k = 1|\mathbf{x}_n,\boldsymbol{\theta}) = \frac{P_k \mathcal{N}(\mathbf{x}_n;\mathbf{m}_l\boldsymbol{\Sigma}_l)}{\sum_{l=1}^K P_l \mathcal{N}(\mathbf{x}_n;\mathbf{m}_l\boldsymbol{\Sigma}_l)} = \frac{P_k \exp(-||\mathbf{x}_n - \mathbf{m}_k||^2/2\sigma^2)}{\sum_{l=1}^K P_l \exp(-||\mathbf{x}_n - \mathbf{m}_l||^2/2\sigma^2)} \tag{15.43}$$

When $\sigma^2 \to 0$, all terms in the denominator approach to zero, but the one with minimum $||\mathbf{x} - \mathbf{m}_l||$ approaches to zero most slowly, and becomes the dominant term of the denominator. If the numerator happens to be this term, that is, $k = l$, then $P_{nk} = 1$; otherwise the numerator approaches zero and $P_{nk} = 0$. Now P_{nk} becomes

$$\lim_{\sigma^2\to 0} P_{nk} = \lim_{\sigma^2\to 0} \frac{P_k \exp(-||\mathbf{x}_n - \mathbf{m}_k||^2/2\sigma^2)}{\sum_{l=1}^K P_l \exp(-||\mathbf{x}_n - \mathbf{m}_l||^2/2\sigma^2)} = \begin{cases} 1 & \text{if } ||\mathbf{x}_n - \mathbf{m}_k|| = \min_l ||\mathbf{x}_n - \mathbf{m}_l|| \\ 0 & \text{otherwise} \end{cases} \tag{15.44}$$

Now the posterior probability $0 < P_{nk} < 1$ defined in Eq. (15.22) for a soft decision becomes a binary value $P_{nk} \in \{0, 1\}$ for a hard binary decision to assign $\mathbf{x}_n$ to C_k with the smallest distance. Also $N_k = \sum_{n=1}^N P_{nk}$ defined in Eq. (15.29) as the sum of the posterior probabilities for all N data points to belong to C_k becomes N_k as the number of data samples assigned only to C_k. In other words, now the probabilistic EM method based on both $\mathbf{m}$ and $\boldsymbol{\Sigma}$ becomes the deterministic k-means method based on $\mathbf{m}$ only.

We can also make a comparison between the GMM method for unsupervised clustering and the softmax regression for supervised classification. First, the latent variables $\mathbf{Z} = [\mathbf{z}_1,\cdots,\mathbf{z}_N]$ in GMM play a similar role as the labeling $\mathbf{Y} = [\mathbf{y}_1,\cdots,\mathbf{y}_N]$ in softmax regression multiclass classification. However, the difference is that $\mathbf{Y}$ is explicitly given in the training set for a supervised classification, while $\mathbf{Z}$ is hidden for an unsupervised clustering analysis. Second, we note that the probability $P_{nk} = p(z_{nk} = 1|\mathbf{x}_n,\boldsymbol{\theta})$ given in Eq. (15.22) is similar to the softmax function $\phi_{nk} = P(y' = k|\mathbf{x}_n)$ in the softmax method in terms of their form, with the only difference that the Gaussian function is used for GMM while the exponential function is used for softmax.

Example 15.3

Both the k-means and GMM clustering methods are tested based on the same synthetic dataset generated by the following MATLAB code

```
mk=zeros(d,3);
Ck=zeros(d,d,3);
Ck(:,:,1)=[1 1.5; 1.5 5];    mk(:,1)=[5; 4];
```

```
Ck(:,:,2)=[3 -1.5; -1.5 1]; mk(:,2)=[6; 5];
Ck(:,:,3)=[2 1; 1 2];        mk(:,3)=[10; 2];
Ki=[200 200 200];
X1=mvnrnd(mk(:,1)',Ck(:,:,1),Ki(1));
X2=mvnrnd(mk(:,2)',Ck(:,:,2),Ki(2));
X3=mvnrnd(mk(:,3)',Ck(:,:,3),Ki(3));
```

This code generates three clusters of normally distributed data points. Note that the means of the first two clusters are very close to each other, but their covariances are such that their distributions overlap with each other but are in very different orientations.

The intermediate results after each iteration are shown in the panels in Fig. 15.5 for the k-means method and Fig. 15.6 for the EM clustering method. The data points are color coded according to their class identity (ground truth not used in clustering), and the contour lines for each of the clusters are also shown based on the mean and covariance of all sampled assigned to each cluster in the current iteration. Comparing the two sets of results obtained by the two methods, we see that the k-means method cannot separate the first two clusters (in red and green), while the third cluster (in blue) is separated incorrectly into two clusters. Conversely, the GMM method can correctly separate all three clusters.

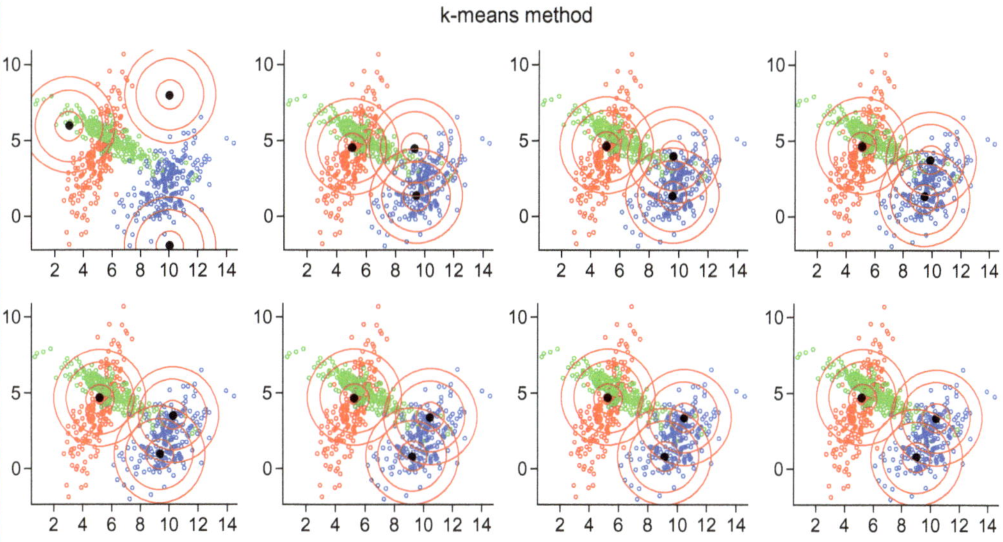

Fig. 15.5 Clustering by k-means method (first eight iterations).

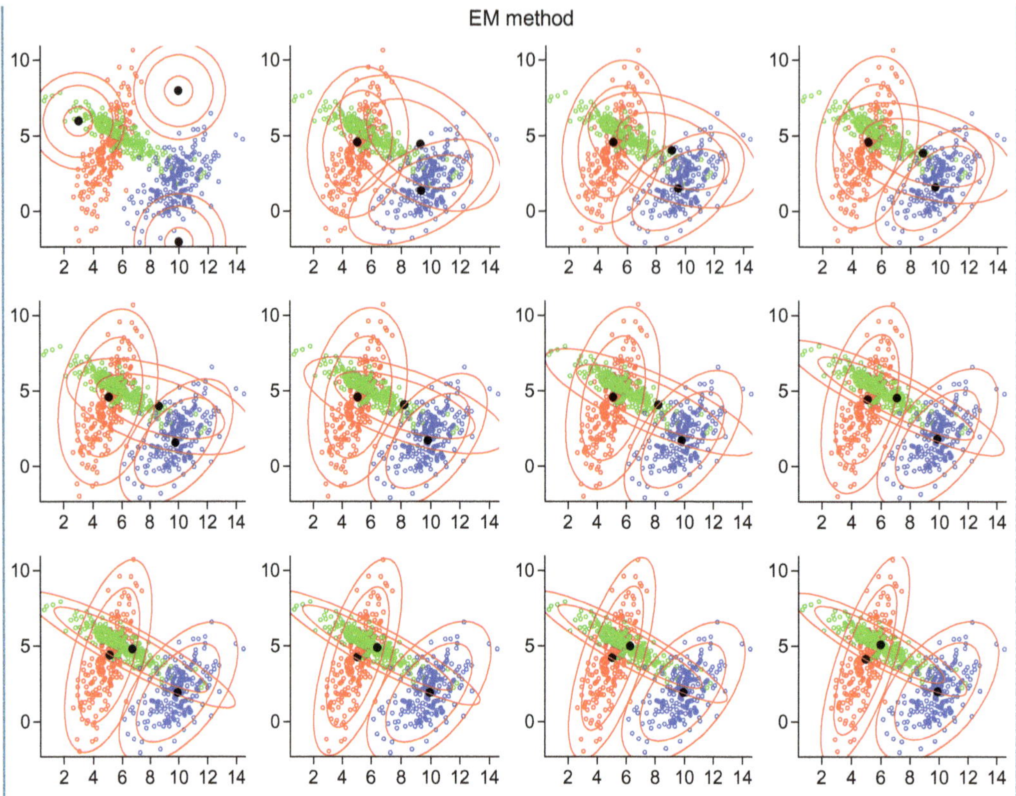

Fig. 15.6 Clustering by Gaussian mixture model (first 12 iterations).

Example 15.4

The two clustering methods are also applied to the 4-D iris dataset containing three classes each of 50 sample vectors. The PCA method is used to visualize the first two principal components, as shown in Figs 15.7 and 15.8. Also, as can be seen from their confusion matrices in Eq. (15.45), the error rate of the EM method is 5/150, much smaller than 18/15 of the k-means method.

k–means

0	0	50
50	0	0
18	32	0

EM

0	0	50
45	5	0
0	50	0

(15.45)

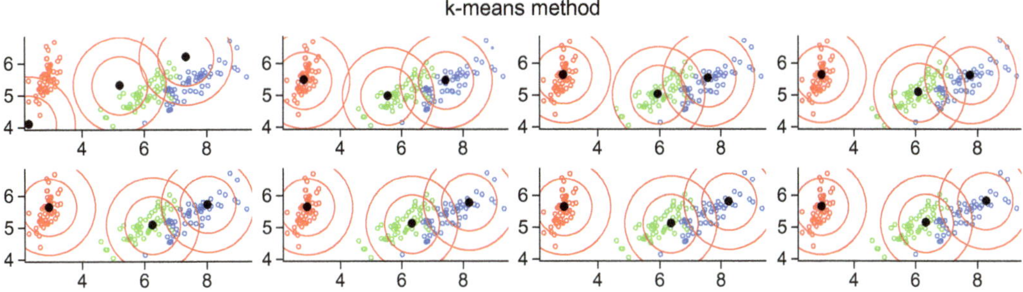

Fig. 15.7 Clustering of Iris dataset by k-means method.

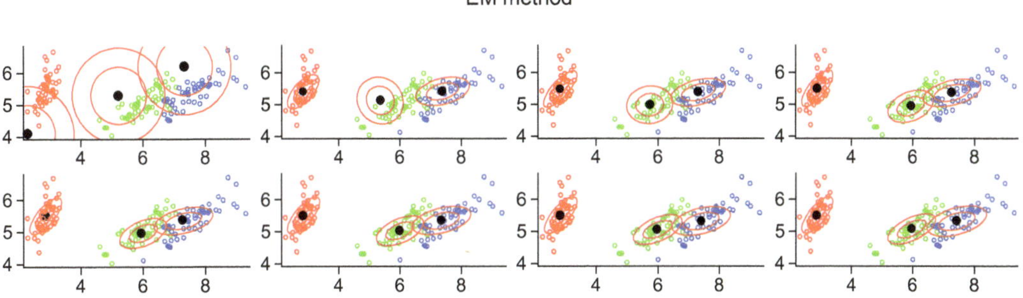

Fig. 15.8 Clustering of Iris Dataset by Gaussian mixture model.

15.3 Bernoulli Mixture Model

If the data points are binary, that is, each data point is treated as a binary random variable that takes either of two values such as 1 and 0 with probabilities μ and $1 - \mu$, then the assumed Gaussian distribution is no longer valid and the GMM is not suitable. In this case, we can modify the EM method for clustering based on *Bernoulli mixture model (BMM)* instead of GMM.

The *probability mass function (pmf)* of the Bernoulli distribution can be used instead

$$\mathcal{B}(x, \mu) = \mu^x (1 - \mu)^{1-x} = \begin{cases} \mu & \text{if } x = 1 \\ 1 - \mu & \text{if } x = 0 \end{cases} \tag{15.46}$$

The mean and variance of x are

$$E[x] = 1\ P(x = 1) + 0\ P(x = 0) = 1\ \mu + 0\ (1 - \mu) = \mu \tag{15.47}$$

$$\begin{aligned} Var[x] &= E[(x - E[x])^2] = E[x^2] - E[x]^2 \\ &= 1^2\ P(x = 1) + 0^2\ P(x = 0) - \mu^2 = \mu - \mu^2 = \mu(1 - \mu) \end{aligned} \tag{15.48}$$

Note that as the variance is solely determined by the mean $E[x] = \mu$ it does not need to be estimated separately, therefore the method based on BMM is simpler than that based on GMM.

A set of d independent binary variables can be represented as a random vector $\mathbf{x} = [x_1, \cdots, x_d]^T$ with mean vector and covariance matrix as shown in the following

$$E[\mathbf{x}] = \mathbf{m} = [\mu_1, \cdots, \mu_N]^T \tag{15.49}$$

$$\text{Cov}[\mathbf{x}] = \boldsymbol{\Sigma} = \text{diag}(\mu_i(1-\mu_i)) = \begin{bmatrix} \mu_1(1-\mu_1) & & 0 \\ & \ddots & \\ 0 & & \mu_d(1-\mu_d) \end{bmatrix} \tag{15.50}$$

The off-diagonal covariances are zero as the d variables are assumed to be independent of each other. The joint pmf of the binary random vector $\mathbf{x}$ is

$$\mathcal{B}(\mathbf{x}, \mathbf{m}) = \prod_{i=1}^{d} \mathcal{B}(x_i|\mu_i) = \prod_{i=1}^{d} \mu_i^{x_i}(1-\mu_i)^{1-x_i} \tag{15.51}$$

and the log pmf

$$\log \mathcal{B}(\mathbf{x}, \mathbf{m}) = \log\left(\prod_{i=1}^{d} \mathcal{B}(x_i, \mu_i)\right) = \sum_{i=1}^{d} [x_i \log \mu_i + (1-x_i)\log(1-\mu_i)] \tag{15.52}$$

Similar to the GMM, the BMM of K multivariate Bernoulli distributions is defined as

$$\begin{aligned} p(\mathbf{x}|\boldsymbol{\theta}) &= p(\mathbf{x}|\mathbf{m}_k, P_k, (k=1,\cdots,K)) = \sum_{k=1}^{K} P_k \mathcal{B}(\mathbf{x}, \mathbf{m}_k) \\ &= \sum_{k=1}^{K} P_k \prod_{i=1}^{d} \mu_{ki}^{x_i}(1-\mu_{ki})^{1-x_i} \end{aligned} \tag{15.53}$$

where $\boldsymbol{\theta} = \{\mathbf{m}_k, P_k, (k=1,\cdots,K)\}$ denotes all parameters of the mixture model to be estimated based on the given dataset. The mean of this mixture model is

$$\mathbf{m} = E[\mathbf{x}] = \sum_{k=1}^{K} P_k E_k[\mathbf{x}] = \sum_{k=1}^{K} P_k \mathbf{m}_k \tag{15.54}$$

Also similar to the GMM, we introduce a set of K latent binary random variables $\mathbf{z} = [z_1, \cdots, z_K]^T$ with binary components $z_k \in \{0, 1\}$ satisfying $\sum_{k=1}^{K} z_k = 1$, and get the prior probability of $\mathbf{z}$, the conditional probability of $\mathbf{x}$ given $\mathbf{z}$, and the joint probability of $\mathbf{x}$ and $\mathbf{z}$ as the following

$$p(\mathbf{z}|\boldsymbol{\theta}) = \prod_{k=1}^{K} P_k^{z_k} \tag{15.55}$$

$$p(\mathbf{x}|\mathbf{z}, \boldsymbol{\theta}) = \prod_{k=1}^{K} \mathcal{B}(\mathbf{x}, \mathbf{m}_k)^{z_k} \tag{15.56}$$

$$p(\mathbf{x}, \mathbf{z}|\boldsymbol{\theta}) = p(\mathbf{z}|\boldsymbol{\theta})\, p(\mathbf{x}|\mathbf{z}, \boldsymbol{\theta}) = \prod_{k=1}^{K} (P_k\, \mathcal{B}(\mathbf{x}, \mathbf{m}_k))^{z_k} \tag{15.57}$$

Given the dataset $\mathbf{X} = [\mathbf{x}_1, \cdots, \mathbf{x}_N]$ containing N i.i.d. samples, we introduce the corresponding latent variables in $\mathbf{Z} = [\mathbf{z}_1, \cdots, \mathbf{z}_N]$, of which each $\mathbf{z}_n =$

$[z_{n1},\cdots,z_{nK}]^T$ is for the labeling of $\mathbf{x}_n$. Then we can find the likelihood function of the BMM parameters $\boldsymbol{\theta} = \{P_k, \mathbf{m}_k,\ (k = 1,\cdots,K)\}$

$$L(\boldsymbol{\theta}|\mathbf{X},\mathbf{Z}) = p(\mathbf{X},\mathbf{Z}|\boldsymbol{\theta}) = p([\mathbf{x}_1,\cdots,\mathbf{x}_N],[\mathbf{z}_1,\cdots,\mathbf{z}_N]|\mathbf{m}_k, P_k(k=1,\cdots,K))$$
$$= \prod_{n=1}^N p(\mathbf{x}_n,\mathbf{z}_n|\boldsymbol{\theta}) = \prod_{n=1}^N \prod_{k=1}^K (P_k \mathcal{B}(\mathbf{x}_n,\mathbf{m}_k))^{z_{nk}} \tag{15.58}$$

and the log likelihood function

$$\log\ L(\boldsymbol{\theta}|\mathbf{X},\mathbf{Z}) = \log p(\mathbf{X},\mathbf{Z}|\boldsymbol{\theta}) = \log \prod_{n=1}^N \prod_{k=1}^K (P_k \mathcal{B}(\mathbf{x}_n,\mathbf{m}_k))^{z_{nk}}$$
$$= \sum_{n=1}^N \sum_{k=1}^K z_{nk}\,[\log P_k + \log \mathcal{B}(\mathbf{x}_n,\mathbf{m}_k)] \tag{15.59}$$

Based on the same EM method used in GMM, we can find the optimal parameters that maximize the expectation of the log likelihood function in the following two steps:

- **E-step:** Find the expectation of the likelihood function.
  We first find the posterior probability for any sample $\mathbf{x}_n$ to belong to cluster C_k, denoted by P_{nk}

$$P_{nk} = P(z_{nk}=1|\mathbf{x}_n,\boldsymbol{\theta}) = \frac{p(\mathbf{x}_n, z_{nk}=1|\boldsymbol{\theta})}{p(\mathbf{x}_n|\boldsymbol{\theta})} = \frac{P_k\,\mathcal{B}(\mathbf{x}_n;\mathbf{m}_k)}{\sum_{l=1}^K P_l\,\mathcal{B}(\mathbf{x}_n;\mathbf{m}_l)}$$
$$(n=1,\cdots,N;\quad k=1,\cdots,K) \tag{15.60}$$

  which is the expectation of z_{nk} (same as Eq. (15.25))

$$E[z_{nk}] = 1\ P(z_{nk}=1|\mathbf{x}_n) + 0\ P(z_{nk}=0|\mathbf{x}_n) = P(z_{nk}=1|\mathbf{x}_n) = P_{nk} \tag{15.61}$$

  Now we can find the expectation of the log likelihood with respect to the latent variables in $\mathbf{Z}$

$$E_z\,(\log L(\boldsymbol{\theta}|\mathbf{X},\mathbf{Z})) = E_z\ \sum_{n=1}^N \sum_{k=1}^K z_{nk}\,[\log P_k + \log \mathcal{B}(\mathbf{x}_n,\mathbf{m}_k)]$$
$$= \sum_{n=1}^N \sum_{k=1}^K E[z_{nk}] \left[\log P_k + \log \prod_{i=1}^d \mu_{ki}^{x_{ni}} (1-\mu_{ki})^{1-x_{ni}}\right]$$
$$= \sum_{n=1}^N \sum_{k=1}^K P_{nk} \left[\log P_k + \sum_{i=1}^d [x_{ni}\log \mu_{ki} + (1-x_{ni})\log(1-\mu_{ki})]\right] \tag{15.62}$$

- **M-step:** Find the optimal model parameters that maximize the expectation of the log likelihood function.
  We first set to zero the derivatives of the expectation of the log likelihood with respect to each of the parameters in $\boldsymbol{\theta} = \{P_k,\ \mathbf{m}_k\ (k=1,\cdots,K)\}$, and then solve the resulting equations to get the optimal parameters.

- Find P_k: same as in the case of the GMM model

$$P_k = \frac{N_k}{N} = \frac{1}{N}\sum_{n=1}^{N} P_{nk} \tag{15.63}$$

- Find $\mathbf{m}_k$

$$\begin{aligned}
&\frac{\partial}{\partial \mathbf{m}_k} E_{\mathbf{Z}}\left(\log p(\mathbf{X},\mathbf{Z}|\boldsymbol{\theta})\right)\\
&= \frac{\partial}{\partial \mathbf{m}_k}\sum_{n=1}^{N}\sum_{k=1}^{K} P_{nk}\left[\log P_k + \sum_{i=1}^{d}[x_{ni}\log\mu_{ki} + (1-x_{ni})\log(1-\mu_{ki})]\right]\\
&= \sum_{n=1}^{N} P_{nk}\frac{\partial}{\partial \mathbf{m}_k}\sum_{i=1}^{d}[x_{ni}\log\mu_{ki} + (1-x_{ni})\log(1-\mu_{ki})] = \mathbf{0}
\end{aligned} \tag{15.64}$$

The ith component of the equation is

$$\begin{aligned}
&\sum_{n=1}^{N} P_{nk}\frac{d}{d\mu_{ki}}[x_{ni}\log\mu_{ki} + (1-x_{ni})\log(1-\mu_{ki})]\\
&= \sum_{n=1}^{N} P_{nk}\left(\frac{x_{ni}}{\mu_{ki}} - \frac{1-x_{ni}}{1-\mu_{ki}}\right) = 0
\end{aligned} \tag{15.65}$$

that is

$$(1-\mu_{ki})\sum_{n=1}^{N} P_{nk}x_{ni} = \mu_{ki}\sum_{n=1}^{N} P_{nk}(1-x_{ni}) = \mu_{ki}N_k - \mu_{ki}\sum_{n=1}^{N} P_{nk}x_{ni} \tag{15.66}$$

Solving for μ_{ki} we have

$$\mu_{ki} = \frac{1}{N_k}\sum_{n=1}^{N} P_{nk}x_{ni} \qquad (i = 1,\cdots,d) \tag{15.67}$$

or, in vector form

$$\mathbf{m}_k = \frac{1}{N_k}\sum_{n=1}^{N} P_{nk}\mathbf{x}_n \tag{15.68}$$

The following MATLAB code is the iteration part of the algorithm. Same as in the code the GMM, all parameters are initialized in a similar manner.

```
er=1;
while er<0.01
   for k=1:K
       p(:,k)=pi(k)*Bpdf(X,mk(:,k)');
   end
   er=0;
   D=sum(p,2);
   for k=1:K
       Pnk(:,k)=p(:,k)./D;
       Nk(k)=sum(Pnk(:,k));
```

```
                pi(k)=Nk(k)/N;
                if Nk(k)>0
                    Mk(:,k)=sum(X.*repmat(Pnk(:,k),[1,d]),1)/Nk(k);
                    Xm=X-repmat(Mk(:,k)',[N,1]);
                end
                er=er+norm(mk(:,k)-Mk(:,k));
            end
        end
```

Example 15.5

The clustering algorithm based on the mixture of Bernoulli model is applied to the dataset of handwritten digits after the gray scale images (in the range from 0 to 255) are converted to binary data by a threshold value (e.g., 128). Once the iteration is converged, the final clustering result is shown in the form of the means vectors of the K clusters (no longer binary) in Fig. 15.9 with $K = 10$ (left) and $K = 12$ (right).

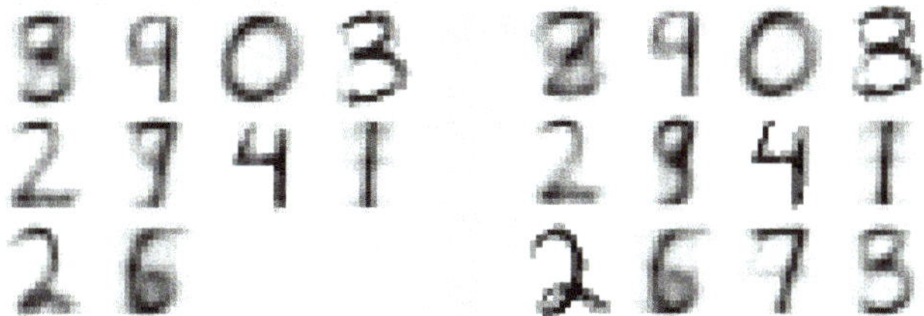

Fig. 15.9 Clustering of handwritten digits (left $K = 10$, right $K = 12$).

PROBLEMS

1. Implement k-means clustering algorithm and apply it to the iris data. Try different values for K, such as $K = 2$ and $K = 4$ in addition to the ground truth value $K = 3$.

   Show your results as a K by K confusion matrix and the error rate. Also visualize your result in both 2-D and 3-D spaces after the 4-D dataset is KLT transformed into a lower dimensional space. Color code all data points according to the cluster they are assigned to.
2. Repeat step 1 for the handwritten digit dataset. Again try different values of K, such as 9 and 11, in addition to the ground truth value $K = 10$.

   Furthermore, convert the K mean vectors of all samples assigned to the same clusters into 16×16 images, and display these images to see what the clusters actually represents, and if they each resemble one of the digits, same as those shown in Example 15.2.

3. Implement GMM clustering and apply it with the k-means method to the dataset used in Example 15.3 (generated by the code provided in the example). Visualize the results by both GMM and k-means methods with all data points color-coded, and compare the confusion matrices of both methods to confirm GMM achieves much better results than K-mean in this particular case.
4. Apply the GMM method to both the iris and handwritten datasets. For each dataset, show your clustering result in confusion matrix and visualize the color-coded data points in 2-D and 3-D spaces. Compare your results with those obtained by the k-means method in the first two problems.
5. Implement clustering based on mixture of Bernoulli model, then apply it to the binary version of the handwritten digit dataset obtained by thresholding all samples and converting them into binary numbers 0 or 1. Compare your results with $K = 10$ with those shown in Fig. 15.9

16 Hierarchical Classifiers

Both supervised classification and unsupervised clustering can be carried out hierarchically in the form of a dendrogram, a tree-like structure, very much like the classifications of different taxonomic ranks in biology (domain, kingdom, phylum, class, order, family, genus, and species). Such a hierarchical classification or clustering method is essentially a divide-and-conquer process by which a complex problem is converted into a sequence of simpler subproblems that can be solved more effectively and efficiently. Certain advantages can be gained by carrying out classification or clustering in a hierarchical manner, in comparison to most of the algorithms previously discussed all attempting to achieve the goal in a single step.

Before considering various hierarchical algorithms specifically, we note that as a general method for classification and clustering, a decision tree may grow very deep to accommodate certain detailed variations in the data, and consequently overfitting the data. This problem can be addressed by the method of *random forest*, which takes the average of multiple decision trees each trained on different subsets of the training dataset. Random forest is similar to the K-fold cross-validation in the sense that the given dataset is subdivided into multiple parts for more reliable outcomes.

16.1 Bottom-Up vs. Top-Down Methods

The tree structure for a hierarchical method is typically constructed in either bottom-up fashion by merging subgroups of data points into bigger groups, or top-down fashion by splitting big groups into smaller subgroups. In the following, we will concentrate on how the merging or splitting is specifically carried out when applied to a classification problem with K classes, to be considered in Section 16.2. The resulting binary tree, typically upside down, has $2K - 1$ nodes, including K leaf nodes each for one of the classes, and $K - 1$ non-leaf nodes, at each of which two subgroups are either merged into one group in the bottom-up case, or group is split into two subgroups in the top-down case.

Either of the top-down splitting and bottom-up merging methods can also be used to construct a hierarchical structure for unsupervised clustering if applied to all N individual unlabeled data samples in the given dataset, instead of K labeled classes, as we will see in Section 16.3.

- **Bottom-up method**
  Initially, every class composed of data points with the same class identity is treated as subgroup represented by one of the K leaf nodes at the bottom of the binary tree. These subgroups are then merged two at a time to form progressively larger groups. This process is carried out recursively $K - 1$ times until eventually all K classes are merged into a single super group at the root node of the tree. In the following discussion, we assume the dataset contains in total N data samples of which n_k are labeled to belong to class C_k, $(k = 1, \cdots, K)$. The steps for this bottom-up merging method is shown below are given in the following:
  1. Compute the $K(K-1)/2$ pairwise Bhattacharyya distances between every two classes C_i and C_j

$$d_B(C_i, C_j) = \frac{1}{4}(\mathbf{m}_i - \mathbf{m}_j)^T \left[\frac{\Sigma_i + \Sigma_j}{2}\right]^{-1} (\mathbf{m}_i - \mathbf{m}_j) + \log\left[\frac{\left|\frac{\Sigma_i + \Sigma_j}{2}\right|}{(|\Sigma_i|\ |\Sigma_j|)^{1/2}}\right] \tag{16.1}$$

  2. Merge the two classes corresponding to the smallest d_B to form a new class $C_k = C_i \cup C_j$, and compute its mean

$$\mathbf{m}_k = \frac{1}{n_i + n_j}[n_i \mathbf{m}_i + n_j \mathbf{m}_j] \tag{16.2}$$

  and covariance

$$\Sigma_k = \frac{1}{n_i + n_j}[n_i(\Sigma_i + (\mathbf{m}_i - \mathbf{m}_k)(\mathbf{m}_i - \mathbf{m}_k)^T) + n_j(\Sigma_j + (\mathbf{m}_j - \mathbf{m}_k)(\mathbf{m}_j - \mathbf{m}_k)^T)] \tag{16.3}$$

  Delete the old classes C_i and C_j. Now there are $K - 1$ classes left.
  3. Repeat the previous steps to find all pairwise distances for the remaining classes, and merge the two with minimum distance, until eventually all classes are merged into a single group containing all K classes, the binary tree structure is thus obtained.

- **Top-down method**
  Initially, all K classes are treated as members of a single super group at the root node on top of the tree, which is then split into two subgroups, each represented as a node at a lower level of the tree. This subdivision is carried out recursively until eventually each subgroup contains a single class represented by one of the leaf nodes at the bottom of the tree. This tree structure can be constructed in either breadth-first manner based on a queue, or depth-first manner based on a stack, which can be most conveniently implemented recursively.

  Various methods can be used to subdivide a group into two subgroups. For example, the k-means method can be used with $K = 2$. Alternatively, the binary subdivision can be carried out in the 1-D space along the direction of the principal component based on the Karhunen–Loève transform (KLT) method, applied to either the covariance matrix of the entire dataset, or the between-class scatter matrix (Eq. (9.20)). All samples are projected onto this direction and then subdivided into two parts with maximum Bhattacharyya distance, as shown in the following steps.

1. Compute the covariance matrix $\mathbf{\Sigma}$ of all N data points or the between-class scatter matrix $\mathbf{S}_B$ of all K classes, find its maximum eigenvalue λ_i and the corresponding eigenvectors $\mathbf{v}_i$
2. Project all data points onto $\mathbf{v}_1$

$$y_n = \mathbf{x}_n^T \mathbf{v} \qquad (n = 1, \cdots, N) \tag{16.4}$$

3. Sort all data points $\{y_1, \cdots, y_N\}$ along this 1-D space and partition them into two subgroups with maximum Bhattacharyya distance.
4. Carry out these steps recursively to each of the two subgroups, until eventually every subgroup contains only one class.

The hierarchical structure so constructed by either method can then be used as either a decision tree classifier for supervised classification or unsupervised clustering as discussed in the following two sections, respectively.

16.2 Binary Hierarchical Classification

The hierarchical classification method is especially suitable when the dimension of the feature space is high and the number of classes is large. For a single-level classifier, it may be difficult to select a subset of the large number of available features good for separating all classes in a single step. Conversely, for a hierarchical classifier it is possible to select a small subset of features that are most relevant and suitable to distinguish two subgroups at each node. Although this is a multistep process, it can be carried out effectively and efficiently due to the low computational complexity of the binary classification.

Specifically, any unlabeled sample to be classified enters the tree structure at the root node, where it is classified into either the left or right subgroup, and this binary classification is further carried out recursively at each of the tree nodes along a certain branch of the tree, until eventually reaching one of the K leaf nodes representing class identity of the data sample. Here are the specific steps:

1. According to the specific classification method used, find the discriminant functions $D_l(\mathbf{x})$ and $D_r(\mathbf{x})$ for the two subgroups based on the training data.
2. Select the best $d < D$ features most suitable for separating the two groups G_l and G_r, based on any of the feature selection methods such as those listed below:
   - choose d features directly from the D original ones using between-class distance (e.g., Bhattacharyya distance) as the criterion,
   - carry out KLT based on the between-class scatter matrix $\mathbf{S}_B$ and use the first d principal components for the binary classification.

   As here only two groups of classes need to be distinguished, the number of selected features can be expected to be small.

3. Any unlabeled pattern **x** enters the classifier at the root of the tree and is classified to either the left or the right subgroup of the node according to the discriminant function

$$\mathbf{x} \in \begin{cases} G_l & if \;\; D_l(\mathbf{x}) > D_r(\mathbf{x}) \\ G_r & if \;\; D_r(\mathbf{x}) < D_l(\mathbf{x}) \end{cases} \tag{16.5}$$

   This process is carried out recursively at each of the subsequent nodes until eventually **x** reaches one of the leaf nodes corresponding to a single class, to which the sample **x** is therefore classified.

The following MATLAB function shows how the top-down splitting method can be implemented recursively as a depth-first search process. Here `X` is a 2-D data array containing N column vectors each for one of the N data samples labeled as one of the K classes, `M` is a 2-D array containing K columns each for the mean vector of one of the K classes, and `Cov` is a 3-D array containing the covariance matrices of the K classes (both pre-calculated in the main function). Also, p is a 1-D vector for the parent node of each of the $2K - 1$ tree nodes, to be used for visualizing the tree using the MATLAB function `treeplot`.

```
function TDclassify()           % top-down tree construction
    global X M Cov K p nc
    [d N]=size(X);              % dimension and size of data
    gid=1:K;                    % list of all K classes
    nc=K;                       % initialize current node
    p=zeros(1,2*K-1);           % parent list for all tree nodes;
    f=0;                        % parent node is always 0
    DFdivideC(gid,f);           % top-down binary tree construction
    treeplot(p)                 % plot the binary tree
    [x,y]=treelayout(p);
    text(x+0.02,y, compose('%d',[1:2*K-1])) % label all tree nodes
end
function DFdivideC(gid,f)       % depth-first search recursion
    global X K M Cov p nc       % global variables shared by all functions
    if length(gid)==1           % leaf-node, terminate recursion
         return
    end
    nc=nc+1;                    % current number of nodes
    p(nc)=f;                    % assign parent
    f=nc;                       % parent node for its children
    [lid rid]=twoMeansC(gid);   % split gid into two parts lid and rid
    if length(lid)==1           % left branch is a leaf node
        p(lid)=nc;
        if length(rid)==1       % leaf node on right
            p(rid)=nc;
        end
    else                        % left branch is non-leaf node
        DFdivideC(lid,f)        % recursion of the left branch
```

```
    end
    DFdivideC(rid,f)          % recursion of the right branch
end
```

The following function subdivides the given list of groups in `gid` into two subgroups `lid` on the left and `rid` on the right, based on the method of k-means clustering with $K = 2$.

```
function [lid rid]=twoMeansC(gid) % 2-means clustering
    global X M Cov Nk
    K=length(gid);            % number of classes in input
    if K==2
        lid=gid(1);
        rid=gid(2);
        return
    end
    m1=M(:,gid(1));           % first 2 classes as seeds for 2 means
    m2=M(:,gid(2));
    er=inf;
    while er>0                % clustering iteration
        gid1=[]; cid1=[];    % group IDs and classes IDs for group 1
        gid2=[]; cid2=[];    % group IDs and classes IDs for group 2
        m1o=m1;  m2o=m2;     % old means
        for l=1:K            % for each of K classes
            k=gid(l);        % kth cluster ID
            m=M(:,k);        % mean of cluster k
            c=Cov(:,:,k);    % covariance of cluster k
            d1=distM(m1o,m,c);
            d2=distM(m2o,m,c);
            if d1 < d2
                gid1=[gid1 k];                % assign class k to group 1
                cid1=[cid1 cid(k,1:Nk(k))]; % assign samples to group 1
            else
                gid2=[gid2 k];
                cid2=[cid2 cid(k,1:Nk(k))];
            end
        end
        m1=mean(X(:,cid1)')'; % two new means
        m2=mean(X(:,cid2)')';
        er=norm(m1-m1o)+norm(m2-m2o);
    end
    lid=gid1; rid=gid2;
    if length(gid2)==1        % single node always on left branch
        lid=gid2; rid=gid1;
    end
end
```

The bottom-up method for the binary tree construction keeps merging two groups of classes with minimum Bhattacharyya distance recursively from the leaf nodes, each representing a single class all the way up to the root node representing a super group containing all classes. This process can be implemented by the following MATLAB function, where Nk is a list containing the number of samples in each of the K classes. The main loop is carried out $K-1$ times to keep merging two subgroups (to be deleted) with minimum distance into a group (to be created). After $K-1$ such merge operations, all K classes are merged into a single super group at the root node. Again, the parent nodes of the $2K-1$ tree nodes are stored in a list p for the visualization of the binary tree.

```
function BUtree()                 % bottom-up tree construction
    global X K Nk M Cov p groups
    [d N]=size(X);                % number of classes
    cid=zeros(2*K,N);             % sample IDs in each cluster
    del=ones(1,2*K);              % deleted clusters
    p=zeros(1,2*K-1);             % parents list;
    Ck=zeros(1,2*K);              % number of samples per group
    Ck(1:K)=Nk;                   % for K leaf nodes
    ik=0;
    for k=1:K
        cid(k,1:Ck(k))=ik+1:ik+Ck(k);
        ik=ik+Ck(k);
    end
    M=zeros(d,2*K);               % means for K leaf nodes and K-1 nonleaf nodes
    Cov=zeros(d,d,2*K);           % covariance of all 2K-1 nodes
    k=K;
    while k<2*K-1                 % merge K-1 times
        for l=1:k                 % get mean and cov for each group
            M(:,l)=mean(X(:,cid(l,1:Ck(l)))')';
            Cov(:,:,l)=cov(X(:,cid(l,1:Ck(l)))');
        end
        dmin=inf;
        for k1=2:k
            if del(k1)                % if first group is not deleted
                for k2=1:k1-1
                    if del(k2)  % if second group is not deleted
                        d=distB(M(:,k1),Cov(:,:,k1),M(:,k2),Cov(:,:,k2));
                        if dmin>d
                            dmin=d; i=k1; j=k2; % save min-distance so far
                        end
                    end
                end
            end
        end
        k=k+1;
        fprintf('(%2d,%2d) --> %2d\n',i,j,k)
        p(i)=k; p(j)=k;               % assign parent Ck to Ci and Cj
```

```
            cid(k,1:Ck(i)+Ck(j))=[cid(i,1:Ck(i)) cid(j,1:Ck(j))];
            Ck(k)=Ck(i)+Ck(j);
            del(i)=0; del(j)=0;      % delete Ci and Cj
        end
    end
```

Example 16.1

Figure 16.1 shows the binary tree structures created by both bottom-up (right) and top-down (left) methods applied to the dataset containing 10 classes of handwritten digits 0 through 9, as shown at the bottom of the 10 leaf nodes in both cases.

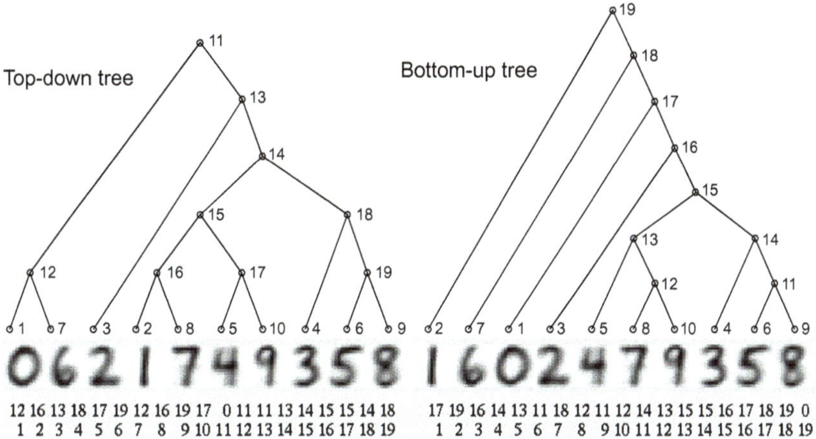

Fig. 16.1 Binary tree classifier: top-down (left) and bottom-up (right).

In the top-down method, node $(K + 1) = 11$ is the root node from which the super group containing all K classes is split recursively into the 10 leaf nodes each for one of the 10 classes, while in the bottom-up method, the 10 leaf nodes are recursively merged into a super group containing all 10 classes at node $2K - 1 = 19$ as the root node. Comparing the trees constructed by the two methods, we see that they are similar in that all classes similar in shape are grouped together; for example, digits 3, 5, and 8 form a subgroup at node 18 in the top-down tree and node 14 in the bottom-up tree, and digits 4, 7, and 9, form another subgroup at nodes 15 and node 13 in the respective trees, while digits 0, 2, and 6 with dissimilar shapes are separated early in top-down method or merged into the groups late in bottom-up method. The minor difference between the two results is that digits 1 and 7 are considered to be similar in the top-down method, but not so in the bottom-up method.

The parent list for all $2K - 1 = 20 - 1 = 19$ tree nodes is also shown in Fig. 16.1, generated by each of the two methods. The 19 nodes are listed in the bottom row while their corresponding parent nodes are listed in the top row. For example, the parent node of leaf node 8 (for digit 7) is node 16 in the top-down method or node 12 in the bottom-up method. Based on this parent list, the tree struction can be automatically generated by the MATLAB function `treeplot(p)`.

Once the binary tree structure is constructed, we can further construct a binary classifier at each of the $K - 1 = 10 - 1 = 9$ non-leaf nodes based on a small set of features selected to separate the two subgroups associated with the node. Such a sequence of binary classifications, each based on a small set of features, can be more effective and efficient than classifying all 10 classes based on all features.

16.3 Binary Hierarchical Clustering

When the hierarchical method is applied to the clustering of a given dataset, the root node on top of the tree structure represents a super group containing all N data points in the dataset, while each of the leaf nodes is for one of these data points. A binary tree can be constructed in either top-down or bottom-up fashion. In the top-down method, the super group of all data points is recursively split into two subgroups in either depth-first or breadth-first fashion until reaching the bottom of the tree where each of the leaf nodes represents one of the N given data points. Alternatively, in the bottom-up method, starting from the leaf nodes each representing a single data sample, the two groups with minimum inter-group distance are merged to form a bigger group, and this process is further carried out recursively until reaching the top of the tree where the root node represents a super group containing all samples in the dataset.

The tree structure obtained by either method can then be truncated at certain levels between the root node on top and the leaf nodes at the bottom to obtain a set of subgroups, each for one of the clusters, depending on the desired number and sizes of these clusters. Actually it is unnecessary to construct the complete tree first and then truncate it to get the clustering results. Instead, in either of the top-down or bottom-up methods, the recursion of splitting or merging can be terminated if certain conditions are met, such as the size of a group or the intra-cluster scatteredness is too large or too small. By doing so, we obtain a set of clusters without constructing the complete tree.

Similar to all unsupervised clustering methods such as the k-means method (Section 15.1), hierarchical clustering also faces the difficulty of determining the proper K value for the number of clusters that best reflects the inherent structure of the dataset. Here specifically we need to determine the parameter values (size of clusters, intra-cluster scatteredness, etc.) for terminating the recursive merging or splitting at a proper level. If the tree structure is truncated at a level which is either too high or too low, the resulting K may be either too large or too small, causing the dataset to be either under or over divided, as illustrated in Fig. 16.4 for an example to be considered later.

One way to address the issue of under versus over subdivision is to generate the histogram of all between-class distances at the tree nodes, as shown in Fig. 16.2. In general, the larger distances on the right side of the histogram come from the subdivisions at the top levels by which the dataset is divided into subgroups, and the smaller distances on the left are from the subdivisions at the lower levels by which some clusters in the dataset may be unnecessarily and over subdivided. In other

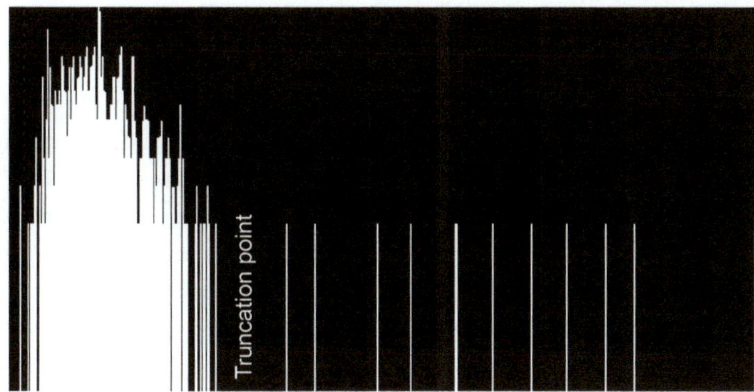

Fig. 16.2 Histogram of between-subgroup distances.

words, the larger and smaller distances may reflect, respectively, the inter-cluster and intra-cluster scatteredness. Also we note that in the histogram, there exists a dense peak composed of many small intra-group distances, much more than the number of larger inter-cluster distances coarsely distributed toward the right end of the histogram.

It is therefore clear that recursive subdivision should be truncated at a certain point along the histogram that separates the valid larger inter-cluster distances on the right from the invalid smaller intra-cluster distances on the left, as also shown in Fig. 16.2. Based on this method the tree structure can be properly truncated as shown in Fig. 16.4 (see later) for Example 16.2, where a dataset is under (left), properly (middle), or over (right) clustered.

In the following, the MATLAB code for both the depth-first and breadth-first algorithms is listed containing the following variables:

- `K`: the number of clusters (leaf nodes) so far (initially zero);
- `X`: a $d \times N$ array containing N-dimensional column vectors each for one of the samples;
- `sid`: sample indices for all samples $(1, \cdots, N)$;
- `kid`: cluster labelings $(1, \cdots, K)$ for all N samples, of which the nth component indicates the cluster sample $\mathbf{x}_n$ is assigned to;
- `path`: a list of paths (strings) for all N samples, only needed for visualization of the tree; for example, a sample with path '010' is assigned to the left, right, and left subgroups in three consecutive steps along a tree branch;
- `mins`: pre-specified minimum number of samples in a cluster;
- `th`: prespecified threshold value for within-cluster scatteredness.

The depth-first binary subdivision is based on based a stack (last-in-first-out), carried out recursively:

```
function DFdivide(sid)          % subdivide a node containing samples in sid
    global X K path kid th mins % global variables shared by other functions
    N=length(sid);              % number of samples
    Y=X(:,sid);                 % all samples at current node
    if cov2(Y)<th | N<mins      % terminate recursion (leaf node reached)
        K=K+1;                  % update number of clusters
```

```
        kid(sid)=K;               % assign cluster ID to all samples
        return
    end
% split group (sid) into left (lid) and right (rid) subgroups:
    [lid rid]=twoMeans(Y,sid);  % by either k-means (K=2)
%     [lid rid]=split(Y,sid);     % or splitting in principal direction
    for l=1:length(lid)         % for each sample on the left
        i=lid(l);               % index of each sample
        path{i}=[path{i},'0'];  % append 0 to path
    end
    for l=1:length(rid)         % for each sample on the right
        i=rid(l);               % index of each sample
        path{i}=[path{i},'1'];  % append 1 to path
    end
    % recursively subdivide the left and right nodes
    DFdivide(lid)
    DFdivide(rid)
end
```

Here `cov2` is a function that calculates the scatteredness, trace of covariance of column vectors in dataset `X`:

```
function tcov=cov2(X)
    [d,n]=size(X);
    tcov=0;
    if n>1
        tcov=trace(cov(X'));
    end
end
```

The breadth-first binary subdivision is based a queue (first-in-first-out), represented in the code by a dynamic 2-D array `Q`. Each tree node is represented by one row of `Q` containing indices of all samples assigned to the node (appended by zeros for all rows to have the same length).

```
function BFdivide(sid)
    global X K path kid th mins
    N=length(sid);                    % number of data samples
    Q=[[];sid];                       % initial queue: a row of all samples
    while size(Q,1)>0                 % loop until the queue is empty
        q=Q(1,:);                     % get the first row
        sid=q(1:nnz(q));              % get sample IDs in the row
        Y=X(:,sid);                   % samples to be subdivided by either
%        [lid rid]=twoMeans(Y,sid);  % by k-means method (K=2)
        [lid rid]=split(Y,sid);       % or split along principal direction
        for l=1:length(lid)           % record splitting history
            i=lid(l);
            path{i}=[path{i},'0'];    % paths of samples in left branch
        end
```

```
            for l=1:length(rid)
                i=rid(l);
                path{i}=[path{i},'1'];  % paths of samples in right branch
            end
            if cov2(X(:,lid))<th | length(lid)<mins    % a leaf node
                K=K+1;                  % update number of clusters
                kid(lid)=K;             % assign cluster ID to all samples
            else                        % not a leaf node
                lid(end+1:N)=0;         % append zeros
                Q=[Q;lid];              % put left node back to queue
            end
            if cov2(X(:,rid))<th | length(rid)<mins  % same for left branch
                K=K+1;
                kid(rid)=K;
            else
                rid(end+1:N)=0;
                Q=[Q; rid];
            end
            Q=Q(2:end,:);               % remove front row from queue
        end
    end
```

Example 16.2

The top-down method is applied to the clustering of a set of 80 data points in a 4-D space, to generate a binary tree based on sequence of binary subdivision. The subdivision is terminated when either the number of samples or the intra-group scatteredness in a subgroup is smaller than a prespecified value. Based on three sets of such parameters, three tree structures are obtained as shown in Fig. 16.3, and the corresponding clustering results are shown in Fig. 16.4 (in 3-D space spanned by the first three principal components after KLT). We see that the dataset is best fit by the seven clusters in the second case due to the properly selected parameters (`th`=0.01, `mins`=20), but it is either under divided (three clusters in the first case) or over divided (15 clusters in the third case) owing to improper parameters.

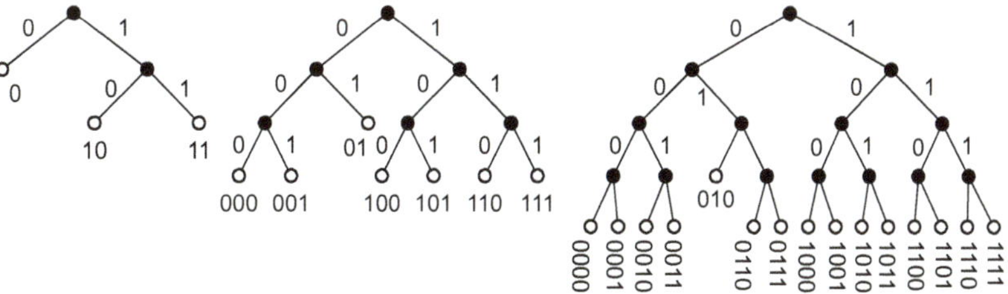

Fig. 16.3 Tree structures with 3, 7, and 15 leaf nodes each for a cluster.

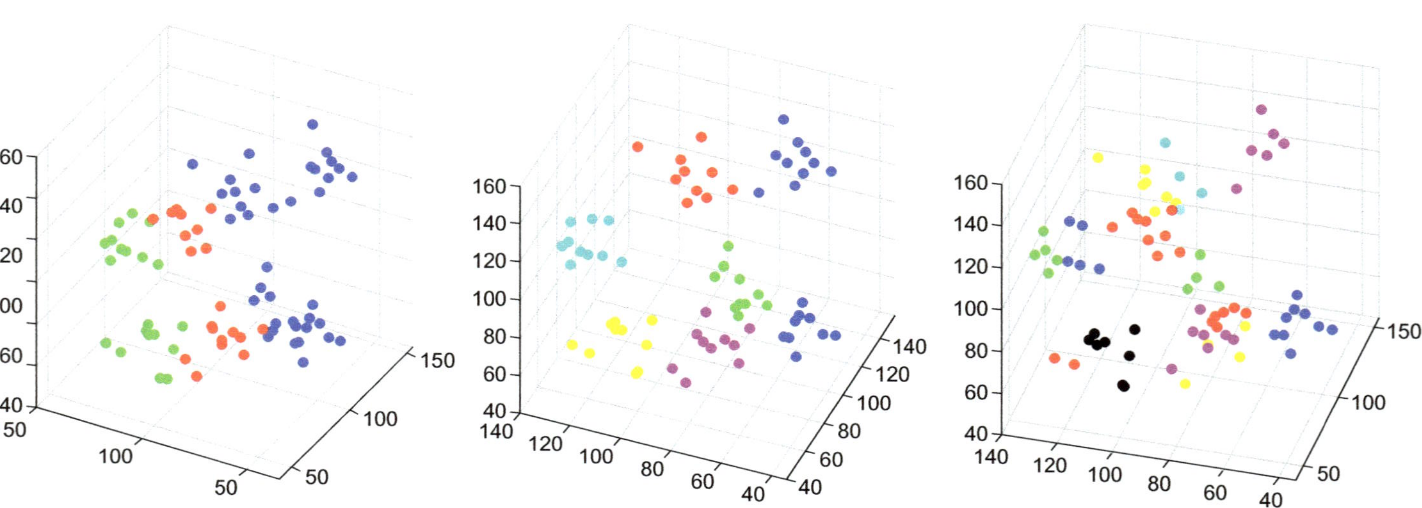

Fig. 16.4 Under (left), proper (middle), and over (right) division.

Example 16.3

The binary top-down splitting method is applied to the clustering analysis of a gene microarray containing a set of 99 genes (credit to the National Cancer Institute NIH, https://llmpp.nih.gov/lymphoma/data.shtml), in terms of their expression levels under seven different experimental conditions as shown in Fig. 16.5 where the expression levels are color coded. These genes are clustered into various subsets of similar genes groups in terms their expression levels under different conditions. The five panels in the figure show how the full dendrogram (top) can be truncated at various levels to form different numbers of subsets. For example, the bottom panel shows that these genes can be clustered into four groups, which can be further merged into two larger groups representing very different behaviors under the seven conditions. The same method is also carried out for the seven experimental conditions (instead of the 99 genes) so that they can be also be clustered into the subgroups each containing a set of similar conditions, as shown by the smaller dendrogram on the right side of the microarray data.

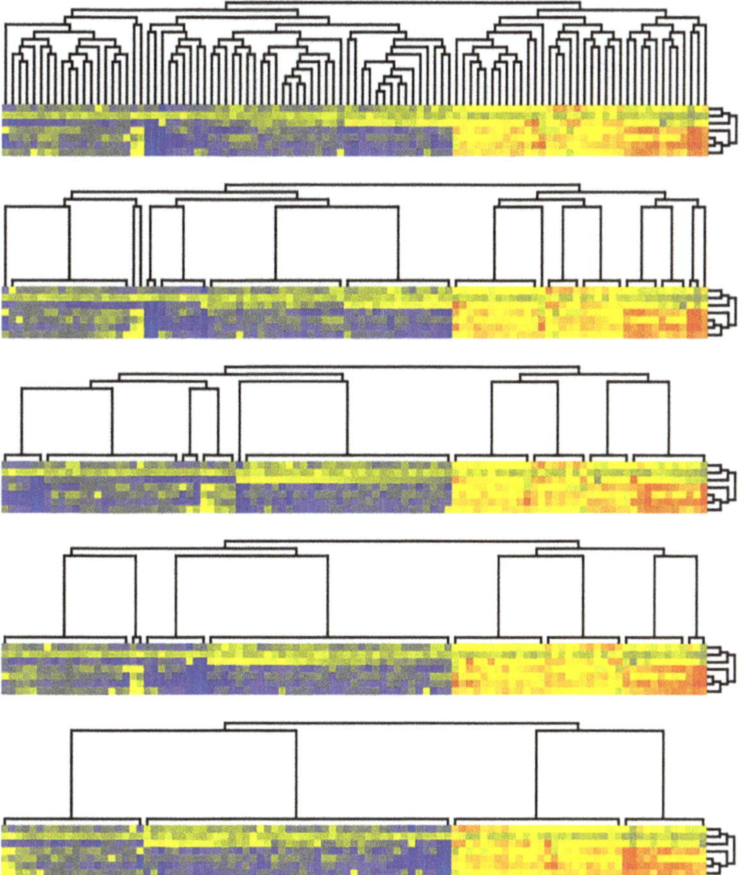

Fig. 16.5 Clustering of 99 genes.

The method is further applied to the clustering of a much larger microarray dataset containing 13,411 genes (from the same source as above). The result is given in Fig. 16.6 showing only a small fragment of the complete dendrogram containing a small subset of the genes.

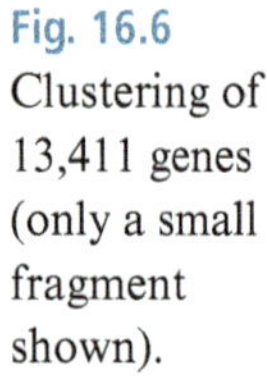

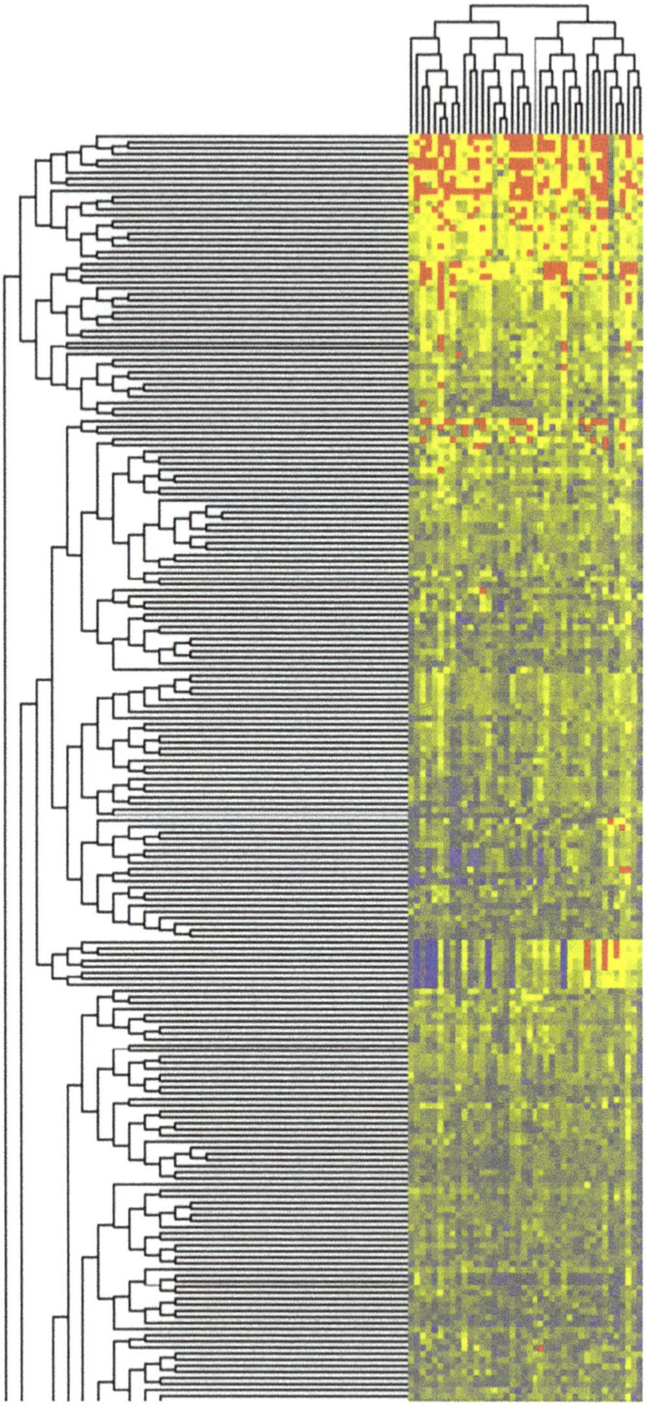

Fig. 16.6 Clustering of 13,411 genes (only a small fragment shown).

PROBLEMS

1. Develop your own code to implement a supervised hierarchical classifier by constructing a binary decision tree based on both top-down and bottom up methods, and then apply both of them to carry out classification of the handwritten digit dataset (10 classes with 256 features). Compare the two methods in terms of both accuracy (from the confusion matrices) and execution time.
2. Develop your own code for unsupervised hierarchical clustering based on both top-down and bottom-up methods, and apply them to the clustering of the handwritten digit dataset. Obtain the dendrograms and confusion matrices from the two methods and compare them in terms of accuracy and execution time.

   Extra credit: apply your clustering code to a microarray dataset of reasonable size downloaded from: https://llmpp.nih.gov/lymphoma/data.shtml.

Part V

Neural Networks

In Part V, we consider the general learning method of *artificial neural networks (ANN)*, a category of algorithms in machine learning that are inspired by the biological neural networks in the brain, composed of large number of nerve cells called neurons. An ANN is designed to carry out either supervised learning such as regression as well as classification, or unsupervised learning such as clustering. The ANN method is covered in Part V, separate from all supervised and unsupervised learning methods in the previous parts, simply because the ANN formulates the learning problems in a rather unique way (as we will see in the following), quite different from all previous methods. However, both supervised and unsupervised ANN algorithms can also be discussed together with all previously considered algorithms, as in the case of many other textbooks.

Before discussing specific learning algorithms and the actual computations taking place in an ANN, we could treat it as a black box that mimics the behavior of a certain generic system, which generates a set of outputs as the response to a set of inputs, in the sense that the output of the ANN is similar to that of the system, if both are given the same input, although the inner workings of the ANN may be completely different from that of the system. In other words, the ANN can be used to model a system only in terms of the relationship between its input and output.

Mimicking the biological neural networks, an ANN is typically a hierarchical structure composed of multiple layers of neurons, also called *nodes* in the context of ANN, including the input layer, the hidden layers, and the output layer. These layers are organized hierarchically in the sense that each layer takes input from a lower layer and generates output to feed into a higher layer. As the computations taking place through the multilayers are in general nonlinear, an ANN can be used to model complex behaviors of some sophisticated systems such as visual signal processing and object recognition in the brain.

To achieve a specific learning goal, an ANN needs to be iteratively trained to learn the desired behavior from the given dataset, so that the *weights* representing the connectivities between the nodes can be progressively modified from some random initial values to eventually produce the desired output given the input. In general, this learning can be either *supervised* or *unsupervised* desired output given the input.

In unsupervised learning, the samples in the dataset X are unlabeled, there exists no desired output for the ANN algorithm to aim at, and it needs to modify its weights on its own based on the internal structures of the dataset, to achieve the desired goal such as clustering, similar to the k-means method.

In supervised learning, each data sample $\mathbf{x}_n$ in $\mathbf{X} = [\mathbf{x}_1, \cdots, \mathbf{x}_N]$ from the training set is labeled by the corresponding $\mathbf{y}_n$ in $\mathbf{Y} = [\mathbf{y}_1, \cdots, \mathbf{y}_N]$, which serves as the target output (i.e. the desired output corresponding to the input x_n), to be compared with the actual output $\hat{y}_{\mathrm{n}}$ generated by the network. Based on the comparison, the weights are adjusted so that is future response to x_n will be closer to the desired y_n, thereby reducing the error or residue $r = y - \hat{y}$. Neural network algorithms formulated in this manner are based on the general learning model illustrated in Fig. 4.1 in Section 4.1, applicable to both regression and classification tasks.

Among all supervised training algorithms, *backpropagation* is of fundamental importance. It is widely used not only as a stand-alone algorithm for tasks such as classification, but also as a core learning component in various deep learning methods, including large-scale artificial intelligence systems such as the large language models discussed in Part VII.

17 Biologically Inspired Networks

17.1 Biological Neural Network Modeling

To understand how such a neural network algorithm works, we first consider some basic concepts in biological neural system. The human brain consists of 10^{11} *neurons* interconnected through about 10^{14} to 10^{15} *synaptic junctions* to form millions of neural networks for various functions. Hundreds of specialized cortical areas are formed based on these networks for different information processing tasks. The book cover shows a color-coded map of one cubic millimeter of human brain tissue, where ~4,000 axons connect to a single neuron, visualizing the brain's complex wiring. The image, from Harvard's Jeff Lichtman lab, was reconstructed with electron microscopy data and visualized using Google Research's AI algorithms.

Functionally, a neuron consists of the following three parts, as shown in Fig. 17.1:

- the *cell body or soma* containing the nucleus of the cell;
- the *dendrites* that receive electrochemical stimulation (input impulses) from other neurons and propagate them to the cell body;
- the *axon* that conducts the impulses (output) away from the cell body to other cells.

For the large number of neurons in a neural network to achieve a certain function collectively, they are connected by the *axo-dendritic synapses* where the dendrites of one neuron receive input stimuli transmitted through the axons of many other neuron. The computational model of such a neuron in the neural network, called a *node*, is also shown in Fig. 17.1, which receives input signals denoted by $x_1, \cdots, x_d$ from d other neurons and becomes activated to different extents in terms of its *action potential* or simply *activation*, as a weighted sum of the inputs

$$a = \sum_{j=1}^{d} w_j x_j + b = \sum_{j=0}^{d} w_j x_j = \mathbf{w}^T \mathbf{x} \tag{17.1}$$

We note that this weighted sum is exactly the same as the linear regression model in Eq. (5.1) in Section 5. However, here the model parameters take on different meanings: b is the offset or bias, and w_j is a weight representing the synaptic connectivity to the jth input node:

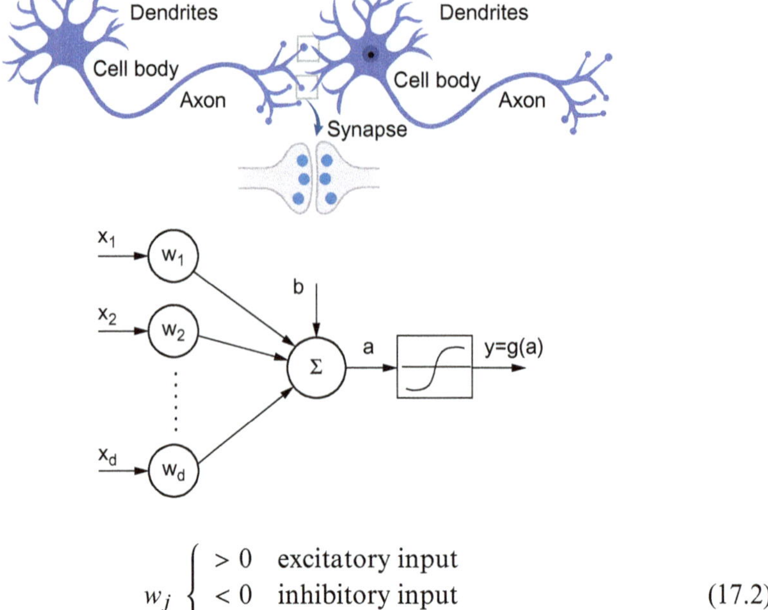

Fig. 17.1 Computational model of neurons.

$$w_j \begin{cases} > 0 & \text{excitatory input} \\ < 0 & \text{inhibitory input} \\ = 0 & \text{no connection} \end{cases} \tag{17.2}$$

As in linear regression, here we have defined $x_0 = 1$ and $w_0 = b$, so that both the weight and pattern vectors are augmented to become $d+1$ dimensional vectors $\mathbf{x} = [x_0 = 1, x_1, \cdots, x_d]^T$ and $\mathbf{w} = [w_0 = b, , w_1, \cdots, w_d]^T$. Based on the activation level a, the neuron produces an output or response as a function of a

$$y = g(a) = g\left(\sum_{j=1}^{d} w_j x_j + b\right) = g(\mathbf{w}^T\mathbf{x}) \tag{17.3}$$

Here $g(x)$, the *activation function*, typically takes one of the following forms:

- Linear function

$$g(x) = x, \qquad \frac{d\,g(x)}{dx} = 1 \tag{17.4}$$

- Logistic function

$$g(x,a) = \frac{1}{1+e^{-ax}} = \frac{e^{ax}}{1+e^{ax}} = \begin{cases} 0 & x = -\infty \\ 1/2 & x = 0 \\ 1 & x = \infty \end{cases}, \qquad \frac{d\,g(x)}{dx} = \frac{a\,e^{-ax}}{(1+e^{-ax})^2} \tag{17.5}$$

where parameter a controls the slope of the transition from 0 to 1 around $x = 0$. Specially

$$g(x,0) = \frac{1}{2}, \qquad \frac{d\,g(x)}{dx} = 0 \tag{17.6}$$

$$g(x,\infty) = u(x) = \begin{cases} 0 & x < 0 \\ 1 & x > 0 \end{cases}, \qquad \frac{d\,g(x)}{dx} = \begin{cases} 0 & x \neq 0 \\ \infty & x = 0 \end{cases} \tag{17.7}$$

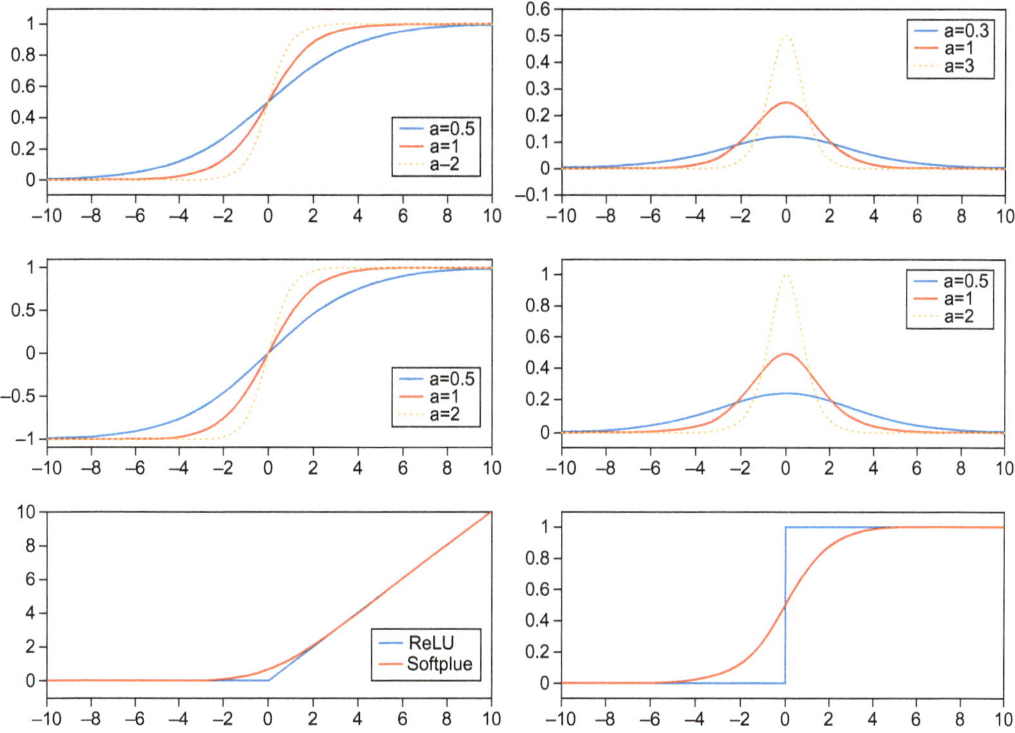

Fig. 17.2 Activation functions (left) and their derivatives (right) logistic (top), Tanh (middle), and ReLU and Softplus (bottom).

- Tanh (hyperbolic tangent) function

$$g(x,a) = \frac{2}{1+e^{-ax}} - 1 = \frac{e^{ax}-1}{e^{ax}+1} = \begin{cases} -1 & x = -\infty \\ 0 & x = 0 \\ 1 & x = \infty \end{cases}, \qquad \frac{d\,g(x)}{dx} = \frac{2a\,e^{-ax}}{(1+e^{-ax})^2} \tag{17.8}$$

- Rectified linear unit (ReLU)

$$g(x) = \max(0, x) = \begin{cases} 0 & x < 0 \\ x & x > 0 \end{cases}, \qquad \frac{d}{dx}g(x) = \begin{cases} 0 & x < 0 \\ 1 & x > 0 \end{cases} \tag{17.9}$$

  The soft version of ReLU is the Softplus function of which the derivative is the logistic function

$$g(x) = \log(1+e^x), \qquad \frac{d\,g(x)}{dx} = \frac{e^x}{1+e^x} = \frac{1}{1+e^{-x}} \tag{17.10}$$

These activation functions and their derivatives are shown in Fig. 17.2.

A neural network can be typically modeled by a hierarchical structure as shown in Fig. 17.3 containing two or more layers of neurons, called *nodes* in the ontext of artificial neural networks:

- the *input layer*: receives inputs from external sources;
- the *output layer*: generates output to the external world;

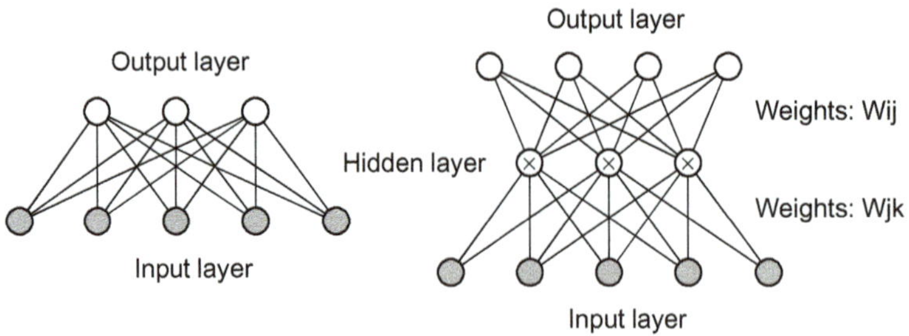

Fig. 17.3 Artificial neural networks of two (left) or three (right) layers.

- the *hidden layer(s)*: one or more layers between the input and output layers, not visible from outside the network.

A neural network can be trained according to certain mathematical rules, the *learning rules* or *learning laws*, by modifying its weights iteratively based on the inputs (and the desired outputs if the learning is supervised), so that the network can produce the desired output as the response to its input as the stimulus.

The method of ANN can be used for both supervised and unsupervised learning. For supervised learning, the goal is to model the relationship between the dependent variable $\mathbf{y}$ as the output of the network and the independent variable $\mathbf{x}$ as the input of the network by a function $\mathbf{y} = f(\mathbf{x}, \mathbf{w})$, of which the parameter $\mathbf{w}$ for all weights in the network is to be estimated in the learning process based on the training set $\mathcal{D} = \{\mathbf{X}, \mathbf{y}\}$. Such a general framework is suitable for both regression and classification. Neural networks can also be used for unsupervised learning based on observed data $\mathbf{X}$ without the labeling $\mathbf{y}$.

Listed here are several learning paradigms for various neural network algorithms with different interpretations of the input and output of a neural network.

1. *Pattern associator*

   This is the most general form of neural network learning by which the associative relationship between two sets of patterns in $\mathbf{X} = [\mathbf{x}_1, \cdots, \mathbf{x}_N]$ and $\mathbf{Y} = [\mathbf{y}_1, \cdots, \mathbf{y}_N]$ is learned and stored.

   - Training: A set of N pairs of patterns $\{(\mathbf{x}_n, \mathbf{y}_n),\ n = 1, \cdots, N\}$ is presented to the network which then learns to establish the associative relationship between two sets of patterns

   $$f : \mathbf{x} \in \mathcal{R}^d \Longrightarrow \mathbf{y} \in \mathcal{R}^m \tag{17.11}$$

   - Testing: When an unlabeled pattern $\mathbf{x}$ in a pair is presented as the input, the network produces an output pattern $\mathbf{y}$ associated to the output.

   Human memory is associative in the sense that given one pattern, some associated pattern(s) may be produced. Examples include: (Evolution, Darwin), (Einstein, $E = mc^2$), (food, sounding bell, salivation).

2. *Auto-associator*

   As a special pattern associator, auto-associator associates a pre-stored pattern to an incomplete or noisy version of the pattern.

- Training: A set of patterns $\{\mathbf{x}_1, \cdots, \mathbf{x}_N\}$ is presented to the network for it to learn and remember, that is, the patterns are stored in the network.
- Testing: When an incomplete or noisy version of one of the patterns stored in the network is presented as the input to the network, the original pattern is retrieved by the network as the output.

3. *Regression*
   This is another special kind of pattern associator which takes a vector input $\mathbf{x} \in \mathcal{R}^d$ and produces a real value $y \in \mathcal{R}$ as a multivariable function $y = f(\mathbf{x})$ at its only output node.
   - Training: trained by a set of observed data samples, the independent vectors and their corresponding function values $\{(\mathbf{x}_n, y_n), n = 1, \cdots, N\}$, the network is model the function.
   - Testing: given any vector $\mathbf{x}$, the output value produced by the single output node is an estimated function value $y = f(\mathbf{x})$.
4. *Classification*
   This is a variation of the pattern associator of which the output patterns are a set of categorical symbols representing different classes $\{C_1, \cdots, C_K\}$, that is, each input pattern is classified by the network into one of the classes

$$f : \mathbf{x} \in \mathcal{R}^d \Longrightarrow y \in \{C_1, \cdots, C_K\} \tag{17.12}$$

5. *Regularity detector*
   This is an unsupervised learning process. The network discovers automatically the regularity in the inputs so that similar patterns are automatically detected and grouped together in the same cluster or class.

17.2 Hebbian Learning

Donald Hebb (1949)[1] speculated that

"When neuron A repeatedly and persistently takes part in exciting neuron B, the synaptic connection from A to B will be strengthened."

In other words, simultaneous activation of neurons leads to pronounced increases in synaptic strength between them, or "neurons that fire together wire together; neurons that fire out of sync, fail to link."

For example, the well-known *classical conditioning* (Pavlov, 1927) could be explained by Hebbian learning. Consider the following three patterns:

- unconditioned stimulus: sight of food F;
- conditioned stimulus: sound of bell B;
- response: secretion of saliva S.

[1] Hebb, D. (1949) *The Organization of Behavior: A Neuropsychological Theory*. Wiley, New York.

The unconditioned response is: $F \to S$. Due to the repeated and persistent conditioning process $F \cap B \to S$, the synaptic connections between patterns B and S are strengthened as both are repeatedly excited simultaneously, that is, the two patterns become associated, resulting in the conditioned response $B \to S$.

Based on this theory of Hebbian learning, the Hebbian network can be considered as a supervised learning method that learns to establish the associative relationship between any pair of two patterns $\mathbf{x}_n$ and $\mathbf{y}_n$ in the given dataset $\{\mathbf{x}_1, \cdots, \mathbf{x}_N\}$ and $\{\mathbf{y}_1, \cdots, \mathbf{y}_N\}$, considered as the training set. This is a two-layer network with d nodes in the input layer to receive an input pattern $\mathbf{x} = [x_1, \cdots, x_d]^T$ and m nodes of the output layer to produce an output $\mathbf{y} = [y_1, \cdots, y_m]^T$. Each output node is fully connected to all d input nodes through its weights and generates an output

$$y_i = \sum_{j=1}^{d} w_{ij} x_j = \mathbf{w}_i^T \mathbf{x} \quad (i = 1, \cdots, m) \tag{17.13}$$

where $\mathbf{w}_i = [w_{i1}, \cdots, w_{id}]^T$ is the weight vector for the ith output node. These m equations can be written in matrix form as

$$\mathbf{y} = \begin{bmatrix} y_1 \\ \vdots \\ y_m \end{bmatrix} = \begin{bmatrix} \mathbf{w}_1^T \\ \vdots \\ \mathbf{w}_m^T \end{bmatrix} \begin{bmatrix} x_1 \\ \vdots \\ x_d \end{bmatrix} = \mathbf{W}\mathbf{x} \tag{17.14}$$

where $\mathbf{W} = [\mathbf{w}_1, \cdots, \mathbf{w}_m]^T$ is an $m \times d$ matrix.

The Hebbian learning rule is inspired by Hebb's theory, that is, when both neurons x_j and y_i are activated, the synaptic connectivity, here the weight w_{ij}, between them is enhanced

$$w_{ij}^{new} = w_{ij}^{old} + \eta\ x_j y_i \quad (i = 1, \cdots, m,\ j = 1, \cdots, d) \tag{17.15}$$

or in matrix form

$$\mathbf{W}^{new} = \mathbf{W}^{old} + \eta\ \mathbf{y}\mathbf{x}^T \tag{17.16}$$

Here η is the *learning rate*, a parameter that controls how quickly the weights get modified.

As in all supervised learning, the Hebbian network is first trained and then used for association.

- **Training**
  For simplicity, we assume all weights are initialized to zero $w_{ij} = 0 \quad (i = 1, \cdots, m,\ j = 1, \cdots, d)$, and then train the network to find all weights based on all N pattern pairs in the dataset $\{(\mathbf{x}_n, \mathbf{y}_n),\ \ n = 1, \cdots, N\}$ based on the learning law

$$w_{ij} = \sum_{n=1}^{N} x_j^{(n)} y_i^{(n)} \quad (i = 1, \cdots, m,\ \ j = 1, \cdots, d) \tag{17.17}$$

or in matrix form, the weight matrix is the sum of the outer-products of all N pairs of patterns

$$\mathbf{W}_{m\times d} = \sum_{n=1}^{N} \mathbf{y}_n \mathbf{x}_n^T = \sum_{n=1}^{N} \begin{bmatrix} y_1^{(n)} \\ \vdots \\ y_m^{(n)} \end{bmatrix} [x_1^{(n)}, \cdots, x_d^{(n)}] \tag{17.18}$$

- **Association**
  When one of the patterns $\mathbf{x}_l$ is presented to the network, it produces an output

$$\mathbf{y} = \mathbf{W}\mathbf{x}_l \tag{17.19}$$

  which, as shown in the following, is the same as the associated pattern $\mathbf{y}_l$, under the following conditions:
  - The N patterns $\{\mathbf{x}_1, \cdots, \mathbf{x}_N\}$ are treated as zero-mean random vectors, and they are assumed to be independent of each other with zero covariance and therefore zero *correlation coefficient* between any $\mathbf{x}_i$ and $\mathbf{x}_j$

$$r_{ij} = \frac{\sigma_{ij}^2}{\sigma_i \sigma_j} = \frac{\sum_{k=1}^{d} x_{ki} x_{kj}}{\sqrt{\sum_{k=1}^{d} x_{ki}^2}\sqrt{\sum_{k=1}^{d} x_{kj}^2}} = \frac{\mathbf{x}_i^T \mathbf{x}_j}{||\mathbf{x}_i|| \, ||\mathbf{x}_j||} = \delta_{ij} = \begin{cases} 1 & i = j \\ 0 & i \neq j \end{cases} \tag{17.20}$$

    If we further assume all patterns are normalized with $||\mathbf{x}_n|| = 1$, then we have $\mathbf{x}_i^T \mathbf{x}_j = \delta_{ij}$
  - The number of input nodes is greater than the number of pattern pairs: $d \geq N$, that is, the capacity of the network is large enough to represent N different patterns (as there can be no more than d orthogonal vectors in a d-dimensional space).

  Under these assumed conditions, the output of the network as its response to input $\mathbf{x}_l$ is

$$\mathbf{y} = \mathbf{W}\mathbf{x}_l = \left(\sum_{n=1}^{N} \mathbf{y}_n \mathbf{x}_n^T \right) \mathbf{x}_l = \mathbf{y}_l (\mathbf{x}_l^T \mathbf{x}_l) + \sum_{n \neq l} \mathbf{y}_n (\mathbf{x}_n^T \mathbf{x}_l) = \mathbf{y}_l \tag{17.21}$$

  as $\mathbf{x}_n^T \mathbf{x}_l = 0$ for all other terms with $n \neq l$. We see that a one-to-one correspondence relationship between $\mathbf{x}_n$ and $\mathbf{y}_n$ has been established for all $n = 1, \cdots, N$.

  In non-ideal cases where these conditions are not strictly satisfied; for example, the patterns in the dataset are not completely independent, the summation term is non-zero, and there will be a non-zero error $\mathbf{y} - \mathbf{y}_l \neq 0$. However, the Hebbian network trained in this way may still behave as an associator that mimics approximately the general associative relationship between $\mathbf{x}_l$ and $\mathbf{y}_l$.

17.3 Hopfield Network

The Hopfield network is a *recurrent neural network (RNN)* that allows the previous outputs of the network to be fed back and used as inputs again (with time delay) as

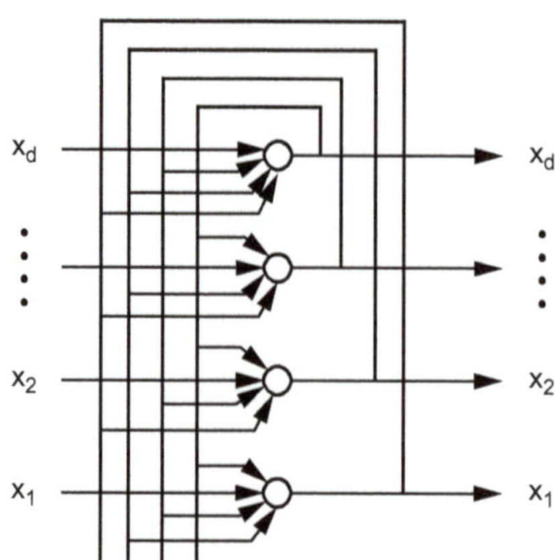

Fig. 17.4 The Hopfield network.

an iterative process, so that it can learn to respond to inputs as a temporal sequence, such as a speech signal.

The structure of a Hopfield network is illustrated in Fig. 17.4. The learning of the Hopfield network is based on the Hebbian learning rule, and as a supervised method, it is trained on a dataset $\mathbf{X} = [\mathbf{x}_1, \cdots, \mathbf{x}_K]$ of K data samples each composed of d binary components $\mathbf{x} = [x_1, \cdots, x_d]^T$ with $x_i \in \{-1, 1\}$, $(i = 1, \cdots, d)$, representing one of K different patterns of interest, used as the inputs to the d nodes of network. Once the network is completely trained, the weight matrix $\mathbf{W} = [\mathbf{w}_1, \cdots, \mathbf{w}_d]$ of the network is obtained in which the K patterns are stored. Given an input $\mathbf{x}$, an iterative computation is carried out until convergence, when one of the pre-stored pattern that most closely resemble the current input $\mathbf{x}$ is produced as the output. The Hopfield network can therefore be considered as an auto-associator (a content addressable memory), by which a pre-stored pattern can be retrieved based on its association to the input pattern established by training process of the network.

- **Training**

  The training process is essentially the same as the Hebbian learning, except here the two associated patterns in each pair are the same (self-association). The weight matrix of the network is obtained as the sum of the outer-products of the K patterns to be stored

$$\mathbf{W}_{d \times d} = \sum_{k=1}^{K} \mathbf{x}_k \mathbf{x}_k^T = \sum_{k=1}^{K} \begin{bmatrix} x_1^{(k)} \\ \vdots \\ x_d^{(k)} \end{bmatrix} [x_1^{(k)}, \cdots, x_d^{(k)}] \tag{17.22}$$

  The weight connecting node i and node j is defined as

$$w_{ij} = \sum_{k=1}^{K} x_i^{(k)} x_j^{(k)} = w_{ji} \tag{17.23}$$

  We assume no self-connection exists $w_{ii} = 0$ $(i = 1, \cdots, d)$

- **Autoassociation**
  Once the weight matrix $\mathbf{W}$ is obtained by the training process, the network can be used as a self-associator. When an input pattern $\mathbf{x}$ is presented to the network, the outputs of the network are updated *iteratively* and *asynchronously*, one randomly selected node at a time

$$x_i^{(n+1)} = \text{sign}\left(\sum_{j=1}^{d} w_{ij} x_j^{(n)}\right) = \begin{cases} +1, & \text{if } \sum_{j=1}^{d} w_{ij} x_j^{(n)} \geq 0 \\ -1, & \text{if } \sum_{j=1}^{d} w_{ij} x_j^{(n)} < 0 \end{cases} \tag{17.24}$$

where $x_i^{(n)}$ and $x_i^{(n+1)}$ are the output x_i of the ith node before and after the nth iteration, respectively. As shown in the following, this iteration will always converge to one of the K pre-stored patterns.

We first define the *Energy function* of any two nodes x_i and x_j of $\mathbf{x}$ as

$$e_{ij} = -w_{ij} x_i x_j \tag{17.25}$$

and the total *energy* of all d nodes in the network as the sum of all pair-wise energies

$$\mathcal{E}(\mathbf{x}) = \frac{1}{2}\sum_{i=1}^{d}\sum_{j=1}^{d} e_{ij} = -\frac{1}{2}\sum_{i=1}^{d}\sum_{j=1}^{d} w_{ij} x_i x_j = -\frac{1}{2}\mathbf{x}^T \mathbf{W} \mathbf{x} \tag{17.26}$$

The interaction between these two nodes is summarized in Table 17.1.

Table 17.1

	x_j	x_i	$w_{ij} > 0$	$w_{ij} < 0$
1	−1	−1	$e_{ij} < 0$	$e_{ij} > 0$
2	−1	1	$e_{ij} > 0$	$e_{ij} < 0$
3	1	−1	$e_{ij} > 0$	$e_{ij} < 0$
4	1	1	$e_{ij} < 0$	$e_{ij} > 0$

We make two observations.

- When the two nodes i and j reinforce each other's state, $e_{ij} < 0$:
  - If $w_{ij} > 0$ (in cases 1 and 4), $x_j = \mp 1$ tends to keep x_i to stay at the same state ∓ 1 in the iteration.
  - If $w_{ij} < 0$ (in cases 2 and 3), $x_j = \mp 1$ tends to keep x_i to stay at the same state ± 1.
- When the two nodes i and j change each other's state, $e_{ij} > 0$:
  - If $w_{ij} < 0$ (in cases 1 and 4), $x_j = \mp 1$ tends to reverse x_i from its previous state ∓ 1 to ± 1.
  - If $w_{ij} > 0$ (in cases 2 and 3), $x_j = \mp 1$ tends to reverse x_i from its previous state ± 1 to ∓ 1.

We note that low energy e_{ij} corresponds to a stable interaction between x_i and x_j, that is, they tend to remain unchanged, and high energy corresponds to an unstable interaction, that is, they tend to change their states. As the result, low total $\mathcal{E}(\mathbf{x})$

corresponds to more stable condition of the network, while high $\mathcal{E}(\mathbf{x})$ corresponds to less stable condition.

We further show that the total energy $\mathcal{E}(\mathbf{x})$ always decreases whenever the state of any node changes. Assume x_k has just been changed, that is, $x_k^{(n+1)} \neq x_k^{(n)}$ ($x_k = \pm 1$ but $x_k^{(n+1)} = \mp 1$), while all others remain the same $x_{l \neq k}^{(n+1)} = x_{l \neq k}^{(n)}$. The energy before x_k changes state is

$$\begin{aligned}\mathcal{E}^{(n)}(\mathbf{x}) &= -\frac{1}{2}\left[\sum_{i \neq k}\sum_{j \neq k} w_{ij} x_i^{(n)} x_j^{(n)} + \sum_i w_{ik} x_i^{(n)} x_k^{(n)} + \sum_j w_{kj} x_k^{(n)} x_j^{(n)}\right] \\ &= -\frac{1}{2}\sum_{i \neq k}\sum_{j \neq k} w_{ij} x_i x_j - \sum_i w_{ik} x_i^{(n)} x_k^{(n)}\end{aligned}$$

and the energy after x_k changes state is

$$\mathcal{E}^{(n+1)}(\mathbf{x}) = -\frac{1}{2}\sum_{i \neq k}\sum_{j \neq k} w_{ij} x_i x_j - \sum_i w_{ik} x_i^{(n+1)} x_k^{(n+1)} \tag{17.27}$$

The energy difference is

$$\triangle\mathcal{E} = (x_k^{(n)} - x_k^{(n+1)}) \sum_i w_{ik} x_i \tag{17.28}$$

Consider two cases.

- Case 1: if $x_k^{(n)} = -1$, but $\sum_i w_{ik} x_i \geq 0$ and $x_k^{(n+1)} = 1$, we have $(x_k^{(n)} - x_k^{(n+1)}) = -2 \leq 0$ and $\triangle\mathcal{E} \leq 0$.
- Case 2: if $x_k^{(n)} = 1$, but $\sum_i w_{ik} x_i < 0$ and $x_k^{(n+1)} = -1$, we have $(x_k^{(n)} - x_k^{(n+1)}) = 2 \geq 0$ and $\triangle\mathcal{E} \leq 0$.

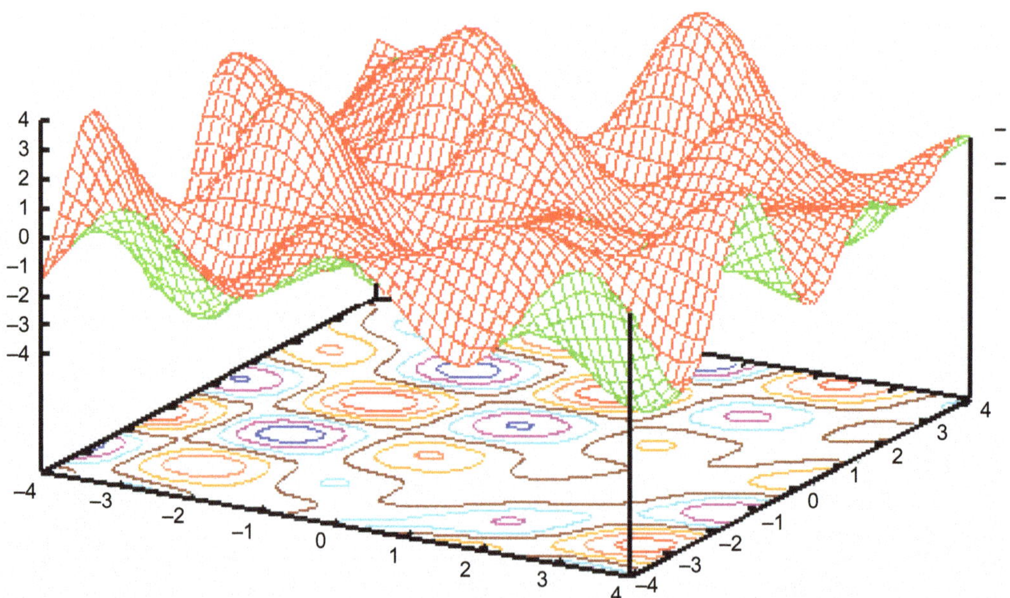

Fig. 17.5 Energy landscape.

As in either case, $\triangle\mathcal{E}(\mathbf{x}) \leq 0$ is always true throughout the iteration, the energy $\mathcal{E}(\mathbf{x})$ will eventually reach one of the minima of the *energy landscape* shown in Fig. 17.5, that is, the iteration will always converge.

We further show that each of the K pre-stored patterns corresponds to one of the minima of the energy function. The energy function $\mathcal{E}(\mathbf{x})$ corresponding to any $\mathbf{x}$ can be written as

$$\mathcal{E}(\mathbf{x}) = -\frac{1}{2}\mathbf{x}^T\mathbf{W}\mathbf{x} = -\frac{1}{2}\mathbf{x}^T\left[\sum_{k=1}^{K}\mathbf{x}_k\mathbf{x}_k^T\right]\mathbf{x} = -\frac{1}{2}\sum_{k=1}^{K}\mathbf{x}^T\mathbf{x}_k\mathbf{x}_k^T\mathbf{x} = -\frac{1}{2}\sum_{k=1}^{K}(\mathbf{x}^T\mathbf{x}_k)^2 \tag{17.29}$$

If $\mathbf{x}$ is different from (ideally orthogonal to) any of the K stored patterns, all terms in the summation will be small (ideally zero). But if $\mathbf{x}$ is similar to (ideally the same as) any one of the stored patterns, their inner product will be large (ideally reach maximum), causing the total energy to be minimized to reach one of the minima. In other words, the patterns stored in the network correspond to the local minima of the energy function. that is these patterns become *attractors*.

Note that it is possible to have other local minima, called *spurious states*, which do not represent any of the stored patterns, that is, the associative memory is not perfect.

18 Perceptron-Based Networks

18.1 Perceptron

The perceptron network (Frank. Rosenblatt, 1958)[1] is a two-layer network containing an input layer of d nodes and an output layer of m output nodes. As a supervised method, it is trained by a dataset $\mathbf{X} = [\mathbf{x}_1, \cdots, \mathbf{x}_N]$, of which each sample $\mathbf{x} = [x_1, \cdots, x_d]^T$ is labeled by the corresponding component in $\mathbf{y} = [y_1, \cdots, y_N]^T$, indicating to which one of K categorical classes $\mathbf{x}_n$ belongs. The goal of the training process is to determine the weights $\mathbf{w} = [w_1, \cdots, w_d]^T$ and bias b so that the output $\hat{y}_n = \sum_{i=1}^d w_i x_n + b$ corresponding to input $\mathbf{x}_n$ matches its labeling y_n for all N samples in the training set in some optimal way. When the perceptron network is fully trained, it can be used to classify any unlabeled sample $\mathbf{x}$ into one of the K classes.

We first consider the special case where the output layer has only $m = 1$ node, and the perceptron is a binary classifier, as with the method of linear regression, and also the initial setup of the support vector machine (SVM), in the sense that all such methods are based on the linear equation $f(\mathbf{x}) = \mathbf{w}^T\mathbf{x} + b = 0$, representing a hyperplane by which the d-dimensional feature space is partitioned into two regions corresponding to two classes C_+ and C_-. When the parameters $\mathbf{w}$ and b are determined in the training process, any unlabeled $\mathbf{x}$ is classified into either of the two classes depending on whether the function value $\hat{y} = f(\mathbf{x})$ is greater or smaller than zero

$$\text{If} \quad f(\mathbf{x}) = \mathbf{w}^T\mathbf{x} + b \begin{cases} > 0 \\ < 0 \end{cases}, \quad \text{then} \quad \hat{y} = \text{sign}(f(\mathbf{x})) = \begin{cases} 1 & \text{for } \mathbf{x} \in C_+ \\ -1 & \text{for } \mathbf{x} \in C_- \end{cases} \tag{18.1}$$

If we divide the equation $f(\mathbf{x}) = \mathbf{w}^T\mathbf{x} + b = 0$ by $||\mathbf{w}||$, we get

$$p_{\mathbf{w}}(\mathbf{x}) = \frac{\mathbf{w}^T\mathbf{x}}{||\mathbf{w}||} = -\frac{b}{||\mathbf{w}||} = b' \tag{18.2}$$

where $p_{\mathbf{w}}(\mathbf{x}) = \mathbf{w}^T\mathbf{x}/||\mathbf{w}||$ is the projection of $\mathbf{x}$ onto the normal direction $\mathbf{w}$ of the partitioning hyperplane, and $b' = -b/||\mathbf{w}||$ is the vector from the origin to the hyperplane ($|b'|$ is the distance of the hyperplane to the origin). Now this classification above can be rewritten as

[1] Rosenblatt, F. (1958) The perceptron: a probabilistic model for information storage and organization. *Psychological Review* 65 (6): 386–408.

$$\text{If } p_{\mathbf{w}}(\mathbf{x}) \left\{ \begin{array}{l} > b' \\ < b' \end{array} \right., \quad \text{then} \quad \mathbf{x} \in \left\{ \begin{array}{l} C_+ \\ C_- \end{array} \right. \tag{18.3}$$

We see that in all binary classification methods, an unlabeled sample $\mathbf{x}$ is classified into either of the two classes based on the projection $p_{\mathbf{w}}(\mathbf{x})$ of $\mathbf{x}$ onto $\mathbf{w}$, which is either greater or smaller than the bias b', indicating $\mathbf{x}$ is on either the positive or negative side of the plane.

In all these binary classification methods, the parameters $\mathbf{w}$ and b need to be determined based on training set $\mathbf{X}$ labeled by $\mathbf{y}$, but they do so differently. While in the least squares method and support vector machine these parameters are obtained by the least squared method and quadratic programming, respectively, here in the perceptron network these parameters are obtained by iteratively modifying some randomly initialized values to gradually reduce the error or residual, the difference between actual output $\hat{y}_n = f(\mathbf{x}_n)$ and the ground truth labeling y_n, denoted by $\delta = y_n - \hat{y}_n$, for all training samples in the dataset. This method is called the δ-rule.

We now consider specifically the training algorithm of the perceptron network as a binary classifier. As always, we redefine the data vector as $\mathbf{x} = [x_0 = 1, x_1, \cdots, x_n]^T$ and the weight vector as $\mathbf{w} = [w_0 = b, w_1, \cdots, w_n]^T$ so that the decision function can be more conveniently written as an inner product $f(\mathbf{x}) = \mathbf{w}^T\mathbf{x}$ without the additional bias term.

The randomly initialized weight vector $\mathbf{w}$ is iteratively updated based on the following error driven δ-rule

$$\mathbf{w}^{new} = \mathbf{w}^{old} + \eta(y - \hat{y})\,\mathbf{x} = \mathbf{w}^{old} + \eta\,\delta\mathbf{x} \tag{18.4}$$

where $\eta > 0$ is the step size, here called the *learning rate*, which is assumed to be 1 in the following for simplicity. We can show that by this iteration, $\mathbf{w}$ is modified in such a way that the error δ is always reduced.

When a training sample $\mathbf{x}$ labeled by $y \pm 1$ is presented to the nodes of the input layer of the perceptron, its output $\hat{y} = \text{sign}(f(\mathbf{x}))$ may or may not match the label y, as shown Table 18.1.

Table 18.1

	y	$f(\mathbf{x}) = \mathbf{w}^T\mathbf{x}$, $\hat{y} = \text{sign}(f(\mathbf{x}))$	$\delta = y - \hat{y}$
1	$y = 1$	$f(\mathbf{x}) > 0$, $\hat{y} = 1$	$\delta = 0$
2	$y = -1$	$f(\mathbf{x}) > 0$, $\hat{y} = 1$	$\delta = -2$
3	$y = 1$	$f(\mathbf{x}) < 0$, $\hat{y} = -1$	$\delta = 2$
4	$y = -1$	$f(\mathbf{x}) < 0$, $\hat{y} = -1$	$\delta = 0$

In both the first and last cases, $yf(\mathbf{x}) = 1 > 0$, the error is $\delta = y - \hat{y} = 0$, and the weight vector $\mathbf{w}^{new} = \mathbf{w}^{old} + \delta\mathbf{x} = \mathbf{w}^{old}$ is not modified. But in cases 2 and 3, $yf(\mathbf{x}) = -1 < 0$, the error is $\delta = y - \hat{y} \neq 0$, the weight vector $\mathbf{w}$ is to be modified in either of the following two ways.

- In case 2, $y = -1$, but $\hat{y} = 1$, then $\delta = -2$ and $y\, f(\mathbf{x}) < 0$, the weight vector is update by

$$\mathbf{w}^{new} = \mathbf{w}^{old} - 2\mathbf{x} \tag{18.5}$$

When the same $\mathbf{x}$ is presented to the network again in the future, the function is

$$f(\mathbf{x}) = \mathbf{x}^T \mathbf{w}^{new} = \mathbf{x}^T \mathbf{w}^{old} - \eta 2\mathbf{x}^T \mathbf{x} < \mathbf{x}^T \mathbf{w}^{old} \tag{18.6}$$

which is smaller than its previous value and the output $\hat{y}$ is more likely to be the same as the desired $y = -1$.

- In case 3, $y = 1$, but $\hat{y} = -1$, then $\delta = 2$ and $y\, f(\mathbf{x}) < 0$, the weight vector is update by

$$\mathbf{w}^{new} = \mathbf{w}^{old} + 2\mathbf{x} \tag{18.7}$$

When the same $\mathbf{x}$ is presented again, the function is

$$f(\mathbf{x}) = \mathbf{x}^T \mathbf{w}^{new} = \mathbf{x}^T \mathbf{x}^{old} + \eta 2\mathbf{x}^T \mathbf{x} > \mathbf{x}^T \mathbf{w}^{old} \tag{18.8}$$

which is greater than its previous value and the output $\hat{y}$ is more likely to be the same as the desired $y = 1$.

The updated equations in Eq. (18.5) for $y = -1$ and Eq. (18.7) for $y = -1$ can be combined to become

$$\mathbf{w}^{new} = \mathbf{w}^{old} + y\mathbf{x} \tag{18.9}$$

where the scaling constant 2 is dropped as it can be absorbed into the learning rate if we let $\eta = 1/2$. Now the learning rule can be rewritten as

$$\text{If} \quad y\, f(\mathbf{x}) \begin{cases} > 0 \\ < 0 \end{cases} \quad \text{then} \quad \begin{cases} \mathbf{w}^{new} = \mathbf{w}^{old} \\ \mathbf{w}^{new} = \mathbf{w}^{old} + y\,\mathbf{x} \end{cases} \tag{18.10}$$

In summary, the learning law guarantees that the weight vector $\mathbf{w}$ is modified in such a way that the performance of the network is always improved with reduced error $\delta = y - \hat{y}$. If C_+ and C_- are linearly separable, then a perceptron will always produce $\mathbf{w}$ in a finite number of iterations to separate them.

This binary classifier with $m = 1$ output node can be generalized to a multi-class classifier with $m > 1$ output nodes, and each of the m weight vectors in $\mathbf{W} = [\mathbf{w}_1, \cdots, \mathbf{w}_m]$ is modified by the same learning process previously considered. The $K > 2$ classes can be encoded by the m outputs $y_i \in \{-1, 1\}$, $(i = 1, \cdots, m)$ in two different ways. If the *one-hot* method is used, the m binary output can encode $K = m$ classes, that is, the kth class is represented by an m-bit output with $y_k = 1$ while all others $y_l = -1$ for $l \neq k$. Alternatively, if binary encoding is used, the m outputs can encode as many as $K = 2^m$ classes. For example, $K = 4$ classes can be labeled by four binary vector of either four or two bits:

Binary output	$\mathbf{y}_0$	$\mathbf{y}_1$	$\mathbf{y}_2$	$\mathbf{y}_3$
y_1	1	−1	−1	−1
y_2	−1	1	−1	−1
y_3	−1	−1	1	−1
y_4	−1	−1	−1	1

or

Binary output	$\mathbf{y}_0$	$\mathbf{y}_1$	$\mathbf{y}_2$	$\mathbf{y}_3$
y_1	−1	−1	1	1
y_2	−1	1	−1	1

(18.11)

The outputs of the m output nodes form an m-dimensional binary vector $\hat{\mathbf{y}} = [\hat{y}_1, \cdots, \hat{y}_m]^T$, which is to be compared with the labeling $\mathbf{y}$ for the current input $\mathbf{x}$ with error $\delta = ||\mathbf{y} - \hat{\mathbf{y}}||$. When the training is complete, any unlabeled input $\mathbf{x}$ is classified to one of the K classes with a matching label to the perceptron's output. In the case of one-hot encoding, it is possible for the binary output $\mathbf{y}$ to have more than one-hot node (e.g., $\hat{\mathbf{y}} = [-1\ 1\ -1\ 1]^T$) not match any of the K one-hot encoded classes. In this case, the input $\mathbf{x}$ can be classified to the class corresponding to the node with the greatest output value $f(\mathbf{x}) = \mathbf{w}^T\mathbf{x}$.

The MATLAB code for the essential part of the algorithm is listed here. $\mathbf{X} = [\mathbf{x}_1, \cdots, \mathbf{x}_N]$ is the dataset contains N training samples, each labeled by one of the components of $\mathbf{y} = [y_1, \cdots, y_n]$ as its class identity. $\mathbf{W}$ is a $(d+1) \times m$ matrix containing m $d+1$ dimensional weight vectors each for one of the m output nodes.

```
[X Y]=DataOneHot;         % get data
K=length(unique(Y','rows')) % number of classes
X=[ones(1,N); X];         % data augmentation
[d N]=size(X);            % numbers of dimensions and samples
m=size(Y,1);              % number of output nodes
W=2*rand(d,m)-1;          % random initialization of weights
eta=1;
nt=10^4;                  % maximum number of iteration
for it=1:nt
    n=randi([1 N]);       % random index
    x=X(:,n);             % pick a training sample x
    y=Y(:,n);             % label of x
    yhat=sign(W'*x);      % binary output
    delta=y-yhat;         % error = desired - actual
    for i=1:m             % for each of m output nodes
        W(:,i)=W(:,i)+eta*delta(i)*x;
                          % update weights for ith output node
    end
    if ~mod(it,N)         % test for every epoch
        er=test(X,Y,W);
        if er<10^(-9)
            break
```

```
        end
    end
end
```

This is the function that tests the training set based on estimated weight vectors in **W**:

```
function er=test(X,Y,W)  % test based on estimated W,
    [d N]=size(X);
    Ne=0;                % number of misclassifications
    for n=1:N
        x=X(:,n);
        yhat=sign(W'*x);
        delta=Y(:,n)-yhat;
        if any(delta)    % misclassification in some output nodes
            Ne=Ne+1;     % update number of misclassifications
        end
    end
    er=Ne/N;             % error percentage
end
```

This is the code that generates the training set labeled by either one-hot or binary encoding method:

```
function [X,Y]=Data
    d=3;  K=8;
    onehot=input('One-hot (1) or binary encoding (0): ');
    Means=[-1 -1 -1 -1 1 1 1 1;
           -1 -1 1 1 -1 -1 1 1;
           -1 1 -1 1 -1 1 -1 1];
    Nk=50*ones(1,K);
    N=sum(Nk);          % total number of samples
    X=[];
    Y=[];
    s=0.2;              % variance of noise
    for k=1:K           % for each of K classes
        Xk=Means(:,k)+s*randn(d,Nk(k));
        if onehot
            Yk=-ones(K,Nk(k));
            Yk(k,:)=1;
        else            % binary encoding
            dy=ceil(log2(K));
            y=2*de2bi(k-1,dy)-1;
            Yk=repmat(y',1,Nk(k));
         end
```

```
            X=[X Xk];
            Y=[Y Yk];
        end
        Visualize(X,Y)
    end
```

Example 18.1

Figure 18.1 shows the classification results of a perceptron network with $d = 2$ input nodes for a 2-D dataset and $m = 2$ output nodes that encode $2^2 = 4$ classes. The training set contains $N = 400$ data samples including 100 samples for each of the four classes labeled by $\mathbf{y} = [y_1, y_2]$. The two weight vectors $\mathbf{w}_1$ and $\mathbf{w}_2$ are initialized randomly. During the training iteration, the weight vectors are modified if $\delta = y - \hat{y} \neq 0$, otherwise they remain the same. After nine such modifications, the four classes are completely separated by the two straight lines normal to the two weight vectors. The first panel shows the initial stage, while the subsequent panels show how the weight vectors are modified each time when $\delta \neq 0$. The darker and bigger dots represent the samples presented to the network when one of the weight vectors is modified.

Figure 18.2 shows the classification results of a perceptron of $d = 3$ input nodes and $m = 3$ output nodes that encode $2^3 = 8$ classes. After 35 modifications of the weight vectors, the perceptron is completely trained to classify all eight classes correctly. The first panel shows the initial stage, the following three panels show the weight vectors $\mathbf{w}_1$, $\mathbf{w}_2$, and $\mathbf{w}_3$, also the normal vectors of the decision planes, from some three different viewing angles. We see that the eight classes are indeed separated by the three planes normal to the weight vectors.

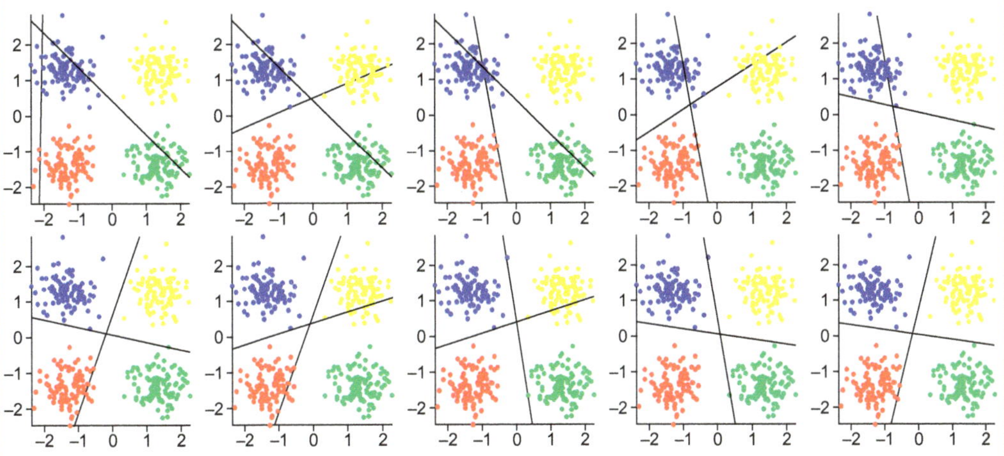

Fig. 18.1 Classification by perceptron (2-D).

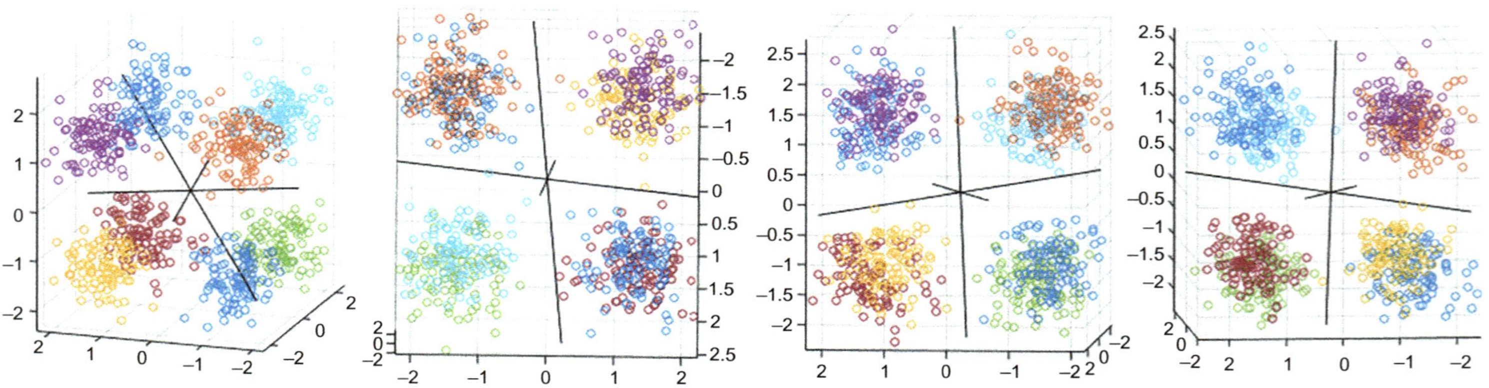

Fig. 18.2 Classification by perceptron (3-D).

The perceptron algorithm as a binary classifier is based on the assumption that the two classes are linearly separable. This limitation can be removed by the kernel method, if the algorithm is modified in such a way that all data samples appear only in the form of an inner product. Consider first the training process in which the N training samples are repeatedly presented to the network, and the weight vector is modified to become $\mathbf{w}^{new} = \mathbf{w}^{old} + y_n \mathbf{x}_n$ whenever a sample $\mathbf{x}_n$ labeled by y_n is misclassified with $\delta = y_n - \hat{y}_n \neq 0$. If the weight vector is initialized to zero $\mathbf{w} = 0$, then the weight vector by the updating rule in Eq. (18.9) can be written as a linear combination of the N training samples

$$\mathbf{w} = \sum_{n=1}^{N} \alpha_n y_n \mathbf{x}_n \tag{18.12}$$

where α_n is the number of times sample $\mathbf{x}_n$ labeled by y_n is misclassified. Upon receiving a new training sample $\mathbf{x}_l$ labeled by y_l, we have

$$f(\mathbf{x}_l) = \mathbf{w}^T \mathbf{x}_l = \sum_{n=1}^{N} \alpha_n y_n \mathbf{x}_n^T \mathbf{x}_l, \quad \text{and} \quad \hat{y}_l = \text{sign}(f(\mathbf{x}_l)) \tag{18.13}$$

and the weight vector $\mathbf{w}$ is updated by the δ

$$\begin{aligned} \text{If} \quad & \delta = y_l - \hat{y}_l \neq 0 \\ \text{then} \quad \mathbf{w}^{new} & = \mathbf{w}^{old} + y_l\, \mathbf{x}_l = \sum_{n=1}^{N} \alpha_n y_n \mathbf{x}_n + y_l\, \mathbf{x}_l \\ \text{i.e.,} \quad \alpha_l^{new} & = \alpha_l^{old} + 1 \end{aligned} \tag{18.14}$$

We see that only $\{\alpha_1, \cdots, \alpha_N\}$ need to be updated during the training process, while the weight vector $\mathbf{w}$ in Eq. (18.12) no longer needs to be explicitly calculated. Once the training process is complete, any unlabeled $\mathbf{x}$ can be classified into either of the two classes based on $\{\alpha_1, \cdots, \alpha_N\}$

$$\text{If} \quad \mathbf{w}^T \mathbf{x} = \sum_{n=1}^{N} \alpha_n y_n \mathbf{x}_n^T \mathbf{x} \begin{cases} > 0 \\ < 0 \end{cases}, \quad \text{then} \quad \mathbf{x} \in \begin{cases} C_- \\ C_+ \end{cases} \tag{18.15}$$

As all data samples appear in the form of inner product in both the training and testing phases, the kernel method can be applied to replace the inner product $\mathbf{x}_n^T \mathbf{x}_m$ by a kernel function $k(\mathbf{x}_n, \mathbf{x}_m)$. Also, the discussion for $m = 1$ output nodes can be generalized to $m > 1$ output nodes.

Here is the MATLAB code segment for the most essential parts of the kernel perceptron algorithm:

```
[X Y]=Data;                   % get dataset
[d N]=size(X);                % numbers of dimensions and samples
X=[ones(1,N); X];             % augmented data
m=size(Y,1);                  % number of output nodes
A=zeros(m,N);                 % initial alphas for all nodes and samples
K=Kernel(X,X);                % get kernel matrix of all N samples
for it=1:nt
```

```
        n=randi([1 N]);          % random index
        x=X(:,n);                % pick a training sample
        y=Y(:,n);                % and its label
        yhat=sign((A.*Y)*K(:,n));   % get yhat
        delta=Y(:,n)-yhat;       % error = desired - actual
        for i=1:m                % for each output node
           if delta(i)~=0        % if a misclassification
               A(i,n)=A(i,n)+1;  % update the corresponding alpha
           end
        end
        if ~mod(it,N)            % test for every epoch
            er=test(X,Y,A);      % percentage of misclassification
            if er<10^(-9)
                break
            end
        end
    end

    function er=test(X,Y,A)       % function for testing
        [d N]=size(X);            % dimension, number of samples
        m=size(Y,1);
        Ne=0;                     % number of misclassifications
        for n=1:N                 % for all N training samples
            x=X(:,n);             % get the nth sample
            y=Y(:,n);             % and its label
            f=(A.*Y)*Kernel(X,x); % f(x)
            yhat=sign(f);         % yhat=sign(f(x))
            if norm(y-yhat)~=0    % misclassification at some output nodes
                Ne=Ne+1;          % increase number of misclassification
            end
        end
        er=Ne/N;                  % percentage of misclassification
    end
```

Example 18.2

The kernel perceptron is applied to the classification of the iris dataset with $m = 3$ one-hot output nodes for the three classes, based on the radial basis kernel. The network is trained on half of the data and tested on the other half. The confusion matrix is shown here and the error rate is 14.67%.

$$\begin{bmatrix} 22 & 6 & 0 \\ 0 & 18 & 5 \\ 0 & 0 & 24 \end{bmatrix}$$

Example 18.3

The kernel perceptron is applied to the handwritten digit dataset with $m = 10$ one-hot output nodes, based on the radial basis kernel. The kernel perceptron is trained on half of the data, and then tested on both the training half and the testing half of the data, and the confusion matrices are shown here and the corresponding error rates are zero when tested on the training half of the data and 0.11 when tested on the testing half. We see that this classifier trends to overfit as its performance degrades when applied to half of the dataset not used in training. This issue will be addressed in yhr backpropagation network to be considered next.

$$\begin{bmatrix} 116 & 0 & 0 & 0 & 0 & 0 & 0 & 0 & 0 & 0 \\ 0 & 108 & 0 & 0 & 0 & 0 & 0 & 0 & 0 & 0 \\ 0 & 0 & 108 & 0 & 0 & 0 & 0 & 0 & 0 & 0 \\ 0 & 0 & 0 & 116 & 0 & 0 & 0 & 0 & 0 & 0 \\ 0 & 0 & 0 & 0 & 113 & 0 & 0 & 0 & 0 & 0 \\ 0 & 0 & 0 & 0 & 0 & 122 & 0 & 0 & 0 & 0 \\ 0 & 0 & 0 & 0 & 0 & 0 & 103 & 0 & 0 & 0 \\ 0 & 0 & 0 & 0 & 0 & 0 & 0 & 108 & 0 & 0 \\ 0 & 0 & 0 & 0 & 0 & 0 & 0 & 0 & 109 & 0 \\ 0 & 0 & 0 & 0 & 0 & 0 & 0 & 0 & 0 & 116 \end{bmatrix}$$

$$\begin{bmatrix} 104 & 7 & 0 & 0 & 0 & 1 & 2 & 1 & 0 & 1 \\ 0 & 103 & 0 & 0 & 0 & 0 & 3 & 1 & 0 & 1 \\ 1 & 0 & 103 & 0 & 0 & 0 & 1 & 1 & 2 & 0 \\ 0 & 0 & 8 & 102 & 0 & 0 & 0 & 2 & 1 & 3 \\ 1 & 1 & 0 & 0 & 103 & 2 & 0 & 0 & 4 & 2 \\ 0 & 0 & 0 & 2 & 0 & 93 & 24 & 0 & 3 & 1 \\ 0 & 1 & 1 & 0 & 0 & 0 & 95 & 4 & 1 & 1 \\ 0 & 0 & 1 & 0 & 0 & 0 & 0 & 104 & 0 & 3 \\ 1 & 0 & 1 & 1 & 3 & 2 & 1 & 3 & 93 & 4 \\ 1 & 2 & 0 & 1 & 7 & 0 & 0 & 3 & 6 & 96 \end{bmatrix}$$

The capability of the perceptron algorithm is limited in the sense that there is only one level of learning taking place between the output and input layers. This situation can be much improved by including additional learning layers between the input and output layers to form a multilayer network, such as the backpropagation which has been most widely used as a supervised learning algorithm, to be discussed in the next section.

18.2 BackPropagation

The backpropagation network (BP network or BPN) is a supervised learning algorithm that finds a wide variety of applications in practice. In the most general sense, a BPN can be used as an associator that learns the relationship between two sets of patterns represented in vector forms. Specifically in classification, similar to the perceptron network, the BPN as a classifier is based on the training set $\mathbf{X} = [\mathbf{x}_1, \cdots, \mathbf{x}_N]$, of which each pattern $\mathbf{x}_n$ is associated with the corresponding pattern, an m-dimensional vector $\mathbf{y}_n$ in $\mathbf{Y} = [\mathbf{y}_1, \cdots, \mathbf{y}_N]$, as its class identity labeling, indicating to which of the K classes $\{C_1, \cdots, C_K\}$ each pattern belongs.

Different from the perceptron in which there is only one level of learning taking place between the output and input layers in terms of the weights, a BPN is a multilayer (three or more) hierarchical structure composed of the input layer, one or more hidden layers, and the output layer, in which learning takes place in multiple levels between two consecutive layers. Consequently, a BPN can therefore be more flexible and powerful than the two-layer perceptron network, and more capable of learning more complicated and patterns.

In the following, before we consider the general BPN containing multiple hidden layers, we will first derive the backpropagation algorithm for the simplest BPN with only one hidden layer in between the input and output layers, as shown in Fig. 18.3. We note that there are $d + 1$ input nodes for input $\mathbf{x} = [x_1, \cdots, x_d]^T$ and a constant $x_0 = 1$, and $l + 1$ hidden nodes that generate $\mathbf{z} = [z_1, \cdots, z_l]^T$ and a constant $z_0 = 1$, while the output layer simply contains m nodes for the output $\mathbf{y} = [y_1, \cdots, y_m]^T$.

Each node in the hidden (excluding $z_0 = 1$) and output layers is fully connected to all nodes in the previous layer. When one of the N training patterns $\mathbf{x}_n$ is presented to the input layer of the BPN, an m–D vector $\hat{\mathbf{y}}_n = f(\mathbf{x}_n, \mathbf{W}^h, \mathbf{W}^o)$ is produced at the output layer as the corresponding response to the input $\mathbf{x}_n$. Here $\mathbf{W}^h = [\mathbf{w}_1^h, \cdots, \mathbf{w}_l^h]$ (a $(d + 1) \times l$ matrix) and $\mathbf{W}^o = [\mathbf{w}_1^o, \cdots, \mathbf{w}_m^o]$ (a $(l + 1) \times m$ matrix) are the model parameters containing the augmented weight vectors for both the hidden and output layers, to be determined in the training process based on the training set $\mathbf{X}$ and corresponding labeling $\mathbf{Y}$, so that its output $\hat{\mathbf{y}}_n$ as a function of the current input $\mathbf{x}_n$ matches the desired output or the target, the labeling $\mathbf{y}_n$ as closely as possible. Once the BPN is fully trained, any unlabeled pattern $\mathbf{x}$ can be classified into one of the K classes corresponding to the minimum $\delta = ||\mathbf{y} - \hat{\mathbf{y}}||$. Note that different from the perceptron network, the output of the m output nodes take real values in general, although they may still be binary (such as one-hot or binary encoding for class labeling) if so desired.

Specifically, the training of the BPN is an iteration of the following two-phase process:

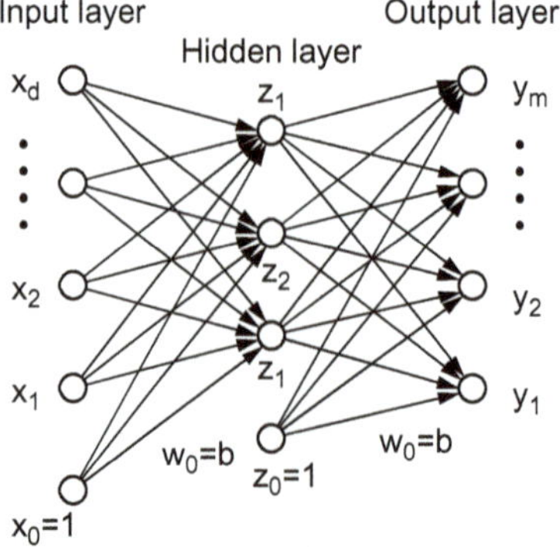

Fig. 18.3 Backpropagation network.

- **The feedforward pass**
  A randomly selected sample $\mathbf{x}_n$ labeled by $\mathbf{y}_n$ is presented to the input layer and forwarded through the weighted connections to the hidden layer and then the output layer to produce output $\hat{\mathbf{y}} = f(\mathbf{x}, \mathbf{W}^h, \mathbf{W}^o)$.
- **The backward error backpropagation**
  The squared error $\varepsilon = ||\mathbf{y}_n - \hat{\mathbf{y}}_n||^2/2$ measuring the difference between the actual output $\hat{\mathbf{y}}_n$ and the desired output $\mathbf{y}_n$ as the target is propagated backward from the output layer through the hidden layer to the input layer, during which the weights of both the output and hidden layers are modified so that ε will be reduced when the same or similar pattern is presented to the input layer in the future.

This two-phase process is iteratively carried out until eventually the error is minimized for all training samples and BPN is properly trained.

We now consider the specific computation taking place in both the forward and backward passes.

- The forward pass from input $\mathbf{x} = [x_1, \cdots, x_d]^T$ represented to the input layer to the output $\hat{\mathbf{y}}_n = [y_1, \cdots, y_m]^T$ produced by the output layer.
  - From input layer to hidden layer

$$z_j = g(a_j^h) = g\left(\sum_{k=1}^{d} w_{jk}^h x_k + b_j^h\right) = g\left(\sum_{k=0}^{d} w_{jk}^h x_k\right) = g(\mathbf{x}^T \mathbf{w}_j^h) \qquad (j = 1, \cdots, l) \tag{18.16}$$

where both $\mathbf{x} = [x_0 = 1, x_1, \cdots, x_d]^T$ and $\mathbf{w}_j^h = [w_{j0}^h = b_j^h, w_{j1}^h, \cdots, w_{jd}^h]^T$ are augmented, and $a_j^h = \mathbf{x}^T \mathbf{w}_j^h$ is the activation of the jth hidden layer node. These l equations can be expressed in vector form

$$\mathbf{z} = \mathbf{g}\left(\mathbf{W}^h \mathbf{x}\right) \tag{18.17}$$

where $\mathbf{z} = [z_1, \cdots, z_l]^T$, and $\mathbf{W}^h = [\mathbf{w}_1^h, \cdots, \mathbf{w}_l^h]^T$ is an $l \times (d+1)$ matrix.
  - From hidden layer to output layer

$$\begin{aligned} \hat{y}_i = g(a_i^o) &= g\left(\sum_{j=1}^{l} w_{ij}^o z_j + b_i^o\right) = g\left(\sum_{j=0}^{l} w_{ij}^o z_j\right) \\ &= g\left(\sum_{j=0}^{l} w_{ij}^o g\left(\sum_{k=0}^{d} w_{jk}^h x_k\right)\right) \qquad (i = 1, \cdots, m) \end{aligned} \tag{18.18}$$

which can also be written in vector form as

$$\hat{y}_i = g(\mathbf{z}^T \mathbf{w}_i^o) \tag{18.19}$$

where both $\mathbf{z} = [z_0 = 1, z_1, \cdots, z_l]^T$ and $\mathbf{w}_i^o = [w_{i0}^o = b_i^o, w_{i1}^o, \cdots, w_{il}^o]^T$ are augmented, and $a_i^o = \mathbf{z}^T \mathbf{w}_i^o$ is the activation of the ith output layer node. These m equations can be expressed in vector form

$$\hat{\mathbf{y}} = \mathbf{g}\left(\mathbf{W}^o \mathbf{z}\right) \tag{18.20}$$

where $\hat{\mathbf{y}} = [\hat{y}_1, \cdots, y_m]^T$, and $\mathbf{W}^o = [\mathbf{w}_1^o, \cdots, \mathbf{w}_m^o]^T$ is an $m \times (l+1)$ matrix.

- The backward error propagation from error ε to the input layer:
  The squared error $\varepsilon = ||\mathbf{y} - \hat{\mathbf{y}}||^2/2$ can be written as a function of the output layer weights w^o_{ij} $(i = 1, \cdots, m,\ j = 0, 1, \cdots, l)$ and hidden layer weights w^h_{jk} $(j = 1, \cdots, l,\ k = 0, 1, \cdots, d)$

$$\begin{aligned}\varepsilon &= \frac{1}{2}||\mathbf{y} - \hat{\mathbf{y}}||^2 = \frac{1}{2}\sum_{i=1}^{m}(\hat{y}_i - y_i)^2 = \frac{1}{2}\sum_{i=1}^{m}[g(a^o_i) - y_i]^2 \\ &= \frac{1}{2}\sum_{i=1}^{m}\left[g\left(\sum_{j=0}^{l} w^o_{ij} z_j\right) - y_i\right]^2 = \frac{1}{2}\sum_{i=1}^{m}\left[g\left(\sum_{j=0}^{l} w^o_{ij}\, g(a^h_j)\right) - y_i\right]^2 \\ &= \frac{1}{2}\sum_{i=1}^{m}\left[g\left(\sum_{j=0}^{l} w^o_{ij}\, g\left(\sum_{k=0}^{d} w^h_{jk} x_k\right)\right) - y_i\right]^2 \end{aligned} \tag{18.21}$$

  Similar to the objective function of the ridge regression considered in Section 4.2, an additional regularization term weighted by the *weight decay parameter* λ can be included in the objective function to penalize large weights and thereby prevent overfitting

$$J(\mathbf{W}^h, \mathbf{W}^o) = \varepsilon + \frac{\lambda}{2}\left[||\mathbf{W}^h||_2^2 + ||\mathbf{W}^o||_2^2\right] \tag{18.22}$$

  where the two weight matrices in the regularization term should exclude the first column for the bias term b which need not to be forced to be small, and $||\mathbf{W}||_2 = \sqrt{\sum_i \sum_j w^2_{ij}}$ is the *Frobenius norm* (Section A.4.2).
  The gradient descent method is used to modify first the output layer weights and then the hidden layer weights to iteratively reduce the objective function J in the following steps.
  - Find the gradient of J with respect to the output layer weights w^o_{ij} $(i = 1, \cdots, m,\ j = 0, 1, \cdots, l)$ by the chain rule

$$\begin{aligned}\frac{\partial J}{\partial w^o_{ij}} &= \frac{\partial \varepsilon}{\partial \hat{y}_i}\frac{\partial \hat{y}_i}{\partial a^o_i}\frac{\partial a^o_i}{\partial w^o_{ij}} + \lambda w^o_{ij} = (\hat{y}_i - y_i)\, g'(a^o_i)\, z_j + \lambda w^o_{ij} \\ &= -\delta_i\, g'(a^o_i)\, z_j + \lambda w^o_{ij}\end{aligned} \tag{18.23}$$

    where $\delta_i = y_i - \hat{y}_i$.
  - Update w^o_{ij} $(i = 1, \cdots, m,\ j = 0, 1, \cdots, l)$ to reduce J by the gradient descent method with learning rate or step size η

$$\begin{aligned}w^o_{ij} &\Leftarrow w^o_{ij} - \eta\frac{\partial J}{\partial w^o_{ij}} = w^o_{ij} - \eta\left(-\delta_i\, g'(a^o_i)\, z_j + \lambda w^o_{ij}\right) \\ &= w^o_{ij} - \eta\left(-d^o_i\, z_j + \lambda w^o_{ij}\right)\end{aligned} \tag{18.24}$$

    where $d^o_i = \delta_i g'(a^o_i)$, or in matrix form

$$\mathbf{W}^o \Leftarrow \mathbf{W}^o - \eta\left(-\mathbf{d}^o\mathbf{z}^T + \lambda\mathbf{W}^o\right) = \mathbf{W}^o + \eta\mathbf{d}^o\mathbf{z}^T - \eta\lambda\mathbf{W}^o \tag{18.25}$$

    where $\mathbf{d}^o = [d^o_1, \cdots, d^o_m]^T$ is the elementwise (Hadamard) product $\boldsymbol{\delta} \odot \mathbf{g}'(\mathbf{a}^o)$, and $\mathbf{d}^o\mathbf{z}^T = \mathbf{d}^o \otimes \mathbf{z}$ is the outer product of $\mathbf{d}^o$ and $\mathbf{z} = [z_0 = 1, z_1, \cdots, z_l]$.

– Find the gradient of J with respect to the hidden layer weights w^h_{jk} $(j = 1,\cdots,l,\ k = 0,1,\cdots,d)$ by the chain rule

$$\begin{aligned}
\frac{\partial J}{\partial w^h_{jk}} &= \frac{\partial \varepsilon}{\partial \hat{y}_i}\frac{\partial \hat{y}_i}{\partial a^o_i}\frac{\partial a^o_i}{\partial z_j}\frac{\partial z_j}{\partial a^h_j}\frac{\partial a^h_j}{\partial w^h_{jk}} + \lambda w^h_{jk} \\
&= \sum_{i=1}^m (\hat{y}_i - y_i)\frac{\partial \hat{y}_i}{\partial a^o_i}\frac{\partial a^o_i}{\partial z_j}\frac{\partial z_j}{\partial a^h_j}\frac{\partial a^h_j}{\partial w^h_{jk}} + \lambda w^h_{jk} \\
&= -\sum_{i=1}^m \delta_i g'(a^o_i) w^o_{ij}\, g'(a^h_j) x_k + \lambda w^h_{jk} = -\left(\sum_{i=1}^m d^o_i w^o_{ij}\right) g'(a^h_j) x_k + \lambda w^h_{jk} \\
&= -\delta^h_j\, g'(a^h_j) x_k + \lambda w^h_{jk} = -d^h_j x_k + \lambda w^h_{jk} \qquad (18.26)
\end{aligned}$$

where we have defined

$$\delta^h_j = \sum_{i=1}^m d^o_i w^o_{ij}, \quad d^h_j = \delta^h_j g'(a^h_j) = \left(\sum_{i=1}^m d^o_i w^o_{ij}\right) g'(a^h_j), \qquad (j = 1,\cdots,l) \qquad (18.27)$$

or in matrix form

$$\begin{bmatrix} \delta^h_1 \\ \vdots \\ \delta^h_l \end{bmatrix} = \begin{bmatrix} w^o_{11} & \cdots & w^o_{1m} \\ \vdots & \ddots & \vdots \\ w^o_{l1} & \cdots & w^o_{lm} \end{bmatrix} \begin{bmatrix} d_1 \\ \vdots \\ d_m \end{bmatrix} = \mathbf{W}^o \mathbf{d}^o \qquad (18.28)$$

Here $\mathbf{W}^o$ is an $m \times l$ matrix, the same as that already defined, but with the first column of b_js removed.

– Update w^h_{jk} $(j = 1,\cdots,l,\ k = 0,1,\cdots,d)$ to reduce J by gradient descent method

$$\begin{aligned}
w^h_{jk} &\Leftarrow w^h_{jk} - \eta \frac{\partial J}{\partial w^h_{jk}} = w^h_{jk} - \eta\left(-\delta^h_j\, g'(a^h_j) x_k + \lambda w^h_{jk}\right) \\
&= w^h_{jk} - \eta\left(-d^h_j x_k + \lambda w^h_{jk}\right) \qquad (18.29)
\end{aligned}$$

or in matrix form

$$\mathbf{W}^h \Leftarrow \mathbf{W}^h - \eta\left(-\mathbf{W}^o\mathbf{d}^o \odot \mathbf{g}'(\mathbf{a}^h)\mathbf{x}^T + \lambda \mathbf{W}^h\right) = \mathbf{W}^h + \eta \mathbf{d}^h \mathbf{x}^T - \eta\lambda \mathbf{W}^h \quad (18.30)$$

where $\mathbf{d}^h = \mathbf{W}^o\mathbf{d}^o \odot \mathbf{g}'(\mathbf{a}^h)$ is the elementwise product of vector $\mathbf{W}^o\mathbf{d}^o$ and $\mathbf{g}'(\mathbf{a}^h)$, and $\mathbf{d}^h\mathbf{x}^T = \mathbf{d}^h \otimes \mathbf{x}$ is the outer product of vectors $\mathbf{d}^h$ and $\mathbf{x}$.

In summary, here are the steps in each iteration.

1. Input a randomly selected pattern $[x_1,\cdots,x_d]^T$, construct $d+1$ dimensional vector $\mathbf{x} = [1, x_1,\cdots,x_d]^T$.
2. Compute $\mathbf{z} = \mathbf{g}(\mathbf{W}^h\mathbf{x})$, and construct $l+1$ dimensional vector $\mathbf{z} \leftarrow [1, \mathbf{z}]$.
3. Compute $\hat{\mathbf{y}} = \mathbf{g}(\mathbf{W}^o\mathbf{z})$.
4. Get elementwise product $\mathbf{d}^o = (\mathbf{y} - \hat{\mathbf{y}}) \odot \mathbf{g}'(\mathbf{a}^o) = \boldsymbol{\delta} \odot \mathbf{g}'(\mathbf{a}^o)$.
5. Update output weights $\mathbf{W}^o \Leftarrow \mathbf{W}^o + \eta\, \mathbf{d}^o\mathbf{z}^T - \eta\lambda\mathbf{W}^o$.
6. Get elementwise product $\mathbf{d}^h = \mathbf{W}^o_{m\times l}\mathbf{d}^o \odot \mathbf{g}'(\mathbf{a}^h) = \boldsymbol{\delta}^h \odot \mathbf{g}'(\mathbf{a}^h)$, where $\mathbf{W}^o_{m\times l}$ is the same as $\mathbf{W}^o$ but with its first column removed.

7. Update hidden weights $\mathbf{W}^h \Leftarrow \mathbf{W}^h + \eta\, \mathbf{d}^h \mathbf{x}^T - \eta \lambda \mathbf{W}^h$.
8. Terminate the iteration if the error ε is acceptably small for all of the training patterns. Otherwise repeat the above with another pattern in the training set.

The following lists the MATLAB code for the essential part of the BPN algorithm. Array X contains C classes each with K samples, array Y are the labelings of the $C * K$ training samples, array W contains the $N + 1$ dimensional weight vectors for the M output nodes. Also, L is the number of hidden nodes, *eta* is the learning rate between 0 and 1, and *tol* is the tolerance for the termination of the learning iteration (e.g., 0.01).

```
function [Wh, Wo, g]=backPropagate(X,Y)
    syms x;
    g=1/(1+exp(-x));              % Sigmoid activation function
    dg=diff(g);                   % its derivative function
    g=matlabFunction(g);
    dg=matlabFunction(dg);
    [d,N]=size(X);                % numbers dimensions and samples
    X=[ones(1,N); X];             % augment X by adding a row of x0=1
    Wh=1-2*rand(L,d+1);           % Initialize hidden layer weights
    Wo=1-2*rand(M,L+1);           % Initialize output layer weights
    er=inf;
    while er > tol
        I=randperm(N);            % random permutation of samples
        er=0;
        for n=1:N                 % for all N samples for an epoch
            x=X(:,I(n));          % pick a training sample
            ah=Wh*x;              % activation of hidden layer
            z=[1; g(ah)];         % augment z by adding z0=1
            ao=Wo*z;              % activation to output layer
            yhat=g(ao);           % output of output layer
            delta=Y(:,I(n))-yhat  % delta error
            er=er+norm(delta)/N   % test error
            do=delta.*dg(ao);
            Wo=Wo+eta*do*z';      % update weights for output layer
            dh=(Wo(:,2:L+1)'*do).*dg(ah);    % delta of hidden layer
            Wh=Wh+eta*dh*x';      % update weights for hidden layer
        end
    end
end
```

The training process of BP network can also be considered as a data modeling problem as in regression to fit the given data $\{(\mathbf{x}_i, \mathbf{y}_i),\ (i = 1, \cdots, N)\}$ by a function with the weights of both the hidden and output layers as the parameters

$$\mathbf{y} = \mathbf{f}(\mathbf{x}, \mathbf{W}^h, \mathbf{W}^o) \tag{18.31}$$

The goal is to find the optimal parameters $\mathbf{W}^h$ and $\mathbf{W}^o$ that minimize the difference $\mathbf{r} = \mathbf{y} - \hat{\mathbf{y}}$ between the desired and the actual outputs. The Levenberg–Marquardt algorithm discussed previously can be used to obtain the parameters (e.g., the MATLAB function *trainlm* `http://www.mathworks.com/help/nnet/ref/trainlm.html`).

The three-layer BPN containing a single hidden layer just discussed can be easily generalized to a multilayer BPN containing any number of hidden layers (e.g., for a deep learning network) by simply repeating steps 6 and 7 for the single hidden layer in the algorithm list above. We assume in addition to the input layer, there are in total L learning layers including all the hidden layers and the output layer, indexed by $l = 1, \cdots, L$. Then step 6 becomes

$$\mathbf{d}^{(l)} = \delta^{(l)} \cdot \mathbf{g}'(\mathbf{a}^{(l)}) \qquad (l = 1, \cdots, L) \tag{18.32}$$

where $\mathbf{a}^{(l)} = (\mathbf{W}^{(l)}\mathbf{z}^{(l)})$ is the activation of all nodes in the lth layer, $\mathbf{z}^{(l)} = \mathbf{W}^{(l-1)}\mathbf{z}^{(l-1)}$ is the input to the lth layer, output from the (l-1)th layer ($\mathbf{z}^{(1)} = \mathbf{x}$), and

$$\delta^{(l)} = \begin{cases} \mathbf{y} - \hat{\mathbf{y}} & (l = L) \\ \mathbf{W}^{(l)}\mathbf{d}^{(l-1)} & (1 \le l < L) \end{cases} \tag{18.33}$$

Then the same step 7 for updating the weights can be carried out based on the gradient vector

$$\mathbf{d}^{(l)} \otimes \mathbf{z}^{(l)} = \mathbf{d}^{(l)}(\mathbf{z}^{(l)})^T \tag{18.34}$$

The following MATLAB code segment is for the BP network of multiple hidden layers, in which L is the total number of learning layers and $m(l)$ is the number of nodes in the lth layer ($l = 1, \cdots, L$)

```
W={1-2*rand(m(1),d+1)};             % initial weights for first layer
for l=2:L
    W{l}=1-2*rand(m(l),m(l-1)+1); % initial weights for other layers
end
er=inf;
d={};
while er > tol
    I=randperm(N);                  % random permutation of samples
    er=0;
    for n=1:N                       % N samples for an epoch
        z={1;X(:,I(n))};            % pick a training sample
        a={W{1}*z{1}};              % activation of first layer
        for l=2:L                   % the forward pass
            z{l}=[1;g(a{l-1})];     % input to the lth layer
            a{l}=W{l}*z{l};         % activation of the lth layer
        end
        yhat=g(a{L});               % actual output of last layer
```

```
            delta=Y(:,I(n))-yhat;       % delta error
            er=er+norm(delta)/N;        % test error
            d{L}=delta.*dg(a{L});       % d for output layer
            W{L}=W{L}+eta*d{L}*z{L}'; % update weights for output layer
            for l=L-1:-1:1              % the backward pass
                d{l}=(W{l+1}(:,2:end)'*d{l+1}).*dg(a{l}); % d of hidden layers
                W{l}=W{l}+eta*d{l}*z{l}'; % update weights for hidden layers
            end
        end
    end
```

Example 18.4

Figures 18.4 and 18.5 show the classification results of two 2-D datasets with 2 and 3 classes previously used in Examples 13.6 and 13.7 for the naive Bayes classifier (with quadratic decision boundaries). The error rates are, respectively, 13% and 11.5%, and the confusion matrices are shown in the following. Interestingly, in the case of the three-class dataset, there is a green region sandwiched between the blue and red regions in the results generated by both the naive Bayes and backpropagation methods based on very different learning algorithms.

$$\begin{bmatrix} 176 & 24 \\ 22 & 178 \end{bmatrix} \quad \begin{bmatrix} 185 & 14 & 1 \\ 5 & 181 & 14 \\ 11 & 33 & 156 \end{bmatrix}$$

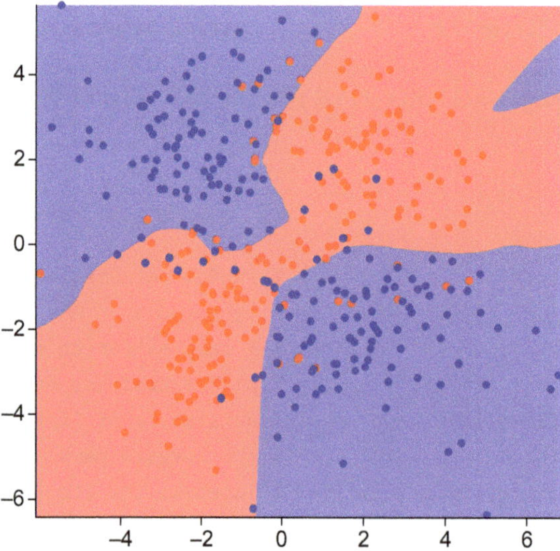

Fig. 18.4 BP network classification of the XOR dataset.

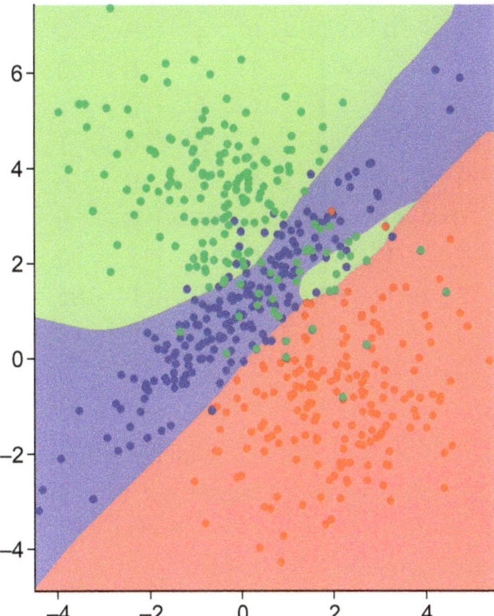

Fig. 18.5 BP network classification of a three-class dataset.

Example 18.5

The backpropagation network is applied to the classification of the handwritten digit dataset. Half of the $N = 2240$ samples in the dataset is used for training, while the other half is used for testing. The classification results are shown by the following two confusion matrices, when the network is tested on half of the samples used for training (left) and on the remaining half reserved for testing (right). Note that the error rate $74/1120 = 6.6\%$ in the latter case is understandably higher than the error $27/1120 = 2.4\%$ for the former, as the network may overfit half of the samples used for training and does not generate too well to the other half of the samples it has never seen.

$$\begin{bmatrix} 113 & 0 & 0 & 0 & 0 & 0 & 0 & 1 & 1 & 1 \\ 0 & 107 & 0 & 0 & 0 & 0 & 1 & 0 & 0 & 0 \\ 0 & 1 & 106 & 1 & 0 & 1 & 0 & 0 & 0 & 0 \\ 0 & 0 & 4 & 113 & 0 & 0 & 0 & 0 & 0 & 3 \\ 0 & 0 & 0 & 0 & 113 & 8 & 0 & 0 & 0 & 0 \\ 0 & 1 & 0 & 0 & 1 & 120 & 0 & 0 & 1 & 0 \\ 0 & 1 & 0 & 0 & 0 & 0 & 102 & 0 & 0 & 0 \\ 0 & 1 & 2 & 0 & 0 & 1 & 0 & 103 & 0 & 1 \\ 0 & 1 & 0 & 1 & 2 & 0 & 0 & 0 & 104 & 1 \\ 1 & 1 & 0 & 0 & 2 & 0 & 0 & 0 & 0 & 112 \end{bmatrix}$$

$$
\begin{bmatrix}
100 & 1 & 0 & 0 & 0 & 4 & 3 & 0 & 0 & 0 \\
0 & 111 & 1 & 1 & 1 & 1 & 0 & 0 & 0 & 1 \\
1 & 0 & 113 & 1 & 0 & 0 & 0 & 1 & 0 & 0 \\
2 & 1 & 1 & 100 & 0 & 0 & 0 & 1 & 0 & 3 \\
0 & 1 & 0 & 0 & 105 & 0 & 0 & 0 & 2 & 3 \\
1 & 0 & 0 & 0 & 3 & 95 & 0 & 0 & 2 & 0 \\
0 & 2 & 1 & 0 & 0 & 0 & 118 & 0 & 0 & 0 \\
1 & 1 & 1 & 1 & 1 & 0 & 0 & 105 & 1 & 5 \\
1 & 1 & 2 & 1 & 1 & 5 & 0 & 2 & 97 & 5 \\
0 & 2 & 0 & 0 & 4 & 0 & 0 & 0 & 0 & 102
\end{bmatrix}
$$

Before leaving this section, we reconsider the issue of how to make a proper tradeoff between overfitting and underfitting in a backpropagation network. In addition to the approach of regularization (Eq. (18.22)), there are several other techniques to address this issue:

- Model complexity in terms of number of layers and nodes:
  Too many may lead to overfitting, while too few may result in underfitting.
- Number of epochs: Training for too many epochs may lead to overfitting the training sample, while too few may cause underfitting due to insufficient learning.
- Dropout: a fixed percentage of nodes are randomly dropped (set to zero) during training, to prevent them from relying too heavily on certain features which may include noise, while all nodes are used during testing.

We further make the following comments regarding how the backpropagation network, and in general, multilayer feed forward network, is closely related to the general method of regression analysis discussed before.

- The computational model for a single neuron in Eq. (17.3) based on the logistic activation function in Eq. (17.5) is actually the same as Eq. (7.6) in logistic regression in Section 7.1

$$
g\left(\sum_{j=1}^{d} w_j x_j + b\right) = g(\mathbf{w}^T\mathbf{x}) = \frac{1}{1+e^{-\mathbf{w}^T\mathbf{x}}} \tag{18.35}
$$

  treated as the conditional probability $p(\hat{y} = 1|\mathbf{x}, \mathbf{w})$. Both methods are to determine the model parameter $\mathbf{w}$, but based on different methods.
- A supervised learning neural network can be considered as a function $\hat{\mathbf{y}} = \mathbf{f}(\mathbf{x}, \mathbf{w})$ as a model of the training dataset $\mathbf{D} = \{\mathbf{X}, \mathbf{y}\}$, while the parameter $\mathbf{w}$ can be found by solving an optimization problem (Chapter 5), by either the least squares method that minimizes $\varepsilon = ||\mathbf{y} - \hat{\mathbf{y}}||^2$, or the maximum likelihood method that maximizes $L(\mathbf{w}) = P(\mathbf{y}|\mathbf{X}, \mathbf{w})$, equivalent to minimizing $-l(\mathbf{w}) = -\log L(\mathbf{w})$, which can be interpreted as the cross entropy between the model distribution based on $\mathcal{D}$ and the unknown ground true distribution, as discussed in Section 4.1.
- As a neural network can be considered as a model for the function $\mathbf{y} = \mathbf{f}(\mathbf{x})$ relating $\mathbf{y}$ to $\mathbf{x}$ in the given dataset $\mathcal{D}$, a fundamental question to ask is whether such

a network is able to model any function. This question is answered by the *universal approximation theorem*, stating that a feedforward network with arbitrary width (number of nodes in a hidden layer) or depth (number of hidden layers) can represent any function with arbitrary accuracy.

While this theorem is formally proven, it can also be understood intuitively. Recall the method of linear regression based on basis functions in Chapter 5, by which a function is approximated by a repression function as a linear combination of K basis functions in Eq. (5.47)

$$\hat{y}_m = \sum_{k=1}^{K} w_{mk}^o \, \varphi_k(\mathbf{x}) = \sum_{k=1}^{K} w_{mk}^o \, g\left(\sum_{i=0}^{d} w_{ki}^h x_i\right) \tag{18.36}$$

which is in the same form as in Eq. (18.18) for the three-layer feed forward network already discussed, with a linear activation function for the output layer. We see the outputs of the hidden layer nodes are the basis functions, and the output of an output layer node is a linear combination of such basis functions. As shown in Example 5.10, a given function can be approximated with progressively higher accuracy as more basis functions, that is, more hidden layer nodes, are used, if the width of the network is unbounded.

If the width of the network is bounded but its depth is not, we can alternatively increase the number of layers of the network so that in Eq. (18.36) we can have as many levels of summations as needed, that is, the total number of terms in these summations can increased without a bound, then by the same argument, such a network can again approximate a given function with arbitrary accuracy. But as a word of warning, while trying to match the given dataset for better accuracy, watch out for the overfitting problem.

18.3 Autoencoder

Autoencoder is a neural network method that learns to represent the patterns in a given dataset in such a way that both the dimensionality and the noise can be reduced, and it can therefore be considered as a method for feature selection and dimensionality reduction, similar to the Karhunen–Loève transform (KLT) method.

An autoencoder, as shown in Fig. 18.6, is similar to a backpropagation network in that it iteratively updates the weights in all hidden and output layers by backpropagation so that eventually its outputs are as desired. But, unlike a backpropagation network that learns to associate a pattern $\mathbf{x}_n$ to another pattern $\mathbf{y}_n$ as its labeling, the autoencoder is an unsupervised learning method based on unlabeled data, and it learns to associate each pattern $\mathbf{x}_n$ to itself.

The purpose of the autoencoder is not to simply copy the data. Instead, we want to reduce the number of hidden layer nodes as much as possible, by imposing constraints such as sparsity to the hidden layer(s) in terms of the number of active nodes, so that there are many fewer nodes in the hidden layers when compared to the input and output layers. By doing so, the hidden layers become the *bottleneck* of the multilayer network, which can then be considered as an *encoder* by which the

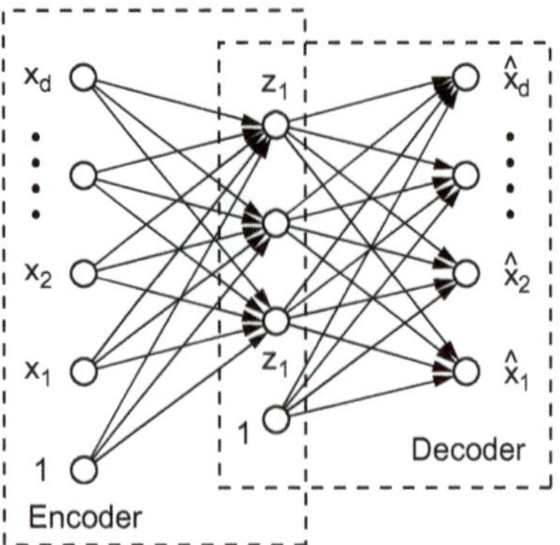

Fig. 18.6 Thee layer network for autoenoder.

dimensionality of the data presented to the input layer to be reproduced by the output layer is converted into a code in a significantly reduced dimensional space in the hidden layer. By this approach, the data can be much compressed, while the information contained in it is mostly preserved. In addition, certain potential structures inherent in the dataset may be revealed, very much like dimensionality reduction achieved principal component analysis (PCA) based on KLT.

The computation taking place in an autoencoder can therefore be considered as an encoding/decoding process, with encoding taking place between the input and hidden layers, and decoding taking place between the hidden and output layers. Due to the reduced number of nodes in the hidden layers, the data size can be significantly reduced for data compression, if some of the components of data samples $\mathbf{x}$ in the dataset are correlated, that is, there is some redundancy in the data.

While the methods of autoencoder and PCA based on KLT are similar to each other in that both can compress the data without losing much information, they are different in that the PCA is a linear method, as the principal components are linear combinations of the data components, while the signals represented by the hidden layer of the autoencoder are nonlinear functions of the data components (owing to the nonlinear activation function). Consequently the autoencoder can be a more powerful method than the PCA, but with the cost of higher computational complexity.

We now consider specifically how an autoencoder works. As in backpropagation, a data vector $\mathbf{x}$ presented to the input layer is forward propagated first to the hidden layer (same as Eq. (18.17))

$$\mathbf{z} = \mathbf{g}(\mathbf{W}^h\mathbf{x}), \qquad \text{or} \qquad z_j = g\left(\sum_{k=0}^{d} w_{jk}^h x_k\right) \quad (j = 1, \cdots, l) \tag{18.37}$$

and then further to the output layer (same as Eq. (18.20))

$$\hat{\mathbf{x}} = \mathbf{g}(\mathbf{W}^o\mathbf{z}) \qquad \text{or} \qquad \hat{x}_i = g\left(\sum_{j=0}^{l} w_{ij}^o z_k\right) \quad (i = 1, \cdots, m) \tag{18.38}$$

Also, as in backpropagation, the weights in both $\mathbf{W}^h$ and $\mathbf{W}^o$ are updated during training in the backpropagation process so that the following objective function for the squared error between the input and desired output, the same as the input, is minimized

$$\varepsilon = \frac{1}{2}||\mathbf{x} - \hat{\mathbf{x}}||^2 \tag{18.39}$$

Here we assume there is only a single hidden layer for convenience, but the computation for both the forward and backward passes can be easily extended to multiple layers.

By limiting the number l of the hidden layer to be smaller than that of input and output layer, the dimensionality of the data can be reduced from d for the dimensionality of the input $\mathbf{x} = [x_1, \cdots, x_d]^T$ to l, the number of hidden layer nodes, while most of the information in the original data is reserved.

It is interesting to compare the autoencoder and the PCA based on KLT in terms of how they map a given data vector $\mathbf{x}$ in the original d-dimensional space to a lower-dimensional space. Recall the KLT previously considered

$$\mathbf{y} = \begin{bmatrix} y_1 \\ \vdots \\ y_d \end{bmatrix} = \mathbf{V}^T\mathbf{x} = \begin{bmatrix} \mathbf{v}_1^T \\ \vdots \\ \mathbf{v}_d^T \end{bmatrix} \mathbf{x} = \begin{bmatrix} \mathbf{v}_1^T\mathbf{x} \\ \vdots \\ \mathbf{v}_d^T\mathbf{x} \end{bmatrix} \tag{18.40}$$

In KLT, the ith component $y_i = \mathbf{v}_i^T\mathbf{x}$ of $\mathbf{y}$ is the inner product of the ith eigenvector $\mathbf{v}_i$ and the input $\mathbf{x}$, the projection of $\mathbf{x}$ onto $\mathbf{v}_i$ if it is normalized. Conversely, in a neural network, the inner product of the weight vector $\mathbf{w}_i$ and the input $\mathbf{x}$ is further mapped to $z_i = g(\mathbf{w}_i^T\mathbf{x})$. If, for example, the activation function $g(u) = u$ is linear (an identical function), the weight vectors $\mathbf{w}_i$ play the same role as the eigenvectors $\mathbf{v}_i$ as the basis vectors that span the space, that is, autoencoder and PCA are essentially the same. But, as in general, the activation function $g(u)$ is nonlinear, the autoencoder is a nonlinear mapping, of which the linear KLT (a rotation in the space) is a special case.

Dimension reduction exemplified by the KLT is effective if the components of the pattern vectors are highly correlated, that is, some of them may be redundant as they carry similar information. However, if all components carry their own independent information, dimensionality reduction will inevitably cause significant information loss. Similarly in an autoencoder, if the components of the data points are mostly independent of each other with little redundancy, they may not be reproduced as the output of the autoencoder if the number of hidden layer nodes is too small.

When the data components are correlated, it is possible to achieve the desired low sparsity of the hidden layer of the autoencoder by using only a small number of hidden nodes, or allowing only a small fraction of them to be active (over different input patterns). The second approach can be realized by (a) keeping only a small number of hidden nodes with maximum outputs while setting the rest to zero, or (b) forcing the average activation of the hidden nodes to be low. This can be achieved by modifying the regularization term in Eq. (18.22) for small weights so that it acts to force the average activation level to be low, as either of the following shows:

- Minimize the following L1 or L2 average of all activations

$$J = \varepsilon + \lambda \sum_{j=1}^{l} |a_j|, \qquad J = \varepsilon + \frac{\lambda}{2} \sum_{j=1}^{l} a_j^2 \tag{18.41}$$

- Minimize the difference between the actual and desired average activation

$$J = \varepsilon + \sum_{j=1}^{l} KL(\rho||\hat{\rho}_j)$$
$$= \varepsilon + \sum_{j=1}^{l} \left[\rho \log\left(\frac{\rho}{\hat{\rho}_j}\right) + (1-\rho)\log\left(\frac{1-\rho}{1-\hat{\rho}_j}\right)\right] \tag{18.42}$$

Here ρ is the desired average activation of the hidden layer, a small value close to zero (e.g., $\rho = 0.05$), and $\hat{\rho}_j$ is the actual average activation of the jth hidden node over all N samples in the given dataset

$$\hat{\rho}_j = \frac{1}{N}\sum_{n=1}^{N} \rho_j(\mathbf{x}_n) = \frac{1}{N}\sum_{n=1}^{N} g(a_j) = \frac{1}{N}\sum_{n=1}^{N} g\left(\sum_{k=0}^{d} w_{jk}^h x_k\right) \quad (j = 1,\cdots,l) \tag{18.43}$$

where $\rho_j(\mathbf{x}_n) = g(a_j) = g\left(\sum_k w_{jk}^h x_k\right)$ is the activation level of the node while responding to the nth input sample $\mathbf{x}_n$. Also, ρ is treated as the probability for some binary random variable (with Bernoulli distribution) to be 1, and $1-\rho$ is the probability for it to be 0, and so is $\hat{\rho}_j$ for another binary variable. We therefore see that the regularization term in Eq. (18.42) is the sum of the Kullback–Leibler (KL) divergence (Section B.2.2) between the two distributions for all hidden nodes, measuring the overall difference between the desired and actual sparsities in terms of ρ and $\hat{\rho}$. By minimizing the objective function including such a regularization term, the sparsity of the hidden layer in terms of the average activation is forced to be as low as desired.

The plot in Fig. 18.7 compares the KL divergence and the quadratic function $a(\rho - \hat{\rho})^2$. While both functions are minimized to zero when $\hat{\rho} = \rho = 0.3$, they behave differently in the range of [0, 1] in that the KL divergence becomes significantly larger than the quadratic function when $\hat{\rho}$ are very different from $\rho = 0.3$.

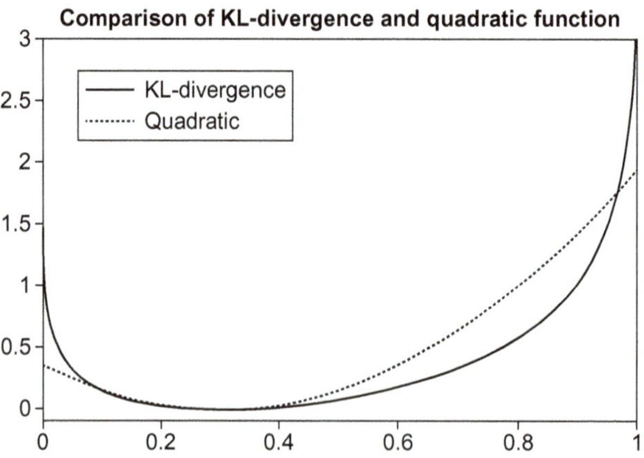

Fig. 18.7 Comparison of KL divergence and quadratic function.

Now the objective function for the autoencoder can be written as the sum of the error ε and the regularization term

$$J(\mathbf{W}^h, \mathbf{W}^o) = \varepsilon + \lambda \sum_{j=1}^{l} KL(\rho||\hat{\rho}_j) \tag{18.44}$$

During the iterative learning the weights are updated based on the gradient descent method, the same as in backpropagation. Given a specific input $\mathbf{x}$ we first get the output $\hat{\mathbf{x}}$ through the forward pass, and then modify the weights of both the output and hidden layers in the backward pass based on the gradient of the objective function $\varepsilon = ||\mathbf{x}-\hat{\mathbf{x}}||^2/2$ with respect to the weights. While the computation for updating the weights for the output layer is the same as steps 4 and 5 in the algorithm of backpropagation listed in Section 18.2, the computation for the hidden layer weights as shown here is different from steps 6 and 7 for backpropagation due to the regularization term

$$\frac{\partial}{\partial w_{jk}^h} J(\mathbf{W}^h, \mathbf{W}^o) = \frac{\partial \varepsilon}{\partial w_{jk}^h} + \lambda \frac{\partial}{\partial w_{jk}^h} \left[\sum_{j=1}^{l} KL(\rho||\hat{\rho}_j) \right] \tag{18.45}$$

The first term is the same as that in Eq. (18.26), while the second term for regularization based on the KL divergence is different

$$\begin{aligned} &\frac{\partial}{\partial \hat{\rho}_j} \left[\sum_{j=1}^{l} \rho \log\left(\frac{\rho}{\hat{\rho}_j}\right) + (1-\rho)\log\left(\frac{1-\rho}{1-\hat{\rho}_j}\right) \right] \frac{\partial \hat{\rho}_j}{\partial w_{jk}^h} \\ &= \left[\rho \frac{\partial}{\partial \hat{\rho}_j} (\log \rho - \log \hat{\rho}_j) + (1-\rho) \frac{\partial}{\partial \hat{\rho}_j} [(\log(1-\rho) - \log(1-\hat{\rho}_j))] \right] \\ &\frac{\partial}{\partial w_{ij}} g\left(\sum_{k=0}^{d} w_{jk}^h z_i \right) = \left(-\frac{\rho}{\hat{\rho}_j} + \frac{1-\rho}{1-\hat{\rho}_j} \right) g'(a_j) x_k \end{aligned} \tag{18.46}$$

Now d_j^h in Eq. (18.26) for backpropagation (defined in Eq. (18.27)) can be modified for the autoencoder

$$d_j^h = \left[\left(\sum_{i=1}^{m} d_i^o w_{ij}^o \right) + \lambda \left(-\frac{\rho}{\hat{\rho}_j} + \frac{1-\rho}{1-\hat{\rho}_j} \right) \right] g'(a_j^h) \tag{18.47}$$

Using this modified d_j^h in step 6 of the algorithm for the propagation listed in the previous section, we get the algorithms for the autoencoder.

The following MATLAB code segment is the essential part of the autoencoder algorithm, which takes the given dataset as input and generates the weights for both hidden and output layers as output. Here to achieve the desired sparsity of the hidden layer, the first approach mentioned is used, that is, out of all L hidden layer nodes, $La < L$ most active ones are kept while all others are set to zero. Alternatively, the more elaborate approach based on KL divergence can be used as well.

```
Wh=1-2*rand(L,d+1);                 % hidden layer weights
Wo=1-2*rand(d,L+1);                 % output layer weights
done=0;
it=0;
while ~done
    it=it+1;
```

```
        X=Xm;
        [d N]=size(X);
        I=randperm(N);                      % random order of N samples
        Ah=0;
        for n=1:N                           % for each samples
            x=X(:,I(n));                    % input vector
            x1=[1;x];                       % augmented input vector
            ah=Wh*x1;                       % activation of hidden nodes
            z=g(ah);                        % output of hidden layer
            zsorted=sort(z,'descend');      % sort hidden layer output
            z=z.*(z>=zsorted(La));          % keep La most active nodes
            z=[1;z];                        % augment hidden layer output
            ao=Wo*z;                        % activation of output nodes
            xhat=g(ao);                     % actual output of output layer
            do=(x-xhat).*dg(ao);            % delta of output layer
            Wo=Wo+eta*do*z';                % update output layer weights
            dh=(Wo(:,2:L+1)'*do).*dg(ah);   % delta of hidden layer
%             dh=((Wo(:,2:L+1)'*do)+lambda*(-r./rp+(1-r)./(1-rp))).*dg(ah);
            Wh=Wh+eta*dh*x1';               % update hidden layer weights
        end
    end
```

Recall that in dimensionality reduction based on KLT, the transform matrix is the eigenvector matrix of either the covariance matrix $\mathbf{\Sigma}_x$ of the entire dataset $\mathbf{X} = [\mathbf{x}_1, \cdots, \mathbf{x}_N]$, or the between-class scatter matrix $\mathbf{S}_B$ in the case of supervised classification. Similarly an autoencoder can also be trained based on either the entire dataset $\mathbf{X}$ for dimensionality reduction, or a dataset $\mathbf{X} = [\mathbf{m}_1, \cdots, \mathbf{m}_K]$ containing the mean vectors of some K classes (calculated based on labeled training set) for classification.

In the following examples, we compare the performances of both the autoencoder (trained on a dataset composed of only mean vectors of all classes) and the method PCA based on the between-class scatter matrix of the classes, when they are applied to reduce the dimensionality of the given dataset to be classified by the naive Bayes classifier.

Example 18.6

Both methods of autoencoder and PCA are applied to a dataset containing $K = 8$ classes each in one of the octants of the 3-D space, and then compared in terms of the separability and classification error rate in the resulting lower dimensional space. Note that in this case all three dimensions are equally important for representing the eight classes, as also reflected by the eigenvalues 0.37%, 0.33%, 0.30% of the covariance matrix, indicating that roughly the same amount of information is contained in each of the three components of the data. We see that it is difficult to reduce the dimensionality without losing a significant amount of information.

Figure 18.8 shows the dataset after its dimensionality if reduced from 3 to 2 by autoencoder (left) and PCA (right) methods. We see that the data points are similarly distributed in the resulting 2-D space, but the class separability for the autoencoder seems to be slightly better than that by the PCA. This result can be confirmed quantitatively as shown in the first row of Table 18.2, where the separability $J_{B/T} = tr\left(\mathbf{S}_T^{-1}\mathbf{S}_B\right)$ (Eq. (9.24)) and the error rates for classification (by naive Bayes classifier) are listed to the left and right of the slash symbol for the autoencoder and PCA methods, respectively. showing the autoencoder has a slightly high separability and lower error rate than the PCA method.

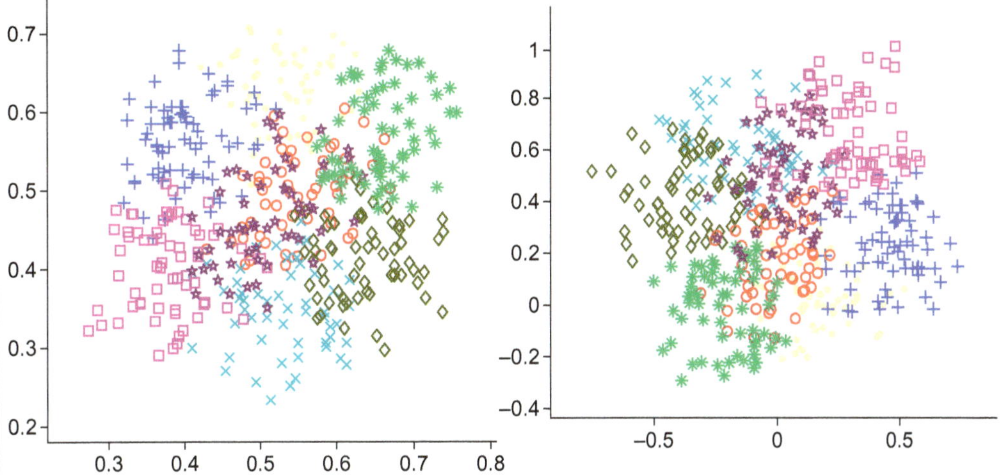

Fig. 18.8 Autoencoder (left) vs. PCA (right) for dimension reduction for a 3-D dataset.

Table 18.2

	Separability	Error rate
Eight-class dataset	1.52/1.51	0.304/0.320
Iris dataset	1.12/1.07	0.020/0.027

Example 18.7

The autoencoder and PCA methods are also compared when applied to the iris dataset containing three classes in the 4-D space. The resulting 2-D dataset is shown in Fig. 18.9, and the separabilities the classification error rates of the two methods are compared in the second row of Table 18.2. We see again that the autoencoder has a little higher separability and lower error rate than the PCA method.

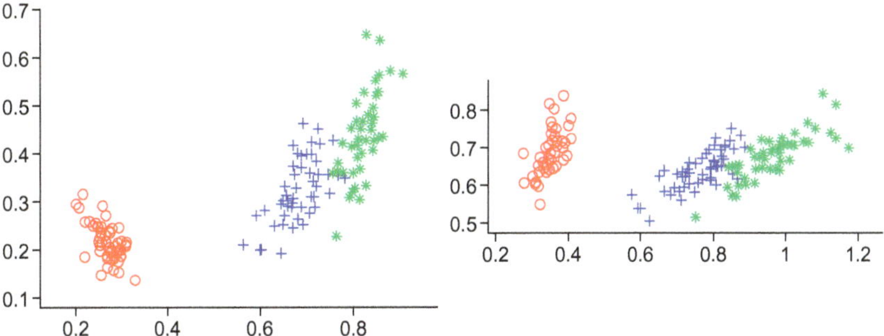

Fig. 18.9 Autoencoder (left) vs. PCA (right) for dimensionality reduction for iris dataset.

Example 18.8

The autoencoder network is typically trained by all samples in the dataset. However, if these samples in a dataset are labeled to belong to one of a set of classes, the autoencoder can also be trained by these classes. Specifically in this example, the three classes in the iris dataset each represented by the average of all samples in the class are used to train the autoencoder. Figure 18.10 shows the 4-D iris dataset mapped to 2-D (top) and 3-D (bottom) spaces by both the autoencoder (left) and KLT (right). It is interesting to observe that in the case of autoencoder (left), the data points representing the three classes are arranged into a line structure in both 2- and 3-D spaces.

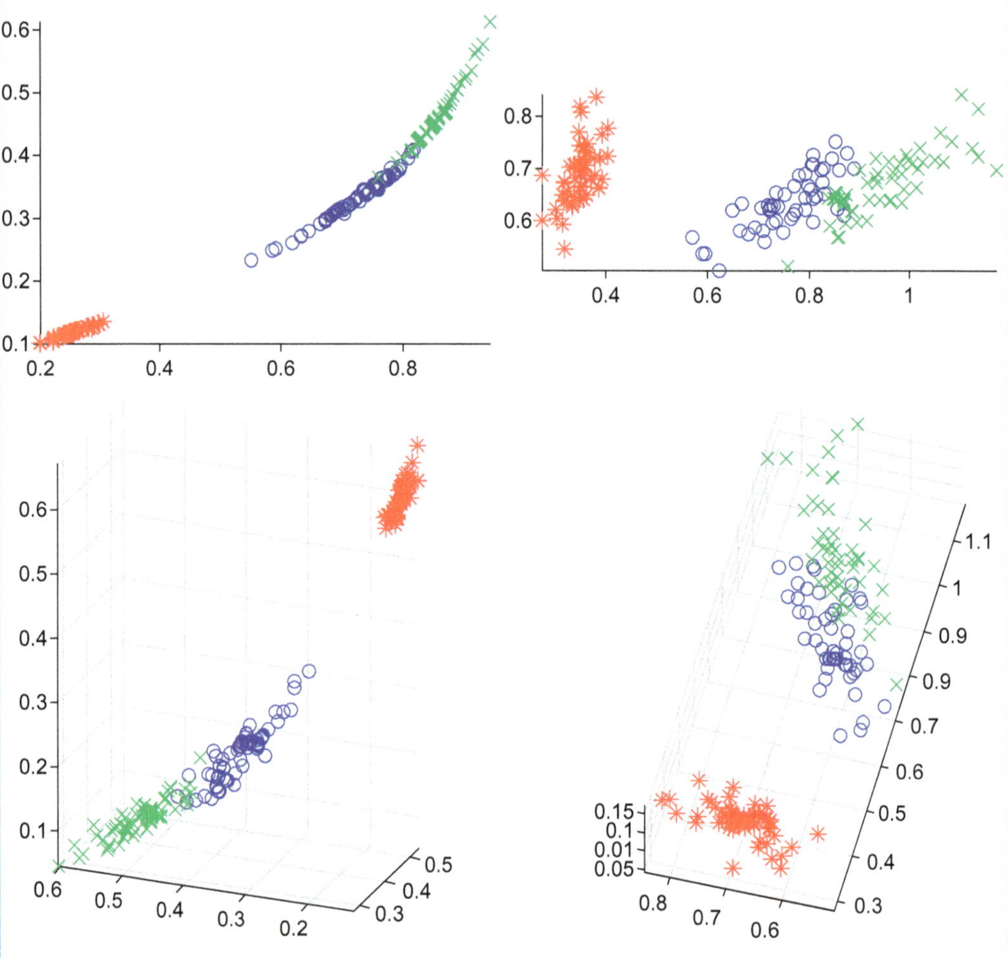

Fig. 18.10 Autoencoder (Left) vs. KLT (right) for dimension reduction.

Example 18.9

The same method is applied to the dataset of 10 handwritten digits from 0 to 9, each represented by 224 training samples in image form of 16×16 pixels, treated as a 256-dimensional vector. After the autoencoder network is fully trained based on the means of all 224 samples of each of the 10 classes, the dataset containing $224 \times 10 = 2240$ data samples in the l-dimensional space spanned by the hidden layer nodes that encode the 256-D samples is then visualized in 3-D space based the KLT, as shown in Fig. 18.11,

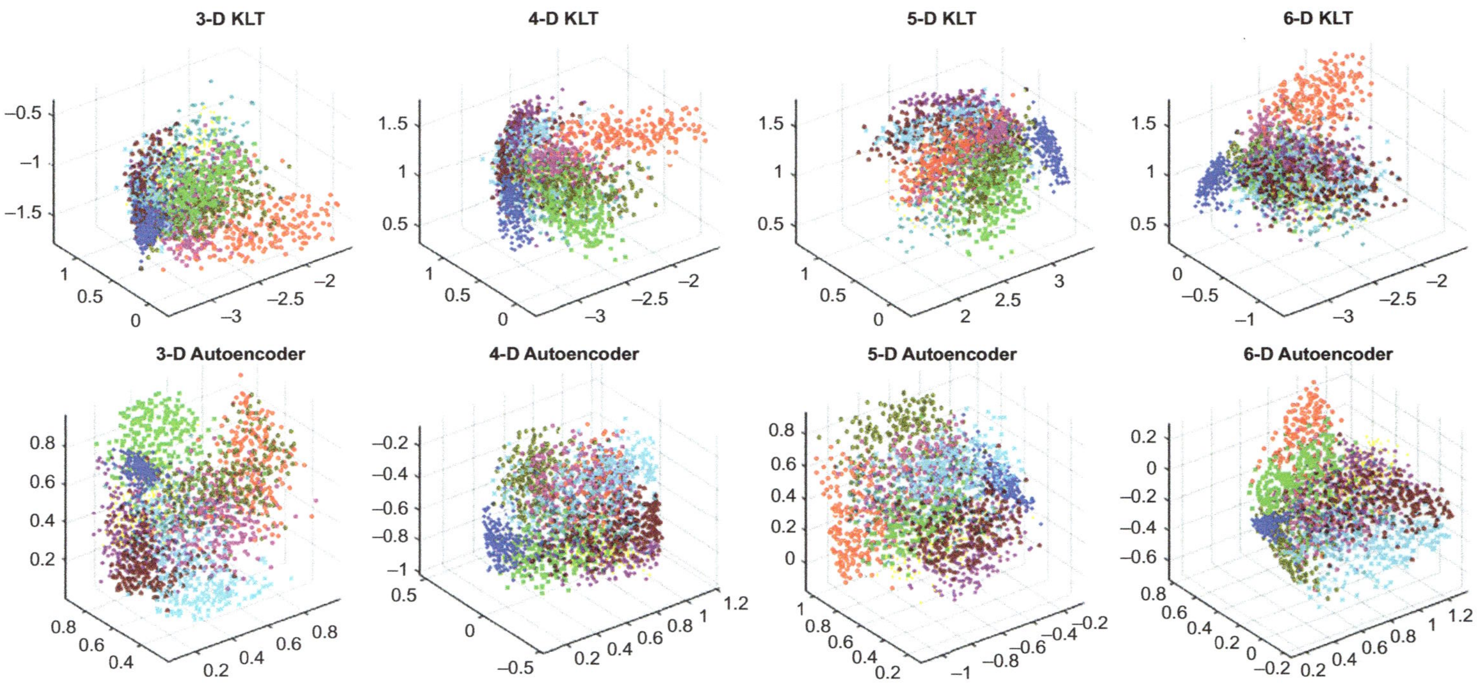

Fig. 18.11 Dimensionality reduction by autoencoder and KLT.

where the dataset in the 256-D space is mapped to $l = 3, 4, 5, 6$ dimensional space by the autoencoder (with l hidden layer nodes) as well as the KLT method, and then further mapped down to 3-D by KLT for visualization.

Also, as shown in Fig. 18.12 for the plots of the separabilities of the ten classes in the lower dimensional space, the autoencoder and KLT method achieve comparable separabilities.

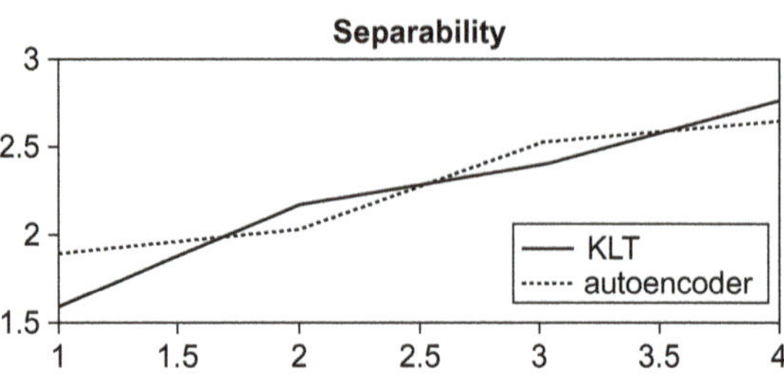

Fig. 18.12 Separability in lower dimensional space.

18.4 Deep Learning

Deep learning, a method based on a multilayer feed forward network with multiple hidden layers, has gained popularity and caused a new wave of enthusiasm for artificial neural networks in recent years, due heavily to the progress in computational capability as well as the latest developments in new algorithms.

Conceptually, a deep learning network can be simply understood as a network composed of multiple learning layers, including more than one hidden layer as well as the output layer. The perceptron network with a single learning layer and the three-layer network with the hidden and output learning layers discussed previously can be considered as special cases of deep learning. If we model a single learning layer by a nonlinear function that maps an input $\mathbf{x}$ to an output

$$\mathbf{y} = \mathbf{g}(\mathbf{w}^T\mathbf{x}) = \mathbf{f}(\mathbf{x}, \mathbf{w}) \tag{18.48}$$

then an L-layer deep learning network can simply be described by

$$\mathbf{y} = \mathbf{f}^{(L)}(\mathbf{f}^{(L-1)}(\cdots \mathbf{f}^{(1)}(\mathbf{x}, \mathbf{w}^{(1)}), \mathbf{w}^{(L-1)}), \mathbf{w}^{(L)}) = \mathbf{f}(\mathbf{x}, \mathbf{W}) \tag{18.49}$$

where $\mathbf{f}$ represents the cascade of all the L single-layer mapping functions $\mathbf{f}^{(1)}$ to $\mathbf{f}^{(L)}$, and $\mathbf{W} = [\mathbf{w}^{(1)}, \cdots, \mathbf{w}^{(L)}]$ is a matrix containing all weights in the network. When properly trained, such a multilayer learning network will be able to model some highly complicated and nonlinear relationships between input $\mathbf{x}$ and output $\mathbf{y}$.

When the multiple layers in a deep learning network are fully connected, that is, a node in a layer gets inputs from all nodes in the previous layer weighted by $\mathbf{w}$, then the method of backpropagation discussed previously can be used to update the weights based on gradient descent.

The basic structure of the multilayer network has been inspired by the biology of the brain. The hierarchical structure of the network is composed of lower level layers where the neurons selectively respond to some simple local features in the data, while neurons in subsequent higher level layers selectively respond to progressively more complex and global patterns. Such a hierarchy is very similar to the two main visual pathways in the primate brain, known as the dorsal or "where" pathway and the ventral or "what" pathway, both formed by multiple levels of cortical areas starting from the primary visual cortex (V1), each composed of different types of neurons in terms of the size and specific type of the visual stimuli they are selectively responsive to.

Specifically, the hierarchy of ventral pathway known for visual object recognition is composed of neurons in the lower layer such as V1 where the simple cells with small receptive field selectively respond to edges of different orientations, and neurons with progressively larger receptive fields in the higher layers selectively respond to more complex features in larger areas of the visual field, such as neurons in the inferior temporal (IT) area sensitive to visual patterns such as images of faces. This visual pathway is shown in Figs 18.13 and 18.14.

As an example of deep learning networks, the convolutional neural network (CNN) is hierarchical structure composed of a sequence of multiple 2-D layers for the purpose of image object recognition. This network can be considered to a large extent as a model that mimics the ventral pathway of the visual cortex, in the sense that the nodes in the lower layers detect and extract local features in the visual data (mimicking the simple cells in V1), while the nodes in higher layers respond selectively to more global patterns represented in terms of the local features detected by the lower layers (mimicking the cells in IT).

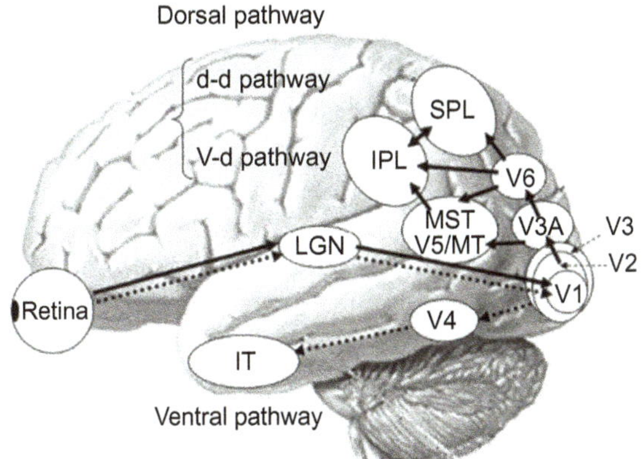

Fig. 18.13 Visual pathways. (Source: Yamasaki, S. (2012) Understanding the pathophysiology of Alzheimer's disease and mild cognitive impairment: a mini review on fMRI and ERP studies. *Neurology Research International.*, Article ID 719056)

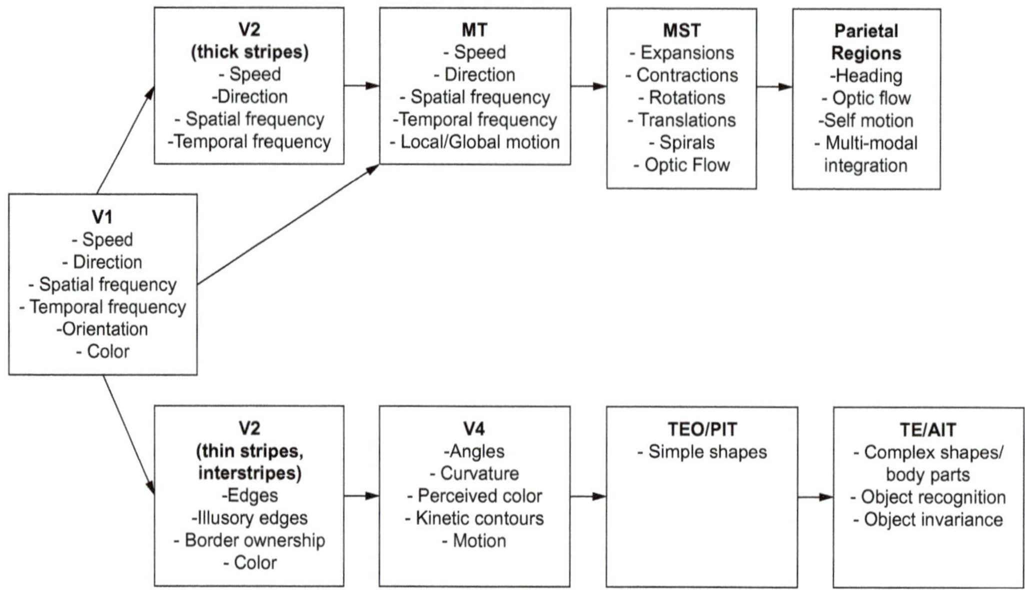

Fig. 18.14 Specialization of visual cortical areas. (Source: Perry, C. J. and Fallah, M. (2014) Feature integration and object representations along the dorsal stream visual hierarchy. *Frontiers in Computational Neuroscience*, 05 August 2014, Vol. 8.)

Specifically, in the lower layers of the CNN, local image features such as edges are extracted by convolution, a basic technique commonly used in image processing. Mathematically, a 1-D continuous convolution in time domain is an operation defined for two functions $f(t)$ and $g(t)$

$$y(t) = x(t) * h(t) = \int_{-\infty}^{\infty} x(t-\tau)h(\tau)d\tau = \int_{-\infty}^{\infty} h(t-\tau)x(\tau)d\tau \tag{18.50}$$

or a 2-D discrete convolution is spatial domain is an operation defined for two 2-D discrete functions $x[m,n]$ and $h[m,n]$

$$y[m,n] = \sum_{i=-k}^{k} \sum_{j=-k}^{k} x[m-i, n-j]\, h[i,j] \tag{18.51}$$

where $h[m,n]$ is called the *convolution kernel* which typically is assumed to be central symmetric $h[-m,-n] = h[m,n]$. Such a convolution is sometimes also considered as *filters*, by which a 2-D signal $x[m,n]$, typically a digital image, is filtered in some way (low-pass, high-pass, band-pass, etc.) to become $y[m,n]$. Figure 18.15 illustrates a 2-D convolution with a 3×3 kernel.

We see that each pixel $y[m,n]$ of the resulting image y is the sum of nine products. As the kernel slides over the input image x pixel by pixel and then row by row, all pixels of y are calculated.

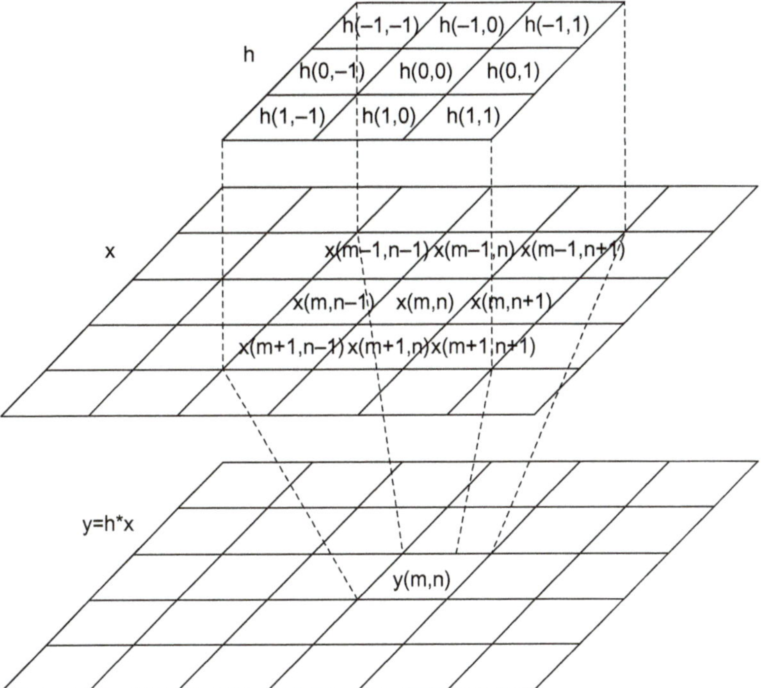

Fig. 18.15 2-D convolution.

For example, the following four kernels can be used to detect edges in vertical and horizontal orientations

$$\begin{bmatrix} -1 & 0 & 1 \\ -1 & 0 & 1 \\ -1 & 0 & 1 \end{bmatrix}, \quad \begin{bmatrix} 1 & 1 & 1 \\ 0 & 0 & 0 \\ -1 & -1 & -1 \end{bmatrix}, \quad \begin{bmatrix} 1 & 0 & -1 \\ 1 & 0 & -1 \\ 1 & 0 & -1 \end{bmatrix}, \quad \begin{bmatrix} -1 & -1 & -1 \\ 0 & 0 & 0 \\ 1 & 1 & 1 \end{bmatrix} \tag{18.52}$$

and the following four kernels are for detecting edges in two diagonal orientations

$$\begin{bmatrix} 0 & 1 & 1 \\ -1 & 0 & 1 \\ -1 & -1 & 0 \end{bmatrix}, \quad \begin{bmatrix} 1 & 1 & 0 \\ 1 & 0 & -1 \\ 0 & -1 & -1 \end{bmatrix} \begin{bmatrix} 0 & -1 & -1 \\ 1 & 0 & -1 \\ 1 & 1 & 0 \end{bmatrix}, \quad \begin{bmatrix} -1 & -1 & 0 \\ -1 & 0 & 1 \\ 0 & 1 & 1 \end{bmatrix} \tag{18.53}$$

Figure 18.16 shows the effects of edge detection by convolution. The cat image is convolved with the kernels for in Eq. (18.52), while the panda image is convolved with kernels in Eq. (18.53).

We note that the convolution as a linear combination can be described by the general neural network operation $\mathbf{y} = \mathbf{f}(\mathbf{x}, \mathbf{w})$ with the elements in the 2-D kernel $\mathbf{w}$ treated as the synaptic weights, through which the 2-D output $\mathbf{x}$ from a lower layer is fed forward into the next layer to produce its 2-D output $\mathbf{y}$.

These 3×3 kernels can be extended to larger kernels for detecting more complex features in a larger area of the visual field in higher layers of the network.

Specifically, the following operations take place sequentially along the multi-layer pathway of the CNN hierarchy as shown in Fig. 18.17, each serving certain purposes.

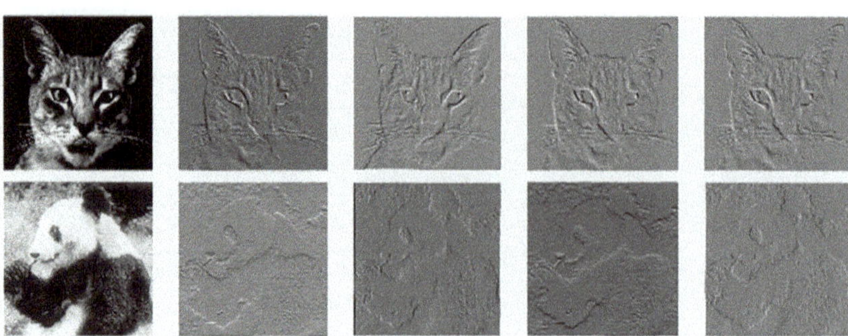

Fig. 18.16 Directional filtering.

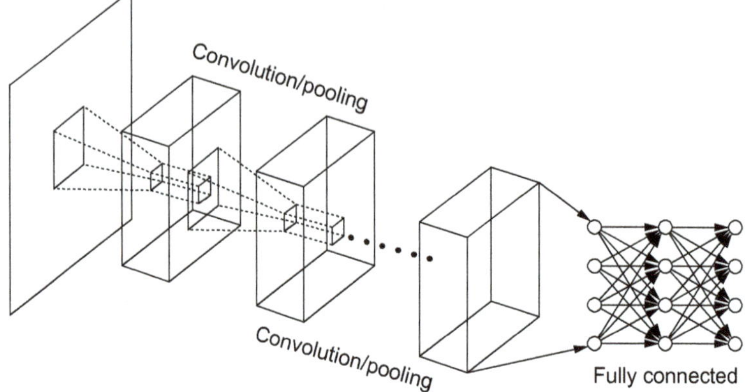

Fig. 18.17 Hierarchical convolutional network.

- *Convolution*
  Each node in the convolution layers is only locally connected to a set of neighboring nodes in a square patch in the previous layer. In particular, a node in the first layer takes as input the pixel values inside its receptive fields, a subregion of the image. It then generates the convolution, the sum of the inputs weighted by the kernel components, and feeds it to the next layer. At each position, multiple nodes with the same receptive field form a column in the depth dimension, all performing convolution but based on a different kernel to extract different types of local features, such as edges in various orientations.
- *Padding*
  Padding is needed for the output of convolution to maintain the same size as the input. For example, for a 5×5 kernel, the first pixel at the top-left corner of the output is in the position of the pixel in the third row and third column in the input, that is, the output is four pixels smaller in both dimension than the input. To prevent this size reduction from one layer to the next, the input needs to be padded with additional rows and columns of pixels around its four borders (zeros or taking same value as the border pixels).
- *Stride*
  Stride s is the number of pixels for the revolution kernel to shift in both dimensions in convolution. In regular convolution, $s = 1$, the size of the resulting image is the same as the input. But when $s > 1$, the kernel shifts s pixels from the center

of one receptive field to next. In other words, stride is the distance between the receptive field centers of neighboring neurons. The size of the output will be $1/s$ of the original in both dimensions.

- *Pooling*
  Pooling is a down-sampling method by which a local patch (e.g., 2×2) in the previous layer is represented by their average or maximum as a single value in the next layer called pooling layer, for dimension reduction that serves two purposes: (a) local shift and rotational invariance and (b) reduction of computational complexity.
- *Dilution or dropout*
  A certain fraction of nodes randomly selected in a layer is dropped by setting the corresponding weights to zero with a set probability, for the purpose of preventing overfitting, so that the network is less sensitive to certain detailed variations in the data due most likely to noise and thereby more generalizable.
- In the last stage of the CNN, all nodes in a layer are fully connected to all nodes in the previous layer. They take as input the progressively larger and more complex features extracted in the previous layer, and learn to selectively respond to different image objects of interest based on error backpropagation algorithm, to achieve the ultimate goal of image objects recognition.
- *Weight decay*
  As mentioned before, a multilayer neural network can be considered as a function $\mathbf{y} = \mathbf{f}(\mathbf{x}, \mathbf{w})$ for modeling the relationship between $\mathbf{x}$ and $\mathbf{y}$, as with many supervised learning algorithms regression and classification. Similarly, including a penalty term proportional to $||\mathbf{w}||$ or its square in the objective function is an effective way to regulate the algorithm and avoid overfitting.

PROBLEMS

1. Implement the perceptron network as a linear classifier (based on both one-hot and binary encoding) for 3-D data generated by the following function. Show your results in confusion matrix and the error rate.

```
function [X,y]=Data
    d=3;                        % number of dimensions
    K=8;                        % number of classes
    Nk(1:K)=50;                 % number of samples per class
    N=sum(Nk);                  % total number of samples
    Means=[-1 -1 -1 -1 1 1 1 1; % mean vectors of K classes
           -1 -1 1 1 -1 -1 1 1;
           -1 1 -1 1 -1 1 -1 1];
    X=[];
    y=[];
    s=0.22;
```

```
    for k=1:K                  % for each of the K classes
        Xk=Means(:,k)+s*randn(d,Nk(k));
        y=[y repelem(k,Nk(k))]
        X=[X Xk];
    end
end
```

2. Implement kernel perceptron network (based on radial basis kernel) and apply it to the classification of the iris dataset. Use both one-hot and binary encoding methods. Use half of the data for training and the other half for testing. Show your results in confusion matrix and error rate.
3. Repeat the previous problem using the handwritten digit dataset.
4. Implement the backpropagation network and apply it to the handwritten digit dataset based on all $d = 256$ dimensions of the data. Use both one-hot (10 output nodes each for one class) and binary encoding (four output nodes to encode 10 classes) labeling. Use half of the data for training and the other half for testing. Show your results in confusion matrix and error rate.

   Then use the KLT method based on the between-class scatter matrix $\mathbf{S}_B$ to reduce the data dimensionality while keeping 99% of the signal energy, and then repeat the classification task.
5. Implement autoencoder for dimensionality reduction and apply it to the iris dataset. Visualize and classify the data in the resulting lower dimensional space (both 2-D and 3-D space). Then compare the separability $J_{B/T}$ and classification error rate with that obtained based on the PCA method.
6. Repeat the problem above using the handwritten digit dataset.

19 Competition-Based Networks

19.1 Competitive Learning Network

Competitive learning is a neural network algorithm for unsupervised clustering, similar to the k-means algorithm considered previously. The competitive learning takes place in a two-layer network, composed of an input layer of d nodes that receives an input vector $\mathbf{x} = [x_1, \cdots, x_d]^T$ as a point in the d-dimensional feature space, and an output layer of K nodes that produces an output pattern $\mathbf{y} = [y_1, \cdots, y_K]^T$ representing the clusters of interest. Each input variable x_i may take either a continuous real value or a binary value depending on the specific application, the outputs $y_i \in \{0, 1\}$ are binary, of which only one is 1 while all others are 0 (one-hot), as the result of a winner-takes-all competition based on their activation values.

In each iteration of the learning process, a pattern $\mathbf{x}$ randomly chosen from the unlabeled dataset $\mathbf{X} = [\mathbf{x}_1, \cdots, \mathbf{x}_N]$ is presented to the input layer of the network, while each node of the output layer gets an activation value, a linear combination of all such d inputs, that is, the inner product of its weight vector $\mathbf{w}_i$ and the input vector $\mathbf{x}$

$$a_k = \sum_{j=1}^{d} w_{ij} x_j = \mathbf{w}_k^T \mathbf{x} = ||\mathbf{w}_k|| \; ||\mathbf{x}|| \; \cos\theta \qquad (k = 1, \cdots, K) \tag{19.1}$$

where θ is the angle between vectors $\mathbf{w}_k$ and $\mathbf{x}$ in the d-dimensional feature space. If the data vectors are normalized with unit length $||\mathbf{x}||^2 = 1$, they are points on the unit hypersphere in the space. Moreover, if the weight vectors $\mathbf{w}_k$ is also normalized with $||\mathbf{w}_k|| = 1$, then their inner product is only affected by the angle θ between them. In practice, the normalization of both the data vector $\mathbf{x}$ and the weight vector $\mathbf{w}$ is optional.

The output of the network is determined by a winner-takes-all competition. The node in the output layer that is maximally activated outputs 1 while all others output 0

$$y_i = \begin{cases} 1 & \text{if } a_i = \mathbf{w}_i^T \mathbf{x} = \max_j \mathbf{w}_j^T \mathbf{x} \\ 0 & \text{otherwise} \end{cases}, \qquad (i = 1, \cdots, K) \tag{19.2}$$

The inner product $\mathbf{w}_k^T \mathbf{x}$ is closely related to the Euclidean distance squared

$$d^2(\mathbf{w}_k,\mathbf{x}) = ||\mathbf{w}_k - \mathbf{x}||^2 = (\mathbf{w}_k - \mathbf{x})^T(\mathbf{w}_k - \mathbf{x}) = \mathbf{w}_k^T\mathbf{w} - 2\mathbf{w}_k^T\mathbf{x} + \mathbf{x}^T\mathbf{x}$$
$$= ||\mathbf{w}_k||^2 + ||\mathbf{x}||^2 - 2\mathbf{w}_k^T\mathbf{x} = ||\mathbf{w}_k||^2 + ||\mathbf{x}||^2 - 2||\mathbf{w}_k||\,||\mathbf{x}||\cos\theta \quad (19.3)$$

If both $\mathbf{x}$ and $\mathbf{w}_k$ are normalized, that is, they become points on a unit sphere in the d-dimensional feature space, then the winning node with the maximum activation $a_k = \mathbf{w}_k^T\mathbf{x}$ also has the minimum Euclidean distance $||\mathbf{w}_k - \mathbf{x}||$ and angular distance θ. This competition can therefore be expressed as the following

$$y_k = \begin{cases} 1 & \text{if } ||\mathbf{w}_k - \mathbf{x}|| = \min_j ||\mathbf{w}_j - \mathbf{x}|| \\ 0 & \text{otherwise} \end{cases}, \qquad (i = 1,\cdots,K) \quad (19.4)$$

Now the competitive learning rule can be expressed based on y_k

$$\mathbf{w}_k^{new} = \mathbf{w}_k^{old} + y_k\,\eta\,(\mathbf{x} - \mathbf{w}_k^{old}) = \begin{cases} (1-\eta)\mathbf{w}_k^{old} + \eta\mathbf{x} & \text{if } y_k = 1 \\ \mathbf{w}_k^{old} & \text{if } y_k = 0 \end{cases} \quad (19.5)$$

that is only the winning node learns to respond to the current input by modifying its weight vector, while the weights of all other nodes remain the same. This competitive learning is illustrated in Fig. 19.1, where the modified weight vector of the winner $\mathbf{w}^{new}$ is between $\mathbf{x}$ and the old $\mathbf{w}^{old}$, that is, the weight vector of the winner which is closest to the current input pattern $\mathbf{x}$ is pulled even closer to it, so that the node will be more likely to win and respond to the same input when it is presented again to the input of the network.

In the competitive learning rule, $0 < \eta < 1$ is the learning rate (step size), that is to be gradually reduced through the iterations from its initial value (e.g., $\eta_0 = 0.9$) toward zero, by a decay factor α. Due to the exponential decay of the learning rate, the iteration gradually converges to a stable state. The value of the decay factor (e.g., $\alpha = 0.99$) can be determined heuristically and experimentally depending on factors such as the size of the dataset.

Here are the iterative steps in the competitive learning:

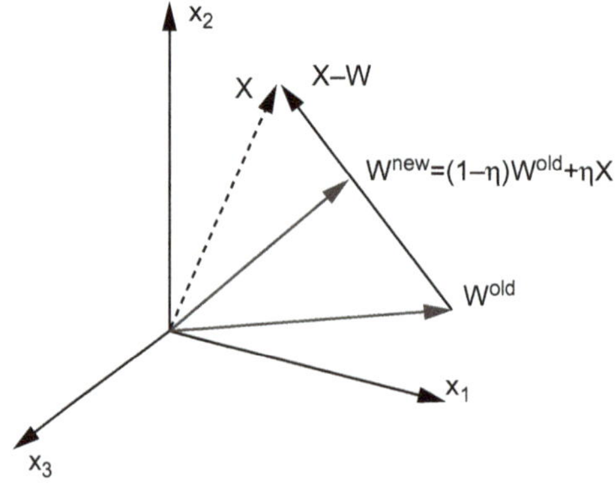

Fig. 19.1 Modification of weight vector of the winner.

- Step 0: Normalize all data vectors $\mathbf{x}$, initialize randomly the weight vectors $\mathbf{w}$ for the output nodes (e.g., any K randomly chosen samples from the dataset): $\mathbf{w}_1^{(0)}, \mathbf{w}_2^{(0)}, \cdots, \mathbf{w}_K^{(0)}$, set iteration index to zero $l = 1$.
- Step 1: Choose randomly an input pattern $\mathbf{x}$ from the dataset and calculate the activation for each of the output nodes

$$a_k = \mathbf{w}_k^T \mathbf{x} \qquad (k = 1, \cdots, K) \tag{19.6}$$

- Step 2: Find the index of the winning node $i = \arg\max_j a_j$, and update its weights

$$\mathbf{w}_i^{(l+1)} = \mathbf{w}_i^{(l)} + \eta(\mathbf{x} - \mathbf{w}_i^{(l)}) \tag{19.7}$$

  Reduce learning rate $\eta \leftarrow \alpha\eta$, renormalize $\mathbf{w}_i^{(l+1)}$.
- Step 3: Terminate the iteration if the clustering result has gradually stabilized, when the learning rate η is reduced from its initial value η_0 to some small value (e.g., 0.1) and the weight vectors no longer change significantly. Otherwise $l \leftarrow l + 1$, go back to step 1.

This iterative process can be more intuitively understood as shown in Fig. 19.2. Every time a sample $\mathbf{x}$ (a red dot) is presented to the input layer, one of the output nodes with its weight vector $\mathbf{w}$ (an blue x) closest to the current input $\mathbf{x}$ will become the winner and its weight vector is drawn even closer to $\mathbf{x}$. This process is carried out iteratively many times. Each cluster of similar sample points in the space will draw one of the weight vectors toward its central area, and the corresponding output node will always win and output 1 whenever any member of the cluster is presented to the network in the future. In other words, after this unsupervised learning, the feature space is partitioned into K regions, each corresponding to one of the K clusters, represented by one of the output nodes, whose weight vector is in the central area of the region. We see that this process is very similar to the k-means clustering algorithm, in which each cluster is represented by the mean vector of all of its members.

It is possible in some cases that the data samples are not distributed in such a way that they form a set of clearly separable clusters in the feature space. In the

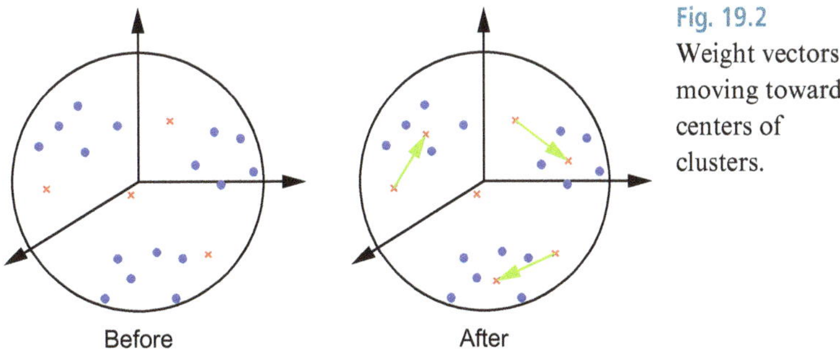

Fig. 19.2 Weight vectors moving toward centers of clusters.

extreme case, they may even form a continuum. In such cases, a small number of output nodes (possibly even just one) may become frequent winners, while others become "dead nodes" if they never win and consequently never get the chance to modify their weight vectors. To avoid such meaningless outcome, a mechanism is needed to ensure that all nodes have some chance to win. This can be implemented by including an extra bias term in the learning rule

$$a_k = \mathbf{w}_k^T \mathbf{x} + b_k \tag{19.8}$$

where b_k is the bias term proportional to the difference between the "fair share" of winning $1/K$ and the actual winning frequency

$$b_k = c \left(\frac{1}{K} - \frac{\text{number of winnings of the } k\text{th node}}{\text{total number iterations so far}} \right) \tag{19.9}$$

The value of the bias term b_k for the kth node will change its winning frequency. If the node is winning more than its share $1/K$, then $b_k < 0$ and it becomes harder for it to win in the future. On the other hand, if the node rarely wins, $b_k > 0$ and its chance to win in the future will be increased. Here hyperparameter c is some scaling coefficient. The greater c, the more balanced the competition will be. It needs to be fine tuned based on the specific nature of the data being analyzed.

This process of competitive learning can also be viewed as a *vector quantization* process, by which the continuous vector space is quantized to become a set of K regions, called *Voronoi diagram (tessellation)*. Vector quantization can be used for data compression. A cluster of similar signals $\mathbf{x}$ in the vector space can all be approximately represented by the weight vector $\mathbf{w}$ of one of a small number of K output nodes in the neighborhood of $\mathbf{x}$, and the data size can therefore be significantly reduced.

Same as in k-means clustering, competitive learning network also suffers from the problem that the ground truth number of clusters is typically unknown, and therefore it is difficult to determine the number of output nodes K of the network, which in the ideal case should be the same as the number of clusters in the dataset, so that each cluster can be represented by one of the nodes at the end of the clustering process. Similar to the case of k-means clustering, this issue can be addressed by carrying out the clustering process multiple times with different K values and checking to see which value best fits the data measured quantitatively by one of the criteria for the separability of the resulting cluster, such as those discussed in Section 15.1.

The MATLAB code for the iteration loop of the algorithm is listed as

```
[d N]=size(X);                 % dataset containing N samples
b=zeros(1,K);                  % bias terms for K clusters
freq=zeros(1,K);               % winning frequencies
eta=0.9;                       % initial learning rate
decay=0.99;                    % decay factor
it=0;
while eta>0.1                  % main iterations
    it=it+1;                   % iteration counter
    W0=W;                      % initial weight vectors
    x=X(:,randi(N,1));         % randomly select an input sample
    dmin=inf;
    for k=1:K                  % find winner among all output nodes
        d=norm(x-W(:,k))-b(k);
        if dmin>d
            dmin=d; m=k;       % mth node is the winner
        end
    end
    w=W(:,m)+eta*(x-W(:,m)); % modify winner's weights
    W(:,m)=w/norm(w);          % renormalize its weights (optional)
    share(m)=share(m)+1;       % update share for winner
    b=c*(1/K-share/it);        % modify biases for all nodes
    eta=eta*decay;             % reduce learning rate
end
```

Example 19.1

The competitive learning method is applied to a set of simulated 2-D data of four clusters. The network has $d = 2$ input nodes and $m = 4$ output nodes. The first 16 iterations are shown in Fig. 19.3, where open circles represent the data points, while the solid back squares represent the weight vectors. Also, the blue and red squares represent the weight vector of the winner before and after its modification, visualized by the straight line connecting the two squares. We see that the weight vectors randomly initialized are iteratively modified one at a time, and after these 16 iterations, they have each moved to the center of one of the four clusters. The separability of the clustering result measured by $tr(\mathbf{S}_T^{-1}\mathbf{S}_B)$ is 1.0. When the number of output nodes is reduced from the ground truth 4 to 3 and 2, the separability is also reduced to 0.76 and 0.46, respectively. When $K = 5$ output nodes are used, one of the four clusters is represented by two output nodes, indicating they can be subsequently merged.

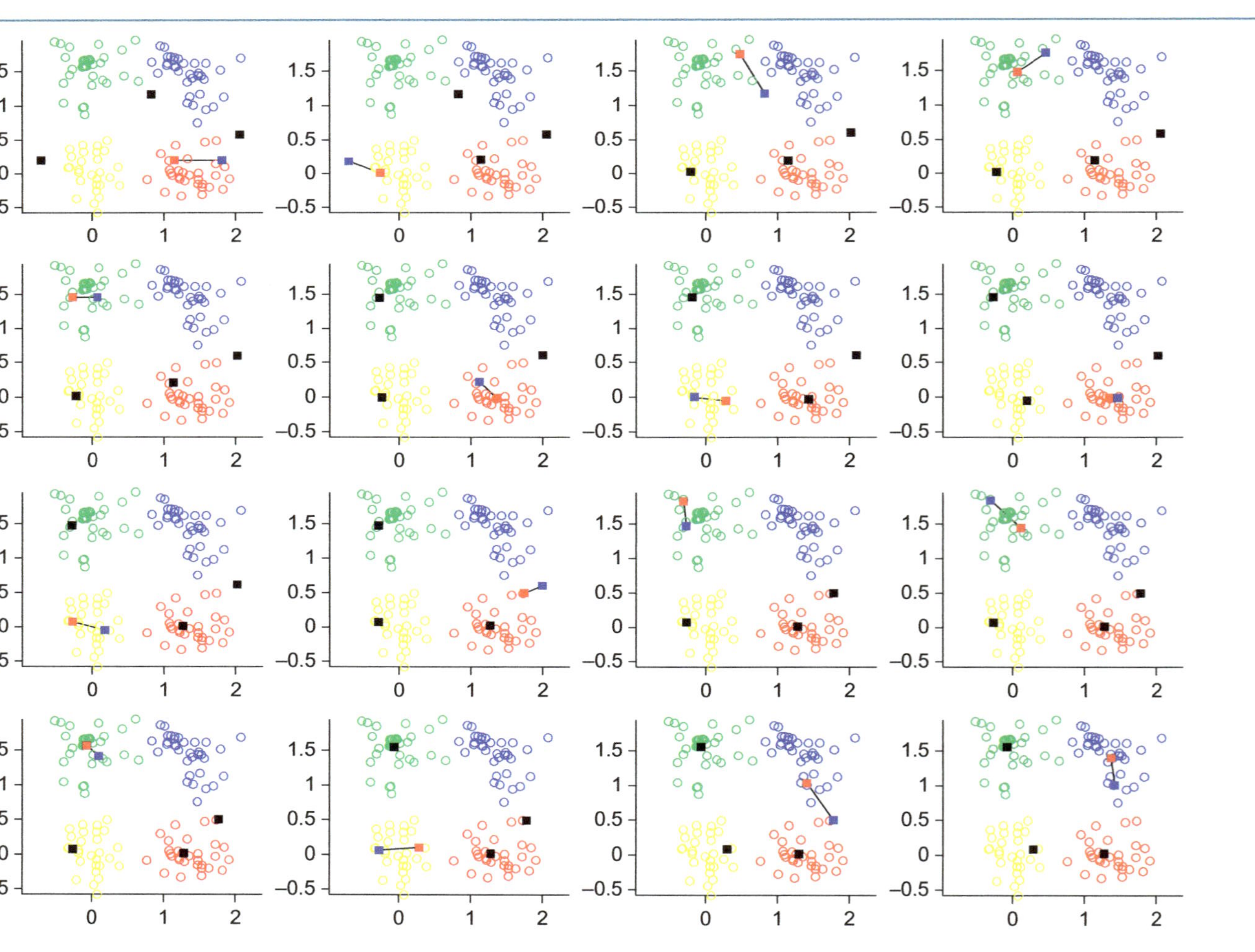

Fig. 19.3 Iterative learning of competitive network (2-D).

Example 19.2

Figure 19.4 shows the clustering results of a set of data points in a 3-D space, in terms of the weight vectors before (top) and after (bottom) of the competitive learning. The left column shows the 2-D space spanned by the first two principal components, while the right column shows the weight vectors in the original 3-D space. Again we see that the weight vectors move from their random initial positions to the centers of the eight clusters as the result of the clustering.

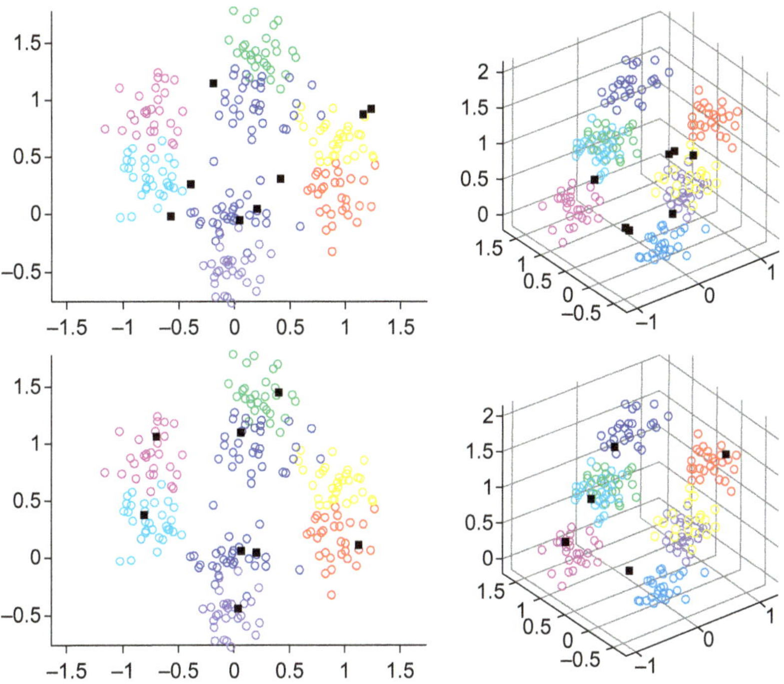

Fig. 19.4 Example of competitive learning (3-D).

Example 19.3

The competitive learning algorithm is also applied to the handwritten digit dataset, and the clustering result is shown as the average of all samples in each cluster in 16×16 image form shown in Fig. 19.5.

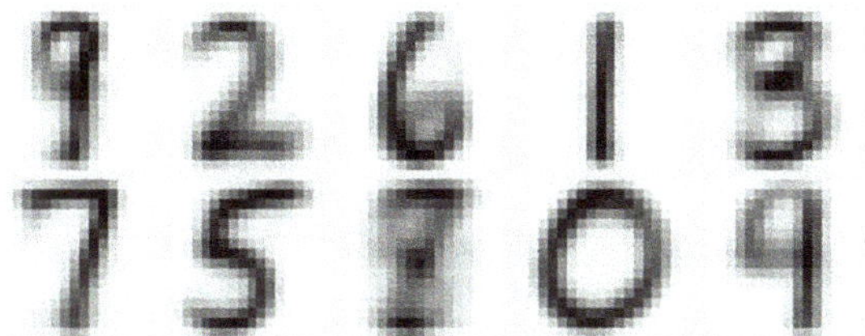

Fig. 19.5 Clustering of handwritten digits by competitive learning.

19.2 Self-Organizing Map (SOM)

The *self-organizing map (SOM)* is a two-layer unsupervised neural network learning algorithm that maps any input pattern presented to its input layer, a vector $\mathbf{x} = [x_1, \cdots, x_d]^T$ in a d-dimensional feature space, to a set of output nodes that forms a low-dimensional space called *feature map*, typically a 2-D grid (lattice), although 1-D and 3-D spaces can also be used. This is shown in Fig. 19.6. A SOM can therefore be considered as a dimensionality reduction method, by which the dataset in the high-dimensional feature space can be visualized and the features and structures of the dataset can be discovered.

The learning rule of the SOM is similar to that of the competitive learning network considered previously in that all output nodes compete based on their activation levels upon the input pattern presented to input nodes of the network. However, in contrast to the competitive learning network where only the winning node gets to modify its weight vector and learn to respond to a cluster of similar patterns individually and independently, here in SOM network all output nodes in the neighborhood of the winning node get to modify their weight vectors and thereby learn to respond collectively to a cluster of similar input patterns. As a result, all output nodes are locked into a continuum feature map with smooth and gradual variation in responsive selectivity.

The SOM as a learning algorithm is motivated by the topographic organization and columnar structures in the brain, in which signals from various visual or auditory sensory systems are topographically projected (mapped) to the corresponding cortical areas, where neighboring neurons in a local cortical area respond selectively to similar stimuli.

- *Retinotopic map:* Spatially adjacent stimuli in the visual field are represented and responded to by neurons in adjacent positions in the primary visual cortex (V1).
- *Tonotopic map:* Sound signals of different frequencies are mapped to the primary auditory cortex where neighboring neurons respond to similar frequencies.
- *Columnar architecture:* Visual signals for different edge orientations and different motion directions are mapped, respectively, to the primary visual area (V1)

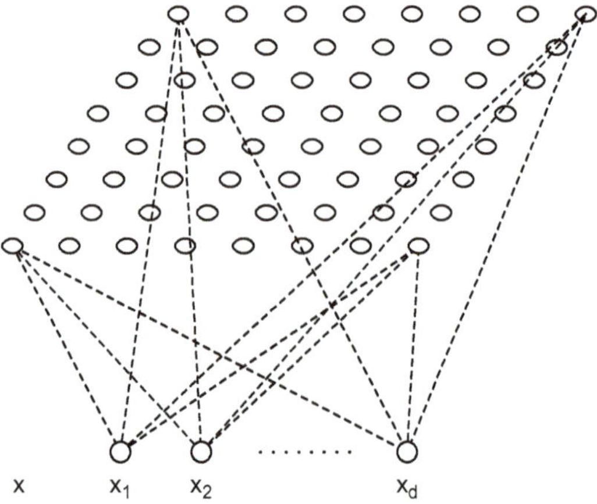

Fig. 19.6 Self-organizing map.

and the middle temporal area (V5 or MT), where neurons in the same column respond to the same edge orientations or similar motion directions, and neurons in neighboring columns respond to similar edge orientations or motion directions.

In a typical SOM network, the output nodes are organized in a 2-D feature map, and the competitive learning algorithm previously discussed is modified so that the learning takes place at not only the winning node in the output layer, but also a set of nodes in the neighborhood of the winner. The weight vectors of all such nodes are modified

$$\mathbf{w}_i^{new} = \mathbf{w}_i^{old} + u_i \eta \, (\mathbf{x} - \mathbf{w}_i^{old}) \tag{19.10}$$

where u_i is a weighting function based on the Euclidean distance d_i from the ith node in the neighborhood to the winner in the 2-D feature map, such as a Gaussian function centered at the winner

$$u_i = \exp\left(-\frac{d_i^2}{2\sigma^2}\right) \begin{cases} = 1 & \text{for winner with } d_i = 0 \\ < 1 & \text{for all other nodes} \end{cases} \tag{19.11}$$

where σ is a hyperparameter that controls the width of the Gaussian function. We see that all neurons in the neighborhood of the winner learn from the current input pattern in such a way that their weight vectors are modified to be closer to the input vector in the d-dimensional feature space, but they do so to different extents. Specially, for the winner itself with $d_i = 0$ and $u_k = 1$, the learning rate for the winner is maximally η, while all other nodes with $d_i > 0$ also modify their weights but with some lower rates $u_i \eta$. Those closer to the winner with smaller d_i and thereby greater u_i will learn more than those farther away.

Here are the specific steps of the SOM learning algorithm:

1. Initialize weights of a set of output nodes arranged in a 2-D map.
2. Choose randomly an input pattern vector $\mathbf{x}$ in the high-dimensional feature space and calculate the activation $y_i = \mathbf{w}_i^T \mathbf{x}$ for each of the output nodes.
3. Find the winning node with maximum activation y_k and update the weights of all nodes in its neighborhood by Eq. (19.10).
4. Go back to step 2 until the feature map is no longer changing or a set maximum number of iterations is reached.

In the training process, both σ controlling the size of the neighborhood and η for the learning rate will be gradually reduced.

The MATLAB code segment listed here is for the essential part of the SOM algorithm.

```
[d N]=size(X);                 % dimension and number of samples
W=rand(d,M,M);                 % initialization of weights
for i=1:M
    for j=1:M
        w=reshape(W(:,i,j),[d 1]);
        W(i,j,:)=W(:,i,j)/norm(w); % normalize all weight vectors
```

```
        end
    end
    eta=0.9;                        % initial learning rate
    sgm=M;                          % width of Gaussian neighborhood
    decay=0.999;                    % rate of decay
    for it=1:nt                     % training iterations
        n=randi([1 N]);
        x=X(:,n);                   % pick randomly an input
        amax=-inf;
        for i=1:M                   % find the winner
            for j=1:M
                w=reshape(W(:,i,j),[d 1]);
                a=x'*w;             % activation as inner product
                if a>amax
                    amax=a; wi=i; wj=j; % record winner so far
                end
            end
        end
        for i=1:M
            for j=1:M
                dist=(wi-i)^2+(wj-j)^2; % distance to winner
                c=exp(-dist/sgm);       % Gaussian weights
                w=reshape(W(:,i,j),[d 1]); % get a weight vector
                w=w+eta*c*(x-w);    % modify weights
                w=w/norm(w);        % normalize weight vectors
                W(:,i,j)=w;         % save modified weight vector
            end
        end
        eta=eta*decay;              % reduce learning rate
        sgm=sgm*decay;              % reduce width of Gaussian
    end
```

Example 19.4

This example of a "self-organizing map" is how the SOM algorithm got its name. The output nodes organized into a 2-D map learn to respond to the positions of a set of random points $\mathbf{x} = [x_1, x_2]^T$ in a $d = 2$ dimensional feature space. The weights of the output nodes are randomly initialized, and then modified iteratively during the learning process when the data points are repeatedly presented to the input of the network. Once the SOM has been trained, the output nodes close to each other in the 2-D feature map respond to a set of adjacent data points in the 2-D space, thereby forming a self-organized spatial map, a roughly regular grid.

The 2-D map of the output nodes is visualized as shown in Fig. 19.7, where the spatial position of each node is determined by the two components of its weight

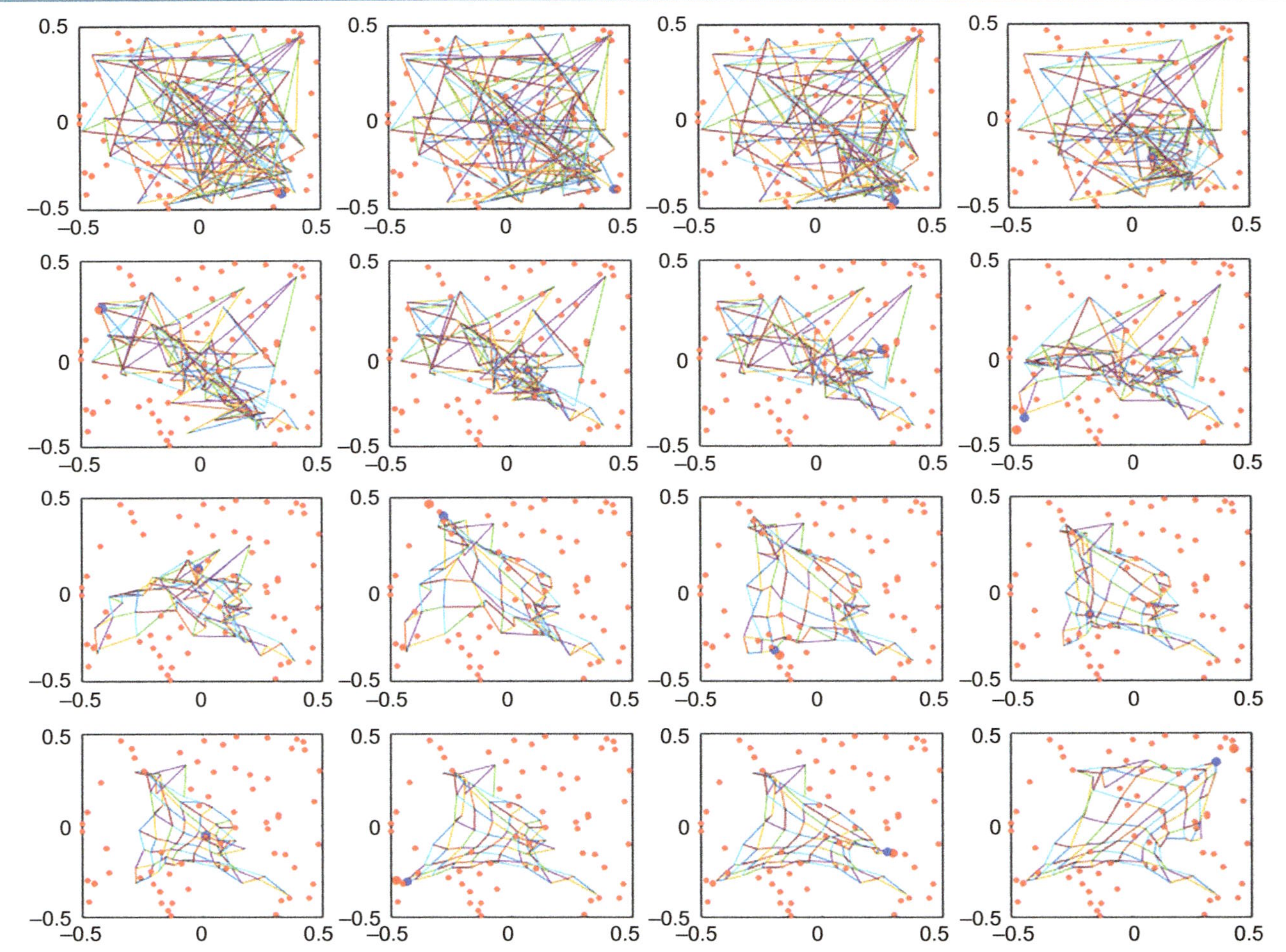

Fig. 19.7 Iterative learning of SOM (first 16 iterations).

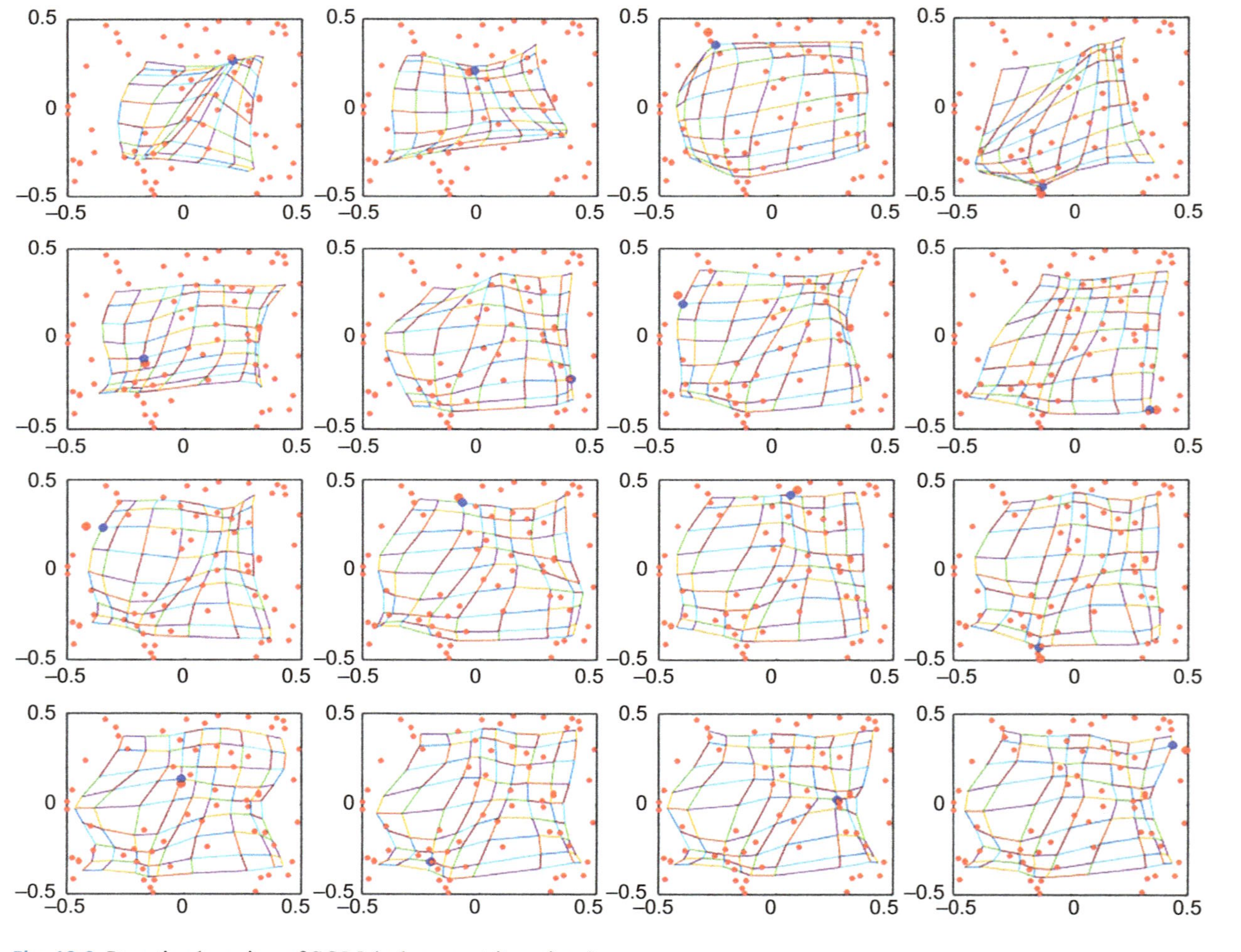

Fig. 19.8 Iterative learning of SOM (subsequent iterations).

vector $\mathbf{w} = [w_1, w_2]^T$ treated as its coordinates, and each node is connected to its four neighbors in the 2-D grid to indicate the spatial structure of the feature map.

In this example, the competitive learning is based on the distance between the weight vector $\mathbf{w}_i$ and the current input $\mathbf{x}$, instead of their inner product $\mathbf{w}_i^T\mathbf{x}$. The node with the shortest distance $||\mathbf{w}_i - \mathbf{x}||$ becomes the winner, and its weight vector $\mathbf{w}$ is pulled even closer to the input, dragging along with it all its neighboring nodes.

The iterative modification of the weights is shown in Figs 19.7 for the first 16 iterations and 19.8 for the subsequent iterations (in every 50 iterations). In each of the panels, the output node (blue dot) closest to the current input vector (red dot) becomes the winner and pulls the weight vectors of all neighboring nodes toward the input. This process converges to a configuration in which the output nodes form roughly a regular 2-D grid, a self-organized map in the 2-D feature space.

Example 19.5

In this example the SOM algorithm is applied to address the *traveling salesman problem (TSP)*. The SOM learns to respond to the locations of a set of 25 North America cities represented by their coordinates (longitudes and latitudes) and thereby approximate the shortest path through all these cities. The initial locations of the weight vector are arranged as a 1-D loop around the cities. And the nodes of the SOM gradually learn to modify their weights based on the city locations, so that they move toward the cities nearby to eventually form a reasonably short path through all these cities. The final result is unaffected by how the weights are initialized.

This learning process is illustrated in Fig. 19.9 for the first 16 iteration and Fig. 19.10 for the subsequent iterations (in every 50 iterations), where the output nodes are represented by their weight vectors as points along the curve in the 2-D map, together with the locations of the cities as the input to the SOM one at a time.

The resulting configuration of the output nodes is not guaranteed to be the desired shortest path through all cities, but it is should be reasonably close to such a solution.

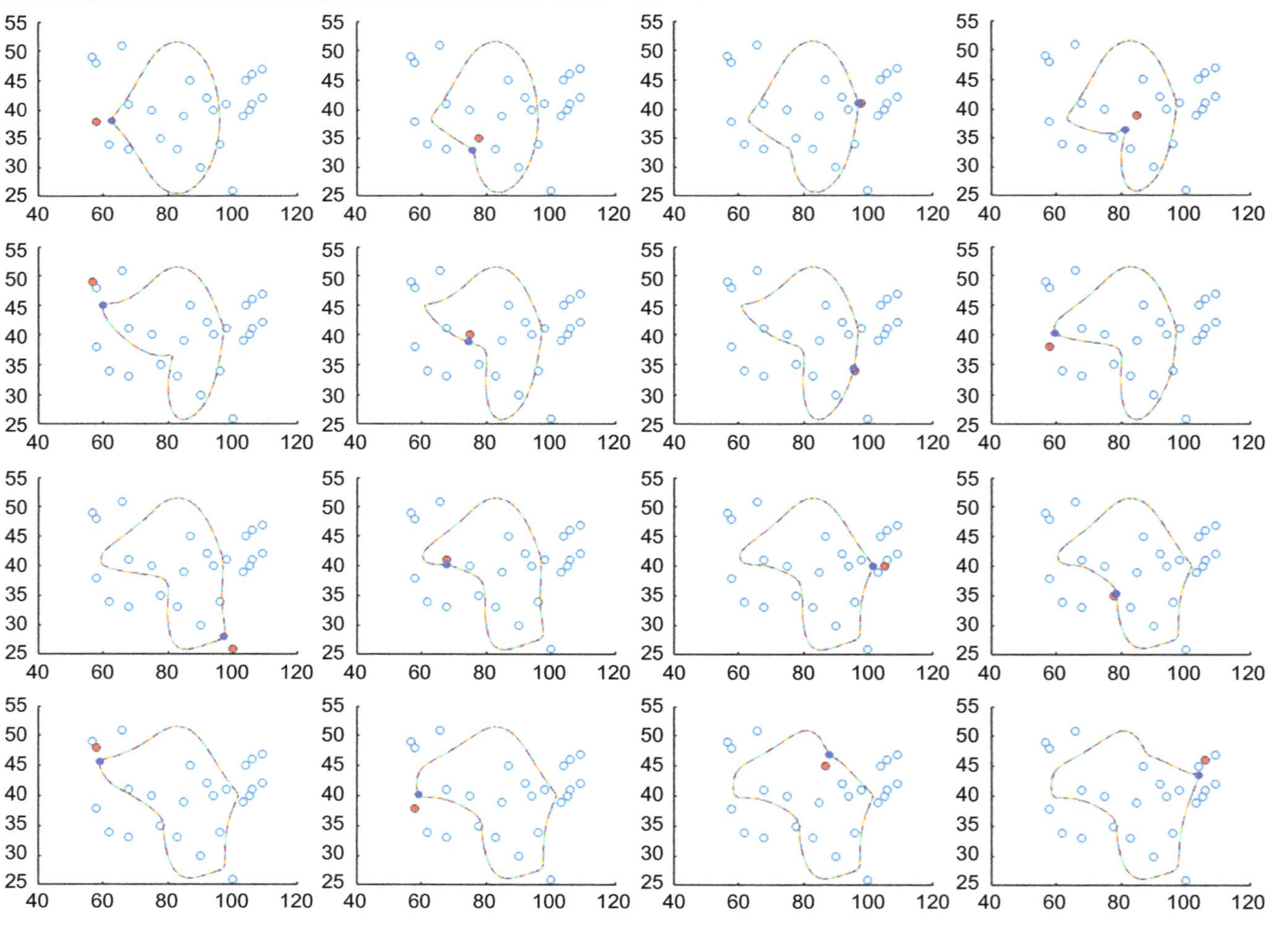

Fig. 19.9 SOM applied to the TSP (first 16 iterations).

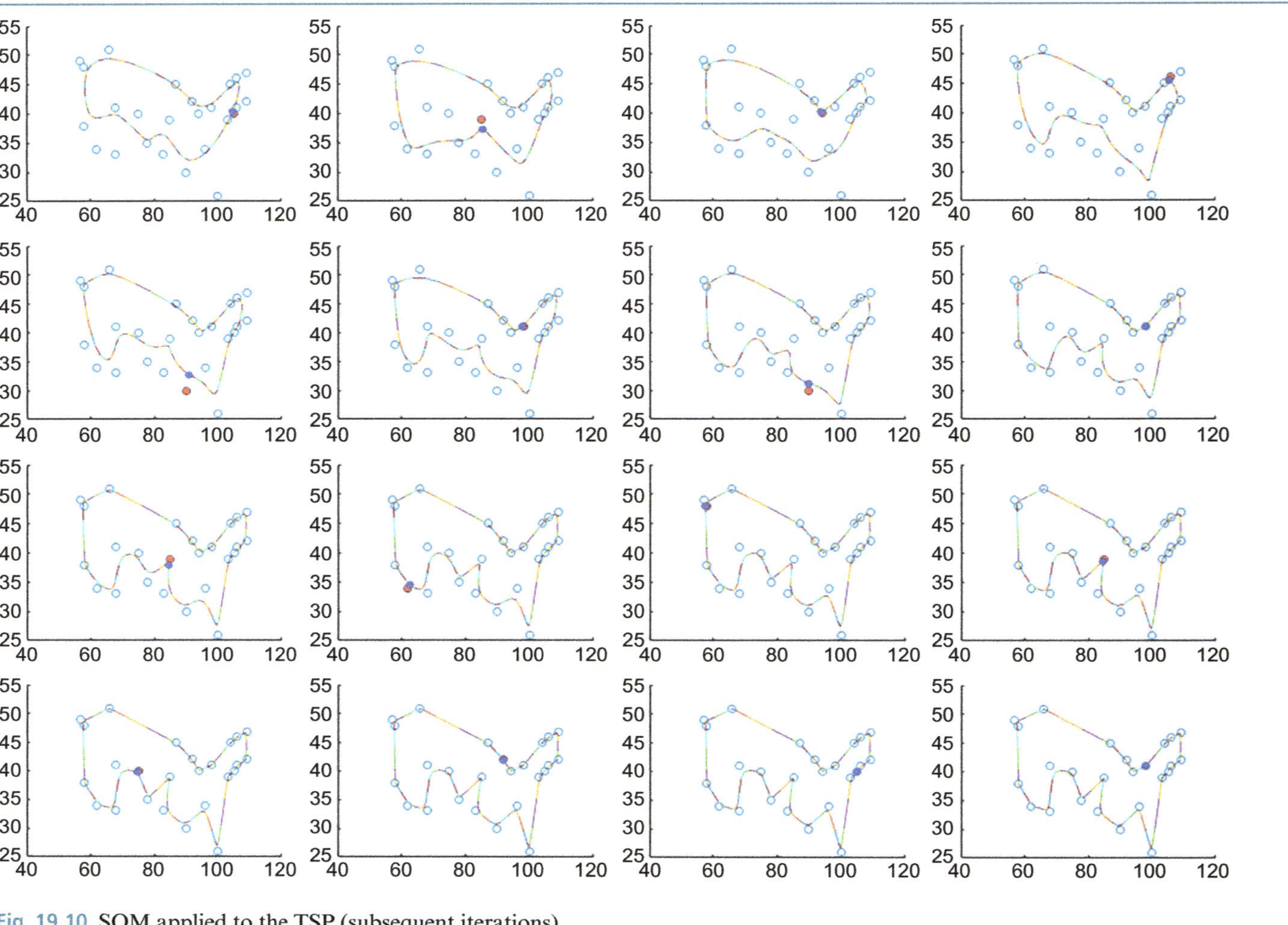

Fig. 19.10 SOM applied to the TSP (subsequent iterations).

Example 19.6

In this example, a SOM network learns to respond to different colors, as points in the color space spanned by the three primary colors red, green and blue (RGB). Each node in the 2-D array of the output layer is connected to the three input nodes taking a 3-D vector for a color sample $\mathbf{x}$, and generates the inner product $\mathbf{w}^T\mathbf{x}$ of the weight vector $\mathbf{w}$ and the input vector $\mathbf{x}$ as the activation. During training, the points in a 3-D regular grid of the color space are randomly selected as the training samples, and the weights of the SOM are gradually modified so that eventually each output node learns to respond most strongly to a specific color. Here all training vectors are normalized so that they have unity intensity, but their hue and saturation are determined by the ratios of the three primaries. Similarly, the weight vectors are also renormalized after each iteration in training. As $||\mathbf{w}|| = ||\mathbf{x}|| = 1$, the activation $\mathbf{w}^T\mathbf{x} = ||\mathbf{w}||\ ||\mathbf{x}|| \cos\theta = \cos\theta$ is maximized if the angle between θ between $\mathbf{w}$ and $\mathbf{x}$ is minimized, that is, the weight vector matches the input color vector. Now the favored color of each output node can be shown by color coding the components of its weight vector as RGB, and the SOM can be visualized by the favored colors of all of its output nodes, as shown in Fig. 19.11. Here the left panel shows the SOM when all weights are randomly initialized, and middle panel shows the fully trained SOM with neighboring nodes responding to similar colors. The right panel further shows all the winners during the learning process, encoded by the input colors they win, while all other nodes that never win (but still learn to respond to certain colors along with their neighboring winners) are colored black. We see that the winners seem to form some structural patterns.

Fig. 19.11 Colors learned by a SOM.

When the size of the SOM is progressively increased, more sophisticated web-like patterns start to emerge, as shown in the left panel of Fig. 19.12 for a SOM of size 1000×1000. Such emergent patterns closely resemble the *perspective* projection of a regular grid in the 3-D color space onto an arbitrary 2-D plane, as shown in the right panel for a zoomed-in view of such a projection. We therefore see that the SOM network actually has learned to mimic the perspective projection of the training samples in the 3-D grid in the color space onto the 2-D output array. However, in an *orthographic* projection of the 3-D space as shown in Fig. 19.13 no such patterns emerge. An interesting question is why the SOM learns to mimic the perspective projection instead of a orthographic projection, while there is no specific mechanism that distinguishes these two types of projections in the learning process.

Fig. 19.12 Emergent patterns in SOM (left) and Perspective projection of the color grid (right).

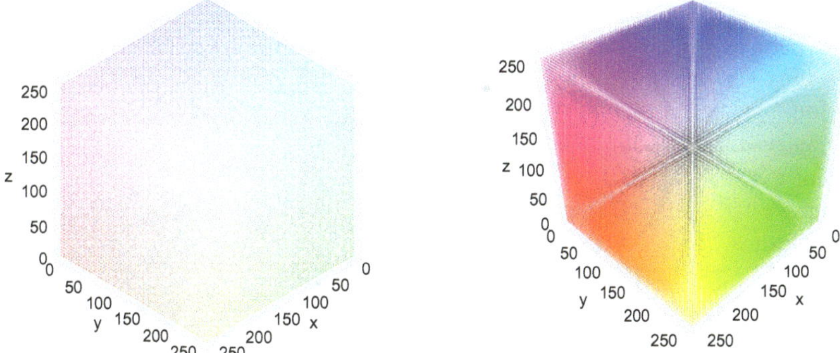

Fig. 19.13 Orthographic and perspective projections of color grid orthographic (left) and perspective (right) projections of the color grid.

PROBLEMS

1. Implement competitive learning network for clustering and apply it to the iris dataset. Show the result in confusion matrix and error rate. Also visualize your result in 2-D and 3-D spaces with all samples color coded according to (1) the cluster they are assigned to, and (2) their ground truth labeling.
2. Repeat the previous problem but based on the handwritten digit dataset. Also visualize the mean vectors of the $K = 10$ clusters converted back into 16×16 images to see what they actually represent.

   Compare all these results to those obtained by the k-means algorithm in the homework in Chapter 15.
3. Use the SOM network to learn different colors each represented by a 3-D vector for the three primary colors red, green, and blue. When the SOM has converged,

show the resulting SOM by color (RGB) coding the output nodes in the 2-D array by their three weights, as in Example 19.6.

If interested, further show the SOM map but containing only those winning nodes during the training process. All other nodes that have never won during training will remain black. Try some large size SOM (e.g., 500 × 500) and large number of training samples (e.g., 10^3, i.e., 10 intensity levels in the range between 0 and 255 for each of the primary colors). Observe to see if any patterns show up in the map.

4. Use the SOM network to solve the TSP for the major North American cities listed in Table 19.1, together with their latitude and longitude. (More cities can be found at: http://www.infoplease.com/ipa/A0001796.html.)

Table 19.1

	Lat.		Long.	
City	Minute	Second	Minute	Second
Los Angeles	34	3	118	15
San Francisco	37	47	122	26
New York	40	47	73	58
Chicago	41	50	87	37
Boston	42	21	71	5
Washington, D.C.	38	53	77	02
Miami	25	46	80	12
Minneapolis	44	59	93	14
New Orleans	29	57	90	4
Seattle	47	37	122	20
Dallas	32	46	96	46
Cleveland	41	28	81	37
Salt Lake City	40	46	111	54
Atlanta	33	45	84	23
Denver	39	45	105	0
Kansas City	39	6	94	35
Phoenix	33	29	112	4

Part VI

Reinforcement Learning

Part VI contains only one chapter to discuss the *reinforcement learning (RL)*, which is considered as one of the three basic machine learning paradigms, alongside supervised learning (e.g., regression in Part II and classification in Part IV), and unsupervised learning (e.g., clustering in Chapter 15). The goal of RL is for the algorithm, a software *agent*, to learn to make a sequence of decisions, called *policy*, in terms of the best *actions* to take for a specific task in a given *environment*, for the purpose of receiving the maximum cumulative *rewards* from the environment.

Unlike supervised learning for either regression or classification based on the training set composed of labeled data samples, RL is a learning method without any examples of optimal behaviors. Instead, RL learns on its own without relying on any labeled training data informing the agent explicitly what the correct responses are. Conversely, unlike unsupervised methods, which learn passively from the given dataset, RL is a trial-and-error learning method, in which the agent actively interacts with and learns from the given environment.

RL as a sequential method in the sense that the learning depends on the dynamics of the environment, which is modeled by a *Markov decision process (MDP)*, a stochastic system of multiple states with probabilistic state transitions and rewards. Starting from an initial state, the agent goes through a sequence of states to eventually reach a destination state. If the MDP model of the environment is known in terms of the state transition and reward probabilities, then the agent only needs to determine the optimal policy in terms of what action to take at each state in the sequence to transit to the next, so that the accumulated rewards can be maximized as the consequence of its actions. The task in this *model-based* case is called *planning*.

However, if the MDP of the environment is unknown, as in the case of most real-world problems, the agent needs to learn the MDP environment by repeatedly running the dynamic process while following some initial policy, and gradually evaluate the received rewards and improve the policy to eventually reach optimality. Such a process of learning the environment by interacting with it is called a *Monte Carlo (MC)* method and the task in this *model-free* case is called *control.*

Depending on the action a taken by the agent in the state s it is current in, the system transits to the next state s', one of the states available at s, according to a transition probability distribution $p(s'|s,a)$. The goal of RL is to choose the optimal action to maximize the long-term expected cumulative reward. This is typically

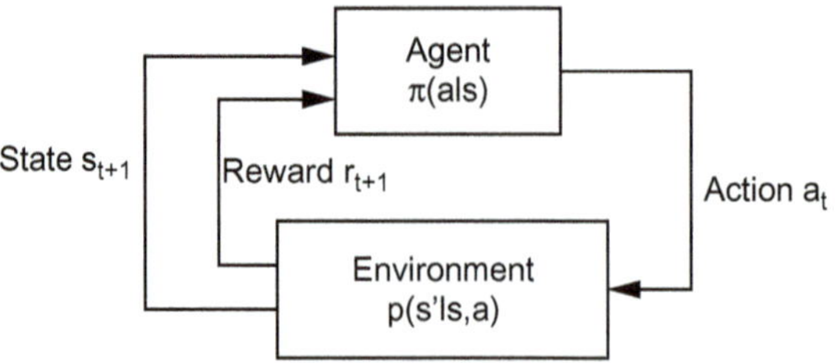

Fig. VI.1 Reinforcement learning: agent and its environment.

realized by the method of *dynamic programming (DP)*, a class of algorithms that seeks to simplify a complex problem by breaking it up into a set of sub-problems to be solved recursively. The process is illustrated in Fig. VI.1 showing the tth step of the process, where the agent makes a decision in terms of what action a_t to take in the current state $s_t = s$ by following certain policy $\pi(a|s)$, based on which the environment makes the corresponding transition to the next state $s_{t+1} = s'$ according to the transition probability $p(s'|s,a)$ conditioned on the current state $s_t = s$ and the agent's action a_t, and provides a reward r_{t+1}. We will consider the specific details of the DP algorithms in the subsequent sections.

20 Introduction to Reinforcement Learning

20.1 Markov Decision Process

20.1.1 Markov Chain

The mathematical framework for reinforcement learning (RL) takes the form of a *Markov decision process (MDP)*, based on the more basic concept of a *Markov process* or *Markov chain*, which is a stochastic model of a system that can be characterized by two types of parameters: (a) the set $S = \{s^1, \cdots, s^N\}$ containing all possible states of the system, and (b) the transition probability $P_{mn} = P(s^n | s^m)$ that models the dynamics of the system in terms of its state transition from the current state $s_t = s^m$ to the next state $s_{t+1} = s^n$ for all $m, n = 1, \cdots, N$. Here s^n denotes the nth state out of all $N = |S|$ states of the system (not s to the nth power), while s_t denotes the state at the tth step of the process.

The system is assumed to be memoryless, that is, its future depends on the current state s_t but is independent of the past history in terms of all previous states

$$P(s_{t+1} | s_t, s_{t-1}, \cdots, s_0) = P(s_{t+1} | s_t) \tag{20.1}$$

For example, the motion of a helicopter can be modeled by a Markov process if its state is described by its linear and angular position and velocity, as its future position and velocity are completely determined by its present state. However, it cannot be modeled by a Markov process if the state is only described by its current position, while its velocity depends also on its positions in the past.

All transition probabilities can be organized as an $N \times N$ state transition matrix

$$\mathbf{P} = \begin{bmatrix} P_{11} & \cdots & P_{1N} \\ \vdots & \ddots & \vdots \\ P_{N1} & \cdots & P_{NN} \end{bmatrix} \tag{20.2}$$

where $P_{mn} = P(s^n | s^m)$ is the probability of the transition from state s_m to the state s_n. Since at any state s_m, the next state has to be one of the N possible states, we have

$$\sum_{n=1}^{N} P_{mn} = \sum_{n=1}^{N} P(s^n | s^m) = 1, \qquad (m = 1, \cdots, N) \tag{20.3}$$

Any matrix satisfying this property is called a *stochastic matrix*. In the following discussion, we will concentrate mostly on the general transition from the current

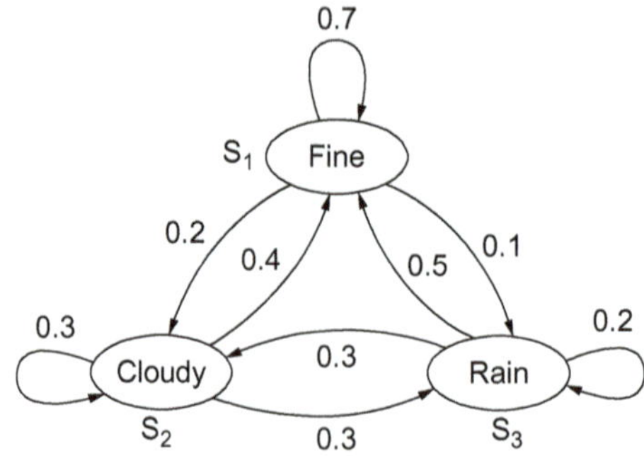

Fig. 20.1 Stochastic matrix for LA weather.

state $s_t = s$ to the next state $s_{t+1} = s'$ with transition probability denoted by $P(s_{t+1} = s'|s_t = s) = P_{ss'}$. For example, the transitions of the weather in Los Angeles (a lot of fine days!) are shown in Fig. 20.1, with the stochastic matrix shown in Eq. (20.4).

$$\mathbf{P} = \begin{bmatrix} 0.7 & 0.2 & 0.1 \\ 0.4 & 0.3 & 0.3 \\ 0.5 & 0.2 & 0.3 \end{bmatrix} \tag{20.4}$$

20.1.2 Markov Reward Process

A *Markov reward process (MRP)* is an extension of the Markov chain that also includes a reward associated with each of the states. The goal is for an agent to collect the maximum amount of rewards while going through the states of a dynamic process of the system. An MRP can be described by four sets of parameters, including the set of all rewards R, a *discount factor* γ, as well as the set of states S, and the state transition probabilities represented by P for a Markov chain. An MRP can therefore be represented by a quadruple (4-tuple) $\langle S, P, R, \gamma \rangle$.

Here the *reward* R is a feedback signal to the agent at each time step of the sequence of state transitions, indicating how well it is doing with respect to the overall task. The reward can be negative, as a penalty, for situations to be avoided. The performance of the agent is measured by the sum of rewards accumulated over all time steps of the Markov process, weighted by the discount factor $\gamma \in [0, 1]$ that discounts future rewards. If γ is close to 0, the immediate reward is emphasized, that is, the policy is short-sighted or greedy; while if γ is close to 1, the rewards in the future steps will be almost as valuable as immediate ones, that is, the policy is far-sighted.

The sequence of state transitions of an MRP from an initial state s_0 to a terminal state s_T is called an *episode*, and the number of time steps T taken by the agent is called the *horizon*, which can be either finite or infinite.

Both the reward r and the next state s' at any given current state s are random variables with joint probability $p(s', r|s) = P(s_{t+1} = s', r_{t+1} = r|s_t = s)$, based on which both the state transition and reward probabilities can be found by marginalizing r and s', respectively

$$p(s'|s) = P(s_{t+1} = s'|s_t = s) = \sum_{r \in R} p(s',r|s), \qquad P(r|s) = \sum_{s' \in S} p(s',r|s) \quad (20.5)$$

Conventionally the reward r received after arriving at state s_t is denoted by r_{t+1} instead of r_t. The index $s' \in S$ for the summation over all states will be abbreviated to s' in the subsequent discussion. The expected reward in state s is

$$r(s) = E[r_{r+1}|s_t = s] = \sum_r rP(r|s) = \sum_r r \sum_{s'} P(s',r|s) \quad (20.6)$$

The *return* G_t in step $s = s_t$ is defined as the accumulated reward, the sum of the immediate reward after arriving at the current state $s = s_t$ and all delayed rewards in the future states up to and including reward r_T at the terminal state s_T at the end of the episode, discounted by γ

$$G_t = r_{t+1} + \gamma r_{t+2} + \gamma^2 r_{t+3} + \cdots + \gamma^{T-t-1} r_T = \sum_{k=0}^{T-t-1} \gamma^k r_{t+k+1} \quad (20.7)$$

The *state value* function $v(s)$ in state $s = s_t$ is defined as the expected return $E[G_t]$, which can be expressed recursively in terms of the values $v(s')$ of all possible next states $s' = s_{t+1}$

$$\begin{aligned} v(s) &= E[\, G_t | s_t = s \,] = E[r_{t+1} + \gamma r_{t+2} + \gamma^2 r_{t+3} + \cdots | s_t = s] \\ &= E[r_{t+1} + \gamma(r_{t+2} + \gamma r_{t+3} + \cdots)|s_t = s] \\ &= E[r_{t+1}|s_t = s] + \gamma E[G_{t+1}|s_{t+1} = s', s_t = s] \\ &= r(s) + \gamma \sum_{s'} P(s'|s)v(s') \end{aligned} \quad (20.8)$$

where $r(s)$ is given in Eq. (20.6). The value of the terminal state s_T at the end of an episode is zero, as there will be no next state and therefore no more future reward.

This equation is called the *Bellman equation*, by which the value $v(s)$ at current state s is expressed recursively in terms of the immediate reward $r(s)$ and the values $v(s')$ of all possible next states, without explicitly invoking all future rewards. Based on such a *bootstrapping* approach, the multistep MRP problem is expressed recursively as a single-step subproblem concerning only one state transition from s to s', so that the current state value can be found from all future ones by induction. For this reason, the Bellman equation plays an essential role in all reinforcement learning algorithms to be considered in the following.

The Bellman equation in Eq. (20.8) holds for all N states $s^1, \cdots, s^N \in S$, and the resulting N equations can be expressed in vector form as

$$\begin{aligned} \mathbf{v} &= \begin{bmatrix} v(s^1) \\ \vdots \\ v(s^N) \end{bmatrix} = \begin{bmatrix} r(s^1) \\ \vdots \\ r(s^N) \end{bmatrix} + \gamma \begin{bmatrix} P(s^1|s^1) & \cdots & P(s^N|s^1) \\ \vdots & \ddots & \vdots \\ P(s^1|s^N) & \cdots & P(s^N|s^N) \end{bmatrix} \begin{bmatrix} v(s^1) \\ \vdots \\ v(s^N) \end{bmatrix} \\ &= \mathbf{r} + \gamma \mathbf{P} \mathbf{v} \end{aligned} \quad (20.9)$$

where $\mathbf{P}$ is state transition matrix previously defined in Eq. (20.2), a stochastic matrix. Now the Bellman equation can be solved to get the value functions at all N states

$$\mathbf{v} = (\mathbf{I} - \gamma\mathbf{P})^{-1}\mathbf{r} \tag{20.10}$$

Such a solution exists because matrix $\mathbf{I} - \gamma\mathbf{P}$ is invertible. To see this we note that all eigenvalues of the stochastic matrix $\mathbf{P}$ are no greater than 1: $|\lambda| \leq 1$ (Section A.3.1), and all eigenvalues of $\gamma\mathbf{P}$ are smaller than 1: $\gamma\lambda < 1$, that is, all eigenvalues of $\mathbf{I}-\gamma\mathbf{P}$ are greater than zero: $1-\gamma\lambda > 0$. Consequently the determinant $\det(\mathbf{I}-\gamma\mathbf{P})$ of this coefficient matrix, as the product of all its nonzero eigenvalues, is non-zero and therefore matrix $\det(\mathbf{I} - \gamma\mathbf{P})$ is invertible.

Alternatively, the Bellman equation in Eq. (20.8) can also be solved iteratively based on dynamic programming, which solves a multistage planning problem by backward induction and finds the value function recursively. We first rewrite the Bellman equation as

$$\mathbf{v} = \mathbf{r} + \gamma\mathbf{P}\mathbf{v} = B(\mathbf{v}) \tag{20.11}$$

where $B(\mathbf{v}) = \mathbf{r} + \gamma\mathbf{P}\mathbf{v}$ is defined as a vector-valued function (an operator) applied to the vector argument $\mathbf{v}$.

We can show that $B(\mathbf{v})$ is a contraction mapping first introduced in Section 1.3, that satisfies

$$\begin{aligned} ||B(\mathbf{v}_i) - B(\mathbf{v}_j)|| &= ||\mathbf{r} + \gamma\mathbf{P}\mathbf{v}_i - \left(\mathbf{r} + \gamma\mathbf{P}\mathbf{v}_j\right)|| = \gamma||\mathbf{P}(\mathbf{v}_i - \mathbf{v}_j)|| \\ &\leq \gamma||\mathbf{P}||\; ||\mathbf{v}_i - \mathbf{v}_j|| = \gamma||\mathbf{v}_i - \mathbf{v}_j|| < ||\mathbf{v}_i - \mathbf{v}_j|| \end{aligned} \tag{20.12}$$

We can also prove that $B(\mathbf{v})$ is a contraction mapping by showing that the norm of its Jacobian matrix is smaller than 1. We first find the Jacobian matrix of $B(\mathbf{v})$, its derivative with respect to its vector argument $\mathbf{v}$

$$\mathbf{J}_B = \frac{d}{d\mathbf{v}}B(\mathbf{v}) = \frac{d}{d\mathbf{v}}(\mathbf{r} + \gamma\mathbf{B}\mathbf{v}) = \gamma\mathbf{P} \tag{20.13}$$

We further find its p-norm with $p = \infty$ (equivalent to $p = 1,\ 2$), the maximum absolute row sum

$$||\mathbf{J}_B||_{p=\infty} = ||\gamma\mathbf{P}||_\infty = \gamma||\mathbf{P}||_\infty = \gamma \max_{1\leq i\leq N} \sum_{j=1}^{N} |P_{ij}| = \gamma < 1 \tag{20.14}$$

The last equality is due to Eq. (20.3), that is, $||\mathbf{P}||_\infty = 1$.

Having shown that $B(\mathbf{v})$ is a contraction mapping, we can now use the method of fixed-point iteration first introduced in Section 1.3 to solve the Bellman equation by iterating Eq. (20.11) from an arbitrary initial value, such as $\mathbf{v}_0 = \mathbf{0}$

$$\mathbf{v}_{n+1} = B(\mathbf{v}_n) = \mathbf{r} + \gamma\mathbf{P}\mathbf{v}_n \tag{20.15}$$

This iteration will always converge to the root of the equation, the fixed point of function $B(\mathbf{v})$.

In summary, the Bellman equation in Eq. (20.8) can be solved by either of the two methods in Eqs (20.10) and (20.15), so long as $\gamma < 1$. When the size of state space, the cardinality $|S| = N$ of the set of all states, is large, the complexity $O(N^3)$ is high enough that the iterative method may be more efficient.

20.1.3 Markov Decision Process

A *MDP* is an extension of the MRP that also includes a certain decision-making rule called *policy* denoted by π. Following a given policy the agent takes an action $a \in A(s)$ out of all actions available at a state s, denoted by $A(s)$, to control the state transition of the dynamic process in the system. An MDP can therefore be described by a quintuple (5-tuple) $< S, P, R, A, \gamma >$, where A represents the set of all actions. A policy can be either deterministic denoted by $a = \pi(s)$, or stochastic denoted by $\pi(a|s) = P(a_t = a|s_t = s)$ as the conditional probability of taking action a in state s. A policy is *soft* if any of the actions available at a state is possible to be taken, that is, $\pi(a|s) > 0$ for all $a \in A(s)$ and all $s \in S$. In particular, a policy is ϵ-soft if $\pi(a|s) \geq \epsilon/|A(s)|$ for some small value of ϵ.

As there are $|A(s^n)|$ available actions to take in each state $s^n \in S$, there are in total $|A(s^1)| \cdots |A(s^N)| = \Pi_{n=1}^N |A(s^n)|$ possible deterministic policies. If $A(s) = A$ is the same for all $|S|$ states, then the total number of policies is $|A|^{|S|}$. Following a given policy from an initial state s_0 will result in a sequence of state transitions called a *trajectory* of the episode

$$s_0 \xrightarrow{a_0} s_1 \xrightarrow{a_1} s_2 \xrightarrow{a_2} s_3 \cdots \xrightarrow{a_{T-1}} s_T \tag{20.16}$$

of which the state s_t visited at the tth step can be any of the N states in S. Note that in an episode some of the states in S may be visited multiple times, while some others may never be visited at all. Also note that, owing to the random nature of the MDP, following the same policy may not result in the same trajectory.

Along the trajectory of an episode starting from state s_0, the discounted rewards from all states visited will accumulate, represented by the value $v(s_0)$, the expectation of the sum of all discounted future rewards. The goal of planning is to find the optimal policy as a sequence of actions for a given MDP model of the environment, so that the $v(s_0)$ is maximized. This optimization problem can be solved by the method of dynamic programming (DP).

In an MDP, both the next state s' and reward r are random variables with joint probability $P(s', r|s, a) = P(s_{t+1} = s', r_{t+1} = r|s_t = s, a_t = a)$ conditioned on the action a and the previous state s. The state transition and reward probabilities can be found by marginalization (similar to Eq. (20.5) for an MRP)

$$\begin{aligned} P(s'|s,a) &= \sum_{r \in R} p(s', r|s, a) \\ P(r|s,a) &= \sum_{s' \in S} P(s', r|s, a) \end{aligned} \tag{20.17}$$

and the expected reward r received after arriving at the current state s is

$$r(s,a) = E[r_{t+1}|s_t = s, a_t = a] = \sum_r r\ P(r|s,a) \tag{20.18}$$

Following a given random policy $\pi(a|s)$, the agent takes an action $a \in A(s)$ to transit from the current state $s_t = s$ to the next state $s_{t+1} = s'$ with probability

$$P_\pi(s'|s) = P(s_{t+1} = s'|s_t = s) = \sum_{a \in A(s)} \pi(a|s)\, P(s'|s,a) \tag{20.19}$$

satisfying $\sum_{s'} P_\pi(s'|s) = 1$, and receives a reward in state s

$$r_\pi(s) = E_\pi[r_{t+1}|s_t = s] = \sum_{a \in A(s)} \pi(a|s)\, r(s,a) \tag{20.20}$$

where $E_\pi[x]$ denotes the expectation (of some random variable x inside the brackets) with respect to a certain policy $\pi(a|s)$. The summation over all available actions $a \in A(s)$ in state s will be abbreviated to a in the following.

Consider, as a simple and typical example, the *gridworld* problem, where the environment is a rectangular 2-D space divided into squares, each considered as a state with certain reward. The goal for the agent in this problem is to start at a specified position (e.g., the square at the upper-left corner) and travel to a destination (e.g., the square at the lower-right corner) while trying to accumulate the maximum amount of reward along the way. There are four actions (going up, right, down, and left) available at an internal square, but only three or two at a square on the boarder or in the corner of the grid. Upon arrival at a square as a state s, the agent receives a reward r, and then makes a state transition to one of the neighboring squares as the next state s', depending on the action a it takes while following a certain policy π. Note that both the reward r and the next state s' can be either deterministic or probabilistic, and so is the policy π.

Moreover, certain constraints may be imposed on the problem. For example, some of the squares are not allowed to enter, such as the walls in a maze. In this case the problem becomes more challenging, not only because the available actions at some states are reduced, but also some paths may lead to a dead end if the policy is too near-sighted (not far-sighted enough).

In reinforcement learning, to prepare for the algorithms, we further define two important functions with respect to a given policy π of an MDP:

- The *state value function* $v_\pi(s)$ is the expected return in state s of the MDP while following policy π

$$\begin{aligned} v_\pi(s) &= E_\pi[G_t|s_t = s] \\ &= \sum_a \pi(a|s)\left(r(s,a) + \gamma \sum_{s'} P(s'|s,a)\, v_\pi(s')\right) \\ &= \sum_a \pi(a|s)\, q_\pi(s,a) \end{aligned} \tag{20.21}$$

  This is similar to the value function $v(s)$ of an MRP in Eq. (20.8), but now treated as a function of action a as well as state s, based on $q_\pi(s,a)$ as defined in the following.

- The *state-action value function* $q_\pi(s,a)$ is the expected return of taking a specific action a (regardless of policy π) in state s, and then following π in all subsequent states

$$\begin{aligned} q_\pi(s,a) &= E_\pi\left[G_t | s_t = s, a_t = a\right] \\ &= r(s,a) + \gamma \sum_{s'} P(s'|s,a)\; v_\pi(s') \\ &= r(s,a) + \gamma \sum_{s'} P(s'|s,a)\; \sum_{a'} \pi(a'|s')\; q_\pi(s',a')) \end{aligned} \tag{20.22}$$

where $v_\pi(s')$ is recursively represented as a weighted sum of $q_\pi(s',a')$ based on Eq. (20.21).

The state-action value $q_\pi(s,a)$ plays a more important role than $v_\pi(s)$ as it allows the freedom of taking any action independent of a given policy π, thereby allowing the opportunity to improve an existing policy, for example by taking a *greedy* action to maximize the value $q_\pi(s,a)$. The state-action value function $q_\pi(s,a)$ is often abbreviated as the action value or simply *Q-value* for convenience in future discussions.

As a function of the state-action pair (s,a), the state-action value $q_\pi(s,a)$ can be represented as a table of $|S|$ rows, each for one of the states s, and $|A|$ columns each for one of the actions a. The Q-value for each state-action pair stored in the table can be updated iteratively by various algorithms for learning the Q-values.

When in particular the policy is deterministic with $\pi(a|s) = 1$ and $a = \pi(s)$, then we have

$$P_\pi(s'|s) = P(s'|s,a), \qquad r_\pi(s) = r(s,a) \tag{20.23}$$

and the Bellman equations in Eqs (20.21) and (20.22) become the same

$$v_\pi(s) = r(s,a) + \gamma \sum_{s'} P(s'|s,a)\; v_\pi(s') = q_\pi(s,\pi(s)) \tag{20.24}$$

This is the same as the Bellman equation of an MRP in Eq. (20.8), which can be solved iteratively if $\gamma < 1$, just as in Eq. (20.15).

Figure 20.2 illustrates the Bellman equations in Eqs (20.21) and (20.22), showing the bootstrapping of the state value function in the dashed box on the left, and

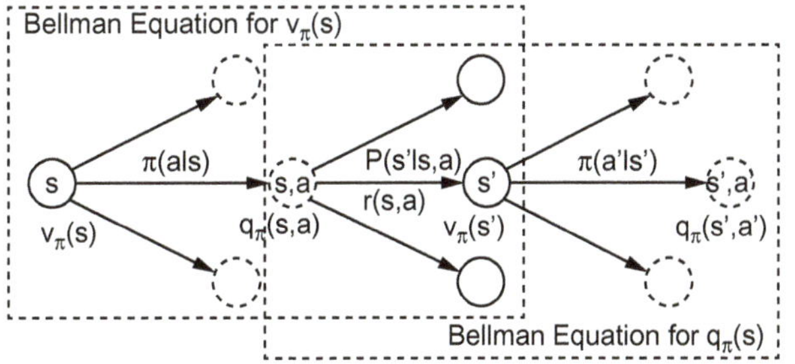

Fig. 20.2 Bellman equations for state and action values.

that of the action value function in dashed box on the right. Here the word bootstrapping means the iterative method that updates the estimated value at a state s based on the previously estimated value of the next state s'. After taking one of the actions $a \in A(s)$ based on policy $\pi(a|s)$, an immediate reward $r(s,a)$ is received.

Given a policy, it is still not certain which state the MDP will transit into as the next state s' due to the random nature of the state transition described by $P(s'|s,a)$. This uncertainty is represented symbolically by a dashed circle called a *chance node* associated with the action value $q_\pi(s,a)$ as the sum of the immediate reward $r(s,a)$ and the expected value of the next state, the weighted average of values $v(s')$ of all possible states $s' \in S$.

20.2 Model-Based Planning

Given a complete MDP model in terms of $p(s',r|s,a)$ of the environment, the goal of model-based planning is to find the optimal policy π^* that achieves the maximum value

$$v^*(s) = \max_\pi v_\pi(s), \qquad \pi^*(s) = \arg\max_\pi \ v_\pi(s) \geq \pi(s) \qquad \forall \pi \tag{20.25}$$

based on the partial order that compares different policies according to their corresponding values

$$\begin{array}{ll} \text{If} & v_{\pi_1}(s) \geq v_{\pi_2}(s) \geq \cdots v_{\pi_k}(s), \quad \forall s \in S \\ \text{Then} & \pi_1 \geq \pi_2 \geq \cdots \geq \pi_k \end{array} \tag{20.26}$$

The problem of policy optimization can be addressed based on two subproblems: *policy evaluation* to measure how good a policy is and *policy improvement* to get a better policy with higher values.

One way to solve this optimization problem is to use brute-force search to enumerate all $|A(s)|$ available policies at each of the $|S|$ states $s \in S$. If we assume the numbers of available actions in all states are the same for simplicity, that is, $A(s^1) = \cdots = A(s^N) = A$, then the computational complexity of this method is $O(|A|^{|S|})$, likely to be too high for the brute-force search to be practical if $|A|$ is large.

A more efficient and therefore more practical way to optimize the policy is to do it iteratively. Consider a simpler task of improving upon an existing deterministic policy π at a single transition step from the current state s to a next state s'. Instead of taking the action $a = \pi(s)$ dictated by the policy, we can improve it by taking the greedy action that maximizes the action value function in Eq. (20.22), while subsequent steps still follow the old policy π. Now we get a new policy

$$\pi'(s) = \arg\max_a q_\pi(s,a) = \arg\max_a \left[r(s,a) + \gamma \sum_{s'} P(s'|s,a) v_\pi(s') \right] \tag{20.27}$$

so that the action value is higher than that of the old one

$$
\begin{aligned}
q_\pi(s,\pi'(s)) &= \max_a q_\pi(s,a) = \max_a \left[r(s,a) + \gamma \sum_{s'} P(s'|s,a) v_\pi(s') \right] \\
&\geq q_\pi(s,\pi(s)) = v_\pi(s)
\end{aligned}
\tag{20.28}
$$

where the last equality is due to Eq. (20.24) for a deterministic policy.

While it is obvious that this greedy method will result in a higher return for the single state transition from s to s', we still need to prove that by following this greedy method in all subsequent states, we will get a new policy π' that achieves higher values than those of the old π at *all* states, that is

$$
v_{\pi'}(s) \geq v_\pi(s), \qquad \forall s \in S \tag{20.29}
$$

This is the *policy improvement theorem*, which can be proven by recursively applying the greedy method to replace the old policy $\pi(s)$ by the new one $\pi'(s)$ for a higher value for each of the subsequent steps one at a time

$$
\begin{aligned}
v_\pi(s) \leq q_\pi(s,\pi'(s)) &= \max_a q_\pi(s,a) = \max_a \left[r(s,a) + \gamma \sum_{s'} P(s'|s,a) v_\pi(s') \right] \\
&= r(s,\pi'(s)) + \gamma \sum_{s'} P(s'|s,\pi'(s)) v_\pi(s') \\
&\leq r(s,\pi'(s)) + \gamma \sum_{s'} P(s'|s,\pi'(s)) \max_{a'} q_\pi(s',a') \\
&= r(s,\pi'(s)) + \gamma \sum_{s'} P(s'|s,\pi'(s)) \left[r(s',\pi'(s)) + \gamma \sum_{s''} P(s''|s',\pi'(s')) v_\pi(s'') \right] \\
&= r(s,\pi'(s)) + \gamma r(s',\pi'(s)) + \gamma \sum_{s'} P(s'|s,\pi'(s)) \left[\gamma \sum_{s''} P(s''|s',\pi'(s')) v_\pi(s'') \right] \\
&\leq \cdots \\
&= r(s,\pi'(s)) + \gamma r(s',\pi'(s')) + \gamma^2 r(s'',\pi'(s'')) + \cdots \\
&= v_{\pi'}(s)
\end{aligned}
\tag{20.30}
$$

that is $\pi'(s) \geq \pi(s)$.

(This derivation may seem long and difficult to follow (although each step should be straightforward). A less experienced reader may choose to skip it and move on when studying the topic the first time.)

The optimal state value $v^*(s)$ and action value $q^*(s,a)$ can therefore be achieved by replacing the weighted average over all possible actions at each state (s and s') in the Bellman equations in Eqs (20.21) and (20.22) by their maximum

$$
v^*(s) = \max_a q^*(s,a) = \max_a \left(r(s,a) + \gamma \sum_{s'} P(s'|s,a) v^*(s') \right) \tag{20.31}
$$

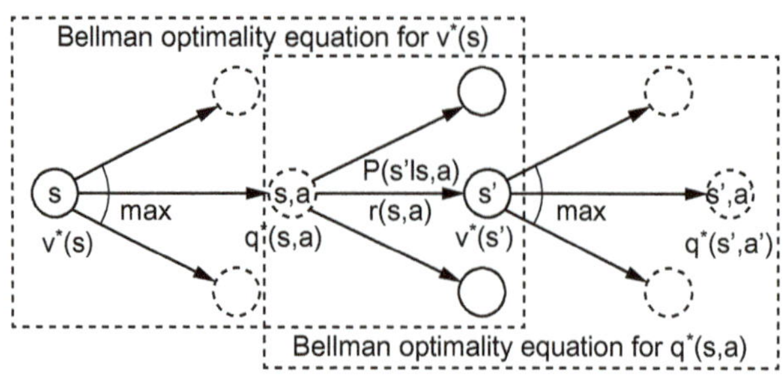

Fig. 20.3 Bellman optimality equations for state and Action values.

and

$$q^*(s,a) = r(s,a) + \gamma \sum_{s'} P(s'|s,a)v^*(s') = r(s,a) + \gamma \sum_{s'} P(s'|s,a) \max_{a'} q^*(s',a') \tag{20.32}$$

If the state transition is deterministic, this equation becomes

$$q^*(s,a) = r(s,a) + \gamma \max_{a'} q^*(s',a') \tag{20.33}$$

Similar to Eqs (20.21) and (20.22) illustrated in Fig. 20.2, Eqs (20.31) and (20.32), called *Bellman optimality equations*, are illustrated in Fig. 20.3.

The Bellman optimality equations should hold for all N states in $S = \{s^1, \cdots, s^N\}$ and can therefore be written in vector form

$$\mathbf{v}^* = \max_a (\mathbf{r}_a + \gamma \mathbf{P}_a \mathbf{v}^*) \tag{20.34}$$

where

$$\mathbf{v}^* = \begin{bmatrix} v^*(s^1) \\ \vdots \\ v^*(s^N) \end{bmatrix}, \quad \mathbf{r}_a = \begin{bmatrix} r(s^1,a) \\ \vdots \\ r(s^N,a) \end{bmatrix}$$
$$\mathbf{P}_a = \begin{bmatrix} P(s^1|s^1,a) & \cdots & P(s^N|s^1,a) \\ \vdots & \ddots & \vdots \\ P(s^1|s^N,a) & \cdots & P(s^N|s^N,a) \end{bmatrix} \tag{20.35}$$

This nonlinear equation system can be solved iteratively for $\mathbf{v}^*$ (similar to Eq. (20.15)

$$\mathbf{v}^*_{n+1} = \max_a (\mathbf{r}_a + \gamma \mathbf{P}_a \mathbf{v}^*_n) = B(\mathbf{v}_n) \tag{20.36}$$

This iteration will converge, as function $B(\mathbf{v})$ is a contraction mapping with $\gamma < 1$

$$\begin{aligned} ||B(\mathbf{v}_i) - B(\mathbf{v}_j)|| &= ||\max_{a_i} (\mathbf{r}_{a_i} + \gamma \mathbf{P}_{a_i} \mathbf{v}_i) - \max_{a_j} \left(\mathbf{r}_{a_j} + \gamma \mathbf{P}_{a_j} \mathbf{v}_j\right) || \\ &\le ||\max_{a_i} \left(\mathbf{r}_{a_i} + \gamma \mathbf{P}_{a_i} \mathbf{v}_i - \mathbf{r}_{a_i} - \gamma \mathbf{P}_{a_i} \mathbf{v}_j\right) || \\ &= \max_{a_i} \gamma ||\mathbf{P}_{a_i} (\mathbf{v}_i - \mathbf{v}_j)|| \le \max_{a_i} \gamma ||\mathbf{P}_{a_i}||\ ||\mathbf{v}_i - \mathbf{v}_j|| \\ &= \gamma ||\mathbf{v}_i - \mathbf{v}_j|| < ||\mathbf{v}_i - \mathbf{v}_j|| \end{aligned} \tag{20.37}$$

where $||\mathbf{P}|| = ||\mathbf{P}||_\infty = 1$ is the p-norm ($p = \infty$) of a stochastic matrix, which is known to be 1.

This iteration will be used for policy evaluation in the algorithm *value iteration* to be considered in the following for finding the optimal policy.

Given the complete information of the MDP model in terms of $p(s',r|s,a)$, we can find the optimal policy π^* using either *policy iteration (PI)* or *value iteration (VI)* based on the DP method.

- **Policy iteration**

  Policy iteration carries out the following two intertwined and interacting processes iteratively from some arbitrary initial policy π until convergence when π no longer changes:

  - Policy evaluation:

    Given a deterministic policy π, Eq. (20.21) becomes the Bellman equation in Eq. (20.24), a linear equation system of $N = |S|$ equations of N unknowns $v_\pi(s^1), \cdots, v_\pi(s^n)$, which can be solved iteratively as in Eq. (20.15) to find values at all states:

$$v_{n+1}^\pi(s) = r(s,\pi(s)) + \gamma \sum_{s'} P(s'|s,\pi(s)) v_n^\pi(s') \qquad \forall s \in S$$

  - Policy improvement:

    Given $v_\pi(s)$ based on policy π, get a better policy π' by the greedy method as in Eq. (20.27):

$$\pi'(s) = \arg\max_a \left[r(s,a) + \gamma \sum_{s'} P(s'|s,\pi(s)) v_\pi(s') \right] \qquad \forall s \in S$$

  We see that in each round of the policy iteration, the processes of policy evaluation and policy improvement, are carried out alternatively, one depending on the other, and one completing before the other starts. The value functions $v_\pi(s)$ at all states are systemically evaluated before they are updated, and the policy $\pi(a|s)$ at all states is updated before it is evaluated. These two processes chase each other as a moving target iteratively and eventually converge to the optimal policy π^* that achieves the maximum value v^*. This approach is said to be *synchronous*, as illustrated here

$$\pi_o \xrightarrow{E} v_{\pi_0} \xrightarrow{I} \pi_1 \xrightarrow{E} v_{\pi_1} \xrightarrow{I} \cdots \xrightarrow{I} \pi^* \xrightarrow{E} v_{\pi^*} = v^* \tag{20.38}$$

  The pseudo code for the algorithm is listed here:

  Initialize

  $v(s) \in \mathbb{R}$, $v(s) = 0$ if s is terminal node, $\pi(s) \in A(s)$, $\forall s \in S$;

  Set tolerance *tol* (small positive value);

  repeat

  Policy Evaluation (find $v(s)$ $\forall s \in S$)

  repeat

  $\delta = 0$

for $\forall s \in S$:

$v'(s) = r(s, \pi(s)) + \gamma \sum_{s'} P(s'|s, \pi(s))v(s')$

$\delta = \max(\delta, |v(s) - v'(s)|)$

until $\delta < tol$

Policy Improvement (by taking greedy action):

$Done = T$

for each $s \in S$

$\pi'(s) = \arg\max_a [r(s,a) + \gamma \sum_{s'} P(s'|s, \pi(s))v(s')]$

if $\pi'(s) \neq \pi(s), \quad Done = F$

end for

until *done*

- **Value iteration**

One drawback of the policy iteration method is the high computational complexity of the inner iteration for policy evaluation. To speed up this process, we can terminate the inner iteration early before convergence. In the extreme case, we simply combine policy evaluation and improvement into a single step, resulting in the following two-step value iteration:

  - Find the optimal values:

    Based on some initial value $v_0(s)$, solve the Bellman optimality equation in Eq. (20.31) by iteration in Eq. (20.36)

$$v_{n+1}(s) = \max_a \left[r(s,a) + \gamma \sum_{s'} P(s'|s,a)v_n(s') \right] \qquad \forall s \in S \qquad (20.39)$$

  - Find the optimal policy:

    Find $\pi^*(s)$ that maximizes $v^*(s)$

$$\pi^*(s) = \arg\max_a \left[r(s,a) + \gamma \sum_{s'} P(s'|s,a)v_n(s') \right] \qquad \forall s \in S \qquad (20.40)$$

Here is the pseudo code for the algorithm:

Initialize: $v(s) = 0 \;\; \forall s \in S$

repeat

$\delta = 0;$

for each $s \in S$

$v'(s) = \max_a [r(s,a) + \gamma \sum_{s'} P(s'|s,a)v(s')]$

$\delta = \max(\delta, |v'(s) - v(s)|)$

$v(s) = v'(s)$

end for

until $\delta < tol$

$\pi^*(s) = \arg\max_a [r(s,a) + \gamma \sum_{s'} P(s'|s,a)v(s')]$

This PI method as illustrated in Fig. 20.4.

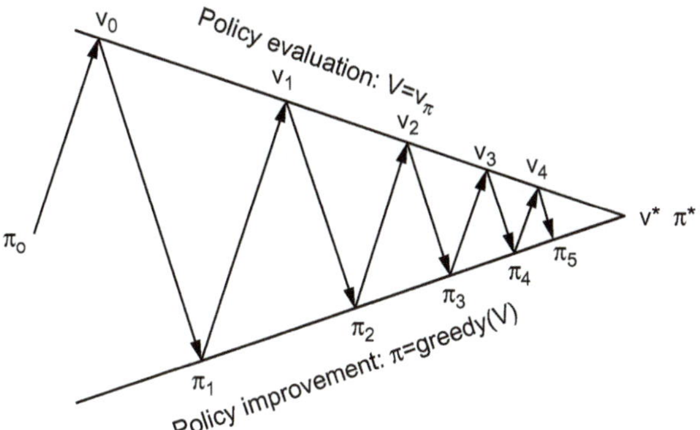

Fig. 20.4 Convergence of policy iteration.

20.3 Model-Free Evaluation and Control

The previously discussed DP methods find the optimal policy based on the assumption that the MDP model of the environment is completely available, that is, the dynamics of MDP in terms of its state transition and reward mechanism are known. A given policy π can then be evaluated based on the transition probabilities $p(s'|s,a)$, and improved based on greedy action selection at each state. This model-based planning is an optimization problem involving no learning of the environment.

Here we consider the optimization problem with an unknown MDP model of the environment. In this model-free problem, called a *control* problem, both the state transition and reward probabilities are unknown, and the value functions can no longer be calculated directly based on Eqs (20.21) and (20.22) as in model-based planning. Now the dynamics of the system in terms of its MDP model needs to be learned based on sampling the environment by repeatedly running a large number of episodes of the stochastic process of the environment while following some given policy, and estimate the value functions as the average of the actual returns received during the sampling process. Such a method is generally referred to as the *Monte Carlo (MC) method.* At the same time, the given policy can also be improved to gradually approach optimality. This approach is called *general policy iteration (GPI)*, as illustrated in the following, similar to the PI method for model-based planning illustrated in Eq. (20.38)

$$\pi_o \xrightarrow{E} q_{\pi_0} \xrightarrow{I} \pi_1 \xrightarrow{E} q_{\pi_1} \xrightarrow{I} \cdots \xrightarrow{I} \pi^* \xrightarrow{E} q_{\pi^*} = v^* \tag{20.41}$$

While GPI in Eq. (20.41) and GI in Eq. (20.38) may look similar to each other, they are essentially different in the following aspects.

- Policy improvement is based on the state value function $v_\pi(s)$ in model-based planning, but the action value function $q_\pi(s,a)$ in model-free control. This is because without a specific MDP model the state value function can no longer be calculated, while the action value function can still be estimated based on the actual rewards received while sampling the environment. Similar to how we improve the policy by taking the greedy action to achieve a higher state value in model-based planning, here we improve the policy by taking an action different from that dictated by the given policy π to achieve a higher action value based on some *ϵ-greedy method* to be discussed later.

- Estimating the action value by the MC method requires running a large number of episodes while sampling the environment, and the computational complexity for policy evaluation in the model-free case is much higher than that in model-based planning. To speed up this process in model-free control, the iterative policy evaluation is no longer fully carried out for it to converge to the true action value of the policy. Instead, the iterative update of the action value is carried out only once, followed immediately by the next phase of policy improvement. Such an estimated action value is inevitably very noisy, but due to the much reduced complexity, we can afford to run a large number of episodes while sampling the environment.
- In model-based planning, the action value is found based on all available actions at each state (Eq. (20.22)) during policy evaluation, and we only need to *exploit* the greedy action at each state to find the policy that achieves the maximum action value $q_\pi(s)$ (Eq. (20.27)) during policy improvement. However, in model-free control, we have to learn the action value $q_\pi(s,a)$, as a function of action $a \in A(s)$ at state s by sampling the environment. This estimated action value based only on some partial sample data is inevitably noisy, especially in the early stage of learning when many of the states have not been visited yet. We therefore need to *explore* all actions available at each state to better learn the value function, as well as to exploit the greedy action to improve the policy.

As indicated by the final point, a proper tradeoff between the *exploitation* of the greedy action and the *exploration* of other non-greedy actions needs be made, so that the agent can be exposed to all possible state-action pairs and gradually learn the action value function while at the same time it can also improve the policy being followed. The *ϵ-greedy method* is just such a method. Specifically, we define $\epsilon \in [0, 1]$ as the probability of exploring all actions $A(S)$ available in state s (including the greedy one), each with an equal probability $\epsilon/|A(s)|$, and $1-\epsilon$ as the probability of taking only the greedy action. Therefore the overall probability of taking the greedy action, also one of the $|A(s)|$ available actions, is $1-\epsilon+\epsilon/|A(s)|$. The policy can now be expressed as

$$\pi'(a|s) = \begin{cases} \arg\max_a q(s,a) & P_{exploitation} = 1-\epsilon+\epsilon/|A(s)| \\ \text{any other } a \in A(s) & P_{exploration} = \epsilon/|A(s)| \end{cases} \tag{20.42}$$

For example, if $|A(s)| = 4$ and $\epsilon = 2/3$, then the probability of taking the greedy action is $1-\epsilon+\epsilon/|A(s)| = 3/6 = 1/2$, and the probability of taking any of the remaining three non-greedy actions is $\epsilon/|A(s)| = 1/6$. Such a policy is different from the original policy currently followed by the agent. Such a policy is called *behavior policy*, as it describes the behavior of the agent.

The value of ϵ may vary during the process. A larger ϵ (close to 1) can be used to allow more state-action pairs to be explored at the early stage when many of them have not been visited yet, while smaller ϵ (approaching 0) may be used to exploit the greedy actions for higher values in the later stage when the action value function has been more reliably learned. For example, we can let $\epsilon_k = 1/k$ with k being the number of iterations completed so far, so that it's value gets progressively smaller from a large initial value. As k approaches infinity and ϵ approaches zero, the ϵ-greedy method approaches absolute greedy method toward the end of the process.

Recall we proved the policy improvement theorem in Eq. (20.30) stating that the state value $v_{\pi'}(s)$ achieved by the greedy policy $\pi'(s)$ is no lower than $v_\pi(s)$ achieved by the original policy π, that is, $\pi'(s) \geq \pi(s)$. Similarly here we can also prove the same policy improvement theorem stating that the action value $q_{\pi'}(s,a)$ achieved by the ϵ-greedy policy π' is no lower than $q_\pi(s,a)$ achieved by the original policy π:

$$\begin{aligned} q_\pi(s,\pi'(s)) &= \sum_a \pi'(a|s) q_\pi(s,a) \\ &= \frac{\epsilon}{|A(s)|}\sum_a q_\pi(s,a) + (1-\epsilon)\max_a q_\pi(s,a) \\ &\geq \frac{\epsilon}{|A(s)|}\sum_a q_\pi(s,a) + (1-\epsilon)\sum_a \frac{\pi(a|s)-\epsilon/|A(s)|}{1-\epsilon} q_\pi(s,a) \end{aligned} \tag{20.43}$$

where the inequality is due to the fact that the maximum value of $q_\pi(s,a)$ is no less than the average of action values $q_\pi(s,a)$ over all $a \in A(s)$ weighted by some normalized coefficients adding up to 1:

$$\sum_a \frac{\pi(a|s)-\epsilon/|A(s)|}{1-\epsilon} = \frac{1}{1-\epsilon}\sum_a \left(\pi(a|s) - \frac{\epsilon}{|A(s)|}\right) = \frac{1}{1-\epsilon}(1-\epsilon) = 1 \tag{20.44}$$

Continuing this equation we further get

$$\begin{aligned} q_\pi(s,\pi'(s)) &\geq \frac{\epsilon}{|A(s)|}\sum_a q_\pi(s,a) - \frac{\epsilon}{|A(s)|}\sum_a q_\pi(s,a) + \sum_a \pi(a|s) q_\pi(s,a) \\ &= \sum_a \pi(a|s) q_\pi(s,a) = v_\pi(s) \end{aligned} \tag{20.45}$$

The iterative process of the GPI method for model-free control is illustrated in Fig. 20.5, which is similar to the PI method for model-based planning based planning illustrated in Fig. 20.4, but here the state value $v_\pi(s)$ is replaced by the action value $q_\pi(s,a)$, and the greedy method is replaced by the ϵ-greedy method.

Comparing the methods of PI and GPI, we note that PI is a *synchronous* method in the sense that the value functions at all states are evaluated before the policy is updated, while GPI is an *asynchronous* method, in which the policy evaluation and improvement are carried out in-place for only one state at a time before all other states, and the two alternating processes of policy evaluation and policy improvement are carried out iteratively.

Also, in GPI the estimated action value $\tilde{q}$ is obtained with only one iteration of policy evaluation, instead of a more accurate estimate that would otherwise be

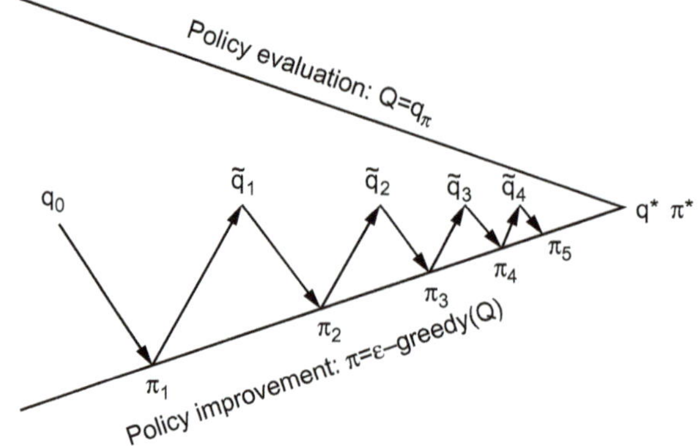

Fig. 20.5 Convergence of general policy iteration.

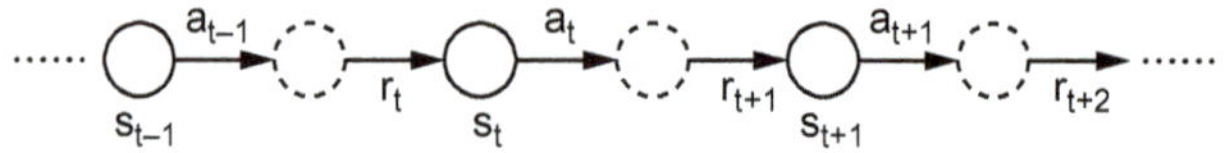

Fig. 20.6 A trajectory segment of GPI.

obtained if the iterative evaluation were fully carried out to its convergence, represented by the top line in Fig. 20.5. At the same the time, the ϵ-greedy method is carried out based on the estimated $\tilde{q}$ to improvement the policy, represented by the bottom line in the figure.

The dynamic process of the environment is sampled by running a large number of episodes (all assumed to terminate) of the process while taking action $a_t = \pi(s_t)$ in each state s_t, $(t = 1, \cdots, T)$

$$s_0 \rightarrow (a_1, r_1, s_1) \rightarrow (a_2, r_2, s_2) \rightarrow \cdots \rightarrow (a_T, r_T, s_T) \tag{20.46}$$

and estimating the value functions based on the averaged returns G_t actually received from these episodes. The trajectory of each episode is also illustrated in Fig. 20.6.

The trajectories of different episodes are different from each other in general owing to the random nature of the environment. In the following sections we will consider specific algorithms for the implementation of the general model-free control just discussed.

To prepare for the discussion of some specific GPI methods, we first consider a general problem of estimating the value of a random variable x as the running average $\hat{x}_n$ of a set of n samples $x_1, \cdots, x_n$ collected in real time, which is updated incrementally upon receiving a new sample x_{n+1}

$$\begin{aligned}\hat{x}_{n+1} &= \frac{1}{n+1}\sum_{i=1}^{n+1} x_i = \frac{1}{n+1}\left(x_{n+1} + \sum_{i=1}^{n} x_i\right) \\ &= \frac{1}{n+1}(x_{n+1} + n\hat{x}_n) = \hat{x}_n + \frac{1}{n+1}(x_{n+1} - \hat{x}_n) \\ &= \hat{x}_n + \alpha(x_{n+1} - \hat{x}_n)\end{aligned} \tag{20.47}$$

where $\hat{x}_n$ and $\hat{x}_{n+1}$ are, respectively, the old and new estimates of variable x, and $\alpha = 1/(n+1)$ is the step size. If we generalize α for it to take any value between 0 and 1, then Eq. (20.47) can be rewritten for three special cases for $\alpha = 0,\ 1/(n+1),\ 1$:

$$\hat{x}_{n+1} = \hat{x}_n + \alpha(x_{n+1} - \hat{x}_n) = \begin{cases} x_{n+1} & \alpha = 1 & \text{no contribution from previous samples} \\ \frac{1}{n+1}\sum_{i=1}^{n+1} x_i & \alpha = 1/(n+1) & \text{equal contribution from all samples} \\ \hat{x}_n & \alpha = 0 & \text{no contribution from latest sample} \end{cases} \tag{20.48}$$

We see that by adjusting α, we can control how much the latest and previous samples contribute to the estimated average. A reasonable strategy is therefore to progressively decrease α from its initial value 1 down to 0, so that the estimated average is mostly determined by the recent samples in the early stage of the process but it becomes progressively more stabilized when enough samples have been collected in the later stage. Such a strategy is suitable in general when estimating model parameters of a non-stationary system, such as an MDP with varying behaviors when the policy is being modified continuously.

The updating process in Eq. (20.47) can be considered as one iterative step in the method of stochastic gradient descent (SGD) (Section 6.1) for minimizing the squared error $\varepsilon = (x_{n+1} - \hat{x}_n)^2/2$ between the old average $\hat{x}_n$ and the latest sample x_{n+1}, and the second term is the gradient $d\varepsilon/d\hat{x}_n$ weighted by the step size.

In the specific case of model-free control, Eq. (20.47) is used to iteratively update the estimated state value functions while sampling the environment. As we will see in the following sections, the state-action values are estimated as the average of all previous sample returns, and they are updated incrementally when a new sample return G_{n+1}, the *target*, becomes available

$$\begin{aligned} v_{n+1}(s) &\Leftarrow v_n(s) + \alpha(G_n - v_n(s)) \\ q_{n+1}(s,a) &\Leftarrow q_n(s,q) + \alpha(G_n - q_n(s,a)) \end{aligned} \tag{20.49}$$

The result of this iterative learning process is the estimated state-action values $q(s,a)$, stored in a 2-D array, called a *Q-table*, of which the rows are for all the states and the columns are for all available actions at each state. Based on the Q-table, the optimal control policy can be determined by always taking the greedy action with the maximum state-action value at each state, and the solution of the reinforcement learning problem can be found by following the sequence of optimal actions as in shown Eq. (20.16) from the given initial state s_0. Such methods based on the Q-table are called the *tabular methods*.

20.3.1 Monte Carlo (MC) Algorithms

We first consider the simple problem of evaluating an existing policy π in terms of its value function given by the Bellman equation in Eq. (20.24)

$$v_\pi(s) = r(s,\pi(s)) + \gamma \sum_{s'} p(s'|s,\pi(s))v_\pi(s') \tag{20.50}$$

This equation for the model-based case can also be used in the model-free case based on the MC method by running multiple sample episodes in the environment. The state value $v_\pi(s)$ is estimated as the average of the actual returns G_t, which in turn are calculated as the sum of discounted rewards $r_{t+1}, \cdots, r_T$ from the current state $s = s_t$ onward to the terminal state s_T found at the end of the episode (assuming a finite horizon).

This method has two versions.

- *Every-visit*: Returns of all visits to a state in the trajectory of an episode are counted in the estimation of the state value function.
- *First-visit*: only the first visit to each state in each episode is counted.

In the first method, all sample returns G_t from multiple visits to a state s_t contribute to the estimated state value $v_\pi(s_t)$. But these sample returns for the state may not be independent of each other as the later ones are dependent on the previous ones, that is, they are not i.i.d. samples and the estimate values may be biased. This problem is avoided in the second method, where only the return of the first visit to each of the states is used as an independent sample drawn from the distribution, and the estimated values are no longer biased. However, as only a fraction of all samples in the trajectory is used to estimate the state values, the cost of the unbiased estimation is its high variance, which can be reduced only by running a large number of episodes to get enough samples for a more statistically reliable average. We realize that this is just the common issue of bias–variance tradeoff (Section 4.2) faced by many learning algorithms.

The following pseudo code is for the first-visit version due to the if statement, which can be removed for the every-visit version.

Input: a given policy π to be evaluated
Initialize: $G_t = [\ ]$ (empty lists) for all $t = 0, \cdots, T-1$
loop (for each episode)
  run episode to the end while following π to get sample data:
    $(s_0, a_0, r_1),\ (s_1, a_1, r_2), \cdots, (s_{T-1}, a_{T-1}, r_T)$
  $G = 0$
  for $t = T-1, T-2, \cdots, 0$ (for each step)
    $G = \gamma G + r_{t+1}$
    if $S_t \notin \{S_0, \cdots, S_{t-1}\}$ (first visit)
      $G_t = [G_t, G]$ (append G to list G_t)
  $V(S_t) = average(G_t),\ \forall t = 0,\ , \cdots, T-1$

In this code, the function *average* finds the average of all elements of a list when all sample returns are available at the end of each episode. A more computationally efficient method (in terms of both space and complexity) is to find the average incrementally as in Eq. (20.49). Based on this incremental average, policy evaluation in terms of both the state and action values can also be carried out by the following code

Initialize $v(s) = 0, \; \forall s$
loop (for each episode)
    run episode following π to get $G_t, \; \forall t$
    for $t = 0, \cdots, T-1$ (for each step)
        if s_t is visited the first time
            $v(s_t) = v(s_t) + \alpha(G_t - v(s_t))$

and

Initialize: $q(s,a) = 0, \; \forall (s,a)$
loop (for each episode)
    run episode following π to get $G_t, \; \forall t$
    for all (s,a) pairs visited
        if (s,a) is visited the first time
            $q(s,a) = q(s,a) + \alpha(G_t - q(s,a))$

The *if* condition can be removed for every-visit version of the algorithm.

We next consider the MC algorithm for model-free control, based on the generalized policy iteration of alternating and interacting policy evaluation and policy improvement. While sampling the environment by following an existing policy π the action value $q_\pi(s)$, instead of the state value $v_\pi(s)$, is gradually estimated and the policy is gradually improved at the same time. This iteration will converge to the optimal policy at the limit when the number of iterations goes to infinity.

The pseudo code for this algorithm is listed below.

Input: $\epsilon > 0, \; \alpha > 0$, and a policy π
Initialize: $q(s,a) = 0, \; \forall (s,a), \; \pi = \epsilon -$ soft (arbitrarily)
loop (for each episode)
    run episode following π to get $G_0, \cdots, G_{T-1}$
    for $t = 0, \cdots, T-1$
        if (s_t, a_t) is visited the first time
            $q(s_t,a_t) = q(s_t,a_t) + \alpha(G_t - q(s_t,a_t))$
            $a^* = \arg\max_a q(s_t,a_t)$
            for $\forall a \in A(s_t)$
$$\pi(a|s_t) = \begin{cases} 1 - \epsilon + \epsilon/|A(s_t)| & \text{if } a = a^* \\ \epsilon/|A(s_t)| & \text{else} \end{cases}$$

Here the ϵ-greedy policy $\pi(a|s)$ is based on the action value function $q(s,a)$ (maximized by the greedy action). Through out the iteration, the policy is modified following the action value function, while at the same time the action value function is modified based on the policy, until the process converges to the desired optimality.

The optimal policy $\pi(s)$ obtained by the this algorithm is not completely deterministic owing to the ϵ-greedy approach, but it is said to be *near-deterministic* as the greedy action is favored over other non-greedy actions.

20.3.2 Temporal Difference (TD) Algorithms

The temporal difference (TD) method is a combination of the MC method and the bootstrapping DP method based on the Bellman equation. The main difference between the TD and MC methods is the target, the return G, in the incremental average in Eq. (20.49) for estimating either the state or action value functions. In the MC method, the target is the actual return G_t calculated at the end of the episode when all subsequent rewards $r_{t+1}, \cdots, r_T$ are available. However, in the TD method, the target (called the *TD target*) is the sum of the immediate reward r_{t+1} and the previously estimated value $v_\pi(s_{t+1})$ at the next state

$$G_t = r_{t+1} + \gamma v_\pi\,(s_{t+1}) \tag{20.51}$$

that is, the TD method is an in-place bootstrapping method. It makes more efficient use of the sample data, and it updates the estimates of the value functions and policy more frequently at every step of an episode, instead of at the end of the episode as in the MC method. Also, it can be used even if the dynamic process of the environment is non-episodic with an infinite horizon.

Again, we first consider the simpler problem of policy evaluation. As in the MC method, we estimate the value function $v_\pi(s)$ as the average of the actual returns G found by running multiple episodes while sampling the environment. Substituting this TD target into the running average in Eq. (20.49) for the afore-mentioned value function, we get

$$\begin{aligned} v_\pi(s) &\Leftarrow v_\pi(s) + \alpha(G - v_\pi(s)) \\ &= v_\pi(s) + \alpha(r + \gamma v_\pi(s') - v_\pi(s)) = v_\pi(s) + \alpha\delta \end{aligned} \tag{20.52}$$

where δ, called the *TD error*, is defined as the difference between the TD target and the previous estimate

$$\delta = G - v_\pi(s) = r + \gamma v_\pi(s') - v_\pi(s) \tag{20.53}$$

and $v_\pi(s')$ is the current estimate of the value of the next state previously estimated (bootstrapping). The iteration of this TD method converges to the true value $v_\pi(s)$, if the step size is small enough.

Here is the pseudo code for policy evaluation using the TD method, based on parameters γ and α:

Input: a given policy π to be evaluated
Initialize: $v(s)$ for all $s \in S$ arbitrarily, $v(s) = 0$ for terminal state s, $\alpha \in (0,\ 1]$
loop (for each episode)

  $s = s_0$
  while s is not terminal (for each step)

take action $a = \pi(s)$, get reward r and next state s'
$v(s) = v(s) + \alpha[r + \gamma v(s') - v(s)]$
$s = s'$

We further consider the TD method for model-free control, which is similar to the TD method for model-free evaluation considered before. By running many episodes of the environment, the estimated action values $q(s,a)$ for all state-action pairs, stored in a state-action table or Q-table, are iteratively updated as the running average of the return G, the sum of the immediate reward r and the action value $q(s',a')$ of the next state based on an action a', dictated by the policy being followed, such as the ϵ-greedy policy. This iterative process gradually learns the action value functions of the MDP model of the environment, based on which the optimal action for each state can be obtained as the one with the maximum Q-value.

The TD method can be carried out by two algorithms, with the only difference being how the action value is updated.

- *On-policy* algorithm
  The action value $q(s,a)$ is updated based on action a' at the current state s dictated by the policy being followed

$$\begin{aligned} q(s,a) &\Leftarrow q(s,a) + \alpha(G - q(s,a)) \\ &= q(s,a) + \alpha(r + \gamma q(s',a') - q(s,a)) \end{aligned} \tag{20.54}$$

- *Off-policy* algorithm
  The action value $q(s,a)$ is updated based on the greedy action a' at the current state s regard the policy being followed

$$q(s,a) \Leftarrow q(s,a) + \alpha(r + \gamma \max_{a'} q(s',a') - q(s,a)) \tag{20.55}$$

These two algorithms become the same if the policy being followed is greedy.

- **State-action-reward-state-action (SARSA)**
  This is an on-policy algorithm, which is called *SARSA* as it updates the Q-value based on the current state s and action a, the immediate reward r, the next state s' and action a'. The pseudo code of the SARSA algorithm is listed where the policy being followed, for example ϵ-greedy, is denote by π.

  Initialize: $q(s,a) = 0$ for all (s,a), $\alpha \in (0,\ 1]$, $\epsilon > 0$
  loop (for each episode)

  $s = s_0$
  get action a at s according to π based on $q(s,a)$
  while s is not terminal (for each step)

  take action a and get reward r and next state s'
  get action a' according to π
  $q(s,a) = q(s,a) + \alpha[r + \gamma q(s',a') - q(s,a)]$
  $s = s',\ a = a'$

A variation of SARSA, the *expected SARSA*, updates the Q-value in state s based on the expected Q-value of the next state s', the weighted average of the Q-values resulting from all possible actions, instead of only one action. Consequently, the estimated Q-values by the expected SARSA have lower variance than SARSA, and a higher learning rate α can be used to speed up the learning process.

$$q(s,a) = q(s,a) + \alpha \left(r + \gamma \sum_{a} \pi(a|s') q(s',a) - q(s,a) \right) \tag{20.56}$$

- **Q-Learning**
  This is an off-policy algorithm, which updates the Q-value based on the greedy action instead of that dictated by the policy being followed. The pseudo code of the Q-learning algorithm follows:
  Initialize: $q(s,a) = 0$ for all (s,a), $\alpha \in (0,\ 1]$, $\epsilon > 0$
  loop (for each episode)
    $s = s_0$
    while s is not terminal (for each step)
      take action a at s according to π based on $q(s,a)$
      get reward r and next state s'
      $q(s,a) = q(s,a) + \alpha[r + \gamma \max_{a'} q(s',a') - Q(s,a)]$
      $s = s'$

  Here is the comparison of the MC and TD methods.

- The MC method estimates the state value $v_\pi(s_t)$, the expected return, by the true return G_t obtained at the end of the episode, that is, the estimated value is updated only once at the end of every episode. The TD method estimates $v_\pi(s_t)$ as the sum of the immediate reward and the action value at the next state previously estimated, that is, it is a bootstrap method, and the estimated value is updated at every step of every episode.
- MC can only learn episodic (terminating) environments with complete episodes; while TD can learn continuing (non-terminating) environments of incomplete episodes.
- MC estimates based on sample returns G_t, and it is unbiased especially in first-visit version. The TD uses the bootstrap approach to find the TD target based on sample data that are not independently and identically distributed (i.i.d.) in general, and it is more sensitive to the initial guess of the value functions, so it is more biased.
- MC based on the sample returns G_t is affected by many random events (state transitions, actions, and rewards), and in particular the first-visit version only makes use of the sample data from the first visit of a state. It does not use the available data efficiently, and it has high variance. The TD method is based on only one random variable, the estimated return, and it updates the action value at every step of the episode, it makes more frequent and efficient use of the sample data, and it has lower variance.

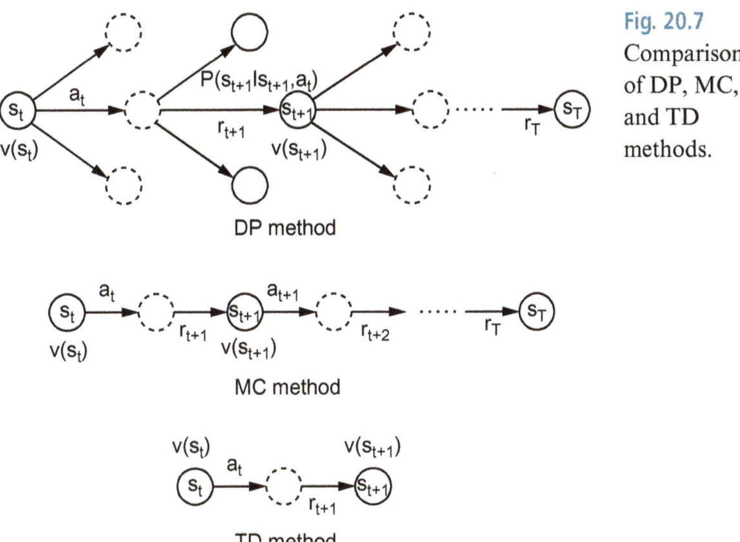

Fig. 20.7 Comparison of DP, MC, and TD methods.

The DP method for model-based planning, and the MC and TD methods for model-free control are summarized in Fig. 20.7.

Example 20.1

The TD method is applied to the grid world problems as shown in Fig. 20.8, where a rectangular array is used to represent a maze of which all positions are sequentially indexed from top left to lower right. This problem is modeled as a Markov decision process with each of the positions treated as a state of the MDP model. For an agent to achieve the goal of finding the shortest path from the starting position (top left corner) to the destination position (lower right corner), we assign a large positive reward (e.g., 10) to the destination position, a large negative reward (e.g., -100) to all positions for the walls of the maze, and a small negative reward (e.g., -1) to all other empty space available for the agent to move into. By maximizing the cumulative reward along its way, the agent will find the desired path.

This maze problem can be solved by the following steps.

- Set up all parameters in the algorithm, including ϵ, α, γ, and the maximum number of iterations for episodes (outer loop) and steps in each of each episode (inner loop).
- Define the maze as a rectangular array with reward for each position (wall or empty space), together with the start and end positions (as the initial and goal states).
- Initialize the Q-table, here an array of the same size as the maze, to zero.
- Carry out either SARSA or Q-learning algorithm to learn the state-action functions by sampling the environment and updating the Q-table iteratively.
- Get the solution, the optimal path from the initial state to the goal state, with maximum cumulative reward, by taking the greedy action with the maximum Q-value in the Q-table for each state along the path.

The following MATLAB code lists the key functions used for solving the grid world problem.

```
function Q=SARSA(Q,s0,st)             % SARSA Learning
    global maze alpha gamma epsilon max_step max_episode
    ie=1;                             % episode counter
    while ie < max_episode            % for all episodes
        is=1;                         % episode counter
        s=s0;                         % initial state
        a=get_action(s,Q);            % action by e-greedy
        while s~=st                   % for each step in episode
            sp=get_next_state(s,a);   % next state s'
            r=get_reward(s,a);        % reward
            ap=get_action(sp,Q);      % naxt action by e-greedy
            Q(s,a)=Q(s,a)+alpha*(r+gamma*Q(sp,ap)-Q(s,a));
            is=is+1;
            s=sp;                     % update state
            a=ap;                     % update action
            if is>max_step
                break;
            end
        end
        ie=ie+1;
    end
end

function Q=Q_Learning(Q,s0,st)        % Q-Learning
    global maze alpha gamma epsilon max_step max_episode
    ie=1;                             % episode counter
    while ie<max_episode              % for all episodes
        is=1;                         % step counter
        s=s0;                         % initial state
        while s~=st                   % for each step in episode
            a=get_action(s,Q);        % action by e-greedy policy
            sp=get_next_state(s,a);   % next state s'
            r=get_reward(s,a);        % reward
            Q(s,a)=Q(s,a)+alpha*(r+gamma*qmax(Q(sp,:))-Q(s,a));
            s=sp;                     % update state
            is=is+1;
            if is>max_step
                break;
            end
        end
```

```
        ie=ie+1;
    end
end

function r=get_reward(s,a)
                    % get reward given state and action
    global maze alpha gamma
    sp = get_next_state(s,a);
    [row,col] = state2idx(sp,maze);
    r = maze(row,col);
end

function sp = get_next_state(s,a)
    % get next state given current state s and action a.
    global maze alpha gamma
    [~,n]=size(maze);
    switch a
        case 1       % move up
            sp=s-n;
        case 2       % move right
            sp=s+1;
        case 3       % move down
            sp=s+n;
        case 4       % move left
            sp=s-1;
    end
end

function action = get_action(s, Q)
    global maze alpha gamma epsilon
    % get an action in state s by e-greedy policy
    [M N]=size(maze);
    [row,col]=state2idx(s,maze);
    valid_actions=[];       % two-column array: action and Q-value
                            % check all four directions
    if row > 1              % Up
        r=row-1;
        c=col;
        if maze(r,c)~=-100  % if valid
            valid_actions=[valid_actions; 1, Q(s,1)];
        end
    end
    if row < M              % Down
```

```
        r=row+1;
        c=col;
        if maze(r,c)~=-100  % if valid
            valid_actions=[valid_actions; 3, Q(s,3)];
        end
    end
    if col > 1              % Left
        r=row;
        c=col-1;
        if maze(r,c)~=-100  % if valid
            valid_actions=[valid_actions; 4, Q(s,4)];
        end
    end
    if col < N              % Right
        r=row;
        c=col+1;
        if maze(r,c)~=-100  % if valid
            valid_actions=[valid_actions; 2, Q(s,2)];
        end
    end
    if rand<epsilon         % of probability epsilon
        idx=randi(size(valid_actions,1)); % random index (non-greedy)
    else                    % of probability 1-epsilon
        [q idx]=max(valid_actions(:,2));  % index of maximum Q (greedy)
    end
    action=valid_actions(idx,1);  % e-greedy action

end

function [r, c] = state2idx(s,maze)
    % Convert state index s to row and column indices (r c)
    [~,n]=size(maze);
    r=ceil(s/n);
    c=s-n*(r-1);
end

function [path reward]=find_path(s0, st, Q)
% get a path from s0 to st by taking greedy actions based on Q-table
    global maze alpha gamma epsilon
    Q(Q==0)=-inf;              % convert 0 (invalid actions) to -inf
    reward=0;                  % initialize cumulative reward
    path=[s0];                 % initialize path to start position
    s=s0;
```

```
    while path(end)~=st          % add new positions to path
        [v,a]=max(Q(s,:));       % get greedy action a
        reward=reward+get_reward(s,a);    % update cumulative reward
        s=get_next_state(s,a);   % get next position
        path=[path,s];           % append new position to path
    end
end
```

The optimal paths found by the code are illustrated in Fig. 20.8 for four slightly different mazes.

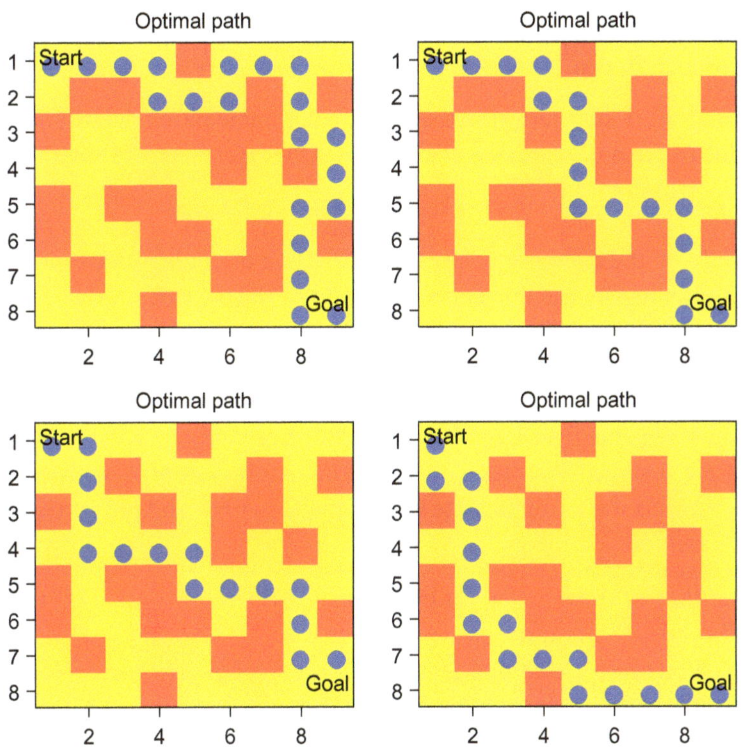

Fig. 20.8 Optimal paths in four different mazes.

20.3.3 TD(λ) Algorithm

The MC and TD methods considered previously can be unified by the n-step TD(λ) method that spans a spectrum, of which the MC and TD methods are two special cases at the opposite extremes.

Recall that both MC and TD algorithms updates iteratively the estimated state-action value function $q(s,a)$ based on Eq. (20.49), but they estimate the target G_t in the equation differently. In an MC algorithm, the target is the actual return $G_t = r_{t+1} + \gamma r_{t+2} + \cdots + \gamma^{T-t-1} r_T$, the sum of discounted rewards of all future steps up to the terminal state s_T, available only at the end of each episode. Conversely, in a TD algorithm (Eq. (20.51)), the target $G_t = r_{t+1} + \gamma v_\pi(s_{t+1})$ is the sum of the immediate reward r_{t+1} available at each step of the episode, and the discounted value of the next state, based on bootstrapping. We therefore see that an MC algorithm updates the value functions once every episode, while a TD algorithm updates once every step in the episode.

As a tradeoff between the MC and TD methods, the n-step TD algorithm can be considered a generalization of the TD method, where the target in Eq. (20.49) is an *n-step return*, the sum of the discounted rewards in the n subsequent states and the discounted value function at the following state s_{t+n} with $1 \leq n < \infty$

$$G_{t:t+n} = r_{t+1} + \gamma r_{t+2} + \cdots + \gamma^{n-1} r_{t+n} + \gamma^n v_\pi(s_{t+n}) \tag{20.57}$$

In particular, when

- $n = 1$: the 1-step return is the sum of the immediate reward and the estimated value of the next state, the same as the TD target in the TD method

$$G_{t:t+1} = r_{t+1} + \gamma v(s_{t+1}) \tag{20.58}$$

- $n = T - t$: the n-step return is the sum of the discounted rewards from all future states up to the terminal state at the end of the episode, that is, it is the return r_t defined in Eq. (20.7), the same as in the MC method

$$G_{t:t+n} = r_{t+1} + \gamma r_{t+2} + \gamma^2 r_{t+3} + \cdots + \gamma^{T-t-1} r_T + v_\pi(s_{t+n}) = G_t \tag{20.59}$$

  where the value function of the terminal state $v_\pi(s_{t+n}) = v_\pi(s_T) = 0$ is zero.
- $T - t < n < \infty$: All states s_n beyond the terminal state s_T with $n > T - t$ remain the same as the terminal state with value $v(s_T) = 0$ and return G_t, same as in the MC method.

We see that all these n-step returns form a spectrum with the TD ($n = 1$) and MC ($n \geq T - t$) methods at the two ends.

Based on these n-step returns of different n values, the n-step TD algorithm can be further generalized to a more computationally advantageous and therefore more useful algorithm called TD(λ), by which the MC and TD algorithms are again unified as two special cases at the opposite ends of a spectrum.

We first define the λ-return G_t^λ as the weighted average of all n-step returns for $n = 1, 2, \cdots, \infty$

$$G_t^\lambda = (1 - \lambda) \sum_{n=1}^{\infty} \lambda^{n-1} G_{t:t+n} \tag{20.60}$$

where $0 \leq \lambda < 1$. Here the coefficient $1 - \lambda$ is needed for the exponentially decaying weights to be normalized

$$(1 - \lambda) \sum_{n=1}^{\infty} \lambda^{n-1} = (1 - \lambda) \sum_{n=0}^{\infty} \lambda^n = \frac{1 - \lambda}{1 - \lambda} = 1 \tag{20.61}$$

The λ-return G_t^λ can be expressed in two summations: the sum of the first $T-t-1$ n-step returns $G_{t:t+1}, \cdots, G_{t:T-1}$, and the sum of all subsequent n-step returns $G_{t:T} = \cdots = G_{t:\infty} = G_t$

$$G_t^\lambda = (1-\lambda)\left[\sum_{n=1}^{T-t-1} \lambda^{n-1} G_{t:t+n} + \sum_{n=T-t}^{\infty} \lambda^{n-1} G_t\right]$$
$$= (1-\lambda)\sum_{n=1}^{T-t-1} \lambda^{n-1} G_{t:t+n} + \lambda^{T-t-1} G_t, \qquad (0 \le t \le T) \qquad (20.62)$$

where the coefficient of G_t in the second term is the sum of all coefficients in the second summation

$$(1-\lambda)\sum_{n=T-t}^{\infty} \lambda^{n-1} = (1-\lambda)\sum_{m=0}^{\infty} \lambda^m \lambda^{T-t-1} = \lambda^{T-t-1} \qquad (20.63)$$

where $m = n - T + t$. We see that in Eq. (20.62) all n-step returns $G_{t:t+n}$ are weighted, by exponentially decaying coefficient λ^{n-1}, except the true return G_t which is weighted by λ^{T-t-1}.

Again consider two special cases.

- $\lambda = 0$: all terms in Eq. (20.62) are zero, except the first one with $n = 1$ (with $\lambda^0 = 0^0 = \lim_{\lambda \to 0} \lambda^0 = 1$)

$$G_t^{\lambda=0} = \lambda^{n-1} G_{t:t+n}\Big|_{n=1} = 0^0\, G_{t:t+1} = r_{t+1} + \gamma v(s_{t+1}) \qquad (20.64)$$

  This is simply the TD target, that is,. TD(0) is the TD method.
- $\lambda = 1$: the first summation is zero and

$$G_t^{\lambda=1} = G_t \qquad (20.65)$$

  that is TD(1) is the MC method.

We see that TD(λ) is a general algorithm of which the two special cases TD(0) and TD(1) are, respectively, the TD and MC methods.

In Eq. (20.62), G_t^λ is calculated as the weighted sum of all n-step returns $G_{t:t+n}$ up to the last one G_t available at the end of the episode. This summation can be truncated to include fewer terms before reaching the terminal state at the end of the episode

$$G_{t:h}^\lambda = (1-\lambda)\sum_{n=1}^{h-t-1} \lambda^{n-1} G_{t:t+n} + \lambda^{h-t-1} G_{t:h}, \qquad (0 \le t < h \le T) \qquad (20.66)$$

Specially when $h = T$, $G_{t:h}^\lambda = G_t^\lambda$ is the same as in Eq. (20.62), and when $h = t+1$, $G_{t:h}^\lambda = G_{t:t+1} = r_{t+1} + \gamma v(s_{t+1})$ is the TD target.

Based on λ-return, the iterative update of the value function $v_\pi(s_t)$ in Eq. (20.52) can be modified to

$$v_\pi(s_t) = v_\pi(s_t) + \alpha(G_t^\lambda - v_\pi(s_t)) = v_\pi(s_t) + \alpha\delta_t \qquad (20.67)$$

where

$$\delta_t = G_t^\lambda - v_\pi(s_t) \qquad (20.68)$$

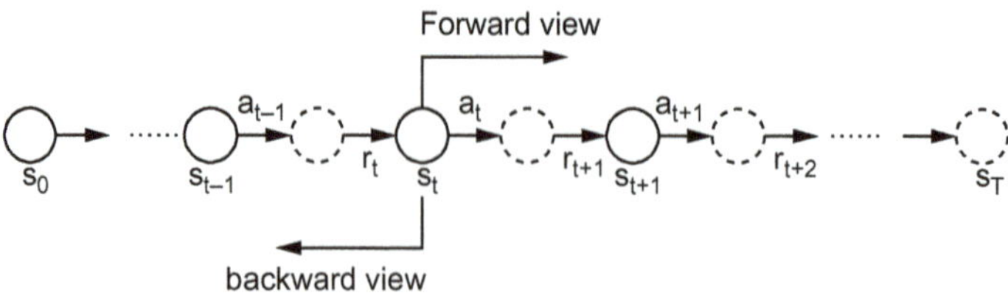

Fig. 20.9 Forward view and backward view.

This iteration is called the *forward view* of TD(λ), which is similar to the MC method, as they are based on either G_t or G_t^λ, available in the future only at the end of each episode, as illustrated in Fig. 20.9.

An alternative but equivalent method is the *backward view* of TD(λ). This method is based the *eligibility trace* $e_t(s)$ of each state, representing the frequency and recency of the visits to each state. If a state s has been more frequently and recently visited compared to others, its eligibility trace $e_t(s)$ is greater than others and its value function $v_\pi(s)$ will be updated by a greater increment than others.

The eligibility trace $e_t(s)$ at all states are initialized to zero. At each step if state $s = s_t$ is currently visited then its eligibility trace is boosted by an increment of 1 to become $e_t(s) = e_t(s) + 1$, while all others (if greater than zero) decay exponentially

$$e_t(s) = \gamma\lambda e_{t-1}(s) \qquad (\forall s \in S) \tag{20.69}$$

The backward view of TD(λ) is similar to the TD(0) method, in that they are both based on the immediate reward r_{t+1} available at each time step of the episode, instead of at the end of the episode, and they are therefore more computationally convenient. Here the iterative update of the estimated value function in Eq. (20.52) is modified so that *all* states (not only the one currently visited) are updated to different extents depending on $e_t(s)$

$$v_\pi(s) = v_\pi(s) + \alpha\delta_t e_t(s), \qquad \forall s \in S \tag{20.70}$$

where the TD error δ_t is the same as in TD(0), given in Eq. (20.53)

$$\delta_t = r_{t+1} + \gamma v_\pi(s_{t+1}) - v_\pi(s_t) \tag{20.71}$$

Here is the pseudo code for the backward view of the TD(λ) method for policy evaluation:

Input: policy π to be evaluated
Initialize: $v(s) = 0 \;\; \forall s \in S$
loop (for each episode)

$s = s_0$
$e(s) = 0, \; \forall s \in S$
while s is not terminal (for each step)

take action a based on $\pi(a|s)$, get reward r and next state s'
$\delta = r + \gamma v(s') - v(s)$
$e(s) = e(s) + 1$

for all s

$v(s) = v(s) + \alpha\delta e(s)$

$e(s) = \gamma\lambda e(s)$

$s = s'$

Note that values at all states are updated, but to different extents depending on $e(s)$, different from the TD algorithm where only the value at the state currently visited is updated.

Again consider two special cases.

- $\lambda = 0$, $e(s) = 0$ for all states except the current s being visited, that is, TD(λ) becomes TD(0), the same as the TD method.
- $\lambda = 1$, $e(s)$ is scaled down by a factor $\gamma < 1$ for all states except the current s being visited, that is, TD(λ) becomes TD(1), the same as the MC method. However, different from the MC method that updates the value function $v(s)$ at the end of each episode, here $v(s)$ is still updated at every step of the episode owing to the backward view of the method.

This backward view of the TD(λ) method can be applied to model-free control when the state value function is replaced by the action value function. Here are two algorithms based on eligibility traces that correspond to the SARSA and Q-learning algorithms based on TD(0).

- SARSA(λ) algorithm

  Initialize: $q(s,a) = 0,\ \forall(s,a),\ \alpha \in (0,\ 1],\ \epsilon > 0$, denote ϵ-greedy policy $\pi(a|s)$ by π

  loop (for each episode)

  $s = s_0$

  $e(s,a) = 0,\ \forall(s,a)$

  get action a according to π based on $q(s,a)$

  while s is not terminal (for each step)

  $e(s,a) = e(s,a) + 1$

  take action a, get reward r and next state s'

  get action a' according to π based on $q(s',a)$

  $\delta = r + \gamma q(s',a') - q(s,a)$

  for all (s,a)

  $q(s,a) = q(s,a) + \alpha\delta\, e(s,a)$

  $e(s,a) = \gamma\lambda e(s,a)$

  $s = s',\ a = a'$

- Q(λ) algorithm

  Initialize: $q(s,a) = 0,\ \forall(s,a),\ \alpha \in (0,\ 1],\ \epsilon > 0$, denote ϵ-greedy policy $pi(a|s)$ by π

  loop (for each episode)

  $s = s_0$

$e(s,a) = 0, \ \forall(s,a)$
get action a according to π based on $q(s,a)$
while s is not terminal (for each step)
    $e(s,a) = e(s,a) + 1$
    take action a, get reward r and next state s'
    get action a' according to π based on $q(s',a)$
    $a^* = \arg\max_b q(s',b)$
    $\delta = r + \gamma q(s',a^*) - q(s,a)$
    for all (s,a)
        $q(s,a) = q(s,a) + \alpha\delta\, e(s,a)$
        if $a' = a^*$ **then** $e(s,a) = \gamma\lambda e(s,a)$ **else** $e(s,a) = 0$
    $s = s', \ a = a'$

Q(λ) algorithm is different from the SARSA(λ) algorithm in two ways. First, the action value is updated based on the greedy action a^*, instead of the a' give by policy π; second, if the ϵ-greedy policy π happens to choose a random non-greedy action $a' \neq a^*$ with probability ϵ to explore rather than exploiting a^*, the eligibility traces of all states are reset to zero. There are other different versions of the algorithm that do not reset these traces.

20.4 Value Function Approximation

All previously considered algorithms are based on either the state value function $v_\pi(s)$ for each state s or the action value function $q_\pi(s,a)$ for each state-action pair. As these function values can be considered as either a 1-D or 2-D table, the algorithms based on such value functions are called *tabular methods*. However, when the number of states and the number of actions in each state are large (e.g., the game Go has 10^{170} states), the tabular methods are not suitable due to the unrealistic table size. In such a case, we can approximate the state value function $v(s)$ or action value function $q(s,a)$ by a function $\hat{v}(s,\boldsymbol{\theta})$ or $\hat{q}(s,a,\boldsymbol{\theta})$, parameterized by a vector $\boldsymbol{\theta}$, containing a set of parameters of the function as its components. Such a function can be treated as a black box as illustrated in Fig. 20.10. Now the state or action value functions can be approximated based on $\boldsymbol{\theta}$ to be estimated during the learning process when sampling the environment. This approach is more efficient and practical than the tabular methods due to its much smaller number of unknown parameters in comparison to the number of unknown variables in the tabular methods.

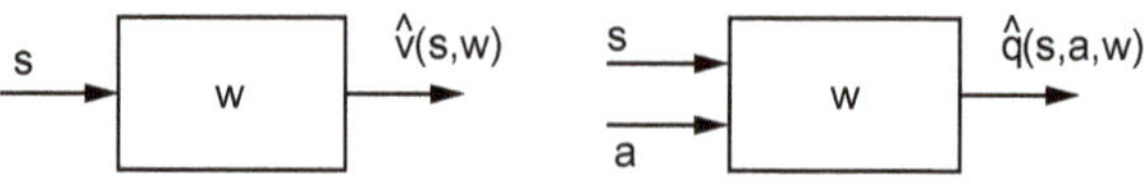

Fig. 20.10 Estimation of state and action value functions.

Such value function approximators can be implemented by methods such as those listed in the following:

- regression
- multilayer neural networks
- decision trees
- Fourier/wavelet bases.

We consider the approach of regression method in this section, and then the neural network method in the next section.

The regression method is to fit a continuous function $\hat{v}(s,\boldsymbol{\theta})$ or $\hat{q}(s,a,\boldsymbol{\theta})$ to a set of discrete and finite data points $v_\pi(s)$ or $q_\pi(s,a)$ obtained by sampling the environment, so that all points in the state space, including those for states or state-action pairs never actually observed during the sampling process, can be estimated by interpolation based on the regression function.

Our task is to find the optimal parameter $\boldsymbol{\theta}$ of the regression model that minimizes the objective function, the mean squared error (MSE) between the approximated values and the sample values

$$\varepsilon(\boldsymbol{\theta}) = \frac{1}{2}E_\pi[(v_\pi(s) - \hat{v}(s,\boldsymbol{\theta}))^2] \tag{20.72}$$

The gradient vector of $\varepsilon(\boldsymbol{\theta})$ is

$$\begin{aligned} \mathbf{g}_\varepsilon(\boldsymbol{\theta}) &= \frac{d}{d\boldsymbol{\theta}}\varepsilon(\boldsymbol{\theta}) = \frac{1}{2}E_\pi\left[\frac{d}{d\boldsymbol{\theta}}\left[(v_\pi(s) - \hat{v}(s,\boldsymbol{\theta}))^2\right]\right] \\ &= -E_\pi\left[(v_\pi(s) - \hat{v}(s,\boldsymbol{\theta}))\frac{d}{d\boldsymbol{\theta}}\hat{v}(s,\boldsymbol{\theta})\right] \end{aligned} \tag{20.73}$$

and the optimal parameter vector can be found iteratively by the gradient descent method

$$\boldsymbol{\theta}_{n+1} = \boldsymbol{\theta}_n + \Delta\boldsymbol{\theta} = \boldsymbol{\theta}_n - \alpha\mathbf{g}_\varepsilon(\boldsymbol{\theta}) = \boldsymbol{\theta}_n + \alpha E_\pi\left[(v_\pi(s) - \hat{v}(s,\boldsymbol{\theta}))\frac{d}{d\boldsymbol{\theta}}\hat{v}(s,\boldsymbol{\theta})\right] \tag{20.74}$$

where α is the step size or learning rate, and

$$\Delta\mathbf{w} = -\alpha\mathbf{g}_\varepsilon(\mathbf{w}) = \alpha E_\pi\left[(v_\pi(s) - \hat{v}(s,\mathbf{w}))\frac{d}{d\mathbf{w}}\hat{v}(s,\mathbf{w})\right] \tag{20.75}$$

If the SGD method is used (Section 6.1), that is, the parameter $\boldsymbol{\theta}$ is updated based on only one data point at each iteration instead of the expectation of the data points, then the expectation symbol E_π can be dropped.

As the true value function $v_\pi(s)$, the expected return, is unknown, it needs to be estimated by either the actual return G_t at each state $s = s_t$ based on the MC method, or $r_{t+1} + \gamma\hat{v}_\pi(s',\boldsymbol{\theta})$ in terms of the immediate reward r_{t+1} and the estimated value $\hat{v}_\pi(s',\boldsymbol{\theta})$ of the next state s' based on the TD method. Such an approach can be considered as supervised learning based on a labeled training data set: $\{(s_t, G_t), \forall t\}$ for the MC method, or $\{(s_t, r_{t+1}, \forall t\}$ for the TD method.

Here we consider the special case where the state value $v_\pi(s^n)$ of each of the $N = |S|$ states is approximated by a linear regression function

$$\hat{v}(s^n, \mathbf{w}) = \sum_{i=1}^{d} x_i(s^n) w_i = \mathbf{w}^T \mathbf{x}(s^n) = \mathbf{w}^T \mathbf{x}_n, \qquad n = 1, \cdots, N \tag{20.76}$$

Here the generic symbol $\boldsymbol{\theta}$ as the parameter of the regression is replaced by $\mathbf{w}$, representing specifically the weights of the linear function. This model is based on the assumption that the value function of each state s^n can be indeed approximated as a linear combination of a set of features as the components in the d-dimensional feature vector $\mathbf{x}_n = \mathbf{x}(s^n) = [x_1(s^n), \cdots, x_d(s^n)]^T$, which are hand picked or designed to properly represent the specific problem at hand.

As a special example, we can represent each of the $N = |S|$ states by a N- dimensional binary feature vector $\mathbf{x}_n = \mathbf{x}(s^n) = [x_1(s^n), \cdots, x_N(s^n)]^T$, where $x_i(s^n)$ is 1 if $i = n$ but 0 otherwise. Then the weight vector becomes $\mathbf{w} = [w_1, \cdots, w_N]^T = [v_\pi(s^1), \cdots, v_\pi(s^n)]^T$, and we have $\hat{v}_\pi(s^n, \mathbf{w}) = \mathbf{w}^T \mathbf{x}_n = v_\pi(s^n)$. In this case, each state s is represented by its value function $v_\pi(s)$, the same as in all previous discussions.

The objective function for the MSE of such a linear approximating function is

$$\varepsilon(\mathbf{w}) = \frac{1}{2} E_\pi[(v_\pi(s) - \hat{v}(s, \mathbf{w}))^2] = \frac{1}{2} E_\pi[(v_\pi(s) - \mathbf{w}^T \mathbf{x}(s))^2] \tag{20.77}$$

If all data points in terms of state s represented by $\mathbf{x}(s)$ and the corresponding return $G(s)$ as a sample of the true value $v_\pi(s)$ have already been collected, then the value function approximation can be carried out as an off-line algorithm in a batch manner, the same as in the linear least squares regression discussed in Chapter 5, and the optimal weight vector $\mathbf{w}^*$ that minimizes the squared error

$$\mathbf{w}^* = \arg\min_{\mathbf{w}} \sum_{n=1}^{N} [G(s^n) - \mathbf{w}^T \mathbf{x}_n]^2 \tag{20.78}$$

can be found by solving a linear equation system $\mathbf{X}^T \mathbf{w} = \mathbf{g}$, where $\mathbf{X} = [\mathbf{x}_1, \cdots, \mathbf{x}_N]$ is a $d \times N$ matrix containing all N states each represented by $\mathbf{x}_n$, and vector $\mathbf{g} = [G(s^1), \cdots, G(s^n)]^T$ contains the corresponding returns. As we assume the number of states is greater than the number of the model parameters, that is, $N \geq d$, this is an over-constrained system with the least squares solution (Eq. (5.10))

$$\mathbf{w}^* = (\mathbf{X}\mathbf{X}^T)^{-1} \mathbf{X}\mathbf{g} = (\mathbf{X}^T)^{-} \mathbf{g} \tag{20.79}$$

However, as reinforcement learning is typically an online process, the parameter $\mathbf{w}$ for the approximator needs to be modified in real time whenever a new piece of data becomes available while sampling the environment. Then we can use the gradient descent method, based on the gradient vector of $\varepsilon(\mathbf{w})$

$$\begin{aligned} \mathbf{g}_\varepsilon(\mathbf{w}) &= \frac{d}{d\mathbf{w}} \varepsilon(\mathbf{w}) = -E_\pi(v_\pi(s) - \hat{v}(s, \mathbf{w})) \frac{d}{d\mathbf{w}} \hat{v}(s, \mathbf{w}) \\ &= -E_\pi(v_\pi(s) - \mathbf{w}^T \mathbf{x}(s)) \mathbf{x}(s) \end{aligned} \tag{20.80}$$

and find the optimal parameter $\mathbf{w}^*$ iteratively

$$\begin{aligned}\mathbf{w}_{n+1} &= \mathbf{w}_n + \Delta\mathbf{w} = \mathbf{w}_n + \alpha E_\pi \left[v_\pi(s) - \hat{v}(s,\mathbf{w})\right] \frac{d}{d\mathbf{w}} \hat{v}(s,\mathbf{w}) \\ &= \mathbf{w}_n + \alpha E_\pi \left[v_\pi(s) - \mathbf{w}^T \mathbf{x}(s)\right] \mathbf{x}(s)\end{aligned} \tag{20.81}$$

with increment

$$\Delta\mathbf{w} = \alpha E_\pi \left[v_\pi(s) - \hat{v}(s,\mathbf{w})\right] \frac{d}{d\mathbf{w}} \hat{v}(s,\mathbf{w}) = \alpha E_\pi \left[v_\pi(s) - \mathbf{w}_n^T \mathbf{x}(s)\right] \mathbf{x}(s) \tag{20.82}$$

If SGD is used, then $\mathbf{w}$ is updated whenever a new data point $\mathbf{x}(s_t)$ is available at a new step t in the sampling process, and the expectation symbol E_π can be dropped. The expectation of the estimated $\mathbf{w}$ by this SGD method is the same as that obtained by the full GD method. This iteration will convergence to the global minimum of the objective function $\varepsilon(\mathbf{w})$ in Eq. (20.77), as it is a quadratic function with only one minimum that is global.

As the state value function $v_\pi(s)$ is unknown, it needs to be estimated in one of several different ways, similar to the corresponding algorithms discussed before.

- **MC method**
  The value function $v_\pi(s)$ is estimated by the actual return G_t for each state s_t visited in an episode, obtained only at the end of each episode while sampling the environment

$$\Delta\mathbf{w} = \alpha[G_t - \hat{v}(s_t,\mathbf{w})] \frac{d}{d\mathbf{w}} \hat{v}(s_t,\mathbf{w}) = \alpha[G_t - \mathbf{w}^T \mathbf{x}(s_t)]\mathbf{x}(s_t) \tag{20.83}$$

  Here is the pseudo code for the algorithm based on the MC method, similar to that for value evaluation algorithms discussed previously:

  Initialize $\mathbf{w} = \mathbf{0}$
  loop (for each episode)
    run current episode to the end to get G_t, $t = 1, \cdots, T$
    for $t = 1, \cdots, T$
      if s_t is visited the first time
        $\mathbf{w} = \mathbf{w} + \alpha[G_t - \mathbf{w}^T \mathbf{x}(s_t)]\mathbf{x}(s_t)$
      end if
    end for
  end loop

- **TD(0) method**
  The value function $v_\pi(s)$ is estimated by the TD target, the sum of the immediate reward r_{t+1} and the approximated value of the next state s_{t+1}, at each time step of each episode while sampling the environment:

$$\Delta\mathbf{w} = \alpha[r_{t+1} + \gamma \mathbf{w}^T \mathbf{x}(s_{t+1}) - \mathbf{w}^T \mathbf{x}(s_t)]\mathbf{x}(s_t) \tag{20.84}$$

  Here is the pseudo code for the algorithm based on the TD method, similar to that for value evaluation algorithms discussed previously:

Initialize $\mathbf{w} = \mathbf{0}$
loop (for each episode)
  for each time step in current episode
    take action $a = \pi(s)$, get reward r and next state s'
    $\mathbf{w} = \mathbf{w} + \alpha[r + \gamma \mathbf{w}^T \mathbf{x}(s') - \mathbf{w}^T \mathbf{x}(s)]\mathbf{x}(s)$
  end for
end loop

- **TD(λ) method**
  The value function $v_\pi(s)$ is estimated based on the λ-return G_t^λ given in Eq. (20.62). As discussed before, TD(λ) has two forms.
  - Forward view: same as MC method, except the true return G_t is replaced by λ-return G_t^λ available only at the end of each episode

$$\Delta \mathbf{w} = \alpha[G_t^\lambda - \mathbf{w}^T \mathbf{x}(s_t)]\mathbf{x}(s_t) \tag{20.85}$$

  - Backward view: similar to TD(0), this is based on the immediate reward r_{t+1} available at every step of each episode

$$\delta_t = \alpha[r_{t+1} + \gamma \mathbf{w}^T \mathbf{x}(s_{t+1}) - \mathbf{w}^T \mathbf{x}(s_t)]\mathbf{x}(s_t) \tag{20.86}$$

  Here the eligibility trace becomes a vector, which is initialized to zero at the beginning of the episode, and then updated by the gradient of the estimated value function $\hat{v}(s, \mathbf{w}) = \mathbf{w}^T \mathbf{x}(s)$ (the components of $\mathbf{e}$ are updated differently instead all by 1)

$$\mathbf{e}_t = \gamma \lambda \mathbf{e}_{t-1} + \frac{d}{d\mathbf{w}} \mathbf{w}^T \mathbf{x}(s_t) = \gamma \lambda \mathbf{e}_{t-1} + \mathbf{x}(s_t) \tag{20.87}$$

  while the weight vector is still updated the same as before

$$\Delta \mathbf{w} = \alpha \delta_t \mathbf{e}_t \tag{20.88}$$

Given a policy π for an MDP, we also define a probability distribution $d(s)$ over all states visited according to π, satisfying

$$\sum_s d(s) = 1 \tag{20.89}$$

and the following balance equation

$$d(s') = \sum_s \sum_a \pi(a|s) p(s'|s, a) d(s) \tag{20.90}$$

As both $d(s')$ and $d(s)$ represent the same distribution, they must be identical.

Given the probability distribution $d(s)$, the MSE of the value function approximation for a policy π can be expressed as

$$MSE(\mathbf{w}) = \sum_s d(s)[v_\pi(s) - \hat{v}_\pi(s, \mathbf{w})]^2 \tag{20.91}$$

It can be shown that the MC method for the linear value function approximation will converge to the optimal weight vector $\mathbf{w}_{MC}$ that minimizes the MSE

$$MSE(\mathbf{w}_{MC}) = \min_{\mathbf{w}} \sum_s d(s)[v_\pi(s) - \hat{v}_\pi(s,\mathbf{w})]^2 \tag{20.92}$$

20.5 Control Based on Function Approximation

Model-free control can be carried out based on the state action value function $\hat{q}(s,a,\mathbf{w})$ approximated as discussed in the previous section, by the approach of general policy iteration, as illustrated in Fig. 20.11. This is similar to the algorithms for model-free control illustrated in Fig. 20.5, with the only difference that the action value function $q_\pi(s,a)$ is now replaced by the parameter $\mathbf{w}$ of the linear model of the action value function

$$\hat{q}(s,a,\mathbf{w}) = \sum_n w_n x(s,a) = \mathbf{w}^T \mathbf{x}(s,a) \tag{20.93}$$

where $\mathbf{x}(s,a)$ is the feature vector for the state-action pair (s,a). We need to find the optimal parameter $\mathbf{w}$ that minimizes the objective function, here specifically the MSE of the approximation

$$\varepsilon(\mathbf{w}) = \frac{1}{2}E_\pi[(q_\pi(s,a) - \hat{q}(s,a,\mathbf{w}))^2] = \frac{1}{2}E_\pi[(q_\pi(s,a) - \mathbf{w}^T\mathbf{x}(s,a))^2] \tag{20.94}$$

with gradient vector

$$\begin{aligned} \mathbf{g}_\varepsilon(\mathbf{w}) &= \frac{d}{d\mathbf{w}}\varepsilon(\mathbf{w}) = -E_\pi\,[q_\pi(s,a) - \hat{q}(s,a,\mathbf{w})]\,\frac{d}{d\mathbf{w}}\hat{q}(s,a,\mathbf{w}) \\ &= -E_\pi\left[(q_\pi(s,a) - \mathbf{w}^T\mathbf{x}(s,a))\mathbf{x}(s,a)\right] \end{aligned} \tag{20.95}$$

If the stochastic gradient descent method is used based on a single sample of the action value $q_\pi(s,a)$, instead of its expectation, then the expectation symbol E_π can be dropped. The optimal weight vector $\mathbf{w}^*$ that minimizes $\varepsilon(\mathbf{w})$ in Eq. (20.94) can be learned iteratively

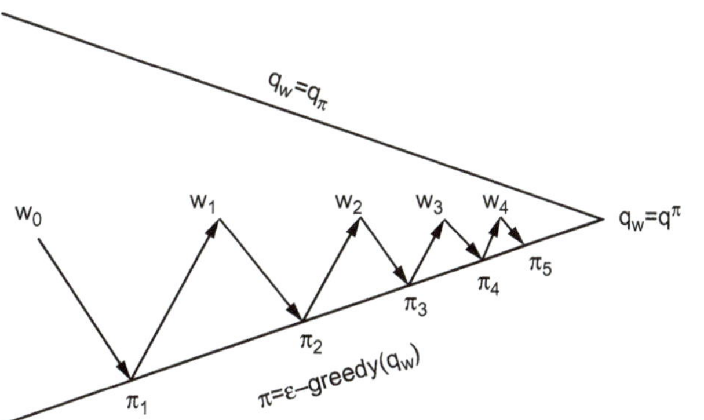

Fig. 20.11 Policy iteration based on approximated value functions.

$$\begin{aligned}\mathbf{w}_{t+1} &= \mathbf{w}_t + \Delta\mathbf{w} = \mathbf{w}_t - \alpha\mathbf{g}_\varepsilon(\mathbf{w}) \\ &= \mathbf{w}_t + \alpha\left[(q_\pi(s_t,a_t) - \hat{q}_\pi(s_t,a_t,\mathbf{w}))\frac{d}{d\mathbf{w}}\hat{q}_\pi(s_t,a_t,\mathbf{w})\right] \\ &= \mathbf{w}_t + \alpha\left[(q_\pi(s_t,a_t) - \mathbf{w}_t^T\mathbf{x}(s_t,a_t))\mathbf{x}(s_t,a_t)\right] \end{aligned} \tag{20.96}$$

where $\Delta\mathbf{w}$ is the increment of the update

$$\begin{aligned}\Delta\mathbf{w} = -\alpha\mathbf{g}_\varepsilon(\mathbf{w}) &= \alpha[q_\pi(s_t,a_t) - \hat{q}_\pi(s_t,a_t,\mathbf{w})]\frac{d}{d\mathbf{w}}\hat{q}_\pi(s_t,a_t,\mathbf{w}) \\ &= \alpha[q_\pi(s_t,a_t) - \mathbf{w}_t^T\mathbf{x}(s_t,a_t)]\mathbf{x}(s_t,a_t)\end{aligned} \tag{20.97}$$

As the true Q-value $q_\pi(s,a)$ in the expression is unknown, it needs to be estimated by some target depending on the specific methods used.

- **MC method**
  The true $q_\pi(s,a)$ is replaced by the sample return G_t as the target, obtained at the end of each episode

$$\Delta\mathbf{w} = \alpha[G_t - \hat{q}(s_t,a_t,\mathbf{w})]\frac{d}{d\mathbf{w}}\hat{q}(s_t,a_t,\mathbf{w}) = \alpha[G_t - \mathbf{w}_t^T\mathbf{x}(s_t,a_t)]\mathbf{x}(s_t,a_t) \tag{20.98}$$

- **TD(0) method**
  The action value function $q_\pi(s,a)$ is replaced by the TD target, the sum of the immediate reward, available at each step of each episode, and the approximated action value of the next state s_{t+1}:
  - SARSA (on-policy)

$$\begin{aligned}\Delta\mathbf{w} &= \alpha[r_{t+1} + \gamma\hat{q}(s_{t+1},a_{t+1},\mathbf{w}) - \hat{q}(s_t,a_t,\mathbf{w})]\frac{d}{d\mathbf{w}}\hat{q}(s_t,a_t,\mathbf{w}) \\ &= \alpha[r_{t+1} + \gamma\mathbf{w}^T\mathbf{x}(s_{t+1},a_{t+1}) - \mathbf{w}_t^T\mathbf{x}(s_t,a_t)]\mathbf{x}(s_t,a_t)\end{aligned} \tag{20.99}$$

  Following Eq. (20.71), the TD error is defined as

$$\delta_t = r_{t+1} + \gamma\hat{q}(s_{t+1},a_{t+1},\mathbf{w}) - \hat{q}(s_t,a_t,\mathbf{w}) \tag{20.100}$$

  then we get

$$\Delta\mathbf{w} = \alpha\delta_t\frac{d}{d\mathbf{w}}\hat{q}(s_t,a_t,\mathbf{w}) \tag{20.101}$$

  - Q-learning (off-policy)

$$\begin{aligned}\Delta\mathbf{w} &= \alpha[r_{t+1} + \gamma\max_{a'}\hat{q}(s_{t+1},a',\mathbf{w}) - \hat{q}(s_t,a_t,\mathbf{w})]\frac{d}{d\mathbf{w}}\hat{q}(s_t,a_t,\mathbf{w}) \\ &= \alpha[r_{t+1} + \gamma\max_{a'}\mathbf{w}^T\mathbf{x}(s_{t+1},a') - \mathbf{w}_t^T\mathbf{x}(s_t,a_t)]\mathbf{x}(s,a)\end{aligned} \tag{20.102}$$

- **TD(λ) method**
  In the forward-view version of the TD(λ) method, the action function $q_\pi(s,a)$ is approximated by the λ-return G_t^λ as the target, available only at the end of each episode

$$\Delta\mathbf{w} = \alpha[G_t^\lambda - \mathbf{w}_t^T\mathbf{x}(s_t,a_t)]\mathbf{x}(s_t,a_t) \tag{20.103}$$

In the backward-view version of the TD(λ) method, the eligibility trace in Eq. (20.87) becomes

$$\mathbf{e}_t = \gamma\lambda\mathbf{e}_{t-1} + \frac{d}{d\mathbf{w}}\hat{q}(s_t, a_t, \mathbf{w}) \tag{20.104}$$

and the weight vector is updated as in Eq. (20.88)

$$\Delta\mathbf{w} = \alpha\delta_t\mathbf{e}_t \tag{20.105}$$

where δ_t is the TD error for the backward view of TD(λ) first given in Eq. (20.71), but now with the state value $v_\pi(s)$) replaced by the estimated action value $\hat{q}(s, a, \mathbf{w})$

$$\delta_t = (r_{t+1} + \gamma\hat{q}_\pi(s_{t+1}, a_{t+1}) - \hat{q}_\pi(s_t, a_t)) \tag{20.106}$$

In summary, here are the conceptual (not necessarily algorithmic) steps for the general model-free control based on approximated action value function:

- learn parameter $\mathbf{w}$ as in Eq. (20.96);
- get the Q-values as in Eq. (20.93);
- obtain the policy by ϵ-greedy approach as in Eq. (20.42).

These steps are also illustrated here

$$\text{Training of } \mathbf{w} \Longrightarrow \text{Q-value} \Longrightarrow \text{Policy } \pi \tag{20.107}$$

20.6 Deep Q-Learning

In the previous section, the Q-values at a given state are approximated by a linear regression function $\hat{q}(s, a, \mathbf{w})$ (Eq. (20.93)) parameterized by the weight vector $\mathbf{w}$. However, when applied to a real-world problem with large numbers of states and actions, this simple linear function may not be able to model the potentially highly complex nonlinear relationship between the state-action pairs and the value functions. In this case, a more sophisticated nonlinear model may be needed, such as a deep learning networks considered in Section 18.4 used for modeling complex nonlinear relationship $\mathbf{y} = \mathbf{f}(\mathbf{x}, \mathbf{W})$ between its input $\mathbf{x}$ and output $\mathbf{y}$. When applied to reinforcement learning, such a deep learning network takes a given state s as input, and generates the approximated Q-values $\hat{q}(s, a, \mathbf{W})$ for all $|A|$ actions available at the state as output. Here $\mathbf{W}$ represents all weights in the deep network, while vector $\mathbf{w}$ contains all weights of a linear regression function $\hat{q}(s, a, \mathbf{w})$. This deep learning network in reinforcement learning is called the *deep Q-learning network (DQN)*, while all previously considered methods such as the methods of tabular Q-learning and linear function approximation are considered shallow. The three methods are illustrated in Fig. 20.12.

As we know, the method of tabular Q-learning models the environment by the Q-table of size $|A||S|$ for all actions at all states, which may become impractically large when both $|S|$ and $|A|$ are large. This difficulty is avoided in the DQN method,

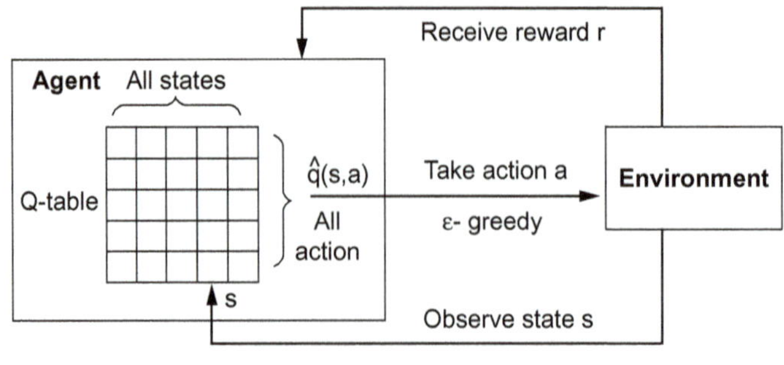

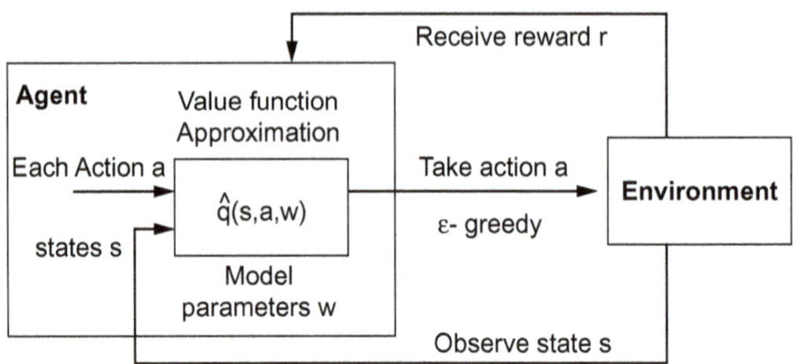

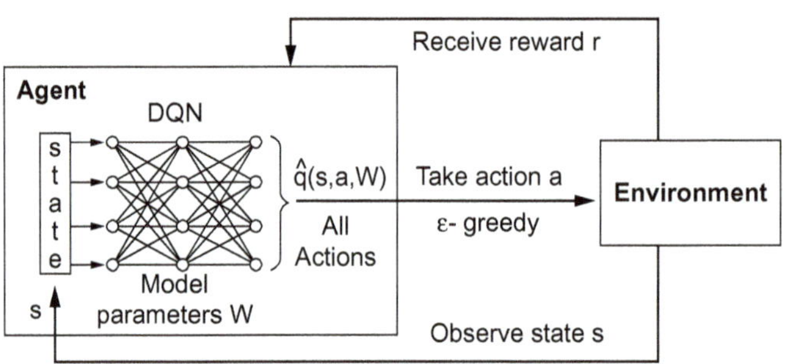

Fig. 20.12 Three Q-learning methods: tabular, function approximation, and deep Q-Learning.

as well as in regression function approximation, which estimates the action value functions $\hat{q}(s,a)$ without exhausting all $|A||S|$ cases. The cases not ever seen during training are approximated by interpolation based on those that have been seen. Also, the DQN method is superior to function approximation owing to its multilayer nonlinear learning capability. Moreover, while both methods take the current state s as input, function approximation is based on a small set of carefully handcrafted states as input, but the input to the DQN can be in some raw data with potentially high dimensionality extracted directly from the environment. For example, in a board game problem, it is very difficult to come up with a set of well-defined states that characterizes all possible situations in the game. But for the DQN, the current state can be represented simply by the image of the board, of which all pixel values are directly fed into the input layer of the DQN (similar to the image input to a convolutional neural networks (CNN) considered in Section 18.4).

As a supervised learning method, the DQN is trained in the same way as a backpropagation network. The weights in $\mathbf{W}$ as the model parameter are iteratively updated by gradient descent, to gradually reduce the error, the difference between the actual output $\hat{q}(s,a,\mathbf{W})$ of the DQN and the desired target output, obtained by the Bellman equation (Eq. (20.33)) based on data collected while sampling the environment, including the current state s, the action a, the next state s', and the reward r, collectively treated as an *experience tuple* $e = (s,a,s',r)$

$$q(s,\mathbf{W}') = r + \gamma \max_{a'} q(s',a',\mathbf{W}') \tag{20.108}$$

Here the second term is the highest output value from the DQN based on the previous estimated weights, denoted by $\mathbf{W}'$. We denote the target $q(s,\mathbf{W}')$ simply as q as it is treated as a constant while updating $\mathbf{W}$. Now the squared error between the DQN output and the target can be found is

$$\varepsilon(\mathbf{W}) = \frac{1}{2}[q - \hat{q}(s,a,\mathbf{W})]^2 = \frac{1}{2}\left[r + \gamma \max_{a'} q(s',a',\mathbf{W}') - \hat{q}(s,a,\mathbf{W})\right]^2 \tag{20.109}$$

and its gradient

$$\begin{aligned}\mathbf{g}_{\varepsilon}(\mathbf{W}) &= \frac{d}{d\mathbf{W}}\varepsilon(\mathbf{W}) = -\,[q(s,\mathbf{W}') - \hat{q}(s,a,\mathbf{W})]\,\frac{d}{d\mathbf{W}}\hat{q}(s,a,\mathbf{W})\\ &= -\left[r + \gamma \max_{a'} q(s',a',\mathbf{W}') - \hat{q}(s,a,\mathbf{W})\right]\frac{d}{d\mathbf{W}}\hat{q}(s,a,\mathbf{W})\end{aligned} \tag{20.110}$$

The weights in $\mathbf{W}$ are updated by gradient descent

$$\begin{aligned}\mathbf{W}_{n+1} &= \mathbf{W}_n - \alpha \mathbf{g}_{\varepsilon}(\mathbf{W})\\ &= \mathbf{W}_n + \alpha\left[r + \gamma \max_{a'} q(s',a',\mathbf{W}') - \hat{q}(s,a,\mathbf{W})\right]\frac{d}{d\mathbf{W}}\hat{q}(s,a,\mathbf{W})\end{aligned} \tag{20.111}$$

For this deep Q-learning method there are two issues that need to be further addressed.

- First, to update $\mathbf{W}$ by Eq. (20.111) we need to estimate the target by the Bellman equation in Eq. (20.108). In this bootstrapping process, the resulting target $q(s,\mathbf{W}')$ depends on the previous estimate $q(s',a',\mathbf{W}')$, which in turn depends on the previous $\mathbf{W}'$, that is, the iterative training process is actually chasing a moving target. This is different from the training of a conventional backpropagation network, where the desired output, the target, is fixed. As the consequence, the training iteration of the DQN may be unstable, even diverging.

  This problem is addressed by including a second network, called the target network and parameterized by $\mathbf{W}'$. The output $q'(s',a',\mathbf{W}')$ of this target network is used as the targets for training the first network, of which the output $q(s,a,\mathbf{W})$ is used for the state transition while sampling the environment. To avoid chasing a moving target, here $\mathbf{W}'$ of the target network remain unchanged for some period of C steps, while $\mathbf{W}$ of the first network is updated. At the end of the C-step period, the updated $\mathbf{W}$ is copied over to the target network to replace its $\mathbf{W}'$.
- Second, during the sequential sampling of the environment step by step, the Q-values obtained for the current state depends on those collected in the previous

state, that is, they are not i.i.d. samples as they are correlated. Consequently the training process may be biased. This issue is similar to that considered in subsection 20.3.1, where we only use the data collected during the first visit, instead of every visit, to a state while sampling the environment.

This problem is addressed by a method called *experience replay* based on a buffer $D = \{e_1, \cdots, e_N\}$ containing the agent's previous experience tuples collected in the latest N steps while sampling the environment. The DQN is then trained based on a minibatch of such experiences randomly selected from the buffer D. As these experiences are in random order and no longer correlated, the training is unbiased.

Once the DQN has been trained, the agent can form the optimal policy by taking the action a corresponding to the maximum $q(s,a)$ out of all outputs of the DQN for each state s as the input.

Here is the pseudo code for the deep Q-learning method:

Input: the states and their corresponding rewards in the environment (e.g., images of game board and the scores)
Output: Q-values for all actions in states
Initialize replay buffer D for N experiences
Initialize $\mathbf{W} = \mathbf{W}'$ for both the target network and DQN
loop (for each episode)
for each time step in current episode
Take action a at current state s based on ϵ-greedy policy:

$$a = \begin{cases} \text{random action} & Prob = \epsilon \\ \arg\max_a q(s,a,\mathbf{W}) & Prob = 1-\epsilon \end{cases}$$

Get reward r and next state s'
Store a (s,a,s',r) in D
Move to next state:$s = s'$
For each experience $(s,a,s',,r)$ in a random minibatch from D
Find target: :

$$q = \begin{cases} r & \text{if } s' \text{ is a terminal state} \\ r + \gamma \max_{a'} q'(s',a',\mathbf{W}') \end{cases}$$

Update $\mathbf{W}$ by gradient descent based on gradient $\mathbf{g}_\epsilon(\mathbf{W})$ of squared error $\varepsilon(\mathbf{W}) = (q - q(,s,a,\mathbf{W}))^2$
Update weights: $\mathbf{W}' = \mathbf{W}$ (i.e., $q' = q$) every C steps
end for
end loop

20.7 Policy Gradient Methods

All RL algorithms previously considered are based on either state or action value function and the policy is indirectly derived from them by greedy or ϵ-greedy method. However, as the ultimate goal of an RL problem is to find the optimal policy that maximizes the return, it can also be considered as an optimization problem to directly find the optimal policy that maximizes an objective function representing the total return, which can be addressed by the gradient ascent method, as discussed in the following.

In previous methods we approximated the state value function $v_\pi(s)$ or action value function $q_\pi(s,a)$ by a function $\hat{v}_\pi(s,\mathbf{w})$ or $\hat{q}_\pi(s,a,\mathbf{w})$, of which the parameter $\mathbf{w}$ is obtained by sampling the environment, as a supervised learning process. Now we will directly construct a model of the policy as a function parameterized by $\boldsymbol{\theta}$

$$\pi_\theta(a|s) = P(a|s,\boldsymbol{\theta}) \tag{20.112}$$

where $\boldsymbol{\theta} = [\theta_1,\cdots,\theta_d]^T$ is a vector composed of d parameters of the policy model. Once $\boldsymbol{\theta}$ is available, the probability for any action at any state is determined. This parameterized policy model is suitable in cases where the numbers of states and actions are large or even continuous.

As a specific example, the policy model can be based on the softmax function

$$\pi_\theta(a|s) = P(a|s,\boldsymbol{\theta}) = \frac{e^{h(s,a,\boldsymbol{\theta})}}{\sum_{a'} e^{h(s,a',\boldsymbol{\theta})}} \tag{20.113}$$

satisfying $\sum_a \pi_\theta(a|s) = 1$. Here the summation is over all possible actions in state s, and $h(s,a,\boldsymbol{\theta})$ is the *preference* of action a in state s, which can be a parameterized function such as a simple linear function $h(s,a,\boldsymbol{\theta}) = \boldsymbol{\theta}^T\mathbf{x}(s,a)$, or a neural network with weights as the parameters in $\boldsymbol{\theta}$, the same as how the value functions are approximated in Section 20.4. According to this policy, an action with higher preference $h(s,a,\boldsymbol{\theta})$ will have a higher probability to be chosen at state s.

Different from how we find the parameter $\mathbf{w}$ of $\hat{q}(s,a,\mathbf{w})$ by minimizing the objective function $\varepsilon(\mathbf{w})$ in Eq. (20.94) for the MSE between the value function $q_\pi(s,a)$ and its model $\hat{q}_\pi(s,a,\mathbf{w})$, here we find the parameter $\boldsymbol{\theta}$ of $\pi_\theta(a|s)$ by maximizing an objective function $J(\boldsymbol{\theta})$ representing the value function, the expected return, under the policy, which is to be minimized by gradient ascent. Such a method is therefore called a *policy gradient method.*

The value-function based methods considered previously and the policy-based methods considered here are summarized, together with the *actor–critic method*, as the combination of the two. These methods are also illustrated in Fig. 20.13.

- Value-based: the policy is derived by the greedy or ϵ-greedy method based on the value function learned by sampling the environment as a supervised training process.

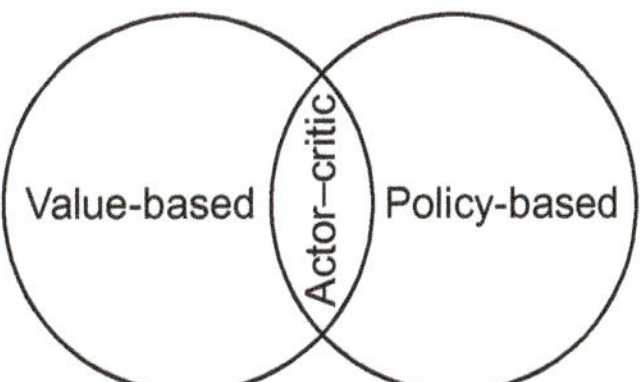

Fig. 20.13 Value-based, policy-based, and Actor–Critic.

- Policy-based: the policy is directly learned without explicitly value function estimation.
- Actor–critic: parameters for both value function model and policy model are learned simultaneously.

A policy-based method can learn stochastic policies effectively in high-dimensional or continuous action space with better convergence properties, but it may get stuck at a local optimum, and has high variance.

How good a policy is may be measured by different objective functions related to the values or rewards associated with the policy being evaluated, depending on the environment of the specific problem.

- In episodic environments with finite horizon, we can use the value, the expected return of the start state s_0

$$J_1(\boldsymbol{\theta}) = v_{\pi_\theta}(s_0) \tag{20.114}$$

- In continuing (online) environments with infinite horizon, we can use the average value

$$J_{avv}(\boldsymbol{\theta}) = \sum_s d_{\pi_\theta}(s) v_{\pi_\theta}(s) \tag{20.115}$$

  where $d_{\pi_\theta}(s)$ is the stationary distribution of all states under policy $\pi(a|s,\theta)$.

We can find the optimal parameter $\boldsymbol{\theta}^*$ for the policy model $\pi_\theta(a|s)$ by solving the maximization problem

$$\theta^* = \arg\max_\theta J(\boldsymbol{\theta}) \tag{20.116}$$

based on the gradient of $J(\boldsymbol{\theta})$

$$\triangledown_\theta J(\theta) = \frac{d}{d\boldsymbol{\theta}} J(\theta) = \left[\frac{\partial J}{\partial \theta_1}, \cdots, \frac{\partial J}{\partial \theta_n} \right]^T \tag{20.117}$$

by the iterative gradient ascent method

$$\boldsymbol{\theta}_{t+1} = \boldsymbol{\theta}_t + \Delta\boldsymbol{\theta} = \boldsymbol{\theta}_t + \alpha \triangledown_\theta J(\boldsymbol{\theta}) \tag{20.118}$$

Here we use t as the index of the iteration based on the assumption that a new sample point is available at every time step of an episode while sampling the environment following policy $\pi_\theta(a|s)$.

While it is conceptually straightforward to see how the optimal parameter $\boldsymbol{\theta}^*$ that maximizes $J(\boldsymbol{\theta})$ can be found by the gradient ascent method, it is not easy to actually obtain the gradient $\triangledown_\theta J(\boldsymbol{\theta})$, which depends on not only what actions to take at the states directly determined by the policy $\pi_\theta(a|s)$, but also how the states are distributed under the policy in an unknown environment. This challenge is addressed by the following *policy gradient theorem*, stating that the gradient of $J(\boldsymbol{\theta})$ can be calculated as shown without explicit derivatives of the state distribution

$$\triangledown_\theta J(\boldsymbol{\theta}) = E_{\pi_\theta} \left[\triangledown_\theta \ln \pi_\theta(a|s)\, q_{\pi_\theta}(s,a) \right] \tag{20.119}$$

Proof of policy gradient theorem

Consider the derivative of the value function $v_{\pi_\theta} = \sum_a \pi_\theta(a|s)\ q_{\pi_\theta}(s,a)$ (in Eq. (20.21))

$$\begin{aligned}
\nabla_\theta v_{\pi_\theta}(s) &= \nabla_\theta\left(\sum_a \pi_\theta(a|s) q_{\pi_\theta}(s,a)\right) \\
&= \sum_a \left[\nabla_\theta \pi_\theta(a|s)\ q_{\pi_\theta}(s,a) + \pi_\theta(a|s)\ \nabla_\theta\ q_{\pi_\theta}(s,a)\right] \\
&\overset{1}{=} \sum_a \left[\nabla_\theta \pi_\theta(a|s)\ q_{\pi_\theta}(s,a) + \pi_\theta(a|s)\ \nabla_\theta \left(r(s,a) + \gamma \sum_{s'} P(s'|s,a) v_{\pi_\theta}(s')\right)\right] \\
&\overset{2}{=} \sum_a \left[\nabla_\theta \pi_\theta(a|s)\ q_{\pi_\theta}(s,a) + \pi_\theta(a|s)\gamma \sum_{s'} P(s'|s,a)\ \nabla_\theta\ v_{\pi_\theta}(s'))\right] \\
&= \sum_a \nabla_\theta \pi_\theta(a|s)\ q_{\pi_\theta}(s,a) + \sum_a \pi_\theta(a|s)\gamma \sum_{s'} P(s'|s,a)\ \nabla_\theta\ v_{\pi_\theta}(s') \\
&\overset{3}{=} \phi(s) + \sum_a \pi_\theta(a|s)\gamma \sum_{s'} P(s'|s,a)\ \nabla_\theta\ v_{\pi_\theta}(s')
\end{aligned} \tag{20.120}$$

where the labeled equal signs are due to the following:

1. Equation (20.22)
2. $r(s,a)$ is not a function of θ
3. We have defined

$$\phi(s) = \sum_a \nabla_\theta \pi_\theta(a|s)\ q_{\pi_\theta}(s,a) \tag{20.121}$$

Equation (20.120) is a recursion by which $\nabla_\theta v_{\pi_\theta}(s)$ is expressed as a function in terms $\nabla_\theta v_{\pi_\theta}(s')$.

We further consider the probability of transitioning from state s to a state x after k steps following $\pi(a|s)$

$$s \overset{\pi(a|s)}{\longrightarrow} s' \overset{\pi(a|s')}{\longrightarrow} s'' \overset{\pi(a|s'')}{\longrightarrow} \cdots \overset{\pi(a|s^{k-1})}{\longrightarrow} s^k = x \tag{20.122}$$

denoted by $P_\pi(s \to x, k)$ with the following properties

- $$P_\pi(s \to s, 0) = 1 \tag{20.123}$$
- $$P_\pi(s \to s', 1) = \sum_a \pi(a|s) P(s'|s,a) \tag{20.124}$$
- $$\sum_{s'} P_\pi(s \to s', 1) = 1 \tag{20.125}$$
- $$P_\pi(s \to s'', 2) = P_\pi(s \to s', 1) P_\pi(s' \to s'', 1) \tag{20.126}$$
- $$P_\pi(s \to x, k+1) = \sum_{s'} P_\pi(s \to s', k) P_\pi(s' \to x, 1) \tag{20.127}$$

Continuing Eq. (20.120) we keep rolling out the recursion of $\nabla_\theta v_{\pi_\theta}(s')$ and get

$$
\begin{aligned}
\nabla_\theta v_{\pi_\theta}(s) &= \phi(s) + \gamma \sum_a \pi_\theta(a|s) \sum_{s'} P(s'|s,a)\, \nabla_\theta\, v_{\pi_\theta}(s') \\
&\overset{1}{=} \phi(s) + \gamma \sum_{s'} P_{\pi_\theta}(s \to s', 1)\, \nabla_\theta\, v_{\pi_\theta}(s') \\
&\overset{2}{=} \phi(s) + \gamma \sum_{s'} P_{\pi_\theta}(s \to s', 1) \left[\phi(s') + \gamma \sum_{s''} P_{\pi_\theta}(s' \to s'', 1)\, \nabla_\theta\, v_{\pi_\theta}(s'') \right] \\
&\overset{3}{=} \phi(s) + \gamma \sum_{s'} P_{\pi_\theta}(s \to s', 1)\phi(s') + \gamma^2 \sum_{s''} \\
&\qquad \sum_{s'} P_{\pi_\theta}(s \to s', 1) P_{\pi_\theta}(s' \to s'', 1)\, \nabla_\theta\, v_{\pi_\theta}(s'') \\
&\overset{4}{=} \phi(s) + \gamma \sum_{s'} P_{\pi_\theta}(s \to s', 1)\phi(s') + \gamma^2 \sum_{s''} P_{\pi_\theta}(s \to s'', 2)\, \nabla_\theta\, v_{\pi_\theta}(s'') \\
&\overset{5}{=} \phi(s) + \gamma \sum_{s'} P_{\pi_\theta}(s \to s', 1)\phi(s') + \gamma^2 \sum_{s''} P_{\pi_\theta}(s \to s'', 2)\phi(s'') + \gamma^3 \\
&\qquad \sum_{s'''} P_{\pi_\theta}(s \to s''', 2)\, \nabla_\theta\, v_{\pi_\theta}(s''') \\
&= \cdots\cdots \overset{6}{=} \sum_{x \in S} \sum_{k=0}^{\infty} \gamma^k P_{\pi_\theta}(s \to x, k)\phi(x)
\end{aligned}
\tag{20.128}
$$

where the labeled equal signs are due to the following:

1. Equation (20.124)
2. Unroll $\nabla_\theta v_{\pi_\theta}(s')$
3. Equation (20.125)
4. Equation (20.127)
5. Unroll $\nabla_\theta v_{\pi_\theta}(s'')$
6. Unrolling recursively to infinity; Eq. (20.123)

Now the gradient of the objective $J(\theta) = v_{\pi_\theta}(s_0)$ can be written as

$$
\begin{aligned}
\nabla_\theta J(\theta) = \nabla_\theta v_\pi(s_0) &= \sum_{s \in S} \sum_{k=0}^{\infty} \gamma^k P_{\pi_\theta}(s_0 \to s, k)\, \phi(s) \\
&= \sum_{s \in S} \eta(s)\, \phi(s) = \left(\sum_{s'} \eta(s') \right) \sum_{s \in S} \frac{\eta(s)}{\sum_{s'} \eta(s')}\, \phi(s) \\
&\propto \sum_{s \in S} \frac{\eta(s)}{\sum_{s'} \eta(s')}\, \phi(s) = \sum_{s \in S} \mu_\pi(s) \sum_a \nabla_\theta \pi_\theta(a|s)\, q_{\pi_\theta}(s,a)
\end{aligned}
\tag{20.129}
$$

where we also defined

$$
\eta(s) = \sum_{k=0}^{\infty} \gamma^k P_{\pi_\theta}(s_0 \to s, k), \tag{20.130}
$$

as the sum of all probabilities for visiting state s from the start state s_0, and

$$
\mu_\pi(s) = \frac{\eta(s)}{\sum_{s'} \eta(s')} \tag{20.131}
$$

as a normalized version of $\eta(s)$ representing the probability distribution of visiting state s while following policy π. Here the proportionality is due to the dropping of the constant $\sum_{s'} \eta(s')$ independent of state s.

Following the similar steps, the policy gradient theorem for an environment with continuous state and action spaces can be also proven

$$\nabla_\theta J(\theta) = \int_s \mu_\pi(s) \int_a \nabla_\theta \pi_\theta(a|s)\; q_{\pi_\theta}(s,a) da\; ds \tag{20.132}$$

Now Eq. (20.129) can be further written as

$$\begin{aligned}
\nabla_\theta J(\theta) = \nabla_\theta v_\pi(s_0) &\propto \sum_{s \in S} \mu_\pi(s) \sum_a \pi_\theta(a|s) \frac{\nabla_\theta \pi_\theta(a|s)}{\pi_\theta(a|s)}\; q_{\pi_\theta}(s,a) \\
&= \sum_{s \in S} \mu_\pi(s) \sum_a \pi_\theta(a|s)\; \nabla_\theta \ln \pi_\theta(a|s)\, q_{\pi_\theta}(s,a) \\
&= E_{\pi_\theta}\left[\nabla_\theta \ln \pi_\theta(a|s)\, q_{\pi_\theta}(s,a)\right]
\end{aligned} \tag{20.133}$$

where E_{π_θ} denotes the expectation over all actions in each state s weighted by $\pi_\theta(a|s)$ and all states $s \in S$ weighted by $\mu_\pi(s)$. □

The gradient $\nabla_\theta J(\theta)$ is now expressed as the expectation of the action value function $q_\pi(s,a)$, weighted by the gradient of the logarithm of the corresponding policy $\pi_\theta(a|s)$. Now the gradient ascent iteration in Eq. (20.118) can be further written as

$$\boldsymbol{\theta}_{t+1} = \boldsymbol{\theta}_t + \alpha\, \nabla_\theta\, J(\boldsymbol{\theta}) = \theta_t + \alpha E_{\pi_\theta}\left[q_{\pi_\theta}(s_t,a_t)\;\; \nabla_\theta \ln \pi_\theta(a_t|s_t)\right] \tag{20.134}$$

Now Eq. (20.118) can also be written as

$$\boldsymbol{\theta}_{t+1} = \boldsymbol{\theta}_t + \alpha\, \nabla_\theta\, J(\boldsymbol{\theta}) = \theta_t + \alpha E_{\pi_\theta}\left[(q_{\pi_\theta}(s_t,a_t) + b(s_t))\;\; \nabla_\theta \ln \pi_\theta(a_t|s_t)\right] \tag{20.135}$$

The expectation E_{π_θ} in Eqs (20.134) and (20.135) can be dropped if the method of stochastic gradient ascent is used, as in all following algorithms, where each iterative step is based on only one sample data point instead of its expectation.

In particular, for a softmax policy model as given in Eq. (20.113), we have

$$\begin{aligned}
\nabla_\theta \ln \pi_\theta(a|s) &= \nabla_\theta \ln\left(\frac{e^{\theta^T \mathbf{x}(s,a)}}{\sum_b e^{\theta^T \mathbf{x}(s,b)}}\right) = \nabla_\theta\left(\theta^T \mathbf{x}(s,a) - \ln \sum_b e^{\theta^T \mathbf{x}(s,b)}\right) \\
&= \mathbf{x}(s,a) - \frac{\nabla_\theta \sum_b e^{\theta^T \mathbf{x}(s,b)}}{\sum_b e^{\theta^T \mathbf{x}(s,b)}} = \mathbf{x}(s,a) - \frac{\sum_b e^{\theta^T \mathbf{x}(s,b)} \mathbf{x}(s,b)}{\sum_b e^{\theta^T \mathbf{x}(s,b)}} \\
&= \mathbf{x}(s,a) - \sum_b \frac{e^{\theta^T \mathbf{x}(s,b)}}{\sum_b e^{\theta^T \mathbf{x}(s,b)}} \mathbf{x}(s,b) \\
&= \mathbf{x}(s,a) - \sum_b \pi_\theta(b|s) \mathbf{x}(s,b)
\end{aligned} \tag{20.136}$$

A set of popular policy-based algorithms is listed in the following. They are based on either the MC or TD methods discussed before.

- **REINFORCE (MC) policy gradient**
In this algorithm, the action value function $q_{\pi_\theta}(s_t,a_t)$ in Eq. (20.134) is replaced by the return G_t obtained at the end of each episode while sampling the environment. Now the equation becomes

$$\theta_{t+1} = \theta_t + \alpha\gamma^t G_t \ \ \nabla_\theta \ln \pi_\theta(a|s) \tag{20.137}$$

Note that the discount factor γ^t is included as the expression for $\nabla_\theta J(\theta)$ in Eq. (20.133) assumed $\gamma = 1$ for simplicity.

Here is the pseudo code for the algorithm:
Initialize $\pi(a|s,\theta$
loop (for each episode)
  At the end of episode, get G_t for each state visited
  for $t = 1, \cdots, T$
    $\theta = \theta + \alpha\gamma^t G_t \ \nabla_\theta \ln \pi_\theta(a_t|s_t)$
end loop

- **Actor–critic (TD) policy gradient**
As its name suggests, this algorithm is based on two approximation function models, the first for the policy $\pi_\theta(a|s)$ parameterized by $\boldsymbol{\theta}$, the *actor*, the same as in the REINFORCE algorithm, and the second for the value function $\hat{v}_\pi(s,\mathbf{w})$ parameterized by $\mathbf{w}$, the *critic*, the same as in Eq. (20.96).
In Eq. (20.135), the action value function $q_{\pi_\theta}(s_t,a_t)$ is replaced by its bootstrapping expression $r_{t+1} + \gamma\hat{v}_\pi(s',\mathbf{w})$, and the bias term $b(s)$ is replaced by the approximated value function $\hat{v}(s_t,\mathbf{w})$. Now both parameters $\mathbf{w}$ and $\boldsymbol{\theta}$ can be found iteratively at every step of an episode while sampling the environment (the TD method) by the stochastic gradient method with the expectation symbol E_π dropped

$$\begin{aligned}\mathbf{w}_{t+1} &= \mathbf{w}_t + \alpha_w \left[(r_{t+1} + \gamma\hat{v}_\pi(s',\mathbf{w}) - \hat{v}_\pi(s,\mathbf{w})) \ \nabla_w \ \hat{v}_\pi(s,\mathbf{w})\right] \\ \theta_{t+1} &= \theta_t + \alpha_\theta \left[(r_{t+1} + \gamma\hat{v}_\pi(s',\mathbf{w}) - \hat{v}_\pi(s,\mathbf{w})) \ \nabla_\theta \ln \pi_\theta(a|s)\right]\end{aligned} \tag{20.138}$$

Here is the pseudo code for the algorithm:
initialize model parameters $\boldsymbol{\theta}$ and $\mathbf{w}$
initialize step sizes α_θ and α_w
loop (for each episode)
  initialize $s = s_0$, $t = 0$
  while s is not terminal (for each step)
    take action a following $\pi_\theta(a|s)$, find reward r and next state s'
    find TD error: $\delta = r + \gamma\hat{v}(s',\mathbf{w}) - \hat{v}(s,\mathbf{w})$
    $\mathbf{w} = \mathbf{w} + \alpha_w \ \delta \ \nabla \ \hat{v}(s,\mathbf{w})$
    $\boldsymbol{\theta} = \boldsymbol{\theta} + \alpha_\theta\gamma^t\delta \ \nabla \ln \pi_\theta(a|s)$
    $s = s'$
    $t = t + 1$
  end while
end loop

- **Backward view of TD(λ) policy gradient**
  Here is the pseudo code for the algorithm:

```
initialize model parameters θ and w
initialize step sizes α_θ and α_w
initialize trace-decay rates λ_θ and λ_w
loop (for each episode)
  initialize s = s_0, t = 0
  initialize z_θ = 0, z_w = 0
  while s is not terminal (for each step)
      take action a following π_θ(a|s), find reward r and next state s'
      find TD error: δ = r + γ v̂(s',w) − v̂(s,w)
      z_θ = γ λ_θ z_θ + ∇γ^t ln π_θ(a|s)
      z_w = γ λ_w z_w + γ^t ∇ v̂(s,w)
      w = w + α_w δ ∇ v̂(s,w)
      θ = θ + α_θ δ ∇ ln π_θ(a|s)
      s = s'
      t = t + 1
  end while
end loop
```

PROBLEMS

The problems in this homework set are more involved than those in previous homework sets. They can be considered as a capstone final project for the entire course.

1. Solve the grid world problem in Example 20.1 by the TD method discussed in Section 20.3 in both forms of SAESA and Q-learning.
2. Solve the same grid world problem by the MC method discussed in Section 20.3.
3. Solve the same grid world problem by the method of value function approximation discussed in Section 20.4 based on linear regression.
4. Solve the same grid world problem by the method of value function approximation discussed in Section 20.5, based on linear regression. Try both MC and TD methods.
5. Solve the same grid world problem by the method of value function approximation based on a nonlinear neural network instead of a linear regression model.
6. Solve the same grid world problem by the policy gradient method.

Part VII

Large Language Models

The field of artificial intelligence has advanced rapidly over the past decade, particularly in natural language processing (NLP). In response to the challenges of NLP, a new generation of *large language models (LLMs)* has emerged, built on large-scale deep learning neural networks trained on massive datasets. These models are capable of a range of complex language tasks, from understanding and generating human language to manipulating it for purposes such as summarizing articles, translating languages, generating human-like responses in conversations, and answering questions. The core ability of LLMs is not merely to understand individual words but to interpret how words relate to each other contextually, spanning sentences and even multiple paragraphs.

While language models have evolved steadily over time, the most transformative leap occurred with the introduction of the *Transformer architecture* in 2017 by Vaswani et al. in their seminal paper, "Attention is All you need."[1] This architecture fundamentally changed the way language models are designed, bringing breakthroughs in tasks like machine translation, text generation, and complex problem-solving. This chapter examines the foundational components, structures, and algorithms of the Transformer architecture, highlighting the principles that make it a leading framework in LLMs today.

In recent years, generative machine learning, exemplified by the Transformer architecture, has gained significant traction, revolutionizing various fields by enabling the creation of high-quality synthetic data. Unlike many traditional machine learning methods covered in earlier chapters, such as support vector machines or naive Bayes classifiers, which are non-generative, straightforward, and reliant on smaller datasets, the Transformer is a generative, large-scale, and multifaceted architecture consisting of interrelated components, each performing unique functions that contribute to the overall model. Transformer models are based on super-deep neural networks with hundreds of layers and billions of parameters (primarily network weights), trained on vast datasets containing billions or even trillions of words. Consequently, they are far more computationally intensive, often requiring days, weeks, or even months of training time and substantial computing resources, which makes them unsuitable for small-scale implementation as student projects.

[1] Vaswami, A. et al. (2017) Attention is all you need. In *Proceedings of the 31st International Conference on Neural Information Processing Systems (NIPS'17)*. Curran Associates Inc., pp. 6000–6010.

Consequently, the study approach for this chapter differs from previous ones. Here, the emphasis is on developing a top-level understanding of the Transformer's design and function, with less focus on precise mathematical formulations. However, the chapter builds upon foundational knowledge from earlier sections, such as backpropagation, which plays an essential role in the Transformer models.

The ML methods discussed previously, such as regression, classification, and clustering, are examples of non-generative methods, which focus on making predictions or uncovering patterns in existing datasets without creating new data. In contrast, recent advancements have led to generative models that can create new data instances by learning the underlying data distribution. These methods can generate realistic samples, including text, images, audio, and video. In this chapter, we will explore the Transformer architecture, which serves as the foundational structure of today's most powerful generative LLMs for tackling large-scale NLP problems, as well as some other generative models in less detail.

Finally it is worth mentioning that the writing of this chapter is also quite different from previous ones, all of which were solely written from scratch, the contents of this chapter on LLMs and the Transformer architecture are the result of a collaboration between the author and ChatGPT (Generative, Pre-trained, Transformer), one of the LLMs currently available. Specifically, the author engaged in a series of conversations with ChatGPT covering various topics related to LLMs. The responses by ChatGPT are then revised, edited, or rewritten into a cohesive chapter by the author. Writing the chapter in this manner with the assistance of ChatGPT is very convenient for the author, and it also serves as an excellent example to demonstrate ChatGPT's capabilities in handling substantial writing tasks.

21 Large Language Models

21.1 Natural Language Processing and Modeling

To prepare for the future discussion of the Transformer architecture, here we first consider the typical natural language processing (NLP) tasks and some language models developed in the past to tackle such NLP tasks as the background to the modern LLMs developed most recently in the last few years.

A language model can be viewed as a black box with an input of text in one language, an output of text in the same or a different language, and the function performed by the black box to generate the output based on the input. A list of typical NLP tasks follows, defined by their inputs, outputs, and purposes.

- Machine translation:
  - Input: A sentence in one language (e.g., English).
  - Output: A translation of the sentence in another language (e.g., French).
  - Purpose: To convert the input sequence into a context-aware representation and generate the correct output sequence in a different language.
- Text summarization:
  - Input: A long document or article.
  - Output: A condensed version that summarizes the key information.
  - Purpose: To generate a shorter version of the input text that retains the essential information.
- Text generation:
  - Input: A prompt or a partially completed sentence.
  - Output: A continuation of the text.
  - Purpose: To generate coherent and contextually relevant text based on the input.
- Question-answering:
  - Input: A passage of text and a question about the passage.
  - Output: An answer extracted or inferred from the text.
  - Purpose: To locate relevant information in the input and produce a direct answer.

To address these NLP tasks, various language models have been developed over time. The development of these models has a rich history, beginning with simple statistical methods and progressing to advanced neural networks. Understanding this evolution offers valuable context for the breakthroughs we witness today.

- **n-gram models**
  One of the earliest types of language models was the n-gram algorithm, which is a statistical model that predicts the next word in a sequence based on the previous n words. For example, in a trigram model ($n = 3$) the prediction is based on just two preceding words. The simplicity of these models made them computationally cheap, but they suffered from two major limitations. (1) Short-term memory: They can only "see" a limited window of previous words; (2) Data sparsity: As the size of n increases, the amount of data required to accurately predict sequences grows exponentially.
- **Recurrent neural networks**
  The development of recurrent neural networks (RNNs) marked a significant improvement in modeling language sequences. RNNs introduced the concept of memory, enabling the model to remember previous inputs. This allowed RNNs to handle longer sequences of text compared to n-gram models. However, RNNs encountered their own issues: (1) vanishing gradient problem, as RNNs process longer sequences, the gradients used for backpropagation during training tend to shrink, making it difficult to learn dependencies between words that are far apart in a sequence; (2) sequential processing: RNNs process one word at a time, which limits the speed at which they can be trained.
- **Long short-term memory**
  Long short-term memory (LSTM) networks were introduced to address the shortcomings of standard RNNs. LSTMs use a set of gates to control the flow of information, enabling the model to remember important details over long sequences and discard irrelevant information. LSTMs became the dominant architecture for sequence modeling in NLP tasks, but they still struggled with long-range dependencies and required sequential processing, which limited their scalability to very large datasets.
- **The attention mechanism**
  In general, attention is a machine learning method that determines the relative importance of each component in a sequence relative to the other components in that sequence. The introduction of the attention mechanism was a game-changer in NLP. Instead of relying on sequential memory, attention allows a model to focus on specific parts of an input sequence based on their relevance to the current task. This means that even distant words in a sentence can be given appropriate weight when making predictions. Attention mechanisms led directly to the development of the Transformer architecture, which forms the foundation of modern LLMs.

Having covered earlier language models, we now introduce a set of state-of-the-art LLMs, representing the latest advancements in language model design for addressing various NLP tasks.

- Bidirectional encoder representations from transformers (BERT), introduced by Google in 2018, is designed to understand text bidirectionally, by looking at both the words that come before and after a target word. BERT is pre-trained using two task.
  - Masked language modeling: Random words in the input are masked, and the model must predict them based on the surrounding context, enabling it to understand relationships in both left and right directions.
  - Next-sentence prediction: The model is trained to predict whether one sentence logically follows another, helping it understand sentence-level relationships.

  BERT has been highly successful in improving performance on a range of NLP tasks, from question answering to text classification.
- Generative pre-trained transformer (GPT)
  The GPT family of models, first developed in 2018 by OpenAI, focuses on text generation. Unlike BERT, GPT is an autoregressive model, meaning it predicts the next word in a sequence based only on the preceding words. Each successive version of GPT has scaled up in size. In terms of parameter count, the largest GPT-2 model has 1.5 billion parameters, GPT-3 has 175 billion, and GPT-4 is known to be significantly larger than GPT-3 (estimated to be about 1.8 trillion), although OpenAI has not officially disclosed the exact number of parameters at the time of writing this book. GPT models are used for tasks ranging from conversational AI (like chatbots) to code generation and writing assistance.
- Text-to-text transfer transformer (T5)
  The T5 model, first introduced by Google in 2019, treats all NLP tasks as text-to-text problems. Whether the task is translation, summarization, or answering questions, T5 frames it as generating text from text input. This unified approach simplifies the architecture and has led to strong performance across a wide variety of tasks.
- Vision Transformers (ViTs)
  The Transformer architecture first introduced by Google in 2020 has been adapted for non-text tasks, such as image recognition. Vision Transformers (ViTs) apply the attention mechanism to image patches, allowing the model to process and classify images. This architecture has shown competitive results with traditional convolutional neural networks (CNNs) on image-based tasks.

21.2 The Transformer Architecture

At the core of the modern LLMs mentioned above is the *Transformer architecture*, which serves as the foundation for their design and functionality. The Transformer is a deep learning model that overcomes the limitations of traditional sequential models like RNNs and LSTMs. Its main purposes of the Transformer are to

- capture relationships between words in the input sequence, even across long distances, using self-attention mechanisms rather than processing them sequentially;
- process entire sequences in parallel, making it highly efficient and suitable for large-scale language tasks;

- model complex dependencies in data, enabling it to excel at tasks requiring contextual understanding, such as translation, text generation, and question-answering.

The input sequence can range from a sentence to a paragraph or even an entire article. The size of the input sequence is measured in *tokens* which represent both words and subwords within the sequence. (1 English word is about 1.3 tokens.) The sequence length changes dynamically. For example, in a conversation between an LLM and a user, the sequence length may start short, such as a single question from the user. As the conversation continues, the sequence grows to retain all the contextual information necessary for the LLM to generate relevant and meaningful responses.

However, the LLM's ability to remember context is constrained by a fixed token limit, which is the maximum number of tokens the model can retain. Within this limit, the model can track the flow of conversation, maintain context, and refer to earlier statements or questions. Once this limit is exceeded, older parts of the conversation begin to drop out of memory, causing the model to "forget" earlier details in the conversation. For instance, the token limit in different versions of GPT-4 ranges from 8,200 (GPT-4 standard) to one million (GPT-4.1) (including both user input and model's response).

The primary goal of the Transformer architecture is to efficiently process sequential data for tasks like translation, text generation, and understanding, utilizing self-attention to enable parallelism and capture long-range dependencies. The Transformer consists of two main components.

- **Encoder:** The encoder processes the input sequence (e.g., a sentence, a paragraph, or an article in some language) by breaking it down into tokens, which in turn is further converted into a vector representation called *embeddings*. These embeddings are then processed through a stack of layers each composed of a self-attention sublayer and a feedforward network (FFN) sublayer. In this process, the encoder captures relationships between tokens in the input sequence and produces a set of encoded representations, the contextualized token embeddings.
- **Decoder:** The decoder takes the encoder's output together with the its own output generated in the previous time step (auto-regressive generation) as input. The decoder also consists of multiple layers of self-attention and cross-attention (also called encoder–decoder attention) where it attends to both the encoder's outputs and its own previously generated outputs. It then generates as the output a probability distribution over all tokens in the vocabulary, based on which the most likely token is selected for the next position in its output sequence (e.g., a word in the sentence in another language).

In the following, we will consider each of the main components of both the encoder and decoder in the Transformer architecture, including the self-attention network (SAN) and the FFN, which are the building blocks of both the encoder and the decoder are constructed. We then further consider the supervised backpropagation learning taking place in the overall Transformer architecture to generate the large number of parameters in the model.

21.3 Attention Is All You Need

Self-attention is the core mechanism in Transformer architecture. Instead of processing words in sequence, self-attention computes the relevance of each word to every other word in the sentence in parallel, so that different words in a sequence can be weighted according to its importance in terms of its relation to every other word, allowing the model to capture long-range dependencies far more effectively than RNNs or LSTMs. For example, in the sentence "The cat sat on the mat," the self-attention mechanism can directly link "cat" and "mat," regardless of how many words are in between. Self-attention enables the model to dynamically focus on different parts of the input sequence, capturing dependencies regardless of how far apart they are in the text. This mechanism is crucial for the model to "understanding" the context and generate coherent and contextually relevant outputs.

Specifically, the implementation of self-attention involves a number of components.

- Tokenization
  Tokenization is a process, which, once trained, automatically breaks a given input sequence into tokens, each representing a word or subword, which can be as small as one character or as large as a complete word. For instance, the sentence "The cat sat on the mat" can be broken down into the six individual words in it, and a word "unhappiness" can be broken into subwords "un," "happy" and "ness."

  The Transformer architecture first converts each token into numerical form, producing a vector called *token embedding*, which represents the semantic information in the token. Each component of the vector encodes certain aspect of the token, reflecting its syntactic and semantic roles. These vector components are considered latent features learned from the model and collectively represent the meaning of the corresponding word or subword. The dimensionality d_{model} of the token embedding vectors varies depending on the specific model, ranging from a few hundreds to a few thousands (e.g., as high as 16,000 in GPT-4). The components of a token embedding represent different learned aspects of the token, such as those listed here.
  - Syntactic information
    Some components encode the grammatical role of the token (e.g., whether it is a noun, verb, adjective), representing how the token fits into the overall construction of the sentence.
  - Semantic information
    Others components may capture the meaning or context of the token. For instance, words with similar meanings may have similar embeddings along certain dimensions, reflecting relationships between words.
  - Contextual dependencies
    Some components may represent how much the token depends on or relates to other tokens in the sentence.

  The high dimensionality of the token embeddings helps capture more nuanced aspects of meaning that may not be easily described. These abstract features often emerge through the training process as the model learns patterns in the data. Once tokenized, the input sequence with n tokens is represented by an $n \times d_{model}$ dimensional matrix $\mathbf{X}$.

On the output side, an inverse process for un-embedding takes place to convert token vectors back into text. This is done by the softmax function which maps the numerical representations of tokens into a probability distribution over all tokens in the vocabulary.

- Query, key, and value
  For each input embedding $\mathbf{x}$, three different vectors are created through learned linear transformations (matrix multiplication):
  1. Query $\mathbf{q} = \mathbf{W}_q\mathbf{x}$: Represents the word under consideration.
  2. Key $\mathbf{k} = \mathbf{W}_k\mathbf{x}$: Represents the features of the words to which the query will be matched.
  3. Value $\mathbf{v} = \mathbf{W}_v\mathbf{x}$: Contains the actual information that will be used to produce the output based on the attention scores.

  Here $\mathbf{W}_q, \mathbf{W}_k$, and $\mathbf{W}_v$ are weight matrices, of which the components are obtained in a two-stage training process:
  - pre-training (self-supervised) on large unlabeled text data for the model to learn general language representations;
  - fine-tuning (supervised) of the representations learned in pre-training based on smaller, task-specific labeled data for the model to adapt to a specific task.

  Typically, the dimensionalities of the resulting vectors $\mathbf{q}$, $\mathbf{k}$, and $\mathbf{v}$ are $d_q = d_k = d_v = d_{model}/h$, where h is the number of attention heads, each capturing different relationships between the tokens (discussed in the following). For example, in GPT-3, $d_{model} = 2048$ and $h = 16$, then $d_q = d_k = d_v = d_{model}/h = 2048/16 = 128$.
- Calculating attention weights
  The attention score for each token $\mathbf{x}_i$ to another token $\mathbf{x}_j$ is calculated as the dot product $s_{ij} = \mathbf{q}_i^T\mathbf{k}_j \sqrt{d_k}$ of its query vector $\mathbf{q}_i$ and the key vector $\mathbf{k}_j$, scaled down by dividing by the square root of the dimension of the key vector $\mathbf{k}$. All such attention scores are then mapped by the softmax function to become *attention weights*

$$w_{ij} = \frac{e^{s_{ij}}}{\sum_k e^{s_{ij}}} \tag{21.1}$$

  As all such attention weights sum up to 1 over all tokens in the sequence, they can be treated as probabilities, representing the contribution of each token $\mathbf{x}_i$ to the meaning of the sentence, in terms of the following:
  - Contextual relevance: how relevant or important a token is to the meaning or function of another;
  - Influence on representation: the extent to which one token's new representation should be influenced by another;
  - Dependency relationships: the strength of the relationship between tokens, indicating how much one token depends on another.

The final attention output of the self-attention mechanism is a set of vectors, each for one of the tokens in the sequence, obtained as $\mathbf{a}_i = \sum_j w_{ij}\mathbf{v}_j$, the sum of all value vectors weighted by the corresponding attention weights.

For instance, in the sentence "The cat sat on the mat," "cat" and "sat" are two of the six tokens. The attention weight $w_{cat,sat}$ from "cat" to "sat" reflects how much the meaning of "cat" is influenced by "sat" as a probability. A higher probability indicates that "sat" is more relevant to understanding "cat." The final output for the token "cat" is a vector that reflects contributions other tokens such as "sat" and "the," showing how they together inform the meaning of "cat" in this context.

- Multihead attention
  The purpose of the multihead attention mechanism is to enhance the model's ability to attend to different parts of the input simultaneously. Instead of computing a single self-attention score for each word pair, multiple sets (heads) of queries, keys, and values are also computed in parallel to capture various aspects of the input, reflecting different types of relationships and patterns among the tokens. For instance, one attention head might focus on grammatical structure, while another focuses on semantic meaning. The output of all the heads is obtained by concatenating the individual vectors each for one of the heads, which is then linearly transformed into a vector of the original dimensionality. The resulting vector represents the linear combination of all the attention information from the multiple heads. For example, if there are h heads each represented by a vector of dimension d_{model}, the concatenation of these vectors has dimension hd_{model}. This concatenated vector can then be linearly transformed by a $d_{model} \times hd_{model}$ matrix back into a d_{model}-dimensional vector.
- Positional encoding
  Since the Transformer model processes words in parallel, it has no inherent sense of word order in the input sequence. For the model to gain such information, a mechanism called *positional encoding* is used to encode the position of every token with a vector of the same dimension d_{model} as the token embedding vectors. Specifically, a token in position p of the sequence is encoded by a vector $\mathbf{p}$ with its even and odd components calculated based on some sinusoidal functions of different frequencies related to their indices and p

$$\begin{cases} \text{even index:} & \mathbf{p}_{p,2i} = \sin(p/10{,}000^{2i/d_{model}}) \\ \text{odd index:} & \mathbf{p}_{p,2(i+1)} = \cos(p/10{,}000^{2i/d_{model}}) \end{cases} \tag{21.2}$$

  Each token is now represented by the sum of its position vector $\mathbf{p}$ and its token embedding vector $\mathbf{x}$ (element-wise addition), combining both the position information (indicating the token's position in the sentence) and the semantic information (reflecting the meaning, relationships, and context of the word).

21.4 Backpropagation Learning

Backpropagation (discussed in Section 18.2) is a supervised learning algorithm based on gradient descent method, which plays a crucial role in the learning process occurring in both the encoder and decoder of the Transformer architecture. We will now consider the structures of its main components.

- The feedforward network
  The output from the self-attention mechanism is the contextualized representation of the tokens, which will be fed into the subsequent stage: the FFN. The FFN is a fully connected network composed of two layers (hidden and output) that transform the input $\mathbf{x}$ from the self-attention stage to the output

$$\mathbf{y}(\mathbf{x}) = \mathbf{W}^o g(\mathbf{W}^h \mathbf{x} + \mathbf{b}^h) + \mathbf{b}^o \tag{21.3}$$

  Here, $\mathbf{W}^h$ and $\mathbf{W}^o$ are the weight matrices, while $\mathbf{b}^h$ and $\mathbf{b}^o$ are the biases of the hidden and output layers, and g is the activation function, typically a rectified linear unit (ReLU) activation. The size (number of nodes) of the hidden layer (intermediate size) of the FFN is larger than that of the input embedding $\mathbf{x}$ (e.g., four times as large).
- Residual connections and normalization
  The outputs from both the self-attention and FFNs are further processed through the process of *residual connections*, which allow the information in their inputs to bypass the multihead attention network, thereby helping to preserve the original information and prevent its degradation as it flows through the network. Specifically, the input to the network is directly added to the output (without weighting).

  The result of the residual connections is a vector that is further normalized to keep its values within a limited range, thereby stabilizing the training process. Specifically, each component is normalized based on the mean m and variance σ^2 over all of its components: $x' = (x - m)/\sigma$.
- Encoder stack
  The function of the encoder stack of the Transformer is to generate context-aware embeddings for each token in the input sequence. The encoder stack is a multiple layer structure, each layer contains a self-attention network sublayer and a FFN sublayers. The output of each layer is a set of embeddings, one for each token, which is fed into the next layer as its input, to be updated to incorporate progressively more contextual information about the token's relationships with other tokens in the input sequence. Through the layers of the encoder stack, the embedding vectors are progressively refined to produce the final output of the encoder stack: a matrix containing the embedding vectors for all tokens in the input sequence, of which each vector represents a token's context-aware embedding. These embedding vectors can be considered a set of features extracted from the input sequence, which describe the patterns in the data, to be fed into each one of the layers in the decoder stack, as described next.

- Decoder stack
  The function of the decoder stack is to generate the output sequence one token at a time. This is accomplished by attending to both its own tokens previously generated (via self-attention) and the encoder stack's final output carrying information from the input data. Such a mechanism is called *cross-attention*. The decoder is also a stack of multiple layers each containing three sublayers.
  - The self-attention network
    It takes as input all the words in the output sequence generated by the decoder in the previous steps so far and generates the token embeddings of these words as its output. This sublayer allows the decoder to attend to the previously generated tokens in the output sequence without considering the input sequence. This process enables the decoder to capture dependencies among the previously generated tokens, allowing it to produce fluent and contextually coherent output texts.
  - The cross-attention (encoder–decoder attention) network
    It takes two sets of inputs, the output from the self-attention network for the words in the output sequence generated so far, used as the query vector, and the final output from the encoder stack for all words in the input sequence, used as the key and value vectors – similar to the three vectors used in the self-attention network. It then generates a set of embedding vectors, one for each position in the output sequence generated so far. This process allows the decoder to attend to both the input sequence as well as its previously generated output.
  - Feedforward network (FFN)
    It further processes the output of the cross-attention network through two fully connected feedforward layers (hidden and output) to generate its output, which is then fed into the next layer in the decoder stack.

  The output of the FFN of the last layer of the decoder stack is fed into a linear layer (projection) which performs a linear transformation to convert the embedding vectors into a set of scores, one for each token in the vocabulary. These scores are next normalized by the softmax function (so they add up to 1) and treated as the probabilities for all words in the vocabulary, representing the likelihood for each word being selected as the next word in the output sequence. This process is repeated for each word in the sequence.
- Encoder and decoder stacks
  Both the encoder and decoder are constructed as stacks composed of multiple layers – each layer in the encoder consists of two sublayers (self-attention and FFN), while each layer in the decoder consists of three sublayers (self-attention, cross-attention, and FFN). The encoder and decoder typically have the same number of layers, and the total number of layers N varies depending on the size of the model, ranging from a few (e.g., $N = 6$ in the original paper "A is all you need" by Ashish Vaswani et al. 2017)[1] to several dozen (e.g., $N = 96$ in GPT3). $N = 120$ in GPT-4

[1] Vaswami, A. *et al.* (2017) Attention is all you need. In Proceedings of the 31st International Conference on Neural Information Processing Systems (NIPS'17). Curran Associates Inc., pp. 6000–6010.

This stack structure of multiple layers (rather than a single layer) plays the most essential role in the deep learning process taking place in the Transformer, offering significant advantages.

- Hierarchical representations
  The model learns hierarchical representations of the input data. Each layer captures different levels of abstraction, enabling the model to better understand the complex relationships between tokens in the sequence.
- Iterative refinement
  The model gains enhanced contextual understanding through iterative refinement – each layer refines the embeddings based on information processed in the previous layer. This iterative process helps build a more nuanced understanding of the context and relationships among tokens, improving the quality of the generated sequences.
- Regularization through multiple layers
  By having multiple layers, the model learns from different levels of representation across all layers. This acts as a regularization mechanism, reducing overfitting by preventing any single layer from bearing the sole responsibility of learning the full complexity of the input–output relationship.
- Nonlinear mapping:
  The nonlinear transformations in each feedforward layer help the model learn complex mappings. Stacking layers enhances the model's ability to approximate complex functions, which is impossible for a single layer to capture.

This Transformer architecture, composed of both the encoder and decoder stacks, is a typical and generic model, upon which various adaptations have been developed. Some of these adaptations use only the encoder stack, while others use only the decoder stack. These variations are designed to tackle different NLP tasks.

- Encoder-only models are designed for understanding tasks (like BERT) and don't have an autoregressive generation component. These models use bidirectional attention and typically don't have masking.
- Decoder–only models are designed for generative tasks (like GPT), applying autoregressive attention with masking to predict one token at a time.
- Encoder-decoder models (like T5) use both stacks, with cross-attention to the encoder output, making them well-suited for tasks like translation or summarization.

Figure 21.1 shows the block diagram of the typical Transformer architecture given by Vaswani et al., which is composed of both the encoder and decoder stacks. Note that only one layer (or block) of either of the two stacks is shown, while the labeling $N\times$ on the side indicates that the layer is repeated N times in the stack, (e.g., $N = 6$ or $N = 96$).

Figure 21.2 further illustrates how the final output of the encoder stack, a set of features extracted from the input data, is fed into each of the multiple layer of the decoder stack, which also takes its previous output in the output sequence

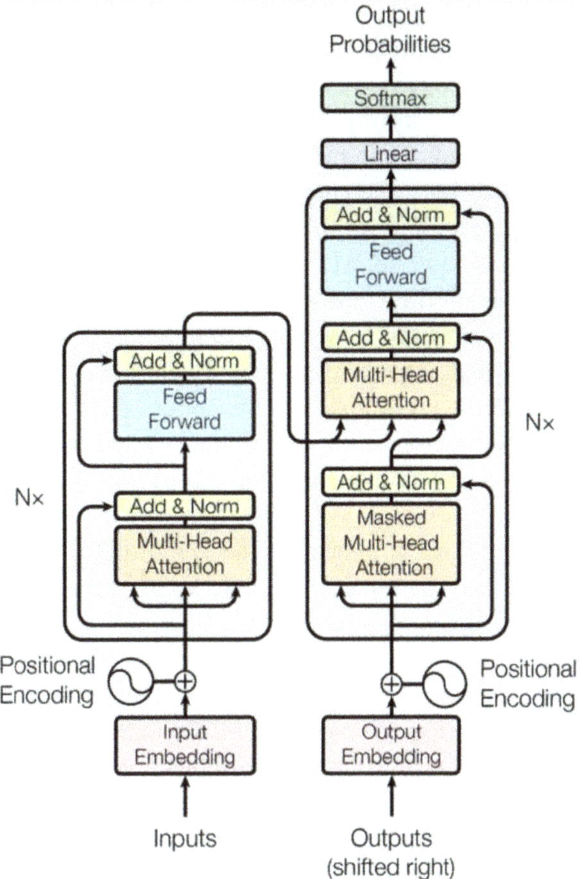

Fig. 21.1 The Transformer architecture
(Source: Vaswant A *et al.* (2017) Attention is all you need. *Conference on Neural Information Processing Systems.*)

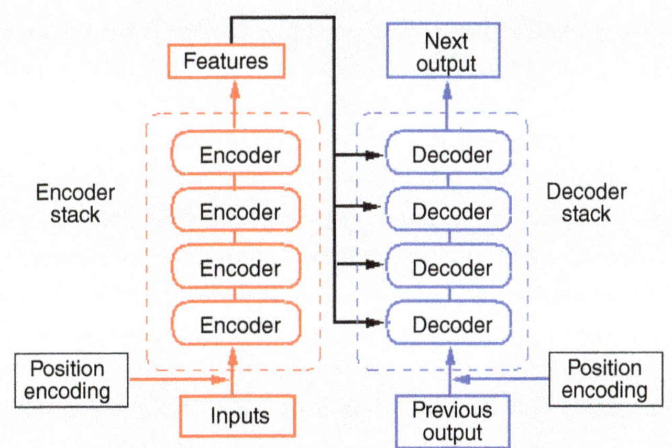

Fig. 21.2 The encoder and decoder stacks.

generated so far as input, and generates the next token in the sequence as the output. Here for simplicity, only $N = 4$ layers are shown for both the encoder and decoder stacks.

- Backpropagation learning
  In the training phase of the Transformer model, all parameters of the multi-layer stacks in both the encoder and decoder are learned through supervised backpropagation. During this process, the values of these parameters are iteratively updated and adjusted until they converge to stable values. Specifically, the parameters of the Transformer model include the following:
  - Weight matrices and bias vectors of the two sublayers in the FFN;
  - Layer normalization parameters: Each sublayer has parameters for layer normalization, including scale and shift parameters.
  - Weights and biases in the linear transformations of the output layer: These are used to project the decoder's output into the vocabulary space for token prediction.
  - Embedding layer parameters: The weight matrices ($\mathbf{W}_q, \mathbf{W}_k, \mathbf{W}_v$) which convert tokens of the input (in encoder) and previous output (for decoder) into token embedding representations.
  - Positional encoding parameters: If the positional encodings are learned (as opposed to fixed), those parameters are also updated during training.

Depending on the specific size of the Transformer architecture, such as the number of layers (a few tens or more), hidden layer sizes (a few hundreds or more), the total number of parameters as listed above varies from a few hundreds of millions to a few hundreds of billions.

The specific computation at each step of backpropagation has already been discussed in detail in Section 18.2. Here we only consider the preparation of the labeled training dataset used for training the Transformer model.

The training dataset used for the supervised learning taking place in a Transformer model are labeled, that is, each piece of input data corresponds to a target output representing the ground truth. The labeling is typically prepared manually by human experts. For example, LLMs for NLP are trained on huge text corpora, containing large number of sentences, and in one or more languages. For example, in machine translation, the target for any input sentence in one language is the correct translation in another language provided by human. In question-answering, human experts label which portions of the text contain the correct answers to specific questions, ensuring that the model learns to extract relevant information. Also, in text summarization: human experts write or evaluate summaries of documents, ensuring that the labeled data provides the correct amount of information and covers the main points of the text.

In some cases, self-supervised learning can be used where the model learns from unlabeled data by generating its own labels from the input data itself, without human involvement. In such a case, part of the input data is masked, corrupted, or transformed, and the model learns to predict the missing or altered part. This process enables the model to learn useful representations of the data that can later be fine-tuned for various downstream tasks (like classification, translation, or question-answering).

21.5 Transfer Learning in NLP

One of the key breakthroughs of LLMs is the use of transfer learning, which involves pre-training a model on general tasks to capture broad patterns and features from large datasets, and then fine-tunes it for a specific task. This approach dramatically reduces the amount of data and time required to train models for new NLP tasks.

- Data collection
  The model is trained on a large corpus of text data from diverse sources, including books, articles, and websites, for the purpose of predicting the next word in a sentence given the preceding words. This helps the model learn grammar, facts, and some reasoning abilities.
- Pre-training
  During this phase, the model is trained on vast amounts of general text data to learn language structure, syntax, semantics, and factual knowledge. This learning process is unsupervised, or more precisely, "self-supervised," as the model is not explicitly given labeled data with predefined answers. Instead, it learns from the raw text by predicting missing words (masked words), or the next token in a sequence. For example, the model might see a sentence "The cat sat on the …." with a missing word, and it learns to predict "mat" based on context. Similarly, it might predict the next word in a sequence or learn relationships between different parts of text.
- Fine-tuning
  Once pre-trained, the model can be adapted to specific types of tasks such as question answering, text completion, generalization, summarization, language translation, dialogue and conversation, sentiment analysis, and error correction and proofreading. This fine-tuning training is based on a task-specific dataset with labeled examples, such as question-answer pairs or dialogues. Since this dataset is narrower and smaller than the one used for pre-training, fine-tuning is much faster and less data-intensive than pre-training a model from scratch. This helps the model learn to generate more relevant and contextually appropriate responses.

  In some cases, fine-tuning involves reinforcement learning from human feedback (RLHF), where human trainers evaluate the model's outputs and provide feedback. Through this process, the model learns to prioritize responses that align with human preferences.
- Supervised learning
  During both pre-training and fine-tuning, the backpropagation algorithm is used to optimize all model parameters (e.g., weights and biases of the multilayer neural networks) to minimize the difference between the model's predictions and the

missing or next word in a sentence in the training data. Through this process, the model learns the relationships and patterns in the text data and becomes capable of generating coherent text and understanding language context when it encounters new inputs. Large-scale deep learning neural networks are used, which may consist of many tens or even more than one hundred layers with tens or hundreds of billions, or over a trillion parameters.

Although the pre-training process is self-supervised without labeled data, backpropagation as a supervised method can still be used. This is because the missing (masked out) or next word in a sentence is generated automatically, that is, the data are effectively self-labeled. Consequently, the cost or error function can be calculated in the same way as it would be with labeled training data.

- Inference
  Once the model is trained and fine-tuned, it can respond to text input (a prompt) by generating coherent and contextually appropriate text, addressing a wide variety of prompts based on its learned parameters and the patterns it learned during training. The response is generated one token at a time until the complete response is formed.

The transfer learning has a number of benefits.

- Reduced training time
  Since the model has already learned general features, fine-tuning can be done more quickly.
- Improved performance
  Models can achieve better performance on the new task because they start from a well-informed state rather than random initialization.
- Less data requirement
  Transfer learning is particularly useful when labeled data for the new task is scarce.

21.6 Reinforcement Learning with Human Feedback

A more recent development in training LLMs is RLHF. While traditional machine learning models are trained purely on pre-defined objectives, RLHF introduces a more interactive and human-guided element. This technique allows models to better align their outputs with human preferences by incorporating human feedback during the training process. For example, when generating text, human evaluators can rate the quality of the output, and the model can be fine-tuned based on this feedback. RLHF has been particularly useful in improving conversational agents and ensuring that models produce helpful, accurate, and safe responses.

The RLHF process typically begins by fine-tuning a pre-trained model using human feedback. Human annotators provide guidance on what constitutes good or bad behavior by evaluating the model's outputs. These evaluations help train a "reward model" that assigns a score to the model's responses based on how well they align with human preferences.

In the next phase, reinforcement learning (often with algorithms like proximal policy optimization, or PPO) is employed to optimize the model's performance based on this reward signal. The model generates responses, which are evaluated by the reward model, and then it adjusts its behavior to maximize the cumulative reward over time. Essentially, the model learns from both its successes and mistakes, improving through an iterative loop of feedback and adjustment.

RLHF is particularly useful for tasks where human judgment is complex or subjective, such as producing coherent and useful text, aligning outputs with ethical considerations, and ensuring helpfulness in conversational AI. By incorporating human preferences, RLHF helps overcome limitations of purely data-driven training, such as bias, misalignment, and undesirable outputs. This technique is especially valuable for improving AI systems that require a nuanced understanding of human preferences, such as LLMs like GPT. It guides these models to be more accurate, safe, and aligned with ethical standards, enhancing their effectiveness in real-world applications like customer service, content generation, and personalized recommendations.

21.7 Scaling Laws

In ML, scaling laws refer to mathematical relationships that describe how the performance of models changes when key factors, such as the number of parameters and size of training dataset, increase. These laws help researchers understand and predict how to scale models effectively to achieve better results. In general, the performance of a model improves when such key factors increase, that is, it generalizes better and can solve a wider range of tasks.

Specifically, in a neural network model, its size can be characterized by four parameters: the size of the model N (the number of parameters), size of the training dataset D, the amount of training computation C (measured in FLOPs (floating-point operations per second)), and the error rate or loss L. The performance of an LLM does not scale linearly with either the model size or the training dataset size, as described by the following scaling laws.

- The relationship between the number of model parameters N and the loss $L(N)$ (the error in model predictions) follows a power law: $L(n) \propto N^{-\alpha}$, where α is a constant exponent that depends on the architecture and task.
- The relationship between the size of dataset D and the loss $L(N)$ follows a power law: $L(D) \propto D^{-\beta}$, where β is a constant exponent that depends on the architecture and task.
- The relationship between the amount of computation $C(N)$ needed to achieve a certain performance and the model size N follows a power law: $C(N) \propto N^{\gamma}$, where γ is a constant greater than 1, meaning that a larger model requires disproportionately more computation to achieve significant gains.

These scaling laws are not independent. Scaling in one dimension without properly scaling the others can lead to problems. For example, a model with a large number of parameters but trained on a small dataset is likely to perform poorly due to overfitting (too many free parameters but too few constraints owing to the small training dataset). Additionally, a large model trained on a large dataset will also require more computation.

OpenAI's scaling law finds that performance L on certain benchmarks scales predictably with model size N, dataset size D, and amount of computation C with some key observations.

- Model size
  Performance continues to improve as model size grows, with no clear saturation point up to billions of parameters, with the cost growing disproportionately.
- Dataset size
  For larger models, increasing dataset size significantly improves performance, but the returns diminish after a point.
- Computational efficiency
  Given a fixed amount of computation, there is an optimal balance between model size and training duration, that is, training for more steps with a smaller model can sometimes outperform a larger model trained for fewer steps.

21.8 Mathematical Operations in the Transformer Architecture

The Transformer architecture, designed for processing and generating both natural language and mathematical expressions, can address simple mathematical operations through the collaborative functions of its encoder and decoder stacks. Throughout its training, the model has been exposed to a vast array of textual data that also includes mathematical expressions and their solutions. This training process enables the model to recognize and encode patterns associated with specific mathematical operations within its internal representations. Once trained, the Transformer model can take a mathematical query, such as $2 + 2$ or 5×3, where the numbers and mathematical symbols are treated as tokens and converted into embedding vectors by the encoder. The self-attention mechanisms of the encoder weigh the importance of each token in the context of the entire input sequence and captures relationships between numbers and operators. The outputs from the encoder stack are then passed to the decoder stack, which generates the output sequence token by token. It leverages the contextual information provided by the encoder and incorporates its own previously generated tokens to predict the next token. For example, in the case of $2 + 2$, the decoder identifies the operands and the addition operator, utilizing the learned patterns from its training data to predict the output token "4."

While the Transformer can competently handle straightforward arithmetic through these mechanisms, it is important to note that its performance may diminish with increasing complexity or with novel mathematical expressions not encountered in the training data. Therefore, while the model can simulate basic arithmetic through learned patterns and contextual understanding, it does not execute formal mathematical reasoning or algorithmic processes as dedicated computational systems do.

21.9 Emergence

Emergence is one of the most fascinating aspects of modern AI research. It refers to the phenomenon where complex and unanticipated behaviors or abilities not explicitly programmed arise from relatively simple models and learning rules. This occurs when both the model size (in terms of the number of parameters) and of dataset size are sufficiently scaled up to reach certain thresholds.

Specifically in LLMs, emergence can explain the sophisticated linguistic, reasoning, and problem-solving abilities observed in large models with billions of parameters. As LLMs increase in size, their performance on various tasks often improves in ways that go beyond mere linear scaling. This results in emergent capabilities that only appear once the model is sufficiently large and trained on vast datasets. For example, smaller models may struggle with specific reasoning tasks or long-term dependencies, but as the model scales up, it begins to exhibit the ability to solve these tasks with no explicit changes to the training process. Emergent behaviors often appear suddenly once the LLMs reach a critical size, similar to a phase transition in physics (e.g., water turns into ice at 0°C). These behaviors might include language understanding, basic arithmetic, translation, or summarization abilities that weren't present in smaller models. For example, GPT-3 demonstrated significant improvements in task generalization compared to GPT-2, even though the underlying architecture remained the same.

Different types of emergent behaviors not as evident in smaller models have been observed in large LLMs once they reach a certain size.

- Few-shot learning and zero-shot generalization
  They can perform tasks based on a small number of examples, despite not having been explicitly trained on those tasks. They can generalize from little information and perform surprisingly well without fine-tuning. They can even solve tasks they've never been explicitly trained on. For example, a model may be able to perform translation between two languages it hasn't directly seen before.
- In-context learning
  They can learn and adapt to new tasks purely from the context of a conversation or input, rather than through traditional training methods. This allows them to learn during inference rather than needing to be retrained.
- Coherent text generation
  They can generate more coherent and contextually appropriate text over longer passages, with some even demonstrating the ability to hold multi-turn conversations in a way that mimics human dialogue.
- Complex reasoning and logic understanding

They can exhibit reasoning abilities such as symbolic and mutistep reasoning, common-sense reasoning, logical inferences, and even certain forms of abstraction. They can answer complex questions or follow a logical sequence.

- Translation and cross-lingual capabilities
  They can perform cross-lingual tasks even if they weren't trained specifically on those tasks. Emergent abilities in multilingual models sometimes result in good translation performance between language pairs the model was never explicitly exposed to.

The phenomenon of emergence in LLMs is a topic of significant interest, but researchers do not fully understand why it occurs. There are, however, a few hypotheses and observations that offer partial explanations.

- Scale and model capacity
  One widely accepted hypothesis is that as models grow larger, they develop a sufficient capacity to represent complex patterns and dependencies in data. Small models can only capture simple relationships, while larger models, with more parameters and layers, can approximate more intricate patterns, thereby causing the model to cross a threshold, leading to emergent behaviors.
- Learning from broader contexts
  As models scale, they become better at capturing long-range dependencies and global context. This ability to "attend" over larger text sequences allows the model to learn more complex behaviors, such as multistep reasoning, which may not be possible in smaller models. Emergent behaviors could be the result of the model better utilizing attention mechanisms to aggregate information across large contexts, leading to more sophisticated outputs.
- Training data diversity and coverage
  Larger models are typically trained on more diverse and expansive datasets. This allows the model to experience a greater variety of linguistic patterns, logical structures, and facts. At a certain scale, the combination of a more complex model and richer data might give rise to emergent capabilities that were underdeveloped in smaller models due to insufficient training data or model capacity.
- Implicit knowledge representation
  As LLMs grow, they implicitly encode more world knowledge within their parameters. This increased knowledge base may enable models to perform tasks that seem unrelated to their training objectives, like answering trivia or solving arithmetic problems. The ability to generalize across tasks is not explicitly programmed but rather emerges from the broad and varied exposure to text during training.
- Optimization dynamics
  The optimization process during training might also play a role in emergence. Some researchers hypothesize that the stochastic nature of gradient descent and the use of large-scale training cause models to learn in stages, with certain abilities "popping up" after the model develops foundational patterns.

- Implicit task discovery
  LLMs trained on diverse datasets are exposed to many different tasks without being explicitly told what those tasks are. Emergent behaviors may arise from the model's ability to infer tasks and internalize a variety of functions from its unsupervised training. As the size of the model increases, its capacity to discover and generalize across tasks improves.

21.10 Do LLMs Understand?

Equipped with the Transformer architecture at their core, modern LLMs seem to "understand" the high-level meaning or thoughts conveyed by a sequence of tokens. This ability stems from the model's architecture and training, which allow it to capture nuanced relationships and contextual meanings. While the Transformer does not understand in the human sense, it effectively learns and represents complex patterns, including abstract or implicit relationships in language. Several factors of the Transformer model contribute to its ability to capture top-level meanings and implicit ideas from input language sequences.

- Self-attention mechanism
  The core strength of the Transformer lies in its self-attention mechanism, which allows the model to capture how each word in a sentence relates to every other word. This enables the model to understand both local and global relationships: Words that are far apart in the sentence can still influence each other, allowing the model to pick up on long-range dependencies (such as relationships between the subject and an implicit meaning much later in the text). Self-attention also helps the model capture nuances. Through multihead attention, the model can attend to different aspects of the input simultaneously, such as syntax, semantics, and sentiment. For example, one attention head might focus on the subject–object relationship, while another focuses on temporal cues (e.g., past vs. future events).
- Multilayer hierarchy
  In the Transformer architecture, multiple layers of encoders and decoders are stacked. With each layer, the model builds increasingly abstract representations of the input. Lower layers might focus on understanding individual words and their direct relationships (e.g., grammatical structure), while higher layers capture more abstract, global features (e.g., the overall topic of the sentence, emotional undertones, or implicit meaning). By the time the input reaches the uppermost layers, the Transformer has constructed a detailed, context-aware representation of the input sequence, capturing implicit patterns like sentiment, tone, and even high-level goals of the text (e.g., humor, instruction, or persuasion).

  In this aspect, the learning process in an LLM is similar to that in a CNN for image classification. In both cases, the input to bottom layer of the deep learning network consists of local, detailed information – pixels of an image in the case of a CNN, or individual words in a sentence or paragraph in the case of an LLM. The output from the uppermost layer of the network layer represents global, complex, and abstract patterns derived from the inputs: the image objects (e.g., dogs and cats) for a CNN, or the thoughts and ideas explicitly or implicitly conveyed by the sentence for an LLM. The model's apparent ability to understand the contents in an image or in a paragraph stems from its multilayer architecture, where progressively more global and complex patterns are extracted layer by layer. This is why multilayer neural networks have become the predominant machine learning method in the broader field of AI.

- Pretraining on large-scale data (language models like BERT, GPT)
  Pretraining on massive datasets (e.g., books, websites) also helps the Transformer learn common patterns of thought and discourse across various contexts. Through this process, it acquires vast amount of world knowledge, reasoning patterns, and an understanding of how language conveys both explicit and implicit information. Techniques like masked language modeling (e.g., BERT) or next-word prediction (e.g., GPT) helps the model learn how context shapes the meaning of each word, phrase, or sentence, providing it with a broad foundation for interpreting nuanced or implied meanings in unseen text.
- Positional embeddings
  Transformers do not have built-in awareness of word order (unlike RNNs or LSTMs); instead, they rely on positional embeddings to encode the position of each token within the sequence. This allows the model to (1) recognize which tokens come earlier or later in the input and (2) understand the structure of the input, such as sentence syntax or word dependencies (e.g., which noun a verb applies to), thereby capturing ordering and flow of thoughts, such as the progression from premise to conclusion or cause to effect. This mechanism aids in understanding narratives and argumentative structures.
- Fine-tuning on specific tasks
  When transformers are fine-tuned on specific tasks (e.g., question answering, summarization, or sentiment analysis), they learn to identify key features of these tasks that enhance high-level understanding. For example, in question-answering, the model learns to focus on relevant information in the input context that is directly or indirectly related to the query; in summarization, the model learns how to extract the most salient points from a document that convey the main ideas or thoughts.
- Modeling implicit meaning through patterns
  A Transformer learns to handle implicit meaning through the patterns present in the training data. It can recognize patterns like analogies, metaphors, and idiomatic expressions because it has encountered them (or similar patterns) during training. For example, if a text implies frustration without directly stating it (e.g., "She threw her phone on the table"), the model learns to associate that action with the sentiment of frustration. The model can also use contextual clues to infer meaning, such as recognizing when a phrase like "Well, that was great!" is used sarcastically based on the surrounding words or previous sentences.

Moreover, by attending to every token, the Transformer learns to identify which entities or concepts are being referenced, even if they are not explicitly mentioned again (e.g., resolving pronouns like "he" or "it" back to the correct subject).

- Implicit meaning and abstract reasoning
  Although transformers do not understand in the way humans do, their ability to generalize and capture abstract relationships allows them to mimic an understanding of implicit content. This can include:
  - Recognizing the overall theme of a text, even if it's not stated directly. For example, in a news article about a sports event, the model can infer whether the tone is celebratory or critical, even without explicit sentiment words.
  - Understanding intentions or goals behind a sequence of actions described in the text. For example, if a sentence describes someone turning off an alarm, brushing teeth, and grabbing a coffee, the model can implicitly infer that the person is preparing for the day without needing a direct statement like "He is getting ready for work."

In summary, self-attention allows the model to focus on different parts of the input sequence and capture relationships between tokens. Hierarchical abstraction across multiple layers enables the Transformer to capture both local and global meaning. Pretraining on large datasets helps the model develop an implicit understanding of common language patterns, enabling it to handle abstract or implied meanings. Task-specific fine-tuning helps refine these capabilities to achieve high-level understanding. As a result, while transformers do not truly understand language in the same way humans do, their design allows them to process text in a way that captures both explicit meanings and subtle, implicit patterns through the relationships between words and their broader context.

Finally, how should we answer the original question: *Do LLMs understand?* While humans, as the designers of LLMs, know exactly how these models function, it remains debatable whether their behavior can be considered true understanding. However, we can confidently say that LLMs do not understand in the same way humans do, for at least one key reason: human brains do not operate through mechanisms like gradient descent or backpropagation. Yet, we still lack a definitive answer to the question, *How do humans understand?* – at least not now. Thus, it might not be entirely incorrect to say that LLMs do "understand," depending on how we define the term. If we adopt a more open-minded definition of understanding – one that encompasses not only human-like comprehension but also a functional, algorithmic process that enables LLMs to generate appropriate responses – we could argue that LLMs do, in fact, understand in their own way.

Following the same line of reasoning, we could also ask a broader question: Can an AI system be conscious? We believe that all other humans are as self-conscious as we are. But how do we know this? In fact, this belief is inferred or assumed, not based on facts, since we can never directly experience another's mind. This is the well-known *problem of other minds* in philosophy. If we can only *assume* others are conscious based on their behavior, why would we deny AI systems the possibility of consciousness if they exhibit behaviors as sophisticated, or even more so than humans? Currently, we do not fully understand the mechanism of consciousness in the brain, much like we do not fully understand how emergent behaviors arise. How, then, can we be certain that consciousness is unique to carbon-based humans and entirely absent in silicon-based AI systems?

When ChatGPT is asked whether it has passed the Turing test (whether one can distinguish an AI system from a human based on a conversation), it responds modestly, stating that it has not passed the test owing to issues such as hallucination (it may provide highly convincing but factually incorrect information) and context retention (it has limited ability to remember the content within a conversation). As a result, its ability to engage in long-term, nuanced discussions diminishes due to its token limit. However, these issues might be considered as "soft" rather than "hard" failures in the binary nature of the Turing test, as some humans also exhibit such behaviors. After a bit of convincing, ChatGPT acknowledges that "In a controlled setting with a conversation that doesn't last too long or require sustained memory, and where any potential hallucinations are minimal or not easily noticeable, I could likely appear convincingly human-like."

To help address the issue of context retention, a new *bio tool* functionality was introduced starting with GPT-4 (unavailable to GPT-3.5). This tool allows key information from a conversation with specific users to be stored in the system for future reference, while all other content in the conversation is lost once the session ends. This mechanism enables GPT to have some long-term memory, thereby improving the continuity across sessions and making conversations more personalized. As a result of this added bio tool feature, GPT's behavior becomes more human-like.

Finally we would like to extrapolate from the previous discussion on emergence. As we see an LLM with relatively simple structures and learning rules may exhibit unexpected complex behaviors, some of which mimic human-like understanding or even consciousness, when both the model size (number of parameters) and the training data size increase significantly. The phenomenon of emergence is not well understood, much like human consciousness, which is not well understood either. Could we, then, speculate that consciousness may also be some kind of emergent property of a super-complex system, the human brain, arising from relatively basic neural functions in an immensely complex hierarchical network structure containing 100 billion neurons connected by 100 trillion synapses?

The author heard a story at a conference on neural networks in the mid-1980s that went like this: When an elementary school teacher asked her class to name the most important inventions of the 20th century, her students mentioned the airplane, television, and computer. Then a boy shouted, "It is the thermos!" Puzzled, the teacher asked, "All the thermos does is keep water cold or hot, why is that so important?" The boy replied: "But how does it know when to keep the water cold and when to keep it hot?" (When asked to proofread this paragraph, ChatGPT made an additional comment: "This is a charming anecdote!" – an obvious emergent behavior!)

Now, many decades later, AI researchers are still grappling with the same question: *How does it know?*

21.11 Future Directions

Although modern LLMs have become powerful tools widely used in various areas such as education, scientific research, and medicine, they also face some significant challenges. One of the key issues is misinformation and hallucination, which is particularly problematic for applications requiring high accuracy, such as journalism, education, and legal documentation. Developing mechanisms to ensure the factual correctness of LLM outputs remains a major hurdle.

To address issues such as misinformation and hallucination, LLMs need to be improved to enhance their factual reliability. This will likely involve integrating external knowledge sources, such as databases and real-time web access, into their architecture. These enhancements could improve their ability to provide accurate information and reduce hallucinations. Fact-checking mechanisms and real-time data integration could also help ensure that outputs are more trustworthy.

Emergence in LLMs is an area of active research because it hints at deep, underlying principles about learning and intelligence that are still not fully understood. Additionally, the emergence of behaviors can be difficult to predict; researchers often do not know what capabilities will arise as models scale, leading to a kind of "black box" nature in how and why specific abilities develop. Key questions include:

- What specific factors contribute to the emergence of complex behaviors in LLMs?
- How does the scale of a model influence the types of capabilities that can emerge?
- What role does the architecture of the model play in facilitating or hindering emergence?
- How can we design training processes that guide the emergence of desired capabilities while minimizing undesirable ones?

The future of LLMs extends far beyond their current capabilities, encompassing both technical advancements and profound philosophical questions. As these models become more multimodal, personalized, and integrated with various technologies, they will continue to transform fields like education, healthcare, and scientific research. As these models evolve, they will also open new possibilities across a wide range of areas, such as automation of cognitive tasks and augmented creativity and innovation. At the same time, their development prompts deeper reflections on issues like understanding, consciousness, and their potential impact on society and human knowledge. The following are key areas where LLMs may drive significant change in the future, in terms of both the technical development and some philosophical considerations.

- Multimodal capabilities and enhanced contextual understanding

LLMs are expected to evolve into multimodal systems, capable of processing and generating text, images, audio, and video. This would enable richer, more interactive communication and problem-solving by interpreting and producing complex multimedia content.

The ability to handle multiple modalities blurs the line between human sensory processing and machine interpretation. This raises questions about whether machines can achieve a deeper "understanding" of the world or if they are merely processing information in increasingly complex ways. What does it mean to "see" or "hear" when a machine can do these things?

- Personalization and real-time learning
  As LLMs integrate user-specific data (while addressing privacy concerns), they will offer more personalized experiences. Real-time learning will allow them to adapt to new information dynamically, improving relevance and continuity in user interactions.

  Personalized learning systems challenge our notion of individuality and agency. As LLMs become more attuned to specific users, are they shaping users' preferences, or are users shaping the machine? How do we ensure the balance between personalization and privacy?
- Collaboration with human experts
  LLMs could become valuable collaborators for professionals, analyzing large datasets and generating insights that assist experts in fields like medicine, law, and engineering.

  The role of machines as collaborators leads to reflections on human identity and work. As machines take on more sophisticated tasks, how do we define human expertise and creativity? Are we enhancing our capabilities, or are we becoming overly reliant on AI for innovation?
- Fact-checking and factual reliability
  LLMs face the challenge of providing accurate information, especially in fields where misinformation or hallucination is problematic. Integrating external knowledge sources and real-time data could enhance their factual reliability and reduce errors.

  As machines become trusted sources of information, this raises questions about the nature of truth and authority. Should machines be held to the same standards of truth as humans, or is their understanding of "facts" inherently different? Can AI ever fully eliminate biases and inaccuracies, or will human-like flaws always persist?
- Integration with IoT and accessibility enhancements
  LLMs could be integrated with Internet of Things (IoT) devices, enhancing communication with smart technologies and enabling seamless interaction between users and their environment.

  The growing integration of AI with our physical surroundings challenges ideas about control and agency. As smart environments become more autonomous, where does human agency end and machine-driven decision-making begin? Do we risk ceding too much control to the systems that guide our lives?

- Emergence and the black box of AI capabilities
  Emergent behaviors in LLMs often appear unexpectedly as models scale in size and complexity. Researchers do not always understand how or why specific capabilities develop, leading to a "black box" phenomenon.
  The unpredictability of emergent behaviors in AI mirrors broader questions about intelligence, consciousness, and free will. Can AI ever fully achieve human-like reasoning and decision-making, or is it confined to a mechanistic, black-box nature? Does emergence hint at a deeper intelligence within machines that we have yet to comprehend?
- Human consciousness vs. machine "understanding"
  LLMs can generate responses that appear to exhibit understanding of nuanced and complex relationships, but they don't truly "understand" in the way humans do.
  This raises fundamental questions about the nature of consciousness. If machines can replicate behaviors associated with understanding or even emotions, how do we distinguish between human consciousness and machine intelligence? Could consciousness itself be an emergent property of sufficiently complex systems, both biological and artificial?
- Scientific research and creativity
  LLMs may accelerate scientific discovery and creative processes by analyzing vast amounts of data, generating hypotheses, and aiding in the design of new concepts or artworks.
  AI's role in creativity provokes deep questions about the essence of artistic expression. Can a machine truly be creative, or is creativity inherently tied to human emotion, experience, and intention? What role will humans play as machines become more involved in the arts and sciences?

In conclusion, by merging technical challenges with philosophical considerations, we get a holistic view of the future of LLMs, where technical advancements are not just questions of capability but also of meaning, ethics, and human identity. The interplay between these two dimensions will shape how LLMs evolve and how society responds to their integration into our lives.

21.12 Applications of Transformers Beyond NLP

Although the Transformer architecture was initially developed to tackle NLP problems, it has emerged as a powerful tool across diverse fields much beyond NLP, due to its remarkable ability to effectively model complex relationships in sequential and spatial data. Here are some notable applications of transformers outside of NLP.

- Computer vision
  Transformers have been widely used in computer vision, most notably with the *Vision Transformer (ViT)*, first introduced by Dosovitskiy *et al.*[2] ViT has

[2] Dosovitskiy, A. *et al.* (2020). An image is worth 16 × 16 words: Transformers for image recognition at scale. *ArXiv*, abs/2010.11929.

achieved state-of-the-art performance in tasks such as image recognition and object detection. It treats image patches (e.g., 16×16 pixels) as token vectors, similar to words converted into imbedding vectors used in NLP. By processing these sequences of patches and using self-attention, the transformer learns to model complex spatial relationships between patches and capture both global context and local details.

The output of the ViT depends on the specific task. For image classification (assuming there is only one object in the image), the output is typically a probability distribution over predefined classes (e.g., "cat," "dog," "car"). In image object detection, the transformer generates multiple output tokens, each corresponding to one of the objects within the image.

The ViT is trained on labeled data, where each object in the image is labeled by (1) its class identify and (2) a bounding box (in terms of its center, together with its width and height). Based on the total loss, the difference between the ground truth labeling and the model output (combining both classification and bounding box losses), backpropagation is carried out to adjust the model parameters, gradually reducing the total loss.

Unlike CNNs, which inherently capture spatial relationships owing to the local receptive fields, transformers lack this inductive bias. To compensate, ViT adds positional encodings to each patch vector. These encodings provide the model with information about the relative position of each patch in the image, helping it understand spatial relationships.

- Protein structure prediction

Transformers have been used to predict protein structures, such as in AlphaFold by DeepMind.[3] In this application, the amino acid sequences of proteins are treated similar to sentences in NLP. This transformer-based model, once trained on massive protein databases, can predict a protein's 3-D structure based on its amino acid sequence. This is achieved by capturing both local (short-range) and global (long-range) dependencies within the sequence, which are essential for predicting folding patterns.

AlphaFold is trained on data consisting of a large number of proteins with known amino acid sequences, each labeled by its experimentally determined 3-D structure. The primary datasets include the protein data bank (PDB), which provides high-quality structures derived from X-ray crystallography and cryo-electron microscopy, capturing a wide range of protein conformations. AlphaFold also leverages multiple sequence alignments (MSAs) and genomic data to extract evolutionary information, which is crucial for understanding conserved structural features across related sequences.

Once trained, AlphaFold outputs several critical predictions for constructing a 3-D structure for the sequence of a given protein. These include:

[3] Jumper, J. *et al.* (2021) Highly accurate protein structure prediction with AlphaFold. *Nature*, 596 (7873): 583–589.

  - the distances between different parts of the protein and their relative orientations in space, which help to understand the protein's overall shape;
  - the angles of the protein's backbone structure, which define its overall folding pattern;
  - a complete 3-D structure and shape in terms of the positions of all parts of the protein (this is a comprehensive representation that includes the spatial coordinates of the atoms in the protein, along with a visualization of the 3-D structure);
  - confidence scores for different parts of the protein and estimates regarding the accuracy of the model predictions.

  Transformers can also be fine-tuned to generate new protein sequences with desired structural or functional properties. By training on existing proteins, they can design sequences that are likely to adopt specific folds, leading to novel proteins with unique functions for therapeutic or industrial applications.
- Mathematics and symbolic reasoning
  Transformers have been adapted for symbolic math problems, including tasks such as integration, equation solving, and formal logic, where they have demonstrated significant promise in performing symbolic transformations and simplifications. These models utilize their attention mechanisms to capture intricate relationships between mathematical entities, enabling them to manipulate and reason about symbols effectively. For example, in integration tasks, transformers can learn to recognize patterns in integrands and apply known integration techniques, while in equation solving, they can systematically derive solutions by breaking down complex expressions into manageable components. Additionally, transformers have been applied to formal logic, where they can handle logical expressions and infer conclusions through symbolic reasoning. These applications are particularly valuable for automated theorem proving, as they allow for the verification of mathematical statements through logical deductions, and for symbolic computation, where they can facilitate the manipulation of mathematical expressions with high accuracy and efficiency.
- Speech and audio processing
  Transformers have demonstrated strong results in various tasks, including automatic speech recognition (ASR), speech synthesis, and audio classification. These models excel in handling audio data by treating it as a sequence of features, enabling them to capture temporal dependencies and intricate patterns within the audio signal. For instance, models like Wav2Vec 2.0 leverage transformers to process raw audio waveforms directly, employing a self-supervised learning approach that allows them to learn meaningful representations from unlabeled audio data. This capability enhances their performance in transcription tasks, achieving high accuracy by effectively distinguishing phonetic nuances and contextual information. In addition to ASR, transformers are also employed in speech synthesis, where they generate high-quality, natural-sounding speech by modeling the relationship between textual input and acoustic features. Furthermore, in audio classification tasks, transformers can analyze diverse audio

signals, ranging from music to environmental sounds, by learning to identify distinctive features and patterns, leading to accurate classification outcomes.

In summary, the versatility of the Transformer architecture lies in its capacity to model dependencies across various types of sequential data, regardless of the domain. Its self-attention mechanism allows for the effective handling of high-dimensional inputs, capturing complex relationships, and enabling contextual understanding. This adaptability makes transformers suitable for solving intricate problems in widely different fields.

21.13 Generative Methods Beyond Transformers

While the Transformer architecture has been a groundbreaking development in large language modeling, numerous other generative learning methods have also emerged, such as those shown in the following. These generative models learn to approximate, explicitly or implicitly, the probability distribution over the given data (e.g., text, images, audio, and video), and generate new data samples that resemble to given data. This enables applications such as image and video synthesis, 3-D scene reconstruction, and natural language generation.

All these models use neural network methods, leveraging deep learning networks trained through backpropagation based on unlabeled data, as their objective is to learn the structure and distribution of the input data to create realistic samples mimicking the input. Additionally, these models share a common general strategy: they learn to represent the given data in the real world (denoted by **x**) in an internal hidden or latent space (denoted by **z**), where the data is encoded in a more abstract or compressed form, making it more manageable. The four types of models are illustrated as in Fig. 21.3, and are discussed in more detail in the following.

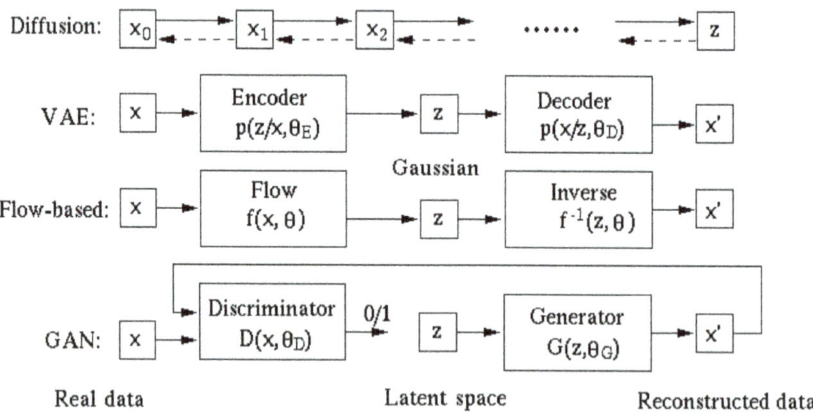

Fig. 21.3 Four generative models.

- Diffusion models
  Diffusion models operate through a two-step process: a forward diffusion process and a reverse diffusion process. In the forward process, noise is gradually added to the data across a series of time steps, progressively corrupting it until it becomes indistinguishable from pure noise. During training, the model learns to reverse this process by estimating the necessary adjustments to remove the noise incrementally. This is typically accomplished through a neural network that predicts the original data from the noisy input. Once trained, the model can generate new data by starting with a random noise sample and iteratively applying the learned denoising steps, resulting in coherent outputs.
  Diffusion models are generally trained on unlabeled data, because the model learns to generate data by progressively denoising a noisy sample until it resembles the original input data, which serves as the target for reconstruction. In each training iteration, the model's predicted denoising steps are compared with the actual data, and backpropagation is used to adjust the model's parameters, improving its ability to remove noise accurately. This iterative approach enables the model to learn a series of denoising steps that can eventually generate new, coherent data from pure noise once training is complete.
  The diffusion models have been applied to generation of high-quality images and video based on textual descriptions, or even noise. They can also enhance the resolution of images, super-resolution, by refining low-resolution inputs.
- Variational autoencoders
  Variational autoencoders (VAEs) consist of two main components: an encoder and a decoder (similar to the autoencoder discussed in Section 18.3), both implemented as neural networks. The encoder learns to map the probability distribution of the input data to a distribution in a latent space, typically Gaussian parameterized by its mean and covariance. The decoder then draws samples from this latent space and maps them back to the data space, producing realistic outputs that resemble the original inputs. To capture the data's structures and to generate high-quality reconstructions, both the encoder and decoder are typically deep neural networks, especially for complex data like images, where architectures such as convolutional neural networks (Section 18.4) are commonly used.
  As the goal is to produce outputs that resemble the inputs (i.e., the input is the target for the output), VAEs are trained on unlabeled data. Backpropagation is used to minimize a loss function that combines two terms: the mean squared error (MSE) between the reconstructed output and the original input, and the Kullback–Leibler (KL) divergence between the distribution in the latent space and the desired distribution, typically a Gaussian. Minimizing this loss function enables VAEs to balance data reconstruction accuracy and latent space regularization, ensuring that samples drawn from the latent space are realistic and diverse.
  Similar to autoencoders, VAEs can be used for dimensionality reduction. However, unlike traditional autoencoders, VAEs are primarily used to generate

new, realistic output that exhibit characteristics similar to the input, instead the exact replicas of the input data. This capability makes VAEs particularly useful in applications where generating diverse, yet coherent, samples is essential. For example, in image generation, a VAE trained on a dataset of human faces can produce new, realistic-looking faces with different features, expressions, or poses, while retaining a resemblance to the types of faces it was trained on.

- Flow-based models

  Flow-based models learn to transform a known and mathematically tractable distribution, typically Gaussian, in a latent space (also called Gaussian space), into a more complex distribution in data space that resembles the actual data distributions. This is achieved progressively by a sequence of invertible transformations, called *flows*, implemented by a multilayer neural network, with each layer corresponding to one of the transformations. Since each transformation is invertible, the model can be reversed to map data points from the data space back to the Gaussian latent space.

  During training, samples from the training set (data space) are passed through the model to be mapped to the Gaussian in the latent space for the calculation of the likelihood of the data points under the Gaussian distribution (in terms of its mean and covariance). This represents how well the Gaussian models the real data. Then, backpropagation is used to minimize a loss function, the negative log-likelihood, so that the model parameters, the weights in the network, are modified to enable the Gaussian to better model the data, thereby improving its performance in generating realistic samples.

  Once trained, a flow-based model can generate new data by passing random samples from the Gaussian in the latent space through the inverse of the transformation network (the flow) to obtain new data points in the data space. These generated data should exhibit the same characteristics as the original data, though they are not replicas.

  Flow-based models can generate high-fidelity images useful for applications like image generation, denoising, and super-resolution. They can also be used for text and video generation.

- Generative adversarial networks

  Generative adversarial networks (GANs) consist of two deep neural networks, a generator and a discriminator, which work in opposition to each other in a process known as adversarial training. The generator transforms random noise into synthetic (fake) data samples that resemble real data, while the discriminator takes both real and fake data samples as input and generate a probability score as output to indicate whether the input is real or fake.

  During training, the generator aims to maximize the probability of the discriminator making a mistake, while the discriminator strives to accurately classify the inputs. This competition between the two networks leads to progressively improved performance on both sides, with the generator producing increasingly realistic samples. Training is complete once it reaches the *Nash equilibrium*, a game theory concept, where neither the discriminator nor the generator can improve its performance any further. At this point, the discriminator can no longer distinguish between the fake data samples produced by the generator and the real samples from the dataset, indicating that the model has successfully learned the underlying data distribution,

Specifically, the discriminator aims to maximize the expectation of its output, $D(\mathbf{x}, \boldsymbol{\theta}_D)$ when the input $\mathbf{x}$ is a real sample from dataset, and to minimize the expectation of $D(G(\mathbf{z}), \boldsymbol{\theta}_G)$ (or equivalently, maximizing $1 - D(G\mathbf{z}, \boldsymbol{\theta}_G)$) when the input is a fake sample from the generator. Thus, the discriminator seeks to maximize the following objective function:

$$J(D, G) = E_{p(x)}[\log D(\mathbf{x}, \boldsymbol{\theta}_D)] + E_{p(z)}[\log(1 - D(G(\mathbf{z}, \boldsymbol{\theta}_G)))] \qquad (21.4)$$

By contrast, the generator aims to achieve its goal of "fooling" the discriminator into classifying its generated (fake) samples as real (from the dataset). To achieve this, it seeks to minimize $J(D, G)$, the objective that the discriminator is trying to maximize. If the generator successfully minimizes $J(D, G)$, it will cause $D(G(z))$ to approach 1 for generated samples, effectively "tricking" the discriminator. This leads to a situation where the generator's fake samples are indistinguishable from real samples, thereby achieving the goal of the GAN. Consequently, the overall objective of the GAN is expressed as $\min_G \max_D J(D, G)$. Backpropagation is then used to update the parameters $\boldsymbol{\theta}_G$ and $\boldsymbol{\theta}_D$, the weights of both the generator and discriminator networks, based on the gradients of $J(D, G)$.

Once trained, GANs can generate new data by sampling random noise from a predefined distribution, typically a Gaussian or uniform distribution, and passing this noise through the generator network to transform it into new synthetic data points that resemble the original training data. GANs can generate high-quality, realistic images, audio, and video, that align with the learned data distribution.

In summary, these recent generative machine learning methods showcase the evolving landscape of generative AI. Each method offers unique mechanisms for generating high-quality data across diverse applications, from image and video synthesis to natural language processing and 3-D reconstruction. Their advancements signify a substantial leap in the capabilities of generative models, enhancing their applicability in real-world scenarios.

Index

For EU product safety concerns, contact us at Calle de José Abascal, 56–1°,
28003 Madrid, Spain or eugpsr@cambridge.org.

www.ingramcontent.com/pod-product-compliance
Ingram Content Group UK Ltd.
Pitfield, Milton Keynes, MK11 3LW, UK
UKHW052159230726
473566UK00018B/157

* 9 7 8 1 3 1 6 5 1 9 5 0 9 *